Physics for Engineering Technology

Physics for Engineering Technology

SECOND EDITION

Alexander Joseph
Professor Emeritus of Physics
Division of Science and Mathematics
John Jay College of Criminal Justice
The City University of New York

Kalman Pomeranz
Professor
Department of Physics
Bronx Community College
The City University of New York

Jack Prince
Professor
Chairman, Department of Physics
Bronx Community College
The City University of New York

David Sacher
Associate Professor
Department of Physics
Bronx Community College
The City University of New York

JOHN WILEY & SONS, New York • Chichester • Brisbane • Toronto • Singapore

Dedications

To my wife, Janet, whose astute criticism is responsible for most of the merit in my work.

K. B. P.

To Marilyn, whose inner strength, courage and unrestrained love are my source of inspiration and creativity.

J. P.

To my sons, Michael and Barry, for their encouragement and criticisms from the students' point of view.

D. S.

Library of Congress Cataloging in Publication Data:

Joseph, Alexander.
   Physics for engineering technology.

   Includes index.
   1. Physics. I. Title.
QC21.2.J67 1977   530'.02'462   76-55696
ISBN 0-471-45075-8

Printed in the United States of America

20 19 18 17 16 15 14 13 12

Preface

We have written this textbook for students of the various engineering technologies that are offered in technical institutes, and community and junior colleges. It is intended for students whose mathematical preparation includes at least one year of elementary algebra and some basic elements of plane geometry. The necessary operational formulas in trigonometry are provided in the text.

Our aim, in a world of rapidly evolving new technological developments, is to provide an understanding of the basic principles of physics and the ways in which these principles are directly utilized in the engineering technologies. During the preparation of the first edition and since its publication, this book has been used by the students of engineering technology at the Bronx Community College. It has also been used as the textbook for a noncredit introductory course for engineering science students who have not had any previous exposure to physics.

The second edition has been prepared with both student and professor in mind. To help the student read the text with greater facility, each chapter begins with a brief overview and a listing of the major learning objectives. The student, therefore, is directed to those parts of the chapter that require his most diligent attention. At the close of each chapter, we have introduced a brief summary of the key terms and equations that have been studied. This is followed immediately by a series of questions aimed at testing the student's recollection of the material he has read. The answers to these questions can generally be found in the body of the chapter material. These questions can be used profitably by the students as a self-testing program.

Following the questions, we have provided the instructor with an increased selection of exercises from which to assign homework or recitation practice. For most chapters, these exercises have been grouped in two sections, A and B. The A group generally represents the basic problems that require minimum mathematical manipulation. The student can successfully solve these problems by recalling an appropriate principle or formula and substituting the given information. The B problems are often more challenging, requiring a sequence of thought processes or mathematical steps to arrive at the correct answer.

In revising much of the original text material, we have concentrated on simplifying the reading and mathematical levels while retaining the rigorously accurate physics that characterized the first edition. To help the student master the problem-solving techniques, we have added many solved examples to the body of the text. In streamlining the coverage, we have reduced the number of chapters from 41 to 39. We have increased and revised the material on modern physics while deleting much of the technical material on hydraulics and motion through fluids. We have also combined the chapters on thermometry and expansion into a single chapter. Much of the material previously discussed in the chapter on electrical conduction in solids, liquids, and gases has been deleted or rewritten into other parts of the book. In anticipation of the probable increase in demand by industry and hospitals for nuclear technicians, the revised chapter on atomic and nuclear physics (Chapter 39) emphasizes the subjects of radioactivity, the nuclear reactor and the laser. To our original treatment of applications in optics we have added sections on the camera, polarization and spectroscopy.

We believe this book can be adapted to serve the needs of students within a wide spectrum of ability levels. The more complex derivations have been set apart typographically. This arrangement enables the instructor to use those portions of the text that are best suited to his students. Answers to selected problems are included in the text, and a set of solutions for all problems is available to the instructor from the publisher.

Although the trend among many authors is to use the metric systems of units exclusively, we have chosen to retain the British system of engineering units wherever present practice in engineering technology still calls for it. Thus we use this latter system extensively in the sections on mechanics and to a lesser degree in the sections on heat. In most other areas the metric systems are used exclusively.

The following are suggested possible sequences of chapters for particular technologies using this book for two semesters. They apply, of course, only to situations in which the students take different physics courses in the various technologies:

1. Mechanical engineering technology: Chapters 1 to 3, 7 to 20, 22, 25 to 27, 29 to 32, 34 to 36, 38, 39.

2. Electrical engineering technology: Chapters 1 to 3, 7 to 13, 15 to 19, 22 to 33, and 39.

3. Construction engineering technology: Chapters 1 to 9, 11 to 13, 15 to 24, 29, and 34 to 38.

4. Chemical engineering technology: Chapters 1, 2, 7 to 9, 11 to 13, 16 to 22, 24, 25, and 28 to 38.

5. Medical laboratory technology: Chapters 1 to 3, 5 to 7, 9 to 12, 14 to 18, 22, and 24 to 39.

6. Nuclear technology: Chapters 1 to 3, 7 to 13, 17 to 22, 25 to 39.

When all technologies must take the same physics sequence, we recommend covering the chapters in order, with the possible exclusion of part or all of Chapters 11, 13, 16, 24, 32, 33, 37, 38, and 39.

The production of a textbook requires the dedication and skill of many competent professionals with expertise in the technical aspects of publishing. We are indebted to the staff at Wiley who orchestrated the myriad of details leading from the manuscript to the published book.

We also wish to acknowledge with gratitude the comments received from those who used the first edition during the past 10 years and to those who reviewed parts of the present book and offered us many helpful suggestions. A special thanks, also, to Rhoda Mark, Leonie Meiselman, Marilyn Prince, and Elsie DeCesare for typing the manuscript.

We hope that the students find in this book much of the excitement that the physicist and technologist experience in their work.

ALEXANDER JOSEPH
KALMAN POMERANZ
JACK PRINCE
DAVID SACHER

We mourn the recent passing of our friend, colleague and co-author, Dr. Alexander Joseph. His devotion to physics education for more than four decades was recognized by all who worked with him. We especially recall his encouragement and important contributions toward the publication of the first edition of this text. We take this opportunity, also, to express our sorrow over the passing of Dr. Morris Meister, first president of Bronx Community College, to whom the first edition was dedicated.

K. B. P.

J. P.

D. S.

Contents

1 Physics and Engineering Technology 1

2 The Composition of Vectors 9

3 Concurrent Forces in Equilibrium 22

4 Nonconcurrent Forces 35

5 The Force of Friction in Equilibrium Problems 52

6 Elastic Forces in Static Structures 66

7 Kinematics and Linear Motion 77

8 Uniformly Accelerated Motion and Projectiles 89

9 Forces and Linear Motion 104

10 Forces and Curvilinear Motion 121

11 Forces and Rotation 143

12 Work and Energy 164

13 Power, Efficiency, and Simple Machines 187

14 Impulse and Momentum 208

15 Simple Harmonic Motion 231

16 Fluid Mechanics 245

17 Thermometry 270

18 Heat 283

19 The First and Second Laws of Thermodynamics 293

20 Thermal Properties of Gases 312

21 Changes of Phase 327

22 Heat Transfer 337

23 Wave Motion 345

24 Intensity and Quality of Sound Waves 363

25 Coulomb's Law 375

26 Electric Fields and Electric Lines of Force 386

27 Electric Potential Energy and Electric Potential 399

28 The Electric Circuit 410

29 Magnetism and Electromagnetism 429

30 Electromagnetic Theory 447

31 Important Applications of Electromagnetic Theory 462

32 Elements of an Alternating Current 478

33 Electronics 497

34 Light and Photometry 506

35 Light Reflection and Mirrors 519

36 Refraction 535

37 Thin Lenses—Optical Instruments 546

38 The Wave Nature of Light 567

39 Modern Physics 581

Appendix 600

Answers to Problems 611

Index 621

1.
Physics and Engineering Technology

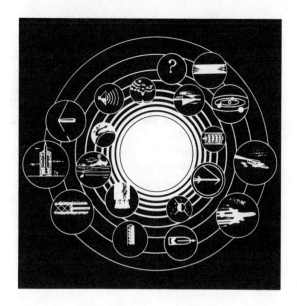

Modern technology leans heavily on physics. Technical progress follows advances in physics and the other basic sciences. It was the increased understanding of electricity and magnetism that led to the electrical power industry, radio, television, radar, and other means of electrical and electronic communication. This applies to electric lighting and heating, too. In medicine today, many of the instruments of diagnosis and therapy are applications of physics, such as X-rays, radioisotopes, electrocardiographs, and electroencephalographs.

Engineering technology is responsible for space flight, rocketry, satellites, and satellite communication systems, which are all applications of physics. The development of modern supersonic aircraft is an obvious application of the principles of physics.

Let us take a look at what physics is. It is often defined as the study of matter, energy, and their transformations. Although we often use "matter" and "energy" as everyday words, the everyday meanings do not provide us with the exact scientific definitions. The physicist in dealing with matter is concerned with the properties of the fundamental particles that make up the atom and with the combinations of these particles that make up the universe.

The concept of energy must be used to understand the combination of these fundamental particles into atoms. Energy is involved in all the engineering technological applications of physics. It is impossible to say exactly when we are dealing only with matter or only with energy, and we shall discover that these two ideas are always closely associated.

This book is divided into six basic areas:

1. Mechanics.
2. Heat.
3. Wave motion.
4. Electricity and magnetism.
5. Light and other electromagnetic radiation.
6. Electronic and nuclear applications.

Mechanics is the foundation for all the other areas of physics. To understand sound, light, and electromagnetic radiation, the physics of wave motion must be studied. The basic laws of the conservation of energy developed in mechanics are needed for the study of heat, sound, wave motion, magnetism and electricity, electromagnetic radiation, and the physics of the nucleus. In other words there is a unity to physics; it is not merely a collection of unrelated topics.

Although this book is concerned with the physics of today, physics is not static; the frontiers of physics change every day. It must be understood, however, that these new developments will probably not change the fundamentals that are studied in this book.

Physics functions to combine our understanding of the world about us into an orderly form that can be intellectually digested. A physicist creates theories based on experimentation in order to understand the nature of the world about

him. The role of the engineering technologist is to further the implementation of these ideas into the development of new products and industrial processes.

Although physics goes back in time to ancient history (see Fig. 1-1, showing the development of electricity) the greatest growth has taken place in the past few centuries. The chief growth in engineering technology has occurred during·the present century and is an extension of the basic discoveries in physics made in the centuries be-

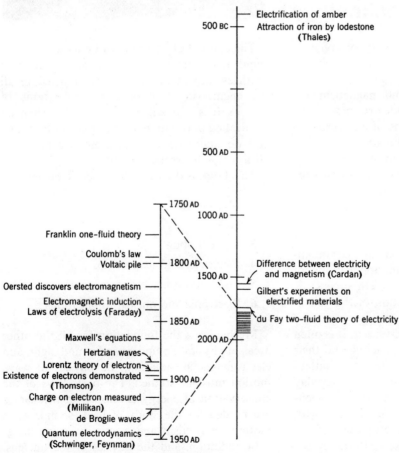

Figure 1-1 "Timetable" showing the development of electricity. (From O. H. Blackwood, W. C. Kelly, and R. M. Bell, *General Physics*, 4th ed., John Wiley & Sons, Inc., New York, 1973.

fore. However, the rate of physical discovery basic to engineering technology is so rapid that it is essential for a solid foundation in physics to be a part of the background of every student of engineering technology.

Since the engineering technologist serves society, it may be useful to understand how basic science, research and development, applications, and the needs of society lead to new products and processes. This complex interaction is illustrated in Fig. 1-2.

Perhaps a good example of the connections between physics and engineering technology is the study of a problem like "sonic boom," a

problem still not completely solved. This is related to the physics of shock waves, a field that permeates many new technological applications, as shown in Fig. 1-3.

Physics and engineering technology both deal almost exclusively with quantities that must be measured. We may, for example, measure a room with a foot rule or a yardstick and learn that it is 12×15 ft. The unit of length used in measuring the room is the foot; this is an old unit that is still commonly used today.

The measurement of distances was developed by the Egyptians to measure land for tax purposes. In ancient business transactions the mea-

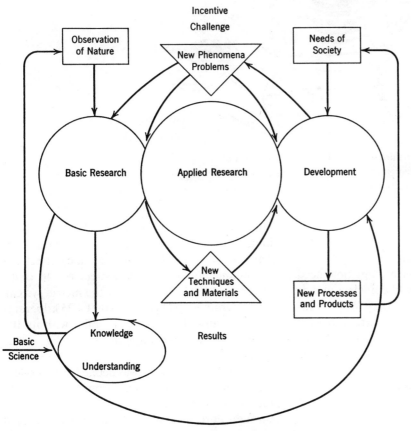

Figure 1-2 Interrelationships in the development of products and processes for society from basic science.

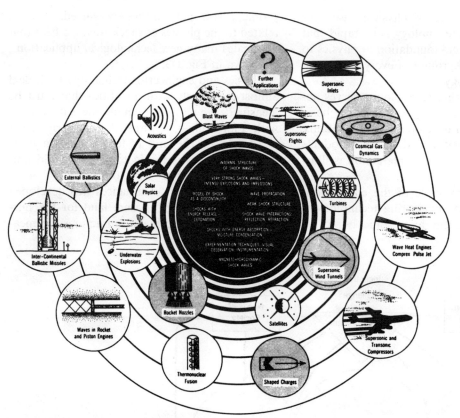

Figure 1-3 Basic research in the nature of shock waves led to a variety of technological applications. (Courtesy Naval Research Advisory Committee.)

surement of weight and volume eventually led to units such as the ounce, the pound, the gill, the pint, the quart, and the gallon.

Physics and engineering technology now need units as small as the diameter of a nucleus and as large as the dimensions of the Milky Way. Transistors, vacuum tubes in TV sets, gyroscopes for controlling the flight of aircraft and rockets, atomic piles (nuclear reactors), and automobile parts must be precisely made if they are to work at all. The manufacturers of these devices require precise measurements based on definite units of measure.

Physicists use the metric system, which was designed in France during the late eighteenth century. It is convenient because it is based on multiples of ten and can be divided into tenths, hundredths, or thousandths. The kilowatt is 1000 watts; a meter is 100 centimeters; a centimeter is 10 millimeters. Thus, 1.46 meters is 146 centimeters or 1460 millimeters. In the metric system, the kilogram is 1000 grams so that 4.25 kilograms is 4250 grams. The American dollar with its 100 cents is a familiar application of the metric system.

At one time the meter was defined as a one-ten-millionth part of the distance from the equator to the North Pole measured along a line passing through Paris. A platinum iridium bar measured at the temperature of melting ice, 0°C, (or 32°F) and marked appropriately served as the standard meter.

Table 1-1 Metric units of length and approximate British equivalents

1 kilometer (km) = 1000 meters or about $\frac{5}{8}$ mile, or 0.62 mile
1 meter (m) = about 1.1 yards or 3.3 feet
1 centimeter (cm) = 1/100 meter = 1/2.54 inch
1 millimeter (mm) = 1/1000 meter = 10^{-3} meter
1 micron (μ) = 1/1,000,000 meter = $1/10^6$ meter or 10^{-6} meter

Later, more accurate measurements of the size of the earth led to the discovery that the original meter was slightly too short. In 1872 the new, correct standard meter (a platinum bar) was made. Exact copies are kept in the Bureau of Weights and Measures in Sèvres near Paris, and other exact copies can be found in the Bureau of Standards in Washington, D.C., London, and the capitals of other nations.

Today the wavelength of the orange-red line in the spectrum emitted by krypton 86 is used to measure the distance between the marks on the standard meter. The wavelengths of light emitted by atoms are not affected by external physical changes, and the wavelength of light is used to provide the most precise standards for the measurement of length.

The yard that you use in everyday life is 3600/3937 of the standard meter. This makes 39.37 inches equal to one meter, and 2.54 centimeters equal to one inch, while one foot equals 30.5 centimeters. Table 1-1 gives the relationships between the metric system and the British system of lengths.

In everyday experiences we do not encounter the quantity *mass*. The scientific meaning of the word differs from the everyday meaning of mass, which makes little distinction between this quantity and weight. Some of the British and metric units for volume and weight are compared in Table 1-2. Weight is the force that gravity exerts on the atoms that make up a substance. An object on the earth weighing 600 lb would weigh only about 100 lb on the moon, where the gravitational attraction is one-sixth that on the earth. However, the mass does not change no matter whether you are on earth or anywhere in space. Mass and the units used for mass will be studied in detail in Chapter 9.

The British system uses the units of foot, pound weight, and second, and is called the fps system or British Engineering System. The oldest metric system, or cgs system, uses the centimeter, gram, and second. In 1904 Giorgi proposed the mks system which uses the meter, kilogram, and second.

In the work in this book, it is assumed that the reader has studied at least two years of high school mathematics (algebra, geometry, and trigonometry). The trigonometric functions needed in this book are presented in Table 1-3. The tables of the values of these trigonometric functions are found in the appendix; it may be wise to review these functions.

Table 1-2 Metric units of weight and volume and British equivalents

1 metric ton = weight of 1000 kilograms = 1.10 short tons
weight of 1 gram (g) = $\frac{1}{454}$ pound
= $\frac{1}{28.4}$ avoirdupois ounce
weight of 1 kilogram (kg) (1000 grams) = 2.20 pounds
1 liter (l) (1000 cubic centimeters) = 1.06 liquid quarts

Table 1-3 Trigonometry of the right and obtuse triangles

Right Triangle

Base is side b; altitude is side a; hypotenuse is side c; angle A is opposite to side a; angle B is opposite side b; angle C is opposite side c.

For the acute angles of a right triangle:

$$\text{sine} = \text{side opposite the angle/hypotenuse}$$
$$\text{cosine} = \text{side adjacent to the angle/hypotenuse}$$
$$\text{tangent} = \text{side opposite the angle/side adjacent to the angle}$$
$$\text{cotangent} = \text{side adjacent to the angle/side opposite the angle}$$
$$\text{secant} = \text{hypotenuse/side adjacent to the angle}$$
$$\text{cosecant} = \text{hypotenuse/side opposite the angle}$$

$$\sin A = a/c$$
$$\cos A = b/c$$
$$\tan A = a/b$$
$$\cot A = b/a$$
$$\sec A = c/b$$
$$\csc A = c/a$$

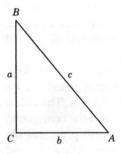

The same equations can be written for angle B using the appropriate sides.

The Pythagorean theorem gives the relationship between the sides a, b, and the hypotenuse c.

$$a^2 + b^2 = c^2$$

Oblique Triangles

$$\frac{a}{\sin A} = \frac{b}{\sin B} = \frac{c}{\sin C}, \text{ or the law of sines.}$$

$$a^2 = b^2 + c^2 - 2bc \cos A, \quad \text{or} \quad \cos A = \frac{b^2 + c^2 - a^2}{2bc},$$

which is the law of cosines.

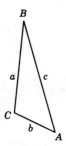

Table 1-3 (*continued*)

Elementary Geometry

The area of a circle

$$A = \pi r^2 = \frac{\pi}{4} d^2$$

where r = the radius and d = the diameter. The circumference of a circle $C = 2\pi r = \pi d$.

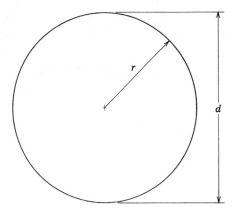

The area of a triangle $A = \frac{1}{2}bh$, where b = the base and h = the altitude.

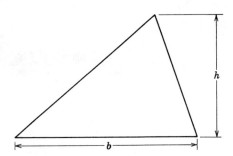

The area of a trapezoid $A = \frac{1}{2}(b_1 + b_2)h$, where b_1 and b_2 are the bases and h = the altitude.

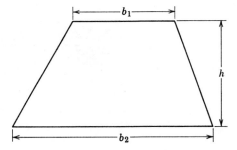

Table 1-3 (*continued*)

The volume of a sphere $V = \frac{4}{3}\pi r^3$, where r = the radius. The surface area of the sphere $A = 4\pi r^2$.

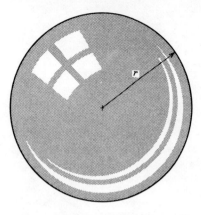

The volume of a cylinder $V = \pi r^2 h$, where r = radius of the base and h = the height.

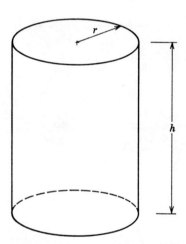

Elementary Algebra

The quadratic equation $ax^2 + bx + c = 0$ where a, b, and c are constants.

$$x = \frac{-b \pm \sqrt{b^2 - 4ac}}{2a}$$

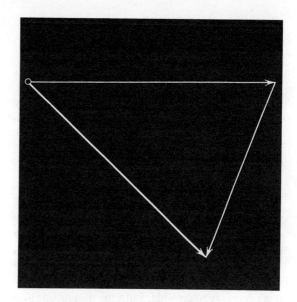

2.
The Composition
of Vectors

Many of the ideas and theories studied in physics involve the concept of *vector* quantities. In this chapter we shall consider the methods used in physics for the addition and subtraction of forces, displacements, and other vector quantities.

LEARNING GOALS

In this chapter the following mathematical skills are developed:

1. *The composition of vectors by the graphical (polygon) method.*
2. *The parallelogram method for the composition of two vectors.*
3. *The trigonometric method for determining the resultant of two vectors.*
4. *The resolution of a vector into components.*
5. *Finding the magnitude and direction of the resultant by the method of resolution into components.*

2-1 SCALAR ADDITION AND VECTOR ADDITION

In studying the rules of arithmetic in elementary school, we learned that

$$30 \text{ cents} + 40 \text{ cents} = 70 \text{ cents}$$

or $30 \text{ apples} + 40 \text{ apples} = 70 \text{ apples}$

These two addition problems represent an addition of scalar quantities. In doing scalar addition, we have to consider only the magnitudes (sizes) of the quantities that are added together. In physics and engineering work, however, we often have to add vector quantities by the process of vector addition. In performing vector additions, it is necessary to consider both the magnitude and the direction of the quantities, as in Fig. 2-1.

vector sum: 30 miles + 40 miles = 50 miles

But it is also possible to combine or add a 40-mile and a 30-mile displacement to obtain a different resultant. We may have

vector sum: 30 miles + 40 miles = 70 miles

as in Fig. 2-2, or

vector sum: 30 miles + 40 miles = 10 miles

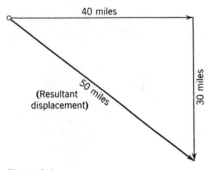

Figure 2-1

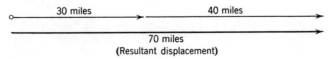

Figure 2-2

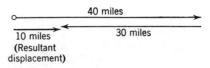

Figure 2-3

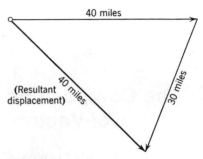

Figure 2-4

as in Fig. 2-3. A fourth possibility is

vector sum: 30 miles + 40 miles = 40 miles

as in Fig. 2-4.

In adding a 40-mile displacement and a 30-mile displacement, the magnitude of the resultant displacement may vary from a minimum of 10 miles up to a maximum of 70 miles. It is clear that the directions must be taken into account when we are combining displacement vectors.

Forces (pushes or pulls) are also vectors, and when forces are added together to find the resultant force, we must consider the direction of each force vector. We shall discover in our study of physics that velocities, accelerations, electric fields, magnetic fields, and many other important quantities are vectors.

But we shall also discover that quantities such as temperature, volume, density, mass, etc., do not have any direction. Such quantities which have only a size (magnitude) are called *scalars*. *Vectors*, however, are quantities that possess both magnitude and direction.

Vectors may be added together by the process of vector addition to find the resultant. The *resultant* is a single vector that will produce the same effect as two or more other vectors. The process of finding the resultant is described as the *composition of vectors*.

2-2 COMPOSITION OF VECTORS BY THE GRAPHICAL (POLYGON) METHOD

In adding vectors by the graphical method, we indicate the magnitude and direction of each vector by an arrow. This arrow points in the direction of the vector; the length of the arrow represents (to scale) the magnitude of the vector.

Example 2-1   A ship travels 100 miles due north on the first day of a voyage. It travels 80 miles to the northeast on the second day, and 130 miles due east on the third day. What is the resultant displacement for the three days?

Solution.   A suitable scale might be 20 miles = 1 cm (see Fig. 2-5). Then

100 miles = 100 ~~miles~~ × (1 cm/20 ~~miles~~) = 5 cm

  80 miles =   80 miles × (1 cm/20 miles) = 4 cm

130 miles = 130 miles × (1 cm/20 miles) = 6.5 cm

By measuring with a ruler, we find from the scale diagram that the arrow for the resultant is 12.2 cm in length. Therefore, the magnitude is 12.2 cm × (20 miles/1 cm) = 244 miles. The direction is found with a protractor, and the angle θ is measured and found to be 50°. The resultant displacement is therefore 244 miles in a direction 50° east of north.   ◀

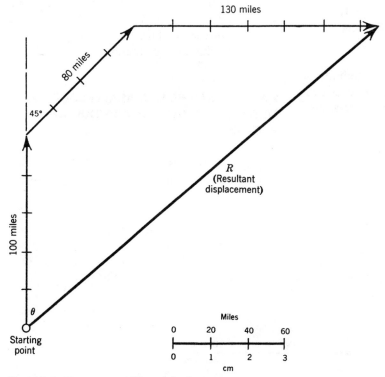

Figure 2-5   The vector addition of displacements.

In Example 2-1 the ship would have had the same displacement if it had sailed 130 miles due east on the first day, and then 80 miles northeast on the second day, and 100 miles north on the third day. The order in which vectors are added does not change the resultant in any way.

The polygon graphical method may be used to find the resultant of all kinds of vectors. In the following problem we use this method to find the resultant of force vectors.

Example 2-2   A 60-lb pull and a 40-lb pull are applied to an object. The angle between these two forces is 120°. Find the resultant force.

Solution.   Using a scale of 1 cm = 10 lb, we find that

$$60 \text{ lb} \times (1 \text{ cm}/10 \text{ lb}) = 6 \text{ cm}$$

and $40 \text{ lb} \times (1 \text{ cm}/10 \text{ lb}) = 4 \text{ cm}$

From Fig. 2-6 we find that $R = 5.3$ cm = 5.3 cm × (10 lb/1 cm) = 53 lb. Moreover, by using a protractor, we find that $\theta = 41°$. The resultant force is therefore a 53-lb force at an angle of 41° with the 60-lb force.   ◄

In using the graphical polygon method for adding vectors, a ruler must be employed to measure lengths and a protractor to measure angles. The steps to be followed are

1. Choose a suitable scale and determine according to this scale the length of the arrow that will represent each vector.

2. Choose an origin or starting point and draw the arrow representing the first vector. This arrow must be of the proper length and point in the appropriate direction.

3. The arrow of the second vector is drawn so that its tail is joined to the head of the first vector. The arrow of the second vector must·also be drawn to scale and must point in the correct direction.

4. If there is a third vector to be added, its arrow is joined to the head of the second vector. This process of joining tail to head is continued until we have drawn to scale all the vectors to be added.

5. The arrow representing the resultant is drawn with its tail at the origin (starting point), and its head is joined to the head of the last vector. The arrow of the resultant indicates the magnitude and direction of the resultant.

If only two vectors are added by this method, the resultant forms the third side of a triangle (a three-sided polygon). We can easily extend this process, however, and use the polygon method to find the resultant of any number of vectors.

2-3 THE GRAPHICAL (PARALLELOGRAM) METHOD FOR THE COMPOSITION OF VECTORS

If only two vectors are being added, it is often convenient to determine the resultant by a graphi-

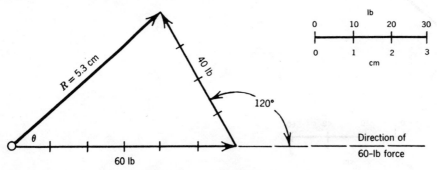

Figure 2-6   The vector addition of forces; polygon method.

cal scheme known as the parallelogram method. The steps in this construction are

1. The two vectors to be added are drawn to scale with their tails meeting at a common point. The two arrows then form two adjoining sides of a parallelogram (see Fig. 2-7).

2. The parallelogram is then completed by drawing or constructing the other two sides.

3. The resultant is represented by the diagonal of the parallelogram, which is included between the two vector arrows.

These three steps in the parallelogram method are illustrated in the following example.

Example 2-3   Solve the problem of Example 2-2 by the parallelogram graphical method.

The 60-lb force and the 40-lb force are drawn to scale in the proper direction. The parallelogram in Fig. 2-7 is completed, and the resultant R is the diagonal. This method gives the same resultant as the polygon method, because the diagonal of the parallelogram divides it into two congruent triangles, and each of these triangles is identical with the triangle of Example 2-2.

2-4 THE TRIGONOMETRIC METHOD

In the trigonometric method the magnitude and direction of the resultant is determined by applying the laws of trigonometry. This method is preferable to the graphical methods of Sections 2-2 and 2-3, since calculations will give a more accurate result than the scale diagrams of the graphical methods.

Example 2-4   Solve the problem of Example 2-2 by the trigonometric method.

Solution.   Using the law of cosines,

$$a = \sqrt{b^2 + c^2 - 2bc \cos A}$$

or

$$b = \sqrt{a^2 + c^2 - 2ac \cos B}$$

or

$$c = \sqrt{a^2 + b^2 - 2ab \cos C}$$

and the table of trigonometric functions (p. 00) we find from Fig. 2-8 that

$$R = \sqrt{(40)^2 + (60)^2 - 2(60)(40) \cos 60°} \text{ lb}$$
$$= 10 \text{ lb} \sqrt{(4)^2 + (6)^2 - 2(6)(4)(0.5)}$$
$$= 10 \text{ lb} \sqrt{16 + 36 - 24} = 10 \text{ lb} \sqrt{28}$$
$$= 52.9 \text{ lb}$$

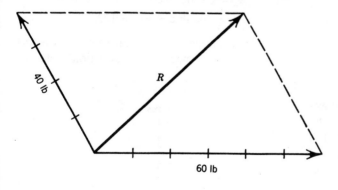

60 lb

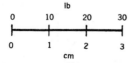

Figure 2-7   The vector addition of forces; parallelogram method.

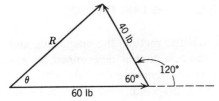

Figure 2-8 The trigonometric method.

The angle, or direction, of the resultant is found by using the law of sines:

$$\frac{a}{\sin A} = \frac{b}{\sin B} = \frac{c}{\sin C}$$

$$\frac{R}{\sin 60°} = \frac{40 \text{ lb}}{\sin \theta}$$

$$\sin \theta = \frac{40 \text{ lb}}{52.9 \text{ lb}} \sin 60° = 0.606$$

$$\theta = 40.9° \qquad \blacktriangleleft$$

The calculation of the resultant by the trigonometric method is very useful, but if we have three or more vectors, we usually employ the method of resolution into components. This method is described in Sections 2-5 and 2-6.

2-5 RESOLUTION OF A VECTOR INTO COMPONENTS

When a letter is used to represent a vector quantity the letter is in boldface type such as **A**. If the magnitude of a vector is indicated it is printed in ordinary type as A. In order to employ the method of resolution into components, we use the following simple definitions:

If A is any vector, we may drop perpendiculars from the head of A onto two perpendicular axes (see Fig. 2-9).

Then $$\frac{A_x}{A} = \cos \theta_x$$

or $$A_x = A \cos \theta_x$$

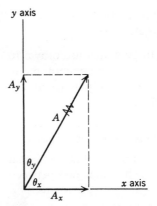

Figure 2-9 Resolution of a vector into components.

and $$\frac{A_y}{A} = \cos \theta_y$$

$$A_y = A \cos \theta_y$$

A_x is the x component and A_y is the y component of the vector A.

Example 2-5 A 200-lb, force acts upward and to the left at an angle of 60° with the horizontal. Resolve this force into an x and a y component.

Solution

$$A_x = -200 \text{ lb} \cos 60° = -100 \text{ lb}$$
$$A_y = 200 \text{ lb} \cos 30° = 173.2 \text{ lb}$$

In this example A_x was negative because it was pointing in the negative x direction (refer to Fig. 2-10). $\qquad \blacktriangleleft$

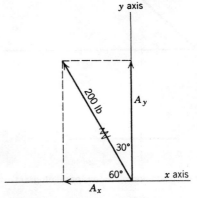

Figure 2-10 Resolution of a vector into components.

	Positive y axis	
upward and toward the left: y positive x negative		upward and toward the right: y positive x positive
Negative x axis	Origin	Positive x axis
	O	
downward and toward the left: y negative x negative		downward and toward the right: y negative x positive
	Negative y axis	

The positive x direction is usually taken as being toward the right, and vectors directed toward the left will, consequently, have a negative x component. The positive y direction is usually taken as being vertically upward, and vectors that point downward have a negative y component. The following chart indicates the signs that are given to the x and y components.

Although the x axis is usually taken to be horizontal and the y axis, vertical, it is sometimes convenient, as in the following problem, to choose axes that are inclined.

Example 2-6   A 100-lb block is resting upon an inclined plane that slopes at an angle of 36.9° with the horizontal. Resolve the 100-lb force of gravity

into a component parallel to the plane and one perpendicular to the plane (see Fig. 2-11).

Solution.   Taking the axes as in the diagram, we find that

x component $= 100$ lb $\sin 36.9° = 100$ lb $\cos 53.1°$
$\qquad\qquad = 60$ lb

y component $= 100$ lb $\cos 36.9° = 80$ lb   ◀

2-6 FINDING THE RESULTANT BY THE RESOLUTION METHOD

If several vectors are to be added, we can find the x component of the resultant by summing up all the x components. We can also find the y component of the resultant by summing all the y components. This method is illustrated in the following example

Example 2-7   Find the resultant in Example 2-2 by the method of resolution into components.

Solution.   The positive direction on the x axis is toward the right in Fig. 2-12. Since B_x is directed toward the left, this x component must be listed as a negative quantity in Table 2-1.   ◀

This result checks with the calculations in Section 2-4, where we employed the trigonometric method to solve the problem. The great advantage of the resolution method, however, is that it can be employed to find the resultant of three or more vectors, as in the next example.

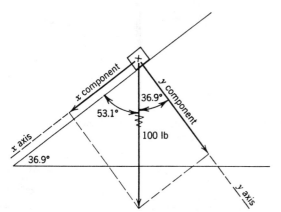

Figure 2-11   Resolution of a vector into components along inclined axes.

Table 2-1

	x	y
Components of $A = 60$ lb	$A_x = 60$ lb	$A_y = 0$
Components of $B = 40$ lb	$B_x = -40$ lb cos 60° $= -20$ lb	$B_y = 40$ lb cos 30° $= 34.6$ lb

$$R_x = 60 \text{ lb} - 20 \text{ lb} = 40 \text{ lb}$$
$$R_y = 0 + 34.6 \text{ lb} = 34.6 \text{ lb}$$
$$R = \sqrt{(34.6)^2 + (40)^2} \text{ lb}$$
$$= 10 \text{ lb} \sqrt{(3.46)^2 + (4)^2} = 10 \text{ lb} \sqrt{12 + 16}$$
$$= 52.9 \text{ lb}$$
$$\tan \theta_x = \frac{34.6 \text{ lb}}{40 \text{ lb}} \qquad \theta_x = 40.9°$$

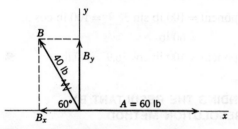

Figure 2-12 Vector sum by resolution in components.

Example 2-8 The following three forces act upon an object. (a) 1000 lb upward and to the right at an angle of 30° with the horizontal, (b) 1500 lb downward and to the right at an angle of 53° below the horizontal, (c) 2000 lb downward and to the left at an angle of 60° below the horizontal. Determine the resultant force.

Solution. The components of the three forces are calculated from Fig. 2-13. See Table 2-2.

The magnitude of the resultant may be calculated from

$$R = \sqrt{(766)^2 + (2432)^2} \text{ lb}$$
$$= 1000 \text{ lb} \sqrt{(0.766)^2 + (2.432)^2}$$
$$= 1000 \text{ lb} \sqrt{0.587 + 5.90}$$
$$= 1000 \text{ lb} \sqrt{6.49} = 2545 \text{ lb}$$

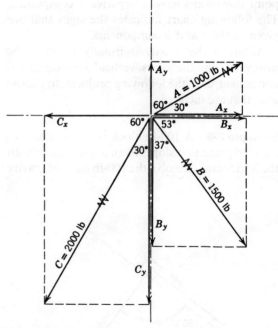

Figure 2-13 Finding the resultant by the method of resolution into components.

Table 2-2

Force	x	y
$A = 1000\ \text{lb}$	$A_x = 1000\ \text{lb cos } 30°$ $= 866\ \text{lb}$	$A_y = 1000\ \text{lb sin } 30°$ $= 500\ \text{lb}$
$B = 1500\ \text{lb}$	$B_x = 1500\ \text{lb cos } 53°$ $= 900\ \text{lb}$	$B_y = -1500\ \text{lb sin } 53°$ $= -1200\ \text{lb}$
$C = 2000\ \text{lb}$	$C_x = -2000\ \text{lb cos } 60°$ $= -1000\ \text{lb}$	$C_y = -2000\ \text{lb sin } 60°$ $= -1732\ \text{lb}$
	$R_x = 866 + 900\ \text{lb} - 1000\ \text{lb}$ $= 766\ \text{lb}$	$R_y = 500 - 1200\ \text{lb} - 1732\ \text{lb}$ $= -2432\ \text{lb}$

The direction of the resultant may be found by noting in Fig. 2-14 that

$$\tan \theta_x = \frac{2432}{766} \qquad \theta_x = 72.5°$$

The resultant of the three forces is 2545 lb acting to the right and downward at an angle of 72.5° below the horizontal. ◀

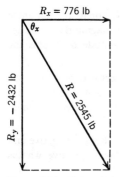

Figure 2-14   Determining the direction of the resultant.

It is very important to note that in determining R_x we had to take the algebraic sum of the x components. Here C_x was taken as -1000 lb in Example 2-8 because it was directed along the negative x axis; B_y and C_y were also negative because these components were in the negative y direction. We may summarize the procedure in the following set of formulas.

$$R_x = A_x + B_x + C_x + \cdots \text{(algebraic sum)}$$
$$R_y = A_y + B_y + C_y + \cdots \text{(algebraic sum)}$$
$$R = \sqrt{R_x{}^2 + R_y{}^2}; \tan \theta_x = \frac{R_y}{R_x}$$

or

$$\sin \theta_x = \frac{R_y}{R}$$

If a slide rule is being used to shorten the computation, it is easier to find $\tan \theta_x$ first and then to find the magnitude of R by using the sine or cosine of θ_x. In Example 2-8, for instance, after finding $\theta_x = 72.5°$ we might have let $\sin 72.5° = 2432$ lb/R and $R = 2432$ lb/sin 72.5° $= 2545$ lb.

2-7 CONCURRENT AND NONCONCURRENT FORCES

The procedures described in this chapter are very useful in finding the resultant of *concurrent forces*. Concurrent forces are forces whose lines of action intersect at a single point. *Noncurrent forces* are forces whose lines of action do not intersect at a common point.

When we have a number of nonconcurrent forces, the resultant force may be calculated by the method of resolution into components. But these forces may also produce a resultant *moment* or *torque* upon the object. To determine the torque (or turning effect) of the forces, we shall apply the *principle of moments*. We shall study some of the very important applications of this principle and of concurrent and noncurrent forces in Chapters 3 and 4.

SUMMARY

Vectors are quantities that possess both magnitude (or size) and direction.

Scalars are quantities that have only a size and do not have a direction.

The *resultant* is a single vector that will produce the same effect as two or more other vectors.

In adding vectors by the graphical (polygon) method, the magnitude and direction of each vector is indicated by an arrow drawn to scale in the correct direction. Starting at the origin, the arrows are drawn so that the tail of the second vector is joined to the head of the first vector, the tail of the third is joined to the head of the second, and so on. The arrow representing the resultant is drawn from the origin to the head of the last vector.

If two vectors are to be added, the graphical (parallelogram) method may be used. The two arrows are drawn to scale with their tails joined. The parallelogram is completed by drawing the other two sides, and the resultant is represented by the diagonal included between the two sides.

When two vectors are combined, the laws of trigonometry may be used to determine the resultant. The magnitude of the resultant may be calculated by using the law of cosines, and the direction of the resultant is found by using the law of sines as in Example 2-4.

If several vectors are to be added, we can find the resultant by resolving all the vectors into x and y components as in Sections 2-5 and 2-6. The x component of the resultant, R_x, is calculated by an algebraic sum of all the x components; the y component of the resultant, R_y, is calculated by an algebraic sum of all the y components. The direction of the resultant is found by the formula

$$\tan \theta_x = \frac{R_y}{R_x}$$

and the magnitude of the resultant by using $R = \sqrt{R_x{}^2 + R_y{}^2}$ or $\sin \theta_x = (R_y/R)$ as in Section 2-6.

QUESTIONS

1. What is the difference between vector quantities and scalar quantities?

2. Give several examples of vector quantities.

3. What does composition of vectors mean?

4. Explain how you would calculate the x component of a vector.

5. How is the y component of a vector calculated?

6. When we add three vectors by the polygon method, the polygon obtained is a figure with four sides. Explain why this is so.

7. When we add two vectors, the magnitude of the resultant depends upon the angle between the vectors.
   (a) At what angle between the vectors will the magnitude of the resultant be a maximum?
   (b) At what angle between the vectors will the magnitude of the resultant be a minimum?

8. Under what conditions will the sum of two vectors give a zero resultant?

9. When we add vector **A** and vector **B**, at what angle between the vectors will the magnitude of the resultant be equal to the magnitude of **A** plus the magnitude of **B**?

10. When we add vector **A** and vector **B**, at what angle between the vectors will the magnitude of the resultant be equal to the difference between the magnitudes of **A** and **B**?

11. If the x component of the resultant is negative and the y component of the resultant is positive, what is the direction of the resultant?

12. For a force that is directed downward and toward the right, what are:
    (a) The sign of the x component.
    (b) The sign of the y component?

13. If the resultant force is downward and toward the left, what are:
    (a) The sign of the x component.
    (b) The sign of the y component?

14. An acute angle is an angle between 0 and 90°. If a vector makes an acute angle with the x axis, what is the sign of the x component?

15. An obtuse angle is an angle between 90 and 180°. If a vector makes an obtuse angle with the y axis, what is the sign of the y component?

16. Suppose that a vector makes an angle of 90° with the x axis. What can we conclude about the x component of this vector?

17. A vector is perpendicular to the y axis. What is the y component?

18. A vector is parallel to the y axis. What is the x component?

19. Suppose that a vector of 100 lb is directed so that it makes an angle of 180° with the x axis. What are
    (a) the y component, and
    (b) the x component?

20. Suppose that a vector makes an angle of less than 45° with the x axis. Which component is greater in magnitude, the x component or the y component?

21. For a given vector suppose that the x and y components are both positive, but that the y component is greater than the x component. The angle between the x axis and the vector must be at least how many degrees?

22. If the x component of the resultant is zero and the y component is negative, what is the direction of the resultant?

PROBLEMS

(A)

1. A 300-lb pull toward the right and a 100-lb pull toward the left act upon an object. Determine the magnitude and direction of the resultant force.

2. Determine the magnitude and direction of the resultant of the following two displacements: 45 miles due north and 60 miles due south.

3. The following three forces act upon an object:
   (a) 90 lb toward the right.
   (b) 160 lb toward the left.
   (c) 50 lb toward the right.
   What is the magnitude and direction of the resultant force?

4. A force of 40 lb vertically upward and a force of 70 lb vertically downward act upon an object.
   (a) What is the magnitude and direction of the resultant force?

(b) What is the magnitude and direction of a force that will make the resultant force zero when it is applied to the object?

5. A small boat will travel 10 miles/hr in still water. Suppose that it travels on a river with a current of 6 miles/hr.
   (a) What is the resultant velocity when the boat heads upstream against the current?
   (b) What is the resultant velocity when the boat travels downstream with the current?

6. A sailboat has the following displacements on two successive days.
   (a) First day: 80 miles due north.
   (b) Second day: 140 miles due east.
   Determine the magnitude and direction of the resultant displacement by the graphical method.

7. A jet airplane has an air speed of 520 knots.
   (a) What is the ground speed of the airplane if it is flying with an 80-knot tailwind?
   (b) What is the ground speed if the airplane is flying into the 80-knot wind?

8. A 100-lb pull acts upward and to the right at an angle of 53° with the horizontal. Resolve this force into an x and a y component.

9. A 200-lb force acts downward and to the right at an angle of 60° with the horizontal. Resolve this force into an x and a y component.

10. (a) Calculate the x component and the y component for each of the two forces in Fig. 2-15.
    (b) What is the x component of the resultant force?
    (c) What is the y component of the resultant force?

Figure 2-15 Composition of two vectors.

11. A 300-lb pull toward the north and a 400-lb pull toward the east act together upon an object.
    (a) What is the magnitude and direction of the resultant?
    (b) What is the magnitude and direction of the force that must be applied to prevent the object from moving?

12. A 200-lb force has an x component of 80 lb.
  (a) What is the angle between the force and the x axis?
  (b) What is the y component of the force?

13. A downward force of 80 lb and a force of 60 lb toward the left act upon an object. What is the magnitude and direction of the resultant force?

14. The following three forces act upon an object:
  (a) 10 lb vertically upward.
  (b) 16 lb vertically downward.
  (c) 8 lb toward the right.
  What is the magnitude and direction of the resultant?

(B)

1. A 100-lb push toward the north and a 200-lb push toward the east act upon an object. Determine the magnitude and direction of the resultant force by the graphical method. Check by using the trigonometric method.

2. A ship that can travel 30 miles/hr in still water is headed due west. A 10-mile/hr current is flowing northeast. Exactly how fast and in what direction will the ship travel?

3. A force of 300 lb vertically and a force of 500 lb horizontally are combined. Determine the magnitude of the resultant and the angle between the resultant and the 500-lb force.

4. An airplane is headed due north at 300 miles/hr, when it encounters a west wind blowing at 60 miles/hr. Calculate the magnitude and direction of the resultant velocity for the airplane.

5. Two 40 lb forces act upon an object. If the resultant is also 40 lb, what is the angle between the two original forces?

6. The following two forces act upon a small object:
  (a) 100 lb horizontally to the left;
  (b) 200 lb to the right at an angle of 37° above the horizontal.
  Find the magnitude and direction of the resultant by the method of resolution into components.

7. A 200-lb pull to the north and a 160-lb pull 110° east of north act upon a fence post. If the fence post does not move, what is the magnitude and direction of the force exerted by the ground on the post?

8. An airplane has an air speed of 300 miles/hr in still air. A 50-mile/hr wind is blowing toward the west.

  (a) How many degrees east of north should the plane be headed if the resultant velocity is to be due north?
  (b) Calculate the magnitude of the resultant velocity.

9. A 5-lb force and an 8-lb force have a resultant of 10 lb.
  (a) What is the angle between the 5-lb force and the resultant?
  (b) What is the angle between the 5-lb and the 8-lb force?

10. A ferry boat can travel 1000 ft/min in still water. If the current in the river is 500 ft/min, at what angle to the river bank must the captain steer in order to head directly across the stream? If the river is 5000 ft wide, how long does it take the boat to travel directly across?

11. A 100-lb and a 60-lb force act upon an object. The angle between the two forces is 30°.
  (a) Determine the magnitude and direction of the resultant by the graphical method.
  (b) Determine the magnitude and direction by trigonometric calculations.

12. The following three forces are acting upon an object:
  (a) 1000 lb horizontally to the right,
  (b) 1500 lb downward and to the left at an angle of 30° below the horizontal, and
  (c) 500 lb upward and to the right at an angle of 53° above the horizontal.
  Find the resultant by the graphical (polygon) method.

13. Find the resultant in Problem 11 by the method of resolution into components.

14. Find the resultant of these three displacements by the method of resolution into components:
  (a) 200 miles 30° east of north.
  (b) 100 miles due west.
  (c) 80 miles due south.

15. An ocean current of 5 miles/hr is flowing in a direction 53° east of north. The captain of a boat that can travel 25 miles/hr in still water wishes to sail due south to reach a harbor 100 miles away.
  (a) In what direction should he sail?
  (b) How long will it take him to reach the harbor?

16. Explorers leave their base and travel the following distances on three successive days.
  (a) First day: 60 miles 50° east of north.
  (b) Second day: 40 miles 90° east of north (due east).
  (c) Third day: 50 miles 120° east of north.
  How far and in what direction must the explorers

travel in order to return to their base on the fourth day?

17. The following four concurrent forces act upon an object:

(a) 100 lb upward and to the right at an angle of 60° above the horizontal.

(b) 200 lb upward and to the left at an angle of 37° above the horizontal.

(c) 300 lb vertically downward.

(d) 150 lb downward and to the right at an angle of 40° below the horizontal.

Determine the horizontal component, the vertical component, and the direction of the resultant.

18. A 1000-lb force acts vertically downward on an object. A second force T acts upward and to the right at an angle of 37° with the horizontal. If the resultant of these two forces is known to be a horizontal force, calculate the magnitude of T and the magnitude of the resultant.

19. In the metric system forces are frequently measured in units called newtons (N). (The newton will be defined in Chapter 9.) The following two forces are applied to an object: $A = 100$ newtons horizontally to the right, and $B = 80$ newtons vertically downward.

(a) What is the magnitude and direction of the resultant of these two forces?

(b) What is the magnitude and direction of a single force which, when added to A and B, will make the resultant zero?

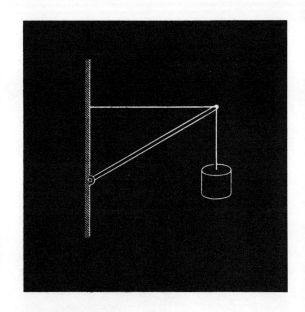

3.
Concurrent
Forces in
Equilibrium

Many of the important problems in physics and engineering involve an understanding of forces in equilibrium. In this chapter we shall learn the first condition for equilibrium and study the application of this rule to the solution of problems involving concurrent forces.

LEARNING GOALS

The reader should be alerted to the following objectives while studying this chapter:

1. *How to draw a free body diagram of concurrent forces.*
2. *Finding the equilibrant of two concurrent forces.*
3. *Using the trigonometric method to calculate an unknown tension or compression.*
4. *Using the resolution method to find unknown forces.*

3-1 STATICS AND EQUILIBRIUM

In studying physics or engineering, we frequently find that the various forces acting upon an object balance or cancel each other so that no motion results. The forces acting on a beam in a building or on a member of a bridge structure must balance each other so that the objects remain at rest. The study of the forces acting upon objects at rest and the conditions under which these objects will remain at rest comprises the very important branch of mechanics called *statics*.

An object at rest will remain at rest if the resultant of all the forces acting upon the object is zero. However, if the object is in motion and the resultant of all the forces is zero, the object will continue to move with its velocity remaining constant in magnitude and direction. Whenever the resultant force system acting upon an object is zero, the object is in *equilibrium*. An object in equilibrium may be at rest or it may be moving with a constant velocity.

3-2 THE EQUILIBRIUM OF CONCURRENT FORCES

If several forces are acting upon an object, it is often possible to produce equilibrium by having an additional force act upon the object. A single force that will just cancel the effects of several forces is called the *equilibrant*.

Example 3-1 A 300-lb force and a 400-lb force act at right angles to each other upon an object. What force must be applied to the object to produce equilibrium?

Solution. The 300-lb and the 400-lb force have the resultant $R = 500$ lb as in Fig. 3-1. The force E is the equilibrant. Here $E = 500$ lb and is opposite in direction to the resultant. ◀

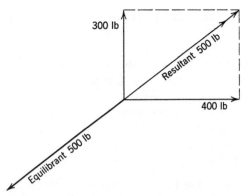

Figure 3-1 The equilibrant is equal to the resultant but opposite in direction.

From the preceding simple problem it is apparent that the equilibrant of several concurrent forces will be a single force which is equal in magnitude to the resultant but opposite in direction.

3-3 THE FIRST CONDITION OF EQUILIBRIUM

If an object is in equilibrium, the resultant of all the forces acting upon the object must be zero. In order for the resultant force to be zero, all the components of the resultant must be zero. If we resolve all the forces into components, the algebraic sum of all the x components must add up to zero. This condition for equilibrium is expressed in mathematical form by writing $\sum F_x = 0$. The symbol $\sum$ is the Greek capital letter sigma; in mathematical formulas it means sum or summation. Since the algebraic sum of the y components and that of the z components must also add up to zero, we have the condition that

$$\sum F_x = 0; \quad \sum F_y = 0; \quad \sum F_z = 0 \quad (3\text{-}1)$$

These three equations are called the first condition for equilibrium. They may be used to calculate unknown forces, as in the following example.

Example 3-2 A 100-lb object is placed upon a smooth inclined plane that is inclined at an angle of 30° with the horizontal. A force parallel to the plane pushes the object up the plane with constant velocity. (a) What force is needed to push the block up the incline? (b) What force is exerted on the block by the plane?

Solution. The force P pushes the block up the plane (see Fig. 3-2). The weight W is the 100-lb force that acts vertically downward. The plane exerts the force N upon the block. This force N is called the normal force because it acts at right angles to the plane (normal means perpendicular). We resolve the 100-lb force into an x and a y component. (These axes are taken parallel and perpendicular to the plane, respectively.) From the laws of equilibrium

$$\sum F_x = 0 \qquad P - 100 \text{ lb sin } 30° = 0$$
$$P = 50 \text{ lb} = \text{the push required}$$
$$\sum F_y = 0 \qquad N - 100 \cos 30° = 0$$
$$N = 86.6 \text{ lb} = \text{the force exerted}$$

on the block by the plane.   ◄

3-4 THE TRIANGLE METHOD IN EQUILIBRIUM PROBLEMS

If several forces are in equilibrium, these forces may be added graphically by the polygon method. Because the resultant force is zero, the arrowhead of the last force vector must coincide with the origin. In other words, concurrent forces in equilibrium must form a closed polygon. If we have three concurrent forces in equilibrium, the polygon is a triangle, and the unknown forces may be found by applying the methods of trigonometry.

The resolution method is used in equilibrium problems if there are more than three forces acting upon an object. Either the resolution method or the triangle method may be employed, however, if there are only three forces. In Fig. 3-2b, P, N, and $W = 100$ lb, the three forces acting upon the the block in Example 3-2 are indicated as forming a closed polygon. The triangle indicates that

$$P = 100 \text{ lb sin } 30° = 50 \text{ lb}$$
$$N = 100 \text{ lb cos } 30° = 86.6 \text{ lb}$$

In order to compare the trigonometric and resolution methods, the following problem is solved by both methods.

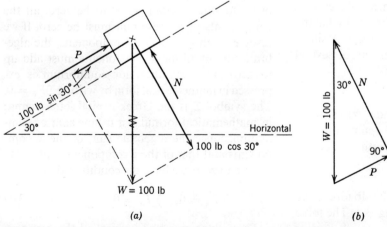

(a)   (b)

Figure 3-2 Block on an inclined plane. (a) Resolution method. (b) Triangle method.

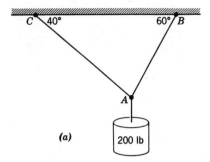

(a)

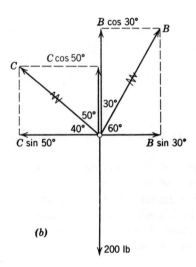

(b)

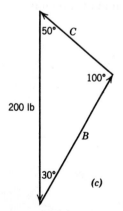

Figure 3-3 (a) and (b) Calculating the unknown tensions.
(c) Three concurrent forces in equilibrium form a closed
triangle.

Example 3-3 A 200-lb weight is supported by
the two cables as in Fig. 3-3a. Calculate the
tension in the two cables.

Solution. RESOLUTION METHOD. The knot at
A may be regarded as an object upon which three
concurrent forces are acting. The diagram in Fig.
3-3b is a free body diagram of the knot. B and C
are the unknown tensions that are resolved into
components. From the equations of equilibrium

$$\sum F_x = 0, \qquad B \sin 30° - C \sin 50° = 0$$

$$B = C \frac{\sin 50°}{\sin 30°} = 1.531C$$

$$\sum F_y = 0, \qquad B \cos 30° + C \cos 50° - 200 \text{ lb} = 0$$
$$(1.531C)(0.866) + (0.643)C = 200 \text{ lb}$$
$$C(1.968) = 200 \text{ lb}$$
$$C = 101.6 \text{ lb}$$

and $\qquad B = 1.531(101.6 \text{ lb}) = 155.6 \text{ lb}$

TRIGONOMETRIC METHOD. We draw B upward
and to the right at an angle of 30° with the vertical.
C is drawn upward and to the left at an angle of
50° with the vertical. The 200-lb weight is indicated as a force acting vertically downward. See
Fig. 3-3c.
From the law of sines,

$$\frac{C}{\sin 30°} = \frac{200 \text{ lb}}{\sin 100°}$$

Since $\qquad \sin 100° = \sin 80°$

$$C = \frac{200 \text{ lb} \sin 30°}{\sin 80°} = 101.6 \text{ lb}$$

Also $\qquad \dfrac{B}{\sin 50°} = \dfrac{200 \text{ lb}}{\sin 100°}$

$$B = \frac{200 \text{ lb} \sin 50°}{\sin 80°} = 155.6 \text{ lb} \blacktriangleleft$$

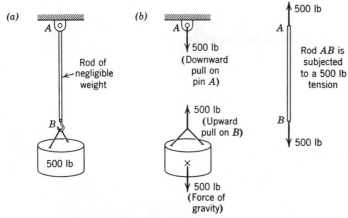

Figure 3-4   An object subjected to a tension will pull inward on its supports.

3-5 TENSION AND COMPRESSION

The rod AB in Fig. 3-4a is subjected to a 500-lb tension. An object under tension tends to be slightly elongated or stretched. As the free body diagram in Fig. 3-4b indicates, an object that is under tension will pull inward on its supports.

In Fig. 3-5 a 500-lb compression is applied to rod AB. An object under compression is subjected to a squeeze that tends to decrease the length of the object. The free body diagram in Fig. 3-5b indi-

cates that an object under compression will push outward on its supports.

In the design of static structures it is necessary to determine the compression or tension that will be produced in different members or parts of the structure. As the following example indicates, this can often be accomplished by applying the first condition for equilibrium.

Example 3-4   Determine the tension in the cable and the compression in the boom in Fig. 3-6.

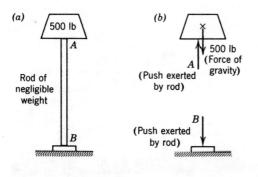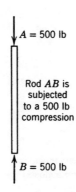

Figure 3-5   A rod subjected to a compression will push outward on its supports.

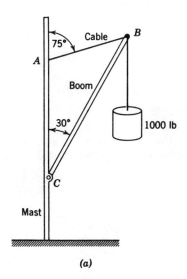

(a)

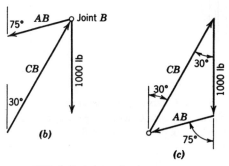

(b)

(c)

AB is indicated as a tension.
CB is indicated as a compression.

Figure 3-6 The boom and mast.

Solution. If the weights of *AB* and *CB* are ignored, we have three concurrent forces acting upon joint *B*. Figure 3-6*b* indicates a free body diagram of joint *B*, where *CB* is under compression so that it pushes on *B* and *AB* is under tension so that it pulls on *B*. Although these two forces may be found by the method of resolution, it is somewhat easier to employ the triangle method of Fig. 3-6*c*.

The third angle of the triangle (opposite the 1000-lb force) must be 45°, since the three angles

must add up to 180°. The law of sines gives

$$\frac{CB}{\sin 105°} = \frac{1000 \text{ lb}}{\sin 45°}$$

$$CB = \frac{1000 \text{ lb } \sin 105°}{\sin 45°}$$

$$= 1000 \text{ lb } \frac{\sin 75°}{\sin 45°} = 1366 \text{ lb}$$

Thus *CB* is subjected to a 1366-lb compression. Also,

$$\frac{AB}{\sin 30°} = \frac{1000 \text{ lb}}{\sin 45°}$$

$$AB = 707 \text{ lb}$$

Hence *AB* is subjected to a 707-lb tension. ◀

SUMMARY

Whenever the resultant force system acting upon an object is zero, the object is in *equilibrium.* An object in equilibrium may be at rest or it may move with a constant velocity.

A single force that will just cancel the effects of several forces is called the *equilibrant.*

In solving equilibrium problems, a free body diagram is drawn indicating all the forces acting upon the object. The forces may then be resolved into components and the sum of the components set equal to zero as in Eq. 3-1. This set of equations is called the first condition of equilibrium.

If an object is in equilibrium and there are only three forces acting upon the object, these forces may be added by the polygon method to form a triangle. The laws of trigonometry may then be used to calculate the unknown forces or angles. This triangle method is an alternate method for solving equilibrium problems with three forces.

A *tension* or a pull tends to stretch an object. An object under tension pulls on its supports.

A *compression* is a force that tends to decrease the length of an object. An object subjected to a compression pushes outward on its supports.

QUESTIONS

1. What is the difference between the resultant and the equilibrant of a force system?

2. Several different forces are acting upon an object that is at rest. What can we conclude about the resultant of these forces?

3. Several different forces act upon an object that is moving. However, we notice that the object is moving with constant velocity. What can we conclude about the resultant of these forces?

4. What is meant by the statement that an object is in equilibrium?

5. What is a free body diagram?

6. What is the difference between the resolution method and the trigonometric method in statics problems?

7. Why is the resolution method applicable to more statics problems than the triangle method?

8. What is the first condition of equilibrium? How is this condition stated in mathematical form?

9. Suppose that an object is acted upon by three forces, *A*, *B*, and *C*, which are in equilibrium. If *A* is acting vertically downward and *B* is directed horizontally toward the right, what must be the direction of *C*?

10. Suppose that an object weighing 100 lb is supported by two ropes and the tension in each rope is 100 lb. Draw a free body diagram of this object and indicate the angle at which the ropes must pull.

11. Explain the difference between tension and compression.

12. Suppose that a heavy object is supported by a rigid rod under compression. How should the arrow representing the compressive force in the rod be drawn in the free body diagram for the object?

13. A diesel engine that weighs 1500 lb is resting on a four-wheeled dolly that weighs 200 lb. Draw a free body diagram of the diesel engine indicating the two forces that act upon it. Then draw a free body diagram of the dolly indicating the upward forces at the four wheels and the two downward forces.

14. A heavy object is supported by a cable under tension and a rod under compression. In drawing the free body diagram of the object, the force of gravity, the tension, and the compression must be indicated. How does the arrow representing the tension force differ from the arrow representing the compression force?

PROBLEMS

(A)

1. The following two forces act upon an object: 80 lb vertically downward; and 60-lb toward the left. What is the magnitude and direction of the equilibrant?

2. The block in Fig. 3-7 weighs 90-lb and the push *P* = 50 lb. The block is maintained in equilibrium by the table. What vertical force is exerted by the table?

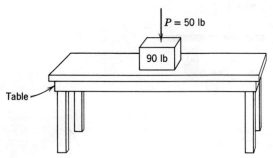

Figure 3-7   Forces in equilibrium.

3. The pull *T* = 100 lb produces a tension of 100 lb in the cable in Fig. 3-8. Calculate the vertical force exerted on the 300-lb block by the floor.

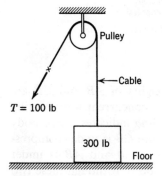

Figure 3-8   Calculate the force exerted by the floor.

4. The pull *P* in Fig. 3-9 maintains the suspended 100-lb weight in equilibrium. Draw a free body diagram of the pulley at *A* and calculate the tension in the rope.

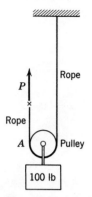

Figure 3-9   Calculate the tension in the rope.

5. (a) Draw a free body diagram of the 100-lb platform in Fig. 3-10.
   (b) What is the compression in the spring?
   (c) What force is exerted by the platform on the man?

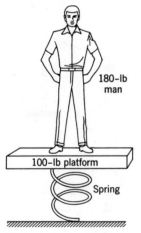

Figure 3-10   Forces in equilibrium.

6. The force $F = 100$ lb acts upon the 200-lb block in Fig. 3-11. The rough floor keeps the block in equilibrium.
   (a) Resolve the 100-lb force into horizontal and vertical components.
   (b) Calculate the vertical and horizontal force exerted on the block by the rough floor.

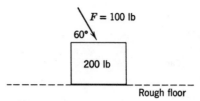

Figure 3-11   The floor exerts a vertical force and a horizontal force.

7. Calculate the tension in the cable in Fig. 3-12.

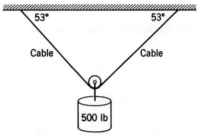

Figure 3-12   Determine the tension in the cable.

8. The force $F = 100$ lb is applied at joint B. Calculate the compression in the two rods in Fig. 3-13.

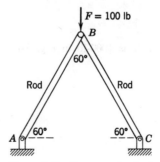

Figure 3-13   Compression in the rods.

9. (a) Draw a free body diagram of joint B in Fig. 3-14. Indicate the tension in the cable, the compression in CB, and the tension in AB in the free body diagram.
   (b) Calculate the compression in CB and the tension in AB.

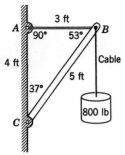

Figure 3-14 Determining the tension and the compression.

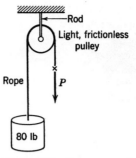

Figure 3-17 Forces on a simple pulley.

10. Determine the tension in ropes A, B, and C in Fig. 3-15.

12. (a) What is the force P in Fig. 3-17 required to support the 80-lb weight?
   (b) Draw a free body diagram of the pulley and calculate the tension in the rod.

13. The acrobat in Fig. 3-18 weighs 150 lb, and he is balanced on the taut wire. Calculate the tension in the wire.

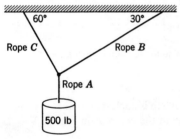

Figure 3-15 Calculate the tensions.

11. The tension in the cable is 1000-lb in Fig. 3-16. Calculate the weight W.

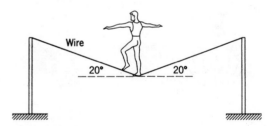

Figure 3-18 Tension in the wire.

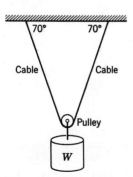

Figure 3-16 Determining the unknown weight.

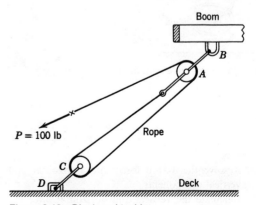

Figure 3-19 Block and tackle.

14. Fig. 3-19 indicates the block and tackle arrangement used on a small sailboat. The force $P = 100$ lb is exerted on the rope by a member of the crew.
  (a) Draw a free body diagram of the pulley at A and calculate the tension in AB.
  (b) Draw a free body diagram of the pulley at C and calculate the tension in CD.

(B)

1. Ropes B and C are of the same length. Determine the tension in ropes A, B, and C of Fig. 3-20.

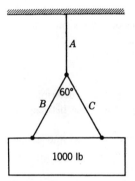

Figure 3-20   Determining the unknown tensions.

2. If the 100-lb weight is suspended as in Fig. 3-21, determine the tension in each of the three chains.

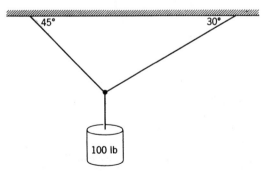

Figure 3-21   Determining the unknown tensions.

3. Ignore the weight of the horizontal strut and calculate the tension in the cable of Fig. 3-22.

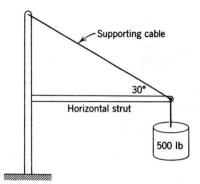

Figure 3-22   Calculate the tension in the supporting cable.

4. An elastic cord is pulled taut between two fixed supports 4 ft apart. A 30-lb force is applied at right angles to the cord at the center of the cord. The center is pulled 1 ft to one side. Calculate the tension in the cord.

5. Push $P = 40$ lb keeps a 100-lb block stationary on the inclined plane of Fig. 3-23. Calculate the force exerted on the block by the inclined plane. (This force has an x component and a y component.)

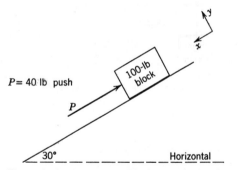

Figure 3-23   A stationary block on an inclined plane.

6. Find the equilibrant for the following two forces:
  (a) 200 lb upward and to the right at an angle of 30° above the horizontal.
  (b) 100 lb downward and to the left at an angle of 53° below the horizontal.

7. Determine the maximum weight if the maximum allowable compression in the strut of Fig. 3-24 is 500 lb.

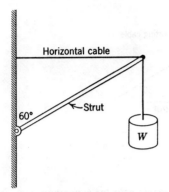

Figure 3-24 Compression in the strut.

8. In Fig. 3-25 A is a horizontal rope. Calculate the tension in A and B.

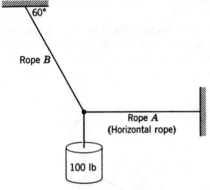

Figure 3-25 Calculating the tension in the supporting ropes.

9. Determine the compression in each of the rods of Fig. 3-26.

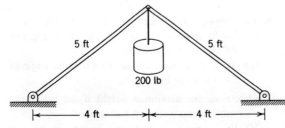

Figure 3-26 Compression in the supporting rods.

10. In Fig. 3-27 A weighs 10 lb. Calculate the tension in a, b, and c, and the weight of B.

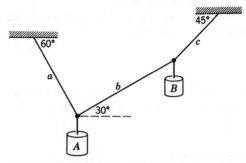

Figure 3-27 Calculating the tension in the supporting ropes.

11. (a) Use the laws of trigonometry to determine the length of the cable and the angle between the cable and the wall in Fig. 3-28.

(b) Ignore the weight of the 10-ft rod and assume $W = 100$ lb.

Determine the tension in the cable.

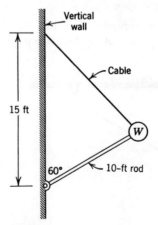

Figure 3-28 Tension in the supporting cable.

12. Draw a free body diagram of joint C (Fig. 3-29) and calculate the internal forces in AC and BC. State whether these forces are tensions or compressions.

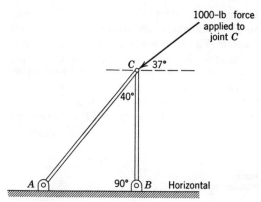

Figure 3-29 Internal forces in supporting members.

13. If $W = 1000$ lb, determine the compression in BC and the tension in AC (see Fig. 3-30).

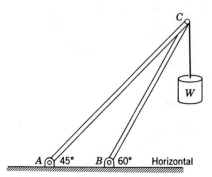

Figure 3-30 Compression and tension in rigid members.

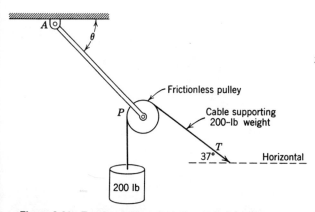

Figure 3-31 Tension in the supporting rod of a pulley.

14. (a) What is the tension T if the pulley in Fig. 3-31 is of negligible weight and is frictionless?
(b) Determine the angle θ and the tension in rod AP (assume that the rod is free to pivot about point A).

15. (a) Calculate the tension T if the pulley at P in Fig. 3-32 is light and frictionless.
(b) Determine θ and the tension in PA if the rod PA is free to pivot about point A.

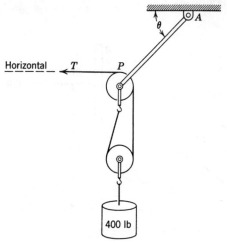

Figure 3-32 Tension in the supporting rod of a double pulley.

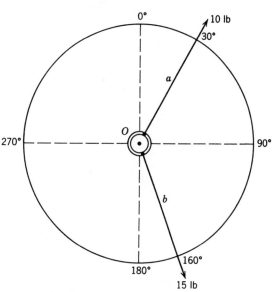

Figure 3-33 Forces on a force table.

16. The diagram (Fig. 3-33) indicates a top view of a force table. Three cords *a*, *b*, and *c* are tied to the ring at *O*. The tension in *a* is 10 lb (at 30°) and the tension in *b* is 15 lb (at 160°). Determine the tension and the direction of rope *c* if the ring at *O* is in equilibrium.

17. The tension in the guy wire of Fig. 3-34 is adjusted so that no horizontal forces act on the vertical member *AB* at pin *A*. Calculate:
    (a) The tension in the guy wire.
    (b) The force upon *AB* at *A*.

18. In the metric system the mass of an object is measured in kilograms. The tension in wire *a* in the diagram (Fig. 3-35) is equal to the force of gravity acting upon the 10-kg object. In Chapter 9 we shall discover that the force of gravity is 9.8 newtons per kilogram of mass. Determine:
    (a) The tension (in newtons) in wire *a*.
    (b) The tension in *b* and angle θ if $P = 40$ newtons.

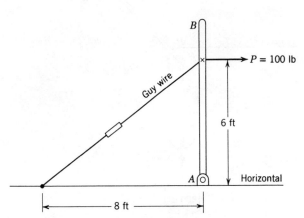

Figure 3-34 Forces on a vertical member.

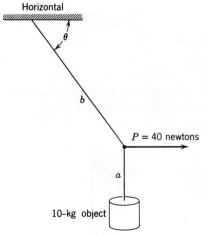

Figure 3-35 Calculating the unknown tensions.

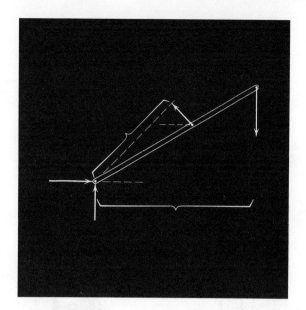

4.
Nonconcurrent Forces

In many cases the forces acting are not concurrent and we must consider the tendency of the forces to produce rotation. In studying nonconcurrent forces in this chapter, we shall learn about the second condition of equilibrium and how it is used in the solution of problems involving nonconcurrent forces.

LEARNING GOALS

In studying this chapter, the reader should keep in mind the following principal learning objectives:

1. *Drawing a free body diagram involving nonconcurrent forces.*
2. *Calculating the moment of forces.*
3. *Calculating the moment of a couple.*
4. *Locating the center of gravity.*
5. *Determining the resultant of nonconcurrent forces.*
6. *Using the first and second condition of equilibrium to find unknown forces.*

4-1 PARALLEL AND NONCURRENT FORCES

Very frequently in our study of physics, we shall discover that the forces acting upon an object do not act on or through a single point in the object. In Fig. 4-1, for example, the three forces acting on the lever form a system of *parallel forces*. It is clear that the lines of action of these parallel

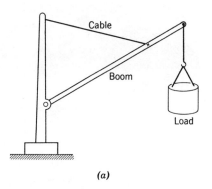

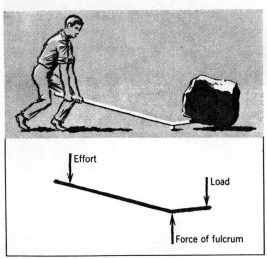

(a)

Figure 4-1 Parallel forces; the lever.

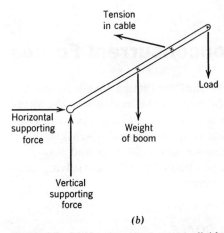

(b)

Figure 4-3 (a) Nonconcurrent, nonparallel forces; the crane. (b) Free body diagram of boom.

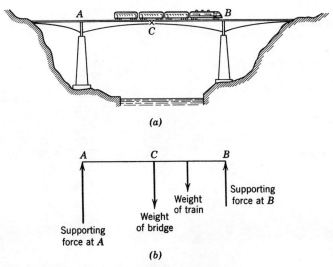

Figure 4-2 Parallel forces on a railroad bridge.

forces do not intersect at a point of concurrence. In Fig. 4-2 we find that the four forces acting upon the bridge also form a system of parallel forces.

Whenever the lines of action of forces do not intersect at a single point, the forces are said to be *noncurrent*. Nonconcurrent forces do not have to be parallel, and in Fig. 4-3 we have an important example of *noncurrent, nonparallel forces*.

The reader should carefully examine the illustrations and the freebody diagrams in Figs. 4-1, 4-2, and 4-3. In order to analyze force systems such as these, we must extend the theory of Chapter 3 and study the principles relating to nonconcurrent forces.

4-2 THE MOMENT OF A FORCE

The turning effect of a force depends upon the size (magnitude) of the force and the lever arm of the force. In Fig. 4-4 the turning effect of the 60-lb force applied to the wrench depends on the perpendicular distance from the line of action of the force to the axis. In Fig. 4-4a the moment of the force is

$$60 \text{ lb} \times 9 \text{ in.} = 540 \text{ lb-in.}$$

or

$$60 \text{ lb} \times \tfrac{3}{4} \text{ ft} = 45 \text{ lb-ft}$$

In Fig. 4-4b the turning effect of the force (the tendency to produce rotation) is greater because the lever arm is longer. The moment is

$$60 \text{ lb} \times 18 \text{ in.} = 1080 \text{ lb-in.}$$

or

$$60 \text{ lb} \times \tfrac{3}{2} \text{ ft} = 90 \text{ lb-ft}$$

In problems involving the moment of a force, we may use the following equation:

$$\text{moment} = \text{force} \times \text{lever arm}$$
$$M = F \times L \tag{4-1}$$

In using Eq. 4-1, it is necessary to recall that the lever arm is the perpendicular distance from the line of action of the force to the axis.

Example 4-1 Calculate the moment of the 100-lb force about point P in Fig. 4-5a.

Solution 1 (Fig. 4-5b)

$$L = 2 \text{ ft} \sin 60° = 2 \text{ ft} \cos 30° = 1.732 \text{ ft}$$
$$M = F \times L = 100 \text{ lb} (1.732 \text{ ft}) = 173.2 \text{ lb-ft}$$

Solution 2 (Fig. 4-5c). The 100-lb force is resolved into two components:

$$100 \text{ lb} \cos 60° = 50 \text{ lb parallel to } PA$$

and

$$100 \text{ lb} \sin 60° = 86.6 \text{ lb perpendicular to } PA$$

The lever arm of the 50-lb component is zero, and this component produces no moment about P. The lever arm is 2 ft for the 86.6-lb component. Therefore,

$$M = F \times L = 86.6 \text{ lb} (2 \text{ ft}) = 173.2 \text{ lb-ft} \blacktriangleleft$$

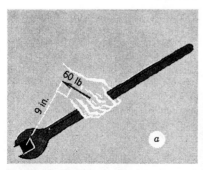
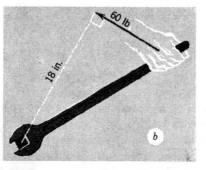

Figure 4-4   The moment of a force.

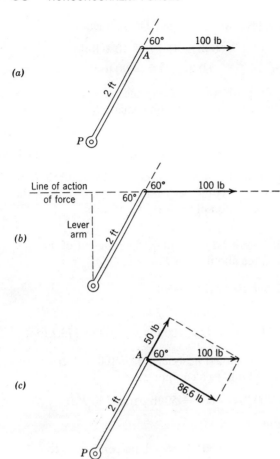

(a)

(b)

(c)

Figure 4-5 Calculating the moment of a force.

4-3 THE RESULTANT OF PARALLEL FORCES

In finding the resultant of two or more parallel forces, we must take into account the moments of the forces. The forces are added together to find the resultant force. The moments must also be added together to calculate the resultant moment, however.

In adding moments together, we must keep track of the sense or sign of the moments. Clockwise moments (moments tending to produce clockwise rotation) are indicated as negative moments. Counterclockwise moments (moments tending to produce counterclockwise rotation) are indicated as positive moments. The resultant

moment is then found as an algebraic sum of the positive and negative moments.

Example 4-2 In Fig. 4-6 a downward force of 40 lb is applied on the right and an upward force of 40 lb on the left. Calculate the resultant of these two forces.

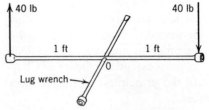

Figure 4-6 The resultant of a couple.

Solution. The resultant force is 40 lb − 40 lb = 0. However, the resultant moment is

$$-40 \text{ lb (1 ft)} - 40 \text{ lb (1 ft)} = -80 \text{ lb-ft}$$

which means 80 lb-ft in the clockwise sense. ◀

The two 40-lb forces in the example formed a *couple*. A couple consists of two equal forces that are opposite in direction but do not have the same line of action. We can calculate the moment of a couple if we multiply the magnitude of one of the forces by the perpendicular distance between the two lines of action.

Example 4-3 Find the resultant of the three forces in Fig. 4-7.

Solution. The resultant force is

$$F_R = \sum F_y = -50 \text{ lb} + 100 \text{ lb} - 60 \text{ lb} = -10 \text{ lb}$$

which means that 10 lb is acting in the negative or downward direction.

To obtain the resultant moment, we can take the moments about A.

$$M_R = \sum M_{(A)}$$
$$= -50 \text{ lb (4 ft)} + 100 \text{ lb (10 ft)} - 60 \text{ lb (12 ft)}$$
$$= -200 \text{ lb-ft} + 1000 \text{ lb-ft} - 720 \text{ lb-ft}$$
$$= 80 \text{ lb-ft}$$

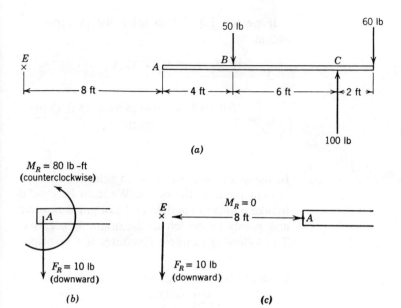

(a)

$M_R = 80$ lb-ft
(counterclockwise)

$F_R = 10$ lb
(downward)

(b)

$M_R = 0$

$F_R = 10$ lb
(downward)

(c)

Figure 4-7   (a) Calculating the resultant of parallel forces. (b) The resultant force system at A. (c) The resultant force system at E.

or 80 lb-ft in the counterclockwise direction. The resultant at A is therefore the force F_R and the moment M_R of Fig. 4-7b. ◀

In the preceding problem the resultant moment would have been different if we had calculated the moments about point B or some other reference axis. In fact, the resultant moment would have been zero if we had calculated the moments about point E, 8 ft to the left of A. See Fig. 4-7c. The resultant could, therefore, have been a 10-lb downward force (and no moment) acting at point E. In many problems with parallel forces we shall express the resultant as a single force through a point for which the resultant moment is zero. This procedure is especially important in problems involving the center of gravity.

4-4 THE CENTER OF GRAVITY

The forces of gravity acting upon the different parts of an object are all directed downward to-

ward the earth. The forces of gravity acting upon an object thus form a system of parallel forces. Using the methods of Section 4-3, we can find the resultant of these forces as a single force called the *weight* of the object. This force acts through a point which is the *center of gravity* of the object.

Example 4-4   Find the center of gravity of the object in Fig. 4-8a.

Solution 1.   The resultant moment about A (see Fig. 4-8b) is

$$M_R = 10 \text{ lb } (0 \text{ ft}) - 20 \text{ lb } (5 \text{ ft}) - 15 \text{ lb } (7 \text{ ft})$$
$$= -205 \text{ lb-ft}$$

The resultant force (total weight) is 45 lb. This force acts through the center of gravity, $\bar{X}$ feet to the right of A, and produces a moment of -205 lb-ft about A. Therefore,

$$-45 \text{ lb } \bar{X} = -205 \text{ lb-ft}$$
$$\bar{X} = 4.56 \text{ ft}$$

The center of gravity is 4.56 ft to the right of A.

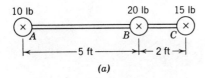

(a)

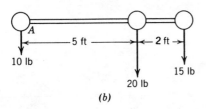

(b)

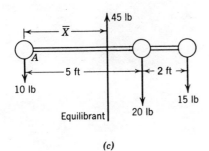

(c)

Figure 4-8 Finding the center of gravity.

Solution 2. In the equilibrium method, Fig. 4-8c, we see that an upward force of 45 lb applied through the center of gravity will support or balance the object. Since there is no unbalanced moment in Fig. 4-8c, we have

$$\sum M_{(A)} = 0$$
$$- 20 \text{ lb } (5 \text{ ft}) - 15 \text{ lb } (7 \text{ ft}) + 45 \text{ lb } (\bar{X}) = 0$$
$$45 \text{ lb } \bar{X} = 205 \text{ lb-ft}$$
$$\bar{X} = 4.56 \text{ ft} \quad \blacktriangleleft$$

It is obvious that solutions 1 and 2 both lead to exactly the same result in the preceding problem. Both methods of solution may be summarized in a single formula.

$$W\bar{X} = W_1 X_1 + W_2 X_2 + W_3 X_3 + \cdots \qquad (4\text{-}2)$$

where W is the total weight.

If we use Eq. 4-2 to solve the example, we obtain

$$\bar{X} = \frac{W_1 X_1 + W_2 X_2 + W_3 X_3}{W}$$

$$= \frac{10 \text{ lb } (0 \text{ ft}) + 20 \text{ lb } (5 \text{ ft}) + 15 \text{ lb } (7 \text{ ft})}{45 \text{ lb}}$$

$$= 4.56 \text{ ft}$$

In using Eq. 4-2, we have to select some convenient point as the origin. We must be careful to take points to the right of the origin as positive and points to the left of the origin as negative. The following example illustrates this principle.

Example 4-5 In Fig. 4-9 locate the center of gravity of the four weights.

Solution. Using point O as the origin, we have

$W_1 = 20 \text{ lb}$	$x_1 = -5 \text{ ft}$	$y_1 = 0$
$W_2 = 10 \text{ lb}$	$x_2 = 5 \text{ ft}$	$y_2 = 0$
$W_3 = 5 \text{ lb}$	$x_3 = 5 \text{ ft}$	$y_3 = 3 \text{ ft}$
$W_4 = 15 \text{ lb}$	$x_4 = 5 \text{ ft}$	$y_4 = -2 \text{ ft}$

Then

$$\bar{X} = \frac{W_1 X_1 + W_2 X_2 + W_3 X_3 + W_4 X_4}{W_1 + W_2 + W_3 + W_4}$$

$$= \frac{20 \text{ lb}(-5 \text{ ft}) + 10 \text{ lb}(5 \text{ ft}) + 5 \text{ lb}(5 \text{ ft}) + 15 \text{ lb}(5 \text{ ft})}{20 \text{ lb} + 10 \text{ lb} + 5 \text{ lb} + 15 \text{ lb}}$$

$$= \frac{-100 \text{ lb-ft} + 50 \text{ lb-ft} + 25 \text{ lb-ft} + 75 \text{ lb-ft}}{50 \text{ lb}}$$

$$= 1 \text{ ft}$$

To find $\bar{Y}$, we use the formula

$$\bar{Y} = \frac{W_1 Y_1 + W_2 Y_2 + W_3 Y_3 + W_4 Y_4}{W_1 + W_2 + W_3 + W_4}$$

$$= \frac{20 \text{ lb } (0) + 10 \text{ lb } (0) + 5 \text{ lb } (3 \text{ ft}) + 15 \text{ lb } (-2 \text{ ft})}{50 \text{ lb}}$$

$$= \frac{-15 \text{ lb-ft}}{50 \text{ lb}} = -\frac{3}{10} \text{ ft}$$

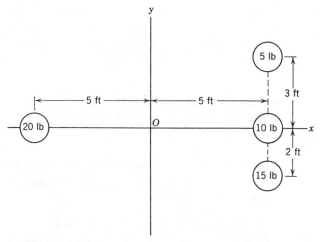

Figure 4-9 Locating the center of gravity of four weights.

The center of gravity is therefore 1 ft to the right of O and $\frac{3}{10}$ ft below O. ◀

In some center of gravity problems the weights are not given, but the diagram indicates the volume of the different parts of the object. If the object is of uniform density, we can assume that the weight of each part is proportional to the volume. This procedure is used in the following problem.

Example 4-6 A 10-ft uniform rod is bent into the L-shaped object indicated in Fig. 4-10. Where is the center of gravity of this object?

Solution. See Fig. 4-10*b*.

$$\bar{X} = \frac{W_1 X_1 + W_2 X_2}{W_1 + W_2} = \frac{6 \text{ units } (3 \text{ ft}) + 4 \text{ units } (0)}{6 \text{ units} + 4 \text{ units}}$$

$$= 1.8 \text{ ft}$$

$$\bar{Y} = \frac{W_1 Y_1 + W_2 Y_2}{W_1 + W_2} = \frac{6 \text{ units } (0) + 4 \text{ units } (2 \text{ ft})}{10 \text{ units}}$$

$$= 0.8 \text{ ft}$$

The center of gravity may be located on Fig. 4-10*b* as 1.8 ft to the right of O and 0.8 ft above O. ◀

In Example 4-6 and in many other cases the center of gravity is a point outside the object.

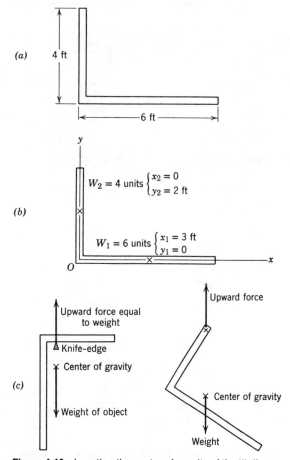

Figure 4-10 Locating the center of gravity of the "L."

Since the entire weight of the rod appears to be concentrated at the center of gravity, the rod in Example 4-6 may be suspended as in Fig. 4-10c. The center of gravity always comes to rest below the point of suspension.

When an object is balanced on a knife-edge, the center of gravity is located along the vertical line through the knife-edge. If the center of gravity is below the knife-edge, the object will be in *stable equilibrium*. However, if the center of gravity is above the knife-edge, the object will be very precariously balanced and the slightest displacement will cause it to fall. The object is then in *unstable equilibrium*. This situation is illustrated in Fig. 4-11a and b. An object is balanced in unstable equilibrium if a very small rotation or displacement will cause the center of gravity to fall. If a small displacement causes the center of gravity

to rise, the object is in stable equilibrium. In Fig. 4-11c, the cone is in neutral equilibrium. The object moves but the center of gravity remains at the same height above the table.

4-5 THE RESULTANT OF NONCURRENT FORCES; THE SECOND CONDITION FOR EQUILIBRIUM

The resultant of nonconcurrent forces may be found by resolving all the forces into components. We then obtain:

1. The x component of the resultant force is ΣF_x (the algebraic sum of all the x components). The y component of the resultant force is ΣF_y (the algebraic sum of all the y components).

2. The resultant moment depends on the reference point chosen as the origin (axis). Taking clockwise moments as negative, and counterclockwise moments as positive, we find the resultant moment as $\Sigma M_{(A)}$ (the algebraic sum of all the moments).

In equilibrium problems involving nonconcurrent forces, the resultant (both force and moment) must be zero. We therefore obtain two conditions for equilibrium.

First condition: $\Sigma F_x = 0$, and $\Sigma F_y = 0$; the resultant force is zero.

Second condition: The moment about any axis must be zero, $\Sigma M_{(A)} = 0$.

In the following example we shall employ both conditions of equilibrium.

Example 4-7 In Fig. 4-2 assume that the railway bridge weighs 100 tons and is 200 ft long. The train weighs 40 tons; its center of gravity is 40 ft from end B. Calculate the weight supported by A and B.

Solution

$$\Sigma F_y = 0$$
$$A - 100 \text{ tons} - 40 \text{ tons} + B = 0$$
$$A + B = 140 \text{ tons} \qquad (4\text{-}3)$$

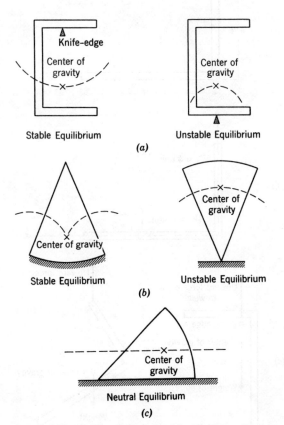

Stable Equilibrium   Unstable Equilibrium

(a)

Stable Equilibrium   Unstable Equilibrium

(b)

Neutral Equilibrium

(c)

Figure 4-11 Stable equilibrium and unstable equilibrium.

Taking moments about A, we have

$$\sum M_{(A)} = 0$$

$$-100 \text{ tons } (100 \text{ ft}) - 40 \text{ tons } (160 \text{ ft})$$
$$+ B(200 \text{ ft}) = 0$$

$$B = \tfrac{1}{2}(100 \text{ tons}) + \tfrac{4}{5}(40 \text{ tons})$$
$$= 82 \text{ tons}$$

We can find A by substituting in (4-3)

$$A + 82 \text{ tons} = 140 \text{ tons}$$

$$A = 58 \text{ tons}$$

We can also find A by taking moments about some other point. If we take moments about an axis through B,

$$\sum M_{(B)} = 0$$

$$-A(200 \text{ ft}) + 100 \text{ tons } (100 \text{ ft})$$
$$+ 40 \text{ tons } (40 \text{ ft}) = 0$$

$$A = \tfrac{1}{2}(100 \text{ tons}) + \tfrac{1}{5}(40 \text{ tons}) = 58 \text{ tons} \quad \blacktriangleleft$$

As a second example of an equilibrium problem involving nonconcurrent forces, we shall consider the boom and mast problem of Fig. 4-12.

Example 4-8 Ignore the weight of the boom in Fig. 4-12. Determine the tension in the cable BD, and the horizontal and vertical thrust on the boom at A.

Solution

$$\sum F_x = 0$$
$$H - T \cos 45° = 0$$
$$H = 0.707T \qquad (4\text{-}4)$$

$$\sum F_y = 0$$
$$T \sin 45° + V - 1000 \text{ lb} = 0$$
$$V = 1000 \text{ lb} - 0.707T \qquad (4\text{-}5)$$

$$\sum M_{(A)} = 0$$
$$-1000 \text{ lb } (20 \text{ ft } \cos 30°) + T \, 12 \text{ ft } \sin 75° = 0$$

$$T = \frac{17{,}320 \text{ lb-ft}}{12 \text{ ft } \sin 75°}$$

$$= 1492 \text{ lb}$$

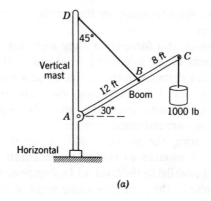

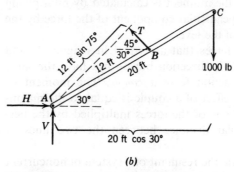

Figure 4-12 (*a*) The boom and the mast. (*b*) Free body diagram of a boom.

From Eq. 4-4,

$$H = 0.707(1492 \text{ lb})$$
$$= 1057 \text{ lb}$$

From Eq. 4-5,

$$V = 1000 \text{ lb} - 1057 \text{ lb}$$
$$= -57 \text{ lb}$$

The minus sign means that the vertical thrust on A is downward. $\quad \blacktriangleleft$

SUMMARY

When the lines of action of a system of forces do not intersect at a single point, the forces are *nonconcurrent*. In the study of nonconcurrent forces we must consider moments or the turning effects of the forces.

The *moment* of a force equals the force times the lever arm of the force. The *lever arm* is the perpendicular distance from the line of action of the force to the axis. Clockwise moments are indicated as negative quantities; counterclockwise moments are indicated as positive.

In calculating the moment of a force, it is sometimes convenient to resolve a force into a component parallel to the lever and a component perpendicular to the lever. The component of the force parallel to the lever produces no turning effects; the moment is calculated by multiplying the perpendicular component of the force by the length of the lever.

Two forces that are equal in magnitude and opposite in direction but do not have the same line of action form a *couple*. The moment or turning effect of a couple is equal to the magnitude of one of the forces multiplied by the perpendicular distance between the two lines of action.

To find the resultant of a system of noncurrent forces, the resultant force is found by the composition of all the forces; the resultant moment is found by the algebraic sum of all the moments taking into account the sign or the sense of each moment. In the English system of units the resultant force is calculated in pounds (lb) and the resultant moment in pound-feet (lb-ft). The resultant moment generally depends upon the reference point chosen as the origin (axis); and in many problems the resultant can be found as a force through a point for which the resultant moment is zero.

The forces of gravity acting upon different parts of any object form a system of parallel forces directed downward. The resultant of these forces is a single force called the weight of the object. This force acts through a point called the *center of gravity* which can be located by applying Eq. 4-2.

When an object is balanced on a knife-edge, the center of gravity is located on a vertical line through the knife-edge. If the center of gravity is below the knife edge, the object is in stable equilibrium. If the center of gravity is above the knife edge, the object is in unstable equilibrium.

In equilibrium problems involving noncurrent forces, the first condition of equilibrium ($\Sigma F_x = 0$ and $\Sigma F_y = 0$) is used together with the *second condition* of *equilibrium* to calculate the unknown forces. The second condition of equilibrium states that the resultant moment about any axis must by zero. This second condition is indicated by the formula $\Sigma M_{(A)} = 0$, where the sum involves the algebraic sum with counterclockwise moments as positive moments and clockwise moments as negative. The first and second condition of equilibrium may lead to simultaneous algebraic equations that can be solved for the unknown forces as in Examples 4-7 and 4-8.

QUESTIONS

1. Explain the meaning of a "nonconcurrent force system."

2. What is the moment of a force?

3. Define "lever arm."

4. Suppose that a force is applied to a wrench so that the force acts at an angle of 60° to the wrench and thereby produces a counterclockwise turning effect. Describe two methods of calculating the moment of the force.

5. What is a couple?

6. How is the moment of a couple calculated?

7. How can we tell whether a moment is a clockwise moment or a counterclockwise moment?

8. What is the center of gravity of an object?

9. What is the difference between stable and unstable equilibrium of an object, and how are these states related to the center of gravity of the object?

10. How is the resultant force and resultant moment calculated for a system of nonconcurrent forces?

11. What is the second condition for equilibrium?

12. In a system of parallel forces it turns out that the resultant force is zero, but the resultant moment is −200 lb-ft. Explain how this could happen.

13. Suppose that the resultant force on an object is 100 lb acting vertically upward. However, the mo-

ment about point A is zero. How could this be possible?

14. In an equilibrium problem the tension in a supporting rod turns out to be -400 lb. What is the physical meaning of the minus sign?

PROBLEMS

(A)

1. The rod AB in Fig. 4-13 is 18 in. long. Calculate the moment in pound-feet about the axis at A if the angle θ is (a) 90°, (b) 60°, (c) 30°, (d) 0°.

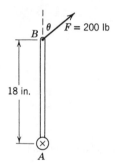

Figure 4-13   The moment of a force.

2. The 80-lb force and the 100-lb force in Fig. 4-14 act upon the member ABC.
   (a) Calculate the resultant moment about A.
   (b) Calculate the resultant moment about B.

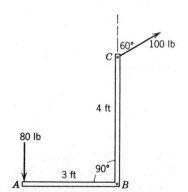

Figure 4-14   Calculating the resultant moment.

3. In Fig. 4-15 calculate the moment about O produced by
   (a) The force $A = 40$ lb.
   (b) The force $B = 30$ lb.

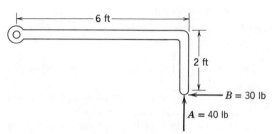

Figure 4-15   Calculating the resultant moment.

4. In Fig. 4-16 the weight of the man at A just balances the weight of the boy at B. Ignore the weight of the plank and calculate the weight of the boy.

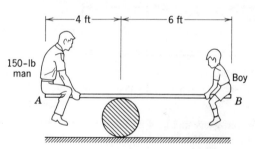

Figure 4-16   The resultant moment is zero.

5. AB is a uniform rod 10 ft long that weighs 40 lb (see Fig. 4-17). Locate the center of gravity of the rod with the two attached weights.

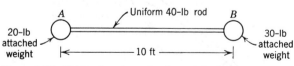

Figure 4-17   The center of gravity of a weighted rod.

6. A uniform pole that is 6 ft long weighs 10 lb. A 20-lb lead weight is attached to one end of the pole. Where is the center of gravity located for the weighted pole?

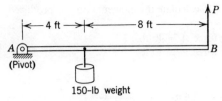

150-lb weight

Figure 4-18 Supporting a pivoted rod.

7. The uniform rod in Fig. 4-18 is 12 ft long and weighs 60 lb.
   (a) Calculate the force P required to support the rod in a horizontal position.
   (b) How much weight is supported by the pivot at A?

8. In Fig. 4-19 the weight $W = 300$ lb. Ignore the weight of the lever and calculate
   (a) the force P required to lift the weight,
   (b) the load supported by the fulcrum.

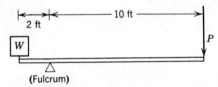

(Fulcrum)

Figure 4-19 Lifting a weight with a simple lever.

9. In the pliers in Fig. 4-20 the forces $P = 50$ lb are applied to the handles. Calculate the compressive force on the small box at A.

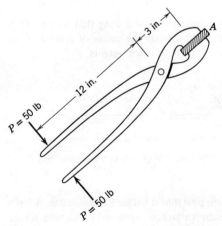

Figure 4-20 Forces applied by the pliers.

10. The claw hammer in Fig. 4-21 is used to pull a nail out of a wooden beam. Calculate the force on the nail.

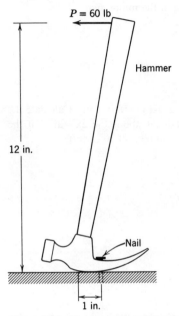

P = 60 lb

Hammer

12 in.

Nail

1 in.

Figure 4-21 The second condition for equilibrium and the claw hammer.

11. The uniform beam ABC in Fig. 4-22 weighs 200 lb and is 10 ft long. The beam is pivoted at A and supported by a vertical force at B.
    (a) Calculate the vertical force at B.
    (b) How much weight is supported at A?

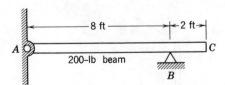

200-lb beam

Figure 4-22 Supporting a heavy beam.

12. In Fig. 4-1 the load is 8 in. from the fulcrum, while the effort is a 100-lb force which is applied 5 ft from the fulcrum. How large a load can be lifted by this effort?

13. Locate the center of gravity for the two weights in Fig. 4-23.

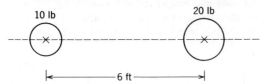

10 lb   20 lb

|← 6 ft →|

Figure 4-23   Locating the center of gravity for two weights.

14. Calculate the moment of the couple that is applied to *AB* in Fig. 4-24.

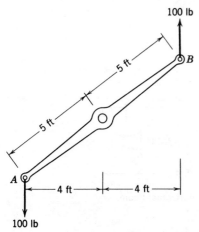

100 lb

B

5 ft

5 ft

A

|← 4 ft →|← 4 ft →|

100 lb

Figure 4-24   Moment of a couple.

15. Calculate the force and the couple that are applied to the 6-ft rigid rod in order to support it at point *A* in Fig. 4-25.

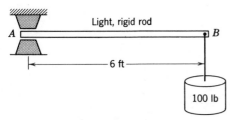

Light, rigid rod

A   *B*

|← 6 ft →|

100 lb

Figure 4-25   Supporting a rod with a force and a couple.

16. In Fig. 4-2 the bridge weighs 200 tons and the train weighs 100 tons. The center of gravity of the train is 50 ft from *B* and 150 ft from *A*.
(a) How much weight is supported by *A*?
(b) How much weight is supported by *B*?

17. The pickup truck in Fig. 4-26 is loaded with 5000 lb of cargo. The center of gravity of the cargo is located at the center of the 8-ft cargo box.

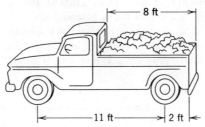

|← 8 ft →|

|← 11 ft →|2 ft|←

Figure 4-26   A loaded pickup truck.

(a) How much of the load is supported by the rear axle?
(b) How much is carried by the front axle?

(B)

1. Find the center of gravity of the three weights in Fig. 4-27.

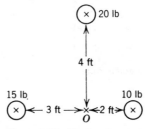

× 20 lb

4 ft

15 lb   10 lb
× ←— 3 ft —→×←2 ft→ ×
O

Figure 4-27   Finding the center of gravity of three weights.

2. (a) Find the resultant force and the resultant moment about *P* produced by the two forces in Fig. 4-28.
(b) Locate the resultant as a single force that will produce the same effect as the two forces *A* and *B*.

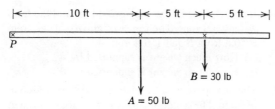

Figure 4-28 The resultant of parallel forces.

3. *AB* is a uniform rigid bar 20 in. long that weighs 2 lb (refer to Fig. 4-29). This bar is balanced upon the knife-edge at *C*. An unknown weight is suspended from the bar at *D* and a 10-lb weight is suspended at *A*. How much does the unknown weight weigh?

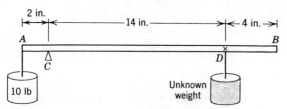

Figure 4-29 Calculating an unknown weight.

4. The horizontal bar *AB* of Fig. 4-30 weighs 100 lb, and its center of gravity is at *C*. The upward push

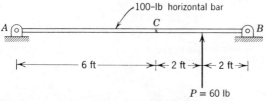

Figure 4-30 Parallel forces in equilibrium.

$P = 60$ lb. What forces act upon the bar at *A* and at *B*?

5. In Fig. 4-31 the uniform horizontal boom *AB* is 6 ft long and weighs 100 lb. Calculate
(a) the tension in the cable *CB*, and
(b) the horizontal and vertical thrust on the boom at *A*.

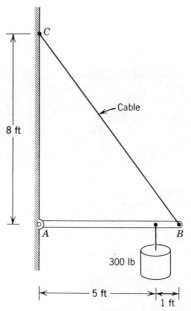

Figure 4-31 The horizontal boom.

6. Refer to Fig. 4-32.
(a) Calculate the resultant moment about point *a*.
(b) Calculate the resultant moment about point *b*.
(c) What are the magnitude, direction, and location of a single force that will produce the same effect as the three forces in the diagram?

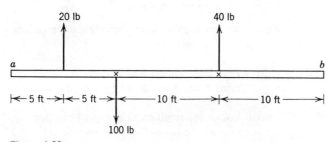

Figure 4-32

7. Forces A and B are applied to the rod in the diagram (Fig. 4-33). Ignore the weight of the rod.
   (a) What third force must be applied to the rod to produce equilibrium? (Determine the x and y components of this force.)
   (b) At what position along the rod must the equilibrant be applied?

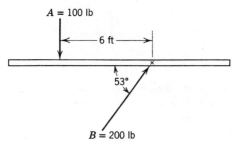

A = 100 lb

Figure 4-33   Determining the equilibrant.

8. Ignore the weight of the boom AC in Fig. 4-34.
   (a) Calculate the tension in the cable BD.
   (b) Determine the horizontal and vertical supporting forces acting upon the boom at A.

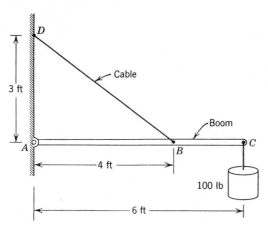

Figure 4-34   The horizontal boom.

9. AB in Fig. 4-35 is a rigid vertical member that weighs 200 lb. Find
   (a) the tension in the guy wire, and
   (b) the horizontal and vertical forces acting upon AB at A.

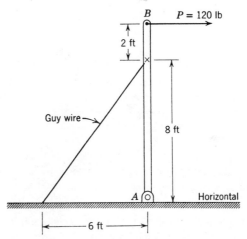

Figure 4-35   Forces on a vertical member.

10. Ignore the weight of the boom ACB of Fig. 4-36. Calculate
    (a) the tension in the cable CD, and
    (b) the horizontal and vertical components of the force acting on ACB at the pivot.

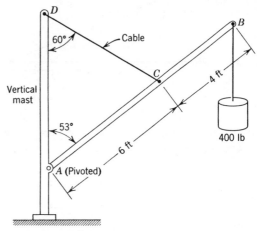

Figure 4-36   The boom and cable.

11. In Fig. 4-37 ABC is a rigid beam 15 ft long that weighs 100 lb. The center of gravity of the beam is 7 ft from end A. Determine the maximum weight W that can be suspended from C if the compression in BD is not to exceed 500 lb.

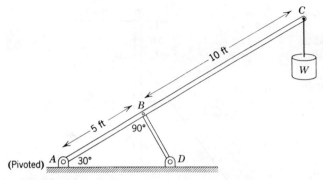

Figure 4-37 The equilibrium of a rigid beam.

12. *AB* is a uniform rod 12 ft long that weighs 100 lb (refer to Fig. 4-38). A 60-lb weight is welded to *AB* at *C*, and an 80-lb weight is attached at *D*.
   (a) Locate the center of gravity of the 240-lb object *ACDB*.

(b) Calculate the tension in the cable.
(c) Determine the force acting on the vertical mast *GH* at point *A*.

13. Tie rod *CD* connects the midpoint of members *AE* and *BE* in Fig. 4-39. Ignore the weight of the mem-

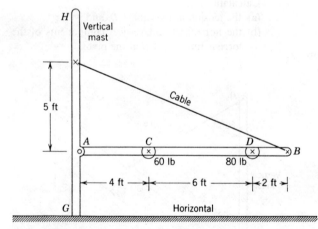

Figure 4-38 Forces in a static structure.

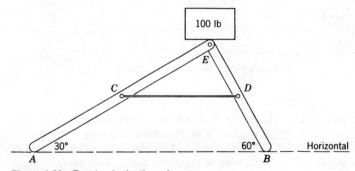

Figure 4-39 Tension in the tie rod.

bers and assume that the frictional forces are negligible at *A* and *B*. Calculate
(a) the vertical forces at *A* and at *B*, and
(b) the tension in the tie rod.

14. The structure in the diagram (Fig. 4-40) is made of wood that weighs 50 pounds per cubic foot.
(a) What is the weight of the object?
(b) Where is the center of gravity?

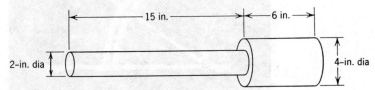

Figure 4-40   Calculating the weight and the center of gravity.

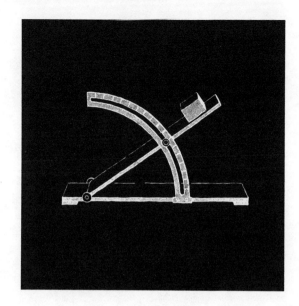

5.
The Force of Friction in Equilibrium Problems

Very frequently, the forces encountered in physics and engineering problems involve forces of *friction*. In this chapter we shall study the forces of friction in equilibrium problems and the simple laws that may be employed to calculate these forces.

LEARNING GOALS

This chapter has five major learning objectives:

1. *How to draw the normal force in free body diagrams.*
2. *How to indicate the friction force in free body diagrams.*
3. *Calculating the coefficient of friction.*
4. *Finding the force of friction in equilibrium problems.*
5. *Calculating the forces necessary to overcome friction and start motion.*

5-1 THE KINDS OF FRICTION

Whenever we attempt to slide one object over another, we find that our efforts are opposed by a force of friction. In order to push a table or slide a box along the floor, we must overcome the force of *sliding friction.* It is, of course, easier to move a heavy object by placing it on wheels or rollers. We must overcome the rolling resistance or the force of *rolling friction* when we roll a wheel or cylinder along the ground.

In moving an airplane through air or a ship through water, the thrust of the engine is opposed by the drag of air resistance or water resistance. Both air resistance and water resistance are examples of *fluid friction.*

In general, we may classify the forces of friction into three main types:

1. Sliding friction.
2. Rolling friction.
3. Fluid friction.

These forces of friction often become very large and troublesome. Friction wastes much power in machines. The frictional resistance or viscosity of heavy oil makes it difficult to pump the cold oil. Friction between moving parts causes them to wear out and often results in the production of excessive amounts of heat. In order to minimize some of these undesirable effects, the designers of machinery have to devise elaborate and expensive lubrication systems. The use of roller bearings or ball bearings also helps to reduce the amount of friction between moving parts. Special cooling devices such as radiators, water circulation pumps, and cooling fans are often employed to remove the heat produced by friction. The designers of automobiles, ships, and aircraft try to produce streamlined body shapes to minimize the amount of fluid friction that acts to resist the forward motion of the body.

Although friction is usually considered to be a wasteful or troublesome force, there are many useful applications for the force of friction. It would be impossible for a person to walk or run on a completely frictionless surface. Nails and screws remain firmly embedded in their holes because of frictional forces. The traction or forward thrust of a locomotive or an automobile depends on the force of friction. When we apply the brakes in an automobile, the brake linings exert a force of friction upon the wheel drums, and friction between the tires and the road is needed to slow or stop the vehicle. The transmission of power by means of fan belts or clutches also depends on the force of friction.

5-2 SLIDING FRICTION

If a block of wood is resting upon the floor and we apply a small force to the block, it will not move or slide forward. The block will not slide because the force of friction is acting to prevent motion. The forces that act upon the block are indicated in a free body diagram in Fig. 5-1, where P is the horizontal push applied to the block; W is the weight of the block and is shown acting through the center of gravity C; and N is the upward push of the floor upon the block. This upward force N is called a *normal force,* because

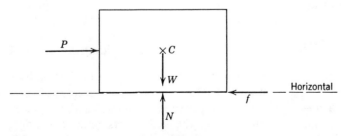

Figure 5-1   Friction on a horizontal plane.

it acts in a direction normal (perpendicular) to the floor. The horizontal push P is equal in magnitude but opposite in direction to the force of static friction f.

If we gradually increase push P, the force of static friction increases by the same amount. Finally, we may reach a value for P that will just serve to start the block moving. The force of static friction when the block is just on the verge of sliding is called the force of *starting friction*.

Example 5-1 A 100-lb packing crate is resting upon the floor of a garage. A horizontal pull is applied to the crate. This pull is increased until we find that a 40-lb pull will just start the crate moving. (a) What is the normal force acting upon the crate? (b) How big is the force of friction when the horizontal pull is 0 lb? (c) What is the force of friction when the horizontal pull is 15 lb? (d) What is the force of starting friction?

Solution. The free body diagram in Fig. 5-1 may be used in this problem. Let $W = 100$ lb. (a) The normal force is equal to the weight; in this problem $N = 100$ lb. (b) If the pull $P = 0$, the force of friction f is also 0. (c) If $P = 15$ lb, we have $\Sigma F_x = 0.15$ lb $- f = 0$, and $f = 15$ lb (d) The force of starting friction is usually written f_s (we note that $f \leq f_s$). When P is increased to 40 lb, the object will just start moving. Therefore, the force of starting friction is $f_s = 40$ lb. ◄

Many experiments indicate that the force of starting friction is proportional to the total normal force that is acting. The force of starting friction also depends on the nature of the two sliding surfaces that are in contact. These two experimental facts are expressed in terms of an algebraic formula

$$f_s = \mu_s N \qquad (5\text{-}1)$$

$$\frac{\text{force of}}{\text{starting friction}} = \frac{\text{coefficient of}}{\text{starting friction}} \times \text{normal force}$$

The *coefficient of starting friction*, μ_s, is a numerical constant that depends on the sliding surfaces. The coefficient of starting friction is also called the coefficient of sliding friction or the

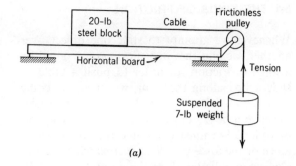

(a)

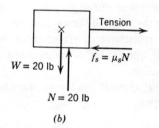

(b)

Figure 5-2 Determining the coefficient of starting friction. (a) The block is just on the verge of moving. (b) Free body diagram of block.

coefficient of static friction. The numerical value for μ_s may be calculated from the results of a simple experiment (see Fig. 5-2 and the following example).

Example 5-2 A 20-lb steel block is placed on a horizontal wooden board and connected to a suspended weight by means of a pulley and cable (refer to Fig. 5-2). When the weight is 7 lb, the block is just on the verge of moving. Calculate the coefficient of starting friction.

Solution. Since $\Sigma F_y = 0$ and $\Sigma F_x = 0$, we find that $N = 20$ lb and $f_s = 7$ lb.

$$f_s = \mu_s N \qquad \text{and} \qquad \mu_s = \frac{f_s}{N} = \frac{7 \text{ lb}}{20 \text{ lb}}$$

$$\therefore \ \mu_s = 0.35 \qquad \text{(for steel sliding on wood)} \ ◄$$

Some values for the coefficient of starting friction for different surfaces are given in Table 5-1. These are approximate values; as the surfaces become worn, the coefficients tend to change. We all know that badly worn or slick tires fail to provide enough friction to give a car good

Table 5-1 The coefficient of starting friction for various surfaces

Surface	Average value for coefficient of starting friction
Wood on wood	0.3
Metals on wood	0.2 to 0.6
Rubber on concrete	0.7
Iron on stone	0.5
Leather on metals	0.4
Steel on steel	0.18
Greased surfaces	0.05

traction. During the "breaking in" period the amount of friction in an engine may decrease. On the other hand, the amount of friction in bearings may increase greatly if the bearing surfaces become worn. Also, it is often observed that damaged brake linings may cause the brakes to "grab" in some cases and to slip in other cases. The coefficient of starting friction for any pair of surfaces may vary quite a bit depending upon the condition of the surfaces.

The values for μ_s given in Table 5-1 are average values for clean, dry surfaces. If the surfaces are covered by a thin layer of grease or moisture, the coefficient of starting friction may decrease by a large amount.

Many experiments with sliding objects indicate that after sliding has started, the force of friction opposing the motion tends to decrease slightly. If an object is sliding, the force of friction acting upon the object is called the *force of kinetic friction* (kinetic means "moving" or "in motion"). Experiments indicate that the force of kinetic friction is slightly less than the force of starting friction, and therefore the *coefficient of kinetic friction* is a little smaller than the coefficient of starting friction. The law for kinetic friction is

$$f_k = \mu_k N \qquad (5\text{-}2)$$

$$\frac{\text{force of}}{\text{kinetic friction}} = \frac{\text{coefficient of}}{\text{kinetic friction}} \times \text{normal force}$$

Example 5-3 A 50-lb block is resting on a horizontal surface. A horizontal force of 30 lb is required to just start the block moving. A con-

stant horizontal force of 26 lb is needed to keep the block moving at a steady speed. Calculate (a) the coefficient of starting friction, and (b) the coefficient of kinetic friction.

Solution. (a) In Fig. 5-1 we let $W = 30$ lb and $P = 30$ lb. Then we easily find that

$$N = 50\text{ lb} \qquad \text{and} \qquad f_s = 30\text{ lb}$$

$$f_s = \mu_s N \qquad \mu_s = \frac{f_s}{N} = \frac{30\text{ lb}}{50\text{ lb}} = 0.60$$

(b) To consider kinetic friction, we must let $P = 26$ lb and apply the laws of equilibrium again.

$$\sum F_x = 0 \qquad 26\text{ lb} - f_k = 0 \qquad f_k = 26\text{ lb}$$

$$f_k = \mu_k N \qquad \mu_k = \frac{f_k}{N} = \frac{26\text{ lb}}{50\text{ lb}} = 0.52 \quad \blacktriangleleft$$

The reader should note that the total normal force is not always equal to the weight of the object. In the following problem the normal force is greater than the weight of the object.

Example 5-4 In order to push a 60-lb bench across the floor, a force is applied that acts downward and to the right at an angle of 30° below the horizontal. The applied force is 40 lb in magnitude and is just great enough to keep the bench moving. Calculate the coefficient of kinetic friction (Fig. 5-3a).

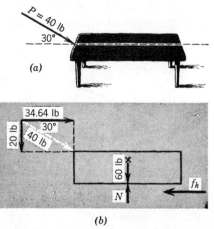

Figure 5-3 (a) Pushing a bench against the force of kinetic friction. (b) Free body diagram.

Solution. The 40-lb force is resolved into a horizontal component 40 lb cos 30° = 40 lb (0.866) = 34.64 lb; and a vertical (downward) component of 40 lb sin 30° = 40 lb (0.500) = 20 lb (Fig. 5-3b).

$$\sum F_y = 0 \quad N - 60 \text{ lb} - 20 \text{ lb} = 0 \quad N = 80 \text{ lb}$$
$$\sum F_x = 0 \quad \quad 34.64 \text{ lb} \quad f_k = 0 \quad f_k = 34.64 \text{ lb}$$

The coefficient of kinetic friction is

$$\mu_k = \frac{f_k}{N} = \frac{34.64 \text{ lb}}{80 \text{ lb}} = 0.43 \quad \blacktriangleleft$$

5-3 FRICTION ON THE INCLINED PLANE

When an object is on an incline or a slope, the normal force must be indicated as a force acting perpendicular to the inclined surface. The weight of the object is a force that acts vertically downward, but the normal force is not vertically upward. The normal force N always acts at an angle of 90° to the inclined surface.

Example 5-5 An inclined plane is sloping at an angle of 20° with the horizontal. A 100-lb block is pushed uphill by a force parallel to the incline. The coefficient of static friction is 0.2. Determine the minimum force needed to start the block up the incline.

Solution. Here P is the push needed to overcome static friction; $W = 100$ lb is drawn vertically down and then resolved into two compo-

nents; N is drawn perpendicular to the plane; and f_s is acting down the plane, since it opposes the motion of the block up the incline (Fig. 5-4).

$$\sum F_y = 0 \quad N - 100 \text{ lb} \cos 20° = 0 \quad N = 94 \text{ lb}$$
$$\sum F_x = 0 \quad P - 100 \text{ lb} \sin 20° - 0.2N = 0$$
$$P - 34.2 \text{ lb} - 0.2(94 \text{ lb}) = 0$$
$$P = 34.2 \text{ lb} + 18.8 \text{ lb} = 53 \text{ lb} \quad \blacktriangleleft$$

On the inclined plane, just as on a horizontal surface, we may describe three different cases of sliding friction:

1. f is less than $\mu_s N$ when the object is stationary.

2. $f_s = \mu_s N$ (force of starting friction) when the object is about to start sliding.

3. $f_k = \mu_k N$ (force of kinetic friction) when the object is moving.

In every case the force of friction is indicated as acting in such a direction as to oppose the sliding motion or the tendency toward motion.

Example 5-6 An inclined plane that rises 6 ft is 10 ft long along the incline. The coefficient of static friction is 0.3 for a 100-lb box, which is on the incline. (a) Determine the minimum force parallel to the plane that will prevent the block from sliding downhill. (b) Determine the force parallel to the plane that will just start the block moving uphill. (c) Determine the force parallel to the plane that will make the force of friction zero.

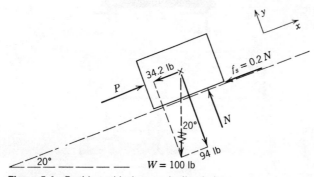

Figure 5-4 Pushing a block up an inclined plane.

Solution. (a) The horizontal distance h in Fig. 5-5 is found by the Pythagorean theorem:

$$h^2 + 6^2 = 10^2$$

$$h = \sqrt{10^2 - 6^2} = \sqrt{100 - 36} = 8 \text{ ft}$$

Here P is the minimum force required to prevent downhill motion. The force of friction $f_s =$

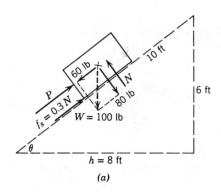

(a)

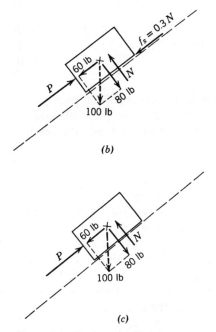

(b)

(c)

Figure 5-5   The force of friction acts to oppose motion. (a) The block tends to slide down and the force of friction acts up the incline. (b) The block tends to move up the plane and the force of friction acts downward along the incline. (c) The force P is just large enough to make the force of friction zero.

$0.3N$ is acting uphill because it opposes the tendency of the block to slide downhill.

The components of $W = 100$ lb may be calculated by employing similar triangles.

$$\frac{W_x}{100 \text{ lb}} = \frac{6 \text{ ft}}{10 \text{ ft}} \qquad W_x = 0.6(100 \text{ lb}) = 60 \text{ lb}$$

$$\frac{W_y}{100 \text{ lb}} = \frac{8 \text{ ft}}{10 \text{ ft}} \qquad W_y = 0.8(100 \text{ lb}) = 80 \text{ lb}$$

We could also have used trigonometric functions to resolve W into components: $\sin \theta = \frac{6}{10} = 0.60$ and $\theta = 37°$;

$$W_x = W \sin \theta = 100 \text{ lb} \sin 37° = 100 \text{ lb} (0.6) = 60 \text{ lb}$$
$$W_y = W \cos \theta = 100 \text{ lb} \cos 37° = 100 \text{ lb} (0.8) = 80 \text{ lb}$$

From the conditions of equilibrium

$$\sum F_y = 0 \qquad N - 80 \text{ lb} = 0, \qquad N = 80 \text{ lb}$$

Then $f_s = 0.3(80 \text{ lb}) = 24 \text{ lb}$

$$\sum F_x = 0 \text{ gives } P - W_x + f_s = 0$$
$$P - 60 \text{ lb} + 24 \text{ lb} = 0$$
$$P = 36 \text{ lb}$$

(b) If we try to push the block uphill, the force of friction will act downhill. Except for this change in the free body diagram, the procedure is similar to that in part (a). We again have

$$N = W_y = 80 \text{ lb}, \qquad f_s = 24 \text{ lb}$$

But now

$$\sum F_x = 0 \text{ gives } P - W_x - f_s = 0$$
$$P - 60 \text{ lb} - 24 \text{ lb} = 0$$
$$P = 84 \text{ lb}$$

(c) From the free body diagrams of Fig. 5-5c it is obvious that the force of friction will be zero if

$$P = W_x = 60 \text{ lb}$$

In the last problem we have proved that the block will remain stationary if P is bigger than 36 lb, but less than 84 lb. For any value of P within these limits the force of friction acting upon the block will be just large enough to prevent motion. The force of friction may be equal to zero pounds or it may increase to a maximum value of $\mu_s N = 24$ lb acting either up or down the plane.

5-4 THE ANGLE OF REPOSE AND THE ANGLE OF UNIFORM SLIP

A block or any other object placed on an inclined plane will not start to slide down the incline if the slope is too small. If the component of the weight parallel to the incline is less than the force of starting friction, the object will remain stationary until we apply an additional downward force. An object tends to remain stationary on a given inclined plane if the slope is less than or equal to the angle of repose.

The *angle of repose* is defined as the steepest slope angle for which an object will remain at rest on a given inclined surface. The angle of repose is determined by the coefficient of static friction. If the coefficient is large, the object will remain stationary on a steep incline and the angle of repose will be large.

Example 5-7 The coefficient of static friction is 0.75 between some rubber tires and the roadway. Determine the steepest incline upon which the automobile will remain at rest if the parking brake has locked the wheels.

Solution. The steepest incline is the angle of repose θ_r (see Fig. 5-6c). At this angle, $W_x = W \sin \theta_r$ equals the force of starting friction, f_s. Since

$$\sum F_y = 0 \qquad N = W \cos \theta_r$$

and $\qquad f_s = \mu_s N = \mu_s W \cos \theta_r$

Then $\sum F_x = 0$ gives $f_s = W \sin \theta_r$ and $\mu_s W \cos \theta_r = W \sin \theta_r$. Canceling W gives

$$\mu_s \cos \theta_r = \sin \theta_r$$

or $\qquad\qquad \mu_s = \dfrac{\sin \theta_r}{\cos \theta_r}$

Since $\sin \theta_r / \cos \theta_r = \tan \theta_r$, we have proved that

$$\mu_s = \tan \theta_r \qquad (5-3)$$

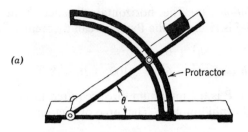

(a)

← Protractor

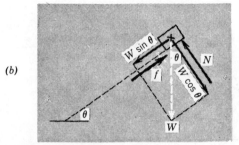

(b)

Figure 5-6 Determining the angle of repose and the angle of uniform slip.

Applying this law to the case of the car, we can calculate the angle of repose:

$$\tan \theta_r = 0.75 \qquad \theta_r = 37° \qquad \blacktriangleleft$$

If the incline in the last problem were made steeper, the component of W perpendicular to the incline would be decreased, and the normal force and the force of starting friction would also be decreased. But if the incline is made steeper, W_x will increase. Consequently, if the slope is made greater than the angle of repose, W_x becomes greater than f_s, and there is an unbalanced force acting to push the object down the incline.

The angle of repose has some important applications in engineering. The slope of an earth enbankment, for example, must be less than the angle of repose. If earth is cut away so that the slope is greater than the angle of repose, a landslide will occur. Piles of coal, gravel, or sand can be heaped up so that the sides are as steep as the angle of repose but no steeper.

Measuring the angle of repose is a convenient laboratory method of determining the coefficient of static friction. An inclined plane that can be pivoted is used so that the slope may be increased or decreased. After some trials, we can determine the steepest angle for which the block will just remain stationary on the plane. This angle will be the angle of repose, and the coefficient of static friction can then be calculated from Eq. 5-3.

The same apparatus that is used to find the angle of repose may also be used to determine the angle of uniform slip (see Fig. 5-6a).

The block on the inclined plane is given a small downward push to overcome starting friction and start the block sliding. If the slope of the incline is less than the *angle of uniform slip*, the block will slow up (decelerate) as it slides. If the angle of the incline is greater than the angle of uniform slip, the block will speed up (accelerate) as it slides. If the angle of the incline is exactly equal to the angle of uniform slip, the block will slide downhill with a constant speed. If we can determine this angle, it is then a simple matter to calculate the coefficient of kinetic friction.

• **Example 5-8**   The angle of repose for a brick on an inclined ramp is 30°. If the brick is started, it will slide with a constant speed when the angle of the incline is 26°. (a) Calculate the coefficient of static friction and (b) Calculate the coefficient of kinetic friction.

(a) From Eq. 5-3

$$\mu_s = \tan \theta_r = \tan 30° = 0.577$$

(b) The angle of uniform slip is $\theta_w = 26°$. From the free body diagram of Fig. 5-6b,

$$\Sigma F_y = 0 \qquad N - W \cos \theta_w = 0$$
$$N = W \cos \theta_w$$
$$f_k = \mu_k N = \mu_k W \cos \theta_w$$

For uniform slip $\Sigma F_x = 0$

$$W \sin \theta_w = f_k$$
$$W \sin \theta_w = \mu_k W \cos \theta_w$$
$$\mu_k = \frac{\sin \theta_w}{\cos \theta_w}$$

or $$\mu_k = \tan \theta_w \qquad (5\text{-}4)$$

Equation 5-4 is a general equation for the angle of uniform slip. In this problem we have $\theta_w = 26°$ and $\mu_k = \tan 26° = 0.488$.

5-5 EQUILIBRIUM PROBLEMS INVOLVING FRICTION

The steps in the solution of equilibrium problems should be familiar by now:

1. Draw a free body diagram indicating all the known and unknown forces.

2. Apply the equations of equilibrium.

3. Solve for the unknowns.

In equilibrium problems that involve the force of friction, it is important to indicate the normal force and the force of friction. The force of friction should be drawn in the direction that opposes the motion or the tendency toward motion.

Example 5-9   A uniform 160-lb ladder is 20 ft long. It is placed against a smooth wall so that the base of the ladder is 12 ft from the wall (see Fig. 5-7a). (a) Determine the force of friction acting on the base of the ladder. (b) Calculate the minimum coefficient of friction if the ladder is not to slip.

Solution.   (a) Since the wall is smooth, there is no force of friction at *B*. Force *H* represents the horizontal thrust of the wall upon the ladder. The height *h* is found by applying the Pythagorean theorem:

$$12^2 + h^2 = 20^2$$
$$h = \sqrt{400 - 144} = 16 \text{ ft}$$

The 160-lb weight is indicated as acting through the center of the ladder. From the condition of equilibrium we have

$$\Sigma M_{(A)} = 0$$
$$-160 \text{ lb}(6 \text{ ft}) + H(16 \text{ ft}) = 0$$
$$H = \tfrac{6}{16}(160 \text{ lb}) = 60 \text{ lb}$$

Since $\Sigma F_x = 0$, we have $f = H = 60$ lb.

(b) To calculate μ_s, we first note that $N = 160$ lb. Then

$$\mu_s = \frac{f_s}{N} = \frac{60 \text{ lb}}{160 \text{ lb}}$$

$$\mu_s = 0.375 \qquad \blacktriangleleft$$

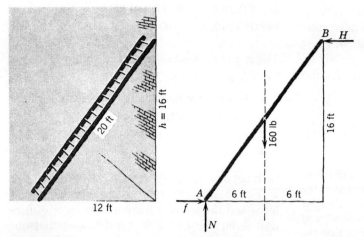

Figure 5-7   Free body diagram of a ladder resting on a vertical wall.

In the following example we have a somewhat more complicated application of the same principles of solution employed in the preceding example.

Example 5-10   A 180-lb man climbs up a uniform 20-ft ladder that weighs 100 lb. One end of the ladder is set against a smooth vertical wall, and the other end is on the ground 10 ft from the base of the wall. The coefficient of starting friction between the bottom of the ladder and the ground is 0.25. How far up along the ladder can the man climb before the ladder starts to slip?

Solution.   We draw a free body diagram indicating that the man has climbed a distance d up the ladder (refer to Fig. 5-8). At end A we let $f = f_s = 0.25N$, since the ladder is on the verge of slipping.

To find the vertical height h, we have

$$10^2 + h^2 = 20^2$$
$$h = \sqrt{400 - 100} = \sqrt{300} = 17.3 \text{ ft}$$

Then
$$\sum F_y = 0$$
$$N - 100 \text{ lb} - 180 \text{ lb} = 0 \qquad N = 280 \text{ lb}$$

Also
$$f_s = 0.25(N) = 70 \text{ lb}$$

Since $\sum F_x = 0$, we have $H = 70$ lb. The lever arm of the 100 lb force is $\frac{1}{2}(10 \text{ ft}) = 5$ ft. The lever arm of the 180 lb force is $\frac{1}{2}d$.

Then $\sum M_A = 0$ gives

$$-100 \text{ lb} (5 \text{ ft}) - 180 \text{ lb} \left(\frac{d}{2}\right) + 70 \text{ lb} (17.3 \text{ ft}) = 0$$

$$-500 \text{ ft-lb} - 90 \text{ lb} \, d + 1211 \text{ ft-lb} = 0$$

$$d = \frac{711}{90} \text{ ft} = 7.9 \text{ ft} \qquad \blacktriangleleft$$

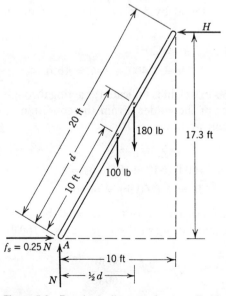

Figure 5-8   Free body diagram of a man climbing a ladder.

Sometimes problems in statics lead to several simultaneous equations that must be solved for the unknowns. As an example of this procedure the reader should study and understand the following.

• **Example 5-11** A 150-lb sled is loaded with 250 lb of freight. The loaded sled is pulled across the ice by means of a tow rope that is held so that it makes an angle of 37° with the horizontal. The coefficient of starting friction is 0.20 and the coefficient of kinetic friction is 0.16. Determine (a) the force needed to start the sled moving and (b) the minimum force needed to keep the sled moving.

Solution. (a) The tension T in the tow rope is resolved into two components as indicated in Fig. 5-9. Then

$$\sum F_y = 0 \qquad N + T \sin 37° = 400 \text{ lb}$$
$$N + 0.6T = 400 \text{ lb} \qquad (5\text{-}5)$$

If we are overcoming static (starting) friction,

$$f = \mu_s N = 0.2N$$
$$\sum F_x = 0 \qquad T \cos 37° = 0.2N$$
$$T(0.8) = 0.2N \qquad N = 4T$$

Substituting in Eq. 5-5, we have

$$4T + 0.6T = 400 \text{ lb}$$
$$T = \frac{400}{4.6} \text{ lb} = 87 \text{ lb}$$

(b) For the case when the sled is already moving, we have

$$f = \mu_k N = 0.16N$$

Then $\sum F_x = 0$ gives $0.16N = T(0.8)$

$$N = 5T$$

Substituting in Eq. 5-5 and solving for T gives

$$5T + 0.6T = 400 \text{ lb}$$
$$5.6T = 400 \text{ lb}$$
$$T = \frac{400}{5.6} \text{ lb} = 71.5 \text{ lb}$$

SUMMARY

Whenever we attempt to slide one object over another, our efforts are opposed by the force of *friction*. The force of friction acts in a direction that opposes the motion or the tendency toward motion.

The force of *static friction* increases when we attempt to push an object. The maximum force of static friction occurs when the object is just on the verge of sliding. This maximum force of friction is called the force of *starting friction*; it is proportional to the normal force acting. The force of starting friction can be calculated by using the *coefficient of starting friction* and the normal force as in Eq. 5-1.

If an object is already sliding, the force of friction is called the force of *kinetic friction*. The force of kinetic friction can be calculated from the normal force by using the coefficient of kinetic friction, μ_k, and Eq. 5-2.

The *angle of repose* is the steepest angle for which a given object will remain at rest on a given inclined surface. The tangent of the angle of repose is equal to the coefficient of static friction as in Example 5-7 and Eq. 5-3. The coefficient of kinetic friction may be calculated by taking the

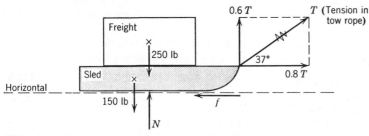

Figure 5-9 Free body diagram of a loaded sled being pulled.

tangent of the angle of uniform slip (Eq. 5-4 and Example 5-8).

QUESTIONS

1. What are the two important factors that determine the force of starting friction?

2. Explain what is meant by the "coefficient of static friction."

3. When is the force of friction less than $\mu_s N$?

4. Under what condition will the force of friction just equal $\mu_s N$?

5. Give an example to illustrate that the normal force is sometimes less than the weight of the object.

6. Under what circumstances will the normal force be greater than the weight of the object?

7. What is the coefficient of kinetic friction?

8. Suppose an inclined ramp is treated so that it becomes very smooth. What effect does this have on the angle of repose?

9. Explain the difference between the angle of repose and the angle of uniform slip.

10. Suppose that a car is traveling toward the right and the motor is applying power to the rear wheels. In what direction does the force of friction act upon the rear wheels? In what direction would friction act if the motor were turned off and the brakes applied?

11. The sled in Fig. 5-8 is pulled by means of an upward-angled tow rope. Suppose that instead of using the tow rope, the sled were to be pushed by a force that is angled downward or, in other words, directed below the horizontal. Explain why the force required to move the sled would be greater in this situation.

12. For the ladder indicated in Fig. 5-7 the normal force and the weight form a couple with a clockwise moment, while the force of friction and the horizontal thrust of the wall form a couple with a counterclockwise moment. Use the second condition for equilibrium to explain why the force of friction needed to support the ladder becomes less when the base of the ladder is moved closer to the wall.

13. When an object is on an inclined ramp, the force of friction is sometimes acting upward along the ramp and sometimes acting downward along the ramp. Explain how to determine the direction of the force of friction.

PROBLEMS

(A)

1. A 100-lb block rests on the floor. The coefficient of friction between the block and the floor is 0.2. How large a force is required to push the block across the floor?

2. In Fig. 5-10 the force P must equal 60 lb to just start the 200-lb block moving. Calculate
   (a) The force of friction.
   (b) The coefficient of friction.

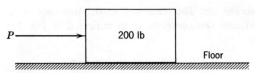

Figure 5-10 Starting a block against the force of static friction.

3. A horizontal push of 30 lb is needed to slide a 90-lb box across the floor. What is the coefficient of friction?

4. Two steel strips are pushed together with a force of 300 lb. Calculate the maximum force of friction between them if the coefficient of friction is 0.2.

5. The coefficient of friction between the road and the tires of a 4000-lb automobile is 0.4. What is the maximum force of friction acting to stop the automobile if the brakes are applied with sufficient force so that the tires are on the verge of skidding?

6. The coefficient of friction between a 200-lb sled and the ice on a lake is 0.06. What is the minimum horizontal force required to push the sled?

7. A loaded railroad car weighs 5 tons. When it is pushed on a horizontal track, a force of 900 lb will keep the car moving with a small, steady velocity. To start the car moving, a force of 1100 lb is required. Calculate
   (a) the coefficient of kinetic friction, and
   (b) the coefficient of starting friction.

8. The coefficient of starting friction for stone blocks on a steel ramp is 0.4. What is the limiting angle of repose?

9. A roof is pitched so that it makes an angle of 35° with the horizontal. If a man is to stand on this roof,

what is the minimum coefficient of friction between his shoes and the roof?

10. What force parallel to the incline is required to push a 200-lb object up a 30° incline if
    (a) the incline is very smooth (frictionless), and
    (b) the incline is rough and the coefficient of friction is 0.25?

11. A concrete block is placed on a straight wooden ramp that is 10 ft long. One end of the ramp rests against the ground and the other end is elevated. When the elevated end is lifted to a height of 4 ft above the ground, the concrete block starts to slide. What is the coefficient of starting friction?

12. A 10-lb object slides down a ramp at constant speed. The ramp is inclined at an angle of 20° with the horizontal.
    (a) What is the magnitude of the force of friction acting upon the object?
    (b) How great a force parallel to the ramp would be required to push the block back up the ramp with constant speed?

13. The two jaws of a pipe wrench are closed down on a pipe that is 2 in. in diameter. The two jaws are clamped together so that they exert a force of 200 lb on the pipe. The coefficient of friction between the jaws and the pipe is 0.90. What is the maximum torque (in pound-feet) that can be applied to the pipe before the wrench starts to slip?

(B)

1. The coefficient of starting friction is 0.4 and the coefficient of kinetic friction is 0.35 between a 100-lb weight and the floor. A horizontal force, P, is applied to the weight. Determine the force of friction under the following conditions:
    (a) The force $P = 20$ lb and the weight is stationary.
    (b) The force P is just large enough so that the weight is on the verge of moving.
    (c) The weight is moving.
    (d) The force $P = 60$ lb and the weight is moving.

2. On the inclined plane in Fig. 5-11 the coefficient of starting friction is 0.40 and the coefficient of kinetic friction is 0.30. No force of friction acts upon the suspended weight W.
    (a) Determine the weight W that will just start the 100-lb block moving uphill.
    (b) Determine the minimum weight W that will keep the 100-lb block moving uphill.

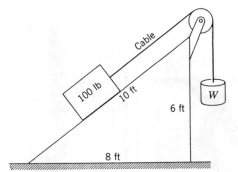

Figure 5-11 Static and kinetic friction on an inclined plane.

(c) What is the minimum weight W that will keep the 100-lb block stationary on the ramp? (*Note*: Static friction will help to keep the weight stationary by acting uphill.)
(d) What is the value of W if the 100-lb weight slides downhill without speeding up or slowing down?

3. A 100-lb ladder is 20 ft long. The center of gravity is 8 ft from the end of the ladder, which is touching the floor, and 12 ft from the end of the ladder, which is resting against a smooth wall. The ladder is set so that it makes an angle of 60° with the floor.
    (a) Calculate the force of friction acting on the ladder at the floor.
    (b) What is the minimum coefficient of friction if the ladder is not to slip?

4. The effective coefficient of friction between a lawn mower and the ground is 0.4. The lawn mower weighs 70 lb. The lawn mower is pushed along a level lawn at constant speed by means of a handle that makes an angle of 53° with the ground.
    (a) Draw a free body diagram of the lawn mower.
    (b) Calculate the force applied along the handle.

5. A packing case 6 ft × 3 ft × 3 ft is placed so that it rests on the 3 ft × 3 ft sq side. The case and its contents weigh 500 lb, and the center of gravity is at the geometrical center of the case (see Fig. 5-12). The coefficient of friction between the box and the floor is 0.35.
    (a) What is the minimum horizontal push needed to slide the box across the floor?
    (b) What is the maximum height h at which the force calculated in (a) may be applied without tipping the case over? (*Note*: When the case is

Figure 5-12 Pushing a packing case. Will it tip over?

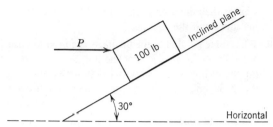

Figure 5-14 Pushing a block up an inclined plane with a horizontal force P.

on the verge of tipping, the normal force and the force of friction act through the edge AA. Use the second condition for equilibrium to calculate h.)

6. The 200-lb cabinet in Fig. 5-13 rests upon four short legs. The center of gravity is at C. The coefficient of friction is 0.50.
   (a) What force P is required to move the cabinet toward the right with a steady speed?
   (b) When the cabinet is being pushed by force P, most of the weight rests upon the two legs at B. Determine the normal force and the force of friction at these two legs.
   (c) How much weight rests upon the two legs at A? Calculate the force of friction on the two legs at A.

7. In Fig. 5-14 the horizontal force P serves to push the 100-lb block up the incline with constant speed. The coefficient of kinetic friction is 0.25.
   (a) Draw a free body diagram of the block and write out the equations of equilibrium.
   (b) Solve the equations of equilibrium to determine the numerical value of the normal force and of P.

8. In the metric system of units, forces are measured in newtons. The force of gravity acting upon object A in Fig. 5-15 is 100 newtons. The coefficient of sliding friction between A and the surface is 0.4. Force P is just great enough to keep A moving with a constant velocity. Determine the normal force and P in case (a) and in case (b).

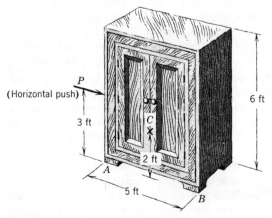

Figure 5-13 Pushing a 200-lb cabinet.

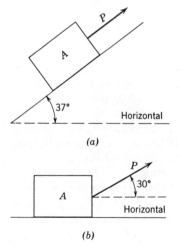

Figure 5-15 Pushing the block.

9. In Fig. 5-16 the coefficient of friction between the brake mechanism at B and the wheel is 0.8. Deter-

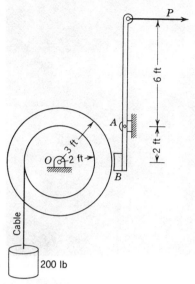

Figure 5-16 Friction in a brake mechanism.

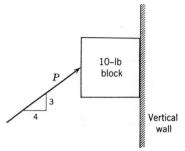

Figure 5-17 Friction on a vertical surface.

mine the minimum force P required to keep the 200-lb weight stationary. (Friction is negligible in the bearings at O and A.)

10. The coefficient of friction is 0.4 between the vertical wall and the 10-lb block in Fig. 5-17.
    (a) What is the force of friction if $P = 15$ lb? What is the normal force?
    (b) What is the normal force and force of friction if $P = 20$ lb?
    (c) What is the minimum force P that will keep the block from sliding down?
    (d) What is the maximum force P that may be applied before the block starts to slide up the wall?

11. The coefficient of starting friction between a uniform wooden beam and the floor is 0.25. The beam is placed so that one end rests upon a smooth vertical wall and the other end rests upon the floor.
    (a) Draw a free body diagram of the beam.
    (b) What is the smallest possible angle between the beam and the floor if the beam is not to slip? [*Hint*: Prove from the free body diagram that the beam is on the verge of slipping when $\tan \theta = 1/(2\mu_s)$.]

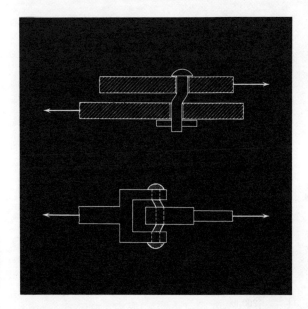

6.
Elastic Forces in Static Structures

In this chapter we shall study the *stress* and *strains* that result when forces are applied to any material or structural member. The elastic deformation of a stressed object is described by *Hooke's law*, which is used very extensively in structural design and engineering.

LEARNING GOALS

This chapter has as its goals the following main ideas:

1. *The calculation of stresses when the forces are known or can be found.*

2. *The application of Hooke's law to the calculation of the spring constant.*

3. *The use of Young's modulus and Hooke's law to calculate the strain.*

4. *The use of the bulk modulus and compressibility to determine the elastic effects produced by pressure.*

5. *The application of Hooke's law to shearing stress and strain.*

6-1 ELASTIC DEFORMATION AND THE ELASTIC LIMIT

In our study of static structures and forces in equilibrium, we have usually assumed that the beams or rods or cables were not broken or stretched out of shape or otherwise deformed by the forces applied to them. It is well known, however, that whenever forces are applied to an object, there is a change in the dimensions and shape of the object. When loads are applied to structural members, these members undergo a temporary or a permanent deformation.

Scientists and engineers have very carefully studied the deformation properties of different structural materials. Many experiments in testing the strength and elastic properties of these materials indicate that the deformation of a structural member tends to pass through two different stages. If the load applied to a member is not too great, the deformation is said to be in the elastic stage. The deformation produced in the object is directly proportional to the forces acting upon the object. If the forces applied to the object are released, there is no permanent deformation and the object returns to its original dimensions. During this elastic stage the internal restoring forces in the material will cause the object to return to its original size when the load is removed.

If the loads applied to an object are too great, the *elastic limit* of the material is exceeded and the deformation enters the plastic (or nonelastic) stage. The material now suffers a permanent deformation, and if the load is released, the object does not return to its original dimensions. If the forces applied are increased beyond the elastic limit, the deformation in a solid object becomes very great and the object may break or shatter. In many processes and operations, precautions must be taken to assure that the applied forces do not produce stresses that exceed the elastic limit of the material.

6-2 HOOKE'S LAW; SPRING CONSTANT

A study of the elastic deformation and the elastic properties of materials usually begins with a consideration of *Hooke's law*. This law was first stated by Robert Hooke during the seventeenth century in connection with his invention of a balance spring for spring-driven clocks and watches. However, this law also applies to elastic forces in tie rods, cables, beams, propellor shafts, etc.

Hooke's law states that the elastic deformation is directly proportional to the force if the elastic limit is not exceeded. An application of this law to the simple case of a coil spring is illustrated in Fig. 6-1. The graph and the table in this figure indicate that the deformation of the spring is directly proportional to the load. From the data we can easily calculate the constant of proportionality. This constant is called the *spring constant* or the *stiffness of the spring*.

$$k \text{(spring constant)} = \frac{\text{load}}{\text{elongation}}$$

$$= \frac{10 \text{ lb}}{5 \text{ in.}} = 2 \text{ lb/in.}$$

The stiffness of the spring is 2 lb/in. We might also have written

$$k = 2 \text{ lb/in.} \times 12 \text{ in./ft} = 24 \text{ lb/ft}$$

Different springs will, of course, have a different spring constant.

When a spring is stretched (within its elastic limit), the spring exerts a *restoring force* that is opposite in direction to the elastic deformation. When a spring is compressed (as in Fig. 6-2c), the spring will again exert a restoring force that is opposite in direction to the elastic deformation. If the elastic deformation is indicated by x, the restoring force is

$$F = -kx \qquad (6\text{-}1)$$

The minus sign indicates that x and F are always in opposite directions.

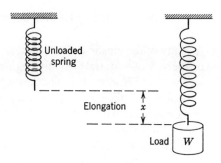

W (pounds)	x (inches)
1	0.5
3	1.5
5	2.5
10	5.0

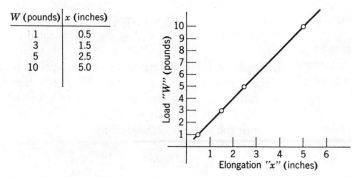

Figure 6-1 Determining the spring constant "*k*."

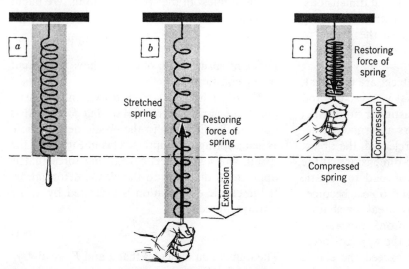

Figure 6-2 The restoring force in a spring. The restoring force is always opposite in direction to the elastic deformation.

6-3 STRESS AND STRAIN

The designers of mechanical structures have found that a very useful statement of Hooke's law may be given in terms of stress and strain. In order to use this form of the law, we must understand the following two definitions:

Stress is the force per unit area.

Strain is the fractional deformation in the object. The use of these definitions is illustrated in the following example.

Example 6-1 A 5000-lb load is suspended from a steel cable that is 10 ft long and 0.5 in.² in cross section. The load stretches the cable by 0.04 in (a) What is the stress? (b) What is the strain?

Solution. (a) Stress $= F/A = 5000$ lb/0.5 in.² $= 10,000$ lb/in.² The stress could also have been calculated in lb/ft². However, lb/ft² is not as useful as lb/in.² in many cases.

(b) Strain $= \dfrac{\text{change in length}}{\text{original length}} = \dfrac{\Delta l}{l}$

$= \dfrac{0.04 \text{ in.}}{120 \text{ in.}} = 3.33 \times 10^{-4}$

Since the strain is a ratio of two lengths, it has no units. ◄

The general statement of Hooke's law of elasticity is

The stress in an elastic object is directly proportional to the strain if the elastic limit of the material is not exceeded.

The constant of proportionality between stress and strain is called the *elastic modulus* of the material. Hooke's law thus states that

$$\text{stress} = \text{elastic modulus} \times \text{strain} \quad (6\text{-}2)$$

We shall study some very important applications of this law in the next section.

6-4 YOUNG'S MODULUS OF ELASTICITY; LENGTH ELASTICITY

The elastic modulus used in calculations of length elasticity is called Young's modulus. This is a very important quantity. Its value for most materials can be found in engineering reference tables. If Young's modulus is known, it is possible to calculate the amount of stretching that will be produced by a given stress.

When a rod or cable of cross section A is subjected to a stretching force F, the stress is F/A. The strain produced will be $\Delta l/l$. If we let Y stand for Young's modulus, Eq. 6-2 becomes

$$\frac{F}{A} = Y\frac{\Delta l}{l} \quad (6\text{-}3)$$

Some typical valves of Young's modulus are shown in Table 6-1.

Table 6-1   Values of Young's Modulus

| Substance | Young's Modulus | | Stress at Elastic Limit | Breaking Stress |
	lb/in.²	dynes/cm²	lb/in.²	lb/in.²
Aluminum, rolled	1×10^7	7.0×10^{11}	25,000	29,000
Gold	1.14×10^7	7.85×10^{11}		25,000
Iron, wrought	2.8×10^7	20×10^{11}	26,000	42–52,000
Lead, rolled	0.24×10^7	1.67×10^{11}		3000
Rubber		0.05×10^{11}		
Steel, annealed	3.0×10^7	20×10^{11}	40,000	64,000

Example 6-2 Measurements indicate that a steel wire 5 ft long is elongated 0.020 in. when a 100-lb weight is suspended from it. The wire is 0.01 in.^{2} in cross section. Calculate Young's modulus from these data.

Solution. From Eq. 6-3

$$Y = \frac{F/A}{\Delta l/l} = \frac{\text{stress}}{\text{strain}}$$

$$\text{stress} = \frac{100 \text{ lb}}{0.01 \text{ in.}^2} = 10^4 \text{ lb/in.}^2$$

$$\text{strain} = \frac{0.020 \text{ in.}}{60 \text{ in.}} = \frac{2}{6000} = \frac{1}{3000}$$

$$Y = \frac{10^4 \text{ lb/in.}^2}{\frac{1}{3000}} = 30 \times 10^6 \text{ lb/in.}^2 \quad \blacktriangleleft$$

As a second example in the use of Eq. 6-3, the reader should carefully study the following.

Example 6-3 Young's modulus for copper is 16×10^6 lb/in.^{2} A 200-lb load is to be suspended from a copper rod that is 3 ft long. The elastic limit for the coper rod is 5×10^3 lb/in.^{2}
(a) What is the minimum diameter for the rod if the elastic limit is not to be exceeded? (b) How much will the minimum-diameter rod be elongated when the load is applied?

Solution. (a) The stress at the elastic limit is

$$\frac{F}{A} = 5.0 \times 10^3 \text{ lb/in.}^2$$

$$A = \frac{F}{5 \times 10^3 \text{ lb/in.}^2} = \frac{200 \text{ lb}}{5.0 \times 10^3 \text{ lb/in.}^2}$$

$$= 40 \times 10^{-3} \text{ in.}^2$$

Using the formula for the area of a circle, we obtain

$$A = \pi r^2 = \frac{\pi d^2}{4} = 40 \times 10^{-3} \text{ in.}^2$$

$$d^2 = \frac{4(4 \times 10^{-2}) \text{ in.}^2}{\pi}$$

$$d = \frac{4}{\sqrt{\pi}} \times 10^{-1} \text{ in.} = 0.225 \text{ in.}$$

(b) From Eq. 6-3

$$\text{stress} = Y \Delta l/l$$

$$\Delta l = \frac{l}{Y} \times \text{stress} = \frac{3 \text{ ft} (5 \times 10^3 \text{ lb/in.}^2)}{16 \times 10^6 \text{ lb/in.}^2}$$

$$= \frac{15}{16} \times 10^{-3} \text{ ft} \times \frac{12 \text{ in.}}{1 \text{ ft}}$$

$$= 0.0113 \text{ in.} \quad \blacktriangleleft$$

The objects in Examples 6-2 and 6-3 were subjected to tensile stresses. A *tensile stress* tends to elongate or stretch an object. The strain produced by a tensile stress may be calculated from Eq. 6-3 if the elastic limit of the material is not exceeded.

The form of Hooke's law given by Eq. 6-3 may also be used in some cases to calculate the strain when an object is subjected to a *compressive stress*. A compressive stress tends to decrease the length of a stiff rod or a supporting column. If the compressive stress is small enough so that the elastic limit of the material is not exceeded and if there is no sideways buckling in the member, then Eq. 6-3 may be used to calculate the strain and the decrease in length.

6-5 BULK MODULUS; VOLUME ELASTICITY

If any material is subjected to pressure on all sides, the volume tends to decrease. The bulk modulus B is the elastic modulus used in calculating the decrease in volume. The stress is the force per unit area or the pressure P. Since ΔV signifies an increase in volume, we shall let $-\Delta V$ stand for the decrease in volume. The strain will therefore be the fractional decrease in volume, or $-\Delta V/V$. Substitution in Eq. 6-2 gives

$$P = B\left(\frac{-\Delta V}{V}\right) \qquad (6\text{-}4)$$

Example 6-4 The cylinder of a large hydraulic press contains 5 gal of oil. The oil is subjected to a pressure of 2400 lb/in.^{2} Calculate the decrease in the volume of the oil (the bulk modulus of the oil is 8×10^4 lb/in.^{2}).

Solution. From Eq. 6-4

$$-\Delta V = \frac{VP}{B} = 5 \text{ gal} \times \frac{2400 \text{ lb/in.}^2}{8 \times 10^4 \text{ lb/in.}^2}$$

$$= 0.15 \text{ gal} \qquad \blacktriangleleft$$

It should be obvious from Eq. 6-4 and from the last example that the bulk modulus is measured in the same units as pressure: force per unit area. In studying the elastic properties of materials, it is sometimes more convenient to use the reciprocal of the bulk modulus or the compressibility k, defined by

$$k = \frac{1}{B} \qquad (6-5)$$

If Eq. 6-5 is substituted into Eq. 6-4, we obtain

$$P = \frac{1}{k}\left(\frac{-\Delta V}{V}\right)$$

or

$$k = \frac{-\Delta V/V}{P} \qquad (6-6)$$

Equation 6-6 indicates that the compressibility is the fractional decrease in volume produced per unit increase in pressure. The compressibility of the oil in Example 6-4, for instance, may be calculated as

$$k = \frac{1}{B} = \frac{1}{8 \times 10^4 \text{ lb/in.}^2}$$

$$= 12.5 \times 10^{-6} \frac{1}{\text{lb/in.}^2}$$

$$= 12.5 \times 10^{-6}(\text{lb/in.}^2)^{-1}$$

This result means that the volume of oil will decrease by 12.5/1,000,000 when the pressure is increased by 1 lb/in.^{2}

Pressure is sometimes measured in units called atmospheres (atm). One atmosphere is equal to normal air pressure at sea level:

1 atmosphere (1 atm) = 14.7 lb/in.^{2}

If the pressure is measured in atmospheres, the compressibility will be given in parts per atmo-

sphere, or atm^{-1}. This unit is illustrated in the next example.

Example 6-5 The bulk modulus of cast iron is given in the tables as 8.7×10^6 lb/in.^{2} Calculate (a) the compressibility in atm^{-1}, (b) the decrease in volume produced when a pressure of 100 atm is applied to an iron block that has a volume of 2 ft^3, and (c) the pressure required to produce a fractional decrease in volume of $\frac{1}{10}$ of 1%.

Solution. From Eq. 6-5,

(a) $k = \dfrac{1}{B} = \dfrac{1}{8.7 \times 10^6 \text{ lb/in.}^2}$

$$= 0.115 \times 10^{-6} \frac{1}{\text{lb/in.}^2}$$

$$\times \frac{14.7 \text{ lb/in.}^2}{1 \text{ atm}}$$

$$= 1.69 \times 10^{-6} \text{ atm}^{-1}$$

(b) From Eq. 6-6,

$$-\Delta V = VPk$$

$$= 2 \text{ ft}^3 (100 \text{ atm}) \times 1.69 \times 10^{-6} \text{ atm}^{-1}$$

$$= 3.38 \times 10^{-4} \text{ ft}^3$$

(c) $\dfrac{-\Delta V}{V} = \dfrac{1}{10}$ of 1% $= 0.001$

$$P = \left(\frac{-\Delta V}{V}\right)\frac{1}{k} = \frac{10^{-3}}{1.69 \times 10^{-6} \text{ atm}^{-1}}$$

$$= \left(\frac{1}{1.69}\right) \times 10^3 \text{ atm}$$

$$= 591 \text{ atm}$$

6-6 SHEAR MODULUS AND SHEARING STRESSES

In Section 6-4 we learned that Young's modulus may be used in calculating the strain caused by a simple tensile or compressive stress. In Section 6-5 it was explained that the bulk modulus may be used to calculate the volume strain if the stress

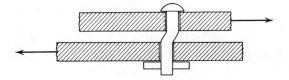

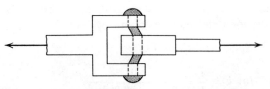

Figure 6-3 Shearing stress.

is a uniform pressure. The elastic modulus called the *shear modulus* is also of great importance in many engineering and design problems.

The bolts in Fig. 6-3 are subjected to a *shearing stress*. If the shearing stress becomes too great, the bolts may be "sheared off." However, if the elastic limit of the material is not exceeded, the shearing strain may be calculated by using Hooke's law and looking up the *shear modulus* in the tables.

The effect of a shearing stress on a steel rod is indicated in Fig. 6-4. The shearing stress is defined as

$$\frac{\text{shearing force}}{\text{area}} = \frac{F}{A}$$

The shearing strain is the fractional deformation, or x/l. If the elastic limit is not exceeded, the stress

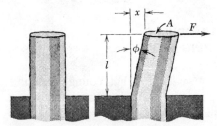

Figure 6-4 Shearing stress and shearing strain on a steel rod.

is proportional to the strain, or

$$\frac{F}{A} = \frac{\eta x}{l} \qquad (6\text{-}7)$$

The constant of proportionality η is the shear modulus of the material.

The diagram indicates that $x/l = \tan \phi$ where ϕ is the angle of shear. Since ϕ is a small angle, the tangent of ϕ is almost exactly equal to ϕ measured in radians. Using the angle of shear, we can rewrite Eq. 6-7 in the form

$$\frac{F}{A} = \eta \phi \qquad (6\text{-}8)$$

Example 6-6 The steel rod in Fig. 6-4 is 1.0 in. high and 0.5 in. in diameter. The shearing force, F, is 6000 lb. The shear modulus of steel is 11.6×10^6 lb/in.^{2} Determine (a) the shearing stress, (b) the deflection x, and (c) the angle of shear.

Solution. (a) Stress $= \dfrac{F}{A}$

$$A = \frac{\pi d^2}{4} = \frac{\pi (0.5 \text{ in.})^2}{4}$$

$$= 0.196 \text{ in.}^2$$

$$\text{stress} = \frac{6000 \text{ lb}}{0.196 \text{ in.}^2}$$

$$= 30.6 \times 10^3 \text{ lb/in.}^2$$

(b) $\dfrac{F}{A} = \dfrac{\eta x}{l}$

$$x = \frac{F}{A} \frac{l}{\eta} = \frac{30.6 \times 10^3 \text{ lb/in.}^2}{11.6 \times 10^6 \text{ lb/in.}^2} 1.0 \text{ in.}$$

$$= 2.64 \times 10^{-3} \text{ in.}$$

(c) $\tan \phi = \dfrac{2.64 \times 10^{-3} \text{ in.}}{1.0 \text{ in.}}$

$$\phi = 2.64 \times 10^{-3} \text{ rad}$$
$$\times (57.3 \text{ degrees}/1 \text{ rad})$$
$$= 0.15 \text{ degree} \qquad \blacktriangleleft$$

The shear modulus is sometimes called the *coefficient of rigidity*; it is used in calculating the strain when a shaft or rod is subject to a twisting or *torsional stress*. Figure 6-5 illustrates the effect of torsion upon a rod. If the elastic limit of the material is not exceeded, the laws of elasticity may be used to calculate the strain (the amount of twisting deformation).

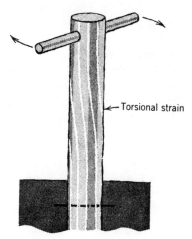

Figure 6-5 Torsion.

Shearing stresses are also involved in the *flexure* or bending of loaded beams (see Fig. 6-6). Structural designers and engineers must calculate the deflections of the beams used in construction projects and other engineering applications. These technical methods are usually studied in specialized engineering and technology programs.

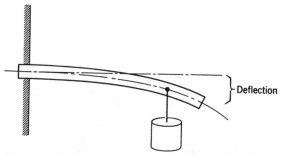

Figure 6-6 Flexure (bending) in a loaded beam.

SUMMARY

Hooke's law states that the elastic deformation is directly proportional to the force if the elastic limit is not exceeded.

The restoring force is opposite in direction and directly proportional to the elastic deformation in a spring. Equation 6-1 is a mathematical statement of Hooke's law applied to a spring. The spring constant or the stiffness k is the constant of proportionality giving the ratio of force to elastic deformation for a given spring. The units for k may be lb/ft or lb/in.

A general form of Hooke's law states that the stress is proportional to the strain if the elastic limit is not exceeded. In other words

$$\text{stress} = \text{elastic modulus} \times \text{strain}$$

where *stress* is the force per unit area and *strain* is the fractional deformation.

The elastic modulus for length elasticity is called *Young's Modulus*, and when used in Hooke's law, we obtain Eq. 6-3.

The elastic modulus for volume elasticity is the *bulk modulus*. Hooke's law takes the form of Eq. 6-4 when pressure causes an object to decrease in volume. The stress in this case is the *pressure*, and the fractional decrease in volume is the strain.

Pressure is sometimes measured in atmospheres:

$$1 \text{ atm} = 14.7 \frac{\text{lb}}{\text{in.}^2}$$

The *compressibility* is the reciprocal of the bulk modulus. The compressibility is the fractional decrease in volume per unit of pressure (Eq. 6-6).

Hooke's law may also be applied to shearing stress and shearing strain, where the *shear modulus* is the appropriate elastic modulus (Eq. 6-7).

QUESTIONS

1. What is an elastic deformation?
2. What is the difference between an elastic deformation and a nonelastic deformation?

3. What is meant by "elastic limit"?

4. Explain the meaning of "spring constant."

5. What is the difference between the stretching force and the restoring force?

6. What is the difference between "stress" and "strain"?

7. Define the term "elastic modulus."

8. Define Young's modulus and explain why the units for Young's modulus must be the same as the units for stress.

9. Suppose that 2000-lb weights are suspended from different steel wires. In which wire will the stress be greatest, and in which wire will the stress be least?

10. When the same stress is applied to different copper rods, the change in length is not the same in all the rods even though Young's modulus is the same. Which rods have the greatest increase in length and which have the least?

11. Define the term *compressibility* and give two different units for measuring the compressibility.

12. Explain what is meant by a shearing force and define shearing stress.

13. What is the difference between a tensile stress and a shearing stress?

PROBLEMS

(A)

1. A 16-lb weight causes a 10-in. elongation in a spring. Calculate the force constant of the spring.

2. An unstretched spring is 8 in. long. When a 10-lb weight is suspended from the spring, the length is 12 in. Calculate the spring constant.

3. How great a push is required to shorten a spring by 4 in.? The spring constant is 30 lb/ft.

4. A 5000-lb load is suspended from a cable 0.5 sq in. in cross-sectional area. Calculate the stress.

5. A steel cable that is 100 ft long carries a load which stretches it by 0.25 in. Calculate the strain.

6. When a tensile stress of 50 lb/in.^{2} is applied to an elastic shock cord, the cord increases in length by 25%. Calculate an effective value for Young's modulus for this material.

7. The maximum allowable stress in a steel cable is 15,000 lb/in.^{2} Calculate the maximum weight that may be suspended from a cable 0.5 in. in diameter.

8. An elastic rod is 5 ft long and has a cross-sectional area of 0.25 in.^{2} When a 1000-lb weight is suspended from the rod, the length increases by 0.016 in. Calculate the stress, the strain, and Young's modulus.

9. The weight of a two-wheel trailer rests upon two leaf springs. The stiffness of each spring is 200 lb/in. How much weight placed on the trailer will cause the trailer body to move down 3 in.? Assume that the elastic limit of the springs is not reached.

10. When a load of 10 tons is applied to a large automotive spring, the spring is compressed a distance of 9 in. Calculate the spring constant in lb/ft and lb/in.

11. An iron post is 4 ft high and has a cross-sectional area of 10 in.^{2} How much does the post contract in length if it supports a weight of 10 tons? Assume that Young's modulus for the iron is 25×10^6 lb/in.^{2}

12. In the metric system, forces may be measured in units called dynes. A force of 1000 dynes causes a spring to stretch 5 cm.
    (a) Calculate the force constant in dynes/cm.
    (b) How large a force is required to shorten the spring by 2 cm?

13. A metal wire that is 30 in. long is placed under tension, which causes the wire to stretch by 0.006 in. Young's modulus for this material is 25×10^6 lb/in.^{2}
    (a) Calculate the strain in this wire.
    (b) Calculate the stress.

(B)

1. Young's modulus for copper is 16×10^6 lb/in.^{2} How much stress in a copper wire will increase the length by $\frac{1}{100}$ of 1%?

2. A 50 lb lead weight is suspended from a steel wire that is $\frac{1}{16}$ in. in diameter and 5 ft long.
    (a) Calculate the stress in the steel wire.
    (b) What is the strain in the wire?
    (c) How much does the wire stretch?
    Assume that the elastic limit is not reached and that $Y = 30 \times 10^6$ lb/in.^{2} for steel.

3. (a) In Fig. 6-7 how much tension is there in the cable?
    (b) How much stretch (elongation) is there in each spring?

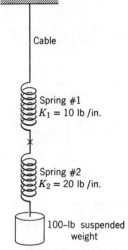

Figure 6-7 The extension of two connected springs.

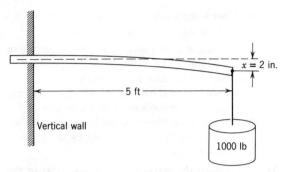

Figure 6-8 Elastic deflection of a loaded beam.

4. A flexible steel measuring tape is 100 ft long. The tape is 0.5 in. wide and 0.02 in. thick. Young's modulus for the steel is 30×10^6 lb/in.^{2} How much of a force will cause the tape to increase in length by 0.1 in.?

5. A stone block 2 ft × 2 ft × 3 ft is hanging from a steel wire 0.5 in. in diameter and 10 ft long. The weight per unit volume of the stone is 300 lb/ft^3.
(a) Calculate the stress in the wire.
(b) How much does the wire stretch?
(Assume that Young's modulus for the steel is 30×10^6 lb/in.^{2})

6. The bulk modulus for water is 31×10^4 lb/in.^{2} How great must the pressure be to cause the volume of a given quantity of water to decrease by $\frac{1}{10}$ of 1%?

7. The weight per unit volume of sea water is 64 lb/ft^3 at normal pressure. The compressibility is 50×10^{-6} atm^{-1}. Deep under the sea the pressure is 10,000 lb/in.^{2}; under these conditions, calculate:
(a) The fractional decrease in volume.
(b) The density.

8. A steel cube, 2 in. on a side, is placed in a vise that applies a compressive force of 300 lb to the cube. Young's modulus for this steel is 30×10^6 lb/in.^{2}
(a) Calculate the stress on the cube.
(b) How much does the thickness of the cube decrease because of the compressive stress?

9. The 1000-lb load causes the end of the beam in Fig. 6-8 to undergo an elastic deflection of 2 in.

($x = 2$ in.) The beam is 5 ft long and 4 in.^{2} in cross section, and the weight of the beam may be neglected.
(a) Calculate the shearing stress.
(b) Determine the force constant in lb/ft for the elastic deflection.

10. The piston in Fig. 6-9 has an area of 10 in.^{2}, and a force $F = 500$ lb is applied to the piston. Assume that no liquid leaks out of the cylinder and the walls are perfectly rigid. The compressibility of glycerine is 22×10^{-6} atm^{-1}.
(a) What is the pressure on the glycerine in atmospheres?

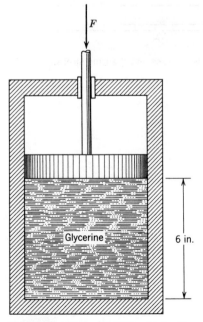

Figure 6-9 Compressibility of a liquid.

(b) What is the fractional decrease in volume?

(c) How far down does the piston move?

11. A hydraulic system contains a type of oil whose bulk modulus is 1.8×10^5 lb/in.^{2} A pressure of 200 lb/in.^{2} is produced in the system.

(a) By what fraction will the volume of oil decrease because of the application of this pressure?

(b) If the original volume of oil is 10 gal, calculate the decrease in volume caused by the application of this pressure.

12. In Fig. 6-10 the maximum allowable stress in the supporting cable is 10,000 lb/in.^{2} Ignore the weight of the horizontal boom.

(a) Calculate the tension in the supporting cable.

(b) What is the minimum allowable cross-sectional area for this cable?

(c) How much will the supported cable be elongated?

Assume that the cable is of minimum thickness and that $Y = 30 \times 10^6$ lb/in.^{2}

13. In Fig. 6-11 the cross section of AC is 0.25 in.^{2}; AC and AB are made of steel. For steel Young's modulus $= 30 \times 10^6$ lb/in.^{2} and shear modulus $= 11.6 \times 10^6$ lb/in.^{2} Design specifications allow a

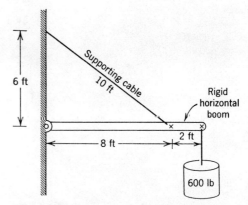

Figure 6-10   Stress in a supporting cable.

maximum tensile stress of 16,000 lb/in.^{2} in AC and a maximum shearing stress of 10,000 lb/in.^{2} in AB. Calculate:

(a) The maximum force P.

(b) The elongation of AC when P is applied to this member.

(c) The minimum cross section of AB.

(d) The deflection in AB.

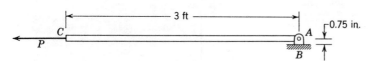

Figure 6-11   Tensile stress and shearing stress.

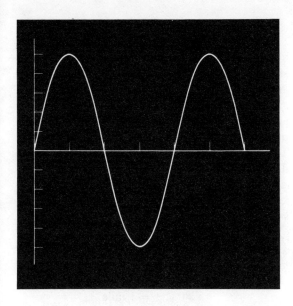

7.
Kinematics and
Linear Motion

The further study of physics will require an analysis of motion and moving objects. In order to understand motion, we shall use the important concepts of *displacement, velocity,* and *acceleration*. Both the English engineering and metric units will be employed in this study.

LEARNING GOALS

In studying the material of this chapter, the reader should be alerted to the following objectives:

1. *The drawing and interpretation of displacement–time diagrams.*
2. *The calculation of instantaneous velocity from the displacement–time diagrams or other data.*
3. *The drawing of velocity–time diagrams.*
4. *The calculation of displacement and acceleration from the velocity–time diagram or data.*
5. *The use of correct units in measuring displacement, velocity, and acceleration.*

7-1 KINEMATICS AND THE DESCRIPTION OF MOTION

The preceding chapters have been directed toward a study of mechanical forces with special emphasis on the branch of mechanics called statics. But it should be obvious to the student that moving parts and moving objects are of great significance in all of science and technology. In order to proceed further into our study of physics, we must learn useful methods for representing, measuring, and analyzing motion. These mathematical techniques for describing motion comprise the branch of mechanics called *kinematics*.

7-2 MOTION AND LINEAR DISPLACEMENT

The *linear displacement* of a moving object refers to a change in location of the object with respect to some suitably chosen reference point. Because change of location involves both a magnitude and direction, linear displacement is a vector quantity; therefore, the vector methods of Chapter 2 may be employed to describe and combine linear displacements.

The usual mathematical symbol used for the linear displacement vector is indicated as s. However, an arrow is written over the letter $\vec{s}$ or the boldface letter **s** is used in printed matter when we wish to discuss the directional and vector characteristic of the linear displacement.

In the English Engineering System of units the magnitude of the linear displacement is measured in feet. The foot is the basic unit of length in this system. However, the other units in Fig. 7-1 are also useful for measuring linear displacements.

Although the English Engineering System is still employed in many branches of engineering and technology, it is being replaced by the metric system of units. In scientific work and in many engineering specializations, the metric system of units has completely replaced the older system. The basic unit for measuring length and linear displacement in the metric system is the *meter* (see Fig. 7-1). The other units in this system (especially the centimeter) are also used in measurement of length.

(a) English Engineering units

$$12 \text{ inches (in.)} = 1 \text{ foot (ft)}$$
$$3 \text{ feet (ft)} = 1 \text{ yard (yd)}$$
$$5280 \text{ feet} = 1 \text{ mile}$$

(b) Metric system of units

$$1000 \text{ millimeters (mm)} = 1 \text{ meter (m)}$$
$$100 \text{ centimeters (cm)} = 1 \text{ meter}$$
$$1000 \text{ meters} = 1 \text{ kilometer (km)}$$
$$\text{milli} = \tfrac{1}{1000} \text{ (one thousandth)}$$
$$\text{centi} = \tfrac{1}{100} \text{ (one hundredth)}$$
$$\text{kilo} = 1000 \text{ (thousand)}$$

To convert from one system to the other,

$$1 \text{ yard} = 0.9144 \text{ meter} = \tfrac{3600}{3937} \text{ meter}$$
$$1 \text{ inch} = 2.540 \text{ centimeters}$$
$$1 \text{ foot} = 30.48 \text{ centimeters}$$
$$1 \text{ mile} = 1.609 \text{ kilometers}$$

Although the magnitude of linear displacement is measured in length units, length and linear displacement are not synonymous. Length (or path length) is a scalar quantity and possesses only

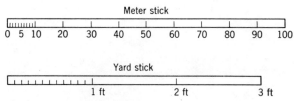

Figure 7-1   Measurement of length.

a magnitude. Linear displacement, a vector quantity, is represented by a magnitude and a direction. Some of the important differences between length and linear displacement are emphasized in the following example.

Example 7-1 An object moves from *A* to *B* along a semicircle (see Fig. 7-2). The radius of the curved path is 3 m. (a) Calculate the path length from *A* to *B*. (b) What is the linear displacement?

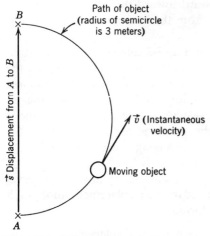

Figure 7-2 Path length and displacement.

Solution. (a) The path length from *A* to *B* is one-half the circumference of a circle of radius 3 m.

$$\text{path length} = \tfrac{1}{2}(2\pi r) = \pi r$$
$$= \pi(3 \text{ m}) = 9.42 \text{ m}$$

(b) The displacement from *A* to *B* is represented by the arrow drawn in the diagram. The direction is upward on the diagram and the magnitude of the linear displacement is $2r = 6$ ms. ◀

7-3 SPEED AND VELOCITY

In describing the motion of an object, it is common practice to state the speed. The *speed* of an object is the rate at which it is traversing its path.

According to this definition,

$$\text{average speed} = \frac{\text{length of path}}{\text{time}} \qquad (7\text{-}1)$$

The units for measuring speed depend on the units used for measuring length and time.

Example 7-2 An object travels a distance of 80 cm in 2.0 sec. Calculate the average speed in (a) centimeters per second, (b) meters per second, (c) kilometers per hour, (d) feet per second, and (e) miles per minute.

Solution

(a) average speed = 80 cm/2.0 sec = 40 cm/sec.

(b) 40 cm/sec × (1 m/100 cm) = 0.40 m/sec.

(c) To convert to km/hr,

$$0.40 \text{ m/sec} \times 1 \text{ km/1000 m}$$
$$\times (3600 \text{ sec/1 hr}) = 1.44 \text{ km/hr}$$

(d) To convert to ft/sec, we have

$$40 \text{ cm/sec} \times (1 \text{ ft/30.48 cm}) = 1.31 \text{ ft/sec}$$

(e) 1.31 ft/sec × (1 mile/5280 ft) × 60 sec/1 min
$$= 1.49 \times 10^{-2} \text{ mile/min} \qquad ◀$$

Students sometimes confuse the terms speed and *velocity*. The two, however, are different. Speed is a scalar quantity, whereas velocity is a vector. The magnitude of the velocity vector is measured in the same units as speed, but there is always a direction associated with velocity.

If the direction of motion changes, there is always a change in velocity, even though the speed of the object may remain constant. If an automobile is traveling on a curved road at a steady speed of 60 miles/hr, the speed is constant. The magnitude of the velocity is also constant (60 miles/hr), but the direction of the velocity vector is not constant and the velocity is changing. In most instances it is more important to study the velocity and its rate of change rather than the speed.

The velocity of an object describes the rate at which its displacement is changing. Velocity is defined as the time rate of change in the displacement. If we let $\mathbf{v}_{av}$ stand for the average velocity (a vector), the definition of average velocity is

$$v_{av} = \frac{s}{t} \qquad (7\text{-}2)$$

Bearing in mind that the displacement and velocity are vectors, it is permissible to write

$$s = v_{av}t \qquad (7\text{-}3)$$

Example 7-3   In Fig. 7-2 the object takes 4 sec to travel from A to B along the curved path. Compute (a) the average speed and (b) the average velocity.

Solution

(a) average speed $= \dfrac{\text{path length}}{\text{time}}$

$$= \frac{9.42 \text{ m}}{4 \text{ sec}} = 2.36 \text{ m/sec}$$

(b) $v_{av} = \dfrac{s}{t} = \dfrac{6 \text{ m}}{4 \text{ sec}} = 1.5 \text{ m/sec}$

The magnitude of the average velocity during the 4-sec interval is 1.5 m/sec. The direction of v_{av} is the same as the direction of the displacement, from A toward B. However, it should be obvious that at the beginning of the 4-sec interval, the object was moving toward the right at 2.36 m/ sec; and the end of the interval it was moving toward the left at 2.36 m/sec. At any given instant during the 4-sec interval the object was moving at 2.36 m/sec in a direction tangent to its curved path.   ◄

In Example 7-3 and in many other cases the average velocity during some time interval is different from the *instantaneous velocity* at a particular instant of time. At any given instant of time the direction of the instantaneous velocity is always tangent to the path of the object and the magnitude of the instantaneous velocity is equal to the speed. In using the formulas of kinematics, the reader must distinguish carefully between the average velocity, indicated as v_{av} or v_{av}, and the instantaneous velocity, indicated as v or v.

7-4 DISPLACEMENT–TIME DIAGRAMS

If an object is moving along a straight-line path, a a very useful and simple method of studying the motion involves the drawing of a displacement–time graph. The table and the graph in Fig. 7-3 illustrate this method. All the linear displacements (distances) in the table are measured with respect to a zero reference point called the origin. The data and the graph indicate that the object is originally 4 m from the origin. The object then moves out to a maximum distance of 9.5 m. It then moves backward toward the origin and reaches a point 2 m from the zero reference position at $t = 20$ sec.

The graph in Fig. 7-3 conveys a lot of information about the motion of the object. When the slope of the graph is very steep, the displacement is changing rapidly, and, consequently, the instantaneous velocity is large. When the slope is small and the graph is nearly horizontal, the displacement is changing slowly and the instantaneous velocity is small.

The average velocity during any time interval can be calculated from a displacement–time graph by using the formula

$$v_{av} = \frac{\text{change in displacement}}{\text{time interval}}$$

or $v_{av} = \dfrac{s_2 - s_1}{t_2 - t_1} \qquad (7\text{-}4)$

If the time interval $t_2 - t_1 = \Delta t$ is made very short, Eq. 7-4 may be used to calculate the slope of the displacement–time graph. This slope is equal to the instantaneous velocity. A formula for finding the slope and the instantaneous velocity is

$$v = \frac{\Delta s}{\Delta t} \qquad (7\text{-}5)$$

(when Δt is made very small).

If the displacement–time graph is very curved, it is usually possible to determine the slope by placing a ruler or straight-edge tangent to the curve at the point where we wish to determine the instantaneous velocity. However, if the curve is

Time (t) (seconds)	0	2	4	6	8	10	12	14	16	18	20
Displacement (s) meters	4.0	5.0	6.0	7.0	9.0	9.5	9.0	8.0	7.0	6.0	2.0

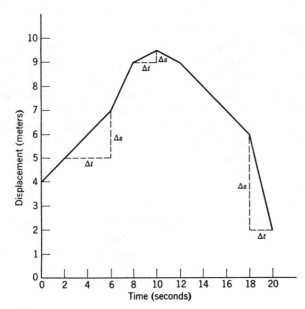

Figure 7-3 A displacement-time graph.

reasonably smooth, the slope may be calculated directly by using Eq. 7-5.

Example 7-4 From the displacement–time diagram in Fig. 7-3, calculate (a) the average velocity during the first 10 sec; (b) the average velocity for the 20 sec of motion; and (c) the instantaneous velocities at $t = 4$ sec, 9 sec, and 19 sec.

Solution. (a) For the first 10 sec we have

$$v_{av} = \frac{s_2 - s_1}{t_2 - t_1} = \frac{9.5 \text{ m} - 4 \text{ m}}{10 \text{ sec}} = 0.55 \text{ m/sec}$$

(b) For the entire 20 sec,

$$s_2 - s_1 = 2 \text{ m} - 4 \text{ m} = -2 \text{ m}$$

$$v_{av} = \frac{s_2 - s_1}{t_2 - t_1} = \frac{-2 \text{ m}}{20 \text{ sec}} = -0.10 \text{ m/sec}$$

The minus sign signifies that the average velocity is in the negative or backward direction.

(c) At $t = 4$ sec we find from the graph

$$v = \frac{\Delta s}{\Delta t} = \frac{7 \text{ m} - 5 \text{ m}}{6 \text{ sec} - 2 \text{ sec}} = \frac{2 \text{ m}}{4 \text{ sec}} = 0.5 \text{ m/sec}$$

At $t = 9$ sec,

$$v = \frac{\Delta s}{\Delta t} = \frac{9.5 \text{ m} - 9 \text{ m}}{10 \text{ sec} - 8 \text{ sec}} = 0.25 \text{ m/sec}$$

At $t = 19$ sec,

$$v = \frac{\Delta s}{\Delta t} = \frac{2 \text{ m} - 6 \text{ m}}{20 \text{ sec} - 18 \text{ sec}} = -2.0 \text{ m/sec}$$

At $t = 19$ sec the slope of the graph is negative, and the minus sign in the instantaneous velocity indicates that the object is moving backward. ◀

In some very important instances the displacement–time graph forms a straight line. Since the slope of a straight line does not change, a straight line on the displacement–time curve indicates that the velocity is constant. If the straight line is also horizontal, the velocity is zero and the object is stationary.

The displacement–time graph is sometimes referred to as the *s–t* curve. In the next section we shall consider a type of graph or curve that is of even greater usefulness in studying motion.

7-5 THE VELOCITY–TIME GRAPH

In Fig. 7-4 a graph has been drawn to indicate how the velocity of an automobile varies with time. The table and the *v–t* curve indicate that the velocity is initially zero, which means that the car is at rest. The car then begins to move and, after 12 sec, the velocity has increased to a maximum of 100 ft/sec. The horizontal portion of the *v–t* curve indicates that the car travels with a constant,

unchanging velocity of 100 ft/sec for 18 sec. The brakes are then applied and the velocity decreases until the car is brought to rest.

A great deal of information can be obtained from a *v–t* curve such as Fig. 7-4. When the slope of the curve is steep, the velocity is changing quickly. When the slope is small, the velocity is changing slowly. If the *v–t* curve is horizontal, the velocity is not changing. The slope of the *v–t* curves indicates the acceleration, or the rate at which the velocity is changing.

Acceleration is defined as the rate of change in the velocity. Both acceleration and velocity are vector quantities. A large acceleration means that the velocity of the object is changing rapidly or abruptly. A small acceleration signifies that the velocity is changing slowly. A negative value for the acceleration occurs when the velocity is decreasing. If the acceleration is zero, the velocity is not changing, and the velocity remains constant in magnitude and direction.

To determine the acceleration from a *v–t* diagram, the slope may be calculated by using the

Time (seconds)	0	2	4	6	8	10	12	15	20	30	31	32	34	35
Velocity (ft /sec)	0	20	40	60	80	90	100	100	100	100	80	60	20	0

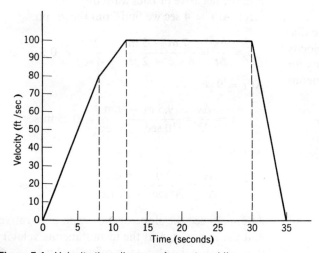

Figure 7-4 Velocity-time diagram of an automobile.

formula

$$a = \frac{\Delta v}{\Delta t} \tag{7-6}$$

(where $\Delta t = t_2 - t_1$ is a small time interval). The units for measuring acceleration depend on the units used to measure velocity and time. In Fig. 7-4 the velocity is measured in ft/sec and the time in seconds so that the acceleration will be measured in ft/sec^2 (feet per second per second). In the metric system the units for acceleration would be m/sec^2 (meters per second per second) or cm/sec^2 (centimeters per second per second).

It should be apparent that the slope of a v–t curve has great physical importance. The area under a v–t diagram is also of significance. The area under any curve is equal to the horizontal dimension multiplied by the effective or average height. But for a v–t curve, the horizontal dimension represents the time and the average height of the curve represents the average velocity. Therefore we have

$$\text{area} = \text{average velocity} \times \text{time}$$
$$= v_{av} \times t = s$$

Obviously, the area under the v–t curve represents the displacement of the object.

Example 7-5 From the data and the v–t diagram in Fig. 7-4 determine (a) the acceleration of the auto during the first 5 sec of its motion, (b) the negative acceleration (deceleration) of the auto as it is stopping, and (c) the distance traveled during the motion.

Solution. (a) The acceleration (slope) is constant for the first 8 sec.

$$a = \frac{\Delta v}{\Delta t} = \frac{20 \text{ ft/sec}}{2 \text{ sec}} = 10 \text{ ft/sec}^2$$

(b) The deceleration is constant during the last 5 sec because the v–t curve forms a straight line.

$$a = \frac{\Delta v}{\Delta t} = \frac{v_2 - v_1}{t_2 - t_1} = \frac{0 - 100 \text{ ft/sec}}{35 \text{ sec} - 30 \text{ sec}}$$
$$= -20 \text{ ft/sec}^2$$

(c) To determine the displacement, we shall calculate the area under the curve. For the first 8 sec, the area is enclosed by a triangle with

$$\text{area} = \tfrac{1}{2}b \times h$$
$$= \tfrac{1}{2}(80 \text{ ft/sec})(8 \text{ sec}) = 320 \text{ ft}$$

For the time interval from 8 to 12 sec, the area forms a trapezoid. The two bases are 80 and 100 ft/sec. The altitude of the trapezoid is 4 sec.

$$\text{area} = \tfrac{1}{2}(b_1 + b_2) \times h$$
$$= \tfrac{1}{2}(80 \text{ ft/sec} + 100 \text{ ft/sec}) \times 4 \text{ sec}$$
$$= 90 \text{ ft/sec} (4 \text{ sec}) = 360 \text{ ft}$$

From the time interval from 12 to 30 sec the velocity is constant and the area under the curve is

$$\text{area} = b \times h = (100 \text{ ft/sec})(18 \text{ sec})$$
$$= 1800 \text{ ft}$$

During the deceleration from 30 to 35 sec the area under the curve

$$\text{area} = \tfrac{1}{2}b \times h$$
$$= \tfrac{1}{2}(100 \text{ ft/sec})(5 \text{ sec}) = 250 \text{ ft}$$

The total displacement is

$$320 \text{ ft} + 360 \text{ ft} + 1800 \text{ ft} + 250 \text{ ft} = 2730 \text{ ft} \blacktriangleleft$$

In the v–t diagram of the last example it was possible to calculate the displacement by using the geometry formulas for the area of a triangle, a trapezoid, and a rectangle. If the area cannot be divided into such simple figures, there are other schemes for estimating the displacement. One method is to draw the v–t curve on graph paper and then count the number of little boxes under the curve. Since each box has the same area, the total area (and the displacement) can then be easily found by multiplication.

SUMMARY

Kinematics is the mathematical description and analysis of motion.

The *linear displacement* of a moving object is a vector quantity that refers to a change in location of the object. The magnitude of a linear displacement may be measured in feet, meters, centimeters and other length units indicated in Fig. 7-1.

Path length is a scalar quantity, and it represents the distance traveled by an object along its path. If the path is curved, the path length is greater than the magnitude of the linear displacement.

The *speed* is the rate at which an object is traversing its path. The average speed equals the path length divided by time.

The velocity of an object is the rate at which the displacement is changing. The *average velocity* during a time interval t is $v_{av} = s/t$, where s is the displacement during the time interval. The average velocity has both magnitude and direction and is a vector.

The *instantaneous velocity* is also a vector and it is equal to $v = \Delta s/\Delta t$ where Δs is the change in displacement during a very short time interval Δt. If an object is traveling in a straight line the instantaneous velocity at any particular time is equal to the slope on the displacement–time graph.

The time rate of change in the velocity is the *acceleration*, i.e., $a = \Delta v/\Delta t$. The acceleration is also a vector quantity. When the velocity–time graph is drawn, the slope on this graph is the time rate of change in the velocity or the acceleration.

Velocity is measured in ft/sec, m/sec, cm/sec, etc. The units for acceleration are ft/sec², m/sec², cm/sec², etc.

QUESTIONS

1. When will the magnitude of the displacement be exactly equal to the path length?

2. Why is the path length from A to B always greater than or equal to the magnitude of the displacement from A to B?

3. An object travels around the circumference of a circle at the constant rate of 5 feet per second. What is the speed of this object? What is the magnitude of the velocity? Why is the velocity of this object not constant?

4. What is the difference between average velocity and instantaneous velocity?

5. Describe the type of motion in which the average velocity is equal to the instantaneous velocity.

6. An object is moving forward at time $t = 4$ sec, but it is moving backward at $t = 6$ sec. How would this appear on the displacement–time curve?

7. How would the situation described in Question 6 be indicated on the velocity–time curve?

8. Describe the velocity–time curve for the following three types of motion:
   (a) the acceleration is zero;
   (b) the object has a constant positive acceleration;
   (c) the object has a constant negative acceleration (a deceleration).

9. How is the instantaneous velocity obtained from the displacement–time curve?

10. How is the displacement obtained from the velocity–time curve?

11. How does the displacement–time curve indicate that an object reverses its direction of motion and starts to move backwards?

12. How does the velocity–time curve indicate that an object reverses its direction of motion and starts to move backward?

13. How would the displacement–time curve indicate that an object has moved and returned back to its starting point?

14. How would the velocity–time curve indicate that an object has moved and returned back to its starting point?

15. What is the difference between "change in velocity" and "acceleration"?

16. For an object traveling with constant speed, the acceleration is not necessarily zero. Explain.

17. When an object is thrown up into the air and it is moving upward, the velocity vector is directed upward. However, the acceleration vector is directed downward. Explain.

PROBLEMS

(A)

1. A runner travels 300 ft in 10 sec. What is his average speed?

2. A jet plane travels 6000 ft along a runway in 20 sec. What is the average speed in the following units?
   (a) ft/sec.
   (b) miles/hr.

3. In 1975 John Walker ran a mile race in 3 min and 49.4 sec. What was his average speed in
   (a) feet per minute,
   (b) feet per second, and
   (c) miles per hour?

4. A world's track record was set by an athlete who ran 20,052 m in 1 hr. What was his average speed in
   (a) kilometers per hour, and
   (b) meters per second?

5. In 1974 F. Bayi set a world record by running 1500 m in 3 min and 32.2 sec. What was his average speed in
   (a) meters per second, and
   (b) feet per second?

6. A supersonic jet plane has been clocked at 1200 miles/hr.
   (a) How long does it take the plane to travel 2 miles?
   (b) How far does the plane travel in 1 sec?

7. A man can run 1000 meters in 2 min and 25 sec. What is his average speed in meters/sec?

8. An athlete can run 100 yd in 10 sec. If he could maintain this same speed, how long would it take him to run a mile?

9. A trainer clocks a race horse at 1 min 50 sec for a distance of 1 mile. What is the average speed in the following units?
   (a) ft/sec.
   (b) miles/hr.

10. Figure 7-5 is the displacement–time curve of an object moving in a straight line with a constant velocity.

   (a) How far does the object travel in the first minute of its motion?
   (b) How far does the object travel in 5 min?
   (c) What is the velocity of the object?

11. The velocity of a high-powered automobile is checked for 10 sec; the data are recorded in Fig. 7-6.

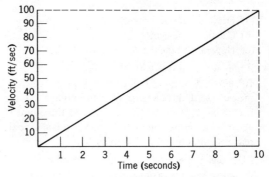

Figure 7-6   Velocity-time graph of an accelerating automobile.

   (a) What is the maximum velocity for the car during the 10 sec?
   (b) What is the displacement during the 10 sec?
   (c) What is the average velocity?
   (d) Calculate the acceleration.

12. Figure 7-7 is the velocity–time diagram of an express train.
   (a) How far does the train travel in 10 sec?
   (b) Calculate the speed of the train in miles per hour.

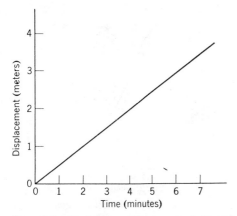

Figure 7-5   Displacement-time graph of object moving with constant velocity.

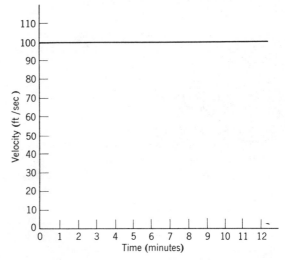

Figure 7-7   Velocity-time diagram of a train.

13. The following data were taken for an accelerating racing car:

Time (sec)	0	2	4	6	8	10
Velocity (ft/sec)	0	25	50	75	100	125

(a) Draw a velocity–time graph for the motion;
(b) Calculate the acceleration;
(c) How far did the car travel in the 10 sec?

14. Radar waves travel with the speed of light 3.0×10^8 m/sec. A beam of radar waves is projected from the earth so that the beam strikes the moon and is reflected back to earth. The distance to the moon is 384,000 km. How long does it take a radar wave to reach the moon and return to earth?

15. Electrons travel 6 cm through a vacuum tube in 2.5 microseconds (a microsecond is a millionth of a second). What is the average velocity of the electrons?

16. An electron travels in a vacuum tube with a constant velocity of 6.5×10^6 m/sec.
(a) How long does it take the electron to travel 35 cm?
(b) How far does the electron travel in 2 microseconds?

17. High energy protons travel in a cyclotron with a speed of 2.95×10^{10} cm/sec.
(a) How long does it take the protons to travel 100 m?
(b) How far do the protons travel in 1 nanosecond? (1 nanosecond = 10^{-9} second.)

18. A satellite travels in an orbit around the earth. It travels 40,000 km in 90 min. What is the average velocity in meters/sec?

19. A rocket is accelerated from 50 m/sec to 150 m/sec in a time of 5 sec. What is the acceleration during these 5 sec?

20. A rocket is initially traveling with a velocity of 200 m/sec. It is then given an acceleration of 10 m/sec² for 8 sec. What is the velocity after the 8 sec of acceleration?

21. A racing car is traveling at 200 ft/sec when the brakes are applied. The velocity then decreases at a uniform rate and the car is brought to rest in 20 sec. Calculate the deceleration (the negative acceleration).

(B)

1. A driver takes a 160-mile automobile trip. He travels at the rate of 45 miles/hr for the first 40 miles. He then gets on a turnpike and drives at 60 miles/hr for the next 100 miles. During the last 20 miles of the trip he is forced to drive at an average speed of 25 miles/hr because of heavy traffic.
(a) How long does the entire trip take?
(b) What is the average speed for the trip?

2. A high-powered car can accelerate from a standstill to 60 miles/hr in 8 sec. Assume that the acceleration is constant.
(a) Draw a v–t diagram. Plot v in ft/sec and t in seconds.
(b) How far did the car travel in the 8 sec?
(c) Calculate the acceleration in ft/sec².

3. A Diesel locomotive is traveling at 100 ft/sec when the brakes are applied. The velocity then decreases at the constant rate of 4 ft/sec² until the locomotive comes to rest.
(a) Draw a velocity–time graph for the decelerating locomotive.
(b) How long does it take the locomotive to stop?
(c) How far does the locomotive travel after the brakes are applied?

4. Figure 7-8 is a velocity–time curve for a decelerating automobile. The brakes were applied at $t = 0$.

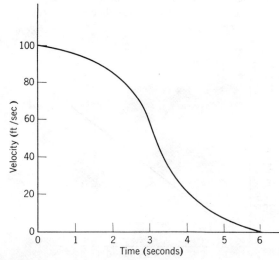

Figure 7-8   Velocity-time curve of a decelerating automobile.

(a) Determine from this figure, as accurately as possible, the distance traveled by the automobile before it came to rest.

(b) Calculate the maximum deceleration during the braking.

5. Figure 7-9 indicates the v–t curve of 15 sec during the motion of a rocket projectile.

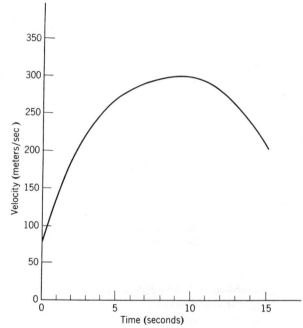

Figure 7-9   Velocity-time curve for a rocket.

(a) What is the initial velocity during this time interval?

(b) What is the maximum velocity?

(c) What is the final velocity?

(d) Determine, as accurately as possible, the distance traveled during the 15 sec.

(e) What is the average velocity during the 15 sec?

6. The marathon foot race is scheduled for a distance of 26 miles 385 yards.

(a) If a runner completes the course in 3 hr and 20 min, calculate his average velocity in ft/sec and in miles/hr.

(b) Suppose that a marathon runner maintains an average velocity of 10 ft/sec. How long does he take to complete the course?

7. The information in Table 7-1 was recorded by a coach during a 400-m trial by a long-distance swimmer.

Table 7-1

Distance from Start (meters)	Time (minutes)	Time (seconds)
50		45
100	1	30
150	2	5
200	2	48
250	3	32
300	4	23
350	5	12
400	5	57

(a) Draw an accurate displacement–time graph for the swimmer.

(b) When did the swimmer achieve his greatest speed?

(c) What was the swimmer's greatest speed?

8. A 100-mile automobile trip is completed in 3 hr. The displacement–time curve is indicated in Fig. 7-10. Determine:

(a) The average velocity during the first 2 hr.

(b) The distance traveled during the third hour.

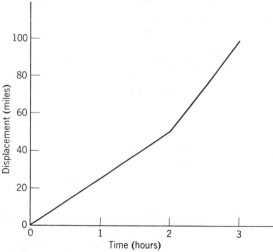

Figure 7-10   Displacement-time diagram of an automobile trip.

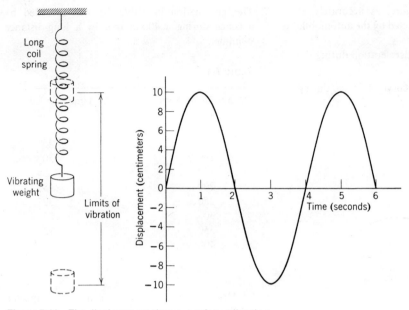

Figure 7-11 The displacement-time curve for a vibrating weight.

(c) The average velocity during the third hour.

(d) The average velocity for the entire trip.

9. The first half of the distance between two towns is traveled at the rate of 20 miles/hr; the second half of the distance is traveled at the rate of 40 miles/hr. What is the average speed for the complete trip? (*Hint*: the average speed is less than 30 miles/hr.)

10. A city bus starts from rest and accelerates at 4 ft/sec² for 10 sec. The bus then travels for 30 sec with a constant velocity. The brakes are then applied and the bus glides to a stop with constant deceleration in 20 sec.

(a) Draw a velocity–time diagram for the 60 sec of the motion.

(b) What was the velocity during the 30-sec interval when the velocity was constant?

(c) How far did the bus travel during:

(i) The first 10 sec?

(ii) The next 30 sec?

(iii) The last 20 sec?

11. The displacement–time graph of a vibrating weight is drawn in Fig. 7-11.

(a) When does the weight have its maximum positive displacement?

(b) At what times is the velocity zero?

(c) Determine the maximum velocity by finding the slope at the times when the velocity is greatest.

12. An object starts from rest and travels with a uniform acceleration of 4 m/sec² for 10 sec. The object then undergoes a constant deceleration of 2 m/sec² until it comes to rest.

(a) Draw a *v–t* diagram for the entire motion.

(b) What is the maximum velocity attained during the motion?

(c) What is the distance traveled?

8.
Uniformly Accelerated Motion and Projectiles

A great many of the important motion problems in physics and engineering involve motion with constant acceleration. This important kind of motion occurs in freely falling objects, the motion of projectiles, and the acceleration of electrons in high voltage electric circuits.

LEARNING GOALS

In studying the material of this chapter, the reader should note the following main learning goals:

1. *Understanding the formulas for displacement and velocity when the acceleration is constant.*
2. *Calculating the position and velocity of freely falling objects.*
3. *Calculating the position and velocity of objects moving with constant acceleration.*
4. *Determining the location and velocity of objects thrown vertically upward.*
5. *Determining the motion of projectiles when air resistance is neglected.*

8-1 UNIFORM ACCELERATION

The principles of kinematics discussed in Chapter 7 can be applied to many kinds of motion. In this chapter we will employ the methods of kinematics to derive the useful formulas needed to study *uniformly accelerated motion*, which is a comparatively simple but very important kind of motion.

Uniformly accelerated motion is motion with a constant acceleration. Objects that are falling through a vacuum have a constant acceleration. Heavy objects that encounter little air resistance as they fall have an acceleration which is almost constant. When we study the principles of electricity, we shall discover that electrically charged particles frequently have a uniformly accelerated motion. According to Newton's laws of motion (which will be discussed in Chapter 9), any object will have a constant acceleration and move with uniformly accelerated motion if the forces acting upon it are constant.

8-2 THE FORMULAS OF MOTION FOR CONSTANT ACCELERATION

The v–t diagram in Fig. 8-1 is a straight line and, therefore, represents the motion of an object that

has a constant acceleration. The constant slope is equal to the acceleration. It is obvious from Fig. 8-1 and from the definition of the slope that

$$a = \frac{v_2 - v_1}{t} \tag{8-1}$$

In some problems v_1 is called the original (or initial) velocity v_0; and v_2 is the final velocity v_f. Equation 8-1 may therefore be written:

$$a = \frac{v_f - v_0}{t}$$

or $\qquad v_f = v_0 + at \tag{8-2}$

Example 8-1   An automobile decelerates from 60 miles/hr to 45 miles/hr in 10 sec. Calculate the acceleration in ft/sec^2 assuming that the acceleration is constant.

Solution.   Since the acceleration is constant, we use Eq. 8-1

$$a = \frac{v_2 - v_1}{t} = \frac{45 \text{ mile/hr} - 60 \text{ miles/hr}}{10 \text{ sec}}$$

$$= -1.5 \text{ mile/hr-sec}$$
$$\times (5280 \text{ ft/1 mile})(1 \text{ hr/3600 sec})$$
$$= -2.2 \text{ ft/sec}^2$$

The minus sign appears in this solution because v_2 is less than v_1, and the acceleration is negative. ◀

The displacement in time t can be easily found in Fig. 8-1. Using the formula for the area of a trapezoid, we can determine that the area under the v–t curve is

$$\frac{(v_2 + v_1)}{2} \times t$$

The displacement in uniformly accelerated motion is therefore

$$s = \frac{(v_f + v_0)}{2} \times t \tag{8-3}$$

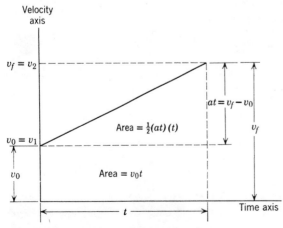

Figure 8-1   The velocity-time curve for uniformly accelerated motion.

The displacement can also be found from the formula

$$s = v_{av} \times t \tag{8-4}$$

Comparing Eqs. 8-4 and 8-3 gives

$$v_{av} = \frac{v_f + v_0}{2} \tag{8-5}$$

Example 8-2   A freely falling stone falls with a constant acceleration of 32 ft/sec². The stone is released from rest and falls for 3 sec. Calculate (a) the velocity after 3 sec of fall, and (b) the distance traveled.

Solution.   (a) $v_0 = 0$, since the stone is released from rest.

$$v_f = v_0 + at$$
$$= 0 + 32 \text{ ft/sec}^2 \text{ (3 sec)}$$
$$= 96 \text{ ft/sec}$$

(b) $v_{av} = \dfrac{v_f + v_0}{2} = \dfrac{96 \text{ ft/sec} + 0}{2}$

$$= 48 \text{ ft/sec}$$

$$s = v_{av} \times t$$
$$= 48 \text{ ft/sec} \times 3 \text{ sec}$$
$$= 144 \text{ ft} \qquad \blacktriangleleft$$

Referring again to Fig. 8-1, we can obtain another useful formula for the displacement. The area under the v–t curve consists of a rectangle and a right triangle. The area of the rectangle is $v_0 \times t$. One side of the right triangle is of length t, whereas the perpendicular side is of length $v_f - v_0 = at$. The area of the triangle is, therefore, $\frac{1}{2}at^2$. The total displacement is

$$s = v_0 t + \tfrac{1}{2}at^2 \tag{8-6}$$

The total displacement would be $v_0 t$ if the object continued to travel with its original velocity. The term $\frac{1}{2}at^2$ gives the additional displacement caused by the acceleration (or the increase in velocity). Either Eq. 8-6 or 8-3 may be used to calculate the displacement in uniformly accelerated motion.

Example 8-3   A torpedo boat is traveling at 10 m/sec. The motors are then turned to full power and the boat accelerates at 1.5 m/sec per sec for 8 sec. How far does the boat travel during this 8-sec period?

Solution.   Substituting $v_0 = 10$ m/sec, $a = 1.5$ m/sec², and $t = 8$ sec in Eq. 8-6, we obtain

$$s = v_0 t + \tfrac{1}{2}at^2$$
$$s = (10 \text{ m/sec})(8 \text{ sec}) + \tfrac{1}{2}(1.5 \text{ m/sec}^2)(64 \text{ sec}^2)$$
$$= 80 \text{ m} + 48 \text{ m} = 128 \text{ m}$$

ALTERNATE METHOD

$$v_f = v_0 + at$$
$$= 10 \text{ m/sec} + (1.5 \text{ m/sec}^2)(8 \text{ sec})$$
$$= 22 \text{ m/sec}$$

$$s = v_{av} \times t = \frac{v_f + v_0}{2} \times t$$

$$= \frac{22 \text{ m/sec} + 10 \text{ m/sec}}{2} \times 8 \text{ sec}$$

$$= 16 \text{ m/sec} \times 8 \text{ sec} = 128 \text{ m} \qquad \blacktriangleleft$$

By combining Eq. 8-3 with Eq. 8-2 and eliminating the time t, we can obtain another formula for uniformly accelerated motion. Solving 8-2 for t gives

$$t = \frac{v_f - v_0}{a}$$

Substituting in 8-3, we obtain

$$s = \frac{(v_f + v_0)}{2} \frac{(v_f - v_0)}{a}$$

Carrying out the multiplication, we find that

$$s = \frac{v_f{}^2 - v_0{}^2}{2a}$$

or $\qquad v_f{}^2 = v_0{}^2 + 2as \tag{8-7}$

Example 8-4   A rifle bullet has a muzzle velocity of 2400 ft/sec. If it acquires this high velocity by being accelerated for a distance of 3 ft in the rifle barrel, and assuming that the motion in the barrel is uniformly accelerated motion, calculate the acceleration.

Solution.   From Eq. 8-7

$$a = \frac{v_f{}^2 - v_0{}^2}{2s} = \frac{(2.4 \times 10^3 \text{ ft/sec})^2 - 0}{2(3 \text{ ft})}$$

$$= 9.6 \times 10^5 \text{ ft/sec}^2$$

ALTERNATE METHOD

$$v_{av} = \frac{v_f + v_0}{2} = \frac{2400 \text{ ft/sec} + 0}{2} = 1200 \text{ ft/sec}$$

$$s = v_{av} \times t \qquad t = \frac{s}{v_{av}} = \frac{3 \text{ ft}}{1200 \text{ ft/sec}}$$

$$t = \frac{1}{400} \text{ sec}$$

Then

$$a = \frac{v_f - v_0}{t} = \frac{2400 \text{ ft/sec} - 0}{\frac{1}{400} \text{ sec}}$$

$$= 9.6 \times 10^5 \text{ ft/sec}^2 \qquad \blacktriangleleft$$

8-3 FALLING OBJECTS AND THE ACCELERATION DUE TO GRAVITY

Many careful measurements indicate that all objects fall with the same acceleration in the absence of air resistance. A piece of lead or a feather will have the same acceleration if they fall in a vacuum. The exact value for the acceleration of a freely falling body varies slightly from place to place on the earth. However, the maximum variation is less than 1%, and for most purposes it is sufficiently accurate to assume that the acceleration due to gravity at or near sea level is

$$g = 32 \text{ ft/sec}^2 = 9.8 \text{ m/sec}^2 = 980 \text{ cm/sec}^2$$

Example 8-5   A stone dropped from a high tower strikes the ground in 3 sec. Assume that air resistance is negligible and calculate (a) the height of the tower, and (b) the velocity of the stone as it hits the ground.

Solution.   (a) The initial velocity $v_0 = 0$ and $t = 3$ sec.

$$s = \tfrac{1}{2}gt^2$$

$$= \tfrac{1}{2}(9.8 \text{ m/sec}^2)(9 \text{ sec}^2) = 44.1 \text{ m}$$

or $s = \tfrac{1}{2}(32 \text{ ft/sec}^2)(9 \text{ sec}^2) = 144 \text{ ft}$

(b) $v = v_0 + gt$

$$= 0 + (9.8 \text{ m/sec}^2)(3 \text{ sec}) = 29.4 \text{ m/sec}$$

or $v = (32 \text{ ft/sec}^2)(3 \text{ sec}) = 96 \text{ ft/sec}$ $\blacktriangleleft$

Experiments with falling weights and projectiles disclose that air resistance increases quite rapidly as the velocity of an object increases. Because of this effect, the speed of objects falling through the atmosphere does not increase indefinitely. After an object has been falling for a comparatively short time, it reaches its terminal velocity, and the object undergoes no further acceleration. This occurs when the air resistance (or air friction) has become large enough to balance the downward pull of gravity. It is obvious when bits of paper fall that the terminal velocity is quickly attained. However, if stones or iron balls are allowed to fall, they may travel a considerable distance without reaching their terminal velocity. If these objects are moving with velocities much less than their terminal velocities, it is permissible to assume that the acceleration is constant and practically equal to the value of "g" given above.

If a projectile is released from a satellite at a very great distance from the surface of the earth, the force of gravity and the acceleration g due to gravity will be less than at sea level. We shall study this effect in greater detail in Chapter 9. It is worth noting, however, that we must travel more than 200 miles above the earth before the value of g decreases by 10%.

Any object projected from points outside the atmosphere will, of course, not encounter any air resistance as it travels through space. The velocity of a projectile entering or re-entering the atmosphere is likely to be much greater than the terminal velocity. Unless some other means is used to slow these projectiles, the force of air resistance and the frictional heating effect will damage or destroy the projectile. In any case, a projectile entering or re-entering the atmosphere from outer space will experience a sudden, nonuniform deceleration.

For objects that fall from rest and have a uniform acceleration, we let $a = g$ in the formulas for uniformly accelerated motion to obtain

$$v_f = gt \qquad \text{(a)}$$
$$v_f{}^2 = 2gs \qquad \text{(b)}$$
$$s = \tfrac{1}{2}gt^2 \qquad \text{(c)}$$
$$v_{av} = \frac{v_f}{2} \qquad \text{(d)}$$
$$s = v_{av} \times t \qquad \text{(e)} \qquad \text{(8-8)}$$

If the object is projected vertically downward with an initial velocity v_0 the formulas are

$$v_f = v_0 + gt \qquad \text{(a)}$$
$$v_f{}^2 = v_0{}^2 + 2gs \qquad \text{(b)}$$
$$s = v_0 t + \tfrac{1}{2}gt^2 \qquad \text{(c)}$$
$$v_{av} = \frac{v_f + v_0}{2} \qquad \text{(d)}$$
$$s = v_{av} \times t \qquad \text{(e)} \qquad \text{(8-9)}$$

Example 8-6   A weight is thrown downward with an initial velocity of 40 ft/sec from the roof of a tall building. The building is 200 ft high. Ignore air resistance and calculate (a) the time it takes for the weight to reach the ground, and (b) the velocity with which the weight strikes the ground.

Solution

(a) $s = v_0 t + \tfrac{1}{2}gt^2$.

$$200 \text{ ft} = 40 \text{ ft/sec } t + \tfrac{1}{2}(32 \text{ ft/sec}^2)t^2$$

(t is in seconds)

$$0 = 16t^2 + 40t - 200$$
$$0 = 2t^2 + 5t - 25$$
$$0 = (2t - 5)(t + 5)$$
$$t = 2.5 \text{ sec}$$

($t = -5$ sec is rejected as physically impossible in this problem).

(b) $v_f = v_0 + gt$

$$= 40 \text{ ft/sec} + 32 \text{ ft/sec}^2 \ (2.5 \text{ sec})$$
$$= 40 \text{ ft/sec} + 80 \text{ ft/sec} = 120 \text{ ft/sec}$$

ALTERNATE METHOD.   Solve for v_f first in

(b) $v_f{}^2 = v_0{}^2 + 2gs$

$$v_f = \sqrt{v_0{}^2 + 2gs}$$
$$= \sqrt{(40 \text{ ft/sec})^2 + 2(32 \text{ ft/sec}^2)(200 \text{ ft})}$$
$$= \sqrt{(1600 + 12{,}800) \text{ ft/sec}}$$
$$= 120 \text{ ft/sec}$$

Then to find the time

$$v_f = v_0 + gt$$
$$t = \frac{v_f - v_0}{g} = \frac{120 \text{ ft/sec} - 40 \text{ ft/sec}}{32 \text{ ft/sec}^2}$$
$$= \frac{80}{32} \text{ sec} = 2.5 \text{ sec} \qquad \blacktriangleleft$$

8-4 OBJECTS THROWN STRAIGHT UPWARD

If an object is thrown straight upward, it reaches a maximum height h which is predetermined by the initial velocity (see Fig. 8-2). On the way up, the speed of the object decreases; on the way down, the speed increases. The acceleration g,

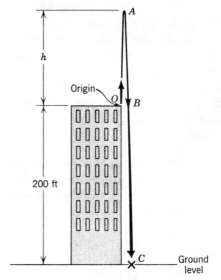

Figure 8-2   The trajectory of an object thrown straight upward.

however, is always constant and directed downward. In order to study this motion mathematically, the point of projection is taken as the origin and the following sign conventions are employed:

1. Displacements above the origin are positive.

2. Displacements below the origin are negative.

3. Velocities upward are positive; downward velocities are negative.

4. The acceleration caused by the downward pull of gravity is downward and therefore negative. The acceleration $a = -32$ ft/sec$^2 = -9.8$ m/sec$^2 = -980$ cm/sec$^2 = -g$.

Substituting $-g$ for a in the equations for uniformly accelerated motion, we obtain

$$v_f = v_0 - gt \qquad (a)$$
$$v_f{}^2 = v_0{}^2 - 2gs \qquad (b)$$
$$s = v_0 t - \tfrac{1}{2}gt^2 \qquad (c)$$
$$v_{av} = \frac{v_f + v_0}{2} \qquad (d)$$
$$s = v_{av} \times t \qquad (e) \qquad (8\text{-}10)$$

Example 8-7 The building in Fig. 8-2 is 200 ft high, and a stone is projected upward with an initial velocity of 40 ft/sec. Ignore air resistance and determine (a) the time it takes for the stone to reach point A, the highest point on its path; (b) the maximum height reached by the stone; (c) the time it takes the stone to reach point B on its path; (d) the velocity of the stone as it passes B; and (e) the velocity of the stone as it strikes the ground.

Solution. (a) When the stone reaches point A, the stone has ceased moving upward but has not yet started to fall down. Therefore $v_f = 0$ and from Eq. 8-10(a)

$$0 = v_0 - gt_h$$

$$t_h = \frac{v_0}{g} = \frac{40 \text{ ft/sec}}{32 \text{ ft/sec}^2}$$

$$= 1.25 \text{ sec}$$

(b) The maximum height may be calculated from Eq. 8-10(b), (c), or (e). In Eq. 8-10(b) let $v_f = 0$ and $s = h$.

$$0 = v_0{}^2 - 2gh$$

$$h = \frac{v_0{}^2}{2g} = \frac{(40 \text{ ft/sec})^2}{2(32 \text{ ft/sec}^2)} = 20(1.25) \text{ ft}$$

$$= 25 \text{ ft}$$

In Eq. 8-10(c) let $t = 1.25$ sec to obtain

$$h = v_0 t - \tfrac{1}{2}gt^2$$
$$= 40 \text{ ft/sec} (1.25 \text{ sec}) - \tfrac{1}{2}(32 \text{ ft/sec}^2) \times (1.25 \text{ sec})^2$$
$$= 50 \text{ ft} - 25 \text{ ft} = 25 \text{ ft}$$

By using Eq. 8-10(d) and (e),

$$v_{av} = \frac{v_f + v_0}{2} = \frac{0 \text{ ft/sec} + 40 \text{ ft/sec}}{2}$$

$$= 20 \text{ ft/sec}$$

$$s = v_{av} \times t = 20 \text{ ft/sec} (1.25 \text{ sec})$$

$$= 25 \text{ ft}$$

(c) The time to reach B may be calculated by letting $s = 0$ in 8.10(c).

$$0 = v_0 t - \tfrac{1}{2}gt^2$$

$$t = \frac{2v_0}{g} = \frac{2(40 \text{ ft/sec})}{32 \text{ ft/sec}^2} = 2.5 \text{ sec}$$

The time to reach B is twice as great as the time to reach A. It is apparent that it takes just as long (1.25 sec) for the weight to rise from the origin to A as it takes to fall back to the origin.

(d) The velocity at B may be obtained from Eq. 8-10(a) or (b). Since $s = 0$ at B

$$v_f{}^2 = v_0{}^2$$

$$v_f = -v_0 = -40 \text{ ft/sec}$$

From Eq. 8.10(a)

$$v_f = v_0 - gt$$
$$= 40 \text{ ft/sec} - 32 \text{ ft/sec}^2 (2.5 \text{ sec})$$
$$= -40 \text{ ft/sec}$$

The minus sign indicates that the velocity at B is downward.

(e) The velocity with which the weight strikes the ground may be calculated from Eq. 8-9(b) if we let $s = -200$ ft.

$$v_f^2 = v_0^2 - 2gs$$
$$= (40 \text{ ft/sec})^2 - 2(32 \text{ ft/sec}^2)(-200 \text{ ft})$$
$$= 1600 \text{ ft}^2/\text{sec}^2 + 12,800 \text{ ft}^2/\text{sec}^2$$
$$= 14,400 \text{ ft}^2/\text{sec}^2$$
$$v_f = -120 \text{ ft/sec} \qquad \blacktriangleleft$$

8-5 PROJECTILE MOTION IN TWO DIMENSIONS

If a bullet or any other object is projected at some angle with the vertical, it is necessary to consider motion in the x or horizontal direction and motion in the y or vertical direction. Some of the principles involved here are suggested by the monkey and hunter puzzle of Fig. 8-3. The hunter has forgotten what he learned in his physics class about the downward acceleration of gravity, and instead of aiming above the monkey, he is aiming directly at the monkey. The unfortunate monkey (who never studied physics) tries to dodge the bullet by releasing his hold on the limb at the same instant that the hunter fires his gun. If the monkey had kept his grip on the limb, the bullet would have passed safely below him. However, the falling monkey and the bullet both had the same downward acceleration; and the downward displacement of the monkey from the limb was exactly equal to the downward displacement of

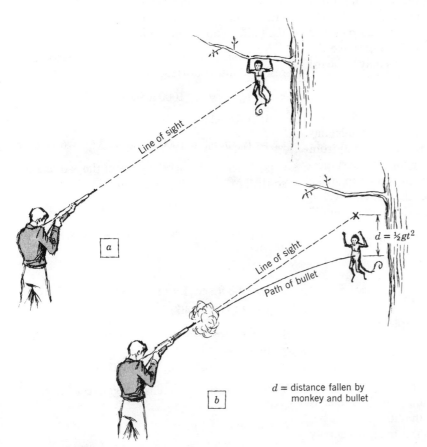

$d =$ distance fallen by monkey and bullet

Figure 8-3 The hunter and the monkey.

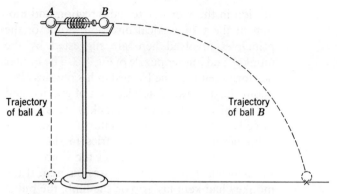

Figure 8-4 An experiment with freely falling objects.

the bullet from the line of sight. This fact may be verified in the laboratory by using a dart gun and a mechanical target which is released by an electromagnet at the instant that the dart gun is fired. A simpler device along the same lines is indicated in Fig. 8-4. The horizontally projected ball and the ball that just drops vertically both have the same vertical acceleration; the balls may be heard striking the floor at the same instant.

Example 8-8 A marksman aims his rifle in a direction that is exactly horizontal. The bullet has a muzzle velocity of 1000 ft/sec and is fired from a point 4 ft above the ground. Ignore air resistance. (a) How long does it take for the bullet to strike the ground? (b) How far does the bullet travel in a horizontal direction before striking the ground?

Solution. (a) The time to fall 4 ft is found from $s = \frac{1}{2}gt^2$.

$$t = \sqrt{\frac{2s}{g}} = \sqrt{\frac{2(4 \text{ ft})}{32 \text{ ft/sec}^2}} = \frac{1}{2} \text{ sec}$$

(b) In $\frac{1}{2}$ sec the horizontal displacement is

$$x = 1000 \text{ ft/sec} \left(\frac{1}{2} \text{ sec}\right) = 500 \text{ ft} \quad \blacktriangleleft$$

Example 8-7 was particularly simple, because the bullet was fired horizontally. If the bullet is fired at some angle θ_0 with the horizontal it becomes necessary to resolve the muzzle velocity v_0 into horizontal and vertical components as in the following problem.

Example 8-9 A bullet with a muzzle velocity of 1000 ft/sec is fired at an angle of 37° with the horizontal. Ignore air resistance and calculate: (a) the position and velocity of the bullet 5 sec after firing, (b) the maximum height reached by the bullet, and (c) the horizontal range.

Solution. In Fig. 8-5b the muzzle velocity is resolved into a horizontal component

$$v_{0x} = v_0 \cos \theta_0 = 1000 \text{ ft/sec} \cos 37° = 800 \text{ ft/sec}$$

and a vertical component

$$v_{0y} = v_0 \sin \theta_0 = 1000 \text{ ft/sec} \sin 37° = 600 \text{ ft/sec}$$

(a) There is no acceleration in the x direction, so the horizontal displacement is

$$x = v_{0x}t = 800 \text{ ft/sec} (5 \text{ sec}) = 4000 \text{ ft}$$

In the vertical direction we have a downward acceleration and

$$y = v_{0y}t - \frac{1}{2}gt^2$$
$$= 600 \text{ ft/sec} (5 \text{ sec}) - \frac{1}{2}(32 \text{ ft/sec}^2)(25 \text{ sec}^2)$$
$$= 3000 \text{ ft} - 400 \text{ ft} = 2600 \text{ ft}$$

In the x direction the velocity remains constant so that

$$v_x = v_{0x} = 800 \text{ ft/sec}$$

In the y direction, there is a negative acceleration and

$$v_y = v_{0y} - gt$$
$$= 600 \text{ ft/sec} - 32 \text{ ft/sec}^2 (5 \text{ sec})$$
$$= 600 \text{ ft/sec} - 160 \text{ ft/sec} = 440 \text{ ft/sec}$$

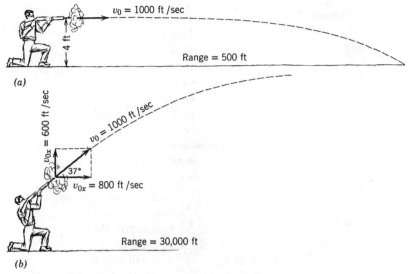

(a)

(b)

Figure 8-5 The trajectory of a bullet. (*a*) Bullet fired horizontally. (*b*) Bullet fired at angle of 37° with horizontal.

(b) The maximum height may be found from

$$v_y{}^2 = v_{0y}{}^2 - 2gy$$

by setting $v_y = 0$. Then

$$h = \frac{v_{0y}{}^2}{2g} = \frac{(600 \text{ ft/sec})^2}{64 \text{ ft/sec}^2}$$

$$= \frac{360,000 \text{ ft}}{64} = 5625 \text{ ft}$$

This could also have been calculated by first finding the time to reach the highest point.

$$v_y = v_{0y} - gt$$

At the highest point, $v_y = 0$ and it is obvious that

$$t_h = \frac{v_{0y}}{g} = \frac{600 \text{ ft/sec}}{32 \text{ ft/sec}^2} = 18.75 \text{ sec}$$

The average velocity in the upward direction is

$$v_{av} = \tfrac{1}{2}(v_f + v_0) = \tfrac{1}{2}(600 \text{ ft/sec} + 0)$$

$$= 300 \text{ ft/sec}$$

Therefore,

$$h = v_{av} \times t$$

$$= (300 \text{ ft/sec})(18.75 \text{ sec}) = 5625 \text{ ft}$$

(c) The time of flight is twice the time required for the bullet to reach its highest point and

$$t_f = 2(t_h) = 2(18.75 \text{ sec}) = 37.5 \text{ sec}$$

Another method of calculating the time of flight is to use

$$y = v_{0y}t - \tfrac{1}{2}gt^2$$

and to let $y = 0$, since the elevation is zero when the bullet returns to the ground. Then

$$0 = v_{0y}t - \tfrac{1}{2}gt^2$$

and

$$t_f = \frac{2v_{0y}}{g} = \frac{2(600 \text{ ft/sec})}{32 \text{ ft/sec}^2} = 37.5 \text{ sec}$$

The horizontal range is

$$R = v_{0x} \times t = 800 \text{ ft/sec} \times 37.5 \text{ sec}$$

$$= 30,000 \text{ ft} \qquad \blacktriangleleft$$

In solving projectile problems the following steps are worth remembering:
(i) Resolve the muzzle velocity into horizontal and vertical components

$$v_{0x} = v_0 \cos \theta_0$$

$$v_{0y} = v_0 \sin \theta_0 \qquad (a)$$

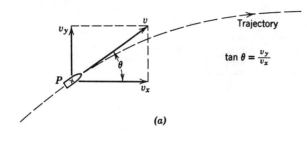

(a)

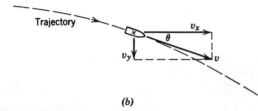

(b)

Figure 8-6 The velocity of a projectile. (a) If $v_y = v_0 \sin \theta_0 - gt$ is positive; v is the resultant velocity and θ is the angle between v and the horizontal. (b) If $v_y = v_0 \sin \theta_0 - gt$ is negative.

(ii) In the x direction the acceleration is zero, and the velocity constant. Therefore we have

$$x = (v_0 \cos \theta_0)t \qquad (b)$$

(iii) In the y direction we are dealing with an object projected straight upward as in Section 8-4. The displacement is

$$y = (v_0 \sin \theta_0)t - \tfrac{1}{2}gt^2 \qquad (c)$$

The velocity in the y direction is

$$v_y = v_0 \sin \theta_0 - gt \qquad (d)$$

Also $\qquad\qquad v_y{}^2 = v_0{}^2 \sin^2 \theta_0 - 2gy \qquad (e)$

(iv) The maximum height may be found from (e) by letting $v_y = 0$. We obtain

$$h = \frac{v_0{}^2 \sin^2 \theta_0}{2g}$$

The time to reach the maximum height may be found from (d) by letting $v_y = 0$. This gives

$$t_h = \frac{v_0 \sin \theta_0}{g}$$

(v) The time required for the projectile to return back to the ground is found by letting $y = 0$ in (c). The time

of flight is easily found to be

$$t_f = \frac{2v_0 \sin \theta_0}{g}$$

The horizontal range is found by multiplying the time of flight t_f by the horizontal velocity. This gives

$$R = \frac{2v_0{}^2 \sin \theta_0 \cos \theta_0}{g} = \frac{v_0{}^2 \sin 2\theta_0}{g}$$

(vi) The direction of the velocity at any time may be found from the diagram in Fig. 8-6.

Some typical projectile trajectories are plotted in Fig. 8-7. It is interesting to note that there are always two angles which give the same range. A target that will be reached by a shell fired at an angle of $\theta_0 = 45° + \alpha$ will also be reached by a shell fired at an angle of elevation of $\theta_0 = 45° - \alpha$.

The trajectories plotted in Fig. 8-7 form curves called parabolas. The theory used to obtain these parabolas ignored the effect of air resistance. Air resistance has the effect of shortening the range, so that the actual trajectory is not a parabola but looks something like the curve of Fig. 8-8.

SUMMARY

For motion with a constant acceleration, the acceleration may be calculated from the formula

$$a = \frac{v_f - v_0}{t}$$

where $v_f =$ the final velocity and v_0 is the original velocity.

The average velocity when the acceleration is constant is given by

$$v_{av} = \frac{v_f + v_0}{2}$$

The displacement in uniformly accelerated motion is calculated from

$$s = v_{av} \times t = \frac{v_f + v_0}{2} \times t$$

or $\qquad\qquad s = v_0 t + \tfrac{1}{2}at^2$

Projectiles without air resistance.
The range is $R = 2v_0^2/g \sin \theta_0 \cos \theta_0$. Using the trigonometry identity $\sin 2\alpha = 2 \cos \alpha \sin \alpha$, we have $R = v_0^2/g \sin 2\theta_0$.

Angle of Projection, θ_0	Height Reached, $h = v_0^2/2g \sin^2 \theta_0$	Range, $R = v_0^2/g \sin 2\theta_0$
90°	$0.500 \, v_0^2/g$	0
80°	$0.485 \, v_0^2/g$	$0.342 \, v_0^2/g$
75°	$0.466 \, v_0^2/g$	$0.500 \, v_0^2/g$
60°	$0.375 \, v_0^2/g$	$0.866 \, v_0^2/g$
45°	$0.250 \, v_0^2/g$	$1.000 \, v_0^2/g$
30°	$0.125 \, v_0^2/g$	$0.866 \, v_0^2/g$
15°	$0.0335 \, v_0^2/g$	$0.500 \, v_0^2/g$
10°	$0.0151 \, v_0^2/g$	$0.342 \, v_0^2/g$
0°	0	0

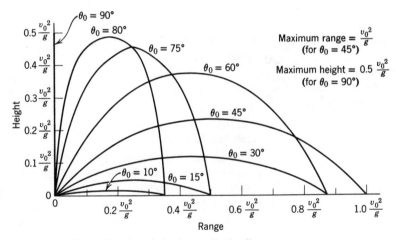

Figure 8-7 The trajectories of short-range projectiles.

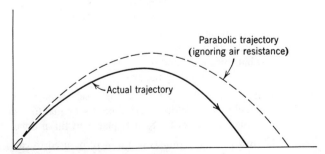

Figure 8-8 The effect of air resistance.

The time can be eliminated from the above formulas to give

$$v_f{}^2 = v_0{}^2 + 2as$$

If the effect of air resistance is neglected (or if an object falls in a vacuum), the motion is uniformly accelerated with the acceleration given by $g = 32$ ft/sec^2 = 9.8 m/sec^2 = 980 cm/sec^2. If an object is dropped from rest and air resistance is negligible, the above formulas for uniformly accelerated motion are used with $v_0 = 0$ and $a = g$.

If an object is thrown downward with an initial velocity v_0 and air resistance is negligible, then the above formulas describe the motion when g is substituted for a.

If an object is thrown upward, the upward direction is taken to be positive and the downward direction is negative. In the absence of air resistance the acceleration is equal to g in magnitude and is downward. Hence the above formulas are used with $a = -g$.

In projectile problems the motion is resolved into a horizontal and a vertical part. Air resistance is usually neglected and in the x or horizontal direction the acceleration is zero. Therefore, we have

$$v_x = v_0 \cos \theta_0$$
$$x = v_0(\cos \theta_0)t$$

In the vertical or y direction, the acceleration is equal to g in magnitude and in the negative or downward direction. Therefore,

$$v_y = v_0 \sin \theta_0 - gt$$
$$y = v_0(\sin \theta_0)t - \tfrac{1}{2}gt^2$$

The time, t_h, for a projectile to reach its highest point is found by setting $v_y = 0$ to yield

$$t_h = \frac{v_0 \sin \theta_0}{g}$$

The maximum height reached is

$$h = v_{av} \times t_h = \frac{1}{2}(v_0 \sin \theta_0) \times t_h = \frac{1}{2}\frac{(v_0 \sin \theta_0)^2}{g}$$

The time t_f for a projectile to return to the ground is found by setting $y = 0$ in the equation

for y. This gives

$$t_f = \frac{2v_0 \sin \theta_0}{g}$$

which is twice the time for the projectile to reach its highest point. The horizontal range is

$$R = v_x \times t_f = (v_0 \cos \theta_0)\frac{(2v_0 \sin \theta_0)}{g}$$

or

$$R = \frac{2v_0{}^2 \sin \theta_0 \cos \theta_0}{g} = \frac{v_0{}^2 \sin 2\theta_0}{g}$$

The velocity components for the projectile velocity may be found by calculating $v_y = v_0 \sin \theta_0 - gt$ and $v_x = v_0 \cos \theta_0$. The magnitude of the velocity is then $v = \sqrt{v_x{}^2 + v_y{}^2}$ and the direction is found from $\tan \theta = v_y/v_x$ as in Fig. 8-6.

QUESTIONS

1. What is meant by the term *uniformly accelerated motion?*

2. Under what conditions will an object have uniformly accelerated motion?

3. An object has a constant velocity of 20 ft/sec. Explain whether this object has or does not have uniformly accelerated motion.

4. How does the velocity–time graph indicate that an object is moving with uniformly accelerated motion?

5. In studying the motion of an object traveling with constant acceleration, the acceleration is found to be -6 m/sec^2. Explain the significance of the minus sign and the meaning of a negative acceleration.

6. How does the v–t graph indicate that an object is moving with a constant negative acceleration?

7. In studying the acceleration of a motor vehicle, the acceleration is found to be 5 miles/hr · sec. Explain what this means and draw a velocity–time graph of the motion labeling the axes.

8. In the formulas of Eq. 8-9, the acceleration is set equal to g, while the acceleration is set equal to $-g$ in the formulas of Eq. 8-10. Explain why this is done.

9. How can the formulas of Eq. 8-10 be used to calculate the maximum height reached by the object?

10. How can the formulas of Eq. 8-10 be used to calculate the time required for an object to reach its maximum height?

11. How can the formulas of Eq. 8-10 be used to calculate the time required for the thrown object to fall back to its starting height?

12. Explain how the equations for uniformly accelerated motion may be used to calculate the time required for a projectile to return back to the ground.

13. How are the equations for uniformly accelerated motion used to calculate the maximum height reached by a projectile?

14. Describe how the time of flight is calculated for a projectile.

15. How is the horizontal range formula obtained?

16. At what angle of projection will a projectile have the maximum horizontal range?

17. At what angle of projection will it take the longest time for a projectile to return to the ground?

18. At what angle should a projectile be fired to reach the maximum possible height?

19. Suppose the muzzle velocity of a gun could be doubled. What effect would this have on the range?

PROBLEMS

(A)

Note: The acceleration of a freely falling body may be taken to be 32 ft/sec^2, or 9.8 m/sec^2, or 980 cm/sec^2 in Problems A and B.

1. An object starts from rest and travels for 3 sec with an acceleration of 5 ft/sec^2. Calculate:
   (a) The distance traveled.
   (b) The final velocity.

2. A heavy metal object is dropped from rest and falls with negligible air resistance for a time of 4 sec. Calculate:
   (a) The distance fallen in feet.
   (b) The distance in meters.
   (c) The final velocity in ft/sec.
   (d) The final velocity in m/sec.

3. An automobile is accelerated from a velocity of 30 miles/hr to a velocity of 60 miles/hr in a time of 5 sec. Assume that the acceleration is constant and calculate:

(a) The distance traveled during the acceleration.
(b) The acceleration in ft/sec^2.

4. A rocket launcher propels a space vehicle from rest on the launch pad with a constant acceleration of 10 m/sec^2. Calculate:
   (a) The distance traveled in 20 sec.
   (b) The velocity at the end of the 20 sec.
   (c) The time required for the vehicle to reach a velocity of 1000 m/sec.

5. An object starts from rest and moves with uniform acceleration. It travels 200 ft in 10 sec.
   (a) Calculate the acceleration.
   (b) What is the velocity at the end of the 20 sec?

6. A motor cyclist can start from rest and in 4 sec attain a speed of 60 miles/hr (88 ft/sec). Assuming that the acceleration is constant, determine:
   (a) The acceleration.
   (b) The distance traveled during the 4 sec of acceleration.

7. An inclined ramp is 10 m long. A ball is released from rest at the top of the ramp, and the ball travels the length of the ramp in 5 sec. The acceleration is constant.
   (a) Calculate the acceleration.
   (b) What was the velocity of the ball when it reached the bottom?

8. A stone is dropped from the top of a tower 100 m high. Ignore air resistance and calculate:
   (a) The time required for the stone to hit the ground.
   (b) The velocity of the stone as it hits the ground.

9. A pitcher is able to throw a baseball vertically to a height of 120 ft. Ignore air resistance and calculate the initial velocity with which the ball was thrown.

10. A target rifle fires a bullet with a muzzle velocity of 1500 ft/sec. The bullet is fired horizontally at a target that is 100 yd away from the rifle. How far below the horizontal line of sight does the bullet fall? (Ignore the effect of air resistance.)

11. In 60 sec of flight the thrust of the rocket motors accelerates a space capsule from rest to a velocity of 3.0 km/sec in the vertical direction. Assume that the motion is uniformly accelerated and calculate:
    (a) The acceleration in meters/sec^2.
    (b) The height reached in 60 sec.

12. A weight is thrown directly upward with an initial velocity of 50 m/sec. If air resistance is negligible, how high will the weight travel?

13. According to some historical reports, Galileo dropped weights from the Leaning Tower of Pisa in order to study the laws for falling objects. The tower is 55 m high. Assume that the effect of air resistance is negligible.
    (a) How long did it take for the weights to reach the ground?
    (b) How fast were the weights traveling when they hit the ground?

14. A gun fires a projectile with a muzzle velocity of 3000 ft/sec. If the projectile is fired vertically, how high does the projectile reach? (Ignore the effect of air resistance.)

15. In testing the ability of a car to accelerate, it is found that a certain model can accelerate from 45 miles/hr to 60 miles/hr in a time of 5 sec. Assume that the acceleration is constant and calculate:
    (a) The acceleration in ft/sec^2.
    (b) The distance traveled during the acceleration.

16. Near the surface of the moon the acceleration of a freely falling object is 5.47 ft/sec^2.
    (a) How far does an object fall in 3 sec near the moon?
    (b) How fast is an object traveling after 3 sec of fall near the moon?

17. A heavy object is released from rest near the earth and falls with negligible air resistance until it reaches a speed of 160 ft/sec.
    (a) How far does it fall?
    (b) How much time is required for the object to reach its final velocity?

18. A lead weight is dropped from rest and falls 60 cm to the ground with negligible air resistance.
    (a) How long does it take to fall the 60 cm?
    (b) How fast is the weight traveling when it strikes the ground?

19. In a drag race one of the automobiles starts from rest and travels 1000 ft in 10 sec with constant acceleration. Calculate:
    (a) The acceleration.
    (b) The maximum velocity (in miles/hr) attained by the automobile.

20. An automobile traveling at 60 miles/hr can be brought to rest by the brakes in a distance of 250 ft. Assume that the deceleration is constant.
    (a) Determine the deceleration in feet per second per second.
    (b) How many seconds is required to stop the automobile?

21. An iron weight is projected vertically upward from the ground with a velocity of 50 m/sec. Ignore air resistance.
    (a) What is the maximum height reached by the weight?
    (b) How fast is the weight traveling when it strikes the ground?
    (c) How long does the weight remain in the air?

(B)

1. A constant acceleration of 10 m/sec^2 is given to an object that is accelerated from an initial velocity of 20 m/sec to a final velocity of 80 m/sec.
    (a) How much time is required for the acceleration?
    (b) How far does the object travel during the acceleration?

2. The plans for a space project require that the space capsule be decelerated from a velocity of 10 km/sec to a velocity of 8 km/sec. The deceleration is to be constant and equal in magnitude to 40 m/sec^2.
    (a) How long should the deceleration last?
    (b) How far does the space capsule travel during the deceleration?

3. A stone is thrown vertically downward from the top of a tall building. Its initial velocity is 40 ft/sec and air resistance is negligible.
    (a) How far does the stone fall in 5 sec?
    (b) How fast is the stone traveling after the 5 sec of fall?

4. A baseball is thrown horizontally from the top of a building 80 ft high. The ball travels a distance of 100 ft in the horizontal direction before it strikes the ground. Ignore air resistance.
    (a) How long did it take for the ball to hit the ground?
    (b) With what horizontal velocity was the ball thrown?
    (c) Determine the magnitude and direction of the resultant velocity as the ball struck the ground.

5. A jet plane becomes airborne at 135 miles/hr. If the acceleration is constant at 5 miles/hr/sec, determine the length of runway required for a takeoff.

6. A mine shaft is 800 ft deep.
    (a) How long does it take a weight released from rest at the top of the shaft to fall to the bottom? (Ignore air resistance.)
    (b) If the weight is thrown downward with an initial velocity of 80 ft/sec, how long does it take to hit bottom?

7. A cannon can shoot its projectile vertically upward to a maximum height of 10,000 m. Ignore air resistance and calculate:
   (a) The muzzle velocity in meters/sec.
   (b) The maximum horizontal range.

8. In an atomic physics experiment protons are decelerated from a velocity of 5.0×10^7 m/sec to 2.0×10^7 m/sec while traveling a distance of 70 cm. The deceleration is constant. Calculate:
   (a) The time it takes the protons to travel the 70 cm.
   (b) The deceleration.

9. A space vehicle returning from a space flight reenters the earth's atmosphere with a speed of 7 miles/sec. Suppose that the retro rockets produce a deceleration of $5 g (1 g = 32$ ft/sec^2) and the vehicle is slowed down to a speed of 2000 ft/sec.
   (a) How long does it take to slow up the vehicle?
   (b) How far does it travel during the deceleration?

10. A projectile is fired with an initial velocity of 250 m/sec at an angle of 30° with the horizontal. Ignore the effect of air resistance. After 10 sec of flight determine:
    (a) The horizontal and vertical displacement.
    (b) The horizontal and vertical components of the velocity, and the direction in which the projectile is traveling.

11. A gun fires a projectile with a muzzle velocity of 2000 ft/sec at an angle of 53° above the horizontal. Assume that the acceleration of gravity, g, is 32 ft/sec^2; ignore air resistance and calculate:
    (a) The maximum height reached by the projectile.
    (b) The horizontal range of the projectile.
    (c) The horizontal and vertical displacement of the projectile after 60 sec.
    (d) The velocity (magnitude and direction) after 60 sec.

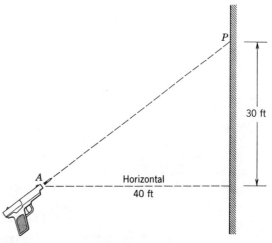

Figure 8-9 Aiming a dart gun.

12. In Fig. 8-9 a dart gun at A is aimed directly toward point P on the vertical wall. The dart has an initial velocity of 100 ft/sec in the direction indicated. Ignore air resistance.
    (a) How long does it take for the dart to strike the wall?
    (b) How far below P does the dart strike the wall?
    (c) Determine the horizontal and vertical components of the velocity of the dart as it strikes the wall.

13. In studying the motion of a baseball that is hit over the fence, some observers note that the ball rises 40 ft above the point where it is struck by the bat. The ball also travels 400 ft horizontally before it gets back to the height at which it left the bat. Ignore the effect of air resistance and determine the speed of the ball after it was struck by the bat.

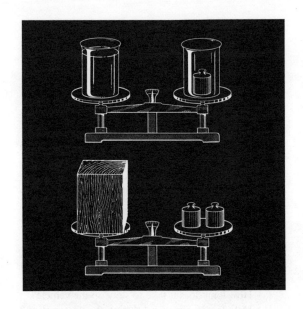

9.
Forces
and Linear Motion

In this chapter we shall study the connection between the forces that act upon an object and the acceleration produced by these forces. We shall base our study upon *Newton's laws of motion* and his *law of gravitational attraction*.

LEARNING GOALS

This chapter has five major objectives:

1. *Learning the unit systems that are employed in measuring forces, accelerations, and masses.*

2. *Being able to calculate the mass of an object from its weight.*

3. *Using the law of universal gravitation to calculate the acceleration of a freely falling object.*

4. *Using Newton's second law of motion to solve problems in dynamics.*

5. *Applying the first and third laws of motion to some important scientific and technical problems.*

9-1 DYNAMICS AND MOTION

In our study of kinematics in Chapters 7 and 8 we considered the motion of objects without inquiring too closely into the factors that caused or influenced the motion. In this chapter we shall expand our knowledge of physics by reviewing the laws of motion which describe the relationship between motion and the forces that produce motion. We shall begin our study of the branch of mechanics called *dynamics*. Dynamics involves a consideration of the forces that act on objects and the motions that are thereby produced.

Most technical and scientific applications of physics are directly concerned with the principles of dynamics. Some of these applications are undoubtedly quite complex and intricate. Nearly all of dynamics, however, is based upon three simply stated laws known as *Newton's laws of motion*. When Sir Isaac Newton formulated these laws during the seventeenth century, he created a firm foundation for the entire science of physics.

9-2 NEWTON'S FIRST LAW OF MOTION; THE LAW OF INERTIA

It is a well-known fact that no object will start moving by itself. Some external (outside) force is always required to start an object moving. Once a material object has started to move, it tends to continue moving unless some force acts to stop the motion. This fundamental property of matter is called *inertia*:

> **An object at rest tends to remain at rest, and an object in motion tends to continue moving in the same direction.**

There are many common situations in which the inertia of natural objects is evident. When we stand in a train or bus that suddenly starts, we experience a tendency to fall backward; our body tends to remain stationary while the motion of the vehicle carries our feet forward. If we are standing in a vehicle that suddenly stops, we note that because of inertia our body tends to continue moving forward. If an automobile or an airplane abruptly changes its speed or direction of motion, the passengers may be thrown from their seats if they are not using their seat belts. Figure 9-1 illustrates a simple parlor trick which is based upon the property of inertia.

Newton was not the first scientist to be aware of the inertia property of material objects. However, *Newton's first law of motion* gives a very

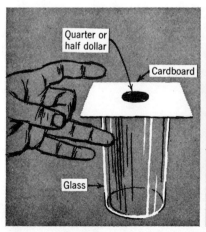

Quarter or half dollar

Cardboard

Glass

Figure 9-1 A parlor trick that illustrates the property of inertia. When the cardboard is flicked, the coin drops into the glass.

clear statement of this attribute of matter:

Every object will remain in a state of rest or in uniform motion in a straight line unless it is acted upon by an unbalanced force.

The phrase "uniform motion in a straight line" means motion with a constant velocity. The term "state of rest" means that the velocity is constant and equal to zero. We may, therefore, shorten the statement of the first law to read:

If no unbalanced force acts upon an object, the velocity will remain constant.

In actual practice the velocity of moving objects rarely remains constant, because forces of friction or other unbalanced forces are usually present. If we could eliminate all the forces acting on an object, this object would continue to move forever with a constant velocity. The laboratory device illustrated in Fig. 9-2 reduces friction to à negligible amount, and this "dry-ice puck" will move across a smooth level table without any sign of slowing up. In outer space, far from any planet, a space capsule or a projectile encounters almost no forces, and these objects will travel great distances at constant velocity with no motive power needed.

9-3 MASS AND MOMENTUM

In Section 9-2 we stated the first law of motion in terms of the velocity of an object. It is useful (and sometimes more accurate), however, to state the laws of motion in terms of *momentum*. The momentum of an object is defined as the product of the mass and the velocity

$$\text{momentum} = \text{mass} \times \text{velocity}$$
$$= mv \qquad (9\text{-}1)$$

The mass of an object refers to the quantity of matter in the object. We can determine the mass of an object by comparing the quantity of matter in the object with a standard mass. It is often possible to use the inertia of an object as an indication of its mass. The greater the mass of an object, the greater its inertia will be. If an experiment indicates that two different objects have the same amount of inertia, the two objects may be said to have the same *inertial mass*.

Figure 9-2   A frictionless dry ice puck. The puck rides upon a layer of carbon dioxide gas.

The fundamental units for measuring mass is the *kilogram* or the *gram*. From the meaning of the prefix *kilo* it is apparent that 1 kilogram = 1000 grams. Originally, the gram and the kilogram were defined in terms of the mass of a standard volume of water at normal temperature (20°C):

> The mass of 1 cubic centimeter of
> pure water is 1 gram.

> The mass of 1000 cubic centimeters of
> pure water is 1 kilogram.

It is worth noting that a volume of 1000 cubic centimeters is also called a *liter*. One liter is a volume that is somewhat larger than the common U.S. quart (1 liter = 1.057 quarts).

Although it is convenient to use the above definitions of the gram and the kilogram, other standard masses have to be used in practical measurements of mass. A common laboratory standard is the 1-kg brass object illustrated in Fig. 9-3. The fundamental standard kilogram is made

of platinum and is carefully stored in the International Bureau of Weights and Measures at Sèvres, France. Very exact copies of this international standard are also kept in the U.S. National Bureau of Standards and in many other countries.

It is obvious from Fig. 9-3 that brass, water, and wood do not have the same density. A kilogram of wood occupies a larger volume than a kilogram of water, and the density of wood is less than the density of water. A kilogram of brass occupies about one-ninth the volume of a kilogram of water; and consequently the density of the brass is about nine times the density of water. An object made of low density material will have a small mass unless its volume is very great. Objects made of high density material feel "heavy," and they have a large mass in proportion to their size.

Density is defined as the mass per unit volume or

$$\text{density} = \frac{\text{mass}}{\text{volume}}$$

The Greek letter ρ (rho) is usually used as the mathematical symbol for density. The definition of density gives us the formula

$$\rho = \frac{m}{V} \qquad (9\text{-}2)$$

It is obvious from the definition of the gram and the kilogram that the density of water is 1 gram per cubic centimeter (1 gm/cm^3) or 1 kilogram per liter (1 kg/l).

Figure 9-3 indicates a very common method of measuring the mass of objects. The indicator of the balance arm is centered when the force of gravity on the object equals the force of gravity on the standard mass. Since the operation of the equal-arm balance depends upon the force of gravity, the mass determined in this way is called the *gravitational mass* of the object. Many experiments have established that the inertial mass and the gravitational mass of an object are equal.

If the object and the balance are moved to a different planet (or to a different place on the earth), the force of gravity will change. Figure

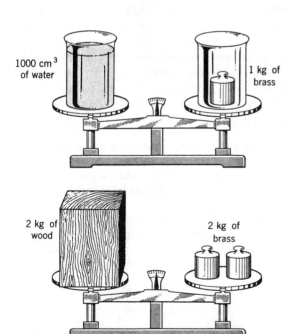

Figure 9-3 Determining the mass of an object with an equal-arm balance.

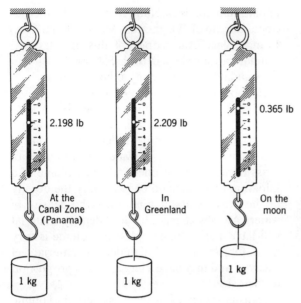

At the Canal Zone (Panama) 2.198 lb

In Greenland 2.209 lb

On the moon 0.365 lb

1 kg 1 kg 1 kg

Figure 9-4 Measuring the pull of gravity with a spring balance.

9-4 indicates how the force of gravity acting upon a 1-kg mass will vary. But the force of gravity acting on the standard masses and the objects will change in the same proportion, and the equal-arm balance will always record the same mass. The mass of an object remains constant even though the weight (and the force of gravity measured by a spring scale) changes.

The mass of an object is a scalar quantity that measures the quantity of matter in an object; but the momentum is a vector that has the same direction as the velocity vector. The definition in Eq. 9-1 indicates that the momentum of an object depends on both the mass and the velocity. Objects with a small mass may have much momentum if the velocity is great.

Under most circumstances the mass of an object remains constant, and the momentum changes when there is a change in the magnitude or the direction of the velocity. Consequently, the law of inertia (first law of motion) may be expressed in terms of momentum as follows:

If no unbalanced forces act upon an object, the momentum will remain constant.

The first law of motion asserts that the momentum of an object remains constant if the resultant force is zero. But in many important cases the resultant force is not zero, and the second law of motion is used to study the motion.

9-4 NEWTON'S SECOND LAW OF MOTION; THE LAW OF ACCELERATION

Whenever a resultant force acts upon an object, it is observed that the momentum undergoes a change. When a small force is applied, the momentum changes slowly. If a large force is applied, the momentum changes rapidly. According to *Newton's second law of motion,*

The time rate of change in the momentum is proportional to the resultant force and in the same direction as the resultant force.

Since the mass of an object usually does not change, the rate of change in the momentum is simply the mass times the rate of change in the velocity. But the time rate of change in the velocity is the acceleration; and the time rate of change in momentum is the mass, m, times the acceleration, $\mathbf{a}$. The second law of motion thus asserts that $m\mathbf{a}$ is proportional to $\mathbf{F}_R$ (the resultant force) if the mass of the object does not change. The second law of motion is also the law of acceleration:

If the mass of an object does not change, the acceleration is directly proportional to the resultant force and in the same direction as the resultant force.

The units of force in the metric system are set up so that the constant of proportionality between $m\mathbf{a}$ and $\mathbf{F}_R$ is equal to unity. The second law may, therefore, be written as an equation

$$\mathbf{F}_R = m\mathbf{a}$$

or
$$F_R = ma \qquad (9\text{-}3)$$

If the same resultant force F_R acts upon different objects, these objects will probably not all have the same acceleration. Equation 9-3 indicates that the objects of small mass will have a large acceleration, and the objects of large mass will undergo a small acceleration. In other words, for a given force the acceleration produced is inversely proportional to the mass of the object being accelerated.

In using Eq. 9-3 we must be careful to measure F_R, m, and a in the appropriate units. If the mass is measured in grams, and the acceleration in centimeters per second per second, then Eq. 9-3 will give the force in units called *dynes*. One dyne is the force that will impart an acceleration of 1 cm/sec² to a 1-gm mass. The dyne is the force unit in the cgs (centimeter-gram-second) metric system.

The cgs system is widely used in scientific and technical work, but there is a second metric system that is of great importance. The mks (meter-kilogram-second) system uses the meter as the unit of length and the kilogram as the unit of mass. If the mass is measured in kilograms and the acceleration in meters per second per second, Eq. 9-3 will give the force in *newtons*. One newton is the force that will impart an acceleration of 1 m/sec² to a 1-kg mass.

Example 9-1 An object has a mass of 5 kg. (a) What is the force required to give this object an acceleration of 3 m/sec²? (b) What is the the force of gravity acting upon this object if $g = 9.8$ m/sec²?

Solution. (a) $F_R = ma = 5$ kg $\times$ (3 m/sec²)
$$= 15 \text{ kg-m/sec}^2$$
$$= 15 \text{ newtons}$$

(b) The force of gravity is usually referred to as the weight W. If the force W is the only force acting, the object would have an acceleration of 9.8 meters/sec². Substituting in the second law, we obtain

$$F_R = ma$$
$$W = mg = 5 \text{ kg } (9.8 \text{ m/sec}^2)$$
$$= 49 \text{ kg-m/sec}^2 = 49 \text{ newtons} \quad \blacktriangleleft$$

This last problem could obviously have been worked in the cgs system, and the force would

have been obtained in dynes. The conversion from newtons to dynes follows very easily from the definition of these units.

$$1 \text{ newton} = 1 \text{ kg} \times 1 \text{ m/sec}^2$$
$$= 1000 \text{ g} \times 100 \text{ cm/sec}^2$$
$$= 100{,}000 \text{ g cm/sec}^2$$
or $\quad 1 \text{ newton} = 10^5 \text{ dynes}$

In the English Engineering System forces are usually measured in pounds of force. The pound is the unit of force that we have employed in our study of statics. Although other force units are sometimes used, we shall employ the pound as the appropriate English force unit in Eq. 9-3. The *slug* is the corresponding unit of mass in the English Engineering System. An object with a mass of 1 slug will be given an acceleration of 1 ft/sec² by an unbalanced force of 1 lb.

Example 9-2 An automobile that weighs 3200 lb is on a level road. (a) What is the mass of this car in slugs? (b) How large a force must be applied to the car to give it an acceleration of 4 ft/sec²? (Ignore friction.)

Solution. (a) The force of gravity acting on the car is 3200 lb. If this were the only force acting, the acceleration would be equal to $g = 32$ ft/sec². Therefore,

$$m = \frac{F_R}{a} = \frac{W}{g} = 3200 \text{ lb}/32 \text{ ft/sec}^2 = 100 \text{ slugs}$$

(b) To produce an acceleration of 4 ft/sec²,

$$F_R = ma = 100 \text{ slugs} \times 4 \text{ ft/sec}^2 = 400 \text{ lb} \quad \blacktriangleleft$$

The relationship between weight and mass used in Problems 9-1 and 9-2 is really only a special form of Newton's second law. The force of gravity (weight) is equal to the mass times the acceleration due to gravity g. This relationship is of such great importance that we shall record it as a separate equation.

$$W = mg$$

and $$m = \frac{W}{g} \qquad (9\text{-}4)$$

Table 9-1

System	Force or Weight	Mass	Acceleration
(1) cgs (metric)	dynes	grams (g)	centimeters per second per second (cm/sec²)
(2) mks (metric)	newtons (N)	kilograms (kg)	meters per second per second (m/sec²)
(3) English Engineering	pounds (lb)	slugs	feet per second per second (ft/sec²)

Equation 9-4 is very frequently employed when the English system of units is used. If the weight (in pounds of force) is divided by g (in ft/sec²), we obtain the mass in slugs. In applying Eqs. 9-3 or 9-4, it is essential that force, mass, and acceleration all be measured within the same unit system. Three unit systems that will be extremely important in further studies are given in Table 9-1.

To obtain a useful relationship between the metric and the English units, we shall consider the standard 1-kg mass in Fig. 9-5. The force of gravity acting on this mass varies with its location in space. However, at a standard location on the earth (sea level, 45° latitude), $g = 9.81$ m/sec² = 32.2 ft/sec². The force of gravity on this mass is $m \times g = 1$ kg $\times$ (9.81 m/sec²) = 9.81 newtons. If this mass is put on a sensitive scale and weighed,

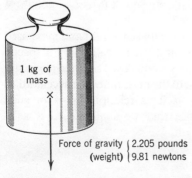

Force of gravity { 2.205 pounds
(weight) { 9.81 newtons

Figure 9-5 The standard kilogram mass.

it is found to weigh 2.205 lb. Therefore,

$$9.81 \text{ newtons} = 2.205 \text{ lb}$$

and

$$1 \text{ newton} = 0.225 \text{ lb}$$

To compare the mass units, we refer to the standard kilogram and calculate $m = W/g =$ 2.205 lb/(32.2 ft/sec²) = 0.0685 slugs. Therefore,

$$1\text{-kg mass} = 0.0685 \text{ slugs}$$

Since the weight of an object (force of gravity) is proportional to the mass, the kilogram and gram are sometimes used to measure weight and force in statics problems. Moreover, the pound is sometimes used to indicate the mass of an object. However, it is advisable to avoid confusion by employing the appropriate force or mass unit from the table just given whenever we use Newton's law of motion (Eq. 9-3).

Equation 9-3 is a vector equation, and in some instances the resultant force $\mathbf{F}_R$ will be known in terms of its x and y components. The x component of the resultant is the sum of all the x force components (ΣF_x); and the y component of the resultant is the sum of all the y force components (ΣF_y). The acceleration vector can also be resolved into components a_x and a_y. The second law of motion may then be written

$$\Sigma F_x = ma_x$$
$$\Sigma F_y = ma_y$$

(9-5)

The first condition of equilibrium (Chapter 3) is just a special case of Eq. 9-5 with the acceleration equal to zero because the resultant force is zero. The first law of motion is also a special case of the second law in which the resultant force (and the acceleration) is zero.

9-5 NEWTON'S THIRD LAW OF MOTION

If a force (a push or a pull) is to be exerted upon an object, something outside the object must exert or apply this force. In other words, whenever a force is exerted upon an object, there must be some other object which is exerting this force. According to *Newton's third law of motion*, these forces always exist in pairs.

> **The force exerted by the first object on the second is equal and opposite to the force exerted by the second object on the first.**

An alternate statement of this law is

> **Action always equals reaction; the forces between two objects are equal in magnitude but opposite in direction.**

Although this is called the third law of motion, this principle of dynamics applies fully to statics problems and to objects at rest. For instance, if a 100-lb concrete block rests on a table, there is a downward force of 100 lb on the table and an upward force of 100 lb on the block. The block weighs 100 lb because the earth attracts it (pulls down on it) with a force of 100 lb. According to the third law, the block also attracts the earth and pulls upward on the earth with a 100-lb force in the opposite direction.

The third law clearly applies to objects that are being accelerated. If we kick a brick forward to give it a forward acceleration, the brick pushes back on our foot with an equal and opposite reaction. If we kick harder, the action increases and the reaction force back on our foot also increases.

The third law predicts some effects that are not too obvious. If an airplane is flying through the air, the earth is exerting a force on the airplane; and the airplane, according to the third law, is pulling on the earth with an equal and opposite force. A space capsule or a projectile falling toward the earth pulls on the earth with a force that is equal and opposite to the force of gravity exerted by the earth. In using the third law, it is important to recall that action and reaction always act upon different objects.

The third law of motion is quite useful in analyzing collisions between objects. This law will be used in a later chapter to study the principles of impact and momentum.

9-6 NEWTON'S LAW OF UNIVERSAL GRAVITATION

Before the time of Newton other scientists had discovered the laws of falling objects. Some of these scientists knew that the earth exerted a force of attraction on all objects; Galileo even understood that the acceleration of all freely falling objects is the same. Newton obviously did not discover gravitation. He was responsible, however, for the inverse square law of universal gravitation. Newton was a brilliant mathematician, and he used the inverse square law of gravitation and the three laws of motion to predict the motion of falling objects and also the movements of the earth and the other bodies in the solar system. Newton's work laid the foundations for the scientific study of both mechanics and astronomy.

The law of universal gravitation states that

> **Any two particles of matter attract each other with a force which is directly proportional to the product of the masses and inversely proportional to the square of the distance between the two particles.**

The formula for the force of attraction between the two masses m_1 and m_2 is

$$F = G\frac{m_1 m_2}{r^2} \qquad (9\text{-}6)$$

where G is the universal gravitation constant and r is the distance between the two masses.

If the two masses are objects of ordinary size, the force of attraction is very small and is difficult to measure. By using a very sensitive torsion balance, however, it is possible to measure the force of attraction between masses in the laboratory. Since the masses m_1 and m_2 and the distance r are easily measured, it is possible to use Eq. 9-6 to calculate G. The results of these experiments are that

$$G = 6.67 \times 10^{-8} \text{ dynes-cm}^2/\text{g}^2$$
$$= 6.67 \times 10^{-11} \text{ newton-m}^2/\text{kg}^2$$
$$= 3.41 \times 10^{-8} \text{ lb-ft}^2/\text{slug}^2$$

Example 9-3   Two 100-kg spheres of lead are placed so that they just touch each other. The density of lead is 11.35 g/cm^3. (a) What is the radius of these spheres? (b) Calculate the force of attraction between the two spheres. (c) Calculate the force of attraction between one of the spheres and the earth.

Solution.   (a) From Eq. 9-2

$$\rho = \frac{m}{V}$$

$$V = \frac{m}{\rho} = \frac{100 \text{ kg}}{11.35 \text{ g/cm}^3} \times (1000 \text{ g}/1 \text{ kg})$$

$$= 8.81 \times 10^3 \text{ cm}^3$$

The volume of a sphere $= \frac{4}{3}\pi r^3 = 8.81 \times 10^3$ cm^3.

$$r^3 = \frac{3(8.81)}{4\pi} \times 10^3 \text{ cm}^3 = 2.10 \times 10^3 \text{ cm}^3$$

$$r = \sqrt[3]{2.10 \times 10^3} \text{ cm} = 12.8 \text{ cm}$$

(b) By using the methods of the calculus, Newton proved that spherical objects attract each other as if their entire mass is concentrated at their centers. The distance between the centers of the two spheres is 2(12.8 cm) = 25.6 cm. Also,

$$m_1 = m_2 = 100{,}000 \text{ g} = 10^5 \text{ g}$$

Substituting in Eq. 9-6, we find that

$$F = G\frac{m_1 m_2}{r^2}$$

$$= 6.67 \times 10^{-8} \frac{\text{dynes-cm}^2}{\text{g}^2} \frac{(10^5 \text{ g})(10^5 \text{ g})}{(25.6 \text{ cm})^2}$$

$$= \frac{6.67 \times 10^2 \text{ dynes}}{(25.6)^2} = 1.02 \text{ dynes}$$

This is comparatively a very small force, since the weight of a postage stamp is about 29 dynes.

(c) The attraction of the earth will be about 2.20 pounds per kilogram. The weight of these lead spheres is about 220 lb. In the cgs system we have

$$W = mg = (10^5 \text{ g})(980 \text{ cm/sec}^2)$$

$$= 9.80 \times 10^7 \text{ dynes}$$

This is nearly 100 million times greater than the attractive force between the two spheres.   ◀

The constant G in the law of gravitation is quite different from g, the acceleration of a freely falling object. There is, however, a mathematical relationship between the two that can be derived from the law of gravitation. Let M stand for the mass of the earth and let W indicate the force of gravity acting upon a mass m which is near the earth. Substituting these symbols in Eq. 9-6 gives

$$W = G\frac{Mm}{r^2}$$

where r is the distance from the center of the earth. But $g = W/m$; and if we divide both sides of this last equation by m we obtain

$$g = G\frac{M}{r^2} \qquad (9\text{-}7)$$

Equation 9-7 has been used by scientists to calculate M, the mass of the earth. This equation will be useful to us when we study earth satellites and satellite motion in a later chapter.

Equation 9-7 indicates clearly that g decreases as the distance r increases. As we move away from the earth, the acceleration due to gravity decreases. Moreover, the weight of objects decreases, since $W = mg$ and m remains constant. It is necessary, however, to go to a great height above the surface of the earth before this decrease in g and in the force of gravity becomes very noticeable.

9-7 LINEAR MOTION AND THE SECOND LAW

Newton's three laws of motion and the law of gravitation are used to solve many important problems in physics and in engineering mechanics. In this section we shall study some applications involving linear acceleration. We shall consider other uses of these laws in later chapters.

In these problems it is essential to draw a clear free body diagram of the object or objects in motion, and to indicate *all* the forces. In applying the equations for the second law of motion, we must be careful to note that the *resultant force* is equal to $m \times a$. If there are several forces acting, it is necessary to calculate F_R before using the equations of Newton's second law.

Example 9-4 A 6-kg block is pushed along a level table. The coefficient of friction is 0.25. How large a force is required to accelerate the object

with uniform acceleration from rest to a velocity of 20 m/sec in a time of 5 sec?

Solution. The appropriate free body diagram is drawn in Fig. 9-6, where P is the force required and W is the weight in newtons.

$$N = 6(9.8)\ \text{N} = 58.8\ \text{N}$$

and

$$f = 0.25(N) = 14.7\ \text{N}$$
$$F_R = P - f = P - 14.7\ \text{N}$$
$$a = \frac{v_F - v_0}{t} = \frac{20\ \text{m/sec} - 0}{5\ \text{sec}} = 4\ \text{m/sec}^2$$

Substituting in the second law of motion, we obtain

$$F_R = ma$$
$$P - 14.7\ \text{N} = 6\ \text{kg}\ (4\ \text{m/sec}^2)$$
$$P - 14.7\ \text{N} = 24\ \text{N}$$
$$P = 38.7\ \text{N} \qquad \blacktriangleleft$$

As a second example of this method we shall consider a problem in English Engineering units.

Example 9-5 A 500-lb sled slides down a snowy slope that is 400 ft long. The slope makes an angle of 37° with the horizontal, and the coefficient of friction between the surface and the sled is 0.2. (a) Calculate the acceleration. (b) How fast is the sled traveling when it reaches the bottom of the slope?

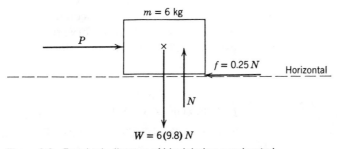

Figure 9-6   Free body diagram of block being accelerated.

Solution. The free body diagram in Fig. 9-7 indicates all the forces in pounds. We calculate the mass of the sled

$$m = \frac{W}{g} = \frac{500 \text{ lb}}{32 \text{ ft/sec}^2} = 15.6 \text{ slugs}$$

To find N, we note that the acceleration in the y direction is zero. From Eq. 9-5

$$\sum F_y = ma_y = 0$$
$$N - 400 \text{ lb} = 0, \qquad N = 400 \text{ lb}$$

Also, $\quad f = \mu N = 0.2(400 \text{ lb}) = 80 \text{ lb}$

The acceleration in the x direction is found from

$$\sum F_x = ma_x$$
$$300 \text{ lb} - 80 \text{ lb} = 15.6 \text{ slugs} \times a_x$$
$$a_x = 220 \text{ lb}/15.6 \text{ slugs} = 14.1 \text{ ft/sec}^2$$

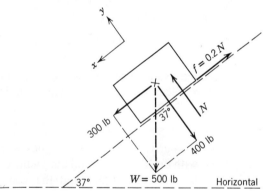

Figure 9-7   Free body diagram of an object sliding down an incline.

(b) Since the sled starts from rest $v_0 = 0$ and

$$v_f{}^2 = 2as$$
$$v_f = \sqrt{2as} = \sqrt{2(14.1 \text{ ft/sec}^2)(400 \text{ ft})}$$
$$= 106 \text{ ft/sec} \qquad \blacktriangleleft$$

In some applications of Newton's law it may be necessary to draw several free body diagrams. The following problem illustrates this point.

Example 9-6   A 20-ton locomotive is pulling a 60-ton freight train on a level track (see Fig. 9-8*a*). Frictional forces between the locomotive and the track and the freight cars and the track are 50 lb per ton and remain constant. (a) How much forward thrust must be applied to the drive wheels of the locomotive to produce an acceleration in the train of 2 ft/sec^2? (b) What is the tension in the connecting linkage between the locomotive and freight cars during this acceleration?

Solution. (a) The mass of the locomotive is

$$m = \frac{W}{g} = \frac{20 \text{ tons}}{32 \text{ ft/sec}^2} \times \frac{2000 \text{ lb}}{1 \text{ ton}}$$
$$= 1250 \text{ slugs}$$

The mass of the freight cars is

$$m = \frac{60 \text{ tons}}{32 \text{ ft/sec}^2} \times \frac{2000 \text{ lb}}{1 \text{ ton}}$$
$$= 3750 \text{ slugs}$$

From Fig. 9-8*b*, the free body diagram of the entire train ($m = 5000$ slugs),

$$F_R = ma$$
$$P - 4000 \text{ lb} = 5000 \text{ slugs} (2 \text{ ft/sec}^2)$$
$$P = 14,000 \text{ lb}$$

is the forward thrust.

(b) To find the tension in the linkage, the free body diagram of Fig. 9-8*c* or 9-8*d* may be used. In drawing these diagrams, we must indicate the correct amount of friction on the locomotive and on the freight cars. The forward thrust P acts upon the drive wheels of the locomotive. The tension in the linkage T pulls forward on the freight cars but pulls backward on the locomotive. From Fig. 9-8*c*

$$F_R = ma$$
$$T - 3000 \text{ lb} = 3750 \text{ slugs} (2 \text{ ft/sec}^2)$$
$$T = 3000 \text{ lb} + 7500 \text{ lb} = 10,500 \text{ lb}$$

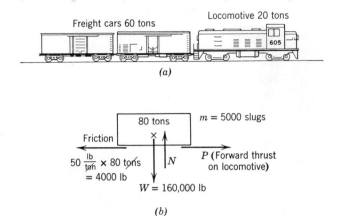

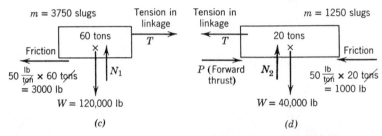

Figure 9-8   (a) The acceleration of a train. (b) Free body diagram of the entire train. (c) Free body diagram of the freight cars. (d) Free body diagram of the locomotive.

As a check we may refer to Fig. 9-8d

$$F_R = P - 1000 \text{ lb} - T$$
$$= 14,000 \text{ lb} - 1000 \text{ lb} - T$$
$$= 13,000 \text{ lb} - T$$
$$F_R = ma$$
$$13,000 \text{ lb} - T = 1250 \text{ slugs} (2 \text{ ft/sec}^2)$$
$$13,000 \text{ lb} - T = 2500 \text{ lb}$$
$$T = 10,500 \text{ lb} \quad \blacktriangleleft$$

An interesting case in which two objects are tied together is illustrated in Fig. 9-9. This comparatively simple device is called an Atwood's machine, and it may be used in a student laboratory experiment to determine g. When the masses are released from rest m_2 (the larger mass) moves

downward and m_1 moves upward. The net accelerating force is the difference between the two weights:

$$F_R = W_2 - W_1 = m_2 g - m_1 g = (m_2 - m_1)g$$

Since the total mass is $m_2 + m_1$, the second law of motion becomes

$$F_R = ma$$
$$(m_2 - m_1)g = (m_2 + m_1)a \qquad (9\text{-}8)$$

Because m_2 is only a little larger than m_1, the masses move slowly and the acceleration a is much less than g. We can use a stopwatch or some other device to measure the time for m_2 to fall a distance s. Then the acceleration a may be calculated from the formula $s = \frac{1}{2}at^2$. When this value for a, and the value for the masses m_2 and

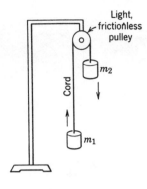

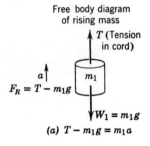

Free body diagram
of rising mass

(a) $T - m_1g = m_1a$

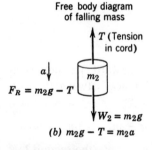

Free body diagram
of falling mass

(b) $m_2g - T = m_2a$

Adding (a) and (b) together T
drops out to give:
$(m_2 - m_1)g = (m_2 + m_1)a$
(Eq. 9-8)

Figure 9-9 Atwood's machine.

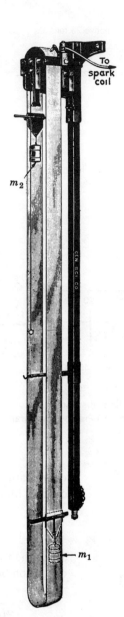

m_1, is substituted in Eq. 9-8, the acceleration due to gravity g may be easily calculated.

Friction in the pulley bearing causes some error in this experiment, but surprisingly good values for g may be determined without too much trouble or expensive equipment.

SUMMARY

Dynamics is the branch of mechanics that studies the forces which act upon objects and the motion produced by these forces.

Newton's *first law of motion* or the law of *inertia*

states that an object at rest will remain at rest and an object in motion will continue to move with constant velocity so long as the net or resultant force on the object is zero.

Newton's *second law of motion* asserts that the resultant force on an object is equal to the product of the mass and the acceleration: $F_R = ma$.

The *mass* of an object refers to the amount of matter in an object. The *weight* of an object is the force of gravity acting on an object and the weight depends upon where the object is located.

According to the second law of motion, we have (see Table 9-1)

Force in dynes = mass in grams × acceleration in cm/sec^2.

Force in newtons = mass in kilograms × acceleration in m/sec^2.

Force in pounds = mass in slugs × acceleration in ft/sec^2.

If the weight of an object is given in pounds, the mass in slugs may be calculated from $W = m \times g$ or $m = W/g$.

Newton's *third law of motion* asserts that the force exerted on an object by a second object is always accompanied by an equal and opposite force (the reaction force) exerted on the second object by the first object.

Newton's *law of universal gravitation* states that the force of gravity between two objects is directly proportional to the product of the masses and inversely proportional to the square of the distance (Eq. 9-6).

The universal gravitational constant G is different from the acceleration of a freely falling object, g. From the law of gravity and the second law of motion we have

$$g = \frac{W}{m} = \frac{GMm/r^2}{m}$$

or

$$g = \frac{GM}{r^2}$$

where M is the mass of the earth.

QUESTIONS

1. What is studied in the branch of mechanics called dynamics?

2. Explain the meaning of the term *inertia*.

3. What is the difference between the mass of an object and its momentum?

4. Is it possible for an object made of material with a small density to have large mass? Explain your answer.

5. A 2-kg object made of wood must have the same mass as a 2-kg object made of lead. What differences would one notice between the wooden object and the lead object?

6. What is the difference between weight and mass?

7. Explain what is meant by the statement that the mass of an object remains constant even though the weight changes.

8. How is the mass of an object related to its density?

9. What are three different units that may be used to measure the mass of an object?

10. What are three different units used in measuring forces?

11. How is the mass of an object in slugs calculated if we know its weight in pounds?

12. When the same force is applied to object A and object B, we find that the acceleration of object B is three times as great as the acceleration of object A. What can we conclude about the masses of the two objects?

13. Suppose that the resultant force acting on a given object is doubled. What effect would this have on the mass? What effect would this have on the acceleration?

14. Compare a pound of feathers and a pound of lead. Which has the greater weight? Which has the greater mass? Which has the greater volume? Which has the greater density?

15. If we move away from the center of the earth, g decreases. What happens to the value of G?

16. What is the difference between g and G?

17. If we travel from the earth to the moon, our mass would not change. However, we would weigh much less on the moon. Explain why this happens.

18. Suppose that the distance between two masses is doubled. What effect does this have on the force of gravitational attraction between the two masses?

19. What is meant by the statement that the force of gravity follows an inverse square law?

20. Suppose that the mass of an object is known in kilograms and the acceleration is measured in centimeters per second per second. How could we calculate the resultant force in newtons? How could we calculate the resultant force in dynes?

21. When an object is sent up in a high altitude balloon the weight of the object decreases. How could this effect be explained by using Eq. 9-7?

22. Suppose that a man pushes a table toward the right with a force of 30 lb. What is the magnitude and direction of the reaction force? On what object does the reaction force act?

PROBLEMS

(A)

1. At a point where $g = 32$ ft/sec$^2 = 9.8$ m/sec^2 determine the mass of an object that weighs
   (a) 64 lb,
   (b) 200 newtons, and
   (c) 2500 dynes.

2. A large machine has a mass of 100 slugs.
   (a) What is the force of gravity on the object if it is weighed at the equator where $g = 32.0878$ ft/sec^2?
   (b) How much does the weight of this object increase if it is moved to the north pole where $g = 32.2577$ ft/sec^2?

3. An object weighs 100 lb on the earth. Compute its mass in slugs.

4. A cubic foot of pure water weighs 62.5 lb on the earth.
   (a) Compute the mass in slugs of a cubic foot of water.
   (b) Calculate the density of water in slugs/ft^3.

5. Compute the weight (on the earth) of the objects whose mass is
   (a) 5 slugs.
   (b) 10 kilograms.
   (c) 30 grams.

6. An object whose mass is 12 slugs is given an acceleration of 5 ft/sec^2. What is the resultant force?

7. An object whose mass is 20 kg is to be given an acceleration of 4 m/sec^2. What resultant force is necessary to produce this acceleration?

8. A resultant force of 100 lb produces an acceleration of 20 ft/sec^2 when it is applied to an object.
   (a) What is the mass of this object?
   (b) How much does this object weigh where $g = 32$ ft/sec^2?

9. A space traveler who weighs 160 lb on the earth is planning to go to the moon where the acceleration of gravity is 5.32 ft/sec^2.
   (a) What is the man's mass on the earth?
   (b) What is his mass on the moon?
   (c) What is his weight on the moon?

10. Electrons have a mass of 9.11×10^{-31} kg. In a special vacuum tube design an electrical force of 2.0×10^{-13} newtons acts upon each electron. What is the acceleration of these electrons?

11. An 8-lb block is traveling at an initial velocity of 10 ft/sec. A constant force then acts upon this object and increases the velocity to 30 ft/sec in 5 sec.
    (a) What is the acceleration?
    (b) Calculate the accelerating force.
    (c) How far did the block travel in the 5 sec?

12. A short wire is attached to an 80-lb weight. The wire is then pulled upward with a force of 100 lb. What is the acceleration?

13. A light automobile weighs 2400 lb.
    (a) What is the mass of this automobile?
    (b) What resultant force is needed to give the automobile an acceleration of 8 ft/sec^2?

14. A resultant force of 100 newtons acts upon a 10-kg mass. Determine the acceleration.

15. A 10-lb block of wood is accelerated from rest to a velocity of 40 ft/sec in 5 sec.
    (a) Calculate the average acceleration.
    (b) What is the mass of the block?
    (c) What is the resultant force acting upon the block?

16. What resultant force is required to impart an acceleration of 400 cm/sec^2 to a 50-g object?

17. A 40-lb block slides down an incline with constant acceleration. It travels 20 ft in 4 sec starting from rest.
    (a) What is the acceleration?

(b) Calculate the resultant force acting to accelerate the block.

18. A 10-kg object is sliding on a horizontal surface. A force acts upon the object reducing the velocity from 12 m/sec to 2 m/sec in 5 sec.
(a) What is the deceleration?
(b) Calculate the force that was acting.
(c) How far did the mass travel in the 5 sec?

19. When we stand on the surface of the earth, we are at a distance of 4000 miles from the center of the earth and $g = 32$ ft/sec². What will g be equal to at a point 8000 miles from the center of the earth?

(B)

1. A push of 200 lb is just sufficient to keep a 2400-lb automobile traveling along a level road at a slow steady speed. Determine the acceleration if the push is increased to 300 lb.

2. An 80-lb block is resting on a rough wooden floor. A horizontal push of 20 lb gives the block an acceleration of 6 ft/sec².
(a) How great is the force of friction that is opposing the motion?
(b) Calculate the coefficient of friction.

3. At a point where $g = 980$ cm/sec² a 10-g weight is dropped. Because of air resistance the actual acceleration is 800 cm/sec².
(a) What is the resultant force in dynes acting on the object?
(b) What is the force of air resistance?

4. What force in newtons is required to push a 10-kg mass up a 30° incline with steady speed? Assume friction is negligible and $g = 9.8$ m/sec².

5. A block of steel is 4 cm × 5 cm × 3 cm. The density of steel is 7.8 g/cm³.
(a) Calculate the mass of the steel block.
(b) Determine the force of gravity acting on the block.
(Assume $g = 980$ cm/sec².)

6. A 4000-lb automobile traveling at 60 miles/hr can be brought to rest by the brakes in a distance of 200 ft.
(a) What is the average deceleration in ft/sec²?
(b) What is the average braking force?

7. A model rocket motor produces 500 lb of thrust. The rocket weighs 200 lb and it is launched vertically from a launch pad on the earth. Calculate the acceleration.

8. A space capsule has a mass of 1500 kg. What is the steady thrust required to accelerate this vehicle from a speed of 7 km/sec to 10 km/sec in a time of 10 min?

9. In a proton accelerator used in nuclear physics, protons are accelerated by a constant force from a velocity of 2×10^7 m/sec to a velocity of 8×10^7 m/sec, while they travel a distance of 20 m. The mass of a proton is 1.67×10^{-27} kg. Calculate the accelerating force.

10. A 160-lb man is standing on a spring scale in an elevator. Determine the reading on the scale if:
(a) The elevator is traveling downward with a steady velocity of 10 ft/sec.
(b) The elevator is accelerating upward with an acceleration of 8 ft/sec².
(c) The elevator is accelerating downward with an acceleration of 8 ft/sec².

11. A loaded freight elevator weighs 6400 lb. It starts from rest and in 2 sec of acceleration attains its maximum upward speed of 10 ft/sec. Assume that the acceleration is constant during this 2-sec interval.
(a) What is the tension in the cables during the upward acceleration?
(b) What is the tension in the cables after the final velocity is attained?

12. A 10-kg mass is released from rest and slides down an incline that makes an angle of 37° with the horizontal. The coefficient of friction is 0.4 on the incline.
(a) Calculate the force of friction acting to oppose the motion.
(b) What is the acceleration?

13. According to the technical tables, a cubic foot of steel weighs 490 lb. In the design of a mechanical device, a solid steel cylinder 2 ft long and 6 in. in diameter will be subjected to a resultant force of 200 lb.
(a) Calculate the weight of the steel cylinder.
(b) What is the acceleration?

14. A 6-kg mass and an 8-kg mass are used in an Atwood's machine (see Fig. 9-9).
(a) What is the acceleration of the masses?
(b) Calculate the tension in the connecting cord.

15. A 10-kg mass is suspended from a wire. Determine the tension in the wire under the following circumstances (assume that $g = 9.8$ m/sec²):
(a) The mass is stationary.
(b) The mass is moving upward with a constant velocity of 5 m/sec.

(c) The mass is moving downward with a constant velocity of 5 m/sec.

(d) The mass is accelerating upward with an acceleration of 8 m/sec².

(e) The mass is accelerating downward with an acceleration of 8 m/sec².

16. A bullet that weighs 0.25 ounces is fired from a rifle with a muzzle velocity of 2000 ft/sec. It acquires this muzzle velocity by being accelerated down a rifle barrel that is 30 in. long. Calculate:

(a) The average acceleration.

(b) The average force acting upon the bullet while it is traveling down the barrel.

17. An automobile of weight W is traveling on a level road with velocity v. The coefficient of friction between the tires and the road is μ.

(a) If the brakes are applied, prove that the maximum deceleration is μg.

(b) Prove that the minimum distance in which the car can stop is $v^2/2\mu g$.

18. An object is placed on an inclined plane that makes an angle of θ with the horizontal.

(a) Assuming that the incline is smooth (frictionless), draw a free body diagram and prove that the acceleration is $g \sin \theta$.

(b) Assuming that the coefficient of friction is μ, draw a free body diagram and prove that the acceleration is $g(\sin \theta - \mu \cos \theta)$.

19. The earth is very nearly spherical in shape. The average radius of the earth is 6370 km, and $g = 9.8$ m/sec² at the surface. The universal gravitational constant is $G = 6.67 \times 10^{-11}$ N-m²/kg². Calculate the mass M of the earth from this data.

20. The weights in Fig. 9-10 are released from rest. Friction is negligible on the suspended 8-lb weight, whereas the coefficient of kinetic friction on the surface is 0.4.

(a) What is the acceleration of the two weights?

(b) Calculate the tension in the cord.

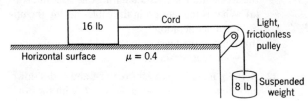

Figure 9-10   A system of accelerating weights.

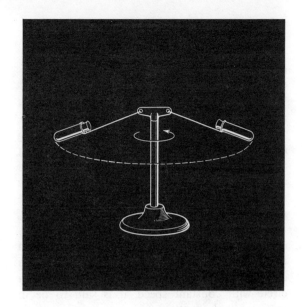

10. Forces and Curvilinear Motion

The laws of motion may be used to study the motion of objects traveling in circles or other curved paths. A *centripetal force* is necessary to force the object into a curved path, and we shall find that these objects are undergoing a *centripetal acceleration*.

LEARNING GOALS

In studying motion on a curved path in this chapter, the reader should pursue the following learning goals:

1. *Be able to calculate the velocity for an object in uniform circular motion.*

2. *Using the formulas for centripetal acceleration to calculate the centripetal acceleration.*

3. *Using the second law of motion to calculate the centripetal force on a moving object.*

4. *Understanding the difference between centripetal and centrifugal force.*

5. *Applying the second law of motion to problems in circular motion involving banking, motion on a vertical circle, and motion on a horizontal circle.*

6. *Understanding the difference between centripetal acceleration and tangential acceleration.*

10-1 CIRCULAR MOTION AND CURVILINEAR MOTION

Whenever an object travels in a path that is not a straight line, the object is in *curvilinear motion*. A railroad train or an automobile traveling around a curve is in curvilinear motion. A rocket or a satellite traveling around the earth follows a curvilinear path through space. Nearly every machine and mechanical device contains some moving parts that are in curvilinear motion.

A very common type of curvilinear motion is *circular motion*. If an object is following a path that is a circle or a part of a circle, the object is in circular motion. When an object travels around a circle with constant speed, it is said to be in *uniform circular motion*.

10-2 THE ACCELERATION IN UNIFORM CIRCULAR MOTION

In uniform circular motion the speed of the moving object remains constant, but the velocity changes. The magnitude of the velocity vector does not vary, but the direction of this vector is continuously changing. As the object travels around the circular path in Fig. 10-1, the velocity

that is $\mathbf{v}_1$ at position A becomes the velocity $\mathbf{v}_2$ at position B. The vector triangle in Fig. 10-1 indicates that

$$\begin{aligned} \mathbf{v}_1 + \Delta\mathbf{v} &= \mathbf{v}_2 \\ \Delta\mathbf{v} &= \mathbf{v}_2 - \mathbf{v}_1 \end{aligned} \tag{10-1}$$

The vector $\Delta\mathbf{v}$ is the change in velocity as the object moves from A to B.

The vector triangle in Fig. 10-1 is an isosceles triangle, and its sides are proportional to triangle OAB. The proportion between corresponding sides is

$$\frac{\Delta v}{v} = \frac{\Delta s}{r}$$

Since $\Delta s = $ velocity $\times$ time $= v\,\Delta t$, this proportion becomes

$$\frac{\Delta v}{v} = v\frac{\Delta t}{r}$$

Multiplying both sides of this equation by $v/\Delta t$ gives the result that

$$\frac{\Delta v}{\Delta t} = \frac{v^2}{r} \tag{10-2}$$

The quantity $\Delta v/\Delta t$ is the rate of change of velocity; we should recognize this as the acceleration.

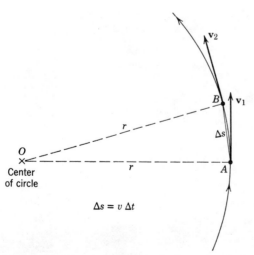

The magnitude
of $\mathbf{v}_2$ and $\mathbf{v}_1$
is v

Figure 10-1 The change in velocity during uniform circular motion.

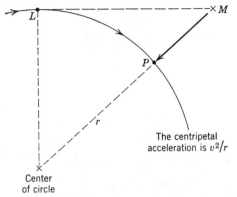

Figure 10-2   An object in circular motion is accelerated toward the center.

Since the change in velocity Δv is directed toward the center of the circle, the acceleration is also directed toward this point. It is easy to see from Fig. 10-2 that an object traveling in a circle is actually accelerating or falling toward the center of the circle. The object at position L would have traveled in a straight line to point M if it were not falling toward the center. The object fell a distance MP in order to remain on the circular path. This acceleration toward the center is called the *centripetal acceleration*. According to Eq. 10-2, the centripetal acceleration is

$$a_c = \frac{v^2}{r} \qquad (10\text{-}3)$$

Example 10-1   An object is traveling at a constant speed of 10 ft/sec in a circular path that is 4 ft in diameter. Calculate the centripetal acceleration.

Solution.   The radius is 2 ft.

$$a_c = \frac{v^2}{r} = \frac{(10 \text{ ft/sec})^2}{2 \text{ ft}} = \frac{100 \text{ ft}^2/\text{sec}^2}{2 \text{ ft}}$$

$$= 50 \text{ ft/sec}^2 \qquad \blacktriangleleft$$

When a moving part of machinery or any moving object is traveling in a circle, the magnitude of the velocity frequently has to be calcu-

lated. The radius of the circle can be measured, and it is often possible to determine the time for one revolution. The magnitude of the velocity may then be calculated as in the following example.

Example 10-2   A flywheel governor is traveling at constant speed in a circle with a radius of 20 cm. One revolution is completed in 0.2 sec. Calculate (a) the velocity, and (b) the centripetal acceleration.

Solution.   (a) In 0.2 sec the object travels once around the circumference, which is a distance of $2\pi r = 2\pi(20 \text{ cm}) = 40\pi$ cm. The magnitude of the velocity is

$$\frac{\text{distance}}{\text{time}} = \frac{40\pi \text{ cm}}{0.2 \text{ sec}} = 200\pi \text{ cm/sec}$$

$$\text{(b) } a_c = \frac{v^2}{r} = \frac{(200\pi \text{ cm/sec})^2}{20 \text{ cm}}$$

$$= 2000\pi^2 \text{ cm/sec}^2$$

$$= 19.7 \times 10^3 \text{ cm/sec}^2 \qquad \blacktriangleleft$$

In order to remember the method used in Example 10-2, we can derive a useful formula for the velocity v. The *period* is the time for one complete revolution and is usually designated by the letter T. The velocity is then the circumference $(2\pi r)$ divided by the period or

$$v = \frac{2\pi r}{T} \qquad (10\text{-}4)$$

Substituting Eq. 10-4 in Eq. 10-3, we can obtain another useful expression for the centripetal acceleration.

$$a_c = \frac{v^2}{r} = \frac{1}{r}\left(\frac{2\pi r}{T}\right)^2$$

$$a_c = \frac{4\pi^2 r}{T^2} \qquad (10\text{-}5)$$

Equation 10-5 can be used to calculate the centripetal acceleration in uniform circular motion when the radius and the period are known.

In many engineering and technical applications involving uniform circular motion, the rate of rotation is given in revolutions per minute or revolutions per second. This quantity is called the *frequency*. The frequency is defined as the number of revolutions per unit time.

Example 10-3   The frequency of rotation for a flywheel is 600 rev/min. Determine the period in seconds.

Solution.   The abbreviation for revolutions per minute is rev/min.

$$f = 600 \text{ rev/min} \times (1 \text{ min/60 sec})$$
$$= 10 \text{ rev/sec}$$

This unit for frequency may also be written 1/sec or $\sec^{-1}$ so that $f = 10 \sec^{-1}$. Since the frequency is the rate of revolution, the time for one revolution is

$$T = \frac{1 \text{ rev}}{10 \text{ rev/sec}} = \frac{1}{10} \text{ sec} \quad \blacktriangleleft$$

In solving Example 10-3 we used an important relationship between frequency (revolutions per second) and period (seconds per revolution). The reader can easily verify that

$$T = \frac{1}{f}$$

and

$$f = \frac{1}{T} \quad (10\text{-}6)$$

Using Eq. 10-6 to eliminate T from Eq. 10-4, we find that

$$v = 2\pi rf \quad (10\text{-}7)$$

Also, Eq. 10-5 can be written in terms of the frequency

$$a_c = 4\pi^2 f^2 r \quad (10\text{-}8)$$

Example 10-4   A flywheel on a gasoline motor is 18 in. in diameter. If the flywheel rotates at 3000 rev/min, calculate the velocity and the acceleration of a point on the rim.

Solution

$$r = \tfrac{1}{2}(\text{diameter}) = \tfrac{1}{2}(18 \text{ in.})(1 \text{ ft/12 in.})$$
$$= \tfrac{3}{4} \text{ ft}$$
$$v = 2\pi rf$$
$$= 2\pi(\tfrac{3}{4} \text{ ft})(3000 \text{ rev/min})$$
$$= 4500\pi \text{ ft/min}$$

This can be changed into ft/sec if we wish, and

$$v = (4500\pi \text{ ft/min})(1 \text{ min/60 sec})$$
$$= 75\pi \text{ ft/sec} = 236 \text{ ft/sec}$$

To get the acceleration in the usual units of ft/sec², we let

$$f = 3000 \text{ rev/min } (1 \text{ min/60 sec}) = 50 \text{ rev/sec}$$
$$= 50 \sec^{-1}$$

Then

$$a_c = 4\pi^2 f^2 r = 4\pi^2(50\tfrac{1}{\text{sec}})^2(\tfrac{3}{4} \text{ ft})$$
$$= 3\pi^2(2500) \text{ ft/sec}^2$$
$$a_c = 74 \times 10^3 \text{ ft/sec}^2$$

The calculation for a_c could also be done with Eq. 10-3.

$$a_c = \frac{v^2}{r} = \frac{(236 \text{ ft/sec})^2}{0.75 \text{ ft}}$$
$$= \frac{(2.36)^2 \times 10^4 \text{ ft/sec}^2}{0.75}$$
$$= 74 \times 10^3 \text{ ft/sec}^2 \quad \blacktriangleleft$$

An object fastened to the rim of the rotating flywheel in Example 10-4 would be subjected to a very great acceleration. Since 32 ft/sec² = 1 g, this acceleration may be described as

$$a_c = 74 \times 10^3 \text{ ft/sec}^2 \frac{1 \, g}{32 \text{ ft/sec}^2}$$
$$= 2.3 \times 10^3 \, g$$

This acceleration is about 2300 g's. An object loosely fastened to the rim of the flywheel would probably fly off on a tangent to the circular path and move with constant velocity according to Newton's first law. In order to keep an object traveling on the circular path with such a great centripetal acceleration, a very large *centripetal force* must be applied to the object.

10-3 THE CENTRIPETAL FORCE

According to the law of inertia, an object will travel in a straight line if there is no resultant force acting upon it. If an object is to travel in a curved path, there must be a force acting on the object that pushes or pulls it into this path. The centripetal acceleration must be caused by a force that accelerates the object toward the center of the curve. It is obvious from the second law of motion that the centripetal force equals the mass times the centripetal acceleration. We have learned three different forms for a_c, and this enables us to write three useful equations for the centripetal force:

$$F_c = \frac{mv^2}{r} \qquad (10\text{-}9)$$

$$F_c = m\left(\frac{4\pi^2 r}{T^2}\right) \qquad (10\text{-}10)$$

$$F_c = m(4\pi^2 f^2 r) \qquad (10\text{-}11)$$

We can also obtain three other equations for F_c if we let $m = W/g$.

Example 10-5 The two masses in Fig. 10-3 are set rotating about the vertical *AA* axis at 10 rev/sec. Each mass is 2 kg. (a) What is the resultant force acting on each mass? (b) What is the tension in the rods?

Solution. (a) The rigid rod exerts a normal (upward) push to counteract the weight *W*. The resultant force is a pull toward the center of the circle.

$$F = ma_c = m4\pi^2 f^2 r$$
$$= (2 \text{ kg})4\pi^2(10\tfrac{1}{\text{sec}})^2(0.2 \text{ m})$$
$$= 8\pi^2(20) \text{ kg-m/sec}^2$$
$$= 1578 \text{ newtons}$$

The same resultant force acts on the other mass.

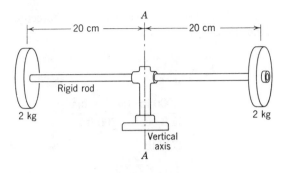

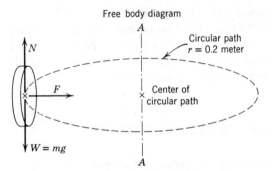

Figure 10-3 Masses traveling in a circular path.

(b) According to the third law, there is an outward pull or tension in the rod of 1578 newtons. ◀

As a second example of the centripetal force problem, we shall consider a problem in English engineering units.

Example 10-6 A 4000-lb automobile is driven at a steady speed of 45 miles/hr around a curve in the highway. The curve has a radius of 200 ft. What is the resultant force acting upon the vehicle?

Solution. Recalling that 60 miles/hr = 88 ft/sec, we find that

$$v = 45 \text{ miles/hr} \times \frac{88 \text{ ft/sec}}{60 \text{ miles/hr}} = \frac{3}{4}(88 \text{ ft/sec})$$

$$= 66 \text{ ft/sec}$$

The resultant force is inward toward the center of the curve and equal to

$$F_c = \frac{W}{g}\frac{v^2}{r}$$

$$= \frac{4000 \text{ lb}}{32 \text{ ft/sec}^2}\frac{(66 \text{ ft/sec})^2}{200 \text{ ft}}$$

$$= 2723 \text{ lb} \qquad \blacktriangleleft$$

As Example 10-6 indicates, a large centripetal force is needed to push the automobile into the curved path. This force on the car must be exerted by the road; the road pushes upon the tires, forcing the car to travel in a curved path. In the next section we study the effect of road friction and the theory for the design of banked curves.

10-4 THE BANKING OF CURVES

In driving an automobile around a curve on an unbanked road, the centripetal acceleration must be produced by the force of friction. The force of friction, however, cannot become indefinitely large, for it must be less than or equal to μN. The maximum value of the force of friction determines the maximum possible speed at which a car can round the curve. If the driver exceeds this maximum speed, his car will skid off the road.

Figure 10-4 indicates the free body diagram of an automobile on an unbanked curve. When the driver is traveling at the maximum speed, he is just on the verge of skidding, and $f = \mu N$. Since $N = W$ on the level road, $f = \mu W$ is the resultant force. The maximum speed may then be calculated from the equation of motion for centripetal acceleration

$$F = ma_c$$

$$\mu W = \left(\frac{W}{g}\right)\left(\frac{v^2}{r}\right)$$

$$v^2 = \mu gr \qquad (10\text{-}12)$$

Example 10-7 An old highway has a sharp curve with a radius of curvature of only 100 ft. If the coefficient of friction is 0.5 between the tires and the road, what is the maximum speed in miles per hour at which an automobile may round the curve?

Solution. Substituting in Eq. 10-12, we obtain

$$v = \sqrt{\mu gr} = \sqrt{0.5(32 \text{ ft/sec}^2)(100 \text{ ft})}$$

$$= \sqrt{1600 \text{ ft}^2/\text{sec}^2} = 40 \text{ ft/sec}\frac{60 \text{ mi/hr}}{88 \text{ ft/sec}}$$

$$= 27.3 \text{ miles/hr.} \qquad \blacktriangleleft$$

The curves on modern superhighways usually have a radius of curvature of more than 1000 ft;

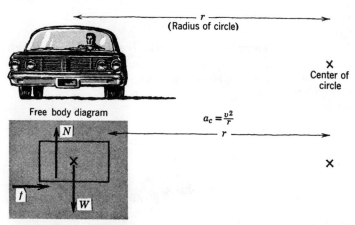

Figure 10-4 An automobile on an unbanked curve.

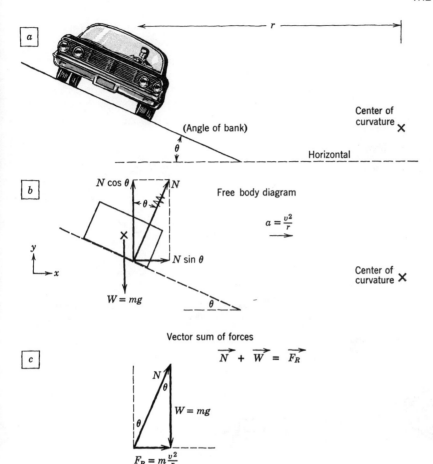

Figure 10-5 An automobile rounding a banked curve without any friction.

thus the curves are very gradual. Also, they are banked so that vehicles traveling at normal road speeds will be able to round the curves without depending on the force of friction.

Figure 10-5a indicates an automobile rounding a banked curve without any frictional force. From the free body diagram (Fig. 10-5b), we can calculate

$$\sum F_y = ma_y = 0$$

(there is no acceleration in the vertical direction).

$$N \cos \theta = W \qquad \text{(a)}$$

$$\sum F_x = ma_x \quad \text{and} \quad a_x = v^2/r$$

$$N \sin \theta = \left(\frac{W}{g}\right)\left(\frac{v^2}{r}\right) \qquad \text{(b)}$$

Dividing Eq. (b) by Eq. (a) gives

$$\frac{N \sin \theta}{N \cos \theta} = \frac{(W/g)(v^2/r)}{W}$$

or

$$\tan \theta = \frac{v^2}{gr} \qquad \text{(10-13)}$$

This result can also be obtained by the method of Fig. 10-5c. There are only two forces acting upon the car, N and W. The vector sum of these two forces produces the horizontal resultant force which is equal to mv^2/r. Using the definition of $\tan \theta$, we find that

$$\tan \theta = \frac{mv^2/r}{mg} = \frac{v^2}{gr}$$

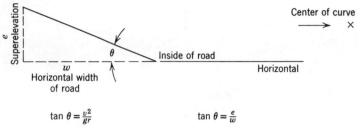

$$\tan \theta = \frac{v^2}{gr} \qquad \tan \theta = \frac{e}{w}$$

Figure 10-6 Superelevation on a banked curve.

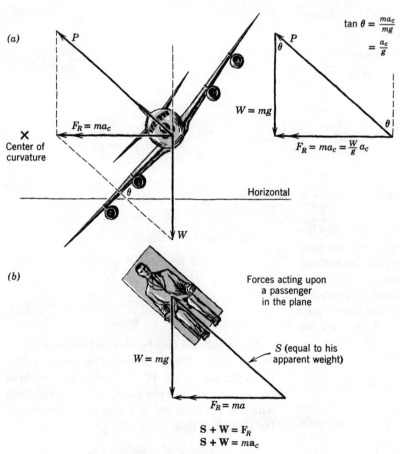

Figure 10-7 An airplane in a horizontal curve.

Example 10-8 A new highway has a curve for which the radius of curvature is 3000 ft. What is the proper angle of banking if automobiles are to round the curve at 60 miles/hr?

Solution. Substituting in Eq. 10-13, we find that

$$\tan \theta = \frac{(88 \text{ ft/sec})^2}{32 \text{ ft/sec}^2 (3000 \text{ ft})} = 0.0807$$

$$\theta = 4.62°$$

Instead of finding θ, it is sometimes more useful to determine the superelevation of the outside of the curve (see Fig. 10-6). For a road that has four 12-ft traffic lanes and a 2-ft dividing wall, $w = 50$ ft and

$$e = w \tan \theta$$

$$= 50 \text{ ft } (0.0807) = 4.04 \text{ ft} \qquad ◀$$

Equation 10-13 gives the speed at which a car can round the curve with no friction. Since some friction is present, even on wet or icy roads, vehicles sometimes succeed in traversing the curve at greater speed. However, these higher speeds may result in a dangerous skid.

Equation 10-13 is also used to calculate the angle of banking on railroad tracks, based upon an average speed for the trains. If the trains travel at the speed used in the calculation, the tracks will have to exert only a normal force on the train as it rounds the bend. However, when a train travels at a lower or a higher speed, the tracks have to resist a thrust at right angles to the road bed.

If an airplane travels in a horizontal circle without any sideslip, Eq. 10-13 gives the angle of banking for the airplane (see Fig. 10-7). Because of the high speed of aircraft, the centripetal acceleration may be quite large, and θ may be a rather large angle. Note that $\tan \theta = a_c/g$ is equal to the number of g in the centripetal acceleration during the banked, horizontal turn. The airplane in the diagram is banked at 53° so that $\tan \theta = \frac{4}{3}$ and the acceleration is about 1.33 g.

The force P acting upon the wings and body of the airplane pushes it around the curve. From the vector triangle in Fig. 10-7a it is apparent

that P will become much greater than the weight of the plane if the acceleration is several g. This force acting upon the wings and control surfaces may produce structural damage. The passenger will also suffer discomfort (Fig. 10-7b).

10-5 THE CONICAL PENDULUM AND THE GOVERNOR

The device in Fig. 10-8 is called a conical pendulum and it forms part of the governors that are used on some engines. The position of the rotating weight is determined by the rate of rotation of the shaft. The position of the weight may be used to regulate the opening and closing of a fuel valve on the engine.

It is obvious from the vector triangle in Fig. 10-8 that $\tan \theta = a_c/g$ and $g \tan \theta = a_c$. Using the expression for a_c given by Eq. 10-8, and noting

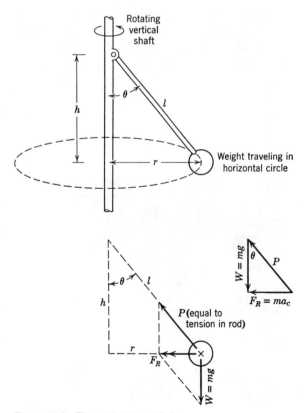

Figure 10-8 The conical pendulum.

that $\tan \theta = r/h$, we find that

$$\frac{gr}{h} = 4\pi^2 f^2 r$$

$$\frac{g}{h} = 4\pi^2 f^2$$

and

$$f = \frac{1}{2\pi} \sqrt{\frac{g}{h}} \qquad (10\text{-}14)$$

Equation 10-14 indicates that h will become smaller (the weight will move up) if the frequency of rotation is increased. It appears, however, that the rod can never get to be exactly horizontal no matter how fast the shaft rotates. From the vector triangle we observe that the tension in the shaft, P, will increase as the centripetal acceleration and angle θ get bigger. Before θ can become exactly 90°, the tension will become great enough to break the rod.

Example 10-9   A small 2-lb weight is attached to a light connecting rod and used as a conical pendulum. The rod is 18 in. long and the shaft is rotating at 90 rev/min. (a) What is the angle θ between the vertical shaft and the rod? (b) Calculate the tension in the rod.

Solution.   (a) $f = 90$ rev/min $\times$ 1 min/60 sec = 1.5 rev/sec. From Eq. 10-14 we obtain

$$h = \frac{g}{4\pi^2 f^2} = \frac{32 \text{ ft/sec}^2}{4\pi^2 (1.5\frac{1}{\text{sec}})^2}$$

$$= 0.36 \text{ ft}$$

From Fig. 10-8 we note that

$$\cos \theta = \frac{h}{l} = \frac{0.36 \text{ ft}}{1.5 \text{ ft}} = 0.24$$

$$\theta = 76.1°$$

(b) Using proportional triangles in Fig. 10-8, we find that

$$\frac{P}{W} = \frac{l}{h}$$

$$P = W \frac{l}{h} = 2 \text{ lb} \frac{(1.5 \text{ ft})}{(0.36 \text{ ft})}$$

$$= 8.34 \text{ lb} \qquad \blacktriangleleft$$

10-6 TRAVELING IN A VERTICAL CIRCLE

Stunt motorcycle drivers are able to ride upside down around the loop in Fig. 10-9 if they are

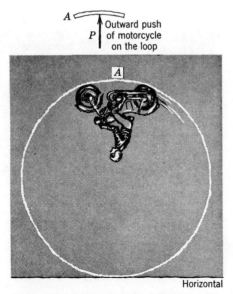

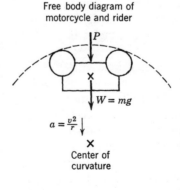

Figure 10-9   Riding the vertical loop.

going fast enough. The minimum speed at which the motorcycle may be traveling as it passes the top of the loop is called the critical velocity v_c.

Example 10-10   In Fig. 10-9 the diameter of the loop is 10 m. The mass of the motorcycle and driver is 200 kg. (a) What is the force with which the motorcycle presses against the loop at A if the speed is 8 m/sec? (b) What is the minimum speed at A if the motorcycle is not to fall away from the loop?

Solution.   (a) Figure 10-9 indicates a free body diagram of the forces that produce the centripetal acceleration (let $r = 5$ m).

$$F_R = ma$$

$$P + mg = \frac{mv^2}{r}$$

$$P = \frac{mv^2}{r} - mg$$

$$= \frac{200 \text{ kg} (8 \text{ m/sec})^2}{5 \text{ m}} - 200 \text{ kg} (9.8 \text{ m/sec}^2)$$

$$= 2560 \text{ kg-m/sec}^2 - 1960 \text{ kg-m/sec}^2$$

$$= 600 \text{ newtons}$$

(b) When the driver is traveling at the minimum speed, the motorcycle is on the verge of falling away from the wall; it just ceases to push outward on the wall and $P = 0$. Therefore,

$$mg = \frac{mv^2}{r}$$

$$v^2 = gr$$

$$v = \sqrt{(9.8 \text{ m/sec}^2)(5 \text{ m})} = \sqrt{49 \text{ m}^2/\text{sec}^2}$$

$$= 7 \text{ m/sec} \qquad \blacktriangleleft$$

It is apparent from Example 10-10 that the critical velocity is

$$v_c = \sqrt{gr} \qquad (10\text{-}15)$$

If an object is guided into a vertical circle by a string or by traveling on the inside of a loop, the velocity at the top of the circle is greater than or equal to v_c. However, if the object is held up by some rigid support, it may travel through the vertical circle with a velocity less than v_c.

Example 10-11   A 10-lb weight travels in a vertical circle as in Fig. 10-10. If the rod r is 2 ft long and the weight travels with a constant speed of 5 ft/sec determine the tension in the rod at (a) and at (b).

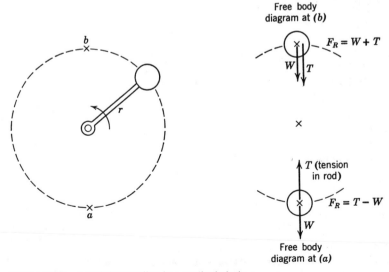

Free body
diagram at (b)

$F_R = W + T$

Free body
diagram at (a)

Figure 10-10   A weight traveling in a vertical circle.

Solution. At (a) F_R is directed upward toward the center of the circle, and $F_R = T - W$.

$$F_R = ma_c$$

$$T - W = \left(\frac{W}{g}\right)\left(\frac{v^2}{r}\right)$$

$$T = W + \left(\frac{W}{g}\right)\left(\frac{v^2}{r}\right)$$

$$= 10 \text{ lb} + \frac{10 \text{ lb}}{32 \text{ ft/sec}^2}\frac{(5 \text{ ft/sec})^2}{2 \text{ ft}}$$

$$= 10 \text{ lb} + 10 \text{ lb}\left(\frac{25}{64}\right) = 10 \text{ lb} + 3.91 \text{ lb}$$

$$= 13.9 \text{ lb}$$

At (b) $F_R = W + T$

$$W + T = \left(\frac{W}{g}\right)\left(\frac{v^2}{r}\right)$$

$$T = \left(\frac{W}{g}\right)\left(\frac{v^2}{r}\right) - W$$

$$= 3.91 \text{ lb} - 10 \text{ lb} = -6.09 \text{ lb}$$

The negative tension indicates that the rod is really under compression when the weight is at the top of its path. The rod is under compression because the velocity is less than the critical velocity. The critical velocity in this case is

$$v_c = \sqrt{gr} = \sqrt{32 \text{ ft/sec}^2 (2 \text{ ft})} = 8 \text{ ft/sec} \quad \blacktriangleleft$$

10-7 SATELLITES IN CIRCULAR MOTION

The orbits of natural and man-made satellites are usually elliptical in shape. However, the earth's orbit about the sun is very nearly a circle because the maximum radius and minimum radius differ by only 3%. The orbits of many artificial earth satellites are nearly circular, and with proper control of the launching rockets a satellite can be launched into an orbit that is an almost perfect circle.

In Fig. 10-11 we have drawn a satellite or space capsule of mass m orbiting around a planet of mass M. According to our discussion of the law of gravity in Chapter 9, we know that the force of gravity may be written as mg_0 where $g_0 = GM/r_0^2$ is the value of g at the orbit of the satellite. Since the force of gravity is the only force acting upon the satellite, and the acceleration is v^2/r_0, we have

$$F_R = ma$$

$$mg_0 = mv^2/r_0$$

The velocity of the satellite in orbit is therefore

$$v = \sqrt{g_0 r_0} \qquad (10\text{-}16)$$

In order to use Eq. 10-16 in any calculation, we need to find g_0. This can be conveniently done in terms of the acceleration due to gravity at the surface of the planet. If r_p is the radius of the planet, and g_p is the acceleration due to gravity at the surface of the planet,

$$g_p = \frac{GM}{r_p^2} \quad \text{and} \quad \frac{g_0}{g_p} = \frac{GM/r_0^2}{GM/r_p^2}$$

Then $$\frac{g_0}{g_p} = \frac{r_p^2}{r_0^2} \qquad (10\text{-}17)$$

Equation 10-17 is the mathematical statement of the law that g is inversely proportional to r^2.

• **Example 10-12** A space capsule is launched into a circular orbit that is 600 miles above the surface of the earth. What is the speed of the capsule? Assume that

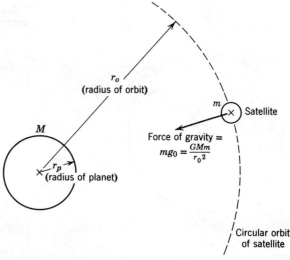

Figure 10-11 The circular orbit of a satellite.

the radius of the earth is 3960 miles and that g at the surface is 32 ft/sec^2

Solution

$$g_p = 32 \text{ ft/sec}^2, \qquad r_p = 3960 \text{ miles}$$

$$r_0 = 3960 \text{ miles} + 600 \text{ miles} = 4560 \text{ miles}$$

$$g_0 = \frac{g_p r_p^2}{r_0^2} = 32 \text{ ft/sec}^2 \frac{(3960 \text{ miles})^2}{(4560 \text{ miles})^2}$$

$$= 24.1 \text{ ft/sec}^2$$

$$v = \sqrt{g_0 r_0} = \sqrt{24.1 \text{ ft/sec}^2 \frac{(4560 \text{ miles})(1 \text{ mile})}{5280 \text{ ft}}}$$

$$= \sqrt{24.1 \frac{(4.56)}{(5.28)} \text{ miles}^2/\text{sec}^2} = 4.56 \text{ miles/sec}$$

$$v = 4.56 \text{ miles/sec} \times 3600 \text{ sec/hr}$$

$$= 16,400 \text{ miles/hr} \qquad \blacktriangleleft$$

Substituting $g_0 = GM/r_0^2$ in Eq. 10-16 gives

$$v = \sqrt{(GM/r_0^2)r_0}$$

$$v = \sqrt{GM/r_0} \qquad (10\text{-}18)$$

Since $\sqrt{GM}$ is a constant for all satellites of the earth, Eq. 10-18 indicates that the velocity v becomes less as the radius of the orbit r_0 increases. The velocity in the circular orbit is inversely proportional to the square root of the radius. It is interesting to note that the mass of the satellite m has no bearing upon the velocity.

The time required to travel around the orbit is the period T of the satellite. If we use Eq. 10-18 and let $v = 2\pi r_0/T$, it turns out that

$$T = \frac{2\pi r_0}{v} = \frac{2\pi r_0}{\sqrt{GM/r_0}} = \frac{2\pi r_0}{\sqrt{GM}}\sqrt{r_0}$$

and

$$T = \frac{2\pi}{\sqrt{GM}} r_0^{3/2} \qquad (10\text{-}19)$$

The period of the satellite is proportional to the three-halves power of the orbital radius, or, in other words, the period squared (T^2) is proportional to the radius cubed (r_0^3). This three-halves power law was discovered for the satellites of the sun (the planets) by Johannes Kepler before Newton was born. However, Kepler developed no scientific theory to explain his discovery. Newton was the first scientist to use the laws of motion and the law of gravitation to account for and predict the motions of the planets.

10-8 WEIGHTLESSNESS AND MOTION

It is obvious from Section 10-6 and from the law of gravitation that a force of gravity acts on an orbiting space vehicle and everything inside the vehicle. Yet we have all seen descriptions and photographs of weightlessness inside space vehicles and satellites. How can objects be weightless if there is a force of gravity?

This confusion arises because the term "weight" is being used to mean two different things:

(i) First meaning: The *weight* of an object is the force of gravity that acts on the object, and is indicated by the force vector $\mathbf{W} = m\mathbf{g}$.

(ii) Second meaning: The *apparent weight* of an object is the force exerted by an object on its supports. If we indicate the supporting forces on the object by $\mathbf{S}$, then the object pushes on its surroundings with a reaction force $-\mathbf{S}$. The apparent weight is the force $-\mathbf{S}$.

• **Example 10-13** A 160-lb man is strapped into a specially designed supersonic test vehicle that can travel up and down like a high speed elevator. What is the true and apparent weight of the man under the following conditions? (Assume that $g = 32$ ft/sec^2.)
(a) The vehicle is traveling upward at a steady speed of 1000 ft/sec. (b) The vehicle is traveling downward at a steady speed of 1000 ft/sec. (c) The vehicle is traveling upward at 1000 ft/sec but is accelerating at 40 ft/sec^2. (d) The vehicle is traveling upward at 1000 ft/sec but is decelerating at 40 ft/sec^2.

Solution. Figure 10-12 indicates the free body diagram that will be used in parts (a), (b), (c), and (d).

$$\mathbf{F}_R = m\mathbf{a}$$

$$\mathbf{W} + \mathbf{S} = m\mathbf{a} \qquad \text{(vector sum)}$$

The apparent weight is the reaction to the supporting force and is equal to $-\mathbf{S}$

$$-\mathbf{S} = \mathbf{W} - m\mathbf{a}$$

$$\text{apparent weight} = \mathbf{W} + (-m\mathbf{a}) \qquad \text{(vector sum)}$$

(a) The acceleration is zero so that

$$\text{apparent weight} = \mathbf{W}$$

$$= 160 \text{ lb downward}$$

(b) Same as (a)

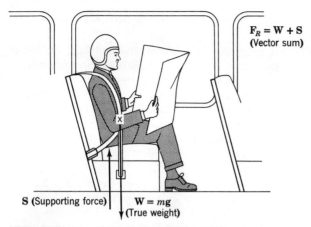

$$F_R = W + S$$
(Vector sum)

S (Supporting force)

$W = mg$
(True weight)

Figure 10-12   A man in an accelerated vehicle.

(c) $\mathbf{a} = 40$ ft/sec^2 upward, $m = 5$ slugs

$(-m\mathbf{a}) = 200$ lb downward

apparent weight $= 160$ lb downward $+ 200$ lb downward

$= 360$ lb downward

The upward acceleration increases the apparent weight.

(d) $\mathbf{a} = 40$ ft/sec^2 downward (because of deceleration)

$(-m\mathbf{a}) = 200$ lb upward

apparent weight $= 160$ lb downward $+ 200$ lb upward

$= 40$ lb upward

During this downward acceleration the man could sit or stand on the roof of the vehicle. A plumb bob in the vehicle would point upward in the direction of the apparent weight.   ◄

This principle of apparent weight also applies to centripetal accelerations. In Fig. 10-7b the apparent weight of the airplane traveler is the reaction to **S**. In centripetal accelerations $(-m\mathbf{a})$ is a vector that is directed opposite to the direction of the acceleration. Hence $(-m\mathbf{a})$ is a vector that points away from the center. In using the principle of apparent weight, it is important to keep track of the direction of the vectors. For all kinds of accelerations, however, we have the principle

apparent weight $= \mathbf{W} + (-m\mathbf{a})$ (vector sum)   (10-20)

When a space capsule is being launched, there is an upward acceleration of several g and the apparent weight of objects is greater than the force of gravity. After the rocket motors discontinue their thrust, the only force acting on the capsule is the force of gravity, and the capsule becomes a free projectile. Since the only force acting is the force of gravity, the acceleration must be g. The apparent weight of any object in the capsule is

$$\mathbf{W} + (-m\mathbf{a}) = m\mathbf{g} + (-m\mathbf{g}) = 0$$

The supporting forces cease to act when the space capsule is falling with the acceleration of gravity. A space traveler would have an apparent weight of zero because he would not exert a force upon the floor of the capsule. All the different parts of our space traveler's body would cease to exert supporting forces on each other, and they would all appear to be weightless. All the objects in the capsule, which are falling with the acceleration of gravity, would fail to exert any force on their surroundings. All these objects would hover or drift slowly within the space craft. This is the state of apparent weightlessness.

For a capsule in a circular orbit, the centripetal acceleration is equal to g and the condition of apparent weightlessness prevails. We also have weightlessness, however, if the capsule is not in orbit but is simply moving through space, in any direction, with the power turned off. This condition of weightlessness can also be achieved by aircraft in the atmosphere if power is used to overcome air resistance and to maintain an acceleration equal to g.

If a space capsule is traveling far from any attracting planet, the force of gravity **W** and the acceleration of gravity **g** become zero. However, if this space capsule is accelerated by its rockets, every object in the capsule will have an apparent weight even though the force of gravity remains zero. The acceleration of the vehicle produces a sort of artificial gravity. Equation 10-20 gives the apparent weight of an object in the accelerating capsule as $-m\mathbf{a}$. However, the apparent weight of these objects is really due to their inertia. They react or "kick back" on their supports, and we interpret this reaction as an apparent weight.

10-9 THE CENTRIFUGE AND "CENTRIFUGAL" FORCE

In the last section it was explained that the acceleration of objects produced a reaction force on the supports. This, of course, is also true for the centripetal acceleration of objects. The reaction force produced by a centripetal acceleration is called the *centrifugal force.*

These definitions will be clearer if we examine the free body diagrams in Fig. 10-13. The centripetal force acts upon the whirling ball and produces the centripetal acceleration. The reaction is the centrifugal force that the athlete feels pulling on his arms. The first law of motion tells us that when the cable is released, the ball will shoot off on a tangent to the circular path.

The centripetal force acts on the accelerated objects and produces the acceleration. The accelerated objects pull upon their supports with a reaction force designated as the centrifugal force. The so-called "centrifugal tendency" that is observed in circular motion is really only the inertia of the objects.

The principle of setting up an "artificial gravity" by means of an acceleration is used in the centrifuge. Centrifuges are extremely useful devices available in many different designs for different applications. They are used in medical laboratories to separate blood cells from the blood plasma and even to separate bacteria cells from a liquid solution. Centrifuges are also widely used in chemical laboratories and manufacturing processes to separate substances of different densities. Modern ultracentrifuges operate at speeds as high as 100,000 rev/min.

A simple laboratory centrifuge is illustrated in Fig. 10-14. When the apparatus is rotated, suspended particles of solids are quickly separated out from the liquid. Dense material will "sink" toward the outer end of the rotating test tube and material of lower density will "float" toward the inner part of the test tube. The apparent weight resulting from the centripetal acceleration may be hundreds or thousands of times greater than the force of gravity.

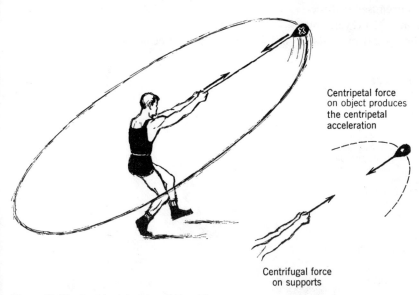

Centripetal force on object produces the centripetal acceleration

Centrifugal force on supports

Figure 10-13 Centripetal and centrifugal force.

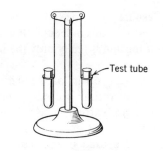

Test tube

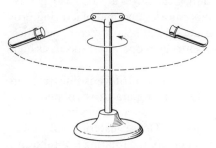

Figure 10-14 A laboratory centrifuge.

10-10 TANGENTIAL ACCELERATION IN CIRCULAR AND CURVILINEAR MOTION

Whenever an object is traveling in a circular path, it has a centripetal acceleration because the direction of the velocity is changing. However, the magnitude of the velocity (the speed) may also

change. The rate at which the magnitude of the velocity is changing is called the *tangential acceleration*. The tangential acceleration has the same direction as the velocity; it is directed along the tangent to the circle (see Fig. 10-15).

The centripetal acceleration is directed toward the center of the circle along the radius and is therefore perpendicular to the tangential acceleration. The centripetal acceleration measures the rate at which the direction of the velocity is changing. The total acceleration must include both the tangential and centripetal components. In uniform circular motion, however, the speed does not change and the tangential acceleration is zero.

Example 10-14 A small cam is fastened to a rotating drum at a distance of 50 cm from the axis of rotation (see Fig. 10-16). The drum is originally at rest, but it is set into rotation so that the speed of the cam increases at the rate of 100 cm/sec². After 3 sec of motion, determine (a) the velocity of the cam, (b) the number of revolutions, and (c) the acceleration of the cam.

Solution. (a) The tangential acceleration is $a_t = 100$ cm/sec². The magnitude of the velocity can be calculated from

$$v = v_0 + a_t t$$
$$= 0 + (100 \text{ cm/sec}^2)(3 \text{ sec}) = 300 \text{ cm/sec}$$

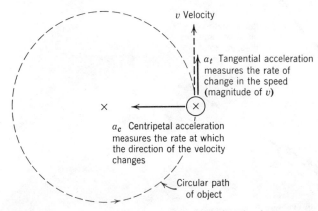

v Velocity

a_t Tangential acceleration measures the rate of change in the speed (magnitude of v)

a_c Centripetal acceleration measures the rate at which the direction of the velocity changes

Circular path of object

Figure 10-15 Tangential and centripetal acceleration in circular motion.

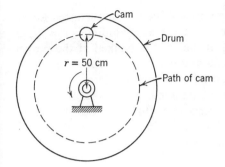

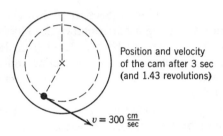

Position and velocity of the cam after 3 sec (and 1.43 revolutions)

$v = 300 \frac{cm}{sec}$

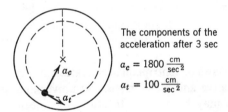

The components of the acceleration after 3 sec

$a_c = 1800 \frac{cm}{sec^2}$

$a_t = 100 \frac{cm}{sec^2}$

Figure 10-16 A cam on a rotating drum.

(b) Because the acceleration is constant, the average speed of the cam is

$$\frac{\text{final speed} + \text{original speed}}{2} = \frac{300 \text{ cm/sec} + 0}{2}$$

$$= 150 \text{ cm/sec}$$

The distance traveled is

average speed × time = 150 cm/sec × 3 sec

$$= 450 \text{ cm}$$

To find the number of revolutions for (b), we calculate the path length for one revolution. The circumference is

$$2\pi r = 2\pi(50 \text{ cm}) = 100\pi \text{ cm}$$

The number of revolutions in the 3 sec is

$$\frac{450 \text{ cm}}{100\pi \text{ cm/rev}} = 1.43 \text{ revolutions}$$

(c) The centripetal acceleration after 3 sec is

$$a_c = \frac{v^2}{r} = \frac{(300 \text{ cm/sec})^2}{50 \text{ cm}} = 1800 \text{ cm/sec}^2$$

The tangential acceleration is constant and equal to 100 cm/sec². ◀

For objects traveling on curved paths that are not circles, it is still possible to resolve the acceleration into two components (see Fig. 10-17). The *tangential acceleration* a_t lies in the direction of the path. The tangential acceleration is the rate of increase in the magnitude of the velocity.

At right angles to the tangential velocity and to the path is the *normal acceleration*, a_N. The normal acceleration is given by the familiar formula

$$a_N = \frac{v^2}{r}$$

where r is the radius of curvature of the curved path.

The figure indicates that a_N is directed toward the center of curvature of the curve. The different portions of the curve may, however, have different centers of curvature and different radii of curvature.

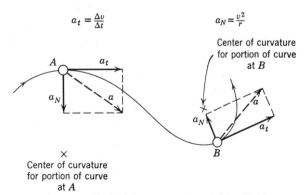

Figure 10-17 Normal acceleration (a_N) and tangential acceleration (a_t) on a curved path.

The normal acceleration measures the rate at which the velocity is changing in direction. The normal acceleration must be zero on a part of the curve that is a straight line. Also, the normal acceleration is the same as the centripetal acceleration if a portion of the curve happens to be a circle.

SUMMARY

If an object is traveling on a curved path, the direction of the velocity is changing even if the magnitude of the velocity (i.e., the speed) remains constant. For any object traveling in a curve, the velocity is changing and, therefore, the object is undergoing an acceleration.

The acceleration toward the center of curvature is called the *centripetal acceleration*; it is calculated from $a_c = v^2/r$.

When an object is traveling in a circle, the magnitude of the velocity is the circumference divided by the period

$$v = \frac{2\pi r}{T}$$

The *period*, T, is the time for one revolution.

The *frequency*, f, is the number of revolutions per unit time and $f = 1/T$. Therefore, $v = 2\pi rf$. Also, the centripetal acceleration can be written in terms of T and f so that

$$a_c = \frac{v^2}{r} = \frac{4\pi^2 r}{T^2} = 4\pi^2 f^2 r$$

In order to force an object to travel in a curved path and impart a centripetal acceleration to the object, there must be a *centripetal force* acting upon the object. The centripetal force is directed toward the center of the curved path and is calculated from

$$F_c = ma_c = \frac{W}{g} a_c$$

When an automobile is driven on an unbanked road, the centripetal force is produced by the force of friction, and the maximum speed for an automobile on the verge of skidding is $v = \sqrt{\mu gr}$ (Eq. 10-12). If the road is banked at the proper angle, the automobile can negotiate the curve without depending upon the force of friction. Because of the banking, the normal force has a horizontal component that provides the centripetal force. The angle of banking can be calculated from

$$\tan \theta = \frac{v^2}{gr}$$

The conical pendulum (Section 10-5) involves a weight traveling in a horizontal circle. The frequency of rotation can be calculated from Eq. 10-14 and the tension in the connecting rod or cord is given by considering the force diagram in Fig. 10-8.

If an object is guided into a vertical circle by a string or by traveling on the inside of a loop, there is a minimum speed called the critical velocity if the object is not to fall away from the circle as it passes the highest point on the circle. If the centripetal force is set equal to mg, the critical velocity is found to be $v_c = \sqrt{gr}$.

The laws of centripetal acceleration and centripetal force and Newton's law of universal gravitation may be applied to study the motion of satellites in circular orbits (Section 10-7).

The acceleration of objects produces a reaction force on the supports. The *apparent weight* of an object is the force exerted by an object on its supports. The *true weight* is the force of gravity on an object. If an object is unaccelerated, the apparent weight of the object is equal to the true weight. The acceleration of an object causes a reaction force on the supports; this is a kind of artificial gravity so that the apparent weight differs from the true weight.

If an object is whirled rapidly in a circular path, the centripetal force acts upon the object and pulls it inward toward the center of the circle. The reaction to the centripetal force is the *centrifugal force*. This centrifugal force is exerted by the object on its supports and acts outward from the center of the circle. The apparent weight resulting from the centripetal acceleration is often much larger than the force of gravity.

If an object is traveling in a curved path, there is always a *normal acceleration* (centripetal acceleration) equal to v^2/r that is directed at right angles to the path and toward the center of curvature of the path. If the magnitude of the velocity (i.e., the speed) is changing, there is also a *tangential acceleration* that is directed along the path and is equal to the time rate of change in the magnitude of the velocity.

QUESTIONS

1. What is meant by the term "uniform circular motion"?

2. Suppose that an object is traveling in a circle at the constant speed of 10 ft/sec. Why can we be certain that the velocity is changing?

3. What is the direction of the centripetal acceleration?

4. Prove that v^2/r always has the units of acceleration.

5. What is the period for an object traveling in a circle?

6. How is the period related to the frequency?

7. What is the definition of frequency of rotation?

8. How would one convert rev/min into rev/sec?

9. Suppose that we wish to calculate the centripetal acceleration in m/sec^2. What units must be used for f in the formula $a_c = 4\pi^2 f^2 r$?

10. A 2000-pound automobile is traveling on an unbanked curve, and the coefficient of static friction is 0.6 between the tires and the road. What is the maximum centripetal force that could be exerted on the automobile to cause it to travel in a curved path?

11. Explain why a vehicle can travel faster around a banked curve than around an unbanked curve.

12. In the conical pendulum of Fig. 10-8, the distance $h = 50$ cm. To what will h be equal if the frequency of the rotating shaft is doubled?

13. A weight travels in a vertical circle as in Fig. 10-10. When the weight is at the top of the circle, the tension may be zero. Under what circumstances will this occur?

14. When a weight travels in a vertical circle as in Fig. 10-10 and passes the highest point in the circle,

the rod may be under compression, or under tension. What determines whether the rod is under compression or under tension?

15. What is centrifugal force and how does it differ from centripetal force?

16. What is the difference between apparent weight and true weight?

17. Under what conditions will the apparent weight and the true weight of object be equal?

18. Under what conditions will the apparent weight of an object be less than the true weight?

19. Under what conditions would the apparent weight of an object be zero?

20. Explain why the apparent weight resulting from centripetal acceleration in a centrifuge may be much greater than the force of gravity.

21. On a rotating wheel, is the centripetal acceleration greater for points near the hub of the wheel or for points near the rim? Explain.

22. Since the earth rotates on its axis, each object on the earth has a centripetal acceleration. Is the centripetal acceleration greater for points near the poles or for points near the equator? Explain.

23. How does Newton's third law of motion explain the cause of centrifugal force?

24. The force of gravity pulls the earth toward the sun and imparts an acceleration of the earth toward the sun. Why does the earth not crash into the sun?

25. How should Newton's second law of motion be used to explain the motion of man-made satellites? What forces the satellites into a curved path?

26. Explain the difference between normal acceleration and tangential acceleration.

27. If an object is undergoing uniform circular motion, the centripetal acceleration is never zero, but the tangential acceleration is always zero. Explain.

28. If an object travels on a straight line, the normal acceleration is always zero. Explain.

PROBLEMS

(A)

1. Calculate the centripetal acceleration of an object that is traveling at 10 ft/sec in a circle 2 ft in radius.

2. An object is traveling at 400 cm/sec in a circle that has a 20-cm radius. Calculate the centripetal acceleration.

3. An automobile travels at a constant speed of 60 miles/hr along a road. If the road curves so that the radius of curvature is 200 ft, calculate the centripetal acceleration.

4. The blades of a fan are 2 ft long and the blades are rotating at the rate of 1200 rev/min. Calculate:
   (a) The frequency in rev/sec.
   (b) The period.
   (c) The centripetal acceleration of a point at the end of the blades.

5. An object travels with a constant speed around a circle that is 3 ft in radius. It completes 6 revolutions each second.
   (a) What is the period for this motion?
   (b) Calculate the speed of the object.
   (c) Calculate the acceleration of the object.

6. An object whose mass is 5 slugs is rotated in a circle that is 10 ft in radius at a speed of 20 ft/sec. Calculate:
   (a) The centripetal acceleration.
   (b) The centripetal force.

7. In a large betatron, electrons travel in a circular path of radius 2 m, with a speed of 2.5×10^7 m/sec.
   (a) Calculate the centripetal acceleration of the electrons.
   (b) Determine the centripetal force on the electrons. (The mass of an electron is 9.11×10^{-31} kg.)

8. A 10-kg mass is traveling at a constant speed in a circle with a radius of 50 cm. Two revolutions are completed each second.
   (a) What is the velocity of the mass?
   (b) What is the centripetal acceleration?
   (c) Calculate the centripetal force acting upon the mass.

9. The flyball on a governor is traveling with uniform speed in a circle that is 20 in. in diameter. The flyball completes 300 revolutions each minute. Determine:
   (a) The speed of the flyball.
   (b) The centripetal acceleration.

10. A child who weighs 80 lb is swinging on a rope that is 16 ft long. What is the tension in the rope if the child is traveling at 20 ft/sec as he swings past the lowest point?

11. The automobile in Fig. 10-18 makes an unbroken U-turn as indicated. If the coefficient of friction is

0.6 between the tires and the road, what is the maximum speed for the automobile not to slip?

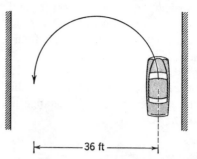

Figure 10-18 Centripetal acceleration during a U-turn.

12. A 100-gram object is whirled around in a horizontal circle 60 cm in diameter so that it completes 20 revolutions in 5 sec. Determine:
   (a) The speed of the object.
   (b) The centripetal acceleration.
   (c) The centripetal force.

13. A merry-go-round in an amusement park carries children in a circle that is 40 ft in diameter. What is the maximum number of revolutions per minute if the acceleration of the children is not to exceed 0.4 g? (*Note*: 1 g = 32 ft/sec².)

14. A force of 20 newtons acts upon a 10-kg mass and forces it to travel in a circular path 8 m in diameter.
   (a) What is the centripetal acceleration?
   (b) How fast is the mass traveling?

15. In a high speed centrifuge a liquid sample is rotated in a circle 40 cm in diameter. How many revolutions per minute are required if the centripetal acceleration is to be 100 g? (*Note*: 1 g = 9.8 m/sec².)

16. On a bobsled run a sharp hairpin turn has a radius of curvature of 60 ft. This curve is banked at an angle of 75°. What is the maximum safe speed for a bobsled on this curve? (Assume that friction is negligible on the icy curve.)

17. An automobile that weighs 3600 lb travels on a level road at a speed of 45 miles/hr. It traverses a curve that is 400 ft in radius.
   (a) Determine the centripetal force acting upon the vehicle.
   (b) What is the minimum coefficient of friction between the tires and the road if the automobile is not to go into a skid?

18. A bridge over a small river has a roadway that is in the shape of an arch. The radius of curvature of the arch is 200 ft. What is the maximum speed at which an automobile can travel across the bridge without leaving the roadway at the top of the arch?

19. The horizontal rod in Fig. 10-19 has a breaking strength of 1000 lb.
    (a) What is the velocity of the 10-lb weight when the rod breaks?
    (b) What is the rate of revolution for the shaft when the rod breaks?

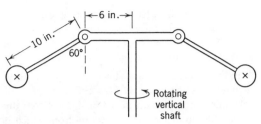

Figure 10-20 Rotating fly weights.

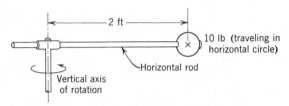

Figure 10-19 A weight traveling in a horizontal circle.

(B)

1. A drum, 18 in. in diameter, is placed on a lathe which rotates it at the rate of 900 rev/min. Calculate:
   (a) The time for one revolution.
   (b) The speed of a point on the rim of the drum.
   (c) The speed of a point 3 in. from the center of the drum.
   (d) The centripetal acceleration of a point on the rim of the drum.

2. A 5-kg mass travels in a vertical circle with constant speed. It completes 3 revolutions per second and the circle is 80 cm in diameter. Calculate:
   (a) The speed in m/sec.
   (b) The centripetal acceleration.
   (c) The tension in the supporting rod when the mass is at the bottom of the circle.
   (d) The tension when the mass is at the top of the circle.

3. A jet plane is traveling at a constant speed of 600 miles/hr. What is the radius of the smallest circle in which this plane can fly if the centripetal acceleration is not to exceed $3 g$?

4. (a) What is the speed of the flyweights in Fig. 10-20?
   (b) Calculate the rate of revolution in rev/min for the shaft.

5. A 4-lb object starts from rest and travels on a horizontal circle of radius 10 ft with a constant tangential acceleration of 2 ft/sec². After 3 sec of motion, determine:
   (a) The velocity of the weight.
   (b) The tangential force acting on the weight.
   (c) The centripetal force.

6. A 100-g iron ball is fastened to a cord 1 m long, and the ball is whirled around in a horizontal circle. The ball completes 10 rev in 12 sec while traveling at a constant speed.
   (a) What is the angle that the cord makes with the vertical?
   (b) Calculate the tension in the cord.

7. A 1-ounce blob of clay is pressed onto the edge of a phonograph record that is 12 in. in diameter. The record is placed on a turntable that rotates at 45 rev/min. Assuming that the clay does not fly off the record, determine:
   (a) The speed at which the clay is traveling.
   (b) The centripetal force acting upon the clay.

8. An earth satellite is launched into a circular orbit that is 600 km above the surface of the earth. Assume that the earth is a sphere with a radius of 6370 km, and that $g = 9.8$ m/sec² at the surface of the earth.
   (a) What is the velocity of the satellite?
   (b) What is the centripetal acceleration of the satellite?
   (c) How long does the satellite take to complete one orbit?

9. A weight travels in a vertical circle that is 4 m in diameter. The velocity at the highest point is 10 m/sec. What should the velocity at the lowest point be if the tension in the rod is to be constant during the rotation?

10. A communication satellite is launched into a circular orbit so that it completes one revolution per day and always remains over the same point on the

equator of the earth. How far above the earth does the satellite have to travel? Assume $g = 32$ ft/sec^2 at the surface of the earth and that the radius of the earth is 3960 miles.

11. A testing procedure on a supersonic rocket plane requires that the plane fly at a constant speed of 1500 miles/hr and complete a 180° turn so that the plane ends up flying in the direction from which it came. The acceleration during the turn is not to exceed 5 g.

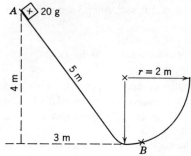

Figure 10-21   Centripetal force on a curved ramp.

(a) What is the minimum distance in miles within which the plane can turn around?
(b) What is the least time required for a 180° turn?

12. The 20-g block is released from rest at A and it slides down the frictionless ramp (see Fig. 10-21).
(a) What is the velocity when the block reaches B?
(b) What force does the block exert upon the curved portion of the ramp at B?

13. The rails on a narrow gauge railroad are 4 ft apart. The trains will traverse a curve of 400 ft radius at an average speed of 30 miles/hr. What is the correct superelevation for the outside rail?

14. See Fig. 10-22a.
(a) What is the tension in AB and BC when the weight is stationary and directly below the horizontal shaft?
(b) What is the maximum frequency of revolution if the tension in AB and BC is not to exceed 100 lb?

15. See Fig. 10-22b.
(a) What is the tension in AB and BC if the members are stationary?
(b) What is the tension in AB and BC if the structure rotates about the vertical axis at a steady rate and completes one revolution every 2 sec?

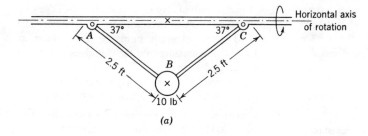

(a)

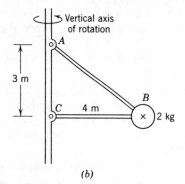

(b)

Figure 10-22   Tension in supporting members.

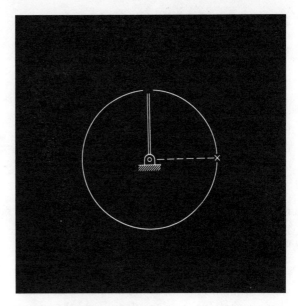

11.
Forces and Rotation

Many of the most important applications of physics and engineering involve a study of rotation and rotating objects. In this chapter we shall study the *angular displacement, angular velocity,* and *angular acceleration* of rotating objects. We shall also study the application of the second law of motion to problems involving *torques* and rotation.

LEARNING GOALS

This chapter has five major learning objectives:

1. *Using the formulas for angular displacement, angular velocity, and angular acceleration.*

2. *Understanding the correct units used in measuring rotational quantities.*

3. *Converting radians per second to revolutions per minute.*

4. *Calculating the moment of inertia in some simple cases.*

5. *Using the second law of motion to calculate the angular acceleration, torque, and moment of inertia.*

11-1 ROTATION AND ROTARY MOTION

Nearly every machine and mechanical device has some parts that rotate about a fixed axis. Flywheels and drive shafts, toothed-gear wheels, the blades on electric fans, and the rotors in electric motors are some examples of objects that describe *rotation* or rotary motion.

During rotation all the parts of the rotating object (except for those points directly on the axis) travel in circular paths, and the fixed axis of rotation passes through the centers of these circles. The linear displacement, velocity, and acceleration is different for these different parts of the rotating object. In order to describe and study the motion of the entire object, the angular displacement, the angular velocity, and the angular acceleration must be considered.

11-2 ANGULAR DISPLACEMENT

The *angular displacement* of an object describes the amount of rotation. If point A on the rotating object moves to point B, then the angle θ is the angular displacement (see Fig. 11-1). We are undoubtedly familiar with the method of measuring this angle in units called degrees. The angle AOB in Fig. 11-1 is a 90-degree angle. The angular displacement of the object is, therefore, 90 degrees (which may also be written 90°).

In engineering or technical applications it is frequently convenient to describe angular dis-

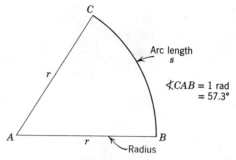
Figure 11-2   Measuring an angle in radians.

placement in revolutions. The angular displacement in Fig. 11-1 is one-fourth of a revolution. The conversion from degrees to revolutions is easily remembered as

$$360° = 1 \text{ rev}$$

In the mathematical study of rotation and in applying the laws of motion to rotation, the most useful unit for measuring angular displacement is the *radian*. An angle of one radian is an angle whose arc is equal in length to the radius (see Fig. 11-2). In general an angle in radians is given by the ratio of arc length s to radius r:

$$\theta \text{ (in radians)} = \frac{s}{r} \qquad (11\text{-}1)$$

To convert from revolutions or degrees into radians is simple if we recall that the arc length for one complete revolution equals the circumference of the circle which is $2\pi r$. The angle in radians for one revolution is therefore

$$\theta = \frac{s}{r} = \frac{2\pi r}{r} = 2\pi$$

The conversion factor for the units is, therefore,

$$1 \text{ rev} = 360° = 2\pi \text{ rad}$$

For purposes of comparison we may convert 1 rad into degrees.

$$1 \text{ rad} = 1 \text{ rad} \times \left(\frac{360°}{2\pi \text{ rad}}\right) = \frac{180}{\pi} \text{ degrees}$$

$$= 57.3° \text{ (to one decimal place)}$$

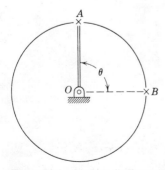
Figure 11-1   Angular displacement.

11-3 ANGULAR VELOCITY; THE RATE OF ROTATION

The *angular velocity* of an object is the rate at which the angular displacement is changing. The angular velocity is used to describe the rate (or speed) of rotation. In most formulas the angular velocity is designated by the Greek letter ω (omega). In many technical and engineering procedures the angular velocity is measured in revolutions per minute (rev/min). In many physics formulas and in mathematical calculations, however, it is necessary to use radians per second (rad/sec) as the unit for angular velocity.

Example 11-1 A rotating pulley completes 10 rev in 4 sec. Determine the average angular velocity in (a) revolutions per second, (b) revolutions per minute, and (c) radians per second.

Solution

(a) $\omega_{av} = 10 \text{ rev}/4 \text{ sec} = 2.5 \text{ rev/sec}$

(b) $2.5 \text{ rev/sec} \times (60 \text{ sec}/1 \text{ min}) = 150 \text{ rev/min}$

(c) $2.5 \text{ rev/sec} \times (2\pi \text{ rad}/1 \text{ rev})$

$\qquad = 5\pi \text{ rad/sec} = 15.7 \text{ rad/sec}$ ◀

The calculation in Example 11-1 gave the average angular velocity during the 4-sec time interval. The formula used was

$$\text{average angular velocity} = \frac{\text{angular displacement}}{\text{time}}$$

$$\omega_{av} = \frac{\theta}{t} \qquad (11\text{-}2)$$

Equation 11-2 may be used in calculations even if the rate of rotation is changing. It is important, however, to distinguish between the average angular velocity ω_{av} and the instantaneous angular velocity ω. The distinction between the average and the instantaneous angular velocity is analogous to that made between the average and the instantaneous linear velocity.

Nearly all the methods used in linear kinematics may be employed in studying rotation. The graph-ical method, in particular, is applicable to rotation and may be used to define the instantaneous value of ω. If a θ–t (angular displacement versus time) graph is drawn, the slope is the rate of change in the angular displacement. If, for example, the angular displacement in revolutions is plotted against the time in seconds, the slope will be in revolutions per second. The slope on the angular-displacement–time graph at any particular time gives the instantaneous velocity at that time.

In the shop or laboratory the average angular velocity of a rotating shaft or flywheel can be determined by using a revolution counter to measure θ in revolutions and a stopwatch to get the time interval for a number of revolutions. Then Eq. 11-2 could be used to calculate ω_{av}. The instantaneous angular velocity at a particular time may be read off from a tachometer calibrated in rev/min or other units of angular velocity.

11-4 THE ANGULAR ACCELERATION

If the instantaneous angular velocity is changing, the rotating object is undergoing an *angular acceleration*. The Greek letter α (alpha) is used as the algebraic symbol for angular acceleration in physics formulas. The angular acceleration is defined as the time rate of change in the instantaneous angular velocity. If the angular velocity is increasing, the angular acceleration is positive. A negative value for α means that the angular velocity is decreasing and the rotational motion is slowing up.

The rotational motion and the angular acceleration of any object can be described by an angular velocity versus time graph of the motion. Figure 11-3, which represents 1 min in the motion of a rotating pulley, is an example of such an angular velocity–time diagram. The analysis of the ω–t diagram proceeds along the same lines as the analysis of the v–t diagram in Chapter 7. The slope of the ω–t diagram represents the rate of change in the angular velocity, or, in other words, the angular acceleration. The area under

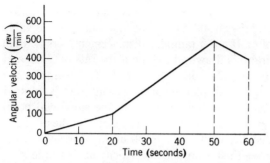

Figure 11-3   The ω-t diagram for the rotation of a pulley.

the ω-t curve is equal to $\omega_{av} \times t$, which is the angular displacement.

Example 11-2   From the ω-t diagram in Fig. 11-3 determine (a) the angular acceleration of the pulley, (b) the number of revolutions during the 60 sec, and (c) the average angular velocity in radians per second.

Solution.   (a) There are three different slopes and three different values for α. During the first 20 sec,

$$\alpha = \frac{\omega_2 - \omega_1}{t_2 - t_1} = \frac{100 \text{ rev/min} - 0}{20 \text{ sec} - 0}$$

$$\alpha = 5 \text{ rev/min-sec}$$

From 20 to 50 sec,

$$\alpha = \frac{\omega_2 - \omega_1}{t_2 - t_1} = \frac{500 \text{ rev/min} - 100 \text{ rev/min}}{50 \text{ sec} - 20 \text{ sec}}$$

$$= 13.3 \text{ rev/min-sec}$$

From 50 to 60 sec the rotation is slowing up.

$$\alpha = \frac{\omega_2 - \omega_1}{t_2 - t_1} = \frac{400 \text{ rev/min} - 500 \text{ rev/min}}{60 \text{ sec} - 50 \text{ sec}}$$

$$= -10 \text{ rev/min-sec}$$

(b)   From 0 to 20 sec the area is

$$A = \tfrac{1}{2}b \times h = \tfrac{1}{2}(100 \text{ rev/min})$$

$$\times 20 \text{ sec} \times \frac{1 \text{ min}}{60 \text{ sec}}$$

$$= 16\tfrac{2}{3} \text{ rev}$$

From 20 to 50 sec the area is

$$A = \tfrac{1}{2}(b_1 + b_2) \times h$$

$$= \tfrac{1}{2}(100 \text{ rev/min} + 500 \text{ rev/min})$$

$$\times 30 \text{ sec} \times \frac{1 \text{ min}}{60 \text{ sec}}$$

$$= 150 \text{ rev}$$

From 50 to 60 sec the area is

$$A = \tfrac{1}{2}(b_1 + b_2) \times h$$

$$= \tfrac{1}{2}(500 \text{ rev/min} + 400 \text{ rev/min})$$

$$\times 10 \text{ sec} \times \frac{1 \text{ min}}{60 \text{ sec}}$$

$$= 75 \text{ rev}$$

The total angular displacement is $16\tfrac{2}{3}$ rev $+$ 150 rev $+$ 75 rev $= 241\tfrac{2}{3}$ rev

(c) $\omega_{av} = \dfrac{\theta}{t} = 241.7 \text{ rev/60 sec} \times 2\pi \text{ rad/1 rev}$

$$= 25.3 \text{ rad/sec} \qquad \blacktriangleleft$$

11-5 UNIFORM ANGULAR ACCELERATION

The mathematical formulas for motion under uniform angular acceleration can be easily derived from ω-t diagrams. These formulas are in the same algebraic form as the equations used in uniform linear acceleration. The only change is that the linear quantities must be replaced by the corresponding angular quantities as Table 11-1 indicates.

In using these formulas for uniform acceleration, we must be careful to use the appropriate units.

Example 11-3   A braking mechanism on a rotating drill is designed to slow the drill from 1000 to 600 rev/min in 5 sec. Assume that the angular acceleration is uniform. (a) What is the angular acceleration in rad/sec^2? How many revolutions are completed during the 5 sec?

Table 11-1

Linear Quantity	Angular Quantity
Linear displacement s	Angular displacement θ
Linear velocity v	Angular velocity ω
Linear acceleration a	Angular acceleration α

Formulas

For uniform linear acceleration	For uniform angular acceleration
$s = v_{av} \times t$	$\theta = \omega_{av} \times t$
$v_{av} = \dfrac{v_f + v_0}{2}$	$\omega_{av} = \dfrac{\omega_f + \omega_0}{2}$
$v_f = v_0 + at$	$\omega_f = \omega_0 + \alpha t$
$s = v_0 t + \frac{1}{2}at^2$	$\theta = \omega_0 t + \frac{1}{2}\alpha t^2$
$v_f^2 = v_0^2 + 2as$	$\omega_f^2 = \omega_0^2 + 2\alpha\theta$

Solution

(a) $\alpha = \dfrac{\omega_f - \omega_0}{t}$

$$= \frac{600 \text{ rev/min} - 1000 \text{ rev/min}}{5 \text{ sec}}$$

$$= -\left(\frac{400}{5}\right) \text{rev/min-sec}$$

$$= -80 \text{ rev/min-sec}$$

$$\times (2\pi \text{ rad/1 rev})(1 \text{ min/60 sec})$$

$$= \frac{-8\pi}{3} \text{ rad/sec}^2 = -8.38 \text{ rad/sec}^2$$

(b) $\theta = \omega_{av} \times t$

$$= \tfrac{1}{2}(1000 \text{ rev/min} + 600 \text{ rev/min}) \times 5 \text{ sec}$$

$$= 4000 \text{ rev-sec/min} \times (1 \text{ min/60 sec})$$

$$= 66\tfrac{2}{3} \text{ rev}$$

ALTERNATE METHOD

$\alpha = -80 \text{ rev/min-sec}$

$\theta = \omega_0 t + \tfrac{1}{2}\alpha t^2$

$$= 1000 \text{ rev/min} \times 5 \text{ sec}$$

$$+ \tfrac{1}{2}(-80 \text{ rev/min-sec})(5 \text{ sec})^2$$

$$= 5000 \text{ rev-sec/min} - 40(25) \text{ rev-sec/min}$$

$$= 4000 \text{ rev-sec/min} \times (1 \text{ min/60 sec})$$

$$= 66\tfrac{2}{3} \text{ rev}$$

◀

11-6 ANGULAR MOTION AND TANGENTIAL VELOCITY

In Fig. 11-4 point P on the rotating object is displaced to P' during a small angular displacement $\Delta\theta$. From the definition of an angle in radians $\Delta s = \Delta\theta\, r$. Dividing both sides by the short time interval Δt gives

$$\frac{\Delta s}{\Delta t} = \frac{\Delta\theta}{\Delta t} r$$

Since Δt is a short time interval,

$$\frac{\Delta s}{\Delta t} = \text{instantaneous linear velocity } v$$

$$\frac{\Delta\theta}{\Delta t} = \text{instantaneous angular velocity } \omega$$

Therefore, we find that the linear velocity of P is

$$v = \omega r \qquad (11\text{-}3)$$

It is important to remember that this equation is based on the definition of a radian. The angular velocity ω in this formula must be in radians per unit time (rad/sec, rad/min, etc.). The entire rotating object and all parts of the object have the same angular velocity.

The magnitude of the linear velocity v depends on the distance of P from the axis of rotation. The linear velocity of points directly on the axis is zero according to Eq. 11-3 because $r = 0$ for these points. The linear velocity v calculated from this equation is often called the tangential velocity, because the velocity vector is tangent to the circular path and at right angles to the radius r.

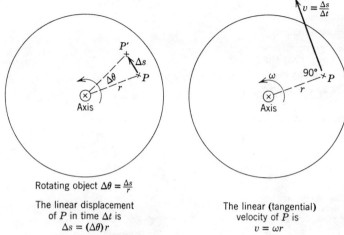

Rotating object $\Delta\theta = \frac{\Delta s}{r}$

The linear displacement
of P in time Δt is
$$\Delta s = (\Delta\theta)r$$

The linear (tangential)
velocity of P is
$$v = \omega r$$

Figure 11-4 Angular and tangential motion.

• **Example 11-4** A grindstone has a radius of 9 in. In a certain grinding operation it is necessary that the rim have a linear velocity of 60 ft/sec. (a) What is the required angular velocity in rev/min? (b) What is the velocity of a point on the wheel 3 in from the axis?

Solution. (a) $r = 9$ in. $= \frac{3}{4}$ ft $\qquad v = \omega r$

$$\omega = \frac{v}{r} = \frac{60 \text{ ft/sec}}{\frac{3}{4} \text{ ft}} = \frac{80 \text{ ft}}{\text{ft-sec}}$$

The ratio ft/ft is equivalent to radians.

$\omega = 80$ rad/sec $\times$ (1 rev/2π rad) $\times$ (60 sec/1 min)

$\qquad = 2400/\pi$ rev/min $= 764$ rev/min

(b) For a point 3 in. from the axis $r = \frac{1}{4}$ ft,

$$v = \omega r = 80 \text{ rad/sec} \times \tfrac{1}{4} \text{ ft} = 20 \text{ ft/sec} \qquad \blacktriangleleft$$

11-7 ROTATION AND LINEAR ACCELERATION

A point on a rotating object is traveling in a circular path and must, therefore, have a centripetal acceleration. Equation 11-3 provides us with a very useful formula for the centripetal acceleration. We have

$$a_c = \frac{v^2}{r} = \frac{(\omega r)^2}{r} = \frac{\omega^2 r^2}{r}$$

or $\qquad a_c = \omega^2 r \qquad\qquad (11\text{-}4)$

In Eq. 11-4 it is necessary to measure ω in rad/time.

• **Example 11-5** Calculate the centripetal acceleration of a point on the rim of the flywheel in Example 11-4.

Solution

$$a_c = \omega^2 r = (80 \text{ rad/sec})^2 \times \tfrac{3}{4} \text{ ft}$$
$$= 4800 \text{ ft/sec}^2$$

ALTERNATE METHOD

$$a_c = \frac{v^2}{r} = \frac{(60 \text{ ft/sec})^2}{\frac{3}{4} \text{ ft}} = 4800 \text{ ft/sec}^2 \qquad \blacktriangleleft$$

If the rate of rotation is changing, ω changes and the magnitude of the linear velocity changes according to Eq. 11-3. The change in v is proportional to the change in ω, since r is a constant. This means that

$$\Delta v = \Delta\omega \, r$$

where Δv is the change in the magnitude of v and $\Delta\omega$ is the change in the magnitude of ω. Dividing both sides of this equation by Δt gives

$$\frac{\Delta v}{\Delta t} = \frac{\Delta\omega}{\Delta t} \times r \qquad\qquad (11\text{-}5)$$

It is apparent that $\Delta\omega/\Delta t$ is the rate of change in angular velocity or the angular acceleration α. From our discussion of circular motion in Chapter 10 we can conclude that $\Delta v/\Delta t$ is the rate of change in the magnitude of v, or the tangential acceleration a_t. Consequently, Eq. 11-5 is equivalent to

$$a_t = \alpha r \qquad (11\text{-}6)$$

In calculations with Eq. 11-6 it is necessary that the radian be used as the unit for angular measure.

• **Example 11-6**   As part of an experiment in space medicine an astronaut is strapped into a seat mounted on a steel framework 3 m from the axis of rotation. The seat starts from rest and is given an acceleration of 1.5 rad/sec^2. After 4 sec of motion determine (a) the number of revolutions made by the man, (b) the angular velocity in rev/min of the steel framework, (c) the linear velocity of the man, and (d) the acceleration of the man.

Solution

(a) $\theta = \omega_0 t + \frac{1}{2}\alpha t^2$

   $= 0 + \frac{1}{2}(1.5 \text{ rad/sec}^2)(4 \text{ sec})^2 = 12 \text{ rad}$

   $= 12 \text{ rad} \times 1 \text{ rev}/2\pi \text{ rad} = 1.91 \text{ rev}$

(b) $\omega = \omega_0 + \alpha t$

   $= 0 + (1.5 \text{ rad/sec}^2)(4 \text{ sec}) = 6 \text{ rad/sec}$

   $= 6 \text{ rad/sec} \times 1 \text{ rev}/2\pi \text{ rad} = 0.955 \text{ rev/sec}$

   $= 0.955 \text{ rev/sec} \times (60 \text{ sec}/1 \text{ min}) = 57.3 \text{ rev/min}$

(c) $v = \omega r = 6 \text{ rad/sec} \times 3 \text{ m} = 18 \text{ m/sec}$

(d) The tangential acceleration is

   $a_t = \alpha r = (1.5 \text{ rad/sec}^2)(3 \text{ m}) = 4.5 \text{ m/sec}^2$

This acceleration is equal to $4.5 \text{ m/sec}^2 \times (1 \text{ } g/9.8 \text{ m/sec}^2)$

or 0.46 g. The centripetal acceleration is

$$a_c = \omega^2 r = (6 \text{ rad/sec})^2(3 \text{ m}) = 108 \text{ m/sec}^2$$

This is equivalent to $108 \text{ m/sec}^2 \times (1 \text{ } g/9.8 \text{ m/sec}^2) =$ 11 g ◀

11-8 THE SECOND LAW OF MOTION FOR A ROTATING OBJECT

In order to analyze the motion of the rotating cam in Fig. 11-5, we first focus our attention on the small shaded portion of the object that is at a distance r from the axis. This portion of the object has a mass m and a tangential acceleration $a_t = \alpha r$. There may be many forces acting on this mass; however, they may be added vectorially to give $\mathbf{F}_R$. This resultant force may then be resolved into the two perpendicular components F_c (centripetal force) and F_t (tangential force). From the second law of motion

$$F_t = ma_t = m\alpha r$$

and $$rF_t = (mr^2)\alpha \qquad (1\text{-}7)$$

The quantity rF_t is the moment of the resultant force about the axis; rF_t is the resultant torque acting upon mass m. If the symbol M is used to indicate torque, Eq. 11-7 may be written

$$M = (mr^2)\alpha \qquad (11\text{-}8)$$

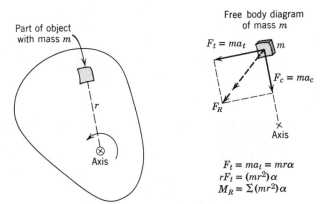

Free body diagram of mass m

$F_t = ma_t$

m

$F_c = ma_c$

F_R

Axis

Part of object with mass m

r

Axis

$F_t = ma_t = mr\alpha$
$rF_t = (mr^2)\alpha$
$M_R = \Sigma(mr^2)\alpha$

Figure 11-5   The second law of motion for a rotating object.

This is the equation of motion for the small shaded portion m of the rotating object. It is obvious, however, that a similar equation can be written for all the other portions of the rotating object. These other portions may have a different mass m and a different radius r, but all portions of the rotating object have the same angular acceleration α. The equations of motion for these different parts are

$$M_1 = (m_1 r_1{}^2)\alpha$$
$$M_2 = (m_2 r_2{}^2)\alpha$$
$$M_3 = (m_3 r_3{}^2)\alpha, \text{ etc.}$$

Summing up all these equations gives

$$M_1 + M_2 + \cdots = (m_1 r_1{}^2 + m_2 r_2{}^2 + \cdots)\alpha \quad (11\text{-}9)$$

The left-hand side of Eq. 11-9 is the *resultant torque* M_R. The sum of all the mr^2 terms is an important physical quantity called the *moment of inertia*. The second law of motion for rotation about an axis may thus be written

$$M_R = I\alpha \quad (11\text{-}10)$$

resultant torque = moment of inertia
$\qquad\qquad\qquad$ × angular acceleration

Equation 11-10 is an extremely important equation having many technical and engineering uses. It is in the same algebraic form as the second law of motion used in studying linear motion. Equation 11-10 states the law:

The angular acceleration of a rotating object is directly proportional to the resultant torque acting upon the object.

In using Eq. 11-10 it is necessary to recall the definition, which was also used in statics, that the torque produced by a force equals the distance from the axis (r) times the perpendicular component of the force (F_t). Also, the resultant torque about the axis of rotation must be the difference between the counterclockwise and the clockwise torques.

The moment of inertia (I) measures the *rotational inertia* of the rotating body. An object with a large moment of inertia possesses a great deal of rotational inertia, and it will require a large torque to produce a comparatively small angular acceleration. On the other hand, objects with a small moment of inertia possess only a small amount of rotational inertia, and these objects may be given a large angular acceleration with a comparatively small torque.

The moment of inertia apparatus in Fig. 11-6 may be used for an experimental verification of Eq. 11-10. A pull P on the rope delivers a torque, rP, to the rod; the moment of inertia may be varied by adjusting the position of the two weights on the rod. The angular acceleration may then be determined by counting the revolutions and using a stopwatch to measure time. It should be noted that the value of α substituted in Eq. 11-10 must be in rad/sec².

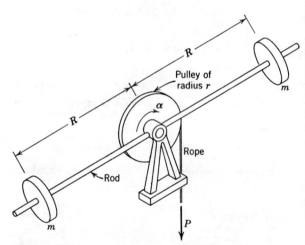

Figure 11-6 Moment of inertia apparatus. Pulley and rod are mounted on the same shaft.

Example 11-7 In the moment of inertia apparatus of Fig. 11-6 the radius of the pulley is 6 in. The two weights are 16 lb each, and they are mounted at a distance of 2 ft from the axis of rotation. When a steady pull P is applied to the rope, the apparatus starts from rest and completes 20 rev in 10 sec with constant angular acceleration.

Table 11-2

System	Force	Torque $(M = r \times F_t)$	Moment of Inertia $(I = \sum mr^2)$	Angular Acceleration (α)
English engineering	pound (lb)	pound-feet (lb-ft)	slug-feet^2 (slug-ft^2)	radians/second^2 (rad/sec^2)
mks (metric)	newton (N)	newton-meter (N-m)	kilogram-meter^2 (kg-m^2)	radians/second^2
cgs (metric)	dyne	dyne-centimeter (dyne-cm)	gram-centimeter^2 (g-cm^2)	radians/second^2

Friction in the bearings is negligible, and the weight of the rod may be considered as negligible by comparison with the two weights. (a) Calculate the angular acceleration. (b) What pull P was needed to maintain this acceleration?

Solution

(a) $\omega_0 = 0$, since the rod started from rest.

$$\theta = \omega_0 t + \tfrac{1}{2}\alpha t^2$$

$$\alpha = \frac{2\theta}{t^2} = \frac{2(20 \text{ rev})}{(10 \text{ sec})^2} = 0.40 \text{ rev/sec}^2$$

(b) The masses are $m = \dfrac{W}{g} = \dfrac{16 \text{ lb}}{32 \text{ ft/sec}^2}$

$$= \frac{1}{2} \text{ slug}$$

$$I = m_1 r_1{}^2 + m_2 r_2{}^2$$
$$= \tfrac{1}{2} \text{ slug } (2 \text{ ft})^2 \times 2$$
$$= 4 \text{ slug-ft}^2$$

In deriving Eq. 11-10 we used $F_R = ma$, which requires that the time be in seconds. Moreover, we let $a_t = \alpha r$, which requires that the angular displacement be in radians. Therefore α must be in rad/sec^2.

$$\alpha = 0.40 \text{ rev/sec}^2 \times 2\pi \text{ rad/1 rev}$$
$$= 0.8\pi \text{ rad/sec}^2$$
$$M_R = I\alpha = 4 \text{ slug-ft}^2 (0.8\pi \text{ rad/sec}^2)$$
$$= 3.2\pi \text{ slug ft/sec}^2 \text{ (ft)} = 3.2\pi \text{ lb-ft}$$

The force P produces this torque with a lever

arm of $r = 6$ in. $= \tfrac{1}{2}$ ft

$$P(\tfrac{1}{2} \text{ ft}) = 3.2\pi \text{ lb-ft}$$
$$P = 6.4\pi \text{ lb} = 20.1 \text{ lb} \qquad \blacktriangleleft$$

In using the torque equation, just as in using $F_R = ma$, we must be certain to employ the appropriate units. Table 11-2 indicates the three systems that will be utilized.

11-9 CALCULATING THE MOMENT OF INERTIA

During the planning and designing of many technical and scientific operations, it often becomes necessary to calculate the moment of inertia of rotating objects. The moment of inertia depends on the mass, the shape, and the size of the rotating objects. It is not always a simple task to determine the moment of inertia from plans or blueprints, but a number of useful methods do exist.

In some simple cases, as in the moment of inertia apparatus of Fig. 11-6, it is possible to calculate the moment of inertia directly from the sum of all the mr^2 terms.

$$I = m_1 r_1{}^2 + m_2 r_2{}^2 + \cdots = \sum(mr^2) \quad (11\text{-}11)$$

This summing procedure is possible if the object consists of a number of separate masses, but this is usually not the case. For rotating objects that consist of a continuous distribution of matter, the sum must be evaluated by using integral calculus. This can be done for many simple figures; the results of these calculations may be found in the engineering handbooks. The moments of inertia for some common objects are indicated in Fig. 11-7. These formulas should not be memorized, but the reader should

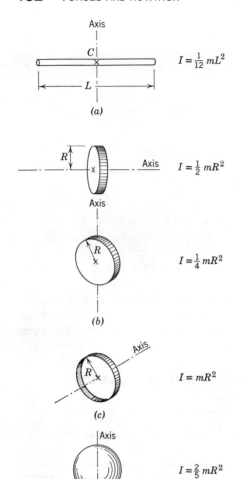

Figure 11-7 The moment of inertia for some common objects. (a) Homogeneous rod. (b) Homogeneous disk or cylinder. (c) Thin-walled wheel or disk. (d) Solid sphere.

understand clearly how they may be used to calculate the moment of inertia.

• **Example 11-8** In Fig 11-8 is a brass disk that is to rotate about the axis through 0. The brass has a weight density of 536 lb/ft³. Calculate the moment of inertia of the disk.

Solution. The mass of one cubic foot of brass is

$$\frac{536 \text{ lb}}{32 \text{ ft/sec}^2} = 16.75 \text{ slugs}$$

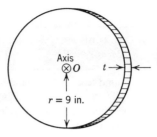

t = thickness = 2 in.

Figure 11-8 A rotating disk.

The density of brass is, therefore, $\rho = 16.75$ slugs/ft³. The volume of the disk is $V = \pi r^2 t$; $r = 9$ in. $= \frac{3}{4}$ ft and $t = 2$ in. $= \frac{1}{6}$ ft.

$$V = \pi(\tfrac{3}{4} \text{ ft})^2(\tfrac{1}{6} \text{ ft}) = \pi(\tfrac{9}{16})(\tfrac{1}{6}) \text{ ft}^3$$
$$= \tfrac{3}{32}\pi \text{ ft}^3$$

The mass is $m = \rho V$

$$= 16.75 \text{ slugs/ft}^3 \ (\tfrac{3}{32}\pi \text{ ft}^3)$$
$$= 4.93 \text{ slugs}$$

For a disk $I = \tfrac{1}{2} mr^2$

$$= \tfrac{1}{2}(4.93 \text{ slugs})(0.75 \text{ ft})^2$$
$$= 1.39 \text{ slug-ft}^2 \qquad \blacktriangleleft$$

In some cases the moment of inertia may be found by using the radius of gyration (k). The radius of gyration squared (k^2) is a convenient way of describing the average or effective value of r^2. The moment of inertia may then be calculated by assuming that the entire mass is concentrated at this distance k from the axis, and

$$I = mk^2 \qquad (11\text{-}12)$$

• **Example 11-9** A flywheel on a stationary engine has a mass of 200 kg. The radius of gyration is 40 cm. Calculate the moment of inertia.

Solution

$$k = 40 \text{ cm} = 0.40 \text{ m}$$
$$I = mk^2 = 200 \text{ kg } (0.40)^2 \text{ m}^2 = 32 \text{ kg-m}^2 \qquad \blacktriangleleft$$

The diagram in Fig. 11-7 indicates the axes passing through the center of mass. In some applications, however, the axis of rotation does not pass through this point. If the axis does not pass through the center of mass of the object, the moment of inertia is larger than the values given in Fig. 11-7. This point is illustrated in Fig. 11-9,

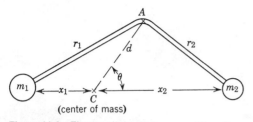

Figure 11-9 The parallel axis theorem. When the axis of rotation does not pass through the center of mass, $I_A = I_c + md^2$.

where C is the center of mass of the object and the axis of rotation passes through A and into the page. We have purposely chosen an object of only two masses so that it will be easy to calculate the moment of inertia about the axis at A:

$$I_A = m_1 r_1^2 + m_2 r_2^2$$

The distance r_1 and r_2 may be found from the law of cosines. By referring to the diagram, it is seen that

$$r_1^2 = x_1^2 + d^2 + 2x_1 d \cos \theta$$

and
$$r_2^2 = x_2^2 + d^2 - 2x_2 d \cos \theta$$

The moment of inertia I_A may then be written as

$$I_A = m_1(x_1^2 + d^2 + 2x_1 d \cos \theta)$$
$$+ m_2[(x_2^2 + d^2 + (-2x_2 d \cos \theta)]$$
$$I_A = (m_1 x_1^2 + m_2 x_2^2) + (m_1 + m_2)d^2$$
$$+ 2d \cos \theta (m_1 x_1 - m_2 x_2)$$

(11-12)

Since C is the center of mass, $m_1 x_1 = m_2 x_2$ and the last term in this equation is zero. The first term in this equation is the moment of inertia (I_C) for an axis through C. If we let $m_1 + m_2 = m$ (the total mass), Eq. 11-12 reduces to a very important theorem in mechanics:

$$I_A = I_C + md^2 \qquad (11\text{-}13)$$

Although this theorem was derived for an object with only two masses, it applies in all cases because any object may be divided into many pairs of masses such as m_1 and m_2. This theorem is called the *parallel axis theorem* or the *transfer theorem*. If we can find the moment of inertia (I_C) for some axis through the center of mass, then Eq. 11-13 may be used to calculate the moment of inertia (I_A) through a parallel axis that does not pass through the center of mass. The distance between the two parallel axes is d in Eq. 11-13.

In practice, most rotating bodies are usually not of simple geometric shape. However, we may imagine that the object is divided into a number of simple parts. I_C is

calculated for each part, and the transfer theorem is then used to obtain the moment of inertia with respect to the axis for each part. The sum of the moments of inertia of the parts equals the total moment of inertia for the object.

• **Example 11-10** A pendulum consists of the 30-g rod and the 100-g disk as in Fig. 11-10. Calculate the moment of inertia for the composite body.

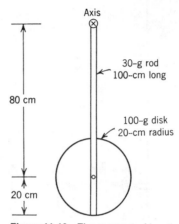

Figure 11-10 The moment of inertia of a composite body.

Solution. For the disk $I_C = \frac{1}{2}mR^2$; $R = 20$ cm and $d = 80$ cm.

$$I_A = I_C + md^2 = \frac{1}{2}mR^2 + md^2$$
$$= m(\tfrac{1}{2}R^2 + d^2)$$
$$= 100 \text{ g} \left[\tfrac{1}{2} \times (20 \text{ cm})^2 + (80 \text{ cm})^2\right]$$
$$= 100 \text{ g} (200 \text{ cm}^2 + 6400 \text{ cm}^2)$$
$$= 660{,}000 \text{ g-cm}^2$$

For the rod, $I_C = \frac{1}{12}mL^2$, and $d = \frac{1}{2}L$, because the distance between the center of mass and A is one-half the length of the rod.

$$I_A = I_C + md^2 = \tfrac{1}{12}mL^2 + m(\tfrac{1}{2}L)^2$$
$$= \tfrac{1}{3}mL^2 = \tfrac{1}{3}(30 \text{ g})(100 \text{ cm})^2 = 100{,}000 \text{ g-cm}^2$$

The total moment of inertia is

$$660{,}000 \text{ g-cm}^2 + 100{,}000 \text{ g-cm}^2 = 760{,}000 \text{ g-cm}^2 \quad \blacktriangleleft$$

In some design problems the moment of inertia of a rotating object (or a scale model) is determined by an experiment. This procedure is illustrated in Fig. 11-11 and the following calculation.

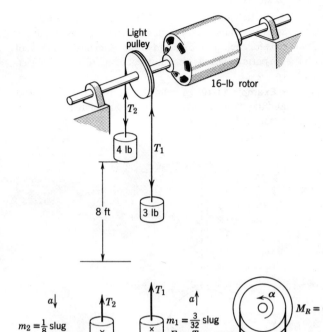

Figure 11-11 Determining the moment of inertia by experiment. Rotor and pulley fixed to shaft that rotates freely in bearings.

• **Example 11-11** In Fig. 11-11 the 4- and 3-lb weights are connected by a rope that passes over the pulley, which is 6 in. in diameter. When the 4-lb weight is released, it falls 8 ft to the ground in 4 sec. Assume that friction in the bearings is negligible, and calculate the moment of inertia and the radius of gyration of the rotor.

Solution. The resultant torque on the rotor is $M_R = T_2 r - T_1 r = (T_2 - T_1)r$. Since $M_R = I\alpha$,

$$(T_2 - T_1)r = I\alpha \qquad \text{(i)}$$

This can be used to calculate I if we determine T_2, T_1, and α. The acceleration of the weights is found from $s = \frac{1}{2}at^2$,

$$a = \frac{2s}{t^2} = \frac{2(8 \text{ ft})}{(4 \text{ sec})^2} = 1 \text{ ft/sec}^2$$

Since the rope does not slip over the pulley $a = a_t = \alpha r$, $r = 3$ in. $= \frac{1}{4}$ ft

$$\alpha = \frac{a}{r} = \frac{1 \text{ ft/sec}^2}{\frac{1}{4} \text{ ft}} = 4 \text{ rad/sec}^2$$

From the equation of motion of the 4-lb weight,

$$F_R = ma, \qquad 4 \text{ lb} - T_2 = \tfrac{1}{8} \text{ slug } (1 \text{ ft/sec}^2)$$
$$4 \text{ lb} - T_2 = \tfrac{1}{8} \text{ lb}$$
$$T_2 = 3\tfrac{7}{8} \text{ lb}$$

From the equation of motion of the 3-lb weight,

$$F_R = ma, \qquad T_1 - 3 \text{ lb} = \tfrac{3}{32}(1 \text{ ft/sec}^2)$$
$$T_1 - 3 \text{ lb} = \tfrac{3}{32} \text{ lb}$$
$$T_1 = 3\tfrac{3}{32} \text{ lb}$$

Substitution in Eq. (i) gives

$$I = \frac{(T_2 - T_1)r}{\alpha} = \frac{(3\tfrac{7}{8} \text{ lb} - 3\tfrac{3}{32} \text{ lb})\tfrac{1}{4} \text{ ft}}{4 \text{ rad/sec}^2}$$

$$= \frac{(\tfrac{25}{32} \text{ lb})\tfrac{1}{4} \text{ ft}}{4 \text{ rad/sec}^2} = \frac{25 \text{ slug-ft}^2}{16(32)} = \frac{25 \text{ slug-ft}^2}{512}$$

$$= 0.049 \text{ slug-ft}^2$$

Since $m = \frac{1}{2}$ slug for the rotor,

$$I = mk^2, \quad k = \sqrt{\frac{I}{m}} = \sqrt{\frac{\frac{25}{512} \text{ slug-ft}^2}{\frac{1}{2} \text{ slug}}}$$

$$k = \sqrt{\frac{25}{256} \text{ ft}^2} = \frac{5}{16} \text{ ft}$$

11-10 ROTATION AND ROLLING

The problems in rotation considered in this chapter have involved rotation about a fixed axis. In many cases, however, the motion of objects involves a combination of rotation and linear motion. The rolling motion of a wheel or a ball is a very common and very important example of a motion that combines rotation and linear motion.

If a wheel rolls without slipping or skidding, the center of the wheel moves forward a distance equal to the circumference of the wheel $(2\pi R)$ whenever the wheel completes one revolution.

The distance moved by the wheel is $2\pi R \times$ the number of revolutions. But 2π times the number of revolutions is the angular displacement in radians. The linear displacement of the center of a wheel that rolls without slipping is therefore given by the familiar formula:

$$s = \theta R \quad (11\text{-}14)$$

with θ in radians. The velocity of the center of a wheel that rolls without slipping is

$$v = \omega R \quad (11\text{-}15)$$

with ω in rad/time. Moreover, the acceleration of the center, if there is no slipping, will be given by

$$a = \alpha R \quad (11\text{-}16)$$

with α in rad/time².

All the points on a rolling wheel do not travel in the same path. The paths followed by different parts of a rolling wheel form rather interesting patterns. The paths traced out in Fig. 11-12 can

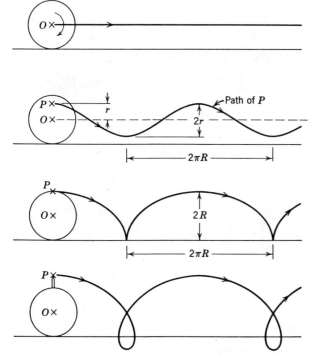

Figure 11-12 Motion during rolling.

be recorded photographically. A time exposure is taken of a wheel that rolls in a darkened room with a glowing flashlight bulb mounted on one point of the wheel. The path of the point is clearly indicated as a light track on the photograph.

The motion of any point on the wheel can be described as the vector sum of two motions: (i) the rotational motion relative to the center, and (ii) the translation (linear motion) of the center. This principle is illustrated in the following problem.

Example 11-12 The wheel in Fig. 11-13 has a velocity of 10 ft/sec and an acceleration of 12 ft/sec^2. Assuming that the wheel does not slip, calculate the velocity and acceleration of points A and B.

Solution. Since the wheel does not slip,

$$v_0 = \omega R$$

$$\omega = \frac{v_0}{R} = \frac{10 \text{ ft/sec}}{2 \text{ ft}}$$

$$\omega = 5 \text{ rad/sec}$$

$$a_0 = \alpha R$$

$$\alpha = \frac{a_0}{R} = \frac{12 \text{ ft/sec}^2}{2 \text{ ft}} = 6 \text{ rad/sec}^2$$

(a) For point A the linear velocity relative to the center is

$$\omega \times r = 5 \text{ rad/sec} \times 2 \text{ ft} = 10 \text{ ft/sec} \leftarrow$$

directed toward the left. When this is combined vectorially with v_0, we find that the velocity of A is zero. The point of the wheel in contact with the ground has a zero velocity if the wheel does not slip. Because the point A has a zero velocity, it is called the instantaneous axis.

The acceleration of A due to rotation relative to O is a tangential acceleration,

$$\alpha r = 6 \text{ rad/sec}^2 (2 \text{ ft}) = 12 \text{ ft/sec} \leftarrow$$

and a centripetal acceleration

$$\omega^2 r = (5 \text{ rad/sec})^2(2 \text{ ft}) = 50 \text{ ft/sec}^2 \uparrow$$

Adding the acceleration a_0 of the center indicates that the resultant acceleration of A is 50 ft/sec^2 $\uparrow$.

(b) For point B the velocity relative to 0 is

$$\omega r = 5 \text{ rad/sec} (1.5 \text{ ft}) = 7.5 \text{ ft/sec} \downarrow$$

Adding $v_0 = 10$ ft/sec $\rightarrow$, we find that the resultant velocity has the magnitude 12.5 ft/sec (see Fig. 11-13b). The acceleration relative to 0 is the tangential acceleration

$$\alpha r = 6 \text{ rad/sec}^2 (1.5 \text{ ft}) = 9 \text{ ft/sec}^2 \downarrow$$

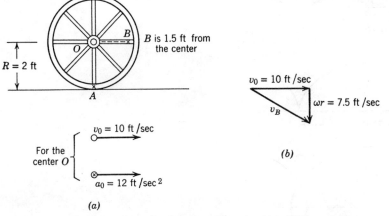

(a)

(b)

Figure 11-13 Velocity and acceleration of a rolling wheel (a). (b) The velocity of point B is $\mathbf{v}_B$.

and the centripetal acceleration

$$\omega^2 r = (5 \text{ rad/sec})^2(1.5 \text{ ft}) = 37.5 \text{ ft/sec}^2 \leftarrow$$

Adding vectorially the acceleration $a_0 = 12 \text{ ft/sec}^2 \rightarrow$, we find that the resultant acceleration has the components

9 ft/sec² ↓   and   25.5 ft/sec² ← ◄

11-11 FORCES AND TORQUES IN ROLLING

When an object is rolling, it is convenient to describe the motion as a combination of two motions:

(i) The acceleration of the center of the mass.

(ii) The rotation about the center of mass.

If a number of forces act upon an object, we may imagine that all the forces act upon the center of mass, and the acceleration of the center of mass may be found from $F_R = ma$. Then the resultant torque of the forces is calculated with respect to an axis through the center of mass. The angular acceleration may be found from $M_R = I\alpha$.

Example 11-13 The wheel in Fig. 11-14 has a radius of gyration of 2 ft. The wheel rolls to the right without slipping. (a) Calculate the angular and linear acceleration. (b) Determine the force of friction that acts to prevent slipping.

Solution. The force of friction (f in the free body diagram) acts to prevent slipping.

$$m = \frac{160 \text{ lb}}{32 \text{ ft/sec}^2} = 5 \text{ slugs}$$

$$F_R = ma$$
$$200 \text{ lb} - f = (5 \text{ slugs})a \quad \text{(i)}$$

The resultant torque about the center of mass is
$M_R = f(3 \text{ ft})$

$$I = mk^2 = 5 \text{ slug }(2 \text{ ft})^2 = 20 \text{ slug-ft}^2$$

$$\alpha = \frac{a}{R} = \frac{a}{3 \text{ ft}}$$

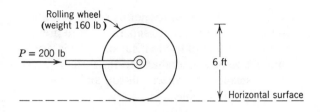

Figure 11-14 The acceleration of a rolling wheel.

Substituting in $M_R = I\alpha$ gives

$$f(3 \text{ ft}) = 20 \text{ slug-ft}^2 \frac{a}{3 \text{ ft}}$$

$$f = \frac{20}{9} \text{ slugs} \times a \quad \text{(ii)}$$

Substituting in Eq. (i), we obtain

$$200 \text{ lb} - (\tfrac{20}{9} \text{ slugs})a = (5 \text{ slugs})a$$
$$200 \text{ lb} = (\tfrac{20}{9} \text{ slugs})a + (5 \text{ slugs})a$$
$$200 \text{ lb} = (\tfrac{65}{9} \text{ slugs})a$$
$$a = \tfrac{9}{65}(200 \text{ lb})/\text{slug} = 27.7 \text{ ft/sec}^2$$

$$\alpha = \frac{a}{3 \text{ ft}} = \frac{27.7}{3} \text{ rad/sec}^2 = 9.2 \text{ rad/sec}^2$$

From Eq. (i),

$$f = 200 \text{ lb} - (5 \text{ slugs})a$$
$$= 200 \text{ lb} - 5 \text{ slugs }(\tfrac{9}{65} \times 200 \text{ ft/sec}^2)$$
$$= 200 \text{ lb} - 138.5 \text{ lb}$$
$$= 61.5 \text{ lb}$$

As a check we may use Eq. (ii),

$$f = \tfrac{20}{9} \text{ slugs }(27.7 \text{ ft/sec}^2) = 61.5 \text{ lb} \quad ◄$$

If the coefficient of friction is small, the wheel in Fig. 11-14 will begin to slip. In order to prevent slipping, 61.5 lb of friction is needed. This means that μ must be greater than $61.5\,\text{lb}/200\,\text{lb} = 0.307$. If μ is smaller, the wheel will roll and slide as in the following problem.

Example 11-14 The coefficient of kinetic friction is 0.1 on the horizontal surface in Fig. 11-14. Calculate the linear and angular acceleration.

Solution

$$f = \mu N = 0.1(200\,\text{lb}) = 20\,\text{lb}$$
$$F_R = 200\,\text{lb} - 20\,\text{lb} = 180\,\text{lb}$$
$$F_R = ma$$
$$a = \frac{F_R}{m} = \frac{180\,\text{lb}}{5\,\text{slugs}} = 36\,\text{ft/sec}^2$$

In this situation α does not equal a/r because of the sliding. The resultant torque about the center of mass is

$$M_R = 20\,\text{lb}\,(3\,\text{ft})$$

Substituting in $M_R = I\alpha$, we obtain

$$20\,\text{lb}\,(3\,\text{ft}) = 20\,\text{slug-ft}^2\,\alpha$$
$$\alpha = 3\,\text{rad/sec}^2 \qquad \blacktriangleleft$$

11-12 TRANSLATION; MOTION WITHOUT ROTATION

When an object is moving with translation (linear motion) only, there is no angular acceleration and no rotation about the center of mass. The resultant torque about the center of mass must be zero. This fact can be used to determine unknown forces in many situations which involve translation only.

Example 11-15 The 4000-lb automobile in Fig. 11-15 has rear wheel drive. The acceleration is 8 ft/sec² in the forward direction. Determine the forward thrust on the rear wheels, and find the weight resting upon the front wheels and the rear wheels. Assume that rolling friction on the wheels and air resistance are negligible.

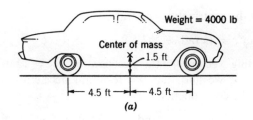
(a)

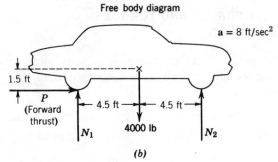
(b)

Figure 11-15 An accelerating automobile moves with translation only.

Solution

$$F_R = \frac{W}{g} \quad (a)$$

$$P = \frac{4000\,\text{lb}}{32\,\text{ft/sec}^2}\,(8\,\text{ft/sec}^2) = 1000\,\text{lb}$$

$$\sum F_y = 0, \qquad N_2 + N_1 = 4000\,\text{lb} \qquad (\text{i})$$
$$M_R = 0$$
$$P(1.5\,\text{ft}) + N_2(4.5\,\text{ft}) - N_1(4.5\,\text{ft}) = 0$$
$$1000\,\text{lb}\,(1.5\,\text{ft}) + (N_2 - N_1)4.5\,\text{ft} = 0$$
$$N_1 - N_2 = 333\,\text{lb} \qquad (\text{ii})$$

Adding Eqs. (i) and (ii) to eliminate N_2 gives

$$2N_1 = 4000\,\text{lb} + 333\,\text{lb}$$
$$N_1 = 2167\,\text{lb}$$

The weight resting on the rear wheels is 2167 lb.

$$N_2 = 4000\,\text{lb} - N_1 = 4000\,\text{lb} - 2167\,\text{lb}$$
$$= 1833\,\text{lb}$$

The weight supported by the front wheels is 1833 lb. $\qquad \blacktriangleleft$

The solution to Example 11-11 is in accord with the common observation that the front of the car rises and the rear sinks lower during an acceleration. This effect is caused by the greater normal force (N_1 in the diagram) which compresses the rear springs. Using the principle that the resultant torque about the center of mass must be zero, the reader should also be able to explain why the front of a car dips during a deceleration.

SUMMARY

The *angular displacement* of an object describes the amount of rotation. The angular displacement may be measured in revolutions, radians, or degrees:

$$1 \text{ revolution} = 360 \text{ degrees} = 2\pi \text{ radians}$$

The rate of rotation is the *angular velocity*, which may be measured in rev/min, rad/sec, or other units that describe the angular displacement per unit time.

The average angular velocity ω_{av} may be calculated from the formula

$$\omega_{av} = \frac{\theta}{t}$$

where θ is the angular displacement during the time t.

The *angular acceleration* α is the time rate of change in the angular velocity. If the angular velocity is plotted on an ω–t curve, the slope of the curve is the angular acceleration and the area under the curve is the angular displacement.

The formulas for uniformly accelerated angular motion are in the same algebraic form as the formulas for uniformly accelerated linear motion. Instead of s, v_{av}, v, and a, the formulas for uniformly accelerated angular motion involve θ, ω_{av}, ω, and α, respectively.

Any point on a rotating object travels on a circular arc, and the distance covered on the circular arc is given by

$$s = \theta \times r$$

when θ is the angular displacement in radians and r is the distance from the axis of rotation. The point traveling on the circular arc also has a *linear* (or *tangential*) *velocity* given by

$$v = \omega \times r$$

where ω is the angular velocity in radians per unit time. For a point on a rotating object, there will be a centripetal acceleration given by

$$a_c = \frac{v^2}{r} = \omega^2 r$$

where a_c is a linear acceleration. If the rate of rotation is changing, there will be an angular acceleration, α, and a point on the rotating object will undergo a *tangential acceleration*.

$$a_T = \alpha \times r$$

where α must be in radians per (time)^2 and a_T is a linear acceleration.

The second law of motion for a rotating object states that the angular acceleration is directly proportional to the resultant torque M_R:

$$M_R = I\alpha$$

where α is in rad/sec^2 and I is the moment of inertia of the object. The *moment of inertia* measures the rotational inertia of the rotating object.

The moment of inertia may be calculated by the formula

$$I = m_1 r_1{}^2 + m_2 r_2{}^2 + \cdots = \sum mr^2$$

Also, we have

$$I = mk^2$$

when m is the total mass and k is the *radius of gyration*. The radius of gyration squared (k^2) is the average or effective value of r^2.

If the axis of rotation does not pass through the center of mass of the rotating object, the moment of inertia, I_A, may be calculated from the *parallel axis theorem*

$$I_A = I_c + md^2$$

where I_c is the moment of inertia for the axis through the center of mass, and d is the distance between the two axes.

For an object of radius R, which rolls without slipping, the linear displacement of the center of the object is given by

$$s = \theta \times R$$

where θ is measured in radians. The linear velocity v and linear acceleration a of the center of the object can then be calculated from

$$v = \omega \times R$$

and
$$a = \alpha \times R$$

If the resultant torque of all the forces is calculated with respect to an axis through the center of mass, the angular acceleration may be found by substituting in the equation $M_R = I\alpha$. If the object moves with translation only, then the resultant torque about the center of mass must be zero.

QUESTIONS

1. What is the meaning of the term *angular displacement*?

2. What units are used in measuring angular displacement?

3. How are revolutions converted into radians?

4. What is meant by *angular velocity*?

5. How is the unit rev/min converted to radians per second?

6. How is the term *angular acceleration* defined?

7. What is the meaning of a negative angular acceleration?

8. How may the tangential velocity of a point on a rotating object be calculated from the angular velocity?

9. For what points on a rotating object is the centripetal acceleration zero?

10. What is the difference between centripetal acceleration and angular acceleration?

11. Explain the difference between centripetal acceleration and tangential acceleration.

12. What quantity describes the rotational inertia of a rotating system?

13. What is the second law of motion for a rotating system?

14. Suppose that the same torque is applied to different objects. How is the angular acceleration related to the moments of inertia of these objects?

15. Explain the meaning of the term *radius of gyration*.

16. Suppose that we take a wheel and remove material from the spokes so that they become very light. What effect does this have upon the radius of gyration of the wheel? Explain your answer.

17. Why does a hollow cylinder have a larger radius of gyration than a solid cylinder of the same diameter?

18. Explain why the point of the wheel that touches the ground has a zero velocity if the wheel does not slip.

19. Suppose a wheel is rolling without slipping. What part of the wheel has the maximum velocity?

20. Draw a free body diagram of a decelerating automobile assuming that the center of mass is halfway between the two axles and some distance above the ground. How can we explain the fact that the normal force on the front wheels of the automobile is greater than the normal force on the rear wheels?

PROBLEMS

(A)

1. A rotating wheel has an angular velocity of 1000 rev/min. How long does it take the wheel to complete one revolution?

2. A rotating drum completes a revolution in $\frac{1}{10}$ sec. Calculate the angular velocity in
   (a) rev/sec.
   (b) rad/sec.

3. The angular velocity of an automotive engine is 3000 rev/min. Calculate the angular velocity in rad/sec.

4. The flywheel on a stationary engine is rotating at the constant rate of 4000 rev/min. Calculate:
   (a) The number of revolutions in 4 sec.
   (b) The angular velocity in rad/sec.

5. A wheel is rotating with a constant angular velocity of 10 rad/sec.
   (a) Compute the angular velocity in rev/min.
   (b) How long does it take to complete 1 rev?

6. A rotating drum completes 10 rev in 2 sec. What is the average angular velocity in
   (a) rev/sec.
   (b) rev/min.
   (c) rad/sec.

7. A rotating flywheel starts from rest with a constant angular acceleration of 5 rad/sec^2. After 10 sec of rotation compute the angular velocity in
   (a) rad/sec.
   (b) rev/min.

8. A rotating flywheel starts from rest and completes 900 rev in 2 min. Calculate the average angular velocity in
   (a) revolutions per second, and
   (b) radians per second.

9. If the flywheel in Problem 8 is uniformly accelerated during the entire 2 min, determine
   (a) the angular acceleration, and
   (b) the final angular velocity.

10. A fan belt is wrapped around a pulley that is 8 in. in diameter. If the pulley rotates at 300 rev/min, what is the linear speed of the fan belt? (Assume that the belt does not slip.)

11. A flywheel 40 cm in diameter is rotating at 10 rev/sec.
    (a) What is the linear velocity of a point on the rim of the flywheel?
    (b) What is the linear velocity of a point 10 cm from the axis of rotation?

12. A wheel is rotating about an axis at the constant rate of 200 rad/sec. For a point 20 cm from the axis of rotation, calculate
    (a) the tangential acceleration, and
    (b) the centripetal acceleration.

13. A torque of 100 newton-meters applied to the rotor of a generator imparts an angular acceleration of 10 rad/sec^2 to the rotor. What is the moment of inertia of the rotor?

14. A sprocket wheel on a bicycle is 6 in. in diameter and is pedaled at the angular velocity of 90 rev/min.
    (a) What is the angular velocity in rad/sec?
    (b) What is the linear velocity of the bicycle chain in ft/sec?

15. A rotating wheel starts from rest and reaches an angular velocity of 400 rev/min in a time of 5 sec. Calculate the angular acceleration in rad/sec^2.

16. A pulley starts from rest with an angular acceleration of 10 rad/sec^2. How long does it take to complete $\frac{1}{4}$ of a revolution?

17. The rotating portion of a generator has a moment of inertia of 200 kg-m^2. What torque must be applied to give this part an angular acceleration of 5 rad/sec^2?

18. In Fig. 11-16 pulley A is rotating at 300 rev/min. The fan belt does not slip.
    (a) What is the linear speed of the fan belt?
    (b) What is the angular velocity of pulley B?

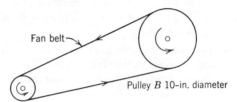

Fan belt

Pulley B 10-in. diameter

Pulley A 4-in. diameter

Figure 11-16   The tangential velocity of a fan belt.

19. A flywheel 4 ft in diameter is mounted on fixed bearings. At what rate must the wheel be rotated so that a point on the rim is subjected to a centrifugal acceleration of 10 g?

20. A part of a machine is rotating at 1200 rev/min. A 2-ounce metal weight is mounted on this rotating part at a distance of 1.5 ft from the axis of rotation.
    (a) Calculate the linear velocity of the metal weight.
    (b) What centripetal force must act to keep the metal weight traveling in its circular path?

21. A rotating flywheel completes 100 rev in 20 sec.
    (a) What is the average angular velocity in rev/sec?
    (b) How long does it take to complete one revolution?

22. The moment of inertia of a drive shaft is 50 kg-m^2. It starts from rest and completes 200 rev in 10 sec. Assume that the angular acceleration is constant. Calculate:
    (a) The average angular velocity.
    (b) The final angular velocity.
    (c) The angular acceleration in rad/sec^2.
    (d) The resultant torque on the shaft.

(B)

1. A propellor starts from rest and reaches an angular velocity of 400 rev/min after completing 100 rev. Calculate the angular acceleration assuming that it is constant during the 100 rev.

2. The rotating drum on a machine is turning at the rate of 600 rev/min. How long does it take the drum to rotate through 15°?

3. A rotating flywheel is 30 cm in radius. The wheel starts from rest and, accelerating uniformly, attains an angular velocity of 100 rad/sec in 20 sec.
   (a) Calculate the angular acceleration of the wheel.
   (b) What is the tangential acceleration of a point on the rim of the wheel?
   (c) What is the centripetal acceleration of a point on the rim of the wheel 15 sec after the wheel starts to rotate?

4. A flywheel on an engine has a moment of inertia of 20 slug-ft^2. What unbalanced torque will accelerate the wheel from rest to an angular velocity of 300 rev/min in 10 sec?

5. A rotating drum has a moment of inertia of 10 slug-ft^2 and is rotating at 600 rev/min. What constant torque must be applied to bring the wheel to rest in 10 sec?

6. A governor rotating at 3000 rev/min is slowed to 1000 rev/min in 30 sec. Assume that the deceleration is uniform. Calculate:
   (a) The deceleration in rad/sec^2.
   (b) The number of revolutions completed during the 30 sec of deceleration.

7. A torque of 100 lb-ft is applied to a flywheel for 5 sec. The flywheel starts from rest and completes 50 rev during the 5 sec of acceleration.
   (a) Calculate the angular acceleration in rad/sec^2.
   (b) What is the angular velocity in rev/min at the end of the 5 sec?
   (c) Calculate the moment of inertia of the flywheel.

8. A propellor weighs 160 lb and has a radius of gyration of 18 in. What torque is required to accelerate this propellor from rest to an angular velocity of 600 rev/min in 10 sec?

9. A rotating flywheel has a moment of inertia of 30 slug-ft^2. A resultant torque of 60 lb-ft is applied to the flywheel.
   (a) Calculate the angular acceleration.
   (b) How long does it take the flywheel to complete 100 rev?
   (c) How long does it take for the flywheel to reach an angular velocity of 600 rev/min?

10. Using the reference data in Fig. 11-7 and the transfer theorem, determine the moment of inertia and the

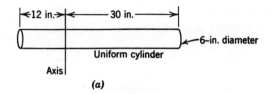

Uniform cylinder

Axis

(a)

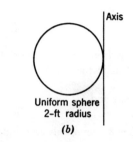

Axis

Uniform sphere
2-ft radius

(b)

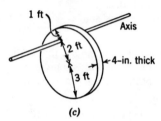

1 ft

Axis

2 ft

4-in. thick

3 ft

(c)

Figure 11-17   Calculating the moment of inertia.

radius of gyration for each of the three objects in Fig. 11-17. Each object has a mass of 10 slugs.

11. A large turbine wheel weighs 800 lb and has a radius of gyration of 2 ft. A frictional torque of 100 lb-ft opposes the rotational motion of the turbine wheel. What torque must be applied to accelerate the wheel from rest to a speed of 600 rev/min in 20 sec?

12. The object in Fig. 11-18 consists of a uniform solid cylinder and a uniform solid sphere. This object is manufactured from a metal alloy that has a density of 10 g/cm^3. Calculate the moment of inertia and the radius of gyration for the axis indicated in the diagram.

13. In Fig. 11-19, $r = 40$ cm and the suspended weight has a mass of 10 kg. The 10-kg mass falls with an acceleration of 6 m/sec^2.
   (a) What is the tension (in newtons) in the cable?
   (b) What is the angular acceleration of the drum?
   (c) Calculate the moment of inertia of the drum.

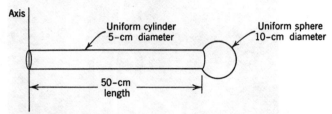

Axis

Figure 11-18   Calculating the moment of inertia of a composite object.

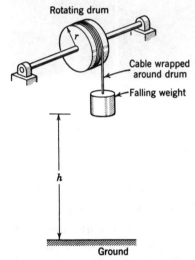

Figure 11-19   Rotation and linear motion.

(a) Calculate the x and y components of linear velocity of point P.

(b) Calculate the x and y components of the acceleration of point P.

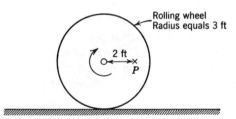

Figure 11-20   Rolling and linear motion.

14. Let the suspended weight in Fig. 11-19 weigh 10 lb and let $r = 2$ ft. The weight is released from rest and falls 8 ft to the ground in 2 sec. Assume that air resistance and friction in the drum bearings are negligible.
  (a) What is the tension in the cable?
  (b) How fast is the weight traveling when it strikes the ground?
  (c) Calculate the moment of inertia of the drum.

15. The wheel in Fig. 11-20 rolls without slipping. The center of the wheel has a linear velocity of 4 ft/sec and a linear acceleration of 15 ft/sec².

16. When the brakes are applied to the automobile in Fig. 11-15a, a braking force of 250 lb acts upon each wheel. (The forward thrust of the motor is zero during the braking.)
  (a) Calculate the deceleration of the automobile.
  (b) Determine the weight resting upon each wheel during the deceleration.

17. In Fig. 11-14 and in Example 11-13, the force $P = 200$ lb acted through the center of the wheel. Suppose instead that the force of 200 lb acts at a height of 4 ft above the ground and, therefore, 1 ft above the center of the wheel. Assume that the wheel does not slip and calculate:
  (a) The angular and linear acceleration.
  (b) The force of friction required to prevent slipping.

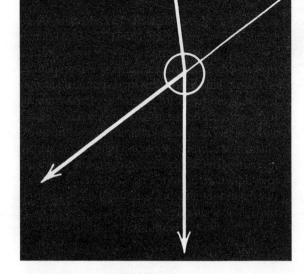

12.
Work and Energy

An understanding of work and energy methods is very basic for any study of mechanics, machinery, or electricity. In this chapter we shall learn the definitions of *work* and *energy* and apply the *conservation of energy* and related principles to the solution of important problems in physics.

LEARNING GOALS

1. *The calculating of work done by a force.*
2. *The use of the work–energy principle to find unknown forces.*
3. *The calculation of kinetic energy.*
4. *The calculation of potential energy.*
5. *The use of the work–energy principle to determine the velocity of a moving object.*
6. *The calculation of the kinetic energy of rotation and the work done by a torque.*

12-1 FORCES AND WORK

The term *work* is a very common word in the English language, and it is used to refer to many kinds of physical and intellectual activity. In scientific and technical operations, however, work has a very precise meaning:

> **Work is done when a force acts upon an object and there is a displacement in the direction of the force, at the point where the force is applied. The work done equals the force times the displacement in the direction of the force.**

This definition of work will be clearer if we refer to Fig. 12-1. Here F is a constant force that acts upon the point of application A; while F is acting, the point of application is displaced to A'. It is evident from the figure that $s \cos \theta$ is the displacement in the direction of the force F. The

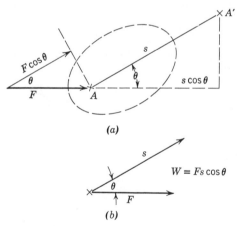

(a)

(b)

Figure 12-1   Work done by a force.

definition of work then gives the formula

$$W = Fs \cos \theta \qquad (12\text{-}1)$$

Because W is used as the symbol for work, we must be very careful to avoid confusing this symbol with weight. One scheme for avoiding confusion is to use the form mg for weight and reserve W for work.

In Eq. 12-1, θ is the angle between the direction of the displacement vector, s, and force vector, $\mathbf{F}$. The figure indicates that $F \cos \theta$ is the component of the force in the direction of the displacement. Since $W = F(s \cos \theta) = (F \cos \theta)s$, we have a second definition of work:

> **The work done by a force is equal to the displacement times the component of the force in the direction of the displacement.**

The units used for measuring work depend on the units used for force and distance. The usual units for work in mechanics problems are given in Table 12-1.

The definitions for these units come directly from the definition of work. One *foot-pound* (ft-lb) is obviously the amount of work done when a one-pound force acts through a displacement of one foot. The *joule* was named after a nineteenth century British physicist, James Prescott Joule, who performed many original experiments to measure the relationship between mechanical work, electrical work, and heat. A joule is the amount of work done when a force of one newton acts for a distance of one meter.

The cgs unit, the *erg*, is used to measure work in atomic and nuclear physics. An erg is the amount of work done when a force of one dyne

Table 12-1

Unit System	Displacement (s)	Force (F)	Work ($W = Fs \cos \theta$)
English engineering	feet	pound	foot-pounds (ft-lb)
mks (metric)	meters	newtons	joules
cgs (metric)	centimeters	dynes	ergs

acts for a displacement of one centimeter. The connection between the erg and the joule can be established in the following way:

$$1 \text{ joule} = 1 \text{ newton} \times 1 \text{ meter}$$
$$= (1 \text{ kg} \times 1 \text{ m/sec}^2) \times 1 \text{ m}$$
$$= (10^3 \text{ g} \times 10^2 \text{ cm/sec}^2) \times 10^2 \text{ cm}$$
$$= (10^5 \text{ dynes}) \times (10^2 \text{ cm})$$
$$= 10^7 \text{ dyne-cm}$$
$$= 10^7 \text{ ergs}$$

One joule is equal to 10 million ergs.

The conversion to English units may be made if we recall that 2.205 pounds = 9.81 newtons and 1 foot = 30.48 centimeters. Then

$$1 \text{ joule} = 1 \text{ newton} \times 1 \text{ meter}$$
$$= 1 \text{ newton} \times \frac{2.205 \text{ lb}}{9.81 \text{ newtons}} \times 100 \text{ cm}$$
$$\times \frac{(1 \text{ ft})}{(30.48 \text{ cm})}$$
$$= 0.737 \text{ ft-lb}$$

A joule is a little less than three-quarters of a foot-pound.

Students sometimes confuse the units of work (force × displacement) with torque (force × lever arm). However, torque and work are quite different ideas. In the definition of torque, the lever arm is measured at right angles to the force. In the definition of work the displacement must be parallel to the force.

It is also important to recognize the difference between force and work. A man supporting a weight on his shoulder is exerting a force, and he certainly feels a sense of muscular fatigue. But unless the man produces some displacement in the direction of his force, he is doing no work. A large force may be applied to an object, but no work will be done unless there is also a displacement in the direction of the force.

12-2 CALCULATING THE NET WORK

The reader will recall from his study of trigonometry that the cosines of obtuse angles (angles larger than 90°) are negative. This means that the work calculated from Eq. 12-1 may be a negative quantity. If the angle between the force and the displacement is more than 90°, work is done *against* the force and the work done is negative. If the angle between the force and the displacement is less than 90°, work is done *by* the force and the work is positive.

In calculating the total or net work done on a system, we must keep track of these signs. The total work is found by an algebraic sum ΣW.

Example 12-1 In Fig. 12-2 the force $P = 200$ lb moves the 160-lb block from A to B. The coefficient of friction is 0.1 on the incline. Calculate the work done by each force acting on the block and determine the net work done.

Solution. The work done by $P = 200$ lb is (see Fig. 12-2c)

$$W_P = 200 \text{ lb} (40 \text{ ft}) = 8000 \text{ ft-lb}$$

The work done by $mg = 160$ lb is

$$W_{mg} = -160 \text{ lb} (30 \text{ ft}) = -4800 \text{ ft-lb}$$

No work is done by N because the displacement has no component in the N direction.

$$N = 0.6(200 \text{ lb}) + 0.8(160 \text{ lb})$$
$$= 120 \text{ lb} + 128 \text{ lb} = 248 \text{ lb}$$
$$f = \mu N = 0.1(248 \text{ lb}) = 24.8 \text{ lb}$$

The work done by f is

$$W_f = -24.8 \text{ lb} (50 \text{ ft}) = -1240 \text{ ft-lb}$$

The net work done is

$$8000 \text{ ft-lb} - 4800 \text{ ft-lb} - 1240 \text{ ft-lb} = 1960 \text{ ft-lb} \blacktriangleleft$$

The net work in Example 12-1 might have been determined in a somewhat simpler fashion if we had calculated the work done by the resultant force. From Fig. 12-2b it is seen that

$$F_R = 0.8(200 \text{ lb}) - 0.6(160 \text{ lb}) - 0.1(248 \text{ lb})$$
$$= 160 \text{ lb} - 96 \text{ lb} - 24.8 \text{ lb} = 39.2 \text{ lb}$$

The work done by F_R equals 39.2 lb (50 ft) = 1960 ft-lb. This result verifies the calculation in

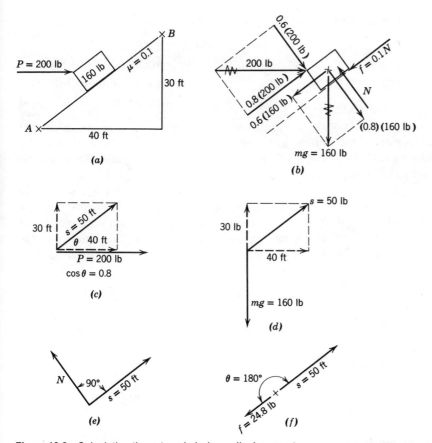

Figure 12-2   Calculating the net work during a displacement.

Example 12-1. It also illustrates an important principle:

If all parts of an object travel with the same velocity, the work done by all the forces acting on the object is equal to the work done by the resultant.

The objects usually encountered in engineering problems are much more complicated than the simple block in Example 12-1. Sometimes, the object in an engineering situation is really a complex arrangement of rods, shafts, wheels, and other moving parts. Different parts of a large-scale object may have different velocities and accelerations. Any object is composed of many small parts, however, and for each of these parts we can draw a free body diagram and write the second law of motion, $F_R = ma$, where F_R, m, and a all refer to the same small part

of the object. If s is the displacement of this little part in the direction of F_R, then the work done on this little part of the object is

$$W = F_R \cdot s = mas \qquad (12\text{-}2)$$

If the acceleration is constant, the quantity as may be written in an equivalent form using the formula for uniformly accelerated motion:

$$v_f{}^2 = v_0{}^2 + 2as$$

$$\frac{v_f{}^2}{2} - \frac{v_0{}^2}{2} = as$$

Substituting this form in Eq. 12-2 we find that the total the total work done on a small part of the object is

$$W = \tfrac{1}{2}mv_f{}^2 - \tfrac{1}{2}mv_0{}^2 \qquad (12\text{-}3)$$

Equation 12-3 was derived by using a formula for uniformly accelerated motion. In the next section we prove that this equation is also true when the acceleration is not constant. Equation 12-3 forms part of a very important principle of mechanics, the kinetic energy principle.

Energy is defined, in general, as the ability to do work. *Kinetic energy* measures the ability of an object to do work because of its state of motion. If all parts of an object are moving with the same velocity, the kinetic energy is, by definition

$$K = \tfrac{1}{2}mv^2 \qquad (12\text{-}4)$$

Kinetic energy is the energy of motion. If different parts of an object have different velocities, the total kinetic energy is the sum of the kinetic energies of all the parts.

Using Eq. 12-3 to sum up the work on parts of a large object, we find that

$$\sum W = \sum(\tfrac{1}{2}mv_f{}^2) - \sum(\tfrac{1}{2}mv_0{}^2)$$
$$= \text{final kinetic energy}$$
$$- \text{original kinetic energy} \quad (12\text{-}5)$$

This equation expresses the *kinetic energy principle*:

The total work done by all the forces equals the increase in the kinetic energy.

Whenever positive work is done on a system or an object, the kinetic energy increases. If negative work is done, the kinetic energy decreases. A moving object can perform work by giving up some or all of its kinetic energy to another object.

Example 12-2   From the data in Example 12-1, calculate the increase in kinetic energy.

Solution.   Since

$$F_R = 39.2 \text{ lb} \quad \text{and} \quad m = \frac{160 \text{ lb}}{32 \text{ ft/sec}^2} = 5 \text{ slugs}$$

the acceleration is

$$a = F_R/m = 39.2 \text{ lb}/5 \text{ slugs} = 7.84 \text{ ft/sec}^2$$
$$v_f{}^2 - v_0{}^2 = 2as = 2(7.84 \text{ ft/sec}^2)(50 \text{ ft})$$
$$v_f{}^2 - v_0{}^2 = 784 \text{ ft}^2/\text{sec}^2$$

The increase in kinetic energy is

$$\tfrac{1}{2}mv_f{}^2 - \tfrac{1}{2}mv_0{}^2 = \tfrac{1}{2}m(v_f{}^2 - v_0{}^2)$$
$$= \tfrac{1}{2}(5 \text{ slugs})(784 \text{ ft}^2/\text{sec}^2)$$
$$= 1960(\text{slug-ft/sec}^2)\text{-ft}$$
$$= 1960 \text{ ft-lb} \qquad \blacktriangleleft$$

In Example 12-1, the work done was calculated to be 1960 ft-lb. As expected, the work done by all the forces is equal to the increase in kinetic energy.

The kinetic energy principle may be used in many cases to solve motion problems that can also be solved by using Newton's second law of motion. In some problems the kinetic energy principle provides a neater and simpler method of solution.

Example 12-3   An 8-g bullet is fired from a rifle that has a barrel 80 cm long. The muzzle velocity of the bullet is 1000 m/sec. Determine the average force acting on the bullet as it travels down the barrel.

Solution.   Using the second law of motion, we first calculate the acceleration from $v_f{}^2 - v_0{}^2 = 2as$. Since $v_0 = 0$.

$$a = \frac{v_f{}^2}{2s} = \frac{(1000 \text{ m/sec})^2}{2(0.8 \text{ m})} = (10^6/1.6)\text{m/sec}^2$$

$$m = 8 \text{ g} = 8 \times 10^{-3} \text{ kg}$$

$$F_R = ma = 8 \times 10^{-3} \text{ kg}\frac{(10^6 \text{ m/sec}^2)}{1.6}$$
$$= 5.0 \times 10^3 \text{ kg-m/sec}^2 = 5.0 \times 10^3 \text{ newtons}$$

ALTERNATE METHOD.   The increase in kinetic energy is

$$\Delta K = \tfrac{1}{2}mv_f{}^2 - \tfrac{1}{2}mv_0{}^2$$
$$= \tfrac{1}{2}(8 \times 10^{-3} \text{ kg})(1000 \text{ m/sec})^2 - 0$$
$$= 4 \times 10^3 \text{ kg-m}^2/\text{sec}^2$$
$$= 4 \times 10^3 \text{ newton-meters}$$
$$= 4 \times 10^3 \text{ joules}$$

$$F_R \cdot s = \Delta K$$

$$F_R = \frac{\Delta K}{s} = \frac{4 \times 10^3 \text{ newton-meters}}{0.8 \text{ meter}}$$

$$= 5.0 \times 10^3 \text{ newtons} \qquad \blacktriangleleft$$

It is obvious from these last two examples that the units for measuring kinetic energy are the same as those for measuring work. Kinetic energy ($\frac{1}{2}mv^2$) is a scalar quantity and must always have a positive magnitude. The reader will recall that momentum (mv) is a vector quantity. Momentum and kinetic energy are different; they are measured in different units, and the two ideas should never be confused.

12-3 CALCULATING WORK FROM THE *F–s* DIAGRAM

In analyzing the performance of an engine and in many other technical operations, it is convenient to calculate the work done by employing an *F–s* (force vs. displacement) diagram. This procedure is illustrated in Fig. 12-3 for a very simple case. The force is a constant upward force equal to the weight *mg* of an object. The work done in a vertical displacement is $F \cdot s = mg(y_2 - y_1)$. The work done is equal to the area under the *F–s* curve.

If the force is not constant, the *F–s* curve is not a straight horizontal line. The curve will slope upward

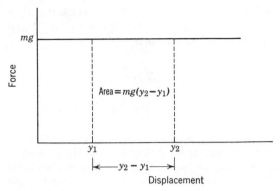

Figure 12–3   The *F–s* curve for a constant force.

as *F* increases, and downward where *F* is decreasing. However, the area under the *F–s* curve will be equal to the average force × displacement. The area is, therefore, always equal to the work done.

In Fig. 12-4 the resultant force is plotted as a function of the displacement. The vertical height of this *F–s* curve is equal to *ma* if all parts of the mass have the same acceleration. The curve is not a straight line because the resultant force and the acceleration are changing. The little rectangular area indicated in the diagram is equal to $(ma_{av}) \times \Delta s$, where a_{av} is the average acceleration

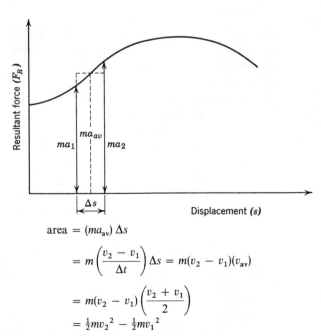

$$\text{area} = (ma_{av})\,\Delta s$$

$$= m\left(\frac{v_2 - v_1}{\Delta t}\right)\Delta s = m(v_2 - v_1)(v_{av})$$

$$= m(v_2 - v_1)\left(\frac{v_2 + v_1}{2}\right)$$

$$= \tfrac{1}{2}mv_2{}^2 - \tfrac{1}{2}mv_1{}^2$$

Figure 12-4   The resultant force (F_R) vs. displacement.

during a small displacement Δs. Now it is clear that $a_{av} = (v_2 - v_1)/\Delta t$ and $\Delta s = v_{av} \Delta t = (v_2 + v_1)/2 \; \Delta t$. Therefore, the small area is equal to

$$\frac{m(v_2 - v_1)}{\Delta t}\frac{(v_2 + v_1)}{2}\Delta t = \frac{m}{2}(v_2 - v_1)(v_2 + v_1)$$

$$= \tfrac{1}{2}m(v_2{}^2 - v_1{}^2)$$

The small area is equal to the change in the kinetic energy during the small displacement. The area under the F–s diagram must, therefore, represent the change in kinetic energy. The work done on an object by the resultant force is equal to the increase in kinetic energy if all parts of the object have the same velocity v. Equation 12-3 is valid even if the acceleration is changing.

12-4 WORK AND POTENTIAL ENERGY

During the motion of an object, work is often done against some mechanical forces which are described as *nonconservative* or *dissipative* forces. Friction is the commonest example of a nonconservative force. When work is done against the force of friction, mechanical energy is dissipated and converted into heat energy. Whenever work is done against a nonconservative force, the negative work represents a loss in mechanical energy.

Example 12-4   A 400-lb cart is located on a level road. Rolling friction in the wheels produces a force of 50 lb. A 150-lb push is applied to the cart and the cart is pushed for 100 yd. (a) How much work is done by the 150-lb push? (b) How much kinetic energy does the cart acquire? (c) How much mechanical work is converted into heat?

Solution.   (a) $W = F \cdot s = 150$ lb (300 ft) = 45,000 ft-lb

(b) The kinetic energy acquired is equal to the work done by the resultant force

$$F_R = 150 \text{ lb} - f = 150 \text{ lb} - 50 \text{ lb}$$
$$= 100 \text{ lb}$$
$$\Delta K = F_R \cdot s = 100 \text{ lb} (300 \text{ ft}) = 30,000 \text{ ft-lb}$$

ALTERNATE METHOD

$$a = \frac{F_R}{m} = \frac{100 \text{ lb}}{400 \text{ lb}/(32 \text{ ft/sec}^2)}$$

$$= \frac{100}{400} \times 32 \text{ ft/sec}^2$$

$$= 8 \text{ ft/sec}^2$$

$$v_f{}^2 - v_0{}^2 = 2as = 2(8 \text{ ft/sec}^2)(300 \text{ ft})$$
$$= 4800 \text{ ft}^2/\text{sec}^2$$

$$\Delta K = \frac{1}{2}m(v_f{}^2 - v_0{}^2)$$

$$= \frac{1}{2}\frac{400 \text{ lb}}{32 \text{ ft/sec}^2}(4800 \text{ ft}^2/\text{sec}^2)$$

$$= 200 \times 150 \text{ ft-lb} = 30,000 \text{ ft-lb}$$

(c) work done by 150-lb force
= increase in kinetic energy + heat

$$45,000 \text{ ft-lb} = 30,000 \text{ ft-lb} + \text{heat}$$

$$\text{heat energy} = 15,000 \text{ ft-lb}$$

ALTERNATE METHOD.   Work done against friction = heat produced = $f \cdot s = 50$ lb (300 ft) = 15,000 ft-lb. ◀

In many instances that are of great practical and theoretical importance, we encounter *conservative* or potential forces. When work is done against conservative forces, the work is not dissipated but is stored so that it can be recovered. When we do work in raising a weight, the work is not lost because the weight can do work as it falls back to the ground. When we do work in winding up the spring on a clock mechanism, this work is also stored, and it can be recovered and used to drive the mechanism. The force of the spring and the force of gravity are two common examples of conservative forces.

When work is done against conservative forces, the work is stored in the form of *potential energy*. Potential energy exists in an object or a system of objects because of the special state, condition, or position of the objects. As the objects return to their ground state level, the potential energy may be converted into energy of motion (kinetic energy) or may be used to perform mechanical work.

Example 12-5   A 2-lb weight is lifted to a height of 8 ft above the ground. (a) How much work is done by the force of gravity? (b) How much potential energy is created? (c) How much kinetic energy will be released if the weight is allowed to fall back to the ground? (d) How much work can the weight deliver to a clock mechanism if the weight is tied to a cable that is hooked up so as to drive the clock?

Solution.   (a) The displacement is up and the force of gravity is down. The work done by the force of gravity is negative.

$$W = -2 \text{ lb (8 ft)} = -16 \text{ ft-lb}$$

(b) Work done against the force of gravity stores 16 ft-lb of potential energy.
(c) The potential energy will be transformed into 16 ft-lb of kinetic energy. This can also be verified by calculating $K = \frac{1}{2}mv^2$.

$$v^2 = 2gs = 2(32 \text{ ft/sec}^2)(8 \text{ ft}) = 512 \text{ ft}^2/\text{sec}^2$$

$$K = \frac{1}{2}(2 \text{ lb}/32 \text{ ft/sec}^2)(512 \text{ ft}^2/\text{sec}^2) = 16 \text{ ft-lb}$$

(d) The weight can deliver 16 ft-lb of work if all the kinetic energy is changed into work. ◀

It is clear from Example 12-5 that the work done against gravity is weight times change in height. If y_1 is the original elevation and y_2 is the final elevation, the increase in gravitational potential energy is

$$\Delta U = mg(y_2 - y_1) \qquad (12\text{-}6)$$

If an object moves down, work is done by the force of gravity and potential energy decreases. If the object moves up, work must be done against the force of gravity and the potential energy increases. The force of gravity is a conservative force, because the work done against gravity in raising a weight is equal to the work done by gravity as the weight descends.

12-5 ELASTIC POTENTIAL ENERGY OF A SPRING

In compressing or stretching a spring, work must be done against the elastic force of the spring. According to Hooke's law this elastic force is given by $F = -kx$ (see p. 67). The force required to stretch or compress the spring is not a constant force but one that increases proportionately to the elongation in the spring if the elastic limit is not exceeded. This is indicated in the F–s diagram of Fig. 12-5, where the stretching force F is plotted against the displacement (elongation) x. The constant slope of the curve is the rate at which F increases with x, and this slope is equal to the spring constant k.

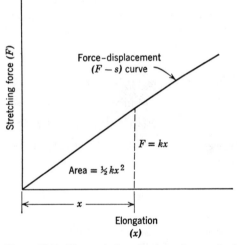

Figure 12-5   The work done in elongating or shortening a spring.

According to Section 12-3 the work done can be calculated by finding the area under the F–s curve. The work done in producing an elongation, x, equals the triangular area under the graph. This area equals $\frac{1}{2}$ base × height $= \frac{1}{2}(kx)(x) = \frac{1}{2}kx^2$. Since the spring has not been stretched beyond the elastic limit, this work has not been dissipated but has been converted into potential energy.

The spring will yield an equal amount of mechanical work as it returns to its original size. The elastic potential energy is, therefore,

$$U = \tfrac{1}{2}kx^2 \qquad (12\text{-}7)$$

In Eq. 12-7 the symbol x indicates the elongation or shortening of a spring with respect to its natural or unstressed length. If the elastic limit is not exceeded, the same potential energy exists whether the spring is elongated or shortened by x units of length.

Example 12-6   A 30-lb weight suspended from a spring elongates the spring 4 in. beyond its natural length. (a) How much gravitational potential energy is lost by the weight when it descends 4 in.? (b) How much potential energy is stored in the spring when it is elongated 4 in. from its natural length? (c) The elastic limit of the spring is 12 in. How much work has to be done to elongate the spring from 4 in. to 12 in.?

Solution

(a) ΔU (gravity) $= mg(y_2 - y_1)$

$$y_2 - y_1 = -4 \text{ in.} = -\tfrac{1}{3}\text{ ft}$$
$$\Delta U = 30 \text{ lb} \left(-\tfrac{1}{3}\text{ ft}\right) = -10 \text{ ft-lb}$$

When the weight descends 4 in., 10 ft-lb of gravitational potential energy is lost.

(b) k (the force constant) $= \dfrac{30 \text{ lb}}{\tfrac{1}{3}\text{ ft}} = 90 \text{ lb/ft}$

$$U \text{ (spring)} = \tfrac{1}{2}kx^2 = \tfrac{1}{2}(90 \text{ lb/ft})(\tfrac{1}{3}\text{ ft})^2$$
$$= 5 \text{ ft-lb}$$

ALTERNATE METHOD.   The final spring force at $x = 4$ in. is 30 lb. The original force at $x = 0$ is 0 lb; $F_{\text{average}} = 15$ lb. The work done against the spring force $= -15$ lb $(\tfrac{1}{3}$ ft$) = -5$ ft-lb.

$$\Delta U = +5 \text{ ft-lb}$$

(c) The work done in elongating the spring $= \Delta U = U_2 - U_1$.

$$\Delta U = \tfrac{1}{2}kx_2{}^2 - \tfrac{1}{2}kx_1{}^2$$
$$= \tfrac{1}{2}k(x_2{}^2 - x_1{}^2)$$

where x_2 is 1 ft and x_1 is $\tfrac{1}{3}$ ft.

$$\Delta U = \tfrac{1}{2}(90 \text{ lb/ft})(1 \text{ ft}^2 - \tfrac{1}{9}\text{ ft}^2)$$
$$= 45 \text{ ft-lb} - 5 \text{ ft-lb} = 40 \text{ ft-lb}$$

ALTERNATE METHOD

$$F_{\text{final}} = kx_2 = 90 \text{ lb/ft } (1 \text{ ft}) = 90 \text{ lb}$$
$$F_{\text{original}} = kx_1 = 90 \text{ lb/ft } (\tfrac{1}{3}\text{ ft}) = 30 \text{ lb}$$

so $F_{\text{av}} = 60$ lb, and displacement $= 1$ ft $- \tfrac{1}{3}$ ft $= \tfrac{2}{3}$ ft.

$$W = F_{\text{av}} \cdot s = 60 \text{ lb } (\tfrac{2}{3}\text{ ft}) = 40 \text{ ft-lb} \quad \blacktriangleleft$$

In Example 12-6 when the weight descended 4 in., we found that 10 ft-lb of gravitational potential energy were lost, but the spring only gained 5 ft-lb of elastic potential energy. There was a loss or decrease of 5 ft-lb of potential energy. Where did this energy go? The work–energy principle discussed in the next section provides an answer to this problem (see Example 12-7).

12-6 THE WORK–ENERGY PRINCIPLE

The kinetic energy principle of Section 12-2 can be extended in a useful way if we employ the concepts of conservative forces and potential energy. The work done can be divided into work done by conservative forces (W_C) and work done by nonconservative forces (W_N). Any force may be considered a nonconservative force if the potential energy function of the force is unknown. The kinetic energy principle can then be written

$$W_C + W_N = \Delta K$$

and $\qquad W_N = \Delta K - W_C \qquad (12\text{-}8)$

The quantity $-W_C$ indicates the work done against conservative forces. But the work done against conservative forces produces an increase in potential energy, ΔU. Equation 12-8 therefore asserts that

$$W_N = \Delta K + \Delta U$$

or $\qquad W_N = K_2 - K_1 + U_2 - U_1 \qquad (12\text{-}9)$

Equation 12-9 expresses an important principle:

The work done by the nonconservative forces acting upon an object or a system of objects is equal to the increase in potential energy plus the increase in kinetic energy.

In many situations it is apparent that the work of the nonconservative forces is zero or negligible. Then Eq. 12-9 becomes

$$K_2 + U_2 = K_1 + U_1 \qquad (12\text{-}10)$$

This equation also expresses an important principle:

The total mechanical energy in a system is the sum of the kinetic energy and the potential energy. The total energy remains constant if no work is done by nonconservative forces.

Example 12-7 The 30-lb weight in Example 12-6 is attached to the unstretched spring and then released from rest. (a) How fast is the weight traveling when it has moved down 4 in.? (b) How much work has to be done to bring the weight to rest with the spring stretched 4 in.?

Solution. (a) Since the weight is originally at rest, $K_1 = 0$. Because the spring is originally unstretched, U_1 (of spring) is also zero; U_1 (of gravity) is zero if we measure all heights from the starting point; K_2 is unknown and will be calculated.

$$U_2 \text{ (spring)} = \tfrac{1}{2}kx^2 = \tfrac{1}{2}(90 \text{ lb/ft})(\tfrac{1}{3} \text{ ft})^2$$
$$U_2 \text{ (spring)} = 5 \text{ ft-lb}$$
$$U_2 \text{ (gravity)} = mgy_2 = 30 \text{ lb } (-\tfrac{1}{3} \text{ ft})$$
$$= -10 \text{ ft-lb}$$
$$K_2 + U_2 = K_1 + U_1$$
$$K_2 + 5 \text{ ft-lb} - 10 \text{ ft-lb} = 0 + 0$$
$$K_2 = 5 \text{ ft-lb}$$
$$\tfrac{1}{2}mv^2 = K_2$$
$$m = \frac{30 \text{ lb}}{32 \text{ ft/sec}^2} = \frac{15 \text{ lb}}{16 \text{ ft/sec}^2}$$

$$v = \sqrt{\frac{2K_2}{m}} = \sqrt{\frac{2(5 \text{ ft-lb})}{\tfrac{15}{16} \text{ lb/ft/sec}^2}}$$
$$= \sqrt{(10)\tfrac{16}{15} \text{ ft}^2/\text{sec}^2} = 3.27 \text{ ft/sec}$$

(b) The weight will oscillate up and down unless nonconservative forces do work. When the weight comes to rest at a distance of 4 in. below its starting point, the final kinetic energy $K_2 = 0$. The final potential energy

$$U_2 = U_2 \text{ (spring)} + U_2 \text{ (gravity)}$$
$$U_2 = 5 \text{ ft-lb} - 10 \text{ ft-lb} = -5 \text{ ft-lb}$$
$$W_N = K_2 - K_1 + U_2 - U_1$$
$$= 0 - 0 + (-5 \text{ ft-lb}) - 0 = -5 \text{ ft-lb}$$

To bring the weight to rest, 5 ft-lb of work must be done against some nonservative force such as friction. ◀

The principle of Eqs. 12-9 and 12-10 is a special form of the law of conservation of energy. In using these equations (as in Example 12-7), care must be taken to calculate the total potential energy (gravitational, elastic, etc.). Since the work–energy principle involves the changes in energy, any convenient reference point may be chosen as the ground level in calculating gravitational potential energy. However, the elongation x in springs must always be measured with respect to the natural (unstretched) length when calculating the elastic potential energy. If several masses are moving independently, it is necessary to calculate the total kinetic energy, as in the following example.

Example 12-8 Assume that friction along the inclined plane and air resistance are negligible for the masses in Fig. 12-6. The two masses are released from rest and the 4-kg mass falls 2 m to the ground. (a) How fast are the masses traveling when the 4-kg mass strikes the ground? (b) How far up the incline does the 5-kg mass travel?

Solution. (a) Use the subscript $_0$ for the original energies and the subscript $_f$ for the final energies.

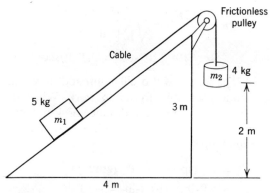

Figure 12-6   Gravitational potential energy in a system of two masses.

$K_0 = 0$ (both masses are at rest)
$U_0 = 0$. We shall measure heights with respect to the original elevations. Since the masses are tied together, they move with the same velocity and

$$K_f = \tfrac{1}{2}(m_1 + m_2)v^2 = \tfrac{1}{2}(5 \text{ kg} + 4 \text{ kg})v^2$$
$$= (4.5 \text{ kg})v^2$$
$$U_f = m_1gh_1 + m_2gh_2$$

The mass $m_2 = 4$ kg falls 2 m so that $h_2 = -2$ m. The mass $m_1 = 5$ kg moves 2 m along the incline and $h_1 = \tfrac{3}{5}(2 \text{ m}) = \tfrac{6}{5}$ m vertically.

$$U_f = 5 \text{ kg } (9.8 \text{ m/sec}^2)(\tfrac{6}{5} \text{ m})$$
$$+ 4 \text{ kg } (9.8 \text{ m/sec}^2)(-2 \text{ m})$$
$$= 58.8 \text{ kg-m}^2/\text{sec}^2 - 78.4 \text{ kg-m}^2/\text{sec}^2$$
$$= -19.6 \text{ kg-m}^2/\text{sec}^2 = -19.6 \text{ joules}$$
$$K_f + U_f = K_0 + U_0$$
$$K_f - 19.6 \text{ joules} = 0 + 0$$
$$K_f = 19.6 \text{ joules} = (4.5 \text{ kg})v^2$$
$$v = \sqrt{\frac{19.6 \text{ kg-m}^2/\text{sec}^2}{4.5 \text{ kg}}}, \qquad v = 2.09 \text{ m/sec}$$

(b) When the 4-kg mass strikes the ground, its kinetic energy is converted into other forms of energy (heat and sound energy). However, the 5-kg mass continues to move upward until its kinetic energy is converted into potential energy.

$$\tfrac{1}{2}m_1v^2 = m_1gh$$
$$h = \frac{v^2}{2g} = \frac{1}{2(9.8 \text{ m/sec}^2)}\frac{19.6 \text{ m}^2}{4.5 \text{ sec}^2}$$
$$= \frac{1}{4.5 \text{ m}} = 0.22 \text{ m}$$

The distance along the incline is s and $h/s = 0.6$, so $s = h/0.6$.

$$s = \frac{0.22 \text{ m}}{0.6} = 0.37 \text{ m}$$

The total distance along the incline is 2.37 m (2 m before m_2 strikes the ground and 0.37 m after).   ◄

It is very useful in engineering and technical problems to divide the work done by the non-conservative forces into two parts:

1. The applied work (W_a) done by some external agency such as a motor.

2. The negative work done by the dissipative forces of friction (W_f).

$$W_a + W_f = K_2 - K_1 + U_2 - U_1$$
and
$$W_a = K_2 - K_1 + U_2 - U_1 - W_f \qquad (12\text{-}11)$$

The term $-W_f$ represents the work done against friction that appears in the form of heat energy. If we designate the amount of heat energy by Q, then $-W_f = Q$ and Eq. 12-11 becomes

$$W_a = K_2 - K_1 + U_2 - U_1 + Q \qquad (12\text{-}12)$$

Equation 12-12 is a very useful statement of the law of conservation of energy as applied to mechanical work. When we study the theory of heat engines and the principles of thermodynamics, we shall recognize that Eq. 12-12 is a special form of the first law of thermodynamics. According to this equation, the principle of work and energy is as follows:

The work done on a mechanical system by an external agency is transformed into mechanical energy (kinetic and potential

energy) and into heat. **The work done by the external agency is equal to the increase in kinetic energy plus the increase in potential energy plus the heat produced by friction.**

In many situations it is apparent that no work is being done on the system by an external agency. Then $W_a = 0$ and Eq. 12-12 becomes

$$U_1 + K_1 = U_2 + K_2 + Q \qquad (12\text{-}13)$$

or

original mechanical energy

$$= \text{final mechanical energy}$$
$$+ \text{ heat produced by friction}$$

Example 12-9   The 10-ton truck in Fig. 12-7 encounters a steady frictional force of 500 lb on the roadway ABC. The truck passes A traveling at 30 ft/sec and it is traveling at 10 ft/sec when it reaches B. The driver then shifts into neutral and coasts from B to C. (a) How much work is done by the motor as the truck travels from A to B? (b) How fast is the truck traveling as it rolls past C?

Solution.   (a) $mg = 20{,}000$ lb, $m = \dfrac{20{,}000 \text{ lb}}{32 \text{ ft/sec}^2}$

Using Eq. 12-12, we find that

$$W_a = \frac{1}{2}m(v_B{}^2 - v_A{}^2) + mg(y_B - y_A) + Q$$

$$= \frac{1}{2} \times \frac{20{,}000 \text{ lb}}{32 \text{ ft/sec}^2}(100 \text{ ft}^2/\text{sec}^2 - 900 \text{ ft}^2/\text{sec}^2)$$

$$+ 20{,}000 \text{ lb} (0.03 \times 1000 \text{ ft})$$
$$+ 500 \text{ lb} (1000 \text{ ft})$$

$$= \frac{10^4(-800)}{32} \text{ ft-lb} + 10^4(60) \text{ ft-lb}$$

$$+ 50 \times 10^4 \text{ ft-lb}$$
$$= -25 \times 10^4 \text{ ft-lb} + 60 \times 10^4 \text{ ft-lb}$$
$$+ 50 \times 10^4 \text{ ft-lb}$$
$$= 85 \times 10^4 \text{ ft-lb}$$

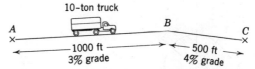

Friction = 500 lb

Figure 12-7   The work done by an external agency.

In traveling from A to B, there was a loss of 250,000 ft-lb of kinetic energy and a gain of 600,000 ft-lb of potential energy; 500,000 ft-lb of work had to be done against friction. As the calculation indicates, the motor had to deliver 850,000 ft-lb of work.

(b) Between B and C the motor does no work. From Eq. 12-13,

$$K_B + U_B = U_C + K_C + Q$$
$$\tfrac{1}{2}mv_B{}^2 + mgy_B = mgy_C + K_C + Q$$

The heat energy produced by friction is

$$Q = 500 \text{ ft } (500 \text{ lb}) = 25 \times 10^4 \text{ ft-lb}$$

It will simplify matters if we take the ground level at C so that $y_B = 0.04(500 \text{ ft}) = 20$ ft and $y_C = 0$. This gives

$$\frac{1}{2}\left(\frac{20{,}000 \text{ lb}}{32 \text{ ft/sec}^2}\right)(10 \text{ ft/sec})^2 + 20{,}000 \text{ lb } (20 \text{ ft})$$

$$= K_C + 25 \times 10^4 \text{ ft-lb}$$

$$3.13 \times 10^4 \text{ ft-lb} + 40 \times 10^4 \text{ ft-lb}$$

$$= K_C + 25 \times 10^4 \text{ ft-lb}$$

$$K_C = 18.13 \times 10^4 \text{ ft-lb}$$
$$K_C = \tfrac{1}{2}mv_C{}^2$$

$$v_C = \sqrt{\frac{2K_C}{m}} = \sqrt{\frac{2(18.13 \times 10^4) \text{ ft-lb}}{\dfrac{20{,}000 \text{ lb}}{32 \text{ ft/sec}^2}}}$$

$$= \sqrt{580.2 \text{ ft}^2/\text{sec}^2} = 24.1 \text{ ft/sec} \qquad \blacktriangleleft$$

12-7   WORK AND ENERGY IN ROTATION

When a force acts on a rotating object, work is done by the tangential component of the force (see Fig. 12-8), F_T. The work done by this force

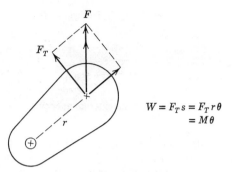

$$W = F_T s = F_T r \theta$$
$$= M \theta$$

Figure 12-8   Work done on a rotating object.

is $F_T s$. If the angular displacement is measured in radians, $s = r\theta$, and the work done is $F_T r\theta$, since $F_T r$ = torque, we have

$$W = M\theta \qquad (12\text{-}14)$$

work in rotation

= torque × angular displacement in radians

Equation 12-14 is easily applied if the torque is constant. If the torque is not constant, it becomes necessary to find the average value for the torque in order to calculate the work done. The graphical method may also be employed in rotation. If a torque versus angular displacement $(M-\theta)$ diagram is constructed, the area under the $M-\theta$ curve equals the work done.

Example 12-10   A gasoline engine of the type used in many automobiles delivers a torque of 200 lb-ft. How much work is done in 10 revolutions?

Solution.   Since θ must be in radians, $\theta =$ 10 rev × 2π rad/rev.

$$\theta = 20\pi \text{ rad}$$
$$W = M\theta = 200 \text{ lb-ft} \times 20\pi \text{ rad}$$
$$= 4000\pi \text{ ft-lb}$$
$$= 12.6 \times 10^3 \text{ ft-lb} \qquad \blacktriangleleft$$

The kinetic energy of a rotating object cannot be calculated from the formula $\frac{1}{2}mv^2$, because different parts of the object have different velocities. However, we can imagine that the object is divided into many small masses,

m_1, m_2, m_3, etc., which have velocities v_1, v_2, v_3, etc. Since ω is the same for the entire object, $v_1 = \omega r_1$, $v_2 = \omega r_2$, $v_3 = \omega r_3$, etc. The total kinetic energy can then be found by summing

$$\begin{aligned} K &= \tfrac{1}{2}m_1 v_1{}^2 + \tfrac{1}{2}m_2 v_2{}^2 + \cdots \\ &= \tfrac{1}{2}m_1 r_1{}^2 \omega^2 + \tfrac{1}{2}m_2 r_2{}^2 \omega^2 + \cdots \\ &= \tfrac{1}{2}(m_1 r_1{}^2 + m_2 r_2{}^2 + \cdots)\omega^2 \end{aligned} \qquad (12\text{-}15)$$

The quantity in the brackets is the moment of inertia, I, which was defined in Chapter 11. Equation 12-15, therefore, states that the total kinetic energy of rotation is

$$K = \tfrac{1}{2}I\omega^2 \qquad (12\text{-}16)$$

The net work done by a constant resultant torque M_R during rotation is $\sum W = M_R \cdot \theta$. Since $M_R = I\alpha$, we have

$$\sum W = I\alpha\theta \qquad (12\text{-}17)$$

Because the resultant torque is constant, the angular acceleration is constant and

$$\omega_f{}^2 = \omega_0{}^2 + 2\alpha\theta$$

$$\frac{\omega_f{}^2}{2} - \frac{\omega_0{}^2}{2} = \alpha\theta$$

When this result is substituted in Eq. 12-17 we obtain

$$\sum W = \tfrac{1}{2}I\omega_f{}^2 - \tfrac{1}{2}I\omega_0{}^2 \qquad (12\text{-}18)$$

net work done in rotation = final kinetic energy

− original kinetic energy

In deriving Eq. 12-18 it was assumed that the work was done by a constant resultant torque. However, by using the M_R versus θ diagram, we can prove that this equation is also true if the resultant torque is not constant. In calculations with Eq. 12-18 the angular velocities ω_f and ω_0 must be expressed in radians per second. This equation is the mathematical statement of the kinetic energy principle for rotation:

The work done by the resultant torque is equal to the increase in kinetic energy of rotation.

Example 12-11   An 800-lb flywheel has a radius of gyration of 2 ft. The wheel is rotating at 600 rev/min. (a) How much work in foot-pounds has to be done to increase the rate of rotation to 900 rev/min? (b) What constant unbalanced torque is needed if the change in angular velocity is to be completed in 100 rev of the wheel?

Solution

(a) $I = mk^2 = \dfrac{800 \text{ lb}}{32 \text{ ft/sec}^2} (2 \text{ ft})^2$

$= 100 \text{ lb-ft-sec}^2 = 100 \text{ slug-ft}^2$

$\omega_f = 900 \dfrac{\text{rev}}{\text{min}} \times \dfrac{1 \text{ min}}{60 \text{ sec}} \times \dfrac{2\pi \text{ rad}}{1 \text{ rev}} = 30\pi \text{ rad/sec}$

$\omega_0 = 600 \dfrac{\text{rev}}{\text{min}} \times \dfrac{1 \text{ min}}{60 \text{ sec}} \times \dfrac{2\pi \text{ rad}}{1 \text{ rev}} = 20\pi \text{ rad/sec}$

The final kinetic energy is

$K_f = \tfrac{1}{2}I\omega_f{}^2 = \tfrac{1}{2}(100 \text{ lb-ft-sec}^2)(900\pi^2 \text{ rad}^2/\text{sec}^2)$

$= 44.4 \times 10^4 \text{ ft-lb}$

The original kinetic energy is

$K_0 = \tfrac{1}{2}I\omega_0{}^2 = \tfrac{1}{2}(100 \text{ lb-ft-sec}^2)(400\pi^2 \text{ rad}^2/\text{sec}^2)$

$= 19.7 \times 10^4 \text{ ft-lb}$

The work done is

$$\Sigma W = K_f - K_0 = 24.7 \times 10^4 \text{ ft-lb}$$

(b) $\theta = 100 \text{ rev} \times \dfrac{2\pi \text{ rad}}{1 \text{ rev}}$

$= 200\pi \text{ rad} = 628 \text{ rad}$

$$\Sigma W = M_R \cdot \theta, \qquad M_R = \dfrac{\Sigma W}{\theta} = \dfrac{24.7 \times 10^4 \text{ ft-lb}}{628 \text{ rad}}$$

$$M_R = 3.93 \times 10^2 \text{ lb-ft} \qquad \blacktriangleleft$$

In analyzing rotation, the work–energy principle of Eqs. 12-12 and 12-13 is often very useful. However, the kinetic energy of the system must include the kinetic energy of rotation, $\tfrac{1}{2}I\omega^2$.

Example 12-12   In Fig. 12-9 the 160-lb drum has a radius of gyration of 1 ft. The cable is wrapped around the drum and when the drum is rotated, the 200-lb weight rises or falls. (a) The cable unwinds smoothly and the 200-lb weight

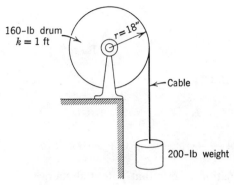

160-lb drum
$k = 1$ ft

$r = 18''$

Cable

200-lb weight

Figure 12-9   Rotating drum and falling weights.

falls a distance of 100 ft to the ground. Friction in the bearings and air resistance are negligible. How fast is the weight traveling when it strikes the ground? (b) A retarding torque is applied to the drum by a brake mechanism. How much heat is produced in the brake mechanism if the weight descends the 100 ft at a slow steady speed?

Solution

(a) $K_0 = 0$ since the weight starts from rest

$U_0 = mgh = 200 \text{ lb} (100 \text{ ft}) = 20,000 \text{ ft-lb}$

$U_F = 0$

$K_F = \tfrac{1}{2}mv^2 + \tfrac{1}{2}I\omega^2$

For the weight:

$$\tfrac{1}{2}mv^2 = \dfrac{1}{2}\left(\dfrac{200 \text{ lb}}{32 \text{ ft/sec}^2}\right)v^2 = 3.13 \dfrac{\text{lb}}{\text{ft/sec}^2} v^2$$

For the drum:

$$I = mk^2 = \dfrac{160 \text{ lb}}{32 \text{ ft/sec}^2} 1 \text{ ft}^2 = 5 \text{ lb-ft-sec}^2$$

$$= 5 \text{ slug-ft}^2$$

Since the cable does not slip, $v = \omega r$ and

$$\omega = \dfrac{v}{r} = \dfrac{v}{18 \text{ in.}} = \dfrac{v}{1.5 \text{ ft}}$$

$$\tfrac{1}{2}I\omega^2 = \tfrac{1}{2}(5 \text{ lb-ft-sec}^2)\left(\dfrac{v}{1.5 \text{ ft}}\right)^2$$

$$= 1.11 \dfrac{\text{lb}}{\text{ft/sec}^2} v^2$$

Therefore

$$K_f = 3.13 \frac{\text{lb}}{\text{ft/sec}^2} v^2 + 1.11 \frac{\text{lb}}{\text{ft/sec}^2} v^2$$

$$= 4.24 \frac{\text{lb}}{\text{ft/sec}^2} v^2$$

$$K_f + U_f = K_0 + U_0$$

$$4.24 \frac{\text{lb}}{\text{ft/sec}^2} v^2 + 0 = 0 + 20{,}000 \text{ ft-lb}$$

(This equation asserts that potential energy lost equals kinetic energy gained.)

$$v^2 = \frac{20{,}000}{4.24} \text{ ft}^2/\text{sec}^2$$

$$v = 68.6 \text{ ft/sec}$$

(b) The work done by the brake mechanism is negative work and results in a loss of mechanical energy. From the work energy principle,

$$W_a = K_f - K_0 + U_f - U_0$$

Since the weight descends at constant speed, $K_f = K_0$, and

$$W_a = U_f - U_0$$

$$= 0 - 20{,}000 \text{ ft-lb} = -20{,}000 \text{ ft-lb}$$

The work done causes a loss of 20,000 ft-lb of mechanical energy and the creation of 20,000 ft-lb of heat energy in the brake mechanism. This effect is also observed in automobile brakes when the kinetic energy of the speeding vehicle is converted into heat energy in the brakes. ◀

When an object is rolling, it possesses both kinetic energy of translation ($\frac{1}{2}mv^2$) and kinetic energy of rotation ($\frac{1}{2}I\omega^2$). In computing the kinetic energy of rotation, I is the moment of inertia with respect to the axis that passes through the center of mass. In computing the kinetic energy of translation, v is the velocity of the center of mass.

If the object is to roll without slipping, friction must act to prevent slipping. However, as long as the point of contact with the surface does not slip, friction will do no work and cause no loss in mechanical energy. But friction does serve to establish an instantaneous axis of rotation at the point of contact with the surface, and some of the mechanical energy will be converted into rotational kinetic energy.

Example 12-13   A 100-g solid sphere starts from rest and rolls down an inclined ramp without slipping. After the sphere has descended a vertical distance of 200 cm, determine (a) the kinetic energy, and (b) the velocity.

Solution

$K_0 = 0$ since the sphere starts from rest

$U_0 = mgh$, where $h = 200$ cm and $U_f = 0$

$K_f = \frac{1}{2}mv^2 + \frac{1}{2}I\omega^2$

For a sphere, the moment of inertia for any axis through the center of mass is $I = \frac{2}{5}mr^2$ (see Fig. 11-7).

$$K_f = \frac{1}{2}mv^2 + \frac{1}{2}(\frac{2}{5}mr^2)\omega^2$$

Since $r\omega = v$, we have

$$K_f = \frac{1}{2}mv^2 + \frac{1}{2}(\frac{2}{5}m)v^2 = \frac{1}{2}mv^2 + \frac{1}{5}mv^2$$
$$= \frac{7}{10}mv^2$$

(a) From the conservation of energy

$$K_f + U_f = K_0 + U_0$$
$$K_f - 0 = 0 + (100 \text{ g})(980 \text{ cm/sec}^2)(200 \text{ cm})$$
$$K_f = 19.6 \times 10^6 \text{ g-cm}^2/\text{sec}^2$$
$$= 19.6 \times 10^6 \text{ ergs}$$

(b) $K_f = U_0,$ $\frac{7}{10}mv^2 = mgh,$ $v^2 = \frac{10}{7}gh$
$v = \sqrt{\frac{10}{7}gh} = \sqrt{\frac{10}{7}(980 \text{ cm/sec}^2)\, 200 \text{ cm}}$
$= \sqrt{(140)(2000) \text{ cm}^2/\text{sec}^2}$
$= \sqrt{28} \times 10^2 \text{ cm/sec} = 529 \text{ cm/sec}$ ◀

12-8 WORK–ENERGY METHODS IN SPACE MECHANICS

According to Newton's law of gravity, the force of gravity at the surface of the earth is $mg_s = GMm/r_s^2$. If a projectile or a space capsule moves away from the surface of the earth, work has to be done against the conservative force of gravity, and the potential energy increases. If the distance from the surface of the earth is not too great, we can assume that the force of gravity is constant and that the potential energy is

$$U = mg_s h \qquad (12\text{-}19)$$

Equation 12-19 can be used in the work–energy solution of many problems involving falling objects.

Example 12-14 A 160-lb parachutist jumps out of a light airplane that is traveling at 120 miles/hr. How fast is the parachutist traveling after 400 ft of free fall? Assume that air resistance is negligible.

Solution. Since we have conservation of energy

$$mg_s h + \tfrac{1}{2}mv_0{}^2 = \tfrac{1}{2}mv_f{}^2$$

Canceling m and solving for $v_f{}^2$ we obtain

$$v_f{}^2 = v_0{}^2 + 2g_s h$$

$$v_0 = 120 \text{ miles/hr} = 2(88 \text{ ft/sec}) = 176 \text{ ft/sec}$$

$$v_f{}^2 = v_0{}^2 + 2g_s h$$

$$= (176 \text{ ft/sec})^2 + 2(32 \text{ ft/sec}^2)(400 \text{ ft})$$

$$= 3.10 \times 10^4 \text{ ft}^2/\text{sec}^2 + 2.56 \times 10^4 \text{ ft}^2/\text{sec}^2$$

$$v_f{}^2 = 5.66 \times 10^4 \text{ ft}^2/\text{sec}^2$$

$$v_f = \sqrt{5.66 \times 10^4 \text{ ft}^2/\text{sec}^2} = 238 \text{ ft/sec}$$

$$= 238 \text{ ft/sec} \times \frac{60 \text{ miles/hr}}{88 \text{ ft/sec}}$$

$$= 162 \text{ miles/hr} \qquad \blacktriangleleft$$

In space mechanics, the magnitude of g decreases and the force of gravity decreases as one moves away from the earth. The work done against the force of gravity may be calculated by drawing an F–s diagram, as in Fig. 12-10. This graph indicates that $F = GMm/r^2$ decreases as one moves away from the earth.

The work done against the force of gravity is equal to the area under this F–s curve. If we consider the small rectangular area of width Δr, the area of this portion = (average height of curve) $\times$ Δr = (average force) $\times$ $(r_1 - r_s)$. The average force in the interval must be less than $GMm/r_s{}^2$ and more than $GMm/r_1{}^2$. Because r_s and r_1 are almost equal, the average force may be approximated as $GMm/(r_s r_1)$. The area is then

$$\frac{GMm}{r_s r_1}(r_1 - r_s) = GMm\left(\frac{1}{r_s} - \frac{1}{r_1}\right).$$

The area of the next small rectangular segment of width $\Delta r = r_2 - r_1$ may be calculated in the same way.

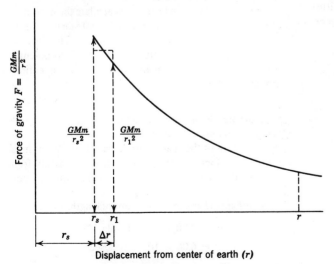

Figure 12-10 The force of gravity decreases with increasing distance from the earth.

This area will be

$$\frac{GMm}{r_2 r_1}(r_2 - r_1) = GMm\left(\frac{1}{r_1} - \frac{1}{r_2}\right)$$

The area of a third rectangular segment will be GMm $(1/r_2 - 1/r_3)$. This procedure can be continued, and the total area from r_s to r can be found by adding up all the rectangular areas. This area is

$$GMm\left[\left(\frac{1}{r_s} - \frac{1}{r_1}\right) + \left(\frac{1}{r_1} - \frac{1}{r_2}\right) + \left(\frac{1}{r_2} - \frac{1}{r_3}\right) + \cdots\right]$$

It is obvious that all the terms except the first and the last will cancel. The total area and work done is

$$GMm\left(\frac{1}{r_s} - \frac{1}{r}\right)$$

This work is equal to the gravitational potential energy for an object at a distance r from the center of the earth. Accordingly we may write the formula

$$U = GMm\left(\frac{1}{r_s} - \frac{1}{r}\right) \qquad (12\text{-}20)$$

Since $g_s = GM/r_s^2$ and $g_s r_s^2 = GM$, we can rewrite Eq. 12-20 in another useful form:

$$U = mg_s r_s^2\left(\frac{1}{r_s} - \frac{1}{r}\right) \qquad (12\text{-}21)$$

The distance r is measured from the center of the earth. Now Fig. 12-11 indicates that $r = r_s + h$, where h is the height of the object above the earth. Substituting this value for r in Eq. 12-21 and doing some algebraic manipulation, we obtain

$$U = mg_s r_s^2\left(\frac{1}{r_s} - \frac{1}{r_s + h}\right)$$

$$= mg_s\left(r_s - \frac{r_s^2}{r_s + h}\right)$$

$$= mg_s \frac{[r_s(r_s + h) - r_s^2]}{r_s + h}$$

$$= mg_s \frac{(r_s^2 + r_s h - r_s^2)}{r_s + h}$$

Therefore, we have a third equivalent equation for the gravitational potential energy.

$$U = mg_s h\left[\frac{r_s}{r_s + h}\right] \qquad (12\text{-}22)$$

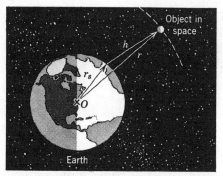

$$U = GMm\left(\frac{1}{r_s} - \frac{1}{r}\right) = mg_s r_s^2\left(\frac{1}{r_s} - \frac{1}{r}\right)$$

$$= mg_s h\left(\frac{r_s}{r_s + h}\right)$$

Figure 12-11 The gravitational potential energy of an object in space.

In using the expressions for the gravitational potential energy, we should recall that mg_s is the weight of an object near the surface of the earth. If the distance from the surface is not very great, then $\dfrac{r_s}{r_s + h}$ is practically equal to unity, and Eq. 12-22 reduces to the simple form of Eq. 12-19. However, in most problems in space mechanics, the potential energy must be calculated by using Eqs. 12-20, 12-21, or 12-22.

These equations can be used to calculate the maximum distance reached by a space capsule that starts from the earth and travels outward into space. The original kinetic energy of the capsule is converted into potential energy; at the maximum height, the energy is entirely potential energy.

• **Example 12-15** (a) A rocket launches a space capsule so that the capsule is ejected with a velocity of 5 miles/sec. How high does the space capsule travel? (b) What must the initial velocity be if the space capsule is to escape from the earth's gravitational field and reach another planet?

Solution. Using miles as a unit of length, we find that

$$g_s = 32 \text{ ft/sec}^2 \times (1 \text{ mile}/5280 \text{ ft})$$

$$= 6.06 \times 10^{-3} \text{ miles/sec}^2$$

For simplicity we will use the approximate value $r_s = 4000$ miles

(a) kinetic energy (original) = potential energy (final)

$$\tfrac{1}{2}mv^2 = mg_sh\left(\frac{r_s}{r_s + h}\right)$$

Before substituting it is convenient to solve this for h:

$$\frac{v^2}{2g_s} = \frac{hr_s}{r_s + h}$$

$$v^2(r_s + h) = 2g_shr_s$$

$$h(v^2 - 2g_sr_s) + v^2r_s = 0$$

$$h(2g_sr_s - v^2) = v^2r_s$$

$$h = \frac{v^2r_s}{2g_sr_s - v^2} \qquad (12\text{-}23)$$

Substituting the values given, we obtain

$$h = \frac{(5 \text{ miles/sec})^2(4000 \text{ miles})}{2(6.06 \times 10^{-3} \text{ miles/sec}^2)(4000 \text{ miles}) - (5 \text{ miles/sec})^2}$$

$$= \frac{25(4000 \text{ miles})}{8(6.06) - 25} = \frac{25(4000 \text{ miles})}{23.48} = 4260 \text{ miles}$$

from the surface of the earth.

(b) If the denominator in Eq. 12-23 becomes very small, h will become very great. The capsule will escape from the earth's gravitational field if h becomes infinitely great. This means that

$$2g_sr_s - v^2 = 0$$

$$v^2 = 2g_sr_s$$

$$v = \sqrt{2g_sr_s} \qquad (12\text{-}24)$$

Equation 12-24 gives the *minimum escape velocity*. Substituting the values given, we obtain

$$v = \sqrt{2(6.06 \times 10^{-3} \text{ miles/sec}^2)(4000 \text{ miles})}$$

$$= \sqrt{48.48 \text{ miles}^2/\text{sec}^2} = 6.96 \text{ miles/sec}$$

SUMMARY

The *work* done by a force equals the force multiplied by the displacement in the direction of the force, i.e.,

$$W = Fs \cos$$

Work may be measured in *foot-pounds, joules,* or *ergs.* (See Table 12-1.)

When work is done on a rotating object by a torque M, the work done is $W = M\theta$, where θ is the angular displacement measured in radians.

If the angle between the force vector and the displacement vector is more than 90°, the cosine is negative and the work done is negative. If the angle between the force and the displacement is less than 90°, the cosine is positive and the work done is positive. According to the *kinetic energy principle*, the total work done by all forces acting on a system equals the final kinetic energy minus the original kinetic energy.

The *kinetic energy* of an object is a measure of the ability of the object to do work because of its state of motion. The kinetic energy of an object which is not rotating is $\tfrac{1}{2}mv^2$. For an object that is rotating, the kinetic energy is $\tfrac{1}{2}I\omega^2$. For an object that is moving with translation and rotating, the total kinetic energy consists of the *kinetic energy of translation* ($\tfrac{1}{2}mv^2$) plus the *kinetic energy of rotation* ($\tfrac{1}{2}I\omega^2$), where v = the velocity of the center of mass and I = the moment of inertia with respect to an axis through the center of mass.

When work is done against *conservative forces* the work is stored in the form of *potential energy*. Potential energy is the energy stored in a system because of the position or state of the objects in the system. The potential energy of an object in a uniform gravitational field is $U = mgh$, where h is the height above some reference level. The elastic potential energy stored in a spring is $U = \tfrac{1}{2}kx^2$, where x is the length the spring is elongated or shortened.

According to the *work–energy principle* the work done by the applied forces on a system equals the increase in kinetic energy plus the increase in potential energy plus the heat produced because of the work done against friction. If frictional forces are negligible and no work is done by applied forces on a system, then the total mechanical energy (potential energy plus kinetic energy) remains constant.

When an object is rolling without slipping, the point of contact with the surface does slip and friction does no work. If friction does no work, then the loss in potential energy equals the gain in kinetic energy. On the other hand, in many problems in mechanics we note that kinetic energy decreases and is converted into potential energy.

QUESTIONS

1. What is the difference between force and work?

2. Why is it possible to exert a large force and still do no work?

3. Under what circumstances will the work done by a force turn out to be a negative quantity?

4. What is meant by the term *kinetic energy*?

5. Suppose that negative work is done on a moving object. What effect does this have on the kinetic energy of the object?

6. What is meant by the term *conservative force*?

7. What is the difference between a conservative force and a nonconservative force?

8. Explain the relationship between conservative forces and potential energy.

9. What is the difference between kinetic energy and potential energy?

10. When does the total mechanical energy in a system remain constant?

11. A 100-lb box is lifted up a vertical distance of 4 ft. In foot pounds,
    (a) what is the increase in potential energy, and
    (b) the work done by the force of gravity?
    Explain why the work done by the force of gravity is negative.

12. A 10-kg object is situated where the acceleration of a freely falling object is 9.8 m/sec². The object falls 2 m with negligible air resistance. What are the force of gravity on this object and the work done by the force of gravity?

13. The force of friction acting upon a 200-lb packing crate is 60 lb when the crate is pushed 3 ft along a horizontal floor. In foot-pounds, what is the work done by the force of friction? Explain why the work done by the force of friction is negative.

14. Explain why the work done by the normal force will be zero for an object that is pushed up an inclined plane.

15. What are two different units used for measuring

work in the metric system? Define these two units in terms of force and displacement.

16. Explain why foot-pound may be used to measure torque and work.

17. How is the conservation of mechanical energy principle used to explain the swinging of a pendulum?

18. How do we calculate the work done by a torque acting on a rotating object?

19. Under what conditions will the total mechanical energy in a system increase?

20. Give an example of a situation in which the kinetic energy of a system is decreasing and the potential energy is increasing.

21. What is the work–energy principle for cases in which applied work is done on a mechanical system?

22. What is the kinetic energy principle? How is this principle stated for rotation?

23. What is meant by "kinetic energy of rotation"?

24. The kinetic energy of rotation is given by $\frac{1}{2}I\omega^2$. Explain the meaning of I and ω. What units must be used in measuring ω?

25. For an object that is rolling without slipping the total kinetic energy is

$$\frac{1}{2}mv^2 + \frac{1}{2}mv^2\left(\frac{k^2}{r^2}\right)$$

Explain how this is obtained from the formulas for translational and rotational kinetic energy.

PROBLEMS

(A)

1. A 200-lb block of concrete is lifted vertically for a distance of 10 ft. How much work is done?

2. A 10-kg mass is lifted vertically at constant speed for a distance of 5 m.
   (a) How great a force (in newtons) is required?
   (b) How much work is done?

3. A force of 6×10^{-8} dynes acts upon an electron passing through a high voltage device. How much work is done if the electron moves a distance of 80 cm in the direction of the force?

4. A 200-lb block of concrete rests on a level roadway. The coefficient of friction between the concrete and the roadway is 0.6. How much work is done in

pushing the block with a constant speed for a distance of 10 ft?

5. A 20-g bullet leaves a rifle barrel with a muzzle velocity of 1000 m/sec. Calculate the kinetic energy in ergs and in joules.

6. A small electric motor delivers a torque of 40 newton-meters. How much work is done during each revolution of the drive shaft?

7. A torque wrench is used to apply a torque of 80 lb-in. How much work in ft-lb is done when the torque is applied through one-half a revolution?

8. A 16-lb bowling ball is a uniform sphere 8 in. in diameter. The ball is rolling without slipping at 12 ft/sec.
   (a) Calculate the kinetic energy of translation.
   (b) Calculate the kinetic energy of rotation.

9. A 500-watt motor does work at the rate of 500 joules/sec. What torque is produced by the motor when the drive shaft is rotating at 10 rev/sec?

10. A force of 100 newtons pushes an object for a distance of 5 m. How much work is done?

11. In a high voltage electrical device, electric forces do 8×10^{-14} joules of work upon electrons that travel 2 m through the device. Calculate the electric force on an electron.

12. The force of friction acting upon a 100-lb table is 40 lb. How much work has to be done to push the table a distance of 5 ft across the floor?

13. How much gravitational potential energy is produced when a 100-lb weight is lifted a distance of 4.5 ft?

14. A mechanical jack is used to lift a 500-kg piece of steel a vertical distance of 40 cm. (Note that $g = 9.80$ m/sec².)
   (a) How great a force in newtons is required to lift the steel?
   (b) How much work in joules is done?
   (c) How much potential energy is created?

15. How much does the potential energy of a mechanical system decrease if a 50-lb weight descends 6 ft?

16. A rocket projectile whose mass is 100 kg is traveling at 400 m/sec. How much kinetic energy is possessed by the projectile?

17. An automotive engine delivers 300 lb-ft of torque to the drive shaft which completes 10 rev/sec. How much work in foot-pounds is done by the engine in 1 sec?

18. A mass of 10 slugs is lifted 30 in. from the floor to the top of a table.
   (a) What force is required to lift this mass where $g = 32$ ft/sec²?
   (b) How much work in ft-lb is done in lifting the mass?
   (c) How much does the potential energy increase?

19. A 10-lb weight falls a distance of 5 ft near the earth. How much kinetic energy is acquired by this weight if air resistance is negligible?

20. Fifty joules of work are done to accelerate a 4-kg object across a rough horizontal surface. The object starts from rest and reaches a velocity of 3 m/sec.
   (a) How much kinetic energy is gained by the object?
   (b) How much work is done against friction?

21. When 400 ft-lb of work are done on a 20-lb object, it moves up vertically a distance of 6 ft. Assume that friction is negligible.
   (a) How much does the potential energy of the object increase?
   (b) Calculate the increase in kinetic energy.
   (c) If the object starts from rest, what is the final velocity?

(B)

1. The force $P = 10$ lb acts upon the 20-lb block in Fig. 12-12 to push the block 10 ft along the horizontal surface from A to B. The coefficient of sliding friction is 0.2.

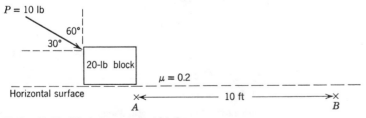

Figure 12-12 Work done against friction.

(a) How much work is done by P?

(b) How much work is done by the force of friction?

(c) If the block starts from rest at A, calculate the kinetic energy when the block reaches B.

2. The 10-kg block in Fig. 12-13 is released from rest at A and slides down the inclined ramp to B. The block is sliding at 10 m/sec when it reaches B.

(a) Calculate the kinetic energy of the block when it reaches B.

(b) How much work was done against friction during the displacement?

(c) Calculate the resultant force acting on the block during the displacement.

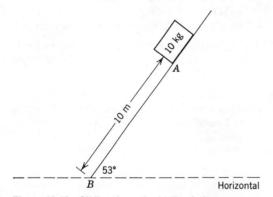

Figure 12-13 Sliding down the inclined plane.

3. A 3200-lb automobile is traveling at 60 miles/hr when the brakes are applied.

(a) How much work must be done by the brakes to slow the automobile down to 30 miles/hr?

(b) If the automobile travels 100 ft in slowing down from 60 miles/hr to 30 miles/hr, what is the average force exerted by the brakes?

4. An 8000-lb truck starts from rest and travels 1000 ft up a 3% grade. The average force of friction acting upon the truck is 100 lb. The truck reaches the top of the grade with a velocity of 20 ft/sec.

(a) How much work was done by the motor?

(b) What was the average thrust delivered by the motor?

5. Figure 12-14 represents a graph of the resultant force versus displacement. The resultant force and the displacement are in the same direction.

(a) How much work is done during the 10-m displacement?

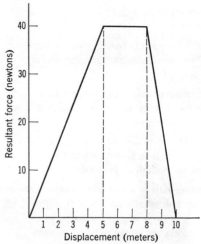

Figure 12-14 The force-displacement diagram for a changing resultant force.

(b) If the kinetic energy is originally zero, what is the kinetic energy after the first 5 m?

(c) What is the final kinetic energy?

6. A pump is used to pump water out of a well into an irrigation ditch 20 ft higher than the water level in the well. The pump delivers 50 ft^3 of water each minute, and the water weighs 62.4 lb/ft^3. How much work is done each minute?

7. A 100-g mass suspended from a coil spring elongates this spring by 20 cm. The elastic limit of the spring is not exceeded.

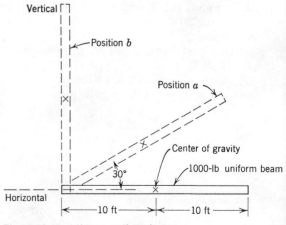

Figure 12-15 Lifting a uniform beam.

(a) Assume that $g = 980$ cm/sec^2 and calculate the spring constant in dynes/cm.

(b) What is the potential energy of the spring when the 100-g mass is suspended from the spring?

8. The uniform beam in Fig. 12-15 is 20 ft long and weighs 1000 lb.

(a) How much work must be done to raise the beam to position a in the figure? (Ignore friction.)

(b) How much work must be done to raise the beam to position b?

9. A spring has a stiffness of 20 lb/in. If 60 ft-lb of work are done in compressing the spring, how much is the spring shortened?

10. A 100-lb force stretches a coil spring by 6 in. The elastic limit is not reached.

(a) How much work is done in stretching the spring the first 3 in.?

(b) How much work is done in stretching the spring the next 3 in.?

11. An 800-lb flywheel is rotating at 300 rev/min. The radius of gyration is 2 ft.

(a) How much work must be done to bring this rotating wheel to rest?

(b) What torque is required to stop the wheel in 100 rev?

12. A 4000-lb automobile starts from rest and coasts down a hill for a distance of 500 ft along the roadway. The vertical height of the hill is 100 ft. The automobile is traveling at 50 ft/sec when it reaches the bottom of the hill.

(a) How much work was done by the frictional forces as the automobile coasted down the hill?

(b) What was the average force of friction acting on the automobile?

13. The object in Fig. 12-16 moves a distance of 10 m from a to b. The three forces A, B, and C in the figure are the only forces that do work upon the object.

(a) How much work is done by each force?

(b) Does the kinetic energy increase or decrease during the displacement? Calculate the change in the kinetic energy.

14. The 2-kg block in Fig. 12-17 is released from rest at A, and slides down the frictionless ramp to B. The block then slides along the horizontal surface. The force of friction on the block causes it to slow up and come to rest at C after traveling a distance of 10 m from B.

(a) How much kinetic energy does the block possess at B?

(b) Calculate the average force of friction acting upon the block as it slides from B to C.

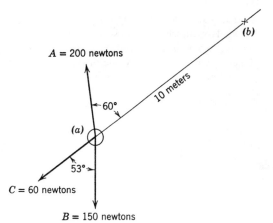

Figure 12-16 The net work done equals the increase in kinetic energy.

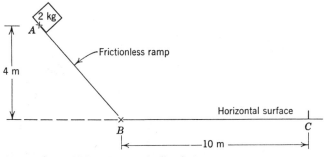

Figure 12-17 Sliding down an inclined ramp.

15. The hollow cylinder in Fig. 12-18 has all its mass concentrated on the rim so that the radius of gyration is equal to r. The cylinder is released from rest at A and it rolls without slipping from A to B.

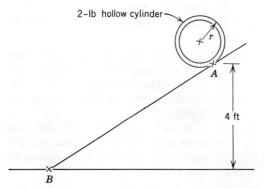

2-lb hollow cylinder

4 ft

A

B

Figure 12-18   Rolling down an incline without slipping.

(a) What is the total kinetic energy of the rolling cylinder when it reaches B?
(b) Calculate the velocity of the cylinder when it reaches B.

16. In order to lift the 200-lb weight in Fig. 12-9, a motor is used to apply a torque of 400 lb-ft to rotate the 160-lb drum. The weight is initially at rest; it starts to rise when the torque is applied. Assume that friction is negligible. After 10 rev determine:
   (a) The work done by the motor.
   (b) The increase in potential energy of the weight.
   (c) The increase in the kinetic energy of the drum and of the weight.
   (d) The velocity of the weight.

17. A space vehicle has a mass of 500 kg. Rocket motors lift this vehicle from the surface of the moon to a height of 200 m above the moon. Calculate the increase in potential energy using the fact that $g = 1.67$ m/sec^2 near the moon.

18. A space capsule weighs 1000 lb on the surface of the earth. Calculate the potential energy of this capsule (in foot-pounds) if it is carried 6000 miles from the surface of the earth.

19. An astronaut and his equipment have a mass of 200 kg. How much work must be done by a rocket to carry this mass away from the earth on an interplanetary voyage? (*Note*: At the surface of the earth $g_s = 9.8$ m/sec^2 and $r_s = 6400$ km.)

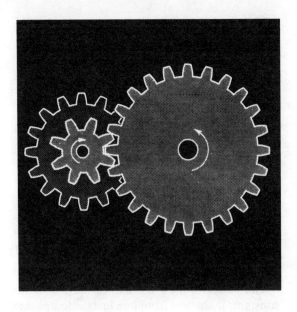

13.
Power, Efficiency, and Simple Machines

In the study of machinery and mechanical engineering the concept of *power* is very widely used. In this chapter we shall study the definition of power and its relationship to work and energy. We shall also use the ideas of *efficiency, mechanical advantage,* and *conservation of energy* in analyzing some important problems in the study of machines and mechanical devices.

LEARNING GOALS

The major goals of this chapter are

1. *The calculation of power in watts and horsepower.*

2. *The use of efficiency in calculating power input and output.*

3. *The calculation of the ideal mechanical advantage for some basic mechanical devices.*

4. *Understanding the relationship between actual and ideal mechanical advantage.*

5. *The use of the work–energy principle and conservation of energy principles in problems involving machines.*

13-1 THE UNITS FOR SIMPLE MACHINES

A small hand-operated mechanical device or a huge motor-driven machine can both do the same amount of work. It is obvious, however, that the large machine will do work at a much faster rate. The large machine is said to have a greater *power output*.

Power is defined as the time rate of doing work. If work W is done in time t, the average power during the time interval is

$$P_{av} = \frac{W}{t} \tag{13-1}$$

The power in most mechanical devices is not constant; it changes from one instant to another. The power output at any instant of time may be found by considering a short time interval Δt during which the rate at which work is done cannot vary appreciably. Then, if ΔW is the work done in the very short time interval Δt, the instantaneous power is

$$P = \frac{\Delta W}{\Delta t} \tag{13-2}$$

The instantaneous power is the rate at which work is done at any instant of time. The units used in measuring power depend on the units used in measuring work and time.

Example 13-1 A 4-ton beam is lifted at the constant velocity of 5 ft/sec by a crane. What is the power output of the crane?

Solution. A force of 4 tons = 8000 lb acts through a displacement of 5 ft in 1 sec. The work done in 1 sec is

$$W = 8000 \text{ lb} \times 5 \text{ ft} = 40{,}000 \text{ ft-lb}$$

Then

$$P = \frac{W}{t} = \frac{40{,}000 \text{ ft-lb}}{1 \text{ sec}} = 40{,}000 \text{ ft-lb/sec} \blacktriangleleft$$

In large machines, the unit used to measure power is frequently the horsepower:

$$1 \text{ horsepower (hp)} = 550 \text{ ft-lb/sec}$$

The power output of this crane may, therefore, also be written as

$$P = 40{,}000 \text{ ft-lb/sec} \times \frac{1 \text{ hp}}{550 \text{ ft-lb/sec}} = 72.7 \text{ hp}$$

The power unit of

$$1 \text{ horsepower} = 550 \text{ ft-lb/sec}$$
$$= 33{,}000 \text{ ft-lb/min}$$

was first established by James Watt during the eighteenth century in using the steam engine. However, in modern engineering and scientific enterprises other units for measuring power are commonly used. In the cgs metric system, work is measured in ergs and power in ergs per second (ergs/sec). In the mks metric system, work is measured in joules and the power unit is joules per second (joules/sec) or *watts*. A power of 1 watt means that 1 joule of work is done in a time of 1 sec.

The conversion factor between watts and horsepower can be found by using the relationship (see Section 12-1) of 1 joule = 0.737 ft-lb. Then

$$1 \text{ watt} = 1 \text{ joule/sec}$$
$$= 0.737 \text{ ft-lb/sec}$$
$$= 0.737 \text{ ft-lb/sec} \times \frac{1 \text{ hp}}{550 \text{ ft-lb/sec}}$$
$$= \frac{0.737}{550} \text{ hp}$$

and

$$1 \text{ hp} = \frac{550}{0.737} \text{ watts}$$

or

$$1 \text{ hp} = 746 \text{ watts}$$

The watt is too small a unit in many engineering applications, so that power is frequently measured in *kilowatts*.

$$1 \text{ kilowatt (kW)} = 1000 \text{ watts}$$

and

$$1 \text{ hp} = 0.746 \text{ kW}$$

13-2 WORK AND POWER

Solving Eq. 13-1 for the work we have

$$W = P_{av} \times t \tag{13-3}$$

According to this equation, a power unit multiplied by a time unit gives a work unit. Because of this fact, work is sometimes measured in watt-seconds (power in watts × time in seconds). It should be obvious that 1 joule = 1 watt-sec. For measuring large amounts of work or energy, the kilowatt-hour (kW-hr) is frequently used.

Example 13-2   A conveyor belt lifts 20 metric tons of ore 100 m up a mine shaft in 30 min. (a) What was the average power output of this machine? (b) How much work was done in kilowatt-hours? (*Note*: 1 metric ton = 1000 kilograms.)

Solution
(a) The work done was equal to the increase in potential energy.

$$W = mgh = 20 \text{ tons} \times \frac{10^3 \text{ kg}}{1 \text{ ton}}$$

$$\times 9.8 \text{ m/sec}^2 \times 10^2 \text{ m}$$

$$= 19.6 \times 10^6 \text{ kg-m}^2/\text{sec}^2$$

$$= 19.6 \times 10^6 \text{ joules}$$

$$P_{av} = \frac{W}{t} = \frac{19.6 \times 10^6 \text{ joules}}{30 \text{ min}} \times \frac{1 \text{ min}}{60 \text{ sec}}$$

$$= \frac{19.6}{18} \times 10^4 \text{ joules/sec} = 1.09 \times 10^4 \text{ watts}$$

$$= 10.9 \text{ kW}$$

(b) $W = P_{av} \times t = 10.9 \text{ kW} \times \frac{1}{2} \text{ hr}$

$$= 5.45 \text{ kW-hr} \qquad \blacktriangleleft$$

A very useful formula for calculating power may be obtained from Eq. 13-2 if we recall that the work done in a short time interval is $\Delta W = F \times \Delta s$. Then Eq. 13-2 gives $P = \Delta W/\Delta t = F \times \Delta s/\Delta t$. Since $\Delta s/\Delta t$ is the velocity v, the

instantaneous power is

$$P = F \times v \tag{13-4}$$

In studying rotating machinery, it is convenient to take the work done in a short time as $\Delta W = M \, \Delta\theta$. If this is substituted into Eq. 13-2, we find that

$$P = \frac{\Delta W}{\Delta t} = M \frac{\Delta\theta}{\Delta t}$$

But $\Delta\theta/\Delta t$ is the angular velocity ω. Therefore, the instantaneous power in rotating machinery is given by

$$P = M\omega \tag{13-5}$$

In using Eq. 13-5, the angular velocity must be in radians per unit time.

Example 13-3   According to the manufacturer's data, the gasoline engine of an automobile delivers 200 lb-ft of torque at 3000 rev/min. (a) Calculate the power output in horsepower. (b) How much work in kilowatt-hours will be done if this engine operates at the rated torque and speed for 2 hr?

Solution
(a) $\omega = 3000 \text{ rev/min} \times 2\pi \text{ rad/1 rev}$

$$= 6000\pi \text{ rad/min}$$

$$P = M\omega = 200 \text{ lb-ft} \times 6000\pi \text{ rad/min}$$

$$= 12\pi \times 10^5 \frac{\text{ft-lb}}{\text{min}} \times \frac{1 \text{ hp}}{33,000 \text{ ft-lb/min}}$$

$$= 114 \text{ hp}$$

(b) $P = 114 \text{ hp} \times \dfrac{0.746 \text{ kW}}{1 \text{ hp}} = 85.0 \text{ kW}$

$$W = P_{av} \times t = 85.0 \text{ kW} \times 2 \text{ hr}$$

$$= 170 \text{ kW-hr} \qquad \blacktriangleleft$$

13-3 WORK OUTPUT AND EFFICIENCY

In all mechanical operations the work or energy supplied to a machine is greater than the useful work output of the machine. Some of the energy input is wasted because of the work that has to be done against frictional forces in the mechanism. Consequently, the work output is always less than the work input. And the power output of a machine is less than the power input.

On many occasions would-be inventors have released proposals for perpetual motion machines that are to perform useful work without any work or energy input. In some of these contrivances work or energy is to be supplied to the device, but the work or energy output, according to the inventors, will be greater than the input. All of these perpetual motion schemes have failed to operate, and most educated people now recognize that these proposals run counter to the law of conservation of energy. When this law is applied to machines, we obtain the following principle:

work input (or energy input)

$$= \text{heat due to friction}$$

$$+ \text{work output (useful work)}$$

If friction is reduced to a minimum, the work output may be almost as large as the work input. The ratio of useful work or work output to the work input is the *efficiency* (eff.). According to this definition we have

$$\text{efficiency} = \frac{\text{work output}}{\text{work input}}$$

or

$$\text{eff.} = \frac{W_o}{W_i} \qquad (13\text{-}6)$$

The efficiency given by Eq. 13-6 is a fraction or a decimal. In many cases it is customary to change the decimal into percent and express the efficiency as a percentage. If practically no work is wasted, the work output is nearly equal to the work input and the efficiency approaches 100%. But in most machines and mechanical devices, the efficiency is much less. In a typical gasoline engine only about one-quarter of the energy released from the fuel is converted into useful work; thus the efficiency of these engines is only about 25%.

Since the power of a machine is equal to the work per unit time, Eq. 13-6 may also be written in terms of power to give

$$\text{efficiency} = \frac{\text{power output}}{\text{power input}}$$

$$\text{eff.} = \frac{P_o}{P_i} \qquad (13\text{-}7)$$

Example 13-4   When a 115 hp motor operates a crane, a 4-ton beam can be raised at the rate of 5 ft/sec. What is the efficiency of the crane?

Solution.   From Eq. 13-4 we calculate the power output:

$$P_o = F \times v = 8000 \text{ lb} \times 5 \text{ ft/sec}$$

$$= 40,000 \text{ ft-lb/sec} \times \frac{1 \text{ hp}}{550 \text{ ft-lb/sec}}$$

$$= 72.7 \text{ hp}$$

$$\text{eff.} = \frac{P_o}{P_i} = \frac{72.7 \text{ hp}}{115 \text{ hp}} = 0.632$$

$$= 63.2\% \qquad ◀$$

13-4 SIMPLE MACHINES AND MECHANICAL ADVANTAGE

No machine can ever create or increase energy. However, machines are sometimes extremely valuable when they convert energy from one form into another. A generator converts mechanical work into electrical energy; an electric motor uses electrical energy to produce mechanical energy; and a diesel engine converts the chemical energy of the fuel into mechanical energy. Although these machines waste a lot of energy, the usefulness of these energy transformations is quite obvious. Machines also produce a great benefit when they enable us to do work by means of a force that has the proper magnitude, direction, and location.

In a simple machine the input work is done by a single applied force, and the machine does work by means of a single output force. The lever, the wedge, the wheel and axle, the pulley, and the screw are examples of simple machines. Complex machines are made of combinations of these simple mechanical devices.

In simple machines the output force F_o is ordinarily not equal to the input force F_i. By using a simple lever, the output force may be made many times greater than the input force. This ratio of output force to input force is called the *actual mechanical advantage* (AMA) of the device.

$$\text{actual mechanical advantage} = \frac{\text{output force}}{\text{input force}}$$

$$\text{AMA} = \frac{F_o}{F_i} \qquad (13\text{-}8)$$

Example 13-5   A 10-lb force is applied to a lever with an actual mechanical advantage of 15. How great a weight can be lifted by this force?

Solution.   The output force is

$$F_o = \text{AMA} \times F_i = 15 \times 10 \text{ lb} = 150 \text{ lb} \blacktriangleleft$$

13-5 THE IDEAL MECHANICAL ADVANTAGE AND EFFICIENCY

Although the output force may be much greater than the input force, a machine cannot make the output work W_o greater than the input work W_i. If s_o is the distance moved by the output force, the output work is $W_o = F_o \times s_o$. And if s_i is the distance moved by the input force, the input work is $W_i = F_i \times s_i$. According to the definition of efficiency,

$$\text{eff.} = \frac{F_o \times s_o}{F_i \times s_i} \qquad (13\text{-}9)$$

If the numerator and denominator in Eq. 13-9 are divided by $s_o \times F_i$, we obtain

$$\text{eff.} = \frac{(F_o \times s_o)/(s_o \times F_i)}{(F_i \times s_i)/(s_o \times F_i)} = \frac{F_o/F_i}{s_i/s_o}$$

or $\qquad \text{eff.} = \frac{\text{AMA}}{s_i/s_o} \qquad (13\text{-}10)$

According to Eq. 13-10, the ratio of input distance to output distance, s_i/s_o, will be equal to

the AMA if the efficiency is 100%. Since efficiencies of 100% are never achieved, the efficiency is a fraction, and the denominator in Eq. 13-10 is greater than the AMA. The ratio in the denominator of Eq. 13-10 is called the ideal mechanical advantage.

$$\text{ideal mechanical advantage} = \frac{\text{input distance}}{\text{output distance}}$$

$$\text{IMA} = \frac{s_i}{s_o} \qquad (13\text{-}11)$$

The actual mechanical advantage (AMA) is always less than the IMA. However, if we could build an "ideal" machine (one with 100% efficiency), then the two mechanical advantages, the actual and the ideal, would be equal. The efficiency of any machine may be calculated in terms of these mechanical advantages if we substitute Eq. 13-11 into Eq. 13-10. This gives

$$\text{eff.} = \frac{\text{AMA}}{\text{IMA}} \qquad (13\text{-}12)$$

The ratio of distances s_i/s_o is equal to the ratio of input and output velocities in any simple machine:

$$\frac{\text{input distance}}{\text{output distance}} = \frac{\text{input velocity}}{\text{output velocity}}$$

$$\frac{s_i}{s_o} = \frac{v_i}{v_o}$$

Because of this fact the IMA is sometimes called the velocity ratio, which is given by the useful formula

$$\text{IMA} = \frac{\text{input velocity}}{\text{output velocity}}$$

$$\text{IMA} = \frac{v_i}{v_o} \qquad (13\text{-}13)$$

Example 13-6   In an automobile bumperjack a 2-ft stroke on the handle causes the jack mechanism to move up one ratchet ($\frac{1}{4}$ in.). It is found that a 40-lb push on the jack handle will lift one wheel on a 4000-lb car. (Assume that 1000 lb of the weight of the car rests on this wheel.) Calculate the (a) IMA, (b) AMA, and (c) efficiency.

Solution

$$\text{IMA} = \frac{s_i}{s_o} = \frac{2 \text{ ft}}{\frac{1}{4} \text{ in.}} \times \frac{12 \text{ in.}}{1 \text{ ft}}$$

$$= 96$$

(b) $\text{AMA} = \dfrac{F_o}{F_i} = \dfrac{1000 \text{ lb}}{40 \text{ lb}} = 25$

(c) eff. $= \dfrac{\text{AMA}}{\text{IMA}} = \dfrac{25}{96} = 26\%$ ◀

13-6 THE IDEAL MECHANICAL ADVANTAGE OF SOME SIMPLE MACHINES

The actual mechanical advantage of a machine cannot be predicted with great accuracy before the machine is constructed and actually placed into operation. However, the actual MA must be less than the ideal MA, and the ideal MA (IMA) can usually be calculated in a rather simple way from the machine design. In Fig. 13-1, for instance, we can calculate the IMA of the lever directly from the diagram. It is obvious from this figure

that we have a proportionality given by

$$\frac{s_i}{s_o} = \frac{\text{effort arm}}{\text{resistance arm}}$$

Therefore, the ideal MA for a lever is given by

$$\text{IMA (lever)} = \frac{\text{effort arm}}{\text{resistance arm}} \qquad (13\text{-}14)$$

The inclined plane or ramp of Fig. 13-2 is another simple machine for which the IMA is easily determined. The load moves up a distance h, while the effort (input force) acts through a

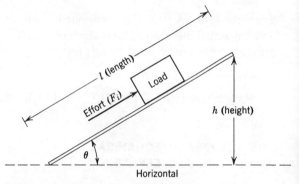

Figure 13-2   The ideal mechanical advantage of an inclined plane.

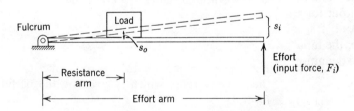

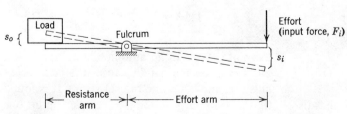

Figure 13-1   The ideal mechanical advantage of a lever.

distance l. Thus we have $s_o = h$ and $s_i = l$. Consequently, the IMA is given by

$$\text{IMA} = \frac{\text{length}}{\text{height}}$$

$$\text{IMA (inclined plane)} = \frac{l}{h} \qquad (13\text{-}15)$$

Since $\sin \theta = h/l$, we can also write

$$\text{IMA (inclined plane)} = \frac{1}{\sin \theta} \qquad (13\text{-}16)$$

The wheel and axle in Fig. 13-3 is a simple machine in which friction can be made low and efficiency quite high. During one revolution, the effort acts through a distance of $s_i = 2\pi R$ and the load travels through a distance of $s_o = 2\pi r$. The IMA is the ratio of the distances, and we find that

$$\text{IMA (wheel and axle)} = \frac{R}{r} \qquad (13\text{-}17)$$

The wheel and axle may also be regarded as a lever with the fulcrum at O. If we look at Fig. 13-3b in this way, it is apparent that the effort arm is R and the resistance arm is r. Therefore, the ideal MA must be R/r as in Eq. 13-17.

The wheel and axle arrangement is sometimes used to increase speed or distance rather than the force. This can be done by applying the effort or input force to the small radius (the axle). The mechanical advantage will be less than one if this is done. The output force exerted on the load will be smaller than the input force, but the load will move with greater speed.

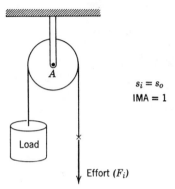

$$s_i = s_o$$
$$\text{IMA} = 1$$

Figure 13-4 The single fixed pulley.

The single fixed pulley in Fig. 13-4 may be considered as a wheel and axle in which $r = R$ and IMA $= 1$. We may also view this single fixed pulley as a lever with the fulcrum at A and with the effort arm equal to the resistance arm. This clearly indicates that the IMA of a single fixed pulley is one. This fact is also obvious from the figure, which indicates that the input distance is equal to the output distance.

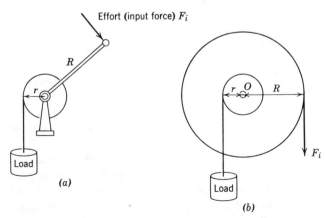

Figure 13-3 The wheel and axle.

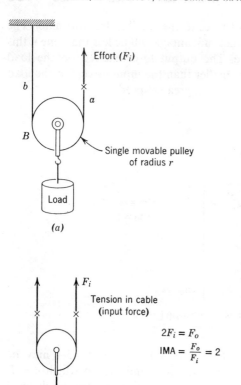

(a)

(b)

Figure 13-5   (a) The single movable pulley. (b) Free body diagram ignoring friction and the weight of the block.

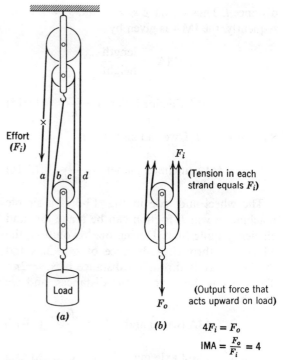

(a)

(b)

$$4F_i = F_o$$
$$IMA = \frac{F_o}{F_i} = 4$$

Figure 13-6   (a) A block and tackle composed of two fixed and two movable pulleys. IMA = 4. (b) Free body diagram ignoring friction and the weight of the block.

The single movable pulley in Fig. 13-5 may be viewed as a lever with the fulcrum at B. This would make the effort arm equal to $2r$ and the resistance arm equal to r for an IMA of two. This may be verified by noticing that the two supporting ropes (rope a and rope b) must each be shortened by 1 ft when the load moves up 1 ft. The effort must pull out 2 ft of rope in order to lift the load a distance of 1 ft. Thus the input distance is twice the output distance in a single movable pulley and the IMA is equal to two. The free body diagram in Fig. 13-5b uses the methods of statics to indicate this important result.

When fixed and movable pulleys are combined as in Fig. 13-6, the device is called a block and tackle. For the block and tackle illustrated in Fig. 13-6, there are four strands of rope supporting the load. It is simple to modify this design of the block and tackle, however, by adding or removing pulley wheels, and the number of supporting strands may be any number greater than two.

If the load in Fig. 13-6 is pulled up a distance of 1 ft, each of the supporting strands a, b, c, and d must be shortened by 1 ft. The effort (input force) must act to pull out 4 ft of rope when the output force acts through a distance of 1 ft. The IMA is, therefore, equal to four. This fact may be verified by means of the free body diagram in Fig. 13-6b.

In Fig. 13-7 the rope is threaded around the two fixed and two movable pulleys so as to produce five supporting strands. Now a, b, c, d, and e must each be shortened by 1 ft. The effort must pull out 5 ft of rope to move the load up a distance

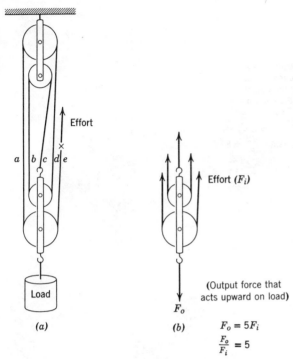

Effort

$a \mid b \mid c \mid d \mid e$

Effort (F_i)

(Output force that acts upward on load)

F_o

Load

(a)

(b)

$F_o = 5F_i$

$\dfrac{F_o}{F_i} = 5$

Figure 13-7 (a) A block and tackle composed of two fixed and two movable pulleys. IMA = 5. (b) Free body diagram ignoring friction and the weight of the block.

of 1 ft; the IMA for this block and tackle is five. This may be verified by means of the free body diagram in Fig. 13-7b.

In general, the IMA for a block and tackle is equal to the number of strands supporting the load. However, in Section 13-7 we shall study some pulley systems in which this method cannot be used to determine the ideal mechanical advantage.

The simple machine illustrated in Fig. 13-8 is a device called a jackscrew. It is easily seen from the figure or by looking at a common screw that the threads represent an inclined plane that is wrapped around a cylinder. When the input force or effort acts through one revolution (a distance of $2\pi R$), the load moves up a distance p. The distance p is equal to the separation between adjacent threads, and this distance is called the pitch of the screw. Since $s_i = 2\pi R$ for $s_o = p$, the IMA is

$$\text{IMA (jackscrew)} = \frac{2\pi R}{p} \qquad (13\text{-}18)$$

Since the pitch is usually a small distance and R is a rather large one, the IMA of a screw can be made very great. Because of friction, the AMA is much less than the IMA. However, friction in the screw also serves a useful purpose in making the screw "self-locking" so that the load will be supported by friction even when the input force F_i is discontinued.

13-7 COMBINATIONS OF SIMPLE MACHINES

The pulley system in Fig. 13-9 is not a simple block and tackle. The load is supported by three different ropes, and the IMA cannot be determined by counting the number of supporting strands. However, the IMA can be determined by examining the diagram. If the load is L and if we ignore friction and the weight of the pulley, then the tension in rope 1 is $L/2$. The tension in rope 2 is $L/4$ and the tension in rope 3 is $L/8$. Consequently, an effort or input force of $L/8$ will

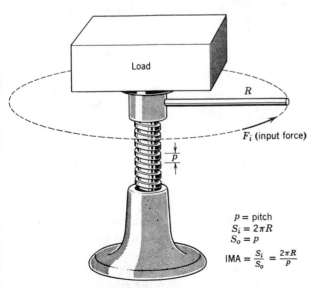

Load

R

F_i (input force)

p

p = pitch
$S_i = 2\pi R$
$S_o = p$

$\text{IMA} = \dfrac{S_i}{S_o} = \dfrac{2\pi R}{p}$

Figure 13-8 The jackscrew.

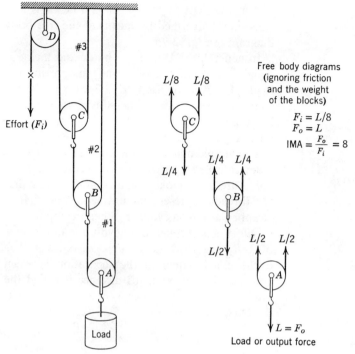

Figure 13-9   A pulley system is a compound machine.

be required to lift or support a load L (if we ignore the weight of the blocks and the effect of friction). This indicates that the IMA of the pulley system is eight.

The IMA of this pulley system can also be determined by noting that the pulley system is a compound machine consisting of four simple machines: one fixed and three movable pulleys. The simple machines are connected so that the output force of the first machine becomes the input force of the second machine; the output force of the second machine becomes the input force of the third machine, etc. When machines are connected in this way, the overall or total MA for the compound machine may be calculated by multiplying together the MA's of the individual machines. The overall mechanical advantage is found by the formula

$$MA = (MA)_1 \times (MA)_2 \times (MA)_3 \times \cdots \quad (13\text{-}19)$$

Example 13-7   Calculate the IMA of the pulley system in Fig. 13-9 by using Eq. 13-19.

Solution

For fixed pulley D: $(IMA)_1 = 1$
For movable pulley C: $(IMA)_2 = 2$
For movable pulley B: $(IMA)_3 = 2$
For movable pulley A: $(IMA)_4 = 2$

The overall

$$\begin{aligned} IMA &= (IMA)_1 \times (IMA)_2 \times (IMA)_3 \times (IMA)_4 \\ &= 1 \times 2 \times 2 \times 2 = 8 \end{aligned} \quad \blacktriangleleft$$

In this last example, Eq. 13-19 was used to calculate the IMA of the compound machine. The same principle can be used to calculate the AMA of the compound machine if the AMAs of the simple machines are used. This procedure is illustrated in the following example.

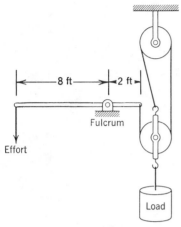

Figure 13-10 The mechanical advantage of a compound machine.

Example 13-8 In Fig. 13-10 the lever has an efficiency of 90% and the block and tackle an efficiency of 50%. Calculate (a) the IMA of the combination, and (b) the AMA of the combination.

Solution

(a) The IMA of the block and tackle is three (three supporting strands). The IMA of the lever is

$$\text{IMA} = \frac{\text{effort arm}}{\text{resistance arm}} = \frac{8 \text{ ft}}{2 \text{ ft}} = 4$$

The overall IMA is $3 \times 4 = 12$.

(b) The AMA of the pulley is equal to the eff. $\times$ IMA

$$\text{AMA} = 0.50(3) = 1.5 \qquad \text{for the pulley}$$

The same relationship is used to find the AMA for the lever.

$$\text{AMA} = 0.90(4) = 3.6 \qquad \text{for the lever}$$

The overall AMA $= (\text{AMA})_1 \times (\text{AMA})_2$
$$= (1.5) \times (3.6) = 5.4 \qquad \blacktriangleleft$$

An important example of a compound machine is the differential pulley or differential hoist in Fig. 13-11. This

machine has a large MA, and it is commonly used to lift automobile engine blocks and other heavy loads. This hoist consists of two wheels mounted on the same shaft, a single movable pulley, and an endless chain. The endless chain passes around the two wheels, as illustrated in Fig. 13-11, and a pull (input force, F_i) is applied to the chain at P.

When the larger wheel of the differential hoist completes one revolution, the input force acts through a displacement of $s_i = 2\pi R$, the circumference of the larger wheel. The rotation of the larger wheel through one complete revolution draws in a length of chain $2\pi R$ at A, and the rotation of the smaller wheel feeds in a length of chain $2\pi r$ at B. The loop of chain AB is thereby shortened in length by an amount $2\pi R - 2\pi r = 2\pi(R - r)$. Since one-half of this length is taken off each side of the loop AB, the load travels up a distance $s_o = \pi(R - r)$. Since the ideal mechanical advantage is equal to s_i/s_0, we have

$$\text{IMA} = \frac{2\pi R}{\pi(R - r)}$$

or $\qquad \text{IMA (differential hoist)} = \frac{2R}{R - r} \qquad (13\text{-}20)$

The result in Eq. 13-20 can be checked by examining the free body diagram in Fig. 13-11b. In this free body diagram, the forces of friction are assumed to be negligible. The hoist is operated with a negligible angular acceleration so that we can apply the laws of statics to the free body diagram. The resultant torque about the bearings at O is zero if there is no angular acceleration, and this gives

$$\sum M = 0$$

$$\left(\frac{L}{2}\right)R = \left(\frac{L}{2}\right)r + F_i R$$

$$\frac{L}{2}(R - r) = F_i R$$

$$\frac{L}{F_i} = \frac{2R}{R - r}$$

Since the forces of friction were assumed to be negligible, the ratio of L/F_i is the mechanical advantage in the absence of friction; but this mechanical advantage is the IMA of the device. We have thus verified Eq. 13-20 by an alternative method.

Equation 13-20 indicates that the IMA of a differential hoist can be very large, especially if r is made almost the same as R. But there is a good deal of friction in the

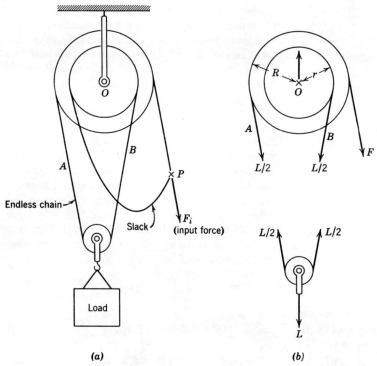

Figure 13-11 The differential pulley (differential hoist).

differential hoist, and this cuts down the efficiency and the actual mechanical advantage. This friction is of some advantage, however, because it enables the hoist to be "self-locking."

13-8 THE TRANSMISSION OF TORQUE

The simple machines and combinations of simple machines discussed in the preceding sections of this chapter are used to transmit and apply forces to loads. In most applications of machinery, however, work is done by means of torque and it is necessary to employ mechanisms for the transmission and application of torques. Transmission belts, gear systems, clutch mechanisms, and propeller shafts are common devices used to transmit and apply torques. In analyzing these mechanisms, we utilize the definitions of work and power given in this chapter and in Chapter 12.

For the work in rotating machinery

work input = torque input
$\qquad$ × input angular displacement
$\qquad$ (in radians)
$$W_i = M_i \times \theta_i \qquad (13\text{-}21)$$

work output = torque output
$\qquad$ × output angular displacement
$\qquad$ (in radians)
$$W_o = M_o \times \theta_o \qquad (13\text{-}22)$$

For the power in rotating machinery we can extend Eq. 13-5 to give

power input = torque input
$\qquad$ × input angular velocity
$\qquad$ (rad/time)
$$P_i = M_i \times \omega_i \qquad (13\text{-}23)$$

and

power output = torque output

× output angular velocity

(rad/time)

$$P_o = M_o \times \omega_o \qquad (13\text{-}24)$$

The efficiency of a rotating mechanical device is calculated, as might be expected, by

$$\text{efficiency} = \frac{\text{work output}}{\text{work input}} = \frac{W_o}{W_i}$$

$$\text{eff.} = \frac{M_o \theta_o}{M_i \theta_i} \qquad (13\text{-}25)$$

It should be obvious that the efficiency can also be calculated by

$$\text{efficiency} = \frac{\text{power output}}{\text{power input}}$$

$$\text{eff.} = \frac{P_o}{P_i}$$

If the numerator and denominator in Eq. 13-25 are both divided by $M_i \times \theta_o$, we obtain another useful formula.

$$\text{eff.} = \frac{M_o/M_i}{\theta_i/\theta_o} \qquad (13\text{-}26)$$

The numerator in Eq. 13-26 is the ratio of torque output to torque input. This quantity is the actual mechanical advantage in rotating machinery:

$$\text{AMA (rotation)} = \frac{M_o}{M_i} \qquad (13\text{-}27)$$

The denominator in Eq. 13-26 is the ratio of input angular displacement to output angular displacement. This ratio is the ideal MA for rotating machinery. Since the angular displacement is directly proportional to the number of revolutions, we have

$$\text{IMA (rotation)} = \frac{\theta_i}{\theta_o}$$

$$= \frac{\text{input revolutions}}{\text{output revolutions}} \qquad (13\text{-}28)$$

Because the angular displacement is directly proportional to the angular velocity in continuously operating machines, we can rewrite Eq. 13-28 in the useful form:

$$\text{IMA (rotation)} = \frac{\omega_i}{\omega_o}$$

$$= \frac{\text{input angular velocity}}{\text{output angular velocity}} \qquad (13\text{-}29)$$

If the output angular velocity is less than the input angular velocity, Eq. 13-29 indicates that there will be an IMA greater than one. All transmission mechanisms that reduce or slow up the rate of rotation have an IMA greater than one. If friction is not too great, these devices produce an output torque that is greater than the input torque. On the other hand, some transmissions are designed to increase rotational speed, and they do this by decreasing the torque output. Some important applications of this principle are discussed in the next section.

13-9 POWER AND EFFICIENCY IN ROTATING MACHINERY

In order to measure the torque output of an engine, the dynamometer in Fig. 13-12 is loosely clamped to the engine shaft or to a drum mounted on the engine shaft. The two half rings of the dynamometer are lined on the inside with brake linings so that the rotating shaft delivers a torque to the dynamometer. This torque may be measured by determining the weight mg at A, which will just serve to keep the arm horizontal. The torque output of the engine is just equal to the balancing torque, mgL, produced by the weight at A. If the angular velocity of the rotating shaft is then measured by a tachometer, it is a simple task to calculate the power output (or brake horsepower) of the engine.

Example 13-9 When the shaft in Fig. 13-12 is rotating at 1000 rev/min, a 10-kg mass at A is just sufficient to keep the arm horizontal. The length of the arm $L = 120$ cm. Calculate the

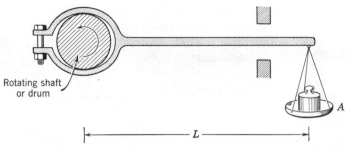

Rotating shaft or drum

A

L

Figure 13-12   A dynamometer for measuring the torque output of an engine.

power output of the engine in kilowatts and in horsepower.

Solution.   The torque output is

$$M_o = mgL$$
$$= (10 \text{ kg}) \times (9.8 \text{ m/sec}^2) \times (1.20 \text{ m})$$
$$= 1.175 \times 10^2 \text{ newton-m}$$

The output angular velocity is

$$\omega_o = 1000 \text{ rev/min} \times \frac{1 \text{ min}}{60 \text{ sec}} \times \frac{2\pi \text{ rad}}{1 \text{ rev}}$$

$$= \frac{\pi}{3} \times 10^2 \text{ rad/sec}$$

The power output is

$$P_o = M_o \times \omega_o$$

$$= 1.175 \times 10^2 \text{ newton-m} \times \frac{\pi}{3} \times 10^2 \text{ rad/sec}$$

$$= 12.3 \times 10^3 \text{ watts} = 12.3 \text{ kW}$$

Since 1 hp = 0.746 kW, we also have

$$P_o = 12.3 \text{ kW} \times \frac{1 \text{ hp}}{0.746 \text{ kW}} = 16.5 \text{ hp} \quad \blacktriangleleft$$

The transmissions illustrated in Fig. 13-13a and b are often used to transmit a torque in rotating machinery. If the fan belt in Fig. 13-13a does not slip, it produces the same effect as the chain and sprocket wheels in Fig. 13-13b. When the input torque M_i is applied to the driving wheel, the chain or belt delivers an output torque M_o to the driven wheel.

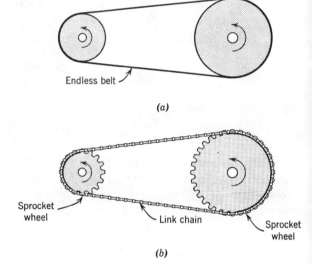

Endless belt

(a)

Sprocket wheel

Link chain

Sprocket wheel

(b)

Figure 13-13   Belt and chain drive transmissions. (a) Belt drive. (b) Chain drive.

When the chain or belt is traveling with a velocity v, and does not slip, the driving wheel has an angular velocity ω_i (input angular velocity) given by the formula $v = \omega_i \times r_i$. The chain or belt is also moving over the driven wheel with the same velocity v, and the output angular velocity may be calculated from the formula $v = \omega_o \times r_o$. It is obvious that

$$\omega_i \times r_i = \omega_o \times r_o$$

and

$$\frac{\omega_i}{\omega_o} = \frac{r_o}{r_i} \qquad (13\text{-}30)$$

But this ratio is just the IMA of the device. Therefore, the ideal mechanical advantage of a belt or chain drive is given by the formula

$$IMA = \frac{\text{radius of driven (output) wheel}}{\text{radius of driving (input) wheel}}$$

$$IMA = \frac{r_o}{r_i} \tag{13-31}$$

Since the ratio of radii is equal to the ratio of diameters, the ideal mechanical advantage may also be calculated from

$$IMA = \frac{\text{diameter of driven (output) wheel}}{\text{diameter of input (driving) wheel}}$$

Since the diameter of sprocket wheels is directly proportional to the number of teeth in the wheel, we have another useful formula:

$$IMA = \frac{\text{number of teeth in driven wheel}}{\text{number of teeth in driving wheel}}$$

Equation 13-30 indicates that the larger wheel in the pair turns more slowly and completes fewer revolutions. If the driving wheel is the larger wheel, and the driven wheel is the smaller one, this type of transmission can be used to increase the angular velocity. But the power output cannot be greater than the power input, and if the transmission is used to increase angular velocity, the torque output will be less than the torque input.

If the driving wheel is the smaller wheel, the output angular velocity will be less than the input angular velocity. This reduction in angular velocity will result in a larger torque output (provided that there is not too much friction).

Example 13-10 In Fig. 13-13a the driving wheel is 8 in. in diameter and the driven wheel is 20 in. in diameter. The power input comes from a 5-hp motor that causes the driving wheel to rotate at 300 rev/min. The efficiency of the transmission is 80%. Calculate the rev/min and the torque delivered to the driven wheel.

Solution. From Eq. 13-30

$$\omega_o = \frac{\omega_i r_i}{r_o} = 300 \text{ rev/min} \times \frac{4 \text{ in.}}{10 \text{ in.}}$$

$$= 120 \text{ rev/min}$$

The power output can be calculated from

$$\text{eff.} = \frac{P_o}{P_i}$$

$$P_o = \text{eff.} \times P_i = 0.80 \times 5 \text{ hp} = 4 \text{ hp}$$

Then

$$P_o = M_o \times \omega_o$$

and

$$P_o = 4 \text{ hp} \times \frac{550 \text{ ft-lb/sec}}{1 \text{ hp}} = 2200 \text{ ft-lb/sec}$$

$$\omega_o = 120 \text{ rev/min} \times \frac{1 \text{ min}}{60 \text{ sec}} \times \frac{2\pi \text{ rad}}{1 \text{ rev}}$$

$$= 4\pi \text{ rad/sec}$$

$$M_o = \frac{P_o}{\omega_o} = \frac{2200 \text{ ft-lb/sec}}{4\pi \text{ rad/sec}} = 175 \text{ lb-ft}$$

ALTERNATE METHOD

$$IMA = \frac{r_o}{r_i} = \frac{10 \text{ in.}}{4 \text{ in.}} = 2.5$$

$$\text{eff.} = \frac{AMA}{IMA}$$

$$AMA = \text{eff.} \times IMA = (0.80)(2.5) = 2.0$$

$$AMA = \frac{M_o}{M_i}$$

$$M_i = \frac{P_i}{\omega_i}$$

$$= \frac{5 \text{ hp} \times \dfrac{550 \dfrac{\text{ft-lb}}{\text{sec}}}{1 \text{ hp}}}{300 \text{ rev/min} \left[(1 \text{ min}/60 \text{ sec}) \times (2\pi \text{ rad}/1 \text{ rev}) \right]}$$

$$= \frac{5(550) \text{ ft-lb/sec}}{5(2\pi) \text{ rad/sec}} = \frac{550}{2\pi} \text{ lb-ft}$$

$$M_o = AMA \times M_i = 2.0 \times \frac{550}{2\pi} \text{ lb-ft} = 175 \text{ lb-ft} \blacktriangleleft$$

The ideal mechanical advantage would be the same if two sprocket wheels meshed directly with each other instead of with the chain. If the sprocket wheels are modified in this way, we obtain the pair of spur gears in Fig. 13-14. The ideal

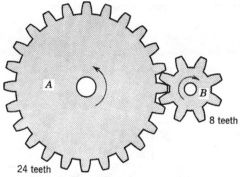

Figure 13-14   Spur gears.

24 teeth

8 teeth

mechanical advantage of the gears would be three if the driving gear were the one with 8 teeth and the driven gear the one with 24 teeth. On the other hand, if the driven gear were the smaller one, the ideal mechanical advantage would be $\frac{1}{3}$ and the gears would produce an increase in angular velocity.

Example 13-11   In Fig. 13-14 a 10-kW electric motor is connected directly to shaft A. This shaft is rotated at 180 rev/min by the motor. The efficiency of the spur gears is 60%. Calculate the power, torque, and rev/min imparted to shaft B.

Solution.   The driven gear has $\frac{1}{3}$ as many teeth and consequently rotates 3 times as fast as the driving gear. The output angular velocity is, therefore, $\omega_o = 3 \times 180$ rev/min $= 540$ rev/min.

$$\text{eff.} = \frac{P_o}{P_i}$$

$$P_o = \text{eff.} \times P_i = (0.60)(10 \text{ kW}) = 6 \text{ kW}$$

The output torque is

$$M_o = \frac{P_o}{\omega_o}$$

$$\omega_o = 540 \text{ rev/min} \times (2\pi \text{ rad/1 rev})$$
$$\times (1 \text{ min/60 sec})$$
$$= 18\pi \text{ rad/sec}$$

$$M_o = 6 \text{ kW}/(18\pi \text{ rad/sec}) \times 10^3 \text{ watts/1 kW}$$

$$= \frac{60}{18\pi} \times 10^2 \text{ newton-m} = 106 \text{ newton-m} \blacktriangleleft$$

Sometimes, two or more pairs of spur gears may be combined to form a gear train, as in Fig. 13-15. The overall mechanical advantage of this compound machine is found by multiplying the mechanical advantage for each pair of spur gears. This principle is illustrated in the following example.

Example 13-12   In the gear train of Fig. 13-15 friction causes a power loss of 30% in each pair of gears. A torque of 100 lb-ft at 300 rev/min is applied to shaft A. (a) What is the overall efficiency of this gear train? (b) What is the angular velocity of C? (c) Calculate the power output and torque output delivered to C.

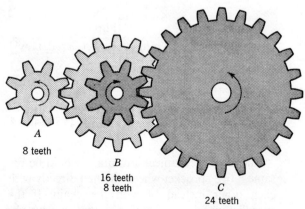

A

8 teeth

B

16 teeth
8 teeth

C

24 teeth

Figure 13-15   A gear train.

Solution

(a) The efficiency of the first pair is 70% and the efficiency of the second pair is also 70%; the overall efficiency $= 0.70 \times 0.70 = 0.49 = 49\%$.

(b) IMA of the first pair is $\frac{16}{8} = 2$; IMA of the second pair is $\frac{24}{8} = 3$

$$\text{Overall IMA} = 2 \times 3 = 6$$

$$\text{IMA} = \frac{\omega_i}{\omega_o}$$

$$\omega_o = \frac{\omega_i}{\text{IMA}} = \frac{300 \text{ rev/min}}{6} = 50 \text{ rev/min}$$

(c) AMA = IMA × eff.

$$= 6 \times 0.49 = 2.94$$

$$\text{AMA} = \frac{M_o}{M_i}$$

$$M_o = M_i \times \text{AMA} = 100 \text{ lb-ft} \times 2.94$$
$$= 294 \text{ lb-ft}$$

To find the power output,

$$P_o = M_o \times \omega_o$$
$$\omega_o = 50 \text{ rev/min} \times 1 \text{ min}/60 \text{ sec}$$
$$\times 2\pi \text{ rad}/1 \text{ rev}$$
$$= \tfrac{5}{3}\pi \text{ rad/sec}$$
$$P_o = 294 \text{ lb-ft} \times \tfrac{5}{3}\pi \text{ rad/sec}$$
$$= 1540 \text{ ft-lb/sec}$$

An alternate procedure for (c) would be to calculate $P_o = $ eff. $\times P_i$, where $P_i = M_i \times \omega_i$. Then the output torque could be found from $M_o = P_o/\omega_o$. The student should do this calculation as a check on the preceding method. ◄

In addition to the spur gears illustrated in Figs. 13-14 and 13-15, many other types of gears are used in the design of machinery and transmission mechanisms. Worm gears, helical gears, and bevel gears are common gear types with many technical applications. The reader will find descriptions of these devices in engineering handbooks and in textbooks on machinery design. The same general principles apply to all these varieties of gears.

The power output is always less than the power input, but the gears may be used to increase torque (with a reduction in angular velocity) or the gears may be designed so as to increase angular velocity (with a reduction in torque).

SUMMARY

Power is the time rate of doing work. The average power in time t is $P_{av} = W/t$ in joules/sec = watts. Also 1000 watts = 1 kilowatt.

In the English Engineering system the power unit is the ft-lb/sec.

The instantaneous power is

$$P = \frac{\Delta W}{\Delta t} = Fv$$

For rotating machinery the instantaneous power is $P = M\omega$, where M is the torque and ω is the angular velocity in radians per unit time.

The work or energy input to a machine is greater than the work output because some work has to be done against frictional forces. The efficiency of a device is defined by

$$\text{eff.} = \frac{\text{work output}}{\text{work input}} = \frac{\text{power output}}{\text{power input}}$$

The efficiency is usually expressed as a percent. Because the output is less than the input, the efficiency is less than 100%.

The actual mechanical advantage (AMA) of a mechanical device is given by

$$\text{AMA} = \frac{\text{output force}}{\text{input force}}$$

The ideal mechanical advantage (IMA) of a device is given by

$$\text{IMA} = \frac{\text{input distance}}{\text{output distance}}$$

and the IMA would equal the AMA if the efficiency were 100%. The AMA is less than the IMA, and we have the efficiency given by the ratio

$$\text{eff.} = \frac{\text{AMA}}{\text{IMA}}$$

For many simple machines the ideal mechanical advantage may be found by finding the ratio of input distance to output distance as in Section 13-6. The mechanical advantage of combinations of simple machines may be calculated by multiplying the mechanical advantages of the individual machines as in Section 13-7.

For rotating machines

$$\text{AMA} = \frac{\text{torque output}}{\text{torque input}}$$

$$\text{IMA} = \frac{\text{input angular velocity}}{\text{output angular velocity}}$$

$$= \frac{\text{input revolutions}}{\text{output revolutions}}$$

The power output is always less than the power input; mechanisms that increase ω, the angular velocity, or the speed of rotation must also cause a reduction in torque.

QUESTIONS

1. In the formula $W = P_{av} \times t$, what is the meaning of P_{av}?

2. How many joules per second are equivalent to one kilowatt?

3. What are two different units for measuring power in the metric system? How are these units defined?

4. The kilowatt-hour is a unit for measuring work or energy. How should kilowatt-hours be converted into joules? How many joules are there in 1 kW-hr?

5. The horsepower-hour may be used to calculate the work done by a large machine. Explain how to calculate the work in horsepower-hours. How could this unit be converted to foot-pounds?

6. If the instantaneous power in watts is to be calculated from the formula $P = Fv$, what units must be used to measure F and v?

7. The instantaneous power for a rotating system may be calculated from the formula $P = M\omega$. What are the units for M and ω if P is to be in watts?

8. Explain what is meant by the efficiency of a mechanical device.

9. Why is the efficiency of a machine always less than 100%?

10. Explain what is meant by the ideal mechanical advantage (IMA) of a machine.

11. The work output for a pulley system is always less than the work input. How can we explain that the force output is greater than the force input?

12. How can the block and tackle in Fig. 13-6a be used to multiply speed instead of force? What would be the ideal mechanical advantage of this modified block and tackle?

13. A long pole is used as a lever. What should be done to assure that this lever has a large ideal mechanical advantage?

14. Define the actual mechanical advantage (AMA) of a mechanical device.

15. Friction in a given mechanical device may be reduced by more effective lubrication. What effect does this have on the IMA of the device? What effect does this have on the AMA? On the efficiency?

16. If a transmission is designed so that the output angular velocity is three times the input angular velocity, what is the ideal mechanical advantage?

17. Suppose that a mechanical linkage is used to increase speed or distance. How will the output force compare with the input force?

18. In a gear system a wheel with many teeth drives a wheel with few teeth. How does the angular velocity of the driving wheel compare with the angular velocity of the driven wheel? What can we conclude about the IMA of this device?

19. Why is it possible for a machine to have a large mechanical advantage but a low efficiency?

20. Explain why a torque converter, which increases the torque, causes a reduction in the angular velocity.

PROBLEMS

(A)

1. A 500-lb weight is lifted by a hoist at a steady speed of 10 ft/sec. What is the power output in horsepower?

2. A 100-kg mass is lifted to a height of 50 m in a time of 20 sec?
   (a) How much work in joules is done?

(b) What is the power output in watts?

3. An electric motor has an efficiency of 70%. The electric power input to the motor is 2000 watts.
  (a) What is the power output in watts?
  (b) What is the power output in horsepower?

4. In the jackscrew of Fig. 13-8 the lever arm length R is 3 ft and the pitch p is $\frac{1}{4}$ in. What is the ideal mechanical advantage of this machine?

5. A 4000-lb elevator is lifted by an electric motor to a height of 150 ft in 12 sec. Neglecting friction, what is the horsepower required to do this work?

6. In Fig. 13-16 a 40-lb pull is applied to the handle of the hammer.
  (a) What is the IMA of this simple machine?
  (b) What force acts upon the nail?

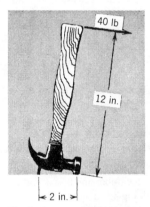

Figure 13-16 The hammer as a simple lever.

7. The coefficient of friction between a 500-lb block and a horizontal surface is 0.4. How much horsepower must be expended to move the block across the surface at the constant rate of 10 ft/sec?

8. An electric motor does 400 joules of work in 10 sec. What is the power output in watts?

9. A 100-lb force pushes an object 40 ft in 10 sec.
  (a) How much work is done?
  (b) What is the power output in ft-lb/sec and in horsepower?

10. A weight lifter can lift a 200-lb weight from the floor to a height of 6 ft in 2 sec.
  (a) How much work is done?
  (b) What is the average power output in ft-lb/sec and in horsepower?

11. A hoist lifts a 80-kg mass at the steady speed of 2 m/sec. Calculate the power output in watts.

12. The pulley wheel in Fig. 13-17 is 16 in. in diameter.
  (a) What is the torque in pound-feet, which is delivered to the pulley?
  (b) If the pulley is rotating at 300 rev/min, what is the power (in foot-pounds per second) delivered to the pulley?

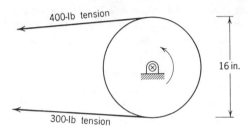

Figure 13-17 Torque applied to a pulley wheel.

13. An iron rod 6 ft long is used as a lever. The weight to be moved is located at one end of the rod and the effort is applied at the other end. Where should the fulcrum be located to give the lever an IMA of 10?

14. An electrically driven pulley system has a power output of 15 kW. It is connected to a cable that lifts a heavy beam at 0.5 m/sec.
  (a) What is the tension in the cable?
  (b) What is the mass of the beam?

15. When 25 kW are used to operate an electric motor, the motor does work at the rate of 12 kW. Calculate the efficiency of the motor.

16. Power is supplied at the rate of 20 hp to a machine that does work with an efficiency of 30%. Calculate the power output in
  (a) horsepower.
  (b) ft-lb/sec.

17. An automotive engine supplies 300 lb-ft of torque at 4000 rev/min. Calculate the power output in
  (a) ft-lb/sec.
  (b) horsepower.

18. A 20-kg mass is lifted to a height of 10 m in a time of 4 sec. Calculate:
  (a) The work output in joules.
  (b) The power output in watts.

19. A diesel engine that operates at 25% efficiency has a power output of 200 hp. Calculate:
    (a) The rate of doing work in ft-lb/min.
    (b) The energy supplied to the engine during each minute of operation.
    (c) The power output in kilowatts.

20. An electric drill draws 1200 watts from the power line. How much electrical energy in kilowatt-hours is consumed when the drill is used for 45 min?

21. Some measurements indicate that the air resistance on an experimental automobile is 500-lb when the automobile is traveling at 60 miles/hr. How much power in horsepower must be expended to overcome this air resistance?

22. A rotating gear mechanism supplies 200 newton-meters of torque when it is rotating at 40 rad/sec. Calculate the power output in watts and in kilowatts.

23. A 5-slug mass is accelerated from a velocity of 10 ft/sec to a velocity of 30 ft/sec in a time of 4 sec.
    (a) Calculate the increase in kinetic energy.
    (b) What is the average power expended in accelerating the mass?

(B)

1. A 4000-lb automobile is accelerated from rest to a speed of 60 miles/hr in 10 sec. Ignore friction and calculate:
   (a) The increase in kinetic energy.
   (b) The work done by the motor.
   (c) The horsepower required.

2. How much horsepower is required to drive a 4000-lb car up a 20% grade at a steady speed of 30 miles/hr? (Ignore friction.)

3. In the wheel and axle of Fig. 13-3, $R = 2$ ft and $r = 8$ in. It is found that a steady 40-lb force on the crank handle will lift a 100-lb load at a constant speed of 2 ft/sec.
   (a) What is the power output of the device?
   (b) What is the AMA?
   (c) What is the IMA?
   (d) Calculate the efficiency.

4. The pulley system of Fig. 13-6 is used to lift a 200-lb load. An 80-lb effort is required, and the weight is lifted 10 ft.
   (a) What is the output work?
   (b) What is the input work?
   (c) How much work is done against friction?

(d) Calculate the efficiency of the pulley system.

5. The manufacturer's data on an electric motor indicates that the power output is 1800 watts at 600 rev/min. What is the torque produced by the motor? Calculate the torque in newton-meters and in pound-feet.

6. When the dynamometer arm in Fig. 13-12 is 6 ft long, a 100-lb weight at A just serves to balance the arm in a horizontal position. The motor shaft is revolving at 900 rev/min.
   (a) What is the torque output of the motor?
   (b) What is the brake horsepower of the motor?

7. A pump lifts 200 gal of water per minute to a height of 100 ft. The efficiency of this pump is 60% and 1 gal of water weighs 8.34 lb.
   (a) What is the power output of the pump?
   (b) What is the horsepower of the motor required to drive this pump?

8. The differential pulley of Fig. 13-11 has $R = 25$ cm and $r = 23$ cm. A 100-kg mass is lifted up a distance of 150 cm in 50 sec. The efficiency of the device is 45%.
   (a) What is the work output in joules?
   (b) What is the power output in watts?
   (c) What is the IMA?
   (d) What is the AMA?
   (e) What input force is needed to lift the weight and overcome friction?

9. A mountain climber and his equipment weigh 200 lb. He is walking up a 40% grade with a velocity of 5 ft/sec. Ignoring friction, calculate the power output in
   (a) foot-pounds per second.
   (b) horsepower.

10. A 50-hp motor drives a conveyor belt that lifts coal out of a barge to a height of 40 ft. If the efficiency of the belt is 40%, how many tons of coal can be lifted in 1 hr?

11. A shaft rotating at 600 rev/min delivers a torque of 200 lb-ft to a gear transmission. The transmission reduces the rotational speed to 200 rev/min and has an efficiency of 60%.
    (a) Calculate the torque output.
    (b) Calculate the power output in horsepower.

12. An inclined plane rises 6 ft, and it is 10 ft long along the incline. A 100-lb weight is pushed from the bottom to the top of the inclined plane and the coefficient of friction is 0.2.

(a) How large a force is required to push the weight up the incline?
(b) What is the work input?
(c) What is the work output?
(d) What is the IMA of the incline?
(e) What is the AMA of the incline?
(f) Calculate the efficiency.

13. A 30-lb input force is applied to the handle of an automobile jack. During one stroke of the jack handle, the input force acts for a distance of 2 ft, and a 3000-lb load is lifted up one-eight of an inch.
    (a) What is the AMA of this jack?
    (b) What is the IMA of this jack?
    (c) What is the output work during one stroke?
    (d) How much work is done against friction during one stroke?
    (e) What is the efficiency of the jack?

14. A turbine generator is driven by water power. The water falls down a height of 100 ft to the turbine blades; 50 ft^3 of water passes through the turbine each second. (Water weighs 62.5 lb/ft^3.)
    (a) What is the power input to the turbine in ft-lb/sec?
    (b) If the turbine converts 60% of the power input into electrical power, calculate the electrical power output in kilowatts.

15. In the pulley system of Fig. 13-18, a 20-lb pull at P is needed to lift the weight of block B and overcome friction. A 300-lb load is then suspended from block B and raised at the rate of $\frac{1}{2}$ ft/sec.
    (a) Calculate the rate (in horsepower) at which the input force must do work.
    (b) What is the efficiency of the system?

16. In the belt transmission of Fig. 13-13 assume that friction in the transmission wastes 20% of the input

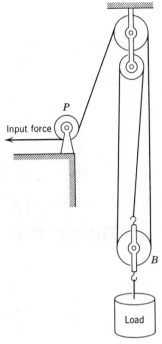

Figure 13-18 The power output and power input of a pulley system.

energy. The driving wheel has a diameter of 4 in. and the driven wheel has a diameter of 12 in. A torque of 60 lb-ft at 300 rev/min is applied to the driving wheel.
    (a) What is the number of revolutions per minute for the driven wheel?
    (b) What is the torque output of the transmission?
    (c) What is the power input to the driving wheel?
    (d) What is the power output of the transmission?

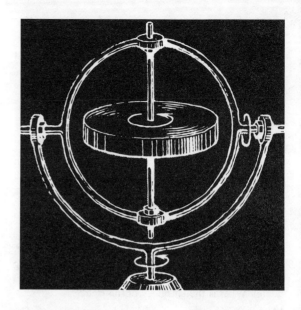

14.
Impulse and Momentum

Many problems in physics and engineering may be solved by applying the *impulse–momentum principle*. In this chapter we shall study the meaning of both impulse and momentum and the *conservation of momentum principle*.

LEARNING GOALS

The major objectives of this chapter are as follows:

1. *Calculating the impulse of a force.*
2. *Using the impulse–momentum principle to calculate unknown forces or the changes in velocity.*
3. *Using the conservation of momentum principle in the study of inelastic collisions.*
4. *Using the conservation of energy and the conservation of momentum to study completely elastic collisions.*
5. *Applying the conservation of angular momentum principle to some problems in rotational motion.*

14-1 THE IMPULSE OF A FORCE

In Chapter 12 we learned that the product of force times displacement represents a very important quantity, the work done by the force. It is also the case that the product of force and time is of great physical significance; this quantity is called the *impulse*, or the *linear impulse*. When a force F acts on an object for a time t, the linear impulse of the force equals the force multiplied by the time. Written as a formula, we find that the definition for linear impulse is

$$\mathbf{J} = \mathbf{F} \times t \qquad (14\text{-}1)$$

The boldface symbols in Eq. 14-1 indicate that the linear impulse J is a vector quantity that has the same direction as the force. If there are several impulses, the net or total impulse is found by the methods of vector addition.

In calculating the impulse when the force F is not a constant force, we must find the average value of F during the time t and then substitute this value into Eq. 14-1. Sometimes, it is convenient to draw a force versus time (F–t) diagram. The area under the F–t curve is equal to the average force multiplied by the time. The area under the F–t curve is, therefore, equal to the impulse of the force.

The units for measuring impulse are very simple to remember:

impulse unit = force unit × time unit

The most useful impulse units are the pound-second (lb-sec), the newton-second (N-sec), and the dyne-second (dyne-sec).

14-2 IMPULSE AND MOMENTUM

The impulse produced by the resultant force during a short time interval Δt is $\mathbf{F}_R \times \Delta t$. According to Newton's second law the resultant force is equal to the time rate of change in momentum or $\mathbf{F}_R = \Delta(m v)/\Delta t$. The impulse of the resultant force is,

therefore,

$$\mathbf{F}_R \times \Delta t = \frac{\Delta(m v)}{\Delta t} \times \Delta t = \Delta(m v)$$

$$= \text{change in linear momentum} \qquad (14\text{-}2)$$

Equation 14-2 expresses a very important principle of mechanics that is called the impulse–momentum principle: The linear impulse acting upon an object is equal to the change in linear momentum or

$$\mathbf{J} = m\mathbf{v}_2 - m\mathbf{v}_1 \qquad (14\text{-}3)$$

impulse = final momentum
— original momentum
(vector difference)

Equation 14-3 may also be written

$$m\mathbf{v}_1 + \mathbf{J} = m\mathbf{v}_2 \qquad (14\text{-}4)$$

original momentum + impulse = final momentum

Equation 14-3 should be compared with the kinetic energy principle of Chapter 12. There we noted that the net work done upon an object equals the increase in kinetic energy. Equation 14-3 asserts that the net impulse upon an object equals the increase in momentum.

Example 14-1   A 100-g block is sliding toward the east at 30 m/sec. What impulse has to be applied to make the block move toward the north at 40 m/sec?

Solution

$$m v_1 = (0.1 \text{ kg})(30 \text{ m/sec})$$
$$= 3 \text{ kg-m/sec toward east}$$
$$m v_2 = (0.1 \text{ kg})(40 \text{ m/sec})$$
$$= 4 \text{ kg-m/sec toward north}$$

Setting up the vector addition (Fig. 14-1) $m\mathbf{v}_1 + \mathbf{J} = m\mathbf{v}_2$, we easily find that the magnitude of J is

$$J = 5 \text{ kg-m/sec} = 5 \text{ kg-m/sec} \times (\text{sec/sec})$$
$$= 5(\text{kg-m/sec}^2) \times \text{sec} = 5 \text{ newton-sec}$$

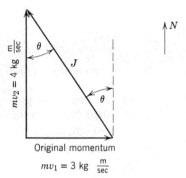

Figure 14-1 Impulse equals change in momentum.

The direction of J is given by $\theta = \text{arc tan } \frac{3}{4} = 37°$. The direction is $37°$ west of north. ◀

With a little imagination the impulse–momentum principle can be extended to a system or group of objects. In Fig. 14-2 we have indicated a system of only two objects that are undergoing a collision. The mass m_A has a velocity $\mathbf{V}_{A1}$ before the collision and velocity $\mathbf{V}_{A2}$ after the collision.

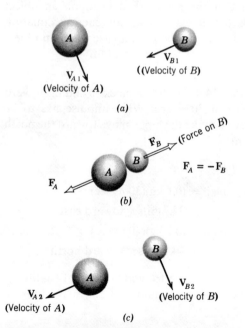

Figure 14-2 Impulse and conservation of momentum during a collision. (a) Before collision; momentum = $m_A\mathbf{V}_{A_1} + m_B\mathbf{V}_{B_1}$. (b) Impulse during collision. (c) After collison; momentum = $m_A\mathbf{V}_{A_2} + m_B\mathbf{V}_{B_2}$.

The mass m_B has a velocity $\mathbf{V}_{B1}$ before the collision and a velocity $\mathbf{V}_{B2}$ after the collision. According to Eq. 14-2 the impulse on object A is

$$\mathbf{F}_A \times t = m_A\mathbf{V}_{A2} - m_A\mathbf{V}_{A1}$$

The impulse on B is

$$\mathbf{F}_B \times t = m_B\mathbf{V}_{B2} - m_B\mathbf{V}_{B1}$$

Adding these last two equations gives

$$(\mathbf{F}_A + \mathbf{F}_B) \times t = m_A\mathbf{V}_{A2} + m_B\mathbf{V}_{B2} \\ - m_A\mathbf{V}_{A1} - m_B\mathbf{V}_{B1} \quad (14\text{-}5)$$

According to Newton's third law of motion $\mathbf{F}_A$ is equal in magnitude to $\mathbf{F}_B$, but the two forces are opposite in direction:

$$\mathbf{F}_A = -\mathbf{F}_B \quad \text{and} \quad \mathbf{F}_A + \mathbf{F}_B = 0$$

This means that Eq. 14-5 can be written

$$m_A\mathbf{V}_{A2} + m_B\mathbf{V}_{B2} = m_A\mathbf{V}_{A1} + m_B\mathbf{V}_{B1} \quad (14\text{-}6)$$

momentum after collision

= momentum before collision

Equation 14-6 is a very important principle called the law of conservation of momentum. This equation applies when two objects collide or exert forces upon each other and no outside forces produce an impulse upon the objects during the collision. The scheme of Eq. 14-6 can be extended to a system of any number of objects if there is no external impulse acting upon the objects. The law of conservation of momentum asserts: The linear momentum of a system of objects remains constant as long as no external impulse acts upon the system.

Example 14-2 A 160-lb man throws a 16-lb bowling ball in a horizontal direction. The ball is thrown at 50 ft/sec. (a) What was the impulse in lb-sec given to the ball? (b) With what velocity would the man move backward if he were on roller skates?

Solution
(a) The mass of the ball is

$$m_B = \frac{16 \text{ lb}}{32 \text{ ft/sec}^2} = \frac{1}{2} \text{ lb-sec}^2/\text{ft} = \frac{1}{2} \text{ slug}$$

The impulse on the ball is

$$J_B = m_B V_{B2} - m_B V_{B1}$$

Since the ball is initially at rest, V_{B1} is zero. Then

$$J_B = m_B V_{B2} = \tfrac{1}{2} \text{ lb-sec}^2/\text{ft} (50 \text{ ft/sec}) = 25 \text{ lb-sec}$$

(b) The impulse applied to the man was

$$J_A = -J_B = -25 \text{ lb-sec}$$

The mass of the man is

$$m_A = \frac{160 \text{ lb}}{32 \text{ ft/sec}^2} = 5 \frac{\text{lb-sec}^2}{\text{ft}} = 5 \text{ slugs}$$

Since the man was initially at rest $V_{A1} = 0$ and

$$J_A = -J_B = m_A V_{A2} - m_A V_{A1}$$

$$-25 \text{ lb-sec} = 5 \frac{\text{lb-sec}^2}{\text{ft}} V_{A2}$$

$$V_{A2} = -5 \text{ ft/sec (in the direction opposite to the ball)}$$

ALTERNATE METHOD. Because the man is on frictionless roller skates, no external impulse is acting and the conservation of momentum may be used:

$$m_A V_{A1} + m_B V_{B1} = m_A V_{A2} + m_B V_{B2}$$

$$0 + 0 = (160 \text{ lb})(V_{A2}) + 16 \text{ lb} (50 \text{ ft/sec})$$

$$V_{A2} = -\frac{16 \text{ lb}}{160 \text{ lb}} (50 \text{ ft/sec}) = -5 \text{ ft/sec} \quad \blacktriangleleft$$

In the conservation of momentum equation (Eq. 14-6) any combination of units may be used, provided only that all the velocities are measured in the same units and that all the masses are given in the same units. However, in the impulse equation (Eq. 14-3) it is necessary to use a consistent unit system. The three systems that will prove most useful are indicated in Table 14-1.

14-3 COMPLETELY INELASTIC COLLISIONS

After a completely inelastic collision the two colliding objects stick together and move with a common velocity. A completely inelastic collision will occur if the two colliding objects have soft or sticky surfaces and adhere to each other after the collision. A blob of clay produces a completely inelastic collision between the metal sphere and the metal cylinder in Fig. 14-3. When a soft ball of putty is dropped, it undergoes a completely inelastic collision with the ground. A serious automobile accident sometimes produces a completely inelastic collision.

When the rifle in Fig. 14-4 is fired, the speeding bullet becomes embedded in the block after the collision. Because the two colliding objects move as a single unit after the impact, this situation is an

Table 14-1

System	Impulse ($F \times t$)	Momentum (mv)
English		
Engineering	lb-sec	slug-ft/sec
mks (metric)	newton-seconds (N-sec)	kilogram-meters second (kg-m/sec)
cgs (metric)	dyne-seconds (dyne-sec)	gram-centimeters second (gm-cm/sec)

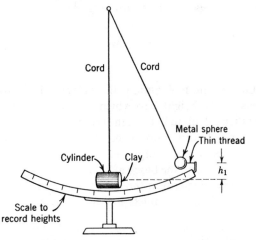

Figure 14-3 Laboratory apparatus for demonstrating conservation of momentum in inelastic collisions. When the thin thread is burned, the metal sphere swings down along a circular arc and makes a completely inelastic collision with the cylinder.

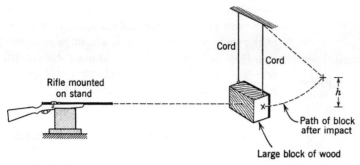

Cord

Cord

Rifle mounted
on stand

h

Path of block
after impact

Large block of wood

Figure 14-4 The ballistic pendulum.

example of a completely inelastic collision. The suspended block is a simple form of the ballistic pendulum, and this apparatus can be used to determine the velocity of the rifle bullet. If the mass of the bullet is m and the velocity v, then the original momentum is mv. After the collision the block (mass M) and the bullet move with a common final velocity. The final momentum is $(m + M)V_f$. Since there is no external impulse, the conservation of momentum gives

$$mv = (m + M)V_f \qquad (14\text{-}7)$$

The final velocity V_f is much less than the bullet velocity v. If V_f can be measured, Eq. 14-7 provides a simple method of determining the velocity of the bullet:

$$v = \frac{m + M}{m} V_f \qquad (14\text{-}8)$$

One method of finding the velocity V_f is to measure the height h to which the block swings and then calculate V_f from the law of conservation of energy. During impact the bullet loses a great deal of its kinetic energy, but the kinetic energy of the bullet and the block after the impact equals the potential energy at the top of the swing. Since the moving mass (bullet plus block) is $m + M$, we have

loss in kinetic energy = gain in potential energy

$$\tfrac{1}{2}(m + M)V_f^2 = (m + M)gh$$

and

$$V_f = \sqrt{2gh} \qquad (14\text{-}9)$$

Substituting Eq. 14-9 into Eq. 14-8 gives

$$v = \frac{m + M}{m} \sqrt{2gh} \qquad (14\text{-}10)$$

Example 14-3 A 20-g bullet is fired into a 5-kg wooden block, and the block swings up a height of 20 cm. (a) Determine the velocity of the bullet. (b) How much heat was produced during the impact?

Solution
(a) $M = 5000$ g, $h = 20$ cm $= 0.2$ m. Substituting in Eq. 4-10

$$v = \frac{20 \text{ g} + 5000 \text{ g}}{20 \text{ g}} \sqrt{2\left(9.8 \,\frac{\text{m}}{\text{sec}^2}\right)(0.2 \text{ m})}$$

$$= \tfrac{502}{2} \sqrt{3.93 \text{ m}^2/\text{sec}^2} = 497 \text{ m/sec}$$

(b) $V_f = \sqrt{2gh} = \sqrt{2(9.8 \text{ m/sec}^2)(0.2 \text{ m})}$

$$= 1.98 \text{ m/sec}$$

The kinetic energy after impact is

$$K_f = \tfrac{1}{2}(m + M)V_f^2 = \tfrac{1}{2}(5.020 \text{ kg})(1.98 \text{ m/sec})^2$$
$$= 9.85(\text{kg-m}^2/\text{sec}^2)$$
$$= 9.85 \text{ N-m} = 9.85 \text{ joules}$$

The kinetic energy before the impact is

$$K_o = \tfrac{1}{2}mv^2 = \tfrac{1}{2}(0.020 \text{ kg})(497 \text{ m/sec})^2$$
$$= 2470 \text{ joules}$$

The conservation of energy gives

$$K_o = K_f + Q$$

where Q is the heat produced during the impact. Then

$$Q = K_o - K_f = 2470 \text{ joules} - 9.85 \text{ joules}$$
$$= 2460 \text{ joules} \quad \blacktriangleleft$$

It is apparent from the last example that much of the original kinetic energy was converted into heat during the impact. Kinetic energy is always lost during inelastic collisions.

In some inelastic collisions the colliding objects are moving in opposite directions before the collision. This introduces a slight complication because one velocity vector must be positive and the other negative. This point is illustrated in the following example.

Example 14-4 A 100-g object moving toward the right at 50 cm/sec collides with a 200-g object moving toward the left at 300 cm/sec. If the two objects stick together, determine the velocity.

Solution

$$m_A = 100 \text{ g} \quad v_A = 50 \text{ cm/sec}$$
$$m_B = 200 \text{ g} \quad v_B = -30 \text{ cm/sec (the left is in the minus direction)}$$

The conservation of momentum for a completely inelastic collision gives

$$m_A v_A + m_B v_B = (m_A + m_B)V_F$$
$$100 \text{ g } (50 \text{ cm/sec}) + 200 \text{ g } (-30 \text{ cm/sec})$$
$$= (300 \text{ g})V_f$$

Dividing by 100 g gives

$$50 \text{ cm/sec} - 60 \text{ cm/sec} = 3V_f$$
$$V_f = -10 \text{ cm/3 sec} = -3.33 \text{ cm/sec}$$

The final velocity is 3.33 cm/sec toward the left. ◀

If the colliding objects are traveling in different directions, the conservation of momentum (Eq. 14-6) must be treated as a vector sum. Trigonometry and vector methods must be employed in calculating the resultant momentum. This procedure is followed in the next example.

Example 14-5 A 4-ton truck traveling toward the east at 30 miles/hr collides with a 3000-lb automobile traveling toward the north at 60 miles/hr. If the two vehicles undergo a completely inelastic collision, determine the final velocity.

Solution

$$m_A \mathbf{v}_{A1} + m_B \mathbf{v}_{B1} = (m_A + m_B)\mathbf{V}_F \text{ (vector sum)}$$
$$m_A v_{A1} = 4 \text{ tons } (30 \text{ miles/hr})$$
$$= 120 \text{ ton-miles/hr (toward the east)}$$
$$m_B v_{B1} = (1.5 \text{ tons})(60 \text{ miles/hr})$$
$$= 90 \text{ ton-miles/hr (toward the north)}$$

The vector sum (resultant momentum) may be easily determined as in the vector triangle of Fig. 14-5:

$$150 \text{ ton-miles/hr} = (m_A + m_B)V_f$$
$$= (5.5 \text{ tons})V_f$$
$$V_f = \frac{150}{5.5} \text{ miles/hr} = 27.3 \text{ miles/hr} \quad \blacktriangleleft$$

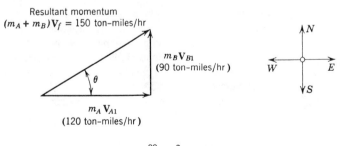

Resultant momentum
$(m_A + m_B)V_f = 150$ ton-miles/hr

$m_B V_{B1}$
(90 ton-miles/hr)

$m_A V_{A1}$
(120 ton-miles/hr)

$$\tan \theta = \frac{90}{120} = \frac{3}{4} \qquad \theta = 36.9°$$

Figure 14-5 Conservation of momentum in an inelastic collision.

14-4 PERFECTLY ELASTIC COLLISIONS

High speed photographs indicate that colliding objects undergo a considerable deformation at the moment of impact (see Fig. 14-6). Work must be done to produce this deformation; usually some of this work is dissipated into heat energy. However, except during completely inelastic collisions, some of the work done during the deformation is converted into elastic potential energy. As the elastic objects return to their original dimensions, they rebound from each other, and the elastic potential energy is converted into kinetic energy.

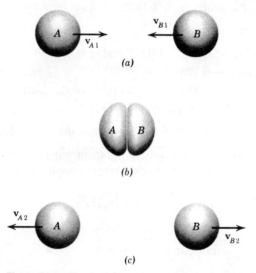

Figure 14-6 Elastic deformation during elastic collisions. (a) Before impact. (b) During impact. (c) After impact.

In completely elastic collisions, the original kinetic energy of the objects is converted into elastic potential energy during the brief time of impact. Upon rebounding from the impact, the potential energy is immediately reconverted into kinetic energy. It is therefore characteristic of a completely (or perfectly) elastic collision that the kinetic energy of the objects before collision equals the kinetic energy after the collision. Any kinetic

energy that is lost by one object will be transferred to the other, and the total kinetic energy will be unchanged in a completely elastic collision.

According to modern theories of heat and matter, colliding objects are composed of tremendous numbers of molecules, and each of these molecules is moving very rapidly. Heat energy is the kinetic energy of the moving molecules. During the impact between objects some kinetic energy is transferred to the molecules; this amount of kinetic energy is lost during the collision. This kinetic energy is dissipated into heat energy, and some kinetic energy is expected to be lost during the collision.

All objects of ordinary size are composed of great numbers of molecules, and completely elastic collisions never occur. In atomic and nuclear physics, however, the colliding particles are composed of atoms or parts of atoms, and completely elastic collisions are rather common.

The collisions between very elastic objects are almost perfectly elastic in some cases; the amount of kinetic energy dissipated into heat is comparatively small. In these cases it is permissible to assume that the collisions are perfectly elastic.

In analyzing completely elastic collisions, we employ two principles of mechanics:

(1) conservation of momentum

$$m_A v_{A1} + m_B v_{B1} = m_A v_{A2} + m_B v_{B2}$$

(2) conservation of kinetic energy

$$\tfrac{1}{2} m_A v_{A1}{}^2 + \tfrac{1}{2} m_B v_{B1}{}^2 = \tfrac{1}{2} m_A v_{A2}{}^2 + \tfrac{1}{2} m_B v_{B2}{}^2$$

Principle 1 is applicable to all collisions (elastic or nonelastic) if there is no external impulse acting upon the objects during the collision. Principle 2, the conservation of kinetic energy, is applicable if there are no external forces doing work upon the objects during the collision and no kinetic energy is dissipated into heat. The conservation of kinetic energy is only applicable for collisions that are perfectly elastic, and this conservation of kinetic energy is used as a definition of perfectly (or completely) elastic collisions.

It is useful to rewrite these equations by transposing the terms in m_A to one side and the terms in m_B to the other side of the equations. The conservation of kinetic energy may then be rewritten (after multiplying through by 2) as

$$m_A(v_{A1}{}^2 - v_{A2}{}^2) = -m_B(v_{B1}{}^2 - v_{B2}{}^2)$$

or

$$m_A(v_{A1} - v_{A2})(v_{A1} + v_{A2})$$
$$= -m_B(v_{B1} - v_{B2})(v_{B1} + v_{B2}) \qquad (14\text{-}11)$$

The conservation of momentum may be written as

$$m_A(v_{A1} - v_{A2}) = -m_B(v_{B1} - v_{B2}) \qquad (14\text{-}12)$$

Since equals divided by equals give equals, we may divide Eq. 14-11 by 14-12 to give

$$v_{A1} + v_{A2} = v_{B1} + v_{B2}$$

or

$$(v_{A2} - v_{B2}) = -(v_{A1} - v_{B1}) \qquad (14\text{-}13)$$

Equation 14-13 summarizes an important principle relating to perfectly elastic collisions. The quantity $v_{A2} - v_{B2}$ is the velocity with which object A is receding from object B after collision. This is the relative velocity of A with respect to B after collision. The quantity $v_{A1} - v_{B1}$ is the relative velocity before collision. According to Eq. 14-13 the relative velocity after an elastic collision is equal to minus the relative velocity before the collision. In other words, the speed with which the objects are receding from each other after the collision is equal to the speed with which they approached each other before they collided.

Problems involving elastic collisions can be solved by using the conservation of kinetic energy and the conservation of momentum. However, in most instances it is simpler to employ the relative velocity principle (Eq. 14-13) and the conservation of momentum.

• **Example 14-6** A 100-g object traveling toward the right at 60 cm/sec collides with a 200-g object traveling toward the left at 80 cm/sec. Calculate the velocity of both objects after collision, assuming the collision to be perfectly elastic.

Solution

$$m_A = 100 \text{ g} \qquad m_B = 200 \text{ g}$$
$$v_{A1} = 60 \text{ cm/sec} \qquad v_{B1} = -80 \text{ cm/sec}$$
$$m_A v_{A1} + m_B v_{B1} = m_A v_{A2} + m_B v_{B2}$$
$$(100 \text{ g})(60 \text{ cm/sec}) + (200 \text{ g})(-80 \text{ cm/sec})$$
$$= 100 \text{ g } v_{A2} + 200 \text{ g } v_{B2}$$

Dividing by 100 g gives

$$60 \text{ cm/sec} - 160 \text{ cm/sec} = v_{A2} + 2v_{B2}$$
$$- 100 \text{ cm/sec} = v_{A2} + 2v_{B2} \qquad (a)$$

Using the relative velocity equation, we obtain

$$-(v_{A1} - v_{B1}) = v_{A2} - v_{B2}$$
$$-60 \text{ cm/sec} + (-80 \text{ cm/sec}) = v_{A2} - v_{B2}$$
$$-140 \text{ cm/sec} = v_{A2} - v_{B2} \qquad (b)$$

Equations (a) and (b) to be solved as simultaneous equations for the two unknowns. Subtracting Eq. (b) from Eq. (a) gives

$$-100 \text{ cm/sec} + 140 \text{ cm/sec} = v_{A2} + 2v_{B2}$$
$$- (v_{A2} - v_{B2})$$
$$40 \text{ cm/sec} = 3v_{B2}$$
$$v_{B2} = 13.3 \text{ cm/sec}$$

The 200 g object rebounds to the right at 13.3 cm/sec. Substituting this result in Eq. (b) gives

$$-140 \text{ cm/sec} = v_{A2} - 13.3 \text{ cm/sec}$$
$$v_{A2} = -127.7 \text{ cm/sec}$$

The 100-g object rebounds to the left at 127.7 cm/sec. ◀

• 14-5 PARTIALLY ELASTIC COLLISIONS

Most collisions are neither completely inelastic nor perfectly elastic. The colliding objects rebound from each other after the collision, but some kinetic energy is dissipated during the impact. The methods of Sections 14-3 and 14-4 cannot be used for these partially elastic collisions. To analyze these collisions, two principles are employed.

1. Conservation of momentum.

2. The restitution equation.

Principle (1) is, of course, the same principle that is used in Sections 14-3 and 14-4. The restitution equation is given by the formula:

$$v_{A2} - v_{B2} = -e(v_{A1} - v_{B1}) \qquad (14\text{-}14)$$

Equation 14-14 is not an exact law of physics but a formula that gives approximately correct results. The quantity e is called the coefficient of restitution, and it depends on the elastic properties of the colliding objects. If $e = 1$, Eq. 14-14 reduces to the relative velocity equation for completely elastic collisions (Eq. 14-13). If $e = 0$, the two final velocities, v_{A2} and v_{B2}, are equal and the two objects stick together in a completely inelastic collision. The values of e may range from 1 (for a completely elastic collision) to 0 (for a completely inelastic collision). Values for e that are greater than zero and less than one indicate partially elastic collisions.

• **Example 14-7** A 10-lb block sliding to the right at 60 ft/sec collides with a 20-lb block sliding to the left at 40 ft/sec. Assume that $e = 0.8$ and determine (a) the final velocities for both blocks, (b) the impulse on the 20-lb block, and (c) the kinetic energy in foot-pounds lost during the collision.

Solution

(a) Conservation of momentum:

$$m_A v_{A1} + m_B v_{B1} = m_A v_{A2} + m_B v_{B2}$$

(10 lb)(60 ft/sec) + 20 lb(−40 ft/sec)

$$= 10 \text{ lb } v_{A2} + 20 \text{ lb } v_{B2} \qquad (a)$$

$$60 \text{ ft/sec} - 80 \text{ ft/sec} = v_{A2} + 2v_{B2}$$

$$- 20 \text{ ft/sec} = v_{A2} + 2v_{B2}$$

From the restitution equation, we obtain

$$-e(v_{A1} - v_{B1}) = v_{A2} - v_{B2}$$

$$-0.8[(60 \text{ ft/sec}) - (-40 \text{ ft/sec})] = v_{A2} - v_{B2} \qquad (b)$$

$$-0.8(100 \text{ ft/sec}) = v_{A2} - v_{B2}$$

Subtracting Eq. (b) from Eq. (a) gives

$$- 20 \text{ ft/sec} + 80 \text{ ft/sec} = (v_{A2} + 2v_{B2}) - (v_{A2} - v_{B2})$$

$$60 \text{ ft/sec} = 3v_{B2} \qquad v_{B2} = 20 \text{ ft/sec}$$

Substituting in Eq. (b) gives

$$- 80 \text{ ft/sec} = v_{A2} - 20 \text{ ft/sec}$$

$$v_{A2} = -60 \text{ ft/sec}$$

(b) impulse $= m_B(v_{B2} - v_{B1})$

$$= \frac{20 \text{ lb}}{32 \text{ ft/sec}^2} [20 \text{ ft/sec} - (-40 \text{ ft/sec})]$$

$$= \frac{20 \text{ lb}}{32 \text{ ft/sec}^2} (60 \text{ ft/sec}) = 37.5 \text{ lb-sec}$$

(c) The original kinetic energy is

$$K_1 = \tfrac{1}{2}m_A v_{A1}{}^2 + \tfrac{1}{2}m_B v_{B1}{}^2$$

$$K_1 = \frac{1}{2}\left(\frac{10 \text{ lb}}{32 \text{ ft/sec}^2}\right)(60 \text{ ft/sec})^2$$

$$+ \frac{1}{2}\left(\frac{20 \text{ lb}}{32 \text{ ft/sec}^2}\right)(40 \text{ ft/sec})^2$$

$$= \tfrac{5}{32}(3600) \text{ ft-lb} + \tfrac{10}{32}(1600) \text{ ft-lb}$$

$$= 562 \text{ ft-lb} + 500 \text{ ft-lb} = 1062 \text{ ft-lb}$$

$$K_2 = \tfrac{1}{2}m_A v_{A2}{}^2 + \tfrac{1}{2}m_B v_{B2}{}^2$$

$$= \frac{1}{2}\left(\frac{10 \text{ lb}}{32 \text{ ft/sec}^2}\right)(-60 \text{ ft/sec})^2$$

$$+ \frac{1}{2}\left(\frac{20 \text{ lb}}{32 \text{ ft/sec}^2}\right)(20 \text{ ft/sec})^2$$

$$= 562 \text{ ft-lb} + 125 \text{ ft-lb} = 687 \text{ ft-lb}$$

loss in KE $= K_1 - K_2$

$$= 1062 \text{ ft-lb} - 687 \text{ ft-lb}$$

$$= 375 \text{ ft-lb} \qquad ◀$$

If an object of mass m_A traveling with velocity v_{A1} collides with the ground or with a wall, we cannot use the conservation of momentum equation unless we take into account the momentum of the earth. Since it is not practical to do this, we usually resort to using the restitution equation with $v_{B1} = 0$ and $v_{B2} = 0$. This equation then becomes

$$v_{A2} = -e v_{A1} \qquad (14\text{-}15)$$

Equation 14-15 applies to the velocity component at right angles to the surface. The velocity parallel to the surface usually does not undergo a change during the collision.

• **Example 14-8** When a golf ball is dropped from a height of 100 cm, it rebounds from the ground to a height of 80 cm. What is the coefficient of restitution?

Solution

$$v_{A1} = \sqrt{2gh_1} \qquad h_1 = 100 \text{ cm}$$

The velocity of rebound is

$$v_{A2} = -\sqrt{2gh_2} \qquad h_2 = 80 \text{ cm}$$

(The minus sign is needed to designate the direction.)

$$e = \frac{-v_{A2}}{v_{A1}} = \frac{\sqrt{2gh_2}}{\sqrt{2gh_1}}$$

$$= \sqrt{\frac{h_2}{h_1}} = \sqrt{0.80} = 0.894 \qquad ◀$$

• 14-6 CONSERVATION OF MOMENTUM IN ROCKETS AND JETS

A rocket is pushed forward when exhaust gases are ejected rearward from the rocket's combustion chamber. The ejected gases usually have a large velocity, and they carry a considerable amount of momentum in the backward direction. Since the total momentum remains constant, the ejection of the hot gases from the combustion chamber must be accomplished by an increase in the forward momentum of the rocket.

If μ is the mass of rocket fuel ejected in one second, and v_e is the velocity of ejection, then μv_e is the momentum transferred into the exhaust during one second of flight. During this same one second interval, the forward momentum must be increased by the same amount. It is obvious that μv_e is the rate at which the forward momentum is increasing. But the rate of increase of momentum must be equal to the force according to Newton's second law of motion. Consequently, the force or thrust is given by the formula

$$\text{thrust} = \mu v_e \qquad (14\text{-}16)$$

• Example 14-9   A rocket that is launched vertically upward from the earth weighs 100,000 lb. Included in this weight is 80,000 lb of fuel. The fuel is burned at a constant rate, and all of it is consumed in 20 sec of flight. The velocity of ejection is 2000 ft/sec. (a) Calculate the thrust of the rocket. (b) Calculate the acceleration when half the fuel has been consumed.

Solution

(a) $\mu = \dfrac{\text{mass of fuel}}{\text{time}} = \dfrac{80{,}000 \text{ lb}/32 \text{ ft/sec}^2}{20 \text{ sec}}$

$= \dfrac{2500 \text{ slugs}}{20 \text{ sec}} = \dfrac{125 \text{ slugs}}{\text{sec}}$

Thrust $= \mu v_e = 125 \text{ slug/sec} \times 2000 \text{ ft/sec}$

$= 250{,}000 \text{ slug-ft/sec}^2 = 250{,}000 \text{ lb}$

(b) When half the fuel is consumed, the weight of the rocket is

$$100{,}000 \text{ lb} - 40{,}000 \text{ lb} = 60{,}000 \text{ lb}$$

The mass of the rocket is $\dfrac{60{,}000 \text{ lb}}{32 \text{ ft/sec}^2}$

The resultant force is $F_R = \text{thrust} - \text{weight}$

$= 250{,}000 - 60{,}000$

$= 190{,}000 \text{ lb}$

$a = \dfrac{F_R}{m} = \dfrac{190{,}000 \text{ lb}}{\dfrac{60{,}000 \text{ lb}}{32 \text{ ft/sec}^2}} = \dfrac{19}{6} \times 32 \text{ ft/sec}^2$

$= 101 \text{ ft/sec}^2$ ◀

A jet engine, like a rocket, produces its forward thrust by ejecting exhaust gases from its combustion chamber. A jet engine, however, draws in large quantities of air to burn the jet fuel. In computing the thrust of a jet, we must add the mass of this air to the mass of the jet fuel.

• Example 14-10   The exhaust of a model jet engine is ejected from the combustion chamber at 100 m/sec. During each second of operation 20 g of jet fuel is burned, and 100 g of air is used in the chemical reaction and ejected with the fuel. Compute the thrust of this model engine.

Solution

$$\mu = 120 \text{ g/sec} = 0.120 \text{ kg/sec}$$

$$\text{thrust} = \mu v_e = 0.120 \text{ kg/sec} \times 100 \text{ m/sec}$$

$$= 12 \text{ kg-m/sec}^2 = 12 \text{ newtons}$$ ◀

14-7 ANGULAR IMPULSE AND ANGULAR MOMENTUM

It should be obvious from the preceding sections of this chapter that the principles of linear impulse and linear momentum are needed for the solution of many problems in linear motion. It is also the case that the ideas of angular impulse and angular momentum are quite useful in the study of rotation. The design of rotating machinery and machine parts, flywheels, and gyroscopes involves applications of the principles of angular impulse and angular momentum.

When a torque M acts upon a rotating object for a time t, there is an angular impulse produced by the torque. Angular impulse is defined by the equation

$$A = M \times t \qquad (14\text{-}17)$$

Table 14-2

System	Torque	×	Time	=	Angular Impulse
English	pound-feet	×	seconds	=	pound-feet-sec (lb-ft-sec)
mks (metric)	newton-meter	×	seconds	=	N-m-sec
cgs (metric)	dyne-centimeter	×	seconds	=	dyne-cm-sec

The units for measuring angular impulse depend on the units used for measuring torque and time. Some useful units for measuring angular impulse are indicated in Table 14-2.

If the torque acting upon an object changes, the angular impulse can be calculated from Eq. 14-17 if we use the average torque that acts during the time t. The angular impulse can also be calculated from a torque–time diagram as in Fig. 14-7. During a short time interval Δt, the angular impulse is

$$\Delta A = M \times \Delta t \qquad (14\text{-}18)$$

But it is obvious from Fig. 14-7 that this is just the area of the little rectangle in the figure. Consequently, the angular impulse during a period of time t is equal to the area under the torque–time diagram.

According to the second law of motion for rotation, the torque M is equal to $I\alpha$. Since $\alpha = \Delta\omega/\Delta t$, we have $M = I\,\Delta\omega/\Delta t$; when this is substituted in Eq. 14-18, we find that

$$\Delta A = M\,\Delta t = \left(I\,\frac{\Delta\omega}{\Delta t}\right)\Delta t = I(\Delta\omega).$$

Here $\Delta\omega$ is the change in angular velocity during a short time interval Δt. During a time t the change in angular velocity is $\omega_f - \omega_0$, and the angular impulse is

$$A = I(\omega_f - \omega_0) \qquad (14\text{-}19)$$

Equation 14-19 represents a very useful principle of mechanics. This equation may be written in another form if we use the definition of angular momentum: the angular momentum of a rotating object is the product of the moment of inertia I and the angular velocity ω. Using this definition, the principle expressed by 14-19 is

$$A = (I\omega)_f - (I\omega)_0 \qquad (14\text{-}20)$$

angular impulse = final angular momentum

− original angular momentum

This is the impulse-momentum principle for rotation or the angular-momentum principle:

The angular impulse on a rotating object or a system of objects is equal to the change in angular momentum.

● **Example 14-11**   (a) Calculate the angular momentum of the rotating system in Fig. 14-8 if it is rotating about zero (0) at 300 rev/min. (b) What angular impulse must be applied to increase the angular velocity to 900

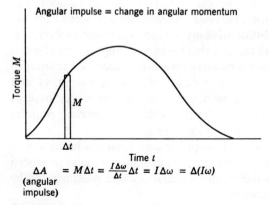

Angular impulse = change in angular momentum

$$\Delta A = M\Delta t = \frac{I\Delta\omega}{\Delta t}\Delta t = I\Delta\omega = \Delta(I\omega)$$
(angular impulse)

Figure 14-7   The torque-time diagram.

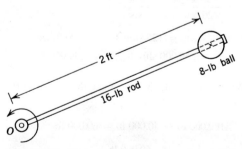

Figure 14-8   The angular momentum of a rotating system.

rev/min? (c) What is the average torque if the impulse is to last for $\frac{1}{4}$ sec?

Solution

(a) The moment of inertia of a rod pivoted at one end is given by

$$I = \tfrac{1}{3}mL^2 \qquad \text{(see Chapter 11)}$$

$$I_{\text{rod}} = \frac{1}{3}\left(\frac{16\text{ lb}}{32\text{ ft/sec}^2}\right)(2\text{ ft})^2$$

$$= \tfrac{2}{3}\text{ slug-ft}^2$$

For the 8-lb ball $I = mr^2 = \dfrac{8\text{ lb}}{32\text{ ft/sec}^2}(2\text{ ft})^2$

$$= 1\text{ slug-ft}^2$$

The total moment of inertia is $\frac{5}{3}$ slug-ft^2. The angular velocity should be in rad/sec if we want to get the torque in the units of pound-feet. Therefore,

$$\omega_0 = 300\text{ rev/min} \times \left(\frac{1\text{ min}}{60\text{ sec}}\right) \times \frac{2\pi\text{ rad}}{1\text{ rev}}$$

$$= 10\pi\text{ rad/sec}$$

The angular momentum is

$$I\omega_0 = \left(\frac{5}{3}\text{ slug-ft}^2\right)(10\pi\text{ rad/sec}) = \frac{50\pi}{3}\frac{\text{slug-ft}^2}{\text{sec}}$$

The "rad" is dropped out of the units because a radian is a ratio. Since slug $=$ lb/ft/sec^2, the last result could also be written as

$$I\omega_0 = \frac{50\pi}{3}\frac{(\text{lb-sec}^2)}{\text{ft}} \times \frac{\text{ft}^2}{\text{sec}} = 52.3\text{ lb-ft-sec}$$

(b) $\omega_f = 900\text{ rev/min} \times \left(\frac{1\text{ min}}{60\text{ sec}}\right) \times \frac{2\pi\text{ rad}}{1\text{ rev}}$

$$= 30\,\pi\text{ rad/sec}$$

$$I\omega_f = \left(\frac{5}{3}\text{ slug-ft}^2\right)(30\pi\text{ rad/sec}) = 50\pi\,\frac{\text{slug-ft}^2}{\text{sec}}$$

$$= 157\text{ lb-ft-sec}$$

$$A = I\omega_f - I\omega_0 = 157\text{ lb-ft-sec} - 52.3\text{ lb-ft-sec}$$

$$= 104.7\text{ lb-ft-sec}$$

(c) $A = M \times t, \qquad M = \dfrac{A}{t}$

$$M = \frac{104.7\text{ lb-ft-sec}}{\frac{1}{4}\text{ sec}} = 419\text{ lb-ft} \qquad \blacktriangleleft$$

In Example 14-11 the moment of inertia of the rotating system did not change. However, this is not always the case, and some of the most interesting applications of the angular momentum principle involve situations in which the moment of inertia is changing. The use of the angular momentum principle in these situations is discussed in Section 14-9.

• 14-8 ANGULAR IMPULSE AND THE CENTER OF PERCUSSION

An angular impulse is applied to the cam in Fig. 14-9 when the rocker arm moves to the right. If the rocker arm is positioned correctly, the cam will start to rotate without delivering any reaction force to the supporting axis. This will occur only if the force F on the cam is set so that the line of action passes through the center of percussion P.

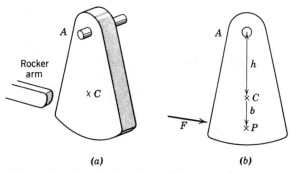

(a)   (b)

Figure 14-9 Angular impulse and the center of percussion.

If the force F acts above or below the center of percussion, a reaction force will act upon the supporting axle. This effect is especially noticeable if the rotating object is a baseball bat or a long pole and the supporting axle is a man's hands. Unless the baseball strikes the bat at the center of percussion, a painful stinging is felt in the hand of the batter. In order to avoid this effect when hammers are used to drive nails, hammers are designed so that the center of percussion is in the hammer head.

In Fig. 14-9 the point A is the axis or center of rotation and the point C designates the center of mass of the rotating object. Since the force F acts through the center of percussion, no accelerating force is exerted on the object by the axis at A. The torque about the center of rotation is $F(h + b)$ and the angular impulse during a short time Δt is $F(h + b)\Delta t$. Since the object starts from rest, the change in angular momentum is $I_A\omega$. Using the principle

$$\text{angular impulse} = \text{change in angular momentum}$$

we have

$$F\,\Delta t(h + b) = I_A\omega \qquad (14\text{-}21)$$

The linear impulse is $F\,\Delta t$, and this must be equal to the increase in linear momentum. For calculating the linear momentum, the entire mass may be assumed to be traveling with the same velocity as the center of mass at C. Since the point C is a distance h from the axis, the velocity of C is ωh. Then the linear impulse principle gives

$$\text{linear impulse} = \text{change in linear momentum}$$
$$F\,\Delta t = m\omega h \qquad (14\text{-}22)$$

When Eq. 14-22 is substituted in Eq. 14-21, we find that

$$m\omega h(h + b) = I_A\omega \qquad (14\text{-}23)$$
$$h(h + b) = \frac{I_A}{m}$$

According to the transfer theorem of Chapter 10

$$I_A = I_C + mh^2$$
$$I_A = mk_C^2 + mh^2 \qquad (14\text{-}24)$$
$$\frac{I_A}{m} = k_C^2 + h^2$$

where k_C is the radius of gyration about an axis through the center of mass. Combining Eq. 14-24 and 14-23, we obtain

$$h(h + b) = k_C^2 + h^2$$
$$hb = k_C^2 \qquad (14\text{-}25)$$

Equation 14-25 may be used to locate the center of percussion if k_C and the location of the axis and the center of mass are known. The symbol h in this formula indicates the distance from the axis to the center of mass, and the quantity b is the distance from the center of mass to the center of percussion.

• **Example 14-12**   The uniform cylinder in Fig. 14-10 forms part of a set of chimes. How far from end B should this cylinder be struck if the blow is not to cause a re-action force on the axis of rotation at A?

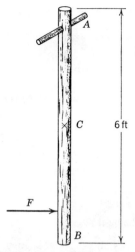

Figure 14-10   Locating the center of percussion.

Solution.   Since the center of mass is at the middle of cylinder $h = \tfrac{1}{2}L$, where $L = 6$ ft and is the length. According to the tables of Chapter 11, $I_C = \tfrac{1}{12}mL^2$. Since $I_C = mk_C^2$, we have

$$mk_C^2 = \tfrac{1}{12}mL^2$$
$$k_C^2 = \tfrac{1}{12}L^2$$

From Eq. 14-25

$$b = \frac{k_C^2}{h} = \frac{\tfrac{1}{12}L^2}{\tfrac{1}{2}L} = \tfrac{1}{6}L = 1\text{ ft}$$

The cylinder should be struck 1 ft below the center of mass (2 ft above B).   ◄

14-9 THE CONSERVATION OF ANGULAR MOMENTUM

If no angular impulse acts on a rotating system, the angular momentum must remain constant. This fact is an obvious conclusion from Eq. 14.20 and from the angular momentum principle. We have, therefore, the law of conservation of momentum.

If there is no angular impulse acting on a rotating system, the angular momentum remains constant.

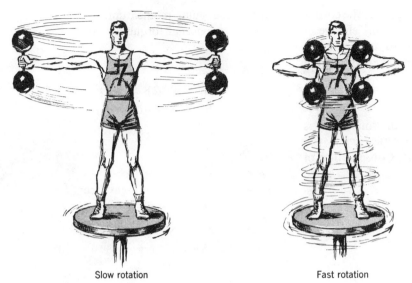

Slow rotation   Fast rotation

Figure 14-11   Demonstrating the conservation of angular momentum.

Many iceskaters and acrobats use the law of conservation of momentum to achieve spectacular effects. If they are turning slowly, they can suddenly increase their rate of rotation (angular velocity) by drawing in their arms and legs, thereby reducing their moment of inertia. This effect is demonstrated by the man on the rotating platform in Fig. 14-11. Because there are no external torques acting on the rotating platform, the angular momentum $I\omega$ does not change. When the man brings the dumbbells up to his chest, he reduces the moment of inertia I, and there is a sudden increase in the angular velocity ω so that the angular momentum ($I\omega$) remains constant. If the man extends his arms, he increases I, thereby causing the angular velocity to decrease. Because the angular momentum remains constant, any increase in the moment of inertia reduces the rate of rotation and any decrease in the moment of inertia increases the rate of rotation.

The conservation of angular momentum can be indicated by means of a simple formula. If no external torques are acting to change the angular momentum of a system, then

original angular momentum

$$= \text{final angular momentum}$$
$$I_0\omega_0 = I_f\omega_f \qquad (14\text{-}26)$$

Example 14-13   The small 200-g mass in Fig. 14-12 completes 10 rev/min on the horizontal surface when the cord is 60 cm long. The cord is then pulled through the hole at A and shortened to 20 cm while the mass is rotating about A.

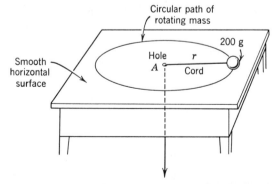

Figure 14-12   Conservation of angular momentum for a rotating mass.

(a) How fast is the mass traveling when the cord is shortened to 20 cm? (b) Calculate the change in angular momentum as the cord is shortened. (c) Calculate the change in kinetic energy as the cord is shortened.

Solution

(a) The moment of inertia of the small mass is mr^2. The original moment of inertia is mr_0^2; the final moment of inertia is mr_f^2. Conservation of angular moment gives

$$I_0\omega_0 = I_f\omega_f$$
$$(mr_0^2)\omega_0 = (mr_f^2)\omega_f$$
$$r_0^2\omega_0 = r_f^2\omega_f$$
$$\omega_f = \frac{\omega_0 r_0^2}{r_f^2}$$

$$= 10 \text{ rev/min} \times (60 \text{ cm}/20 \text{ cm})^2$$
$$= 10 \text{ rev/min} \times 9 = 90 \text{ rev/min}$$
$$= 90 \text{ rev/min} \times (2\pi \text{ rad}/1 \text{ rev})$$
$$= 180\pi \text{ rad/min}$$

$$v_f = \omega_f \times r_f = 180\pi \text{ rad/min} \times 20 \text{ cm}$$

$$= 3600\pi \text{ cm/min} \times \frac{1 \text{ min}}{60 \text{ sec}} = 60\pi \text{ cm/sec}$$

$$= 188 \text{ cm/sec}$$

ALTERNATE METHOD.   Since $\omega = v/r$, the conservation of angular momentum gives

$$(mr_0^2)\frac{v_0}{r_0} = (mr_f^2)\frac{v_f}{r_f}$$

$$r_0 v_0 = r_f v_f$$

$$v_f = \frac{v_0 r_0}{r_f}$$

$$v_0 = \omega_0 \times r_0 = 10 \text{ rev/min} \times \frac{2\pi \text{ rad}}{1 \text{ rev}} \times 60 \text{ cm}$$

$$= 1200\pi \text{ cm/min} \times \frac{1 \text{ min}}{60 \text{ sec}}$$

$$= 20\pi \text{ cm/sec} = 62.8 \text{ cm/sec}$$

$$v_f = (20\pi \text{ cm/sec})(60 \text{ cm}/20 \text{ cm}) = 188 \text{ cm/sec}$$

(b) The angular momentum does not change because no external torque acts on the rotating system.

(c) $K_0 = \frac{1}{2}mv_0^2$
$$= \frac{1}{2}(200 \text{ g})(62.8 \text{ cm/sec})^2$$
$$= 10^2(6.28 \times 10)^2 \text{ g-cm}^2/\text{sec}^2$$
$$= 39.4 \times 10^4 \text{ ergs}$$

$K_f = \frac{1}{2}mv_f^2$
$$= \frac{1}{2}(200 \text{ g})(188 \text{ cm/sec})^2$$
$$= (18.8)^2 \times 10^4 \text{ g-cm}^2/\text{sec}^2$$
$$= 353 \times 10^4 \text{ ergs}$$

$\Delta K = K_f - K_0$
$$= 353 \times 10^4 \text{ ergs} - 39.4 \times 10^4 \text{ ergs}$$
$$= 3.14 \times 10^6 \text{ ergs}$$

The kinetic energy increases because work is done on the system when the cord is pulled through the hole.   ◀

14-10 ANGULAR MOMENTUM AS A VECTOR QUANTITY

Once an object has been set into rotation, it tends to continue rotating unless an external torque acts on the object; the angular momentum remains constant in both magnitude and direction. The directional or vector property of angular momentum becomes quite apparent if we examine the behavior of rotating wheels, tops, and gyroscopes.

Anyone who has ever ridden a bicycle has made use of the vector property of angular momentum. It is almost impossible to balance one-self on a stationary bicycle. But once the wheels are spinning, conservation of angular momentum helps the rider remain upright. The angular momentum of the spinning wheels tends to remain constant in magnitude and direction.

The conservation of angular momentum also keeps a spinning top from falling over on its side. The spin axis of the top remains upright because the angular momentum tends to remain constant in both magnitude and direction. The direction

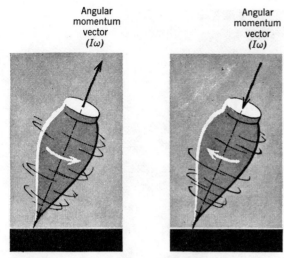

Figure 14-13 The angular momentum vector. The spin is always counterclockwise about the angular momentum vector.

of the spin axis is the direction of the angular momentum vector (see Fig. 14-13).

The vector nature of angular momentum may be demonstrated by performing a simple experiment on the rotating platform (Fig. 14-14). If the man on the platform starts the wheel rotating in a clockwise sense, he and the platform will rotate in a counterclockwise sense because the angular momentum of the system remains constant. The downward angular momentum acquired by the rotating wheel cancels the upward angular momentum acquired by the platform so that the total angular momentum remains constant (and equal to zero). If the man turns the wheel over, he will be changing the direction of the wheel's angular momentum and he will reverse the direction in which the platform is rotating.

Because angular momentum has the properties of a vector, a rotating object tends to maintain constant the direction of its spin axis. If a torque is applied to give the object some additional angular momentum about this axis, the final angular momentum can be found by an algebraic sum:

final angular momentum

= original angular momentum

+ change in angular momentum

However, if the additional angular momentum is not about the original axis, then the additional angular momentum must be added to the original angular momentum by means of a vector sum. This principle will be used in the next section to explain the precessional motion of gyroscopes and tops.

14-11 THE GYROSCOPE

A simple gyroscope is just a wheel with a large moment of inertia that is mounted on gimbal rings (see Fig. 14-15). The wheel can be set into rotation with its spin axis in any given direction. Because of the conservation of angular momentum, the axis tends to maintain a constant direction even when the gyroscope is carried from place to place. If the gyroscope is carried on an airplane and the axis is pointed vertically, the gyroscope can be

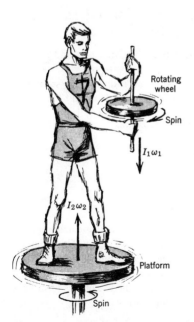

Figure 14-14 Demonstrating the vector nature of angular momentum.

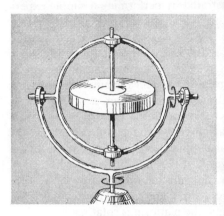

Figure 14-15   A model gyroscope mounted on gimbal rings.

incorporated into an instrument that serves as a banking indicator or an artificial horizon. If the gyroscope axis is pointed in the south-north direction, the gyroscope will continue to point in this direction, and this fact is used in the design of gyrocompasses.

Some very interesting effects can be obtained with the inexpensive model gyroscope of Fig. 14-16. When the gyroscopes are in the position indicated in the diagram, there is an unbalanced torque about the torque axis. However, if the gyroscope wheel is spinning, the unbalanced torque does not cause the gyroscope to fall over. Instead, we observe that the unbalanced torque produces a change in the direction of the angular momentum. The gyroscope spin axis rotates, comparatively slowly, about the vertical precession axis. This type of motion is called precession.

To analyze the precession, the change in the direction of the spin axis during a short time interval Δt is given by $\Delta\phi$ (an angle measured in radians). The torque M produces an angular impulse $M \times \Delta t$ which is equal to the change in angular momentum. The additional angular momentum produced by the torque is angular momentum about the torque axis, which is at right angles to the spin axis. The final angular momentum $(I\omega)_f$ must be found by the vector sum method of Fig. 14-16c. Since the magnitude of the spin angular momentum does not

change, this vector triangle is an isosceles triangle and the apex angle of this triangle in radians is

$$\Delta\phi = \frac{\text{arc length}}{\text{radius}} = \frac{M\,\Delta t}{I\omega\ \sin\theta} \qquad (14\text{-}27)$$

where θ is the angle between the spin axis and the precession axis.

The rate of precession in radians per second is $\Delta\phi/\Delta t = \Omega$ (the symbol Ω is a capital omega.) The rate of precession may be easily calculated if both sides of Eq. 14-27 are divided by Δt to give

$$\Omega = \frac{M}{(I\omega \sin\theta)} \qquad (14\text{-}28)$$

$$\frac{\text{rate of precession}}{\text{(rad/sec)}} = \frac{\text{torque}}{\text{spin angular momentum} \times \sin\theta}$$

Equation 14-28 is a useful equation for calculating the precessional motion of a gyroscope or top. This equation may be used whenever the precessional angular velocity is much smaller than the spin angular velocity. This is usually the case in gyroscopes; the use of this formula is illustrated in the following example.

• **Example 14-14**   The model gyroscope in Fig. 14-16b consists of a rotating 4-lb disk that is 18 in. in diameter.

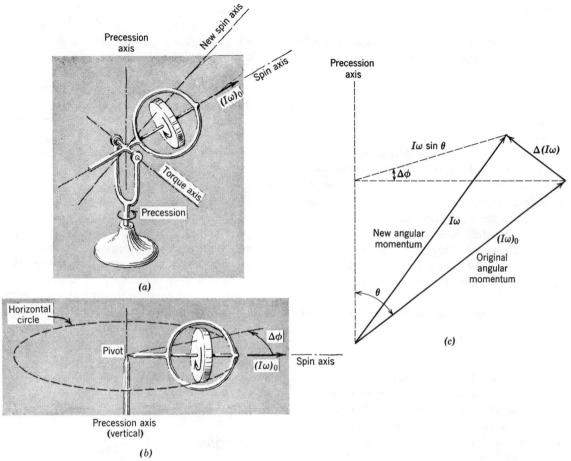

Figure 14-16 Precession.

The framework supporting the disk weighs 2 lb, and the center of gravity of the entire 6-lb instrument is located 20 in. from the pivot. How long does it take for the spin axis to complete 1 rev. about the vertical precession axis when the gyroscope wheel is set rotating at 300 rev/min?

Solution. The torque is

$$M = 6 \text{ lb} \times 20 \text{ in.} \times \frac{1 \text{ ft}}{12 \text{ in.}} = 10 \text{ lb-ft}$$

The radius of the rotating disk is 9 in. $= \frac{3}{4}$ ft. Using the formula for the moment of inertia of a disk (Chapter 11) gives

$$I = \frac{1}{2} mR^2 = \frac{1}{2} \left(\frac{4 \text{ lb}}{32 \text{ ft/sec}^2} \right) \left(\frac{3}{4} \text{ ft} \right)^2$$

$$= \tfrac{9}{256} \text{ lb-ft-sec}^2$$

And

$$\omega = 300 \text{ rev/min} \times \frac{1 \text{ min}}{60 \text{ sec}} \times \frac{2\pi \text{ rad}}{1 \text{ rev}} = 10\pi \text{ rad/sec}$$

The spin angular momentum is

$$I\omega = (\tfrac{9}{256} \text{ lb-ft-sec}^2)(10\pi \text{ rad/sec})$$

$$= 1.102 \text{ lb-ft-sec}$$

Since $\theta = 90°$ and $\sin \theta = 1$, the rate of precession is

$$\Omega = \frac{M}{I\omega} = \frac{10 \text{ lb-ft}}{1.102 \text{ lb-ft-sec}} = 9.07 \text{ rad/sec}$$

The time for 1 rev $= 2\pi$ rad is then

$$\frac{2\pi \text{ rad}}{9.07 \text{ rad/sec}} = 0.692 \text{ sec} \qquad \blacktriangleleft$$

SUMMARY

The *linear impulse* of a force is a vector quantity equal to $F \times t$. The linear impulse of the resultant force is $F_R \times t$. *The impulse–momentum principle* states that the linear impulse acting upon an object is equal to the change in linear momentum.

When objects collide or exert forces on each other and no outside forces produce an impulse on the objects during the collision, then the linear momentum of the system does not change. In other words, we will have *conservation of momentum*: total momentum after the collision = total momentum before the collision.

In a *completely inelastic collision* the two colliding objects stick together and move with a common velocity after the collision. Kinetic energy is lost (converted into heat) during inelastic collisions.

In a *completely elastic collision* the kinetic energy of the objects before the collision equals the kinetic energy after the collision. Conservation of momentum (Eq. 14-6) applies to any type of collision, but it is only in completely elastic collisions that we have conservation of kinetic energy.

The operation of a rocket engine is an application of the conservation of linear momentum. The backward momentum acquired by the exhaust gases from the combustion chamber is equal in magnitude but opposite in direction to the forward momentum acquired by the rocket. The forward thrust on the rocket is equal to the momentum per unit time acquired by the exhaust gases. The thrust may be calculated from Eq. 14-16 as in Example 14-9.

When a torque M acts upon a rotating object for a time t, an angular impulse is produced upon the object. The *angular impulse* $= M \times t$. The *impulse–momentum principle* for a rotating object states that the angular impulse is equal to the change in angular momentum. The *angular momentum* for a rotating object is $I\omega$, where I is the moment of inertia and ω is the angular velocity in rad/sec.

If an angular impulse is applied to an object by a force that exerts a torque about an axis, there will usually be a reaction force on the supporting axis. However, if the force acts through the *center of percussion* (see Fig. 14-9), there will be no reaction on the axis.

If no angular impulse acts upon a rotating system, the angular momentum must remain constant. This is the law of *conservation of angular momentum*. In some situations the moment of inertia changes (see Fig. 14-11). In order for the angular momentum to remain constant, a decrease in the moment of inertia must result in an increase in the angular velocity.

Once an object is set rotating, the angular momentum remains constant unless an external torque acts on the spinning object. Because angular momentum has the properties of a vector, the conservation of angular momentum requires that the spin axis have a constant direction.

In the case of a rotating top or a gyroscope, the gravitational force produces a torque that changes the direction but not the magnitude of the angular momentum. The spin axis of the top or gyroscope changes its direction because of the torque acting, and the rotation of the spin axis about the vertical axis is called *precession*.

QUESTIONS

1. Define the term *linear impulse*.

2. What is the impulse–momentum principle?

3. What units are used to measure impulse?

4. What is the difference between momentum and kinetic energy?

5. Suppose an object starts out moving at 5 m/sec toward the right and after a collision it is moving 5 m/sec toward the left. Has the momentum changed? Has the kinetic energy changed? Explain your answers.

6. Has there been a net impulse acting upon the object in Question 5? Has work been done on the object?

7. What is meant by "a completely inelastic collision"?

8. Under what conditions will the linear momentum of a system of objects remain constant?

9. A system may consist of two rapidly moving objects,

but the momentum of the system may be zero. Explain how this could occur.

10. The momentum after a collision equals the momentum before the collision, but the kinetic energy after the collision is less than the kinetic energy before the collision. Explain why this is generally true.

11. What is a perfectly elastic collision?

12. How is the conservation of momentum used to explain the thrust produced by a rocket engine?

13. Suppose we drop a rubber ball from a height of 4 ft and find that it rebounds to a height of 3 ft. Does conservation of momentum apply to the collision between the ball and the ground? Is the collision perfectly elastic?

14. How could the impulse–momentum principle be used to calculate the force on a golf ball when it is hit by the golf club? Does conservation of momentum apply to the collision between the club and the ball?

15. Suppose the explosion of some gunpowder causes a moving rocket to separate into two parts. Does conservation of momentum apply to this problem? Is there conservation of kinetic energy? Explain your answer.

16. Define *angular impulse*.

17. How is the angular momentum of an object related to the moment of inertia?

18. What is the impulse–momentum principle for a rotating object?

19. What is meant by the law of conservation of angular momentum? Under what conditions is this law valid?

20. Angular momentum is a vector quantity. What is the magnitude of the angular momentum? What is the direction?

21. What is meant by precession?

22. What is the basic principle that explains the operation of the gyrocompass?

23. What is the center of percussion?

24. In Example 14-13 the kinetic energy of the moving mass was calculated by using $K = \frac{1}{2}mv^2$. Why would it have no difference in Example 14-13(c) if the kinetic energy were taken to be $K = \frac{1}{2}I\omega^2$?

25. The tops in Fig. 14-13 have different senses of spin. Describe the direction of precession about the vertical axis for the two cases. (Fig. 14-16 may be used as a guide.)

PROBLEMS

(A)

1. A force of 400 lb acts for $\frac{1}{100}$ sec. Calculate the impulse.

2. A mass of 4 slugs is accelerated from rest to a velocity of 15 ft/sec. What is the impulse acting upon this mass?

3. A 6-kg object is accelerated from a velocity of 10 m/sec to a velocity of 30 m/sec in a time of 4 sec.
   (a) Calculate the change in momentum.
   (b) Calculate the accelerating force.
   (c) Calculate the impulse in newton meters and check whether it equals the changes in momentum calculated in part (a).

4. A torque of 400 lb-ft is applied to a pulley for 0.5 sec. What is the angular impulse?

5. An object with a moment of inertia of 100 slug-ft^2 is rotating with an angular velocity of 40 rad/sec. Calculate the angular momentum.

6. An 800-lb flywheel has a radius of gyration of 15 in. It is rotating with an angular velocity of 600 rev/min. Calculate:
   (a) The moment of inertia in slug-ft^2.
   (b) The angular velocity in rad/sec.
   (c) The angular momentum.

7. A 2-kg object moving at 40 m/sec makes a completely inelastic collision with a stationary 8-kg object. Calculate the joint velocity after the collision.

8. An object whose weight is 80 lb is traveling at 20 ft/sec.
   (a) Calculate the momentum in slug-ft/sec.
   (b) What impulse in lb-sec is required to accelerate this object to 30 ft/sec?

9. A 6-oz baseball traveling at 40 ft/sec is struck with a baseball bat. The ball rebounds from the bat with a velocity of 60 ft/sec.
   (a) Calculate the impulse (in lb-sec) that acts upon the baseball during the impact.
   (b) If the impact lasts for $\frac{1}{100}$ of a second, what is the average force exerted upon the baseball by the bat?

10. A 200-lb football star running at a constant speed runs 80 yd in 10 sec. How large an impulse has to be applied to stop this man?

11. A 20-ton cannon fires a shell that weighs 1000 lb. The shell has a muzzle velocity of 3000 ft/sec.
    (a) What is the recoil velocity of the cannon if it is free to move?
    (b) How much kinetic energy (in foot-pounds) is given to the shell?
    (c) How much kinetic energy is imparted to the recoiling gun?

12. A 10-ton freight car moving at 5 miles/hr bumps into the rear of a 20-ton freight car moving in the same direction at 3 miles/hr. What is the velocity of the two cars if they become coupled together after the collision?

13. A 10-kg block sliding at 20 m/sec collides with a stationary 30-kg block.
    (a) If the two blocks stick together after the collision, calculate the final velocity.
    (b) If the blocks undergo a completely elastic collision, calculate the velocity of each one after the collision.

14. A generator armature has a moment of inertia of 20 kg-m^2. It is accelerated from rest to an angular velocity of 50 rad/(sec. in a time of 10 sec).
    (a) Calculate the final angular momentum.
    (b) What impulse acted upon the armature during the 10 sec?
    (c) Calculate the average torque acting during the time of the acceleration.

15. A 10-lb object moving toward the right at 20 ft/sec collides with a 30-lb object that is not moving.
    (a) Determine the velocity of both objects after the impact if they stick together in a completely inelastic collision.
    (b) Determine the velocity of both objects if the collision is assumed to be perfectly elastic.

16. A 10-g bullet is fired into a 20-kg block of wood that is resting on a smooth floor. The bullet is initially traveling at 300 m/sec and remains imbedded in the block after the impact.
    (a) How fast does the block move after the impact?
    (b) How much kinetic energy is lost during the impact?

17. A 16-lb bowling ball traveling at 10 ft/sec makes a head-on collision with a stationary 2-lb bowling pin. Assume that the collision is completely elastic and determine the velocity of both objects after the collision.

18. In the apparatus in Fig. 14-17 the two balls are attached to strings. The ball at B is released from rest after being raised to a height h above the suspended ball A. Ball B makes a head-on collision with ball A.
    (a) What happens to both balls after the impact, assuming that the collision is completely elastic?
    (b) Suppose that a small bit of putty is placed on ball A so that the two stick together after the impact. How high will the two balls swing?

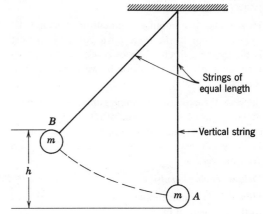

Figure 14-17 Collision between two balls of equal mass.

19. A flywheel with a moment of inertia of 10 slug-ft^2 is accelerated from rest to an angular speed of 20 rev/sec in a time of 10 sec.
    (a) Calculate the angular momentum after 20 sec.
    (b) What angular impulse was applied to the wheel?
    (c) What was the average torque during the 10 sec?

20. A torque of 10 lb-ft is applied to a motor armature for 5 sec. The armature starts from rest and acquires an angular velocity of 30 rad/sec during this time interval.
    (a) Calculate the angular momentum of the armature after 5 sec.
    (b) Determine the moment of inertia of the flywheel.
    (c) What was the kinetic energy of the flywheel after the 5 sec?

(B)

1. A 10-g object traveling at 50 cm/sec toward the right makes a direct (head-on) collision with a 20-g object traveling at 40 cm/sec toward the left.
   (a) If the two objects stick together, determine (1) the final velocity, and (2) the kinetic energy lost in the collision.
   (b) Now assume that the collision is completely elastic. What is the velocity of each object immediately after the impact?

2. A bullet that weighs 0.5 oz is fired horizontally into a 5-lb block. The block is suspended from strings to form a ballistic pendulum. The bullet remains imbedded in the block, and the center of gravity of the block swings up a vertical distance of 6 in. after the impact.
   (a) Calculate the initial velocity of the bullet.
   (b) How much kinetic energy was converted into heat as a result of the impact?

3. A bullet that weighs 0.1 lb is traveling at 2000 ft/sec. The bullet is fired into a 10-lb block of wood. The bullet passes through the block of wood, and the block starts to move at 15 ft/sec.
   (a) What is the final velocity of the bullet after it passes through the wooden block?
   (b) How much kinetic energy is converted into heat during the collision?

4. A small rocket projectile has a mass of 200 kg. During the launching of this projectile, the rocket motor suddenly expels 50 kg of fuel from the projectile. The ejection velocity of this fuel is 900 m/sec. What launching velocity is acquired by the projectile?

5. A 5-kg mass moving with a velocity of 10 m/sec collides head-on with a stationary 10-kg mass. The 10-kg mass starts to move with a velocity of 6 m/sec.
   (a) What is the velocity of the 5-kg mass after the collision?
   (b) How much kinetic energy is converted into heat in the collision?
   (c) Calculate the coefficient of restitution for this collision.

6. A 4-lb object traveling at 80 ft/sec collides directly with an immovable wall. Determine the impulse on the wall and the final velocity of the object in the following cases:
   (a) If the collision is perfectly elastic.
   (b) If the collision is perfectly inelastic.

   (c) If the coefficient of restitution is 0.6.

7. A 20,000 lb airplane is flying horizontally at a steady speed of 300 miles/hr. A 500-lb rocket is then launched from the airplane in a forward direction. The rocket is projected forward from the airplane with a velocity of 2000 ft/sec relative to the plane. By how much does the speed of the plane decrease as a result of the launching?

8. A 160-lb man stands on a turntable that is rotating at 20 rev/min. The radius of gyration for the man is 0.8 ft. By extending his arms, the man can change his radius of gyration to 1.2 ft. How fast will he rotate after extending his arms?

9. A motor armature rotating at 1800 rev/min is brought to rest by friction in a time of 2 min. The moment of inertia of the armature is 5 kg-m^2.
   (a) What angular impulse (in newton-m-sec.) acted to bring the armature to rest?
   (b) Calculate the torque due to friction that acted upon the armature.

10. A 2-ton automobile traveling toward the west at 15 miles/hr collides with an 8-ton truck traveling toward the north at 5 miles/hr. The two vehicles stick together after the collision.
    (a) Determine the magnitude and direction of the combined momentum after the collision.
    (b) What is the common velocity of the two vehicles immediately after the collision?
    (c) If the impact lasts $\frac{1}{10}$ of a second, calculate the average force acting upon the automobile.
    (d) How much kinetic energy is lost during the collision?

11. A flywheel that weighs 40 lb is shaped in the form of a uniform circular disk 4 ft in diameter. If the flywheel is rotating at 3000 rev/min, calculate:
    (a) The moment of inertia in slug-ft^2.
    (b) The angular momentum in slug-ft^2/sec.
    (c) The kinetic energy in ft-lb.

12. A rocket produces 100,000 lb of thrust. The fuel has an ejection velocity of 4000 ft/sec. How many pounds of fuel will be consumed in 1 min of operation?

13. A rocket motor burns 10 kg of fuel per second and ejects it from the combustion chamber, with an ejection velocity of 6000 m/sec. What is the thrust produced by this rocket motor?

14. A two-stage rocket projectile is designed so that a small explosive charge divides the projectile into two parts. The projectile is traveling at 2000 m/sec and

has a mass of 100 kg when 20% of the mass is suddenly ejected forward. The rest of the projectile acquires a backward velocity of 100 m/sec.

(a) Calculate the velocity of the mass that is projected forward.

(b) If the explosion lasts for $\frac{1}{100}$ of a second, what is the average force acting upon the smaller mass?

(c) What is the average force acting upon the larger mass during the explosion?

(d) How much kinetic energy was produced by the explosion?

15. A 4000-lb gun is aimed so that the muzzle makes an angle of 60° with the horizontal. A 100-lb projectile is then fired with a muzzle velocity of 2000 ft/sec. The gun is mounted so that it recoils a distance of 2 ft in the horizontal direction.

(a) Calculate the horizontal recoil velocity of the gun.

(b) How much kinetic energy is acquired by the projectile?

(c) How much kinetic energy is acquired by the gun?

(d) What is the average horizontal force acting upon the gun during the recoil?

(e) If the explosion lasts for $\frac{5}{1000}$ second, calculate the average vertical thrust on the gun platform.

16. A 10-g mass moving at 200 cm/sec toward the right makes a direct collision with a stationary mass. After the collision, the stationary mass moves toward the right, while the 10 g mass rebounds toward the left with a velocity of 100 cm/sec. Frictional forces are negligible, and the heat produced upon impact is also negligible. Determine:

(a) The momentum acquired by the object which was originally at rest.

(b) The kinetic energy acquired by the object that was originally at rest.

(c) The velocity of the second object after the collision.

(d) The mass of the second object.

17. In Fig. 14-18 the two masses are equal and the mass at B is originally at rest. The two masses undergo a completely elastic collision.

(a) Assume that the collision is head-on, so that the two objects are traveling on the line AB after the collision. Prove that the conservation of energy

Figure 14-18   Elastic collision between two equal masses.

and conservation of momentum will be satisfied after the impact if mass A comes to rest ($v_{A2} = 0$) and mass B moves to the right with the original velocity of mass A ($v_{B2} = v_{A1}$).

(b) Assume that the collision is glancing, so that the two masses rebound at an oblique angle to the line AB. Prove from the conservation of kinetic energy that $v_{A1}{}^2 = v_{A2}{}^2 + v_{B2}{}^2$ and that this means that the two final velocities are at right angles to each other after the collision.

18. The uniform disk in Fig. 14-19 is suspended from point A. How far below A should the disk be struck so that there is no horizontal reaction on point A?

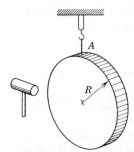

Figure 14-19   Striking a suspended disk at the center of percussion.

19. In Fig. 14-13 assume that the top has a mass of 400 g and the center of gravity is 10 cm from the pivot point on the axis of the top. The radius of gyration of the top is 5 cm, and the top is spinning at 600 rev/sec.

(a) Calculate the moment of inertia and the angular momentum.

(b) If the axis of the top makes an angle of 30° with the vertical, calculate (1) the torque about the pivot produced by the force of gravity, and (2) the rate of precession.

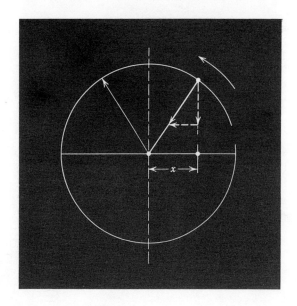

15.
Simple Harmonic Motion

This chapter discusses the special example of the periodic motion of a particle known as *simple harmonic motion* or SHM. SHM is defined and then the general features of the motion are studied leading to expressions for the displacement, velocity and acceleration of the particle. Formulas for the period of the motion are determined for several specific cases of SHM: simple spring, simple pendulum, torsion pendulum and physical pendulum.

LEARNING GOALS

The major learning objectives of this chapter are

1. *To learn the definition of SHM and the vocabulary related to this form of motion.*

2. *To learn the equations for the displacement, velocity and acceleration of a particle in SHM and how to use these equations.*

3. *To understand the definitions of frequency and period for SHM.*

4. *To learn the formulas for the period of the motion for several specific forms of SHM and how to use the formulas.*

5. *To understand how the reference circle can be used to deduce the basic kinematic equations of SHM.*

6. *To learn the graphs of displacement, velocity and acceleration as functions of time for a particle in SHM.*

15-1 SIMPLE HARMONIC MOTION

Suppose that we attach a weight to a helical spring. The spring will stretch and the weight will hang suspended at some equilibrium position (Fig. (15-1a). We assume that the weight does not stretch the spring beyond its elastic limit (see Sec. 6-1). Now let the weight be pulled a short distance below the equilibrium position and released. The weight will execute an oscillatory motion (up and down) about the equilibrium position (Fig. 15-1b).

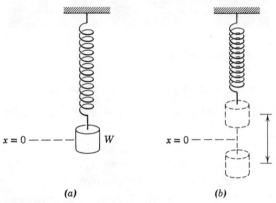

Figure 15-1   Oscillatory motion on a spring (b) about the equilibrium position (a).

This motion is periodic but fundamentally different from the periodic circular motion studied in previous chapters. The acceleration of the weight on the spring is neither constant in magnitude nor in direction, but rather is dependent on the instantaneous displacement of the weight from its equilibrium position.

This general class of periodic motion, of which oscillatory motion of a weight on a spring is one example, is called *simple harmonic motion* (SHM).

> **Simple harmonic motion is that motion which ensues whenever the acceleration of a particle is directly proportional to its displacement from equilibrium, but oppositely directed to the displacement.**

Thus, whenever
$$a = -Cx \qquad (15\text{-}1)$$
where C is a constant of proportionality, the motion that develops will be simple harmonic. The negative sign in Eq. 15-1 indicates that when the displacement x, as measured from equilibrium, is positive, the acceleration will be negative, and when x is negative, the acceleration will be positive.

In addition to the helical spring, other important examples of simple harmonic motion are the simple pendulum, the torsion pendulum, and the physical pendulum. Certain electrical oscillations may also be considered forms of SHM.

Equation 15-1 applies to all forms of simple harmonic motion. In each particular case the constant of proportionality, C, will be related to different physical parameters. We wish to obtain expressions for the velocity and displacement of a particle executing SHM. We shall develop these expressions for the general case of SHM and then indicate how they apply to several specific examples.

In Fig. 15-2, m is a particle moving in a circle of radius R with constant speed v, and P represents the projection of m on the diameter AA'. As m makes one complete revolution, P moves from A to A' and back to A. It has been noted that the circular motion of m is not simple harmonic. However, we shall show that the motion of P is simple harmonic. We shall then derive expressions

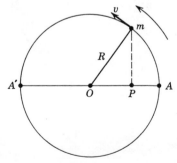

Figure 15-2   Reference circle for simple harmonic motion showing the motion of a particle m along the circumference and the motion of its projection P along a diameter.

for displacement and velocity by observing the motion of P.

The acceleration of m as it moves in the circle is v^2/R, or $R\omega^2$ where ω is the constant angular velocity of the motion. This acceleration is directed toward the center of the circle O (Fig. 15-3) and can be resolved into components in the x and y directions. Thus, as in projectile motion, we can consider the motion of m to consist of two independent, perpendicular motions having the accelerations of their respective components. The motion of P is the x component of the motion of m and has acceleration $R\omega^2 \cos \theta$. However, from Fig. 15-3 it is clear that the displacement of P at any time is

$$x = R \cos \theta \qquad (15\text{-}2)$$

where R is the *amplitude* and represents the maximum displacement in either direction. Further, when x is positive (to the right), $R\omega^2 \cos \theta$ is negative (to the left). Thus

$$a_P = -R\omega^2 \cos \theta = -\omega^2 x \qquad (15\text{-}3)$$

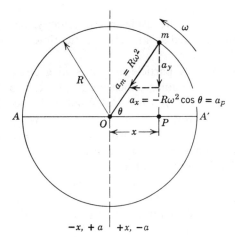

Figure 15-3 Resolution of centripetal acceleration vector into horizontal and vertical components.

Since ω^2 is a constant, the motion of P satisfies Eq. 15-1 and is simple harmonic. Equations 15-2 and 15-3 give expressions for the displacement and acceleration. The velocity of P is obtained by

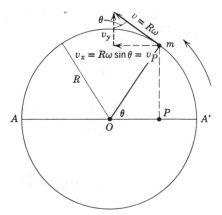

Figure 15-4 The velocity vector resolved into horizontal and vertical components.

taking the x component of the velocity of m. Thus, from Fig. 15-4,

$$v_P = v \sin \theta = R\omega \sin \theta$$

But $\sin \theta = \pm\sqrt{1 - \cos^2 \theta}$. Therefore,

$$v_P = \pm v\sqrt{1 - \cos^2 \theta} = \pm R\omega\sqrt{1 - \cos^2 \theta}$$
$$= \pm\omega\sqrt{R^2 - R^2 \cos^2 \theta}$$

or $\qquad v_p = \pm\omega\sqrt{R^2 - x^2} \qquad (15\text{-}4)$

The $\pm$ sign in the last expression indicates that for a given displacement the particle may either be moving to the left or the right (Fig. 15-5).

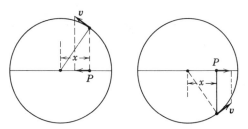

Figure 15-5

Any particle that moves in simple harmonic motion can be treated as if it were the projection of a second identical particle moving in a circle. The circle is referred to as the *reference circle*, and

Table 15-1   Simple harmonic motion relationships

Item	Symbol	Equation
Displacement	x	$x = R \cos \theta = R \cos \omega t = R \cos 2\pi f t$
Velocity	v	$v = -R\omega \sin \theta = -R\omega \sin \omega t$
		$\quad = -R(2\pi f) \sin (2\pi f)t$
		$\quad = \pm \omega \sqrt{R^2 - x^2}$
Acceleration	a	$a = -\omega^2 x = -\omega^2 R \cos \theta$
		$\quad = -\omega^2 R \cos \omega t$
		$\quad = -4\pi^2 f^2 R \cos 2\pi f t$

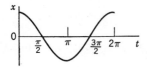

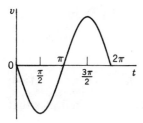

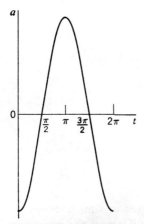

Figure 15-6   Graphs of x, v, and a as functions of t for a particle moving with simple harmonic motion. Graphs show one cycle of the motion.

the angle θ in Eqs. 15-2, 15-3, and 15-4 is measured in this circle. Since the angular velocity is constant, θ may be replaced in the latter equations by ωt. We summarize in Table 15-1. In Table 15-1 we recall that if f is the frequency of the motion and T is the period,

$$\omega = 2\pi f \qquad (15\text{-}5a)$$

and

$$f = \frac{1}{T} \qquad (15\text{-}5b)$$

Figure 15-6 shows the displacement, velocity, and acceleration plotted as a function of the time t.

Example 15-1   A particle m moves in a circle with a constant angular velocity of 2 rad/sec. The radius of the circle is 10 cm. (a) What is the acceleration of m as it moves in the circle? (b) What is the acceleration of the projection of m on a horizontal diameter when the projection, p, is $+5$ cm from the center of the circle? (c) What is the velocity of p when the displacement of p is $+5$ cm? (d) When is the velocity of p maximum? Minimum? (e) When is the acceleration of p maximum? Minimum?

Solution

(a) The acceleration of m as it moves in a circle is its centripetal acceleration. Thus

$$a_m = R\omega^2 = 10(2)^2 = 40 \text{ cm/sec}^2$$

(b) Given the displacement of p, we choose the appropriate form of the acceleration from our chart. Then,

$$a_p = -\omega^2 x = -(2)^2(+5) = -20 \text{ cm/sec}^2$$

(c) From the chart

$$v_p = \pm\omega\sqrt{R^2 - x^2}$$
$$= \pm 2\sqrt{10^2 - 5^2}$$
$$= \pm 2\sqrt{75}$$
$$= \pm 17.4 \text{ cm/sec}$$

(d) If we use the representation

$$v_p = \pm\omega\sqrt{R^2 - x^2},$$

it is clear that the maximum speed of p occurs when $x = 0$, which corresponds to p being at the center of the circle (equilibrium position for p). The minimum speed occurs when $x = \pm R$, which corresponds to p being at the extremities of the path where the velocity changes direction.

(e) From $a_p = -\omega^2 x$ it is evident that $a_p = 0$ when $x = 0$, and a_p is maximum when $x = \pm R$ (since x cannot be larger than R). Thus the acceleration is a maximum when the velocity is minimum, and the acceleration is a minimum when the velocity is maximum. ◀

Example 15-2   A vibrating particle (SHM) has an acceleration of 9 ft/sec² when its displacement is -3 in. What is the period of the vibration?

Solution.   In order to determine the period, it is necessary to know the frequency, f, or the angular velocity, ω. Since we are given the acceleration and displacement, it is straightforward to use Eq. 15-3 to find ω.

Thus

$$a = -\omega^2 x$$
$$9 \text{ ft/sec}^2 = -\omega^2(-\tfrac{1}{4} \text{ ft}) = \tfrac{1}{4}\omega^2 \text{ ft}$$
$$\omega^2 = 36/\text{sec}^2$$

and

$$\omega = 6 \text{ rad/sec}$$

Then, from Eqs. 15-5 it follows that

$$T = \frac{2\pi}{\omega} = \frac{2\pi}{6} = \frac{\pi}{3} \text{ sec}$$ ◀

Example 15-3   A particle moving in simple harmonic motion has a maximum displacement of 10 cm. The velocity of the particle is 120 cm/sec when the particle has a displacement of 8 cm. Find (a) the period of SHM, (b) the maximum velocity, (c) the maximum acceleration, (d) the acceleration when $x = -8$ cm, and (e) the time it takes for the particle to move from $x = 5$ cm to $x = -5$ cm.

Solution

(a) To find the period, we first determine the angular velocity. Since we are given the velocity at a given displacement, we can use

$$v = \pm\omega\sqrt{R^2 - x^2}$$

Substituting the given information, we have

$$120 = \omega\sqrt{10^2 - 8^2}$$
$$120 = \omega\sqrt{100 - 64} = \omega\sqrt{36}$$
$$120 = 6\omega$$
$$\omega = 20 \text{ rad/sec}$$

Then, from Eq. 15-5

$$T = \frac{2\pi}{\omega} = \frac{2\pi}{20} = \frac{\pi}{10} \text{ sec}$$

(b) The maximum velocity is given by

$$v_{\max} = \omega R$$
$$= (20 \text{ rad/sec})(10 \text{ cm})$$
$$= 200 \text{ cm/sec}$$

(c) The maximum acceleration occurs when the displacement is a maximum. Thus

$$a = -\omega^2 x$$
$$a_{\max} = \omega^2 R = (20)^2(10)$$
$$= 4000 \text{ cm/sec}^2 \text{ (magnitude)}$$

(d) The acceleration when $x = -8$ cm is found by using

$$a = -\omega^2 x$$

Thus

$$a = -(20)^2(-8)$$
$$= 3200 \text{ cm/sec}^2$$

(e) To find the time elapsed in moving between two different positions, it is necessary to determine the times at which the particle is located at each of the positions and take the difference between the two times. We obtain the time at which a particle has a particular displacement from the first equation in Table 15-1. Thus

$$\cos \omega t_1 = \frac{x_1}{R}$$

$$\cos (20t_1) = \frac{5 \text{ cm}}{10 \text{ cm}} = \frac{1}{2}$$

Therefore,

$$20t_1 = \cos^{-1}\left(\frac{1}{2}\right) = \frac{\pi}{3} \text{ rad}$$

and $\qquad t_1 = \dfrac{\pi}{60} \text{ sec}$

Similarly, for $x = -5$ cm, we have

$$\cos \omega t_2 = \frac{x_2}{R}$$

$$\cos (20t_2) = -\frac{5 \text{ cm}}{10 \text{ cm}} = -\frac{1}{2}$$

$$20t_2 = \cos^{-1}\left(-\frac{1}{2}\right) = \frac{2\pi}{3} \text{ rad}$$

and $\qquad t_2 = \dfrac{2\pi}{60} \text{ sec}$

Therefore, the time to move from $x_1 = +5$ cm to $x_2 = -5$ cm is

$$t_2 - t_1 = \frac{2\pi}{60} - \frac{\pi}{60} = \frac{\pi}{60} \text{ sec} \qquad \blacktriangleleft$$

15-2 THE VIBRATION OF SPRINGS

In Chapter 6 the reader was introduced to Hooke's law, which states that when a wire is stretched a distance x, which is considerably less than the elastic limit of the wire, the wire exerts a restoring force F, proportional to the displacement x and oppositely directed. Thus

$$F_{\text{restoring}} = -kx \qquad (15\text{-}6)$$

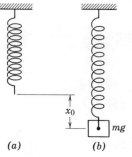

Figure 15-7 (a) The unstretched spring. (b) The spring is stretched an amount x_0 by the weight mg.

where k is the force constant or modulus of the wire. Springs observe Hooke's law also. When a mass m is hung on a spring, the mass will stretch the spring until the restoring force just balances the weight that is pulling the spring down (Fig. 15-7). The amount by which the spring is stretched, x_0, will be determined by the equilibrium condition, $\Sigma \mathbf{F} = 0$. Thus

$$mg + (-kx_0) = 0$$

and $\qquad x_0 = \dfrac{mg}{k} \qquad (15\text{-}7)$

The position, x_0, is the equilibrium position for the motion of the weight on the spring. If the weight is pulled down an additional displacement, x, and released, the extra restoring force exerted by the spring, $-kx$, will be an unbalanced force. This unbalanced force, according to Newton's second law, will give rise to an acceleration a, according to the equation

$$F = -kx = ma$$

or $\qquad a = -\left(\dfrac{k}{m}\right)x \qquad (15\text{-}8)$

Since k and m are constants, the acceleration of a mass m on a spring satisfies the definition of simple harmonic motion with $k/m = \omega^2$. It is important to note that the displacement x in Eq. 15-8 measures the displacement from the equilibrium position of m on the spring and not from the equilibrium position of the spring without m. Hooke's law *does* pertain to the stretching of the spring from *its* equilibrium position, but the *unbalanced force* concerns itself only with the stretching beyond the equilibrium position of *m*.

Since, for the spring, $\omega^2 = k/m$, we have for the period of a spring,

$$T = \frac{1}{f} = \frac{2\pi}{\omega} = \frac{2\pi}{\sqrt{k/m}} = 2\pi\sqrt{\frac{m}{k}} \quad (15\text{-}9)$$

The period does not depend on the amplitude (maximum displacement) of the vibration.

Example 15-4 A helical spring has a modulus $k = 10$ lb/in. A 100-lb weight is attached to the spring. After attaining an equilibrium position, the 100-lb weight is pulled down 3 in. and released. (a) How far does the 100-lb weight stretch the spring? (b) What is the unbalanced force acting on the weight after the spring is stretched the additional 3 in.? (c) What is the maximum acceleration of the weight? (d) What is the velocity of the weight when its displacement is 1 in. above the equilibrium position? (e) How long does it take the weight to go from a displacement of $+1.5$ in. to a displacement of -1.5 in.?

Solution
(a) Equilibrium takes place when the 100-lb weight equals the restoring force. Thus

$$W = kx_0$$

$$x_0 = \frac{W}{k}$$

$$x_0 = \frac{100\ \text{lb}}{10\ \text{lb/in.}} = 10\ \text{in.} = \frac{5}{6}\ \text{ft}$$

(b) The unbalanced force is now given by Eq. 15-6 with the displacement from equilibrium being 3 in. Thus

$$F_{\text{unbalanced}} = -kx$$
$$= (-10\ \text{lb/in.})\ 3\ \text{in.}$$
$$= -30\ \text{lb}$$

(c) The maximum acceleration occurs when $x = \frac{1}{4}$ ft. At that point the unbalanced force is

-30 lb. Then, from Eq. 15-8,

$$ma = -kx = -30\ \text{lb}$$

$$a = \frac{-30\ \text{lb}}{100\ \text{lb}/32\ \text{ft/sec}^2}$$

$$= -\tfrac{960}{100} = -9.6\ \text{ft/sec}^2$$

(d) From our chart

$$v = \pm\omega\sqrt{R^2 - x^2}$$

Here R is the maximum displacement of the weight, which is $\frac{1}{4}$ ft. Thus

$$v = \pm\omega\sqrt{\left(\tfrac{1}{4}\right)^2 - \left(\tfrac{1}{12}\right)^2}$$

Now

$$\omega^2 = \frac{k}{m} = \frac{120}{\frac{100}{32}} = 32(1.2) = 38.4$$

$$\omega = \sqrt{38.4} = 6.2\ \text{rad/sec}$$

Hence

$$v = \pm 6.2\sqrt{\left(\tfrac{3}{12}\right)^2 - \left(\tfrac{1}{12}\right)^2} = \pm 6.2\sqrt{8/(12)^2}$$

$$= \pm\frac{6.2}{12}\sqrt{8} = \pm\frac{(6.2)(2.8)}{12} = \pm 1.45\ \text{ft/sec}$$

(e) From our chart $x = R\cos \omega t$. We have, for the two displacements,

$$x_1 = R\cos \omega t_1$$

$$x_2 = R\cos \omega t_2$$

$$x_1 = +\frac{1.5}{12} = \frac{3}{12}\cos \omega t_1$$

$$\cos \omega t_1 = +\tfrac{1}{2}$$

$$\omega t_1 = \cos^{-1}\left(+\frac{1}{2}\right) = \frac{\pi}{3}$$

since the angle whose cosine is $\frac{1}{2}$ is $60°$ or $\pi/3$ radians. Thus

$$t_1 = \frac{\pi}{3\omega} = \frac{\pi}{3(6.2)} = \frac{\pi}{18.6}\ \text{sec}$$

$$x_2 = -\frac{1.5}{12} = \frac{3}{12}\cos \omega t_2$$

$$\cos \omega t_2 = -\tfrac{1}{2}$$

$$\omega t_2 = \cos^{-1}\left(-\frac{1}{2}\right) = \frac{2\pi}{3}$$

since the angle whose cosine is $-\frac{1}{2}$ is $120°$ or $2\pi/3$ radians. Thus

$$t_2 = \frac{2\pi}{3\omega} = \frac{2\pi}{3(6.2)} = \frac{2\pi}{18.6} \text{ sec}$$

The time it takes the weight to move from x_1 to x_2 is

$$t_2 - t_1 = \frac{2\pi}{18.6} - \frac{\pi}{18.6} = \frac{\pi}{18.6}$$

$$= 0.17 \text{ sec}$$

Note that the time it takes the weight to go from $+3$ to -3, which represents one half-period, is

$$\frac{T}{2} = \frac{1}{2f} = \frac{2\pi}{2\omega} = \frac{\pi}{\omega} = 0.51 \text{ sec}$$

Thus the particle takes less than half the time to cover half the displacement. ◀

Example 15-5 A 100-g mass stretches a spring 80 cm. If the mass is set vibrating on the spring, what will be the period of the motion?

Solution. The force constant of the spring

$$k = \frac{F}{x} = \frac{mg}{x} = \frac{100(980)}{80} = (25)(49) \frac{\text{dyne}}{\text{cm}}$$

Then, from Eq. 15-9,

$$T = 2\pi\sqrt{\frac{m}{k}} = 2\pi\sqrt{\frac{100}{25(49)}} = \frac{4\pi}{7} = 1.80 \text{ sec}$$ ◀

Example 15-6 A helical spring is stretched 20 cm when a certain mass is suspended from the spring. What is the period of oscillation if the mass is displaced downward a short additional distance and released?

Solution. At first glance, this problem does not appear to furnish sufficient information to determine the period, since we are not given the mass of the suspended weight and the spring constant. However, the information given is just enough to determine the ratio m/k which appears in the formula for the period (Eq. 15-9). From

Eq. 15-7

$$x_0 = \frac{mg}{k}$$

or

$$\frac{m}{k} = \frac{x_0}{g}$$

Substituting for x_0 and g, we have

$$\frac{m}{k} = \frac{20}{980} = \frac{1}{49}$$

Then, making use of Eq. 15-9, we find that

$$T = 2\pi\sqrt{\frac{m}{k}}$$

$$= 2\pi\sqrt{\frac{1}{49}}$$

$$= \frac{2\pi}{7} \text{ sec} = 0.90 \text{ sec}$$ ◀

Example 15-7 A spring has a period of vibration of 2 sec when a mass of 400 g is oscillating on the spring. Find the force constant of the spring.

Solution. Using Eq. 15-9, and solving for k, we have,

$$T = 2\pi\sqrt{\frac{m}{k}}$$

$$T^2 = \frac{4\pi^2 m}{k}$$

and

$$k = \frac{4\pi^2 m}{T^2}$$

Substituting the given information yields

$$k = \frac{4\pi^2(400)}{(2)^2}$$

$$= 400\pi^2 \text{ dyne/cm}$$

$$= 3.95 \times 10^3 \text{ dyne/cm}$$ ◀

15-3 SIMPLE PENDULUM

Consider a mass m attached to a thin string or rod of length l and negligible mass. When the string

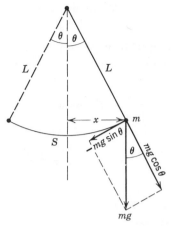

Figure 15-8 Simple harmonic motion in a simple pendulum.

is hung in a vertical position and the mass displaced from equilibrium a short distance and released, the ensuing motion is periodic. It is to be shown that the periodic motion of such a pendulum is simple harmonic. In Fig. 15-8, the pendulum has been displaced by an angle θ from the vertical. The unbalanced force acting on m, tangent to the arc S, is $-mg \sin \theta$. But $\sin \theta = x/l$, where x is the horizontal displacement of m. Thus, from Newton's second law,

$$F = -mg \sin \theta = -\frac{mgx}{l} = ma$$

or $\qquad a = -\frac{gx}{l}$ $\qquad\qquad$ (15-10)

If θ is small, the direction of the unbalanced force, and hence the acceleration, will be opposite to x rather than along the tangent to the arc. Thus Eq. 15-10 satisfies the definition of simple harmonic motion, with $\omega^2 = g/l$. The angular velocity of the pendulum is

$$\omega = \sqrt{\frac{g}{l}} \qquad\qquad (15\text{-}11)$$

and the period

$$T = \frac{1}{f} = \frac{2\pi}{\omega} = 2\pi \sqrt{\frac{l}{g}} \qquad (15\text{-}12)$$

Measuring the period of a simple pendulum of fixed length has been found to be an excellent way of determining g. The period is independent of the angular amplitude of the motion, provided that θ is small. When θ is larger than $10°$, the pendulum's motion is no longer simple harmonic and the period becomes a rather complicated function of the amplitude. The period is also independent of the mass m.

Example 15-8 A simple pendulum, 1 m long, is found to have a period of 2 sec. What is the acceleration due to gravity as determined from this experiment?

Solution. From Eq. 15-12

$$T = 2\pi \sqrt{\frac{l}{g}}$$

$$T^2 = \frac{4\pi^2 l}{g}$$

and $\qquad g = 4\pi^2 \frac{l}{T^2} = 4\pi^2 \times \frac{1 \text{ m}}{4 \text{ sec}^2}$

$$g = 9.87 \text{ m/sec}^2 \qquad\qquad ◀$$

Example 15-9 What is the length (in feet) of a pendulum whose period is π sec?

Solution. From Eq. 15-12

$$T = 2\pi \sqrt{\frac{l}{g}}$$

Solving for l, we find that

$$T^2 = \frac{4\pi^2 l}{g}$$

and $\qquad l = \frac{g}{4\pi^2} T^2$

Substituting the given information, we obtain

$$l = \frac{32}{4\pi^2} \times \pi^2$$

$$= 8 \text{ ft} \qquad\qquad ◀$$

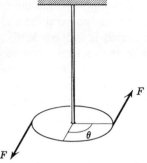

Figure 15-9　Simple harmonic motion in a torsion pendulum. The two forces marked F produce a torque that twists the wire. The restoring torque in the wire is proportional to θ.

15-4 THE TORSION PENDULUM

When a mass m is hung from a wire, twisted as in Fig. 15-9, and released, periodic oscillations that are simple harmonic take place. The restoring torque M is proportional to the angle of twist θ. Thus

$$M = -k'\theta \qquad (15\text{-}13)$$

where k' is a constant that depends on the material from which the wire is made. Since

$$M = I\alpha$$

where I is the moment of inertia of the mass m about an axis through its center, and α the angular acceleration, we have

$$M = I\alpha = -k'\theta$$

and

$$\alpha = -\frac{k'\theta}{I} \qquad (15\text{-}14)$$

Equation 15-14 is identical to the definition of simple harmonic motion, except here expressed in angular variables. Again $\omega^2 = k'/I$, and

$$T = \frac{1}{f} = \frac{2\pi}{\omega} = 2\pi\sqrt{\frac{I}{k'}} \qquad (15\text{-}15)$$

Example 15-10　A solid, homogeneous disk of mass 100 g and radius 10 cm is hung from a wire with torsion constant $k' = 2 \times 10^6$ dyne-cm/rad.

Find the period of the oscillations if the disk is twisted and released.

Solution. The moment of inertia of the disk about an axis through its center is $I = \frac{1}{2}mr^2$. Hence

$$I = \frac{1}{2}(100)(10)^2 = 5 \times 10^3 \text{ g-cm}^2$$

From Eq. 15-15

$$T = 2\pi\sqrt{\frac{I}{k'}} = 2\pi\sqrt{\frac{5 \times 10^3}{2 \times 10^6}} = \frac{2\pi}{\sqrt{400}} = \frac{\pi}{10}$$

$$= 0.31 \text{ sec} \qquad \blacktriangleleft$$

15-5 THE PHYSICAL PENDULUM

When the mass of the rod in a simple pendulum arrangement cannot be neglected, the pendulum is then known as a *physical pendulum* (or compound pendulum). Figure 15-10 shows a physical pendulum suspended about a horizontal axis O. If the pendulum is displaced an angle θ from the vertical, we have

$$\text{resultant torque} = I\alpha$$
$$-mgh\sin\theta = I_0\alpha \qquad (15\text{-}16)$$

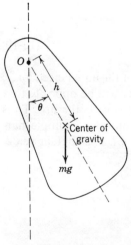

Figure 15-10　The physical pendulum. The weight *mg* produces a torque about the point of suspension O.

where I_0 represents the moment of inertia of the body about the point of suspension. If θ is small, $\sin \theta \approx \theta$, and

$$\alpha = -\frac{mgh}{I_0}\theta \qquad (15\text{-}17)$$

which satisfies the equation for harmonic motion with

$$\omega_p^{\,2} = \frac{mgh}{I_0} \qquad (15\text{-}18)$$

What length of simple pendulum has the same frequency (or period) as this physical pendulum? Equating the two angular frequencies (or their squares), we have

$$\omega_p^{\,2} = \omega_s^{\,2}$$

$$\frac{mgh}{I_0} = \frac{g}{L} \qquad (15\text{-}19)$$

Hence the length of the equivalent simple pendulum is

$$L = \frac{I_0}{mh} \qquad (15\text{-}20)$$

Example 15-11   A physical pendulum consists of a hollow ring of radius 10 cm suspended about a horizontal axis at the circumference. (a) What is the period of the physical pendulum? (b) What is the length of the equivalent simple pendulum?

Solution
(a) We first calculate the moment of inertia about the point of suspension. From the parallel axis theorem, Eq. 11-13,

$$I_0 = I_C + md^2$$
$$= m(10)^2 + m(10)^2$$
$$= 200m$$

where m is the mass of the ring. Then from Eq. 15-18

$$\omega_p^{\,2} = \frac{mgh}{I_0} = \frac{mg(10)}{200m} = \frac{980}{20}$$

$$\omega_p^{\,2} = 49$$

Then the period is given by

$$T^2 = \frac{4\pi^2}{\omega_p^{\,2}} = \frac{4\pi^2}{49}$$

and $\qquad T = \dfrac{2\pi}{7}\ \sec = 0.90\ \sec$

(b) From Eq. 15-20

$$L = \frac{I_0}{mh}$$

$$= \frac{200m}{m(10)}$$

$$= 20\ cm \qquad \blacktriangleleft$$

SUMMARY

Simple harmonic motion occurs whenever the acceleration of a particle is directly proportional to the displacement but oppositely directed: $a = -Cx$, where C is a constant.

The motion of a particle in SHM can be studied by referring to the motion of a particle in a *reference circle*.

The period of oscillation for a mass, m, on a spring of force constant k is

$$T = 2\pi\sqrt{\frac{m}{k}} \qquad \textit{simple spring}$$

The period of oscillation for a simple pendulum of length l is

$$T = 2\pi\sqrt{\frac{l}{g}} \qquad \textit{simple pendulum}$$

The period of oscillation for a torsion pendulum with moment of inertia I and torsion constant k' is

$$T = 2\pi\sqrt{\frac{I}{k'}} \qquad \textit{torsion pendulum}$$

The period of oscillation of a physical pendulum of mass m, whose center of gravity is h units below the point of support, and with moment of inertia I_0 about the point of support is

$$T = 2\pi \sqrt{\frac{I_0}{mgh}} \qquad physical\ pendulum$$

The length of the simple pendulum whose period is the same as this physical pendulum is $L = I_0/mh$.

QUESTIONS

1. Discuss the differences between two classes of periodic motion: circular motion and simple harmonic motion.

2. What are the *two* basic conditions contained in the definition of simple harmonic motion?

3. When a particle moving with simple harmonic motion has a negative displacement, will its velocity be positive, negative, or either?

4. How are the frequency and period related in simple harmonic motion?

5. For a helical spring, what is the value of C in the equation, $a = -Cx$?

6. What is the value of ω^2 for a simple pendulum?

7. Does the period of a simple pendulum depend upon the weight of the pendulum?

8. How does a physical pendulum differ from a simple pendulum?

9. Does the period of a physical pendulum depend upon the total mass of the pendulum or the distribution of the mass? Explain.

PROBLEMS

(A)

1. A 2-lb weight stretches a spring 3 in.
   (a) What is the force constant of the spring?
   (b) What is the natural frequency of vibration of a 64-lb body attached to the spring?

2. A 6-lb body is attached to a long spiral spring. The spring is known to stretch 4 in. for each 2 lb of force exerted on it.
   (a) What is the force constant of the spring?
   (b) Find the period of vibration of the 6-lb body if it is pulled down slightly and released.

3. When an 8-lb weight is attached to a spring, the spring is elongated by 2 in. The weight is pulled down an additional 6 in. and then released.
   (a) What is the force constant of the spring?
   (b) What is the maximum acceleration of the weight?
   (c) What is the speed of the weight as it passes the equilibrium position?
   (d) What is the period of vibration?

4. A spring measures 22 in. when a 10-lb weight is suspended on it, and 28 in. when a 15-lb weight is suspended on it.
   (a) What is the force constant of the spring?
   (b) What was the original length of the spring?

5. A spring has a force constant of 10^4 dynes/cm. A certain weight, when set into vibration on the spring, has a period of oscillation of $\pi/10$ sec. What is the mass of the suspended weight?

6. What is the length of a pendulum whose period is 1 sec?

7. Compute the acceleration of gravity at a place where a simple pendulum, 100 cm long, oscillates with a period of 2 sec.

8. How long is a simple pendulum that makes 25 complete vibrations in 5π sec? (Assume that $g = 980$ cm/sec^2.)

9. A children's swing in a neighborhood playground is 8 ft long. When set into simple harmonic motion, what is the period of the motion?

10. A 9-lb weight executes simple harmonic motion with a period of 4 sec at the end of a helical spring. Determine the period when a 4-lb weight is attached to the same spring.

11. A vibrating particle (SHM) has an acceleration of 6 ft/sec^2 when its displacement is 2 in. What is the period of the vibration?

12. A particle executes simple harmonic motion. When its displacement is 2 cm, its velocity and acceleration are 20 cm/sec and -50 cm/sec^2, respectively.
    (a) What is the frequency of vibration?
    (b) Find the amplitude of vibration.

13. A body of mass 0.4 slug oscillates in simple harmonic motion with a frequency of 10 cycles/sec and an amplitude of 0.25 ft.
    (a) What is the maximum velocity of the body?
    (b) What is the kinetic energy of the body when its displacement is one-half the amplitude?
    (c) What is the maximum elastic potential energy of the body?

14. A mass of 12 g executes simple harmonic motion with an amplitude of 2 cm and a period of 4 sec.
    (a) What is its maximum acceleration?
    (b) What is its maximum velocity?
    (c) What is its acceleration $\frac{1}{2}$ sec after its displacement was a maximum?
    (d) What is its velocity at this instant?
    (e) What is the force acting on the body when its displacement is $\frac{1}{4}$ cm above its equilibrium position?

15. A 2-kg mass vibrates to and fro along a straight path, 18 cm long, in simple harmonic motion. The period of the motion is 4 sec. Determine
    (a) the amplitude of the vibration,
    (b) the maximum velocity and maximum acceleration,
    (c) the velocity and acceleration when the displacement is 5 cm,
    (d) the restoring force when the mass is at the midpoint of its path, and
    (e) the restoring force when the displacement is 7 cm.

16. The prong of a tuning fork vibrates with a frequency of 400 cycles/sec and an amplitude of 4 mm.
    (a) What are the maximum acceleration and velocity?
    (b) What is the velocity when the displacement is 3 mm?

17. A thin rod, 4 ft long, hangs suspended about a horizontal axis through one end.
    (a) Find the period of the motion.
    (b) What is the length of the equivalent simple pendulum?

18. A uniform bar 140 cm long with a mass of 6 kg is suspended from one end. What is the period of oscillation of the physical pendulum?

(B)

1. A helical spring is stretched 25 cm when a given mass is attached to it. What is the period of vibration if the mass is displaced downward slightly and released?

2. A weight W stretches a spring $\frac{1}{2}$ ft. What is the period of vibration of this weight if it is set into vibration with the spring?

3. When a 0.5-kg mass is attached to an unstretched spring and released, the weight falls a maximum distance of 40 cm.
    (a) What is the amplitude of the simple harmonic motion?
    (b) What is the force constant of the spring?
    (c) What is the acceleration of the weight at the lowest point of its motion?

4. What is the period of a simple pendulum, 100 cm long, located on the moon? (The acceleration of gravity on the moon is $\frac{1}{6}$ of its acceleration on earth.)

5. What is the period of vibration of a 2-ft pendulum on Planet X where the acceleration of gravity is 50 ft/sec^2?

6. A large steel ring (6 ft in diameter) is suspended as a physical pendulum on an axis through its circumference (Fig. 15-11). What is the period of vibration of the ring when oscillating in simple harmonic motion? (The moment of inertia of a ring about an axis through its circumference is $2\,mR^2$, where m is the mass of the ring and R is the radius.)

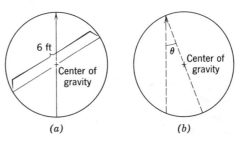

(a)   (b)

Figure 15-11

7. A certain mechanism requires the use of a physical pendulum to keep time. It is essential that the pendulum have a frequency of $5/\pi$ Hz (1 Hz = 1 cycle/sec). If the physical pendulum is a ring, suspended at its circumference, what must be the radius of the ring? (Assume that $g = 980$ cm/sec^2.)

8. A smooth block ($m = 0.2$ kg) is attached to a spring lying on a smooth table (Fig. 15-12). When the block is pulled to the right and released, it begins its motion with an acceleration of 5 m/sec^2. The spring has a stiffness constant of 8 N/m.
    (a) What is the amplitude of the vibrations?
    (b) What is the period of the motion?
    (c) What is the maximum speed of the block?

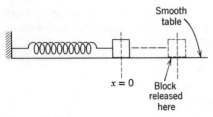

Figure 15-12

9. A spring has a force constant k. The spring is cut into two equal parts.
   (a) What would be the force constant of each part?
   (b) What would be the force constant of a composite spring, constructed from the two parts as shown in Fig. 15-13?

Figure 15-13

10. A 20-g stone sits on a platform that vibrates vertically in simple harmonic motion with an amplitude of 0.05 cm.
    (a) What is the maximum frequency of vibration for which the stone always remains in contact with the platform?
    (b) What is the maximum force exerted by the platform on the stone at this frequency?

11. A mass of 0.5 kg executes simple harmonic motion with an amplitude of 10 cm. The 0.5-kg mass has a velocity of 80 cm/sec when the displacement is -6 cm.
    (a) What is the frequency of the simple harmonic motion?
    (b) What is the maximum velocity?
    (c) What is the maximum acceleration?
    (d) What is the restoring force acting on the mass when the velocity is 60 cm/sec?

12. A 2-lb block is vibrating in simple harmonic motion. The frequency is $2/\pi$ Hz and the amplitude is 4 in.
    (a) What is the maximum acceleration of the block?
    (b) What is the maximum unbalanced force acting on the block?
    (c) How long does it take for the block to make 10 vibrations?

13. A particle moves in simple harmonic motion with a period of 4 sec and an amplitude of 8 in.
    (a) What is the maximum acceleration?
    (b) What is the maximum velocity?
    (c) What is the acceleration $\frac{1}{2}$ sec after the velocity is a maximum?
    (d) What is the displacement $\frac{2}{3}$ sec after the acceleration is a maximum?

14. The particle in Fig. 15-14 is experiencing simple harmonic motion with a period of 2 sec and an amplitude of 20 cm.
    (a) How long does it take the particle to move from $x = 20$ cm to point A, where $x = 10$ cm.
    (b) How long does it take the particle to move from A to point B, where $x = -10$ cm?
    (c) How long does it take the particle to move from B to point C, where $x = 10\sqrt{3}$ cm?

Figure 15-14

15. The piston of an engine moves up and down through a distance of 4.8 in. The engine makes 3000 rev/min. Assuming SHM, determine
    (a) the maximum acceleration of the piston,
    (b) the maximum force on the piston if the piston assembly weighs 1.0 lb.

16. A disk is suspended in a horizontal position by a wire attached to its center. A couple having a moment of 8 newton-meters is required to twist the disk through an angle of 15° in the horizontal plane. When released, the period of the oscillatory motion is $\frac{1}{3}$ sec. Determine the moment of inertia of the disk.

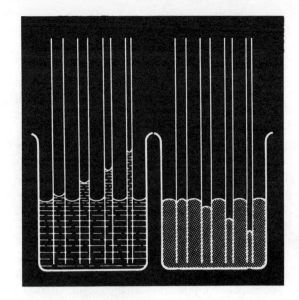

16.
Fluid Mechanics

In this chapter we consider the study of fluids. Fluids are materials that possess the property of flow and includes liquids and gases. After defining the basic concepts of density and pressure, the laws relating to fluids at rest (hydrostatics) and fluids in motion (hydrodynamics) are developed. Some interesting phenomena uniquely associated with fluids (surface tension and capillarity) are discussed at the end of the chapter.

LEARNING GOALS

While you study this chapter be alert to the following major learning objectives:

1. *To learn the definitions of density and pressure and the vocabulary related to fluid mechanics.*

2. *To learn how to calculate fluid pressure by using Pascal's principle.*

3. *To understand how Pascal's principle is utilized in several practical devices.*

4. *To learn the concept of buoyancy and how to use Archimedes' principle in solving problems in fluid statics.*

5. *To learn how to use the equation of continuity and Bernoulli's equation to solve problems in fluid dynamics.*

6. *To recognize how Bernoulli's equation helps to understand several interesting phenomena, such as the curving of a baseball.*

7. *To understand the characteristics of surface tension and capillarity.*

16-1 DENSITY

The term *fluid* refers to materials that flow and includes liquids and gases. In these two physical states, matter has no shape of its own, as it does in the solid state but takes the shape of its container. Liquids have definite volumes at any particular temperature, but the volume of a gas is always the volume of space within which the gas is confined. Thus, if a gas is enclosed in a test tube, its volume is the volume of the test tube. But, if the test tube is opened and the gas is allowed to escape, its new volume, after a short while, will be the volume of the room. The motion of fluids conforms to the same basic laws that have been discussed in conjunction with the motion of solids. Newton's laws and the work–energy principle are used to discover relationships between those variables that describe the state of motion of the fluid. Two of the important variables that are used to describe the state of a fluid are its density and pressure.

The mass density, ρ, is the mass per unit volume,

$$\rho = \frac{m}{V} \qquad (16\text{-}1)$$

The mass density differs from the weight density, which is the weight per unit volume.

$$\text{weight density} = \frac{W}{V} = \frac{mg}{V}$$

or   weight density $= \rho g$

$$(16\text{-}2)$$

We see that the weight density is the product of the mass density and the acceleration of gravity. Unless otherwise indicated, the term *density* shall always refer to the mass density, ρ. The density of a uniform fluid depends upon the temperature and pressure. For liquids, the variations in density with temperature or pressure are very slight compared with similar variations in gases. Table 16-1 gives some densities for common solids, liquids, and gases at standard temperature and pressure (i.e., 0°C and 1-atm pressure). It should be noted from Table 16-1 that the density of liquids is about 1000 times greater than gases, and the density of solids is between 1000 and 10,000 times greater than gases.

We define specific gravity as the ratio of the weight (or mass) density of a substance to the weight (or mass) density of water. Mercury with a density of 13.6 g/cm^3, compared to 1 g/cm^3 for

Table 16-1 Densities for common liquids and gases

Densities (in kg/m^3) at 0°C and 1-atm pressure

Gases		Solids	
Air	1.293	Aluminum	2.70×10^3
Carbon dioxide (CO$_2$)	1.977	Brass (approx)	8.5×10^3
Helium (He)	0.1785	Copper	8.9×10^3
Hydrogen (H$_2$)	0.0899	Cork (approx)	0.24×10^3
Oxygen (O$_2$)	1.429	Gold	19.3×10^3
Nitrogen (N$_2$)	1.251	Ice	0.92×10^3
		Iron (approx)	7.6×10^3
Liquids		Lead	11.3×10^3
Ether	0.73×10^3	Platinum	21.4×10^3
Ethyl alcohol	0.81×10^3	Silver	10.5×10^3
Mercury	13.6×10^3	Wood, elm (approx)	0.57×10^3
Olive oil	0.92×10^3	Wood, white pine	0.42×10^3
Sea water	1.03×10^3	Zinc	7.1×10^3
Water	1.00×10^3		

From R. Resnick and D. Halliday, *Physics for Students and Engineering*, John Wiley & Sons, Inc., New York, 1960, p. 337

water, has a specific gravity of 13.6. Many alcohols have densities near 0.8 g/cm^3 and specific gravities of 0.8. Note that in the engineering system of units, the weight density of water is 62.4 lb/ft^3.

Example 16-1 What is the density of 1000 m^3 of a substance that has a mass of 1.65 × 10^3 kg? Is this substance a solid, liquid, or gas?

Solution The density is the mass per unit volume. From Eq. 16-1

$$\rho = \frac{m}{V}$$

$$= \frac{1.65 \times 10^3 \text{ kg}}{10^3 \text{ m}^3}$$

or $\rho = 1.65$ kg/m^3

Since this value falls more readily among the gaseous densities, the substance is most likely a gas. ◀

Example 16-2 What is the mass of 2 m^3 of olive oil?

Solution. From Table 16-1, we see that the density of olive oil is 0.92 × 10^3 kg/m^3. Using Eq. 16-1, we have

$$\rho = \frac{m}{V}$$

or $m = \rho V$

Substituting the given information, we obtain

$$m = (0.92 \times 10^3 \text{ kg/m}^3)(2 \text{ m}^3)$$
$$= 1.84 \times 10^3 \text{ kg} \quad ◀$$

Example 16-3 Material X has a specific gravity of 12. What is the weight of 11 ft^3 of materal X?

Solution. To find the weight, we must first know the density of material X. Since its specific gravity is 12, its density must be 12 times greater than the density of water, Thus

weight density of X

$$= (\text{specific gravity})(62.4 \text{ lb/ft}^3)$$
$$= (12)(62.4 \text{ lb/ft}^3)$$
$$= 748.8 \text{ lb/ft}^3$$
$$= 750 \text{ lb/ft}^3$$

Now, from the definition of weight density,

$$\text{weight density} = \frac{\text{weight}}{\text{volume}}$$

or $\text{weight} = (\text{weight density})(\text{volume})$
$$= (750 \text{ lb/ft}^3)(11 \text{ ft}^3)$$

and $w = 8250$ lb ◀

Example 16-4 How many times denser is lead than zinc? Platinum than copper?

Solution. To obtain the desired result, we need only form the ratios of the respective densities. Thus, to find out how many times denser lead is than zinc, we form the ratio

$$\frac{\rho_{\text{lead}}}{\rho_{\text{zinc}}} = \frac{11.3 \times 10^3 \text{ kg/m}^3}{7.1 \times 10^3 \text{ kg/m}^3} = \frac{11.3}{7.1} = 1.6$$

Lead is 1.6 times denser than zinc. Similarly, for platinum and copper,

$$\frac{\rho_{\text{platinum}}}{\rho_{\text{copper}}} = \frac{21.4 \times 10^3 \text{ kg/m}^3}{8.9 \times 10^3 \text{ kg/m}^3} = \frac{21.4}{8.9} = 2.4$$

Platinum is 2.4 times denser than copper. ◀

Example 16-5 How many cubic feet of white pine wood are required to make a 1310-lb load?

Solution. From the definition of weight density

$$V = \frac{\text{weight}}{\text{weight density}}$$

The weight density of white pine is its specific gravity (0.42) times the weight density of water. Thus

$$\text{weight density} = (0.42)(62.4 \text{ lb/ft}^3)$$
$$= 26.2 \text{ lb/ft}^3$$

Therefore, the volume required is

$$V = \frac{1310 \text{ lb}}{26.2 \text{ lb/ft}^3}$$
$$= 50 \text{ ft}^3 \quad ◀$$

16-2 HYDROSTATIC PRESSURE AND PASCAL'S PRINCIPLE

The second important variable used in describing the state of a fluid is pressure. The *pressure* exerted against a surface is defined as the force exerted per unit area of the surface. Thus

$$p = \frac{\text{force exerted against surface}}{\text{area of surface}}$$

$$= \frac{F}{A} \qquad (16\text{-}3)$$

Pressure is measured in units of lb/ft^2, N/m^2 or $dynes/cm^2$. Pressure is a scalar quantity and, at any point of a fluid, it is exerted in all directions. If a pressure exists at a point in a fluid and a surface of area A is located at that point, there will be a force on the area, given by

$$F = pA \qquad (16\text{-}4)$$

and this force will be *perpendicular* to the area (Fig. 16-1).

The pressure in a liquid varies with the depth of the liquid. If the liquid is at rest, the pressure is referred to as *hydrostatic pressure*. To determine the relationship between hydrostatic pressure and depth, consider a liquid of density ρ, which is at rest. At a depth h below the surface of the liquid let the pressure be denoted by p. Consider a volume of the liquid as outlined in Fig. 16-2. The pressure on the top surface is denoted by p_0. Since this element of fluid is at rest, the vector sum of all the forces acting on this element must be zero. The forces in the horizontal directions are symmetrically oriented and equal. Thus ΣF_x is clearly equal to zero. In the vertical direction we have three forces acting on our volume element: The pressure above the liquid exerts a downward force on the top surface; the

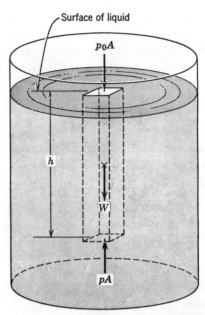

Figure 16-2 Hydrostatic pressure beneath the surface of a liquid. The pressure p_0 is the pressure at the surface of the liquid, which may be caused by the atmosphere or by mechanical means.

weight of our volume element of liquid is a downward force; and the pressure at the depth h exerts an upward force on the element. Applying Newton's first law, we have

$$\Sigma F_y = 0; \qquad p_0 A + W - pA = 0$$

or

$$pA = p_0 A + W$$

This equation can be simplified further if we recall that $W = \rho g V = \rho g(Ah)$, since the volume of our element is the cross-sectional area multiplied by the height. Thus

$$pA = p_0 A + \rho g h A$$

and

$$p = p_0 + \rho g h \qquad (16\text{-}5)$$

Equation 16-5 expresses the relationship between the pressure at a depth h in a liquid and the depth. It indicates that the pressure increases by ρg for each unit of depth below the surface. At the surface, where $h = 0$, the pressure is just p_0. For an open tank this represents atmospheric

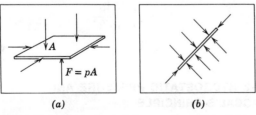

Figure 16-1 The hydrostatic forces exerted by a fluid against a surface are always perpendicular to the surface.

pressure. The quantity $\rho g h$ is the difference between the total pressure p and the atmospheric pressure p_0. This difference is called the *gauge pressure*; almost all pressure gauges are calibrated to measure gauge pressures. The total pressure is called the *absolute pressure*. The gauge pressure, $\rho g h$, is the weight of a column of fluid, h units high and one square unit in cross section. In a similar manner, the atmospheric pressure, p_0, represents the weight of all the air in a column of one square-unit cross section extending from sea level to the top of the atmosphere. If the cross section is one square inch, the weight of all of this air is 14.7 lb. Consequently, atmospheric pressure at sea level is 14.7 lb/in.^{2}, which is equivalent to 1.013×10^5 N/m^2.

Equation 16-5 conveys several important ideas. Note that the equation contains no information about the overall size or shape of the container. The pressure depends *only* on the depth of liquid. This phenomenon was first recognized by the French scientist Blaise Pascal during the eighteenth century. It leads to the "hydrostatic paradox" depicted in Fig. 16-3. All three containers hold different weights of fluid, but the forces exerted on the equal bases for the same height of fluid are equal in all three cases. Another important idea that comes from Eq. 16-5 is that a pressure which is exerted at any point of a confined fluid is transmitted equally throughout the fluid, in all directions. We see this from the presence of the atmospheric pressure in Eq. 16-5. This pressure is exerted at the top of the fluid, yet it is part of the total pressure at any depth h throughout the fluid. This property of a confined fluid, to transmit pressures throughout its volume, is known as Pascal's principle and is employed in many hydraulic systems such as hydraulic brakes, hydraulic jacks, retractable landing gears, recoil mechanisms, and the common barbershop chair. A typical example of the application of Pascal's principle is the hydraulic press.

The hydraulic press consists of one cylinder in which a piston can operate up and down, in addition to a second cylinder of larger diameter which is connected to the first cylinder by means of a pipe (see Fig. 16-4). A closely fitting piston is found in the larger cylinder. If a force of any magnitude is applied to the smaller piston, the pressure produced on the fluid in the smaller cylinder is transmitted to all parts of the fluid. A

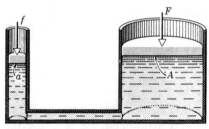

Figure 16-4 The hydraulic press.

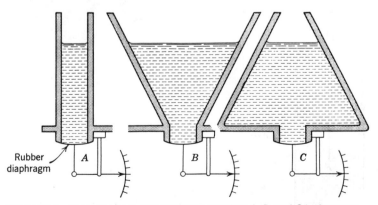

Figure 16-3 Pascal's vases. The pressure at points *A*, *B*, and *C* is the same.

pressure of 1 lb/in.^{2} at the smaller piston will be transmitted to all surfaces of the hydraulic press and to the piston in the larger cylinder. If the area of the large piston is 10 in.^{2}, the total force on the larger piston is 10 lb because 10 in.$^2 \times$ 1 lb/in.^{2} = 10 lb.

The distance that the smaller piston travels is always greater than the distance traveled by the larger one. This is an example of the conservation of energy. If the smaller piston (1 in.^{2}) travels 10 ft, the larger piston (10 in.^{2}) will travel 1 ft. Thus the mechanical advantage obtained is gained at the expense of increasing the distance that the piston must travel. The distance that the larger piston moves is always equal to the distance that the smaller piston moves divided by the ideal mechanical advantage. The actual mechanical advantage is (Fig. 16-5)

$$\text{AMA} = \frac{\text{force output}}{\text{force input}} \quad \text{or} \quad \text{AMA} = \frac{F}{f} \quad (16\text{-}6)$$

The ideal mechanical advantage of a hydraulic press can be determined by the ratio of the area of the large piston to the area of the small piston, i.e.,

$$\text{IMA} = \frac{\text{large piston area}}{\text{small piston area}} = \frac{A}{a} \quad (16\text{-}7)$$

Another method for determining the mechanical advantage of a hydraulic press is to use the ratio of the squares of the diameters or radii of the two pistons:

$$\text{IMA} = \frac{\text{large diameter}^2}{\text{small diameter}^2} = \frac{D^2}{d^2}$$

$$\text{IMA} = \frac{\text{large radius}^2}{\text{small radius}^2} = \frac{R^2}{r^2} \quad (16\text{-}8)$$

The law of machines applies throughout the operation of the hydraulic press. The input work is equal to the effort force applied to the smaller piston multiplied by the distance that the smaller piston travels. The output work is the force on the large piston multiplied by the distance that the large piston travels. Since input work equals output work, effort × distance = resistance × resistance distance, or $f \times s_i = F \times s_0$ (Fig. 16-5).

Many large cannons use a complex hydraulic recoil mechanism. The recoil (the reaction to the explosion that moves the cannon backward) forces a piston to move oil, thus doing work. Since work is done, part of the energy of the recoil is removed. This prevents the cannon mounting or base from moving; only the gun barrel moves. In some cannon the same hydraulic pressure in the recoil mechanism returns the gun to its original loading position.

Jacks to lift trucks and other heavy machines are often of the hydraulic type. A hydraulic jack is simply a hydraulic press (see Fig. 16-6). Almost all chassis lubrication is done by hydraulic pressure. Power steering depends on the same principle (see Fig. 16-7).

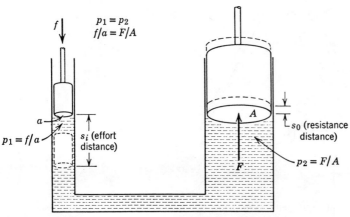

Figure 16-5   Mechanical advantage of the hydraulic press.

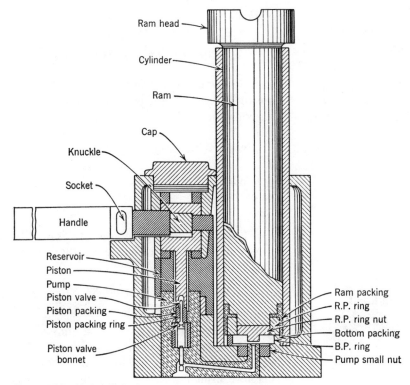

Figure 16-6   Hydraulic jack.

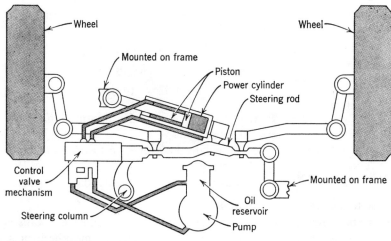

Figure 16-7   Power steering.

Example 16-6 A 4000-lb drum rests on a rectangular base that is 10 ft × 20 ft. What pressure is being exerted on the base by the drum. (Ignore atmospheric pressure.)

Solution. The pressure exerted on the base is the force per unit area that the drum exerts on the base. The total force exerted by the drum is its weight. Consequently,

$$p = \frac{F}{A}$$

$$= \frac{4000 \text{ lb}}{200 \text{ ft}^2} = 20 \text{ lb/ft}^2 \quad \blacktriangleleft$$

Example 16-7 What pressure is exerted by a 110-lb girl who is balancing on her high-heeled shoes which have a combined area of 0.5 in.²?

Solution. Using Eq. 16-3, we find that

$$p = \frac{F}{A}$$

$$= \frac{110 \text{ lb}}{0.5 \text{ in.}^2}$$

$$= 220 \text{ lb/in.}^2$$

This pressure is almost 15 times atmospheric pressure and accounts for the imprints that high-heeled shoes make on asphalt, linoleum, and carpets. It has been estimated that an average young lady on thin high-heels exerts more pressure on the ground than the Empire State Building! ◀

Example 16-8 Salt water has a weight density of 64 lb/ft³. What is the gauge pressure 8 ft below the surface of the Atlantic Ocean?

Solution. From the definition of gauge pressure,

$$p_{\text{gauge}} = \rho g h$$

$$= (64 \text{ lb/ft}^3)(8 \text{ ft})$$

$$= 512 \text{ lb/ft}^2 \quad \blacktriangleleft$$

Example 16-9 A submarine dives to its limit of 400 ft. What is the gauge pressure per square foot on its hull?

Solution

$$p = \rho g h$$

$$= 64 \text{ lb/ft}^3 \times 400 \text{ ft} = 25,600 \text{ lb/ft}^2 \quad \blacktriangleleft$$

Example 16-10 A swimming pool has dimensions of 32 ft × 20 ft and is filled with fresh water to a constant depth of 4 ft.

(a) What is the gauge pressure at the bottom?

(b) What is the absolute pressure at the bottom?

(c) What is the total force exerted on the bottom of the pool?

Solution

(a) The weight density of fresh water is 62.4 lb/ft³. At a depth of 4 ft the gauge pressure will be

$$p_{\text{gauge}} = \rho g h$$

$$= (62.4 \text{ lb/ft}^3)(4 \text{ ft}) = 249.6 \text{ lb/ft}^2$$

$$= 250 \text{ lb/ft}^2$$

(b) The absolute pressure is the sum of the atmospheric pressure and the gauge pressure. The atmospheric pressure is 14.7 lb/in.² or 2120 lb/ft². Therefore, using Eq. 16-5, the absolute pressure is

$$p = p_0 + p_{\text{gauge}}$$

$$= 2120 \text{ lb/ft}^2 + 250 \text{ lb/ft}^2$$

$$= 2370 \text{ lb/ft}^2$$

(c) The total force on the bottom of the pool is given by Eq. 16-4.

$$F = pA$$

where the area of the pool bottom is (32 ft)(20 ft) = 640 ft². Therefore,

$$F = (2370 \text{ lb/ft}^2)(640 \text{ ft}^2)$$

$$= 1,516,800 \text{ lb}$$

$$= 1.52 \times 10^6 \text{ lb} \quad \blacktriangleleft$$

Example 16-11 Using the information of the previous example,

(a) Calculate the average pressure against the walls of the pool.

(b) Calculate the force against the long wall.

Solution

(a) The essential pressure against the walls is the gauge pressure, because the atmospheric pressure is counteracted by the atmospheric pressure against the walls from the outside. Consequently, we are concerned with finding the average value of the gauge pressure, ρgh. Since the gauge pressure varies linearly with the depth, the average value for any interval will be the value at the midpoint of that interval. Therefore, the average pressure against the walls will be the pressure halfway up the wall, or

$$p_{average} = \rho g \times \frac{h}{2} = (62.4 \text{ lb/ft}^3)(2 \text{ ft}) = 124.8 \text{ lb/ft}^2$$

$$= 125 \text{ lb/ft}^2$$

(b) The area of the long wall is $32 \text{ ft} \times 4 \text{ ft} = 128 \text{ ft}^2$. The total force against this wall is found by using Eq. 16-4.

$$F = p_{average}A$$
$$= (125 \text{ lb/ft}^2)(128 \text{ ft}^2)$$
$$= 16,000 \text{ lb}$$
$$= 1.6 \times 10^4 \text{ lb} \qquad \blacktriangleleft$$

Example 16-12 A hydraulic press, like the one pictured in Fig. 16-5, has a small piston area of 4 in.^{2} and a large piston area of 24 in.^{2} Assuming 100% efficiency,

(a) What effort is required to raise a 900-lb block of steel?

(b) Through what distance will the block move when the effort has been exerted through 6 in.?

Solution

(a) For a machine with 100% efficiency the IMA and AMA are equal. Consequently, from Eqs. 16-6 and 16-7 we have

$$\text{IMA} = \text{AMA}$$

$$\frac{A}{a} = \frac{F}{f}$$

$$f = F \cdot \frac{a}{A}$$

$$= (900 \text{ lb})\left(\frac{4 \text{ in.}^2}{24 \text{ in.}^2}\right)$$

$$= 150 \text{ lb}$$

(b) Since the work input and work output must be equal, we have

$$\text{effort} \times \text{effort distance} = \text{resistance}$$
$$\times \text{resistance distance}$$

$$(150 \text{ lb})(6 \text{ in.}) = (900 \text{ lb})(s_0)$$

$$s_0 = \frac{(150 \text{ lb})(6 \text{ in.})}{900 \text{ lb}}$$

$$= 1 \text{ in.} \qquad \blacktriangleleft$$

Example 16-13 A hydraulic press has a small piston of radius 2 in. and a large piston of radius 3 in. It is found that it requires an effort of 100 lb to support a 180-lb resistance.

(a) What is the IMA?

(b) What is the AMA?

(c) What is the efficiency of the press?

Solution

(a) From Eq. 16-8

$$\text{IMA} = \frac{R^2}{r^2} = \frac{(3 \text{ in.})^2}{(2 \text{ in.})^2} = \frac{9}{4}$$

(b) From Eq. 16-6

$$\text{AMA} = \frac{F}{f} = \frac{180 \text{ lb}}{100 \text{ lb}} = \frac{9}{5}$$

(c) The efficiency of a machine is the ratio of the AMA to the IMA. Thus

$$\text{efficiency} = \frac{\text{AMA}}{\text{IMA}} = \frac{\frac{9}{5}}{\frac{9}{4}} = \frac{4}{5}$$

$$= 80\% \qquad \blacktriangleleft$$

16-3 BUOYANT FORCE; ARCHIMEDES' PRINCIPLE

One of the most important effects produced by hydrostatic pressure is the buoyant force which is exerted on all objects that are submerged in fluids. It is this buoyant force that enables a man or a huge steel ship to float on the sea. The buoyant force exerted by the atmosphere keeps balloons and dirigibles floating in the air. If a rock is thrown into a pool of water, the rock will sink, but the water will exert a buoyant force on the rock. If the rock is thrown into a pool of mercury, the buoyant force will be great enough to keep the rock floating. Hydrostatic pressure acts on all objects that are partially or completely immersed in a fluid, and the hydrostatic pressure is the cause of the buoyant force.

The concrete block in Fig. 16-8a weighs 20 lb. However, when the block is submerged in water, it appears to weigh only 15 lb. There is a 5-lb buoyant force acting upward on the block, which causes the apparent weight to be only 15 lb. It should be clear from Fig. 16-8c that the

apparent weight = true weight − buoyant force

The diagram in Fig. 16-9 indicates the hydrostatic forces acting on a submerged object. There are also horizontal forces acting on the object, but the resultant of these forces is zero. There is, however, a net force in the vertical direction. If ρ_l is the density of the fluid, the pressure on the upper surface of the block is $\rho_l g y_1 + p_0$, and there is a downward force of $(\rho_l g y_1 + p_0)A$ acting on this surface (see Fig. 16-9). The pressure on the lower surface is $\rho_l g y_2 + p_0$, and there is an upward force acting on the lower surface of $(\rho_l g y_2 + p_0)A$. The net upward force is

$$(\rho_l g y_2 + p_0)A - (\rho_l g y_1 + p_0)A = \rho_l g(y_2 - y_1)A$$

But $(y_2 - y_1)A$ is the volume of the block that is submerged in the fluid. Therefore, the net upward force exerted by the fluid (the buoyant force) is given by

$$B = \rho_l g V \qquad (16\text{-}9)$$

Since $\rho_l g$ is the weight density (weight per unit volume) of the fluid, $\rho_l g V$ is *the weight of the fluid displaced by the submerged volume of the object.*

Equation 16-9 describes an extremely important principle in fluid mechanics. This principle is named after an early Greek scientist, Archimedes,

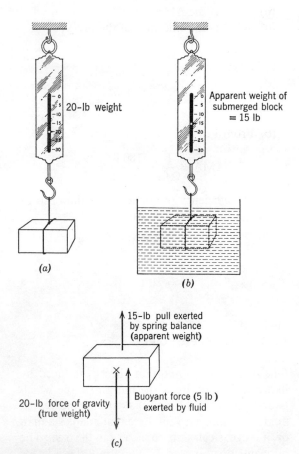

(a)

Apparent weight of submerged block = 15 lb

(b)

15–lb pull exerted by spring balance (apparent weight)

20-lb force of gravity (true weight)

Buoyant force (5 lb) exerted by fluid

(c)

Figure 16-8 A buoyant force acts upward on an object immersed in a fluid, (a) and (b). (c) Free body diagram of a submerged block.

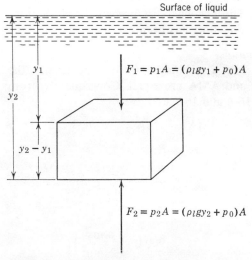

Surface of liquid

$$F_1 = p_1 A = (\rho_l g y_1 + p_0)A$$

$$F_2 = p_2 A = (\rho_l g y_2 + p_0)A$$

Figure 16-9 The hydrostatic forces acting on a submerged object produce the buoyant force.

who was apparently the first scientist to discover and use this principle. Archimedes' principle really consists of two separate parts: (1) If a solid object is partially or completely immersed in a fluid, a volume of fluid is displaced that is equal to the submerged volume of the object. A classic demonstration of this displacement of fluid can be observed by anyone who steps into a bath tub that is already filled to the brim. The volume of water that overflows is equal to the volume submerged in the tub. (2) There is a buoyant force acting on any object that is partially or completely immersed in a fluid. The buoyant force is an upward force that is equal in magnitude to the weight of the displaced fluid.

Example 16-14   A plastic float has a mass of 20 kg. When this mass is floating on salt water, only 10% of its volume is submerged beneath the surface. The specific gravity of the salt water is 1.03. Determine (a) the density of the plastic, (b) the volume of the plastic, and (c) the downward force in newtons required to submerge the entire float beneath the surface.

Solution.   Let the volume of the float be V as in Fig. 16-10. (a) Since the volume of liquid displaced is $0.1V$, the buoyant force is $\rho_l g\,(0.1V)$, and the weight (force of gravity) is $\rho g V$. The den-

sity of the liquid is ρ_l and the density of the plastic float is ρ. Then

$$\rho g(V) = \rho_l g(0.1V)$$

$$\rho = \rho_l \frac{(0.1V)}{V} = 0.1\rho_l$$

The density of the liquid is $(1.03)(1 \text{ g/cm}^3)$ so that

$$\rho = 0.1(1.03 \times 1 \text{ g/cm}^3)$$
$$\rho = 0.103 \text{ g/cm}^3$$

(b) $\rho = \dfrac{m}{V}$ $V = \dfrac{m}{\rho} = \dfrac{20 \times 10^3 \text{ g}}{0.103 \text{ g/cm}^3}$

$$= 19.4 \times 10^4 \text{ cm}^3$$

(c) The buoyant force on the entire block is $B = \rho_l g V$. Calculating this in mks units we have

$\rho_l = 1.03 \times 10^3 \text{ kg/m}^3$ $g = 9.8 \text{ m/sec}^2$
$V = 19.4 \times 10^4 \text{ cm}^3 \times (1 \text{ m}^3/10^6 \text{ cm}^3) = 0.194 \text{ m}^3$

and

$$B = 1.03 \times 10^3 \text{ kg/m}^3 (9.8 \text{ m/sec}^2)(0.194 \text{ m}^3)$$
$$= 1.96 \times 10^3 \text{ kg-m/sec}^2 = 1960 \text{ newtons}$$

The weight of the block is

$$mg = 20 \text{ kg} \times 9.8 \text{ m/sec}^2$$
$$= 196 \text{ newtons}$$

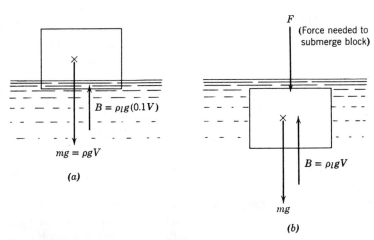

Figure 16-10

From Fig. 16-10 we have

$$F + mg = B$$
$$F = 1960 \text{ N} - 196 \text{ N} = 1764 \text{ N} \quad \blacktriangleleft$$

A very interesting application of Archimedes' principle involves the determination of density and specific gravity. According to some historical writers, Archimedes himself used the principle to calculate the density of the gold in a king's crown.

Example 16-15 For the block in Fig. 16-8 calculate (a) the volume, (b) the density, and (c) the specific gravity. (Take ρg for water as 62.5 lb/ft.^{3})

Solution
(a) The volume may be calculated by noting that $B = \rho_l g V$

$$V = \frac{B}{\rho_l g} = \frac{5 \text{ lb}}{62.5 \text{ lb/ft}^3} = 0.08 \text{ ft}^3$$

(b) The weight density is

$$\rho g = \frac{\text{weight}}{\text{volume}} = \frac{20 \text{ lb}}{0.08 \text{ ft}^3} = 250 \text{ lb/ft}^3$$

The mass density is

$$\rho = \frac{m}{V} = \frac{\frac{20}{32} \text{ slug}}{0.08 \text{ ft}^3} = 7.81 \text{ slugs/ft}^3$$

(c) The specific gravity may be calculated from the definition

$$\text{sp. gr.} = \frac{\text{density of material}}{\text{density of water}} = \frac{250 \text{ lb/ft}^3}{62.5 \text{ lb/ft}^3} = 4.0$$

Alternatively, we may use the definition

$$\text{sp. gr.} = \frac{\text{weight of object}}{\text{weight of equal volume of water}}$$

$$= \frac{20 \text{ lb}}{5 \text{ lb}} = 4 \quad \blacktriangleleft$$

Example 16-16 A block of wood 10 cm on a side floats at the interface between oil and water (Fig. 16-11). Its lower surface is 2 cm below the interface. The density of oil is 0.6 g/cm^3.
(a) What is the mass of the block?

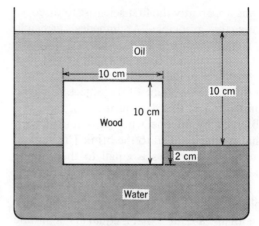

Figure 16-11 A block of wood (10 cm on each side) floating at the interface between oil and water.

(b) What is the gauge pressure at the lower face of the block?

Solution
(a) Since the block is floating, its weight is being supported by the buoyant force. Therefore,

$$mg = B$$

The buoyant force is the weight of the displaced fluid. The block displaces 800 cm^3 of oil and 200 cm^3 of water. Thus

$$B = \rho_{\text{oil}} g \, V_{\text{oil}} + \rho_{\text{H}_2\text{O}} g \, V_{\text{H}_2\text{O}}$$
$$= (0.6)(980)(800) + 1(980)(200)$$

Therefore,

$$m(980) = 0.6(980)(800) + 1(980)(200)$$
$$m = 0.6(800) + 200$$
$$= 680 \text{ g}$$

(b) The gauge pressure 12 cm down is

$$p_{\text{gauge}} = \rho_{\text{oil}} g h_{\text{oil}} + \rho_{\text{H}_2\text{O}} g h_{\text{H}_2\text{O}}$$
$$= 0.6(980)(10) + 1(980)(2)$$
$$= 7840 \text{ dynes/cm}^2 \quad \blacktriangleleft$$

Example 16-17 A block of ice ($\rho = 0.92$ g/cm^3) floats in fresh water. What fraction of the ice's volume lies below the surface of the water?

Solution. In order to float, the weight of the ice must be balanced by the buoyant force. Thus,

$$w_{ice} = B$$

$$\rho_{ice} g V_{ice} = \rho_{H_2O} g V_{H_2O}$$

where the volume of water displaced is equal to the volume of the ice that is submerged and V_{ice} is the total volume of the ice. Therefore,

$$\frac{V_{submerged}}{V_{total}} = \frac{\rho_{ice}}{\rho_{H_2O}} = 0.92$$

Thus, 92% of the ice is *below* the surface. In salt water, where $\rho = 1.03 \text{ g/cm}^3$, 89% of the ice would be below the surface. It is for this reason that icebergs are so potentially dangerous to all sea-going vessels. ◀

16-4 FLUID DYNAMICS

The study of fluids in motion is, in general, a very complex problem. In the special case when the flow of a fluid can be considered as *streamline* or *laminar*, the equations that determine the motion of the fluid are greatly simplified. In streamline flow, every particle passing a particular point in the pipe follows exactly the same path as each preceding particle that passed the same point.

The paths along which the particles move are called *streamlines*. The velocity of a particle will vary along the streamline depending upon the cross section of the pipe. At any given cross section the velocity will be the same for all points if the fluid is nonviscous. A viscous fluid, because of friction between layers of fluid, will have greater velocity in the center of the pipe than at the outside. A fluid will have a streamline flow if its velocity is not too great and its lines of flow are not caused to change direction too abruptly because of bends in the pipe or obstructions. In what follows, it is assumed that the fluid flow is streamline and, in addition, that the fluid is incompressible. The latter condition leads to a constant flow rate throughout the pipe.

The flow rate is the volume of fluid flowing through a cross section of pipe per unit time. The flow rate is represented by the symbol Q. Consider Fig. 16-12. All the fluid in the straight section of pipe having length vt, where v is the velocity of each particle in the straight section, will flow out of this section in time t. The volume of fluid is Avt and therefore the flow rate is

$$Q = \frac{Avt}{t}$$

or

$$Q = Av \qquad (16\text{-}10)$$

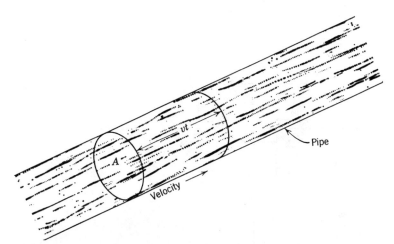

Figure 16-12 The volume of fluid flowing through a section of the pipe in time t is $A \times vt$. Therefore $Q = Av$ is the volume per unit time.

For an incompressible fluid, the flow rate is a constant (Fig. 16-13). If two sections of pipe having different cross sections are joined, the flow rate will be the same in both sections. Since the velocity and cross sections are different in both sections, we have

$$Q = A_1v_1 = A_2v_2 \qquad (16\text{-}11)$$

The last equation indicates that as the cross section increases the velocity decreases, and vice versa. This equation is often referred to as the *equation of continuity.*

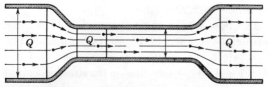

Figure 16-13 In the steady flow of an incompressible fluid the flow rate Q is the same in both the wider and the narrow sections of the pipe.

16-5 BERNOULLI'S EQUATION

The basic equation that describes the flow of a fluid is called Bernoulli's equation and expresses the relation between the pressure, velocity, and elevation of the fluid at different points along the line of flow. Bernoulli's equation is the result of applying the work–energy principle to a sample volume of fluid such as the one depicted in Fig. 16-14. Consider that portion of fluid between cross sections B and C in Fig. 16-14a. At B, the cross-section area is A_1, the pressure, p_1, and the velocity, v_1. At C, the cross section is A_2, the pressure, p_2, and the velocity, v_2. We consider the motion of this section of fluid from its position in Fig. 16-14a to its position in Fig. 16-14b. During this motion, the section at B is moved through a distance l_1 by a force $F_1 = p_1A_1$. The work done *on* the fluid is $F_1l_1 = p_1A_1l_1$. At the same time section C moves a distance l_2 against a force $F_2 = p_2A_2$. Therefore, the work done *by* the fluid is $F_2l_2 = P_2A_2l_2$. Thus the net work done on the fluid in moving it from the position in Fig. 16-14a

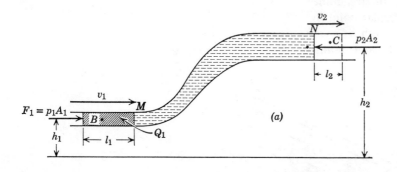

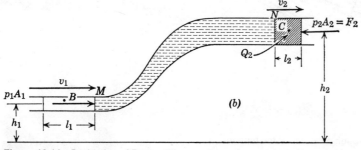

Figure 16-14 Derivation of Bernoulli's equation.

to that of Fig. 16-14*b*, is

$$W = p_1 A_1 l_1 - p_2 A_2 l_2$$

Since the fluid is incompressible, the volume of fluid $A_1 l_1$, contained in section B, is equal to the volume $A_2 l_2$ contained in section C. We call this volume V and recall that the mass, m, of this volume of fluid is equal to ρV, where ρ is the density of the fluid. Then,

$$W = (p_1 - p_2)V = (p_1 - p_2)\frac{m}{\rho}$$

According to the work–energy principle, the net work done on the fluid by forces other than gravity is equal to the increase in its kinetic and potential energy. Since there is no change in the energy status of the fluid in the section between M and N, we need only consider the respective kinetic and potential energies of sections B and C. The increase in kinetic energy is

$$\tfrac{1}{2}mv_2{}^2 - \tfrac{1}{2}mv_1{}^2$$

and the increase in potential energy (measured with respect to an arbitrary reference level) is

$$mgh_2 - mgh_1$$

We have, therefore,

$$(p_1 - p_2)\frac{m}{\rho} = (\tfrac{1}{2}mv_2{}^2 - \tfrac{1}{2}mv_1{}^2) + (mgh_2 - mgh_1)$$

If we divide both sides of the equation by m, multiply by ρ, and rearrange terms, grouping those with the same subscript together, we get

$$p_1 + \tfrac{1}{2}\rho v_1{}^2 + \rho g h_1 = p_2 + \tfrac{1}{2}\rho v_2{}^2 + \rho g h_2 \quad (16\text{-}12)$$

This last equation indicates that the combination of terms, $p + \tfrac{1}{2}\rho v^2 + \rho g h$, has the same value at two arbitrarily chosen points in the fluid. Consequently, this combination of terms must have the same value everywhere in the fluid. Therefore,

$$p + \tfrac{1}{2}\rho v^2 + \rho g h = \text{constant} \quad (16\text{-}13)$$

Equation 16-13 (or 16-12) is Bernoulli's equation for the steady flow of an incompressible fluid with zero viscosity. Equations 16-12 and 16-11 form a system of two simultaneous equations that can be used for solving problems in fluid dynamics.

It should be remembered that in Bernoulli's equation p represents the *absolute* pressure in lb/ft², N/m², or dynes/cm², and ρ is the *mass* density in slugs/ft³, kg/m³, or g/cm³.

Example 16-18   In a Venturi tube the pressure at the inlet (Fig. 16-13) is 30 lb/ft² and the velocity of water flowing through is 10 ft/sec. If at the throat of the Venturi, the pressure drops to 5 lb/ft², what is the velocity of the water in the throat?

Solution.   Since there is no change in height in this problem, we can simply drop $\rho g h$ from both sides of Eq. 16-12.

$$p_1 + \tfrac{1}{2}\rho v_1{}^2 = p_2 + \tfrac{1}{2}\rho v_2{}^2$$

$$30 \text{ lb/ft}^2 + \frac{1}{2}\left(\frac{62.4 \text{ lb/ft}^3}{32 \text{ ft/sec}^2}\right)100 \text{ ft}^2/\text{sec}^2$$

$$= 5 \text{ lb/ft}^2 + \frac{1}{2}\left(\frac{62.4 \text{ lb/ft}^3}{32 \text{ ft/sec}^2}\right)v_2{}^2$$

$$30 \text{ lb/ft}^2 + \tfrac{1}{2}(1.95) \times 100 \text{ lb/ft}^2$$

$$= 5 \text{ lb/ft}^2 + 0.98 \text{ slug/ft}^3 \, v_2{}^2$$

$$30 + 98 = 5 + 0.98 v_2{}^2$$

$$128 = 5 + 0.98 v_2{}^2$$

$$v_2{}^2 = 126 \text{ ft}^2/\text{sec}^2$$

$$v_2 = 11.2 \text{ ft/sec} \quad \blacktriangleleft$$

Example 16-19   Sea water ($\rho = 2 \text{ slugs/ft}^3$) flows through a horizontal pipe of cross-sectional area 2.88 in.² At one section the pipe narrows to a cross-sectional area of 1.44 in.² The pressure difference between the two sections is 0.27 lb/in.² Find the velocity of the water in each section and the flow rate in ft³/min.

Solution.   Let p_1, v_1, refer to the pressure and velocity at a point in the pipe where the cross section is 2.88 in.² and let p_2, v_2 refer to the smaller section. From the equation of continuity,

$$A_1 v_1 = A_2 v_2$$

$$(2.88 \text{ in.}^2)v_1 = (1.44 \text{ in.}^2)v_2$$

or

$$2v_1 = v_2$$

From Bernoulli's equation we know that the pressure is greatest in regions where the velocity is least. Consequently, since $v_2 > v_1$ we must have $p_1 > p_2$. We now turn to Bernoulli's equation

$$p_1 + \tfrac{1}{2}\rho v_1{}^2 = p_2 + \tfrac{1}{2}\rho v_2{}^2$$
$$p_1 - p_2 = \tfrac{1}{2}\rho(v_2{}^2 - v_1{}^2)$$

$$0.27 \frac{\text{lb}}{\text{in.}^2} = \frac{1}{2}\left(2\,\frac{\text{slugs}}{\text{ft}^3}\right)(4v_1{}^2 - v_1{}^2)$$

where we have used $v_2 = 2v_1$. In order to change the pressure difference into lb/ft², we multiply this difference by 144 in.²/ft². Thus

$$(0.27)(144) = 3v_1{}^2$$
$$v_1{}^2 = (0.09)(144)$$
$$v_1 = (0.3)(12) = 3.6 \text{ ft/sec}$$

and
$$v_2 = 2v_1 = 7.2 \text{ ft/sec}$$

We get the flow rate, Q, from $A_1 v_1$ or $A_2 v_2$, since these are equal. In order to express Q in ft³/min, we must express the area in ft² and the velocity in ft/min.

$$A_1 = (2.88 \text{ in.}^2)\left(\frac{1 \text{ ft}^2}{144 \text{ in.}^2}\right) = 0.02 \text{ ft}^2$$

$$v_1 = (3.6 \text{ ft/sec})(60 \text{ sec/min}) = 216 \text{ ft/min}$$

and $\quad Q = A_1 v_1 = (0.02)(216) = 4.32 \text{ ft}^3/\text{min} \quad \blacktriangleleft$

16-6 SOME APPLICATIONS OF BERNOULLI'S EQUATION

Hydrostatics

The basic equation of hydrostatics, Eq. 16-5, is a special case of Bernoulli's equation when the velocity of the fluid is everywhere zero. Consider a fluid at rest in a tank (Fig. 16-15), and apply Bernoulli's principle to points 1 and 2. Let the elevations of these points be measured point 1. Then,

$$p_2 = p_0 \quad \text{(atmospheric pressure)}$$
$$v_1 = v_2 = 0$$
$$h_2 = h$$
$$h_1 = 0$$

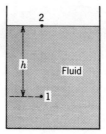

Figure 16-15 Application of Bernoulli's principle to a fluid at rest in a tank open to the atmosphere.

From Bernoulli's equation

$$p_1 + \tfrac{1}{2}\rho v_1{}^2 + \rho g h_1 = p_2 + \tfrac{1}{2}\rho v_2{}^2 + \rho g h_2$$
$$p_1 = p_0 + \rho g h$$

which is Eq. 16-5, from which Pascal's principle follows.

Velocity of Efflux

We consider now the velocity of a fluid as it flows from a small orifice in the side of a large tank (Fig. 16-16). We choose the two points as indicated in the figure and measure our elevations relative to the height of point 1. Both points are open to the atmosphere, so $p_1 = p_2 = p_0$. If the orifice is very small compared with the area at the top of the tank, the level of fluid in the tank

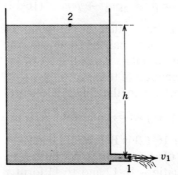

Figure 16-16 The velocity of efflux. The opening at point 1 is assumed small compared with the cross section area of the tank at point 2.

falls very slowly as the fluid empties. Consequently, v_2 will be very small and we assume it to be zero. Then, from Bernoulli's equation,

$$p_0 + \tfrac{1}{2}\rho v_1{}^2 = p_0 + \rho gh$$

and
$$v_1{}^2 = 2\,gh \qquad (16\text{-}14)$$

Equation (16-14), known as *Torricelli's theorem*, indicates that the velocity of efflux is the same as the velocity acquired by an object which falls freely through a height h. If A is the area of the orifice, the volume of fluid per unit time leaving the tank will be $Av_1 = A\sqrt{2\,gh}$.

Curving of a Spinning Baseball

Figure 16-17 shows a top view of a baseball, spinning about a vertical axis, moving toward the right. The air molecules near the surface of the ball are dragged along because of friction. Each of these molecules has a velocity comprised of two parts:

1. A tangential velocity due to the rotation of the ball.

2. A velocity due to the air moving from right to left relative to the ball.

The resultant of these two velocities is less for point a than point b. Therefore, the pressure at a is greater than the pressure at b and the ball will move along the path indicated.

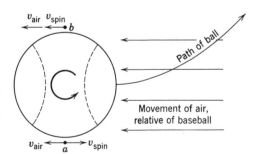

Figure 16-17 The pressure at point a will be greater than the pressure at point b. Therefore, the ball will move along the path shown.

Lift of an Airplane Wing

Figure 16-18 shows the flow pattern of air rushing past the curved wing of an airplane. The velocity of the air above the wing is greater than below. Thus the pressure is greater from below producing a lift on the wing.

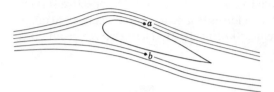

Figure 16-18 The velocity at a is greater than at point b. According to Bernoulli's principle, the pressure will be greater at b than a, producing a lift on the wing.

16-7 SURFACE TENSION

The reader has undoubtedly noticed that water drops have a definite shape, as in Fig. 16-19. Why are drops always shaped this way? Why don't the water molecules just spray out? Instead, we see a

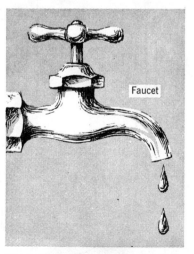

Figure 16-19 Surface tension causes water drops to assume a characteristic tear-drop shape.

definite shape, as if the water were in a special container. Indeed, it has a container. This container is a thin film formed at the surface of the water.

If you try, you can "float" a sewing needle or a thin razor blade on water (see Fig. 16-20). These objects with a density greater than water should sink. Instead they are supported by a thin film called a *surface-tension* film. The effective surface-tension forces can be measured using an arrangement as in Fig. 16-21. The net force required to

lift the square against the surface tension can be measured on the balance.

Another way to measure the surface tension of a film is to make a soap bubble film on a movable wire frame (Fig. 16-22) and hang a small weight from the frame held vertically.

The soap film is very thin. But this thickness is very great compared to molecular size. As the weight "stretches" the film, the inner molecules move into the surface layer. The tension in a surface film is the ratio of the surface force to the distance across which the forces work. The film pulls on both the front and back of the movable wire frame, which is L units long. Thus the surface tension force pulls on an effective length of $2L$. The surface tension, γ, is the force per unit length of wire, or

$$\gamma = \frac{F}{2L} \qquad (16\text{-}15)$$

where F is the total downward force acting on the wire.

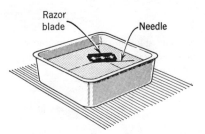

Figure 16-20 A razor blade and a needle can be supported by the surface tension film of water.

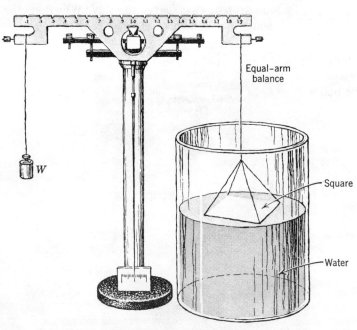

Figure 16-21 Measuring the net force produced by surface tension on a square.

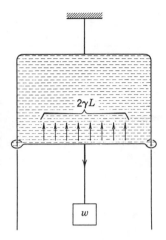

Figure 16-22 There is a film on the front and on the back of the wire frame. Hence, the force produced by surface tension acts on a length $2L$ where L is the length of the wire. Since the force is proportional to the length, $F = \gamma 2L$ and $\gamma = F/2L$.

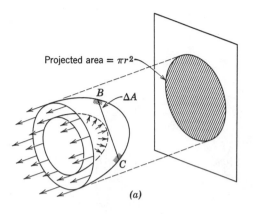

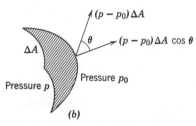

Figure 16-23 (a) The forces on one half of a soap bubble that forms a two-surface film. (b) Component of a force acting on an element of area ΔA to the right (x direction). The component in the y direction at B [in (a)] is balanced by a component in opposite direction at C.

The pressure difference between the inside and outside of a spherical bubble depends upon the surface tension, γ, and the radius of the bubble. Figure 16-23a shows the forces acting on one-half of a soap bubble, which forms a two-surface film. The force to the left is caused by the surface tension pulling, from the left-half of the bubble, over a distance equal to the circumference, $2\pi r$. Since we have a two-surface film, this force is

$$F_{\text{left}} = 2\gamma(2\pi r) = 4\pi r\gamma$$

To determine the force to the right, consider the force on the element of surface area, ΔA. If p represents the internal pressure of the bubble and p_0 is the atmospheric pressure outside, the force on ΔA will be $(p - p_0)\Delta A$ directed outward along the radius. As Fig. 16-23 indicates, the y component of this force is balanced by an oppositely directed y component from a symmetrically placed force. The x component is $(p - p_0)\Delta A \cos \theta$, or the pressure difference multiplied by the projection of the area on a plane perpendicular to the x axis. It follows that the total force to the right will be $(p - p_0)$ multiplied by the total projected area of the hemisphere on the plane perpendicular

to the x axis. This area is just πr^2 (Fig. 16-23). Thus

$$F_{\text{right}} = (p - p_0)\pi r^2$$

and the equilibrium condition requires that

$$(p - p_0)\pi r^2 = 4\pi r\gamma$$

or $$p - p_0 = \frac{4\gamma}{r} \quad \text{(soap bubble)} \quad \text{(16-16)}$$

For liquids that form single surface films, an analysis similar to that of the soap bubble would lead to the result

$$p - p_0 = \frac{2\gamma}{r} \quad \text{(liquid drop)} \quad \text{(16-17)}$$

From Eqs. 16-16 and 16-17 we see that the pressure difference is inversely proportional to the radius of the bubble. Thus the smaller the

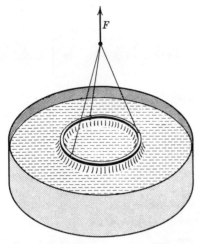

Figure 16-24 Measuring the force produced by surface tension acting on a ring. The surface tension may then be calculated from the equation

$$\gamma = \frac{F}{2L} = \frac{F}{2(2\pi r)}$$

Table 16-2 Experimental values of surface tension

Liquid in Contact with Air	t, °C	Surface Tension, dynes/cm
Benzene	20	28.9
Carbon tetrachloride	20	26.8
Ethyl alcohol	20	22.3
Glycerine	20	63.1
Mercury	20	465
Olive oil	20	32.0
Soap solution	20	25.0
Water	0	75.6
Water	20	72.8
Water	60	66.2
Water	100	58.9
Oxygen	−193	15.7
Neon	−247	5.15
Helium	−269	0.12

bubble, the greater the pressure difference between the inside and outside of the bubble.

In a ring as in Fig. 16-24, we have an alternate method for measuring γ.

The value for surface tension of different liquids varies. Common liquids are listed in Table 16-2.

Example 16-20   What is the pressure difference between the inside and outside of a water drop at 60°C if the radius of the drop is 3 mm?

Solution.   Water drops have a single-surface film, so we can use Eq. 16-17.

$$p - p_0 = \frac{2\gamma}{r}$$

$$= \frac{2(66.2 \text{ dynes/cm})}{0.3 \text{ cm}}$$

$$= 441 \text{ dynes/cm}^2 \qquad \blacktriangleleft$$

Example 16-21   The gauge pressure in a soap bubble is 500 dynes/cm². Find the radius of the bubble.

Solution.   The gauge pressure is just $p - p_0$. Since soap forms a two-surface film, we use Eq. 16-16.

$$p - p_0 = \frac{4\gamma}{r}$$

$$r = \frac{4\gamma}{p - p_0}$$

$$= \frac{4(25 \text{ dynes/cm})}{500 \text{ dynes/cm}^2}$$

$$= 0.2 \text{ cm or } 2 \text{ mm} \qquad \blacktriangleleft$$

Any liquid surface under surface tension forces has the tendency to occupy the minimum area consistent with the boundaries of the container holding the fluid. A drop of liquid has a certain volume. Because of surface tension, the shape of the drop will be spherical, since the spherical surface forms the smallest boundary area for a fixed volume of material. Tiny beads of water on a hard, flat surface quickly form spherically shaped bubbles. Figure 16-25 shows the formation of a spherically shaped drop of milk at the bottom of a vertical tube.

Figure 16-25 Formation of a milk drop.

16-8 CAPILLARITY

When water is placed in an open tube having a small diameter, the water rises as indicated in Fig. 16-26a. Notice that the highest elevation of liquid is in the tube of smallest cross section. In Fig. 16-26b, we see that the mercury in the tubes is depressed below the level of the mercury in the beaker. These effects are described by the term *capillarity*. In the case of water in a glass tube, the molecular attraction between the glass and the water (adhesive forces) is greater than the attraction between different water molecules (cohesive forces). This results in an angle of contact of approximately 0° (Fig. 16-27a). On the other hand, the mercury-to-mercury attraction is

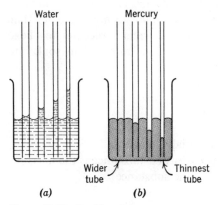

Figure 16-26 Capillary tubes.

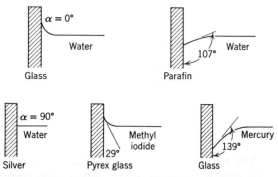

Figure 16-27 Capillary tubes with different contact angles for different tubes and liquids.

greater than the mercury-to-glass attraction producing an angle of contact of approximately 139° (Fig. 16-27b). The height to which a liquid will rise in a capillary tube (like water) or the depth to which a liquid will be depressed (like mercury) is given by

$$h = \frac{2\gamma \cos \alpha}{r\rho g} \qquad (16\text{-}18)$$

where γ is the surface tension, α the angle of contact, r the inside radius of the tube, and ρg the weight density of the liquid.

Wetting agents or detergents are chemicals which, when added to water, change the angle of contact between the water and some surface from an angle greater than 90° to one that is much less than 90°, causing the water to spread out over the surface. Waterproofing agents applied to cloth do just the converse, changing the angle of contact from less than 90° to much greater than 90°, thus causing the water molecules to "bead up" on the cloth surface rather than soak into the garment.

SUMMARY

Fluids are materials that flow.

The *mass density* of a substance is the mass per unit volume of the substance:

$$\rho = \frac{m}{V}$$

The *weight density* of a substance is the weight per unit volume of the substance:

$$\text{weight density} = \frac{mg}{V} = \rho g.$$

Pressure is defined as the force exerted against a surface per unit area of surface:

$$p = \frac{F}{A}$$

The pressure at a depth h beneath the surface of a fluid of mass density, ρ, open to the atmosphere, is

$$p = p_0 + \rho g h$$

The difference between the *absolute pressure* and atmospheric pressure is called the *gauge pressure*. Atmospheric pressure is 14.7 lb/in.^{2} or 1.013×10^5 N/m^2.

Pascal's principle states that the pressure exerted anywhere in a confined fluid is transmitted undiminished throughout the fluid.

Archimedes' principle states that an object immersed in a fluid experiences a buoyant force equal to the weight of the displaced fluid:

$$B = \text{(weight density of fluid)} \\ \times \text{(volume of fluid displaced)}$$

Law of floatation: An object floats when the buoyant force acting on the object is equal to the total weight of the object.

Bernoulli's principle: For a fluid (density ρ) moving in streamline flow, the pressure, velocity, and elevation of the fluid at any two points are related:

$$p_1 + \tfrac{1}{2}\rho v_1{}^2 + \rho g h_1 = p_2 + \tfrac{1}{2}\rho v_2{}^2 + \rho g h_2$$

The *velocity of efflux* from a small hole, h units below the top of an open tank, is given by $v = \sqrt{2gh}$.

The pressure difference between the inside and outside of a spherical bubble depends upon the surface tension, γ, and the radius of the bubble:

$$p - p_0 = \frac{4\gamma}{r} \qquad \text{(soap bubble)}$$

$$p - p_0 = \frac{2\gamma}{r} \qquad \text{(liquid drop)}$$

The height to which a liquid of density ρ and surface tension, γ, will rise in a tube of radius r is

$$h = \frac{2\gamma \cos \alpha}{r\rho g}$$

where α is the angle of contact.

QUESTIONS

1. How is pressure defined?

2. What is the pressure at a depth, h, beneath the surface of a confined fluid of weight density ρg when the pressure of the atmosphere at the surface of the liquid is p_0?

3. What is Pascal's principle? How does it explain the operation of an hydraulic lift?

4. State Archimedes' principle in words. Write an expression for the buoyant force acting on an object submerged in a fluid in terms of the density of the fluid and the volume of the object.

5. Use Archimedes' principle to explain how battleships made of steel are able to float.

6. Explain how submarines are able to submerge and surface.

7. What is meant by the phrase "streamline flow"?

8. How is Bernoulli's equation related to the work–energy principle?

9. Explain how a rotating baseball can be made to curve.

10. How does Bernoulli's equation explain the lifting force developed on the wings of an airplane as it moves through the air?

11. What is surface tension? Capillarity?

12. Distinguish between *adhesion* and *cohesion*.

13. Why does the top surface of mercury in a glass tube appear different from the top surface of water in the same tube?

PROBLEMS

(A)

1. What is the volume of 400 g of mercury (the density of mercury is 13.6 g/cm^3)?

2. What is the volume in gallons of 100 lb of lubricating oil with a specific gravity of 0.80 (1 ft^3 equals 7.48 gal)?

3. What is the density and specific gravity of gasoline if 66 g have a volume of 100 cm^3?

4. The specific gravity of gold is 19.3. What volume is occupied by 1000 kg of gold?

5. The specific gravity of helium at 0°C and 1 atm pressure is 0.18. What mass of helium would occupy a volume of 5000 cm^3?

6. Two cubic feet of water weigh 125 lb. Silver has a specific gravity of 10.5. How much do 2 ft^3 of silver weigh? What volume is occupied by 125 lb of silver?

7. A 4800-lb safe has four legs, each having an area of 0.75 in.^{2} What pressure does the safe exert on the floor beneath the legs?

8. A tank measuring 18 in. long × 12 in. wide × 9 in. high is filled to the top with mercury.
   (a) What is the total force exerted on the base of the tank by the mercury?
   (b) What is the average pressure exerted against the walls?
   (c) What force is exerted against each wall?

9. If a water pressure gauge shows a pressure of 45 lb/in.^{2} at the ground floor, what is the maximum height to which water will rise in the building?

10. Water rises to a maximum height of 128 ft in a certain building. What must be the water pressure at ground level?

11. What is the pressure of the atmosphere on a day that it supports a 78-cm column of mercury?

12. What would be the height of a column of liquid (specific gravity = 8.4) in a barometer on a day when a mercury barometer stands at 76 cm?

13. A hydraulic jack has a large piston with a cross-sectional area of 100 cm^2. The small pump piston has a cross section of 5 cm^2. With a force of 25 newtons on the small piston, what is the total force on the large piston?

14. The larger piston of a hydraulic jack has an area of 48 in.^{2} while the smaller area is 9 in.^{2} What force on the smaller piston will raise an 1800-lb load resting on the larger piston?

15. What is the density of a block that floats in fresh water with $\frac{2}{3}$ of its volume above the waterline?

16. A block floats in salt water with 0.7 of its volume below the surface of the water. What is the density of the block?

17. A barge 100 ft long, 20 ft wide, and 10 ft deep has a hull made of wood $\frac{1}{2}$ ft thick (the specific gravity of wood is 0.8). What is the water level on the outside of the barge when the barge is empty?

18. What is the apparent weight in water of a rock that displaces 16 g of water and has a specific gravity of 5?

19. What is the pressure difference between the inside and outside of a water drop at 20°C if the radius of the drop is 5 mm?

20. Repeat problem 19 for a soap solution.

21. The gauge pressure in a soap bubble at 20°C is 20 dynes/cm^2. Find the radius of the bubble.

22. A liquid drop (single surface film) has a surface tension of 65 dynes/cm and a diameter 0.01 mm. What is the pressure inside the drop?

(B)

1. What is the density of an object that weighs 90 lb in air and 60 lb when submerged in water?

2. What is the specific gravity of an object that weighs 98,000 dynes in air and 95,060 dynes when submerged in water?

3. A metal box weighs 4200 lb and has a specific gravity of 4. How much weight does the box appear to lose when weighed while submerged in water?

4. An alloy of silver (sp. gr. = 10.5) and platinum (sp. gr. = 21.4) weighs 150 lb. When submerged in water, the alloy weighs 140 lb. What is the weight of silver in the alloy?

5. Ice has a specific gravity of 0.917. A 90-lb boy stands on a block of ice, $\frac{1}{2}$ ft thick, so that the ice *just* floats in fresh water. What is the area of the ice block?

6. A cubical block of wood (10 cm on a side) floats as shown in Fig. 16-28. What is the mass of the block if its lower surface is 6 cm below the oil–water interface? The density of the oil is 0.8 g/cm³.

7. A hollow sphere has an inner radius of 8 cm and an outer radius of 10 cm. The sphere floats with half of its volume below the surface of a liquid of specific gravity 0.75. What is the density of the material from which the sphere is made?

8. A hollow sphere (m = 225 g) floats in fresh water with $\frac{1}{5}$ of its volume submerged. What volume of metal (sp. gr. = 9) must be fastened to the bottom of the sphere so that the entire assembly floats just beneath the surface of the water?

9. Repeat Problem 8, but this time find the volume of metal that must be placed *inside* the hollow sphere so that the assembly just floats.

10. A 10-in. diameter pipe has water flowing through it at a velocity of 4 ft/sec. What is the velocity in a 5-in. diameter pipe to which the large pipe is connected?

11. Kerosene flows through a 1-in.-diameter fill line at a velocity of 10 ft/sec. What is the flow in cubic feet per second? What is the flow in gallons per hour? (1 ft³ = 7.48 gal.)

12. The discharge rate through a pipe is 30 ft³/sec.
    (a) What is the cross section at a section where the fluid velocity is 24 ft/sec?
    (b) What is the fluid velocity at a point where the cross section is 36 in.²?

13. Salt water ($\rho g = 64$ lb/ft³) flows in a horizontal tube, such as the Venturi tube of Fig. 16-29. The cross section at the inlet is 2.5 times greater than the cross section at the throat, while the pressure difference is 15.75 lb/in.² Find the velocity of the fluid at the inlet and in the throat.

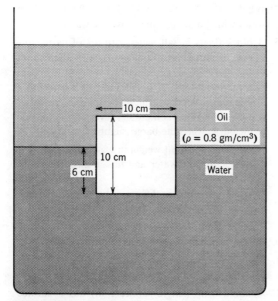

Figure 16-28

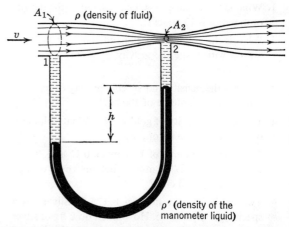

Figure 16-29   A typical Venturi tube. The difference in height between the two mercury columns, *h*, is a measure of the pressure difference between points 1 and 2.

14. A horizontal pipe has cross sections of 0.08 ft^2 and 0.02 ft^2 at the wide and narrow sections, respectively. The flow rate for salt water through the pipe is 0.4 ft^3/sec.
    (a) What is the velocity at the wide and narrow sections of the pipe?
    (b) What is the pressure difference between these sections?

15. The velocity and gauge pressure of salt water at a certain point in a pipeline is 2 ft/sec and 25 lb/in.^{2}, respectively. What is the gauge pressure at a second point 25 ft lower than the first? The cross section at the second point is $\frac{1}{4}$ that of the first point.

16. Water stands in an open tank to a height H. A small hole is made in the wall at a depth y below the surface of the water. At what distance from the foot of the tank will the water strike the ground?

17. Find the velocity of efflux from the tank in Fig. 16-30 through the small hole located 3 ft below the surface of the salt water.

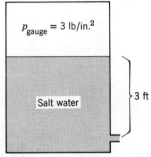

$p_{\text{gauge}} = 3$ lb/in.^{2}

Salt water

3 ft

Figure 16-30 The gauge pressure above the salt water is 3 lb/in.^{2} The height of the water is 3 ft above the outlet pipe.

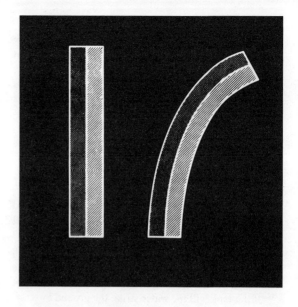

17.
Thermometry

The concept of *temperature* and how temperature is measured will be discussed in this chapter. The effect of temperature changes on the physical size of objects and the strange expansion properties of water are also studied. This chapter is the first of six that treat the subject of heat and thermodynamics.

LEARNING GOALS

The main learning objectives of this chapter are

1. *To understand the definitions of temperature, heat and internal energy.*

2. *To understand what a thermometer is and how it can be calibrated.*

3. *To learn the relationships among the four most often used temperature scales.*

4. *To learn how to calculate changes in the physical size of objects as a consequence of changes in temperature.*

5. *To study the strange expansion properties of water.*

6. *To calculate the thermal stress produced in fixed rods that are heated but are not free to expand.*

17-1 INTERNAL ENERGY

In the preceding chapters we have studied various forms of energy: translational and rotational kinetic energy of rigid bodies and fluids, and gravitational and elastic potential energy. These types of energy were mainly associated with the motion of an object rather than with the individual motions of the molecules in the object. It is well known that even when an object is at rest, each of its molecules is vibrating about some equilibrium point. Such an object has no rotational or kinetic energy as we have defined these quantities previously, since the center of mass of the object is at rest and the object is not rotating. It may have gravitational potential energy, depending on the position at which the reference level for such energy has been chosen. Nevertheless, since the individual molecules are moving, there must be some form of kinetic energy associated with their motion. Further, there are forces between the molecules of an object which hold the object together. These intermolecular forces give rise to a potential energy similar to the types previously studied.

The sum of the kinetic energy and intermolecular potential energy of all the molecules of an object is called the *internal energy* of the object. When two objects are placed in contact, energy will be transferred by molecular collision from the object of higher molecular kinetic energy to the one of lower molecular kinetic energy. This transfer continues until the average kinetic energy of each object is the same. At this point as much energy is being transferred across the surface of contact from body A to body B, as is transferred from B to A. When two objects have the same average kinetic energy, we say they are at the same temperature. Temperature and internal energy are quite different. *Internal energy is the total kinetic and potential energy of the molecules. Temperature is a measure of the average kinetic energy of a molecule.*

It is very possible for an object of low temperature to have more internal energy than an object of higher temperature. Consider two flasks of water, one containing much more water (and therefore many more molecules) than the other. Suppose that the first flask contains water with an average kinetic energy per molecule of 10^{-12} ergs, and the second flask with the greater number of molecules has an average kinetic energy per molecule of 10^{-13} ergs. Clearly, the water with the lower average kinetic energy per molecule, and thus the lower temperature, has a greater total energy because of its larger number of molecules.

There are two fundamental ways of changing the temperature of an object:

1. By doing mechanical work *on* the object.

2. By placing the object in good thermal contact with a second object that is at a different temperature.

Examples of the first method are compressing a gas, hammering a piece of metal, rubbing two solids against each other, or stirring a liquid. In each of these cases work is done on an object resulting in a change in temperature. The relationship between mechanical work and temperature changes is discussed more fully in Chapters 18 and 19. Examples of the second method are heating an object over a flame or mixing hot and cold objects together. The energy that flows between objects of different temperature is called *heat.* Once the heat has been absorbed by the object of lower temperature, this energy becomes part of the internal energy of the object. Thus it is improper to speak of the "heat energy of an object," since heat refers specifically to an energy flow between objects. Once this energy becomes part of an object, it is not possible to tell whether this energy came from a heat flow, as in the second method or from mechanical work, as in the first method. The measurement of heat will be discussed in Chapter 18. In the remainder of this chapter we consider means of measuring the average kinetic energy of molecules (temperature).

17-2 THERMOMETERS

Any substance which has a physical or chemical property that varies with changes in the average kinetic energy of its molecules may be calibrated as a standard for measuring temperature. Most substances expand when heated and contract when cooled. A substance that undergoes such expansions and contractions, uniformly, over a considerable range of average kinetic energies can be calibrated. Gases confined in a fixed volume have uniform pressure variations when their temperature is changed, and therefore can be calibrated. The intensity of the light radiated by an object varies according to a regular pattern as its temperature changes. Hence the temperature of an object may be measured by comparing the intensity of its light spectrum to that of some standard. The electrical resistance of conductors is another property that is temperature-dependent. All of these properties have been utilized to calibrate devices that measure temperature. The various properties are usually sensitive to these temperature changes in different ranges of temperatures. Hence they all play important roles in the development of scientific and technical skills. A device used to measure temperature is called a *thermometer*.

Consider first the property of expansion. Let us put a liquid that expands uniformly into a capillary tube and seal the tube at both ends. Let the height of the original column of liquid be denoted by l_1. Now increase the average kinetic energy of the molecules in the liquid by passing the tube through a hot flame several times. The liquid will expand to a new height, l_2. Suppose, now, we *arbitrarily* assign values to the original and final temperatures, t_1 and t_2. Then, recalling the fact that the liquid expands uniformly with temperature, we can find the appropriate value of the temperature when the column of liquid stands at any other height, l_3, by writing a simple proportion (Fig. 17-1):

$$\frac{l_2 - l_1}{t_2 - t_1} = \frac{l_3 - l_2}{t_3 - t_2} \qquad (17\text{-}1)$$

All the quantities in Eq. 17-1, except t_3, are either measurable or known, so that t_3 can be easily calculated. There are several systems of temperature units, depending on the values of temperature

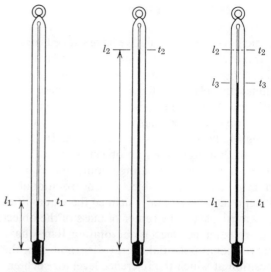

Figure 17-1 The temperature intervals $(t_2 - t_1)$, $(t_3 - t_2)$ are directly proportional to the changes in length $(l_2 - l_1)$, $(l_3 - l_2)$.

assigned to the two arbitrary columns l_1 and l_2. In the Celsius (formerly centigrade) system, the height of a column of liquid mercury when in thermal equilibrium with a mixture of water and ice, under atmospheric pressure, is assigned the value zero degrees Celsius (0°C) and the height of the same column when in thermal equilibrium with water and steam is assigned the value one hundred degrees Celsius (100°C). By placing one hundred equally spaced scratches on the glass tube between 0°C and 100°C, we define the size of a single unit of Celsius temperature. The scratches can then be continued above 100°C and below 0°C as long as we remain within the range of temperatures for which mercury expands uniformly. Now the temperature corresponding to any column of mercury may be read directly off the glass tube without resorting to measuring the height of the column or using Eq. 17-1.

In the Fahrenheit system, the temperature of the water–ice mixture has the value 32°F and the water–steam mixture, 212°F. Thus there are 180 intervals on the Fahrenheit scale where there are 100 intervals on the Celsius scale, and each Fahrenheit degree is $\frac{5}{9}$ of a Celsius degree. We should make it clear that there is an important distinction between the terms "degree Fahrenheit (or Celsius)" and "Fahrenheit (or Celsius) degrees." "Degree Fahrenheit (or Celsius)" refers to a specific scratch on the thermometer, such as 10°F. On the other hand, "Fahrenheit (or Celsius) degree" refers to a temperature interval that may take place anywhere on the scale. Thus 10 F° refers to an interval of 10 units of Fahrenheit temperature irrespective of whether the interval is between 5°F and 15°F or between 95°F and 105°F.

Very often it is necessary to convert Fahrenheit temperatures to Celsius temperatures, and vice versa. From Fig. 17-2 it is clear that

$$\frac{100 - 0}{212 - 32} = \frac{t_c - 0}{t_F - 32}$$

or
$$t_c = \tfrac{5}{9}(t_F - 32) \qquad (17\text{-}2)$$

Solving Eq. 17-2 for t_F, we obtain

$$t_F = \tfrac{9}{5}t_c + 32 \qquad (17\text{-}3)$$

Equations 17-2 and 17-3 are used to convert specific temperatures from one scale to the other. It is not necessary to memorize both equations, since one can easily be obtained from the second. Where and when to insert the parentheses (present in Eq. 17-2 but absent in Eq. 17-3) might cause some confusion. This may be avoided by testing the form of the equation to be used for the water–steam point or water–ice point.

Example 17-1 Convert 68°F to Celsius degrees.

Solution. From Eq. 17-2 we have $t_c = \tfrac{5}{9}(t_F - 32)$. We note that from this equation $t_c = 100$ when $t_F = 212$, and $t_c = 0$ when $t_F = 32$. We are then reasonably safe in assuming the equation has been recalled correctly. Then

$$t_c = \tfrac{5}{9}(68 - 32) = \tfrac{5}{9}(36)$$
$$t_c = 20°C \qquad \blacktriangleleft$$

Example 17-2 Convert an interval of 75 C° to Fahrenheit degrees.

Solution. Since each Fahrenheit degree is $\tfrac{5}{9}$ of a Celsius degree, there are $\tfrac{9}{5}$ as many Fahrenheit degrees as in any corresponding interval of Celsius degrees. Thus

Fahrenheit degree interval
$$= \tfrac{9}{5} \text{ Celsius degree interval}$$
$$\Delta F° = \tfrac{9}{5} \Delta C° = \tfrac{9}{5}(75)$$
$$\Delta F° = 135 \text{ F}° \qquad \blacktriangleleft$$

Scientists have never succeeded in obtaining temperatures colder than $-273.16°C$ or $-459.60°F$ (-273 and -460 to the nearest degree). According to the present theory these are the lowest attainable temperatures. There are two temperature scales that assign the value of zero to these temperatures. These temperature scales are called *absolute scales*, since the lowest attainable temperature is called *absolute zero*. The Kelvin scale defines $-273°C$ as $0°K$, and the Rankine scale defines $-460°F$ as $0°R$. Thus

$$t_K = t_c + 273 \qquad (17\text{-}4)$$
and
$$t_R = t_F + 460 \qquad (17\text{-}5)$$

(see Fig. 17-3).

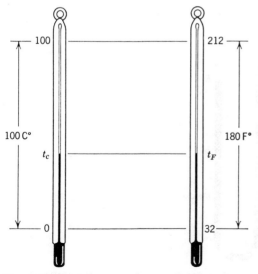

Figure 17-2

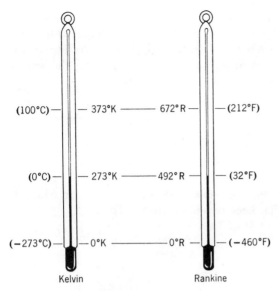

Figure 17-3 The absolute temperature scales.

meter (see Fig. 17-4). When the gas is heated, it expands, forcing the mercury up in side A. By adjusting A, the level of mercury in side B can be brought back to the reference level E at which all readings are taken. By this procedure the pressures are always read at constant volume. From a knowledge of the pressures at the ice and steam points, a proportionality may be written that permits a determination of temperature for other pressures.

The *optical pyrometer* is used for measuring very high temperatures in the white-hot region. The brightness of an object whose temperature is desired is compared to that of a metal filament whose brightness can be adjusted by varying its temperature. When the brightness of the two objects is judged the same, the two are at the same temperature. Tungsten is commonly used as the calibrated filament. The upper limit on the temperature, which can be measured by a tungsten filament, is near the melting point of tungsten.

Example 17-3 The boiling point of sulfur is 444.60°C. What is the boiling point of sulfur on the Kelvin scale?

Solution. Utilizing Eq. 17-4, we have $t_K = t_c + 273$, $t_K = 444.60 + 273$, and so $t_K = 717.60°K$. ◄

The *mercury-in-glass thermometer* is limited in use to a temperature range from $-39°C$ at which temperature mercury freezes to 360°C, where it no longer exhibits the uniform expansion quality required for temperature measurement. This range can be extended on the cold side by using alcohol or pentane in place of mercury, since these liquids have lower freezing points.

The *resistance thermometer* can be used from $-250°C$ to about 1760°C, the melting point of platinum wire, which is generally used as the temperature-sensitive resistor. Because the resistance of a wire can be measured very accurately,

In the *constant-volume* gas thermometer, the changes in pressure of a constant volume of gas are measured by an open-tube mercury mano-

Example 17-4 Copper melts at 1080°C. What is the melting point of copper on the Rankine scale?

Solution. We can obtain Rankine temperatures from Fahrenheit temperatures by using Eq. 17-5. To get the Fahrenheit temperature that

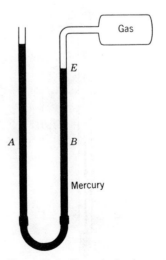

Figure 17-4 A constant-volume gas thermometer.

is equivalent to 1080°C, we use Eq. 17-3. Thus

$$t_F = \tfrac{9}{5}t_c + 32$$
$$= \tfrac{9}{5}(1080) + 32$$
$$= 1944 + 32$$
$$= 1976°F$$

Then, from Eq. 17-5

$$t_R = t_F + 460$$
$$= 1976 + 460$$
$$= 2436°R \qquad \blacktriangleleft$$

17-3 LINEAR EXPANSION

When the temperature of an object is raised, the kinetic energy of its molecules increases. This increase in the mechanical energy of the molecules results in vibrations of larger amplitude about each molecule's equilibrium position. The average distance between adjacent molecules increases, and the object will expand in one, two, or three dimensions. Some objects (i.e., water below 4°C, rubber) contract when their temperature increases. The peculiar behavior of these materials can be traced to the structure and shape of their molecules, but we shall not pursue this part of the theory. We consider, instead, the more usual expansion properties exhibited by most materials.

Thin rods expand principally in one dimension, along their length. Experiments on different-sized rods, at different temperatures, indicate that the change in the length of a rod is proportional to both the original length at the start of the temperature change and the temperature change. That is,

$$\Delta l \sim l_0 \, \Delta t$$

where l_0 represents the length at the initial temperature of the rod and Δt is the temperature change. We transform this proportionality into an equation by introducing a constant of proportionality α, so that

$$\Delta l = l_0 \alpha \, \Delta t \qquad (17\text{-}6)$$

The constant α is called the *coefficient of linear expansion*. From Eq. 17-6, we see that

$$\alpha = \frac{\Delta l / l_0}{\Delta t}$$

which means that α is the fractional change in length per unit change in temperature. The units in which α is expressed are $1/F°$ or $1/C°$, depending upon which temperature scale is used for the temperature change, Δt. Table 17-1 lists the coefficient of expansion for several common materials.

Example 17-5 A steel ruler is 18 in. long at 50°C. What is the length of the ruler at 80°C? What would be the corresponding increase in length for a copper ruler?

Solution. From Eq. 17-6

$$\Delta l = \alpha l_0 \, \Delta t = (18 \text{ in.})(1.2 \times 10^{-5}/C°)(30 \text{ C}°)$$
$$= 0.006 \text{ in.}$$

Table 17-1 Linear coefficients of expansion

Substance	Coefficient (per C°)	Coefficient (per F°)
Aluminum	2.4×10^{-5}	1.3×10^{-5}
Brass	1.8×10^{-5}	1.0×10^{-5}
Copper	1.7×10^{-5}	0.95×10^{-5}
Glass, Pyrex	0.3×10^{-5}	0.17×10^{-5}
Iron	1.2×10^{-5}	0.65×10^{-5}
Lead	3.0×10^{-5}	1.7×10^{-5}
Silver	2.0×10^{-5}	1.1×10^{-5}
Steel	1.2×10^{-5}	0.65×10^{-5}
Zinc	2.6×10^{-5}	1.5×10^{-5}

Therefore, the length of the ruler at 80°C is

$$L = l_0 + \Delta l = 18.006 \text{ in.}$$

For copper,

$$\Delta l = \alpha l_0 \Delta t = (18 \text{ in.})(1.7 \times 10^{-5}/\text{C}°)(30 \text{ C}°)$$
$$= 0.009 \text{ in.} \qquad \blacktriangleleft$$

Linear expansion must be taken into account when bridges are built, railroad tracks laid, or any machinery constructed that will be exposed to both hot and cold weather during the course of its operation. The spaces left between pieces of railroad track allow for the expansion of the track in summer. One end of many bridges is placed on rollers to allow free expansion in hot weather.

17-4 BIMETALLIC STRIP

Suppose we weld together two flat strips of different metals. The strips are initially of equal length and are at the same temperature. Now, we heat this bimetallic strip. The material with the larger coefficient of expansion must expand more. On the other hand, since the two strips are welded together, they must remain in perfect contact. The net effect is to have the strip bend in an arc with the material of larger α on the outside (see Fig. 17-5).

The bimetallic strip is used in the operation of various types of thermostatic control systems, sprinkler systems, and some precision clocks.

17-5 AREA EXPANSION

Flat materials with large surface areas will expand in both length and width, thus increasing their area as a function of temperature. Consider a rectangular strip of metal with linear coefficient of expansion α and dimensions L_0, W_0. The original area of the strip is

$$A_0 = L_0 W_0$$

As heat is applied, the length and width will each expand to new dimensions. We assume that these two expansions are independent of each other and that they obey Eq. 17-6. Thus

(i) $\quad L = L_0 + \Delta L = L_0 + L_0 \alpha \Delta t = L_0(1 + \alpha \Delta t).$

(ii) $\quad W = W_0 + \Delta W = W_0 + W_0 \alpha \Delta t = W_0(1 + \alpha \Delta t).$

The new area (Fig. 17-6) will be the product of (i) and (ii).

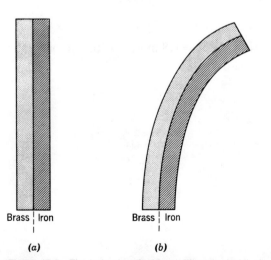

Brass | Iron   Brass | Iron

(a)   (b)

Figure 17-5 The behavior of a bimetallic strip. In (a), the strip at room temperature. In (b), the strip after a sizeable temperature increment.

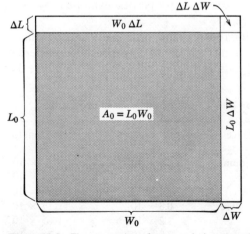

Figure 17-6 The expansion of an area in length and width. The actual increases in the length and width, ΔL and ΔW, represent a very small fraction of their original values. The figure has been drawn out of proportion to provide visual clarity.

$$A = LW = L_0 W_0 (1 + \alpha \, \Delta t)^2$$
$$= L_0 W_0 [1 + 2\alpha \, \Delta t + (\alpha \, \Delta t)^2]$$
$$= A_0 [1 + 2\alpha \, \Delta t + (\alpha \, \Delta t)^2]$$

Since α is very small for most materials, the α^2 term can be neglected with respect to the α term, even for larger temperature increases. Therefore, to sufficient accuracy, we have

$$A = A_0 (1 + 2\alpha \, \Delta t)$$
$$= A_0 (1 + \gamma \, \Delta t) \qquad (17\text{-}7)$$

where $\gamma = 2\alpha$. The *coefficient of areal expansion* is called γ and is just twice the coefficient of linear expansion.

Example 17-6   A flat, round plate of lead of radius 8 in. is at a temperature of 32°F. What is the area of the plate at 212°F?

Solution.   For lead, (see Table 17-1)

$$\gamma = 2\alpha = 3.4 \times 10^{-5}/\text{F}°$$

Then $A = A_0 (1 + \gamma \, \Delta t)$
$$= \pi (8 \text{ in.})^2 [1 + (3.4 \times 10^{-5}) \times 180]$$
$$= \pi (8 \text{ in.})^2 (1.006) = 64.4\pi \text{ in.}^2$$
$$= 202 \text{ in.}^2.$$

Approximately the same answer is obtained if first the new radius r is calculated from Eq. 17-6. Then $A = \pi r^2$ yields the new area.   ◀

17-6 VOLUMETRIC EXPANSION

All liquids and gases, and any type of solid material, expand in their volume when heated. The derivation of the expression for change in volume as a function of temperature change proceeds in the same way as that for area. We shall write down the result without proof and leave the derivation as an exercise. Thus

$$V = V_0 (1 + \beta \, \Delta t) \qquad (17\text{-}8)$$

where for solids the volume coefficient β is three times the linear coefficient. For different liquids, β is given in Table 17-2. All gases have approxi-

Table 17-2   Coefficient of volume expansion for some liquids

Substance	Coefficient (per C°)
Alcohol, ethyl	11.2×10^{-4}
Benzene	12.4×10^{-4}
Ether	16.5×10^{-4}
Glycerin	5.1×10^{-4}
Mercury	1.8×10^{-4}
Water	2.1×10^{-4}

mately the same volume coefficient. The volumetric expansion of ideal gases will be discussed more fully in Chapter 20.

Example 17-7   A Pyrex glass has a volume of 5 liters at 30°C and is filled to the top with glycerin. (a) What is the volume of the glass at 60°C? (b) What is the volume of glycerin at 60°C? (c) How much liquid overflows?

Solution.   (a) The volume coefficient of expansion of Pyrex is $3(0.3 \times 10^{-5})/\text{C}° = 0.9 \times 10^{-5}/\text{C}°$. Then

$$V_p = V_0 (1 + \beta \, \Delta t)$$
$$= 5(1 + 0.9 \times 10^{-5}/\text{C}° \times 30 \text{ C}°) \text{ liter}$$
$$= 5(1 + 0.00027) = 5(1.00027) \text{ liter}$$
$$= 5.0014 \text{ liter}$$

(b) For glycerine β is $5.1 \times 10^{-4}/\text{C}°$; then

$$V_{\text{gly}} = V_0 (1 + \beta \, \Delta t)$$
$$= 5(1 + 5.1 \times 10^{-4}/\text{C}° \times 30 \text{ C}°) \text{ liter}$$
$$= 5(1 + 0.0153) = 5(1.0153) \text{ liter}$$
$$= 5.0765 \text{ liter}$$

(c) Subtracting the new volume of Pyrex from that of the glycerine, we obtain

$$V_{\text{gly}} - V_p = V_{\text{overflow}} = 5.0765 - 5.0014$$
$$= 0.0751 \text{ liter}   ◀$$

During a volumetric expansion, since the mass remains constant, the density of the substance must change. We have (letting d equal the density)

$$m(\text{final}) = m_0(\text{initial})$$

$$dV = d_0 V_0$$

$$d = \frac{d_0 V_0}{V} = \frac{d_0 V_0}{V_0(1 + \beta \, \Delta t)}$$

$$= \frac{d_0}{1 + \beta \, \Delta t} \qquad (17\text{-}9)$$

From Eq. 17-9 we see that the density decreases with an increase in temperature.

Example 17-8 Six-hundred grams of a liquid occupy a volume of 0.5 liter at 0°C. Find the density and volume of the liquid at 70°C if $\beta = 3 \times 10^{-5}/\text{C}°$.

Solution. The original density of the liquid is

$$d_0 = 600 \text{ g}/0.5 \text{ liter} = 1200 \text{ g/liter}$$

$$d = \frac{d_0}{1 + \beta \, \Delta t} = \frac{1200 \text{ g/liter}}{1 + 3 \times 10^{-5}/\text{C}° \, (70 \text{ C}°)}$$

$$= \frac{1200 \text{ g/liter}}{1.0035}$$

$$= 1195.8 \text{ g/liter}$$

$$= 1196 \text{ g/liter} \qquad \blacktriangleleft$$

It should be noted that expansion properties are also functions of pressure. The coefficients of expansion as given in this book are those for which the pressure is held constant throughout the process.

17-7 ANOMALOUS EXPANSION OF WATER

The expansion of water is unusually different from that of most other liquids. Most liquids exhibit expansion properties like those of mercury (see Fig. 17-7a). The density constantly decreases as the temperature increases. Water, on the other hand, exhibits quite a different character (see Fig. 17-7b). The maximum density of water is at 4°C.

Water at 4°C has its minimum volume. When approaching 4°C from below, water first decreases in volume and then increases after passing 4°C. When approaching 4°C from above, water again decreases in volume until 4°C, and then proceeds to increase in volume as it continues on to colder temperatures and eventually turns to ice. Ice occupies a greater volume than water. This is why many water pipes crack when they become frozen during cold winter months.

This anomalous behavior of water is also responsible for the fact that ponds which appear frozen often have water at 4°C at the bottom. As soon as the top layer of liquid cools to 4°C, it becomes denser than the liquid beneath it and

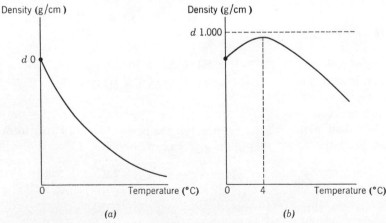

Figure 17-7 Density of mercury (a) and water (b) exhibited as a function of temperature.

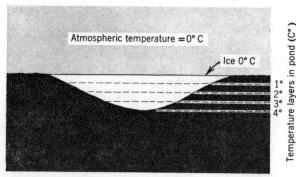

Figure 17-8 Temperature of a pond in the presence of an atmospheric temperature of 0° C.

sinks. This process continues until the entire lake is at 4°C. Now the top layer proceeds to cool further. Cooling and further shifting of layers continue until the layer of maximum density (4°C) is at the bottom and the top surface is at the same temperature as the surrounding atmosphere (see Fig. 17-8).

Although water has its maximum density at 4°C, the density does not differ greatly from unity between 0° and 100°C, and in most problems great error is not introduced by assuming water to have a constant density in the above temperature range (see Table 17-3).

Table 17-3 The density of water for various temperatures between 0 and 100°C

Temperature, °C	Density, g/cm^3
0	0.99987
4	1.00000
5	0.99999
10	0.99973
20	0.99823
40	0.99224
50	0.98807
60	0.98324
80	0.97183
90	0.96534
100	0.95838

17-8 THERMAL STRESS IN RODS

When a thin rod that is constrained from expanding at both ends is heated, the rod will experience a stress due to the forces exerted on it at both ends. If the rod were free to expand, it would increase its length by an amount Δl. Since it does not expand, the supports must be exerting a compressive force at both ends, just great enough to counterbalance the thermal stretching. The force being exerted is given by a perviously discussed equation involving Young's modulus (Eq. 6-3)

$$Y = \frac{F/A}{\Delta l/l}$$

or

$$\frac{F}{A} = \frac{Y \, \Delta l}{l}$$

Here Δl is the amount that the bar would stretch if free, and A is the cross-section area at the ends. From Eq. 17-6

$$\frac{\Delta l}{l} = \alpha \, \Delta t$$

Hence

$$\frac{F}{A} = Y\alpha \, \Delta t \qquad (17\text{-}10)$$

where F/A is the thermal stress present in a constrained rod that has been heated or cooled through a temperature difference Δt.

Example 17-9 A round aluminium rod, rigidly fixed at both ends, is heated through a temperature change of 50°C. Find the thermal stress in the rod. If the rod has a radius of 2 in., find the compressive force exerted on the rod by the supports.

Solution. Young's modulus for aluminum is 10×10^6 lb/in.^{2} and its coefficient of expansion is $2.4 \times 10^{-5}/C°$. From Eq. 17-10

$$\frac{F}{A} = Y\alpha \, \Delta t$$

$$= (10 \times 10^6 \text{ lb/in.}^2)(2.4 \times 10^{-5}/C°)(50 \text{ C°})$$

$$= 1.2 \times 10^4 \text{ lb/in.}^2$$

For a radius of 2 in., the cross-section area is $\pi r^2 = 4\pi$ in.^{2} Therefore,

$$F = (1.2 \times 10^4 \text{ lb/in.}^2)(4\pi \text{ in.}^2)$$

$$= 4.8\pi \times 10^4 \text{ lb}$$

$$= 1.51 \times 10^5 \text{ lb} \qquad \blacktriangleleft$$

SUMMARY

The *internal energy* of an object is the sum of the kinetic energy and potential energy of all the molecules of the object with respect to the center of mass of the object.

Temperature is the measure of the average kinetic energy of a molecule in an object.

The energy that is transferred between objects of different temperatures is called *heat*.

A *thermometer* is a device used to measure temperature. The four common temperature scales are Celsius (formerly centigrade), Fahrenheit, Kelvin, and Rankine. Celsius and Fahrenheit temperatures are related according to the equations:

$$t_c = \tfrac{5}{9}(t_F - 32) \quad \text{or} \quad t_F = \tfrac{9}{5}t_c + 32$$

Celsius and Kelvin temperatures are related according to the equation

$$t_K = t_c + 273$$

Fahrenheit and Rankine temperatures are related according to the equation

$$t_R = t_F + 460$$

According to the present theory, the lowest attainable temperature is $-273.16°C = -459.60°F$.

The *coefficient of linear expansion* of a material is the fractional change in length of an object per unit temperature change:

$$\alpha = \frac{\Delta l/l}{\Delta t} = \frac{\Delta l}{l \, \Delta t}$$

The area and volume of objects also change with temperature:

$$\Delta A = (2\alpha)A_0 \, \Delta t \quad \text{area change}$$
$$\Delta V = (3\alpha)V_0 \, \Delta t \quad \text{volume change (solids)}$$
$$\Delta V = \beta V_0 \, \Delta t \quad \text{volume change (liquids, gases)}$$

Water has its greatest density (smallest volume for a given mass) at 4°C.

Thermal stress is produced in a rod that is heated but is constrained from expanding. The stress depends on Young's modulus, Y, and the temperature change:

$$\frac{F}{A} = Y\alpha \, \Delta t$$

QUESTIONS

1. What is meant by the term *internal energy*?

2. Distinguish clearly between *temperature* and *internal energy.*

3. Which is hotter, a tub filled with water at 50°C or a cup of boiling water? Which do you think has more internal energy?

4. Why is it considered incorrect to speak of the heat contained in steam?

5. Can you determine relative hotness or coldness with your hand (i.e., by touching different objects)? Could you use your hand as a thermometer?

6. Do some library research to find out which thermal equilibrium situations were assigned the temperatures of 0 and 100 in the Fahrenheit system?

7. What is the difference between a Celsius degree and a degree Celsius?

8. Why are the Kelvin and Rankine scales called *absolute* scales?

9. Explain why railroad tracks might buckle if no spaces were left between sections of track.

10. Discuss how a bimetallic strip might be used in a thermostat to regulate the room temperature in your home.

11. Experienced shoppers usually shake canned goods containing liquids before buying them. Why would this be a scientifically sound practice?

12. Examine the ice cubes formed in your home freezer. Why is there a bulge in the center of the cube?

PROBLEMS

(A)

1. Body temperature is normal at 98.6°F. Express this temperature in degrees Celsius.

2. Mercury freezes at a temperature of −39°C. Express this temperature in degrees Fahrenheit.

3. The melting and boiling points of iron are 1535 and 3000°C, respectively. What are the corresponding points in degrees Fahrenheit?

4. The boiling point of dry ice (carbon dioxide) is −112°F. Express this temperature in degrees Celsius.

5. The freezing point of oxygen is −218°C. Convert this temperature to degrees Rankine.

6. Silver boils at 1950°C. Express this temperature in degrees Kelvin.

7. What Rankine temperature corresponds to a temperature of 550°C?

8. At what temperature do the Celsius and Fahrenheit thermometers read the same numerical value?

9. Change the following Fahrenheit temperatures to Celsius temperatures.
   (a) 76°F.
   (b) −195°F.
   (c) 460°F.

10. Convert the following Celsius temperatures to Fahrenheit temperatures.
   (a) 20°C.
   (b) −180°C.
   (c) 410°C.

11. Find the Kelvin temperatures that correspond to each of the following Celsius temperatures:
   (a) −30°C.
   (b) 78°C.
   (c) −40°C.

12. In a certain experiment a volume of water is found to undergo a temperature change of 48°C. What is the equivalent change on
   (a) the Fahrenheit scale, and
   (b) the Kelvin scale?

13. The temperature at the surface of the sun is about 6000°C. Express this temperature in °F, °R, and °K.

14. Change a temperature reading of 870°R to a Celsius temperature.

15. Liquid tungsten boils at a temperature of approximately 6100°K. What is this temperature in °F?

16. Find the increase in length of a 200-ft copper rod when its temperature is changed from 10 to 30°C. What is the new length of the rod?

17. Railroad tracks are laid on a day when the temperature is 58°F. Each steel rail is 40 ft long. What should be the minimum distance between rails if the temperature goes as high as 208°F?

18. A 10-ft rod expands 0.3 in. when its temperature changes from 10 to 90°C. What is the linear coefficient of expansion of the rod?

19. A steel tape is 500 cm long at 5°C. At what temperature will it be 0.5 cm longer?

20. A brass ring has a diameter of 40 cm at a temperature of 50°C. What is the diameter of the ring at 90°C?

(B)

1. A certain hypothetical thermometer is known to depend on the pressure of a certain gas according to the equation:

$$t(°C) = bP − 30°C$$

where b is a constant and P is the absolute pressure of the gas in atmospheres. If the temperature is 100°C when the pressure is 2 atm, what is the temperature when the pressure is 1 atm?

2. A certain hypothetical thermometer is known to depend on the volume of a certain fluid according to the equation:

$$t(°C) = aV + 10°C$$

where a is a constant. If the temperature is 40°C when V, the volume is 5 liters, find the temperature when the volume is 12 liters.

3. At what temperature is the Fahrenheit reading five times the Celsius reading?

4. Prove that $t_K = \frac{5}{9} t_R$, thus showing that the two absolute temperature scales are directly proportional.

5. During a certain thermodynamic process the temperature and volume are related by the equation:

$$TV^{0.4} = b$$

where b is a constant, T is measured in Kelvin degrees, and V is measured in liters. If the temperature is 400°K, when V is 5 liters, find T when V is 10 liters. (If you have trouble here, review logarithms.)

6. A Pyrex glass beaker has a volume of 500 ml at 10°C. What is the volume of the beaker at 30°C?

7. What will be the new volume of 600 l of glycerin when its temperature increases by 20°C?

8. A Pyrex glass beaker is filled to the top with 400 cm³ of mercury at 15°C. How much mercury runs out of the beaker as the temperature is increased to 100°C?

9. Copper has a density of 8.80 g/cm³ at 20°C. Compute the density of copper at 80°C.

10. The ends of a steel rod exactly 0.4 in.² in cross-sectional area are held rigid between two fixed points at a temperature of 40°C. Determine the stress in the rod when the temperature increases to 60°C. Young's modulus for steel is 33×10^6 lb/in.²

11. A rectangular sheet of aluminum measures 6 ft × 8 ft at 48°F. What is the area of the sheet at the temperature of boiling water?

12. A block of metal, in the shape of a rectangular parallelopiped, has its volume increased by 0.3% when the temperature changes by 1000°C. What is the linear coefficient of expansion of the metal?

13. Derive the equation for cubical expansion

$$V = V_0(1 + 3\alpha\, \Delta t)$$

where α is the linear coefficient of expansion.

14. A 1-liter Pyrex glass beaker is filled to the top with mercury when the temperature is 20°C. How much mercury will have run out of the beaker by the time the temperature has reached 60°C?

15. Two hundred cubic centimeters of ethyl alcohol are to be placed in an aluminum cylinder at 20°C. What should be the minimum volume of the container so that no alcohol runs out as the system is heated to 70°C? (Assume no loss of alcohol due to evaporation.)

16. An iron ball of 5-cm diameter is 0.010 mm too large to pass through the hole of a brass ring at a temperature of 40°C. At what temperature, the same for the ball and ring, will the ball just fit through the ring?

17. A clock has a period of 1 sec at a temperature of 20°C. The pendulum consists of a small ball attached to a thin steel rod. What will the period be if the temperature is raised to 30°C?

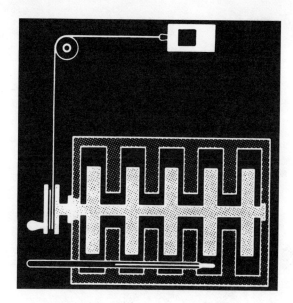

18.
Heat

We have previously defined the internal energy of a body in terms of the kinetic and potential energies of the particles that comprise the body. When the internal energy of an object changes, the object either loses or gains some energy. This energy can be measured, either by measuring the heat flow into or out of the body if it is in contact with an object of different temperature, or by measuring the amount of mechanical work that was done on the object and calculating the equivalent amount of thermal energy. In this chapter we shall discuss methods for calculating the energy of heat flow and the equivalence of mechanical and thermal energy.

LEARNING GOALS

In this chapter you should be alerted to the following major learning objectives:

1. *To recognize that heat is a form of energy transfer.*

2. *To learn the relationship between thermal energy units and mechanical energy units.*

3. *To learn to measure thermal energy changes that result from temperature changes or phase changes.*

4. *To understand the meaning of the term latent heat.*

5. *To learn how to solve problems that involve an interchange of thermal energy among several bodies which are mixed together.*

18-1 MECHANICAL EQUIVALENT OF HEAT

We define two different units for measuring thermal energy. These units are defined in terms of a certain process.

   1. One calorie is the amount of thermal energy required to change the temperature of one gram of water, one Celsius degree.

   2. One British thermal unit (Btu) is the amount of thermal energy required to change the temperature of one pound of water, one Fahrenheit degree.

$$1 \text{ Btu} = 252 \text{ calories}$$

$$1000 \text{ calories} = 1 \text{ kilocalorie}$$

We might now consider the following. If the calorie and Btu are units for thermal energy and thermal energy is related to kinetic and potential energy, then the calorie and Btu should be related to the mechanical units for energy, ergs, joules, and ft-lb. The idea that such a relationship should exist occurred to Joule, and he set about to design an experiment which would take a definite amount of mechanical energy and convert it into a measurable amount of thermal energy. By invoking the law of conservation of energy, the two quantities could be equated, and the relationship between the thermal and mechanical energy units obtained.

In Joule's experiment a weight is attached by means of a cable to a paddle wheel immersed in a vat of water (see Fig. 18-1). When the weight drops, the wheel turns, churning the water in the vat. The loss of potential energy of the weight after it falls, measured in joules, is equal to the gain in the thermal energy of the water (measured in calories). The results of many experiments indicate that

$$4.186 \text{ joules} = 1 \text{ calorie}$$

$$778 \text{ ft-lb} = 1 \text{ Btu}$$

We say that 4.186 joules is the mechanical equivalent of 1 calorie of thermal energy and 778 ft-lb is the mechanical equivalent of 1 Btu of thermal energy.

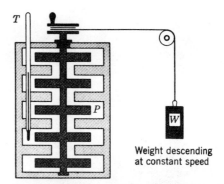

Figure 18-1   The descending weight did work in stirring the water and thus generated heat. One Btu of heat is produced by doing 778 ft-lb of work. (From O. H. Blackwood, W. C. Kelly, and R. M. Bell, *General Physics*, 2nd ed., John Wiley & Sons, Inc., New York, 1962, p. 228.)

Example 18-1   In an experiment to determine the mechanical equivalent of heat, a 200-lb weight fell through a height of 39 ft and raised the temperature of 10 lb of water 1F°. Calculate the mechanical equivalent from this data.

Solution.   We first calculate the loss in potential energy of the weight in foot pounds.

change in potential energy
$$= \text{weight} \times \text{change in height above arbitrary zero level}$$
$$= 200 \text{ lb} \times 39 \text{ ft}$$
$$= 7800 \text{ ft-lb}$$

Next, we calculate the gain in thermal energy of the water in Btu. It takes 1 Btu to raise the temperature of 1 lb of water 1F° or 10 Btu to raise the temperature of 10 lb of water 1F°. Thus

$$10 \text{ Btu} = 7800 \text{ ft-lb}$$

or $\qquad 1 \text{ Btu} = 780 \text{ ft-lb}$ ◀

18-2 SPECIFIC HEAT CAPACITY

From our definition of the calorie, it is clear that, for example, 10 calories of thermal energy will change the temperature of 1 g of water by 10C°.

The same amount of thermal energy, however, will change the temperature of 1 g of glass by about 62.5° and 1 g of copper by about 106° on the Celsius scale. These variations are due to the different specific heat capacities of water, glass, and copper. The *specific heat capacity* of a substance is the amount of thermal energy required to change the temperature of that substance 1C°. From our definition of the calorie, it is clear that the specific heat capacity of water is unity. The specific heat capacities of other substances can be found from experiments that transfer a given amount of thermal energy between water and a known mass of the substance whose specific heat capacity we wish to measure. For example, suppose we drop 10 g of copper at a temperature of 100°C into 10 g of water at a temperature of 20°C. The copper and water interchange energy until an equilibrium temperature is reached. Suppose this temperature is found to be 26.9°C. If we neglect any losses of heat to the surroundings (container walls or atmosphere), the heat gained by the water will be equal to that lost by the copper. How much heat was gained by the water? The temperature changes by 6.9C°, which requires 6.9 cal for each gram of water. For 10 g we have a total heat gain of

$$10 \times 6.9 = 69.0 \text{ cal}$$

This heat was lost by 10 g of copper which underwent a temperature change of 73.1C°. The number of calories required to change 1 g of copper 1C° is then given by

$$\frac{69.0 \text{ cal}}{10 \text{ g} \times 73.1 \text{C}°} = 0.094 \text{ cal/g-C}°$$

The correct values of the specific heat capacities for various substances are listed in Table 18-1.

From the previous discussion we can redefine the specific heat capacity in terms of an equation involving the energy change in the substance, the temperature change of the substance, and the mass of the substance. Thus

$$c = \frac{Q}{m \, \Delta T} \qquad (18\text{-}1)$$

Table 18-1 Specific heats of various substances

Substance	cal/g-C° or Btu/lb-F°
Aluminum	0.22
Copper	0.093
Glass	0.20
Gold	0.03
Ice	0.50
Iron	0.11
Lead	0.03
Mercury	0.033
Methyl alcohol	0.60
Silver	0.056
Steam	0.480
Steel	0.11
Tin	0.055
Turpentine	0.42
Water	1.00
Zinc	0.092

where c = specific heat capacity of the substance, Q = change in the thermal energy of the substance, m = mass of the substance, and ΔT = temperature change of the substance. When Q is measured in calories, m in grams, and ΔT in Celsius degrees, c will be measured in cal/g-C°. If the English system of units is used, Eq. 18-1 is altered so that the weight of the object is used rather than its mass. Then, when Q is in Btu, the weight in pounds, and ΔT in Fahrenheit degrees, c will be in Btu/lb-F°. However, since

$$1\frac{\text{Btu}}{\text{lb-F}°} = 1\frac{252 \text{ cal}}{(453.6 \text{ g})(\frac{5}{9}\text{C}°)} = 1\frac{\text{cal}}{\text{g-C}°}$$

the numerical value of c is independent of the system of units used. Therefore, Table 18-1 does not indicate any specific system of units. Further, although Table 18-1 lists only one value of c for each substance, the specific heat capacity is not a constant for any substance. It varies with temperature and tends towards zero as the absolute zero of temperature is approached. For ordinary temperatures, however, the specific heat capacity of a substance may be considered sufficiently constant without incurring large errors.

The product of the mass of an object and its specific heat capacity is called the *heat capacity*, and represents the amount of thermal energy required to change the temperature of the entire object one degree.

$$\text{heat capacity} = \frac{Q}{\Delta T} = mc = C$$

where $m = $ mass and $C = $ heat capacity of the substance. Thus, if it requires 100 cal to change the temperature of an object 5C°, its heat capacity is 20 cal/C°. If the mass of the object is known to be, for example, 50 g, the specific heat capacity would be $\frac{20}{50} = 0.4$ cal/g-C°.

Example 18-2 When 60 g of a certain metal at a temperature of 80°C is mixed with 20 g of water at 10°C, the final temperature is observed to be 15°C. Determine the specific heat capacity of the metal.

Solution. We equate the heat lost by the metal, the temperature of which decreased, to the heat gained by the water, the temperature of which increased. Thus

$$\text{heat lost} = \text{heat gained}$$

mass of metal × specific heat capacity of metal

$$\times \text{ temperature change of metal}$$

$$= \text{mass of water}$$

$$\times \text{ specific heat capacity of water}$$

$$\times \text{ temperature change of water}$$

$$60(c)(80 - 15) = 20(1)(15 - 10)$$

$$c = \frac{20(5)}{60(65)} = \frac{1}{39} = 0.026 \text{ cal/g-C°}$$

Example 18-3 What is the final temperature when 150 g of copper at 90°C is mixed with 30 g of methyl alcohol at 20°C?

Solution. To solve this problem we begin, once again, with the conservation of energy. Thus

heat lost by copper = heat gained by methyl alcohol

$$(mc\,\Delta T) \text{ for copper} = (mc\,\Delta T) \text{ for methyl alcohol}$$

$$150(0.093)(90 - t) = 30(0.6)(t - 20)$$

$$13.95(90 - t) = 18(t - 20)$$

$$1255.5 - 13.95t = 18t - 360$$

$$1255.5 + 360 = 18t + 13.95t$$

$$1615.5 = 31.95t$$

$$t = \frac{1615.5}{31.95} = 50.6°C \qquad \blacktriangleleft$$

Example 18-4 What is the final temperature when 50 lb of steel at a temperature of 140°F and 10 lb of aluminum at a temperature of 100°F are mixed with 20 lb of water at a temperature of 50°F?

Solution. We shall proceed as we have in the previous examples by equating the thermal energy lost by part of the system to that gained by the rest of the system. We encounter here a serious problem, however. It is clear that the steel loses energy, and that the water gains energy. But whether the aluminum gains or loses energy is not immediately evident. Let us assume that the aluminum loses energy. We then expect the final temperature to lie between 50°F (the initial temperature of the water) and 100°F (the initial temperature of the aluminum). If the final temperature does not lie in this temperature range, we know that our assumption about the behavior of the aluminum was in error, and we have to repeat the problem, assuming that the aluminum gains energy. Thus

heat lost by aluminum + heat lost by steel

$$= \text{heat gained by water}$$

$$10(0.22)(100 - t) + 50(0.11)(140 - t)$$

$$= 20(1)(t - 50)$$

$$220 - 2.2t + 770 - 5.5t = 20t - 1000$$

$$990 - 7.7t = 20t - 1000$$

$$1990 = 27.7t$$

$$t = \frac{1990}{27.7} = 71.8°F$$

Since 71.8°F is in the proper temperature range, it appears that we made the proper choice regarding the behavior of the aluminum. However, let us repeat the problem, assuming that the aluminum gains heat, and see what the final temperature would be.

heat lost by steel = heat gained by aluminum
+ heat gained by water

$$50(0.11)(140 - t) = 10(0.22)(t - 100) + 20(1)(t - 50)$$
$$770 - 5.5t = 2.2t - 220 + 20t - 1000$$
$$770 + 220 + 1000 = 7.7t + 20t = 27.7t$$
$$1990 = 27.7t$$

$$t = \frac{1990}{27.7} = 71.8°F$$

Thus, it appears that regardless of what we assume about the nature of the aluminum's temperature change, we get the same answer.

It should be noted that when expressing the heat gained or lost, ΔT is always written as the positive difference between the high and low temperatures for that object. ◀

Experiments to determine specific heat capacities are usually done with the aid of a calorimeter. The component parts of a calorimeter are depicted in Fig. 18-2. The inner vessel (A) of the calorimeter is the essential element of the equipment, for it is

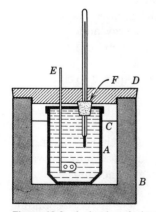

Figure 18-2 A simple calorimeter.

in here that the substance whose specific heat capacity is to be determined is mixed with a substance of known specific heat (usually water). The inner vessel is suspended within the larger outer vessel (B) by a guard ring (C), which is made of asbestos, a poor conductor of heat. The "dead-air" space which separates the inner and outer vessels reduces the loss of heat to the surroundings via convection. The outer surface of the inner and outer vessels should be highly polished to prevent losses by radiation. The transfer of heat by conduction, convection, and radiation will be discussed in Chapter 22. There are holes in the cover (D) for the insertion of a stirrer (E), thermometer, and a rubber stopper (F). When performing an experiment, the apparatus should be covered at all times to prevent interchanges of thermal energy between the air above the calorimeter and the contents of the calorimeter.

Example 18-5 The specific heat capacity of a 60-g block of metal is to be determined. The block is heated to 100°C and placed gently into a calorimeter containing 80 g of water at a temperature of 18°C. The inner vessel and stirrer are made of aluminum and have a combined mass of 50 g. The final temperature of the system is observed to be 24°C. Compute the specific heat capacity of the metal block.

Solution. The initial temperature of the calorimeter (inner vessel) and stirrer is the same as that of the water. Thus the water, calorimeter, and stirrer all gain heat, while the 60-g block loses an equal amount. Or,

heat lost by block

= heat gained by water
+ heat gained by calorimeter
+ heat gained by stirrer

$$60(c)(100 - 24) = 80(1)(24 - 18)$$
$$+ 50(0.22)(24 - 18)$$

We have used the facts that the calorimeter and stirrer are made of the same material and have a combined mass of 50 g and that the specific heat capacity of aluminum is 0.22.

$$4560c = 480 + 66$$

$$c = \frac{546}{4560} = 0.12\ \frac{cal}{g\text{-}C^{\circ}}$$ ◀

18-3 CHANGES OF PHASE

From Eq. 18-1 it would appear that the temperature of an object increases continuously as the object gains more and more heat. Suppose that we put some water at 20°C in a container and heat the system with a strong flame. Place a thermometer in the water so that we can record the rise in temperature of the water. A very curious thing happens. After the thermometer reaches 100°C, it ceases to record any higher temperature until all of the water has turned to steam. If provision is made to hold the steam in the container, and we continue to heat the system, the thermometer will again begin to record increases in temperature. In fact, the temperature of the steam will rise at a faster rate than that of the water.

If, instead of heating the water, we cool the water, we will find an interesting phenomenon taking place at 0°C. The water at 0°C begins to freeze as we remove the heat from it, and the thermometer in the water will not read lower than 0°C until all the water changes to ice (see Fig. 18-3).

Experiments on other objects verify the fact that changes of phase are not accompanied by temperature changes. The heat supplied to a substance while it is changing phase (from a solid to a liquid, or a liquid to a vapor, or a solid to a vapor) is not used to increase the average kinetic energy of the molecules. What then becomes of this thermal energy?

According to the molecular theory of matter, the molecules of a solid are arranged in an ordered crystallike pattern. These molecules can perform only small oscillations about their equilib-

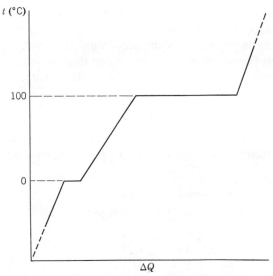

Figure-18-3 A graph showing how the temperature of water changes as thermal energy is added or withdrawn from the water.

rium positions. When the solid changes to the liquid state, the bond between molecules becomes weaker, and the molecules have more freedom in their oscillations, although a partly ordered pattern still exists. In the vapor state the molecules behave, essentially, as free particles and no ordered arrangement is present. Thus the thermal energy supplied during a change in phase manifests itself as a change in the molecular structure of the substance, a potential energy change, rather than as a change in the relative speeds of the molecules. The thermal energy added during a change of state does not, therefore, result in a temperature change.

The number of calories required to change one gram of a substance in the solid state into one gram of the substance in the liquid phase is called the *heat of fusion*. The *heat of vaporization* is the number of calories required to change one gram of a substance in the liquid phase into the vapor state. A similar statement concerning a change directly from the solid to the vapor phase defines the *heat of sublimation*. There are corresponding

Table 18-2 Melting and boiling points and the heats of fusion and vaporization of various substances

Substance	Melting Point (°C)	Heat of Fusion (cal/g)	Boiling Point (°C)	Heat of Vaporization (cal/g)
Alcohol (ethyl)	−114	24	78	204
Bromine	−7	16	60	43
Copper	1080	42	2310	—
Gold	1063	16	2500	—
Helium	−271	—	−268	6
Lead	330	5.9	1170	175
Mercury	−39	2.8	358	71
Nitrogen	−210	6.1	−196	48
Oxygen	−219	3.3	−183	51
Tungsten	3400	—	5830	
Water	0	80	100	540
Zinc	420	24	918	475

definitions for these three quantities in terms of one pound of a substance and the Btu. When a substance goes from the vapor state to the liquid or solid state, or from the liquid to the solid state, it liberates the same amount of heat it absorbed in vaporizing or becoming liquefied.

The heats of vaporization, sublimation, and fusion are called latent,* because their presence is not accompanied by changes in temperature. Latent heats are determined by calorimetry experiments similar to those from which we obtain data about specific heat capacities. For water, ice, steam, the latent heats are

$$\text{latent heat of fusion} = L_f = 80 \text{ cal/g}$$
$$= 144 \text{ Btu/lb}$$
$$\text{latent heat of vaporization} = L_v = 540 \text{ cal/g}$$
$$= 972 \text{ Btu/lb}$$

We do not list a value for the heat of sublimation from ice to steam, since water does not sublimate at ordinary pressures. Values of these latent heats for other substances are listed in Table 18-2.

* "Latent" comes from a Latin word meaning "hidden."

From our discussion, it is clear that the latent heat absorbed or liberated by m grams or pounds of a substance during a change of phase is given by

$$Q = mL \qquad (18\text{-}3)$$

Example 18-6 Ten grams of ice at 0°C are dropped into 50 g of water at 30°C. What is the final temperature of the mixture?

Solution. Using our energy principle, the thermal energy lost by the water, whose temperature decreases, must equal the heat gained by the ice. Part of the heat gained by the ice will change it to water at 0°C and the rest will be used to bring the melted ice to the final temperature. Thus

$$\begin{array}{c}\text{mass of} \\ \text{original water}\end{array} \times \begin{array}{c}\text{specific heat capacity} \\ \text{of water}\end{array} \times \begin{array}{c}\text{temperature change} \\ \text{of original water}\end{array}$$

$$= \begin{array}{c}\text{mass} \\ \text{of} \\ \text{ice}\end{array} \times \begin{array}{c}\text{latent} \\ \text{heat of} \\ \text{fusion}\end{array} + \begin{array}{c}\text{mass of} \\ \text{melted} \\ \text{ice}\end{array} \times \begin{array}{c}\text{specific} \\ \text{heat of} \\ \text{melted ice}\end{array} \times \begin{array}{c}\text{temperature} \\ \text{change of} \\ \text{melted ice}\end{array}$$

$$50(1)(30 - t) = 10(80) + 10(1)(t - 0)$$
$$1500 - 50t = 800 + 10t$$
$$700 = 60t$$
$$t = \tfrac{70}{6} = 11.7°C \qquad \blacktriangleleft$$

Example 18-7   Ten grams of ice at $-5°C$ are mixed with 5 g of steam at $100°C$. Find the final temperature of the mixture.

Solution.   Here part of the heat gained by the ice must first bring the ice to $0°C$ at which point it can be melted. After being melted, the water formed must be brought from $0°C$ to the final temperature. The heat supplied by the steam is liberated in two stages. First, 540 cal/g are liberated in converting the steam to water at $100°C$. The remaining heat is lost when the temperature of the water decreases to its final value. Thus

$$\text{heat lost} = \text{heat gained}$$

$$5(540) + 5(1)(100 - t) = 10(0.5)[0 - (-5)]$$
$$+ 10(80) + 10(1)(t - 0)$$

$$2700 + 500 - 5t = 25 + 800 + 10t$$

$$2375 = 15t, \qquad t = 158°C.$$

Clearly, something is wrong. We assumed that all the steam condenses and all the ice melts. Hence the final composition is water. But water cannot have a temperature greater than $100°C$ under the conditions of the problem. Evidently, not all the steam condenses. Since some steam remains, the final temperature of the mixture must be $100°C$. The problem now is to determine how many grams of steam did condense. We let x represent the number of grams of steam that condense. Using the heat lost-heat gained principle, we have heat lost by condensing steam = heat gained by ice as it is heated from -5 to $0°C$ + heat gained by melting ice at $0°C$ + heat gained by melted ice as it is heated from 0 to $100°C$. Therefore,

$$540x = 10(0.5)[0 - (-5)] + 80(10)$$
$$+ 10(1)(100 - 0)$$

$$540x = 1825$$

$$x = \frac{1825}{540}$$

$$x = 3.4 \text{ g}$$

If the solution for the final temperature for such a problem had yielded a temperature less than $0°C$, it would have indicated that not all of the ice had melted.   ◀

18-4   HEAT OF COMBUSTION

When an object is burned, it gives up thermal energy. The amount of thermal energy released when one gram of a substance is completely burned is called the heat of combustion. In general, only a small percentage of the total heat supplied when a substance is totally burned can be used beneficially. The energy used for personal comfort in private homes and apartment houses is usually supplied by the heat of combustion of coal or oil. In rockets, the heat of combustion from liquid or solid fuels is converted into mechanical energy to give the rockets their necessary thrust. The heats of combustion for various commercial fuels are listed in Table 18-3.

Table 18-3   Heats of combustion of liquid fuels

Fuel	Btu per lb	Calories per gram
Alcohol, fuel or denatured	11,620	6,456
Crude oil, California	18,910	10,506
Kansas	19,130	10,628
Mexico	18,755	10,419
Oklahoma	19,502	10,834
Pennsylvania	19,505	10,836
Texas	19,460	10,811
Wyoming	19,510	10,839
Gas oil	19,200	10,667
Gasoline	20,750	11,528
Fuel oil, California	18,835	10,464
Mexico	18,510	10,283
Mid-continent	19,376	10,764
Furnace oil	19,025	10,569
Kerosene	19,810	11,006

SUMMARY

Thermal energy is usually measured in units of *calories* or *British Thermal Units* (Btu).

One calorie is the thermal energy required to change the temperature of one gram of water, one Celsius degree.

One Btu is the thermal energy required to change the temperature of one pound of water, one Fahrenheit degree.

The specific heat capacity of a substance is the thermal energy change per unit mass per unit temperature change:

$$c = \frac{Q}{m \, \Delta T}$$

When several substances at different temperatures are mixed, the law of conservation of energy requires that the heat lost by all substances whose temperatures decrease is exactly equal to the heat gained by all the substances whose temperatures increase.

During a change of phase, the heat gained or lost by the substance is given by $Q = mL$, where L is the latent heat and m is the mass.

QUESTIONS

1. Which two units are most often used to measure thermal energy? How are these units defined?

2. Describe a situation in which mechanical energy is converted into thermal energy. How could you determine the mechanical equivalent of heat from such a process?

3. Discuss in your own words the significance of the specific heat capacity.

4. Water has a specific heat capacity that is larger than most other common substances. Would it take longer to warm up 2 lb of water from room temperature to 60°F or the same amount of steel through the same temperature change? Explain.

5. Explain how a calorimeter prevents major heat losses during mixture experiments.

6. What is meant by a change of phase? How much heat is required to change one gram of solid lead into molten lead at the melting point?

7. Why is ice used so successfully as a coolant for drinks?

8. What is the heat of vaporization of steam? Explain why a burn caused by 100°C steam is more severe than a burn caused by 100°C water?

9. Why are the heats of fusion, vaporization, and sublimation called latent heats?

PROBLEMS

(A)

1. What is the mechanical equivalent of 2520 calories?

2. What is the mechanical equivalent of 420 Btu?

3. What is the mechanical equivalent of a thermal power of 3000 cal/sec?

4. What is the mechanical power equivalent (in horsepower) of 11,000 Btu/hr?

5. How many calories are given off when 200 g of water cool from 85 to 5°C?

6. How many calories are required to increase the temperature of 1000 g of gold from 0 to 1500°C?

7. Eighty grams of aluminum are warmed from 60 to 120°C. How many calories are required?

8. What is the final temperature when 40 g of water at 10°C are mixed with 100 g of water at 70°C?

9. What is the final temperature when 200 g of copper at 200°C are added to 400 g of water at 10°C?

10. How many Btu are released when 4 lb of steam at 250°F are changed to water at 68°F?

11. How many calories are required to change 40 g of ice at -50°C to water at 50°C?

12. What is the final temperature when 5 g of ice at 0°C are mixed with 25 g of water at 20°C?

13. How many Btu are released when 4 lb of kerosene are completely burned?

14. How many calories are released when 2 kg of California crude oil are completely burned?

(B)

1. A certain engine does work at the rate of 3 hp. What equivalent amount of Btu does the engine give off in 1 hr?

2. Convert the following mechanical energy units to thermal units in the same system as the given units:
   (a) 1764 ft-lb,
   (b) 88 joules, and
   (c) 1.857×10^8 ergs.

3. How much heat is required to melt 10 lb of lead at the melting point?

4. Calculate the amount of thermal energy required to raise 5 kg of gold from 70 to 300°F.

5. How many calories are required to change 10 g of ice at $-10°C$ to steam at $110°C$?

6. A 240-g block of aluminum at a temperature of $145°C$ is dropped into 600 g of water at $20°C$. What is the final temperature of the mixture?

7. Find the final temperature when 600 g of copper at $80°C$ are dropped into a 50-g aluminum calorimeter containing 60 g of water at a temperature of $18°C$.

8. Forty grams of a certain liquid at a temperature of $20°C$ are mixed with 8 g of ice at $0°C$. All of the ice *just* melts (and the final temperature is $0°C$). What is the specific heat capacity of the liquid?

9. In the process of manufacturing a certain iron part for a machine, a 1-lb piece of the iron at $1500°F$ is plunged into 12 lb of water at $50°F$. Find the final temperature of the water and iron.

10. The following data were taken during an experiment designed to determine the heat of vaporization of water:

Mass of aluminum calorimeter	48 g
Mass of water plus calorimeter	112 g
Mass of water, calorimeter, and steam	116 g
Original temperature of water and calorimeter	$15°C$
Original temperature of steam	$100°C$
Final temperature of mixture	$46.7°C$

Determine the heat of vaporization from these data.

11. How much useful heat is obtained when 30 lb of coal are completely burned in a system that is 15% efficient?

12. An oil burner is required to supply 33,000 Btu of useful thermal energy each hour. If the burner is 10% efficient, how much oil must be supplied to the burner each day?

13. In an experiment to determine the heating value of a certain type of coal in Btu/lb, the following data were noted. The temperature of 6 lb of water changed from 72 to $82°F$ when 0.004 lb of the coal was completely burned. Find the heating value of this type of coal from these data.

14. Eight grams of steam at $100°C$ are introduced into a mixture of 200 g of water and 70 g of ice. Find the final temperature and composition of the mixture.

15. Twelve grams of steam are added to a mixture of 20 g of water and 20 g of ice. Find the final temperature and composition of the mixture.

16. A 400-g block of metal at a temperature of $200°C$ is mixed in a calorimeter containing a mixture of 300 g of water and 200 g of ice. All of the ice just melts. Neglecting heat changes in the calorimeter, determine the final temperature of the mixture if the mass of metal is doubled.

17. A 2-g lead bullet traveling at a speed of 250 m/sec strikes a stationary wall and comes to rest.
(a) How many joules of kinetic energy are converted to heat?
(b) Assuming that half of the thermal energy available is kept by the bullet, what will be the increase in temperature of the bullet?

18. An electric heater has a 4000-watt heating element and contains 400 g of water. If the water was originally at $5°C$, how long will it take to bring the water to the boiling point? How much longer will it take to vaporize all of the water?

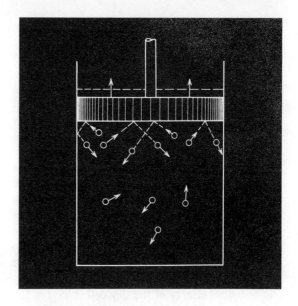

19.
The First and
Second Laws of
Thermodynamics

The two fundamental *laws of thermodynamics* are discussed in this chapter. These laws relate directly to the conversion of thermal energy to mechanical energy or mechanical energy to thermal energy. The operation of *heat engines* and *refrigerators* are subject to these laws and their special consequences.

LEARNING GOALS

The major learning objectives of this chapter are

1. *To understand the significance of the first and second laws of thermodynamics.*
2. *To learn how to calculate the work done on or by a gas.*
3. *To understand the basic operating principles of engines and refrigerators.*
4. *To learn how to calculate the efficiency of any thermodynamic cycle.*
5. *To learn the Carnot cycle and how to calculate its efficiency.*
6. *To understand the important features of several types of engines.*

19-1 THE FIRST LAW

At this point in our study of physics we pause for a moment to consider what we have learned about energy transformations. We know that all energy is conserved. Energy may change its form but never its amount. When a block slides along a rough floor, the kinetic energy of translation is transformed into thermal energy (heat) as the average kinetic energy of the molecules in the block and floor is increased. The electrical energy between the terminals of a dry cell causes electric charges to flow through the filament of a lamp causing the latter to "heat up." The energy in the moving electric charges, obtained from the electric potential energy of the dry cell, is converted into thermal energy in the coils of the toaster, iron, broiler, etc. The flame from a stove burner causes the water in a pot to boil; the energy transferred by radiation from the flame to the pot, and by conduction through the pot, is transferred into thermal energy in the water.

The amount of energy transferred between electrical and mechanical systems, or between mechanical and thermal systems, or any such combination, is determined by the law of conservation of energy. We shall reformulate the law of conservation of energy in terms of the net heat flow into a system, the work done by the system on its surroundings, and the change in the internal energy of the system.

Consider the following examples: (a) A flame under a cylinder closed by a movable piston causes the gas in the cylinder to expand against the piston to a new volume while the temperature of the gas remains unchanged. Here all the thermal energy introduced into the cylinder by the flame has been transformed into mechanical work done on the piston. There has been no change in the internal energy of the gas because there has been no change in the temperature. (b) The electric charges flowing through a wire immersed in some water cause the temperature of the water to increase. In this case, the thermal energy transferred from the wire to the water is transformed into internal energy in the water.

The water does no work because it does not expand against its surroundings. (c) A rotating propeller shaft churns the wine in a vat until the temperature of the wine increases by $10C°$. In this case, the thermal equivalent of the mechanical work done by the rotating shaft is transformed into the change in internal energy of the wine, the latter doing no work on its surroundings.

From these simple examples, and our previous encounter with problems in mechanical energy transformations, we can formulate the following general rule:

The net heat flow into a system is equal to the sum of the thermal equivalent of the work done by the system on its surroundings and the change in the internal energy of the system.

In equation form:

$$Q = W + \Delta U \qquad (19\text{-}1)$$

where Q is the net heat flow into the system, W is the thermal equivalent of the work done by the system on its surroundings, and ΔU is the change in the internal energy of the system. Also, W is considered positive when work is done *by* the system on its surroundings and negative when work is performed *on* the system by the surroundings.

Equation 19-1 is called the *first law of thermodynamics*, and as we have noted it is simply a restatement of the conservation of energy. In Eq. 19-1 Q may be either a mechanical transfer of thermal energy due to work being done on the system, as in example (c) above, or a nonmechanical transfer, as in examples (a) and (b). In very simple terms, Eq. 19-1 suggests the following observation. When an amount of thermal energy is transferred to a system, the system may use this energy in one of three ways. (a) All of the energy is used by the system to do external work ($\Delta U = 0$). (b) All of the energy is used to increase the internal energy of the system ($W = 0$). (c) Part of the energy is used to increase the internal energy of the system; part is used to do work on the surroundings of the system.

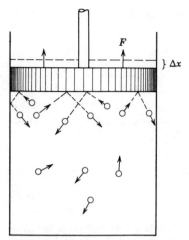

Figure 19-1   Recoil of a piston caused by collisions with gas molecules.

Since we shall be applying many of our ideas about energy changes to gases enclosed in cylinders, it is now necessary to obtain an expression for the work done by a gas when it expands against a movable piston.

Consider a gas at a pressure p, enclosed in a cylinder of initial volume V with a movable piston of cross-sectional area A (see Fig. 19-1.) Suppose the gas expands against the piston at *constant pressure* forcing the piston to move up a distance Δx. The work done on the piston is

$$W = F \, \Delta x$$

where F is the force pushing up on the piston. But,

$$F = pA$$

Therefore, $W = p(A \, \Delta x)$

or $W = p \, \Delta V$ (19-2)

Equation 19-2 says that the work done by a gas in moving a piston at a constant pressure equals the pressure times the change in volume of the gas.

Example 19-1   A gas expands at a constant pressure of 1 atm forcing a piston to move up 10 cm. The cross-sectional area of the piston is 100 cm². How much work is done by the gas?

Express your answer both in liter-atmospheres and in joules. Note from Eq. 19-2 that work can be expressed in units of pressure times volume or liter-atmospheres.

Solution.   The change in volume of the gas is the area of the piston times the distance moved by the piston. Thus

$$\Delta V = A \, \Delta x = 100 \text{ cm}^2 \times 10 \text{ cm}$$
$$= 1000 \text{ cm}^3 = 1 \text{ liter}$$

Therefore,

$$W = p \, \Delta V$$
$$= 1 \text{ atm-1 liter} = 1 \text{ liter-atm}$$
$$= 1.013 \times 10^5 \text{ N/m}^2 \times 1000 \text{ cm}^3 \times 10^{-6} \text{ m}^3/\text{cm}^3$$
$$= 1.013 \times 10^2 \text{ joules} \qquad \blacktriangleleft$$

Example 19-2   A movable piston of 10 in.² cross section weighs 200 lb. A gas expanding against the piston moves the piston 1 in. at a constant velocity. How much work is done by the gas?

Solution.   Since the piston moves at a constant velocity, it must be in equilibrium. Thus the weight of the piston must equal the pressure under the piston times the piston area. Therefore,

$$p = \text{weight of piston/area}$$
$$= 200 \text{ lb/10 in.}^2 = 20 \text{ lb/in.}^2$$

and $W = p \, \Delta V = 20 \text{ lb/in.}^2 \times 10 \text{ in.}^3$
$$= 200 \text{ in.-lb}$$
$$= 16\tfrac{2}{3} \text{ ft-lb} \qquad \blacktriangleleft$$

How do we calculate the work done by an expanding gas if the pressure of the gas varies during the expansion? For such problems we turn to graphical analysis. We assume that we know precisely how the pressure changes with volume so that we can plot a graph of pressure versus volume (see Fig. 19-2). The area of the shaded rectangle is $p_1 \, \Delta V_1$, which represents the work done in expanding by an amount ΔV_1 under a constant pressure p_1. By adding up many of these narrow rectangles, it becomes clear that the

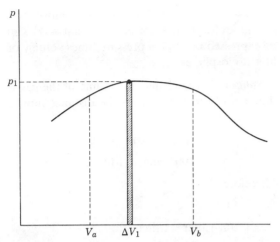

Figure 19-2 Work done by the expanding gas equals the area under the p-v curve.

work done in expanding from a volume V_a to a volume V_b is the total area under the p–V diagram between V_a and V_b.

19-2 SPECIAL CASES OF THE FIRST LAW OF THERMODYNAMICS

We can now consider a few special cases of the first law of thermodynamics.

Isovolumic Process

Consider a gas enclosed in a container that has rigid walls. Then all energy changes in the gas take place at constant volume. Since $\Delta V = 0$, $W = 0$ also, and $Q = \Delta U$.

In an isovolumic process, the thermal energy transferred to the gas is completely converted to internal energy, resulting in an increase in the temperature of the gas.

Adiabatic Process

An adiabatic process is one that allows no thermal energy to be transferred between the gas and its surroundings. The gas is said to be surrounded by *adiabatic* walls. For such a process, from Eq. 19-1

$$Q = 0 \quad \text{and} \quad W = -\Delta U.$$

The work done by the expanding gas is done at the expense of the internal energy. The temperature of the gas, consequently, decreases. If the gas is compressed adiabatically, the temperature of the gas will rise.

Isothermal Process

In an isothermal process the temperature of the gas is kept constant throughout the process. Since the temperature remains fixed, the internal energy cannot change. Thus $\Delta U = 0$, and $Q = W$. All of the thermal energy absorbed by the gas is converted by the gas into work done on the surroundings.

Free Expansion

A free expansion is one during which a system contained within adiabatic walls does no work on its surroundings. For such a process, $Q = 0$ and $W = 0$. Thus $\Delta U = 0$.

Hence, in a free expansion, the internal energy of the gas remains constant and there is no change in the temperature of the gas. Joule performed an experiment aimed at testing this result. A thin membrane separated two chambers of equal volume (see Fig. 19-3). Chamber A is occupied by a gas at a temperature T, while chamber B is evacuated. When the membrane is removed, the gas in A expands into B. The walls are assumed to be properly adiabatic. The results of the experiment indicated that the temperature of the gas did not change significantly after the expansion.

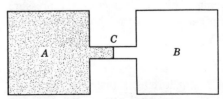

Figure 19-3 When the membrane at C is removed, the gas in A expands into chamber B.

Throttling Process

In a throttling process a gas is expanded by forcing it through a narrow opening between two chambers. A prototype of this experiment is the Joule-Thomson porous-plug experiment.

Consider two chambers separated by a porous plug. As gas diffuses through the plug from chamber A to chamber B, it passes from a region of pressure p_1 to one of pressure p_2. Suppose the process to have reached equilibrium with the walls so that there are no longer any thermal interchanges with the surroundings (see Fig. 19-4). We now interject two pistons, C and D so that the force of C to the right is $p_1 A$, and the force on D to the right is $p_2 A$. Further, assume that the pistons are moving with the same speed as the molecules were without the chamber so that conditions at the plug are not changed in the presence of the pistons. Since $Q = 0$, we have

$$W = -\Delta U$$

Now,

$W = $ work done *by* the gas $-$ work done *on* the gas

$\quad = p_2 V_2 - p_1 V_1$

Therefore,

$$p_2 V_2 - p_1 V_1 = -(U_2 - U_1)$$

or $\qquad U_2 + p_2 V_2 = U_1 + p_1 V_1$

The quantity $U + pV$ is conserved during a throttling process. This combination of quantities occurs often in the study of thermodynamics and is called the *enthalpy*. The enthalpy is gen-

erally denoted by the letter H. Enthalpy changes are very important in isobaric processes. In an isobaric process the pressure is kept constant so that

$$
\begin{aligned}
Q &= W + \Delta U \\
&= p(V_2 - V_1) + (U_2 - U_1) \\
&= (U_2 + pV_2) - (U_1 + pV_1) \\
&= H_2 - H_1 = \Delta H
\end{aligned}
$$

Thus, in an isobaric process, the thermal energy added or subtracted from a system is equal to the change in the enthalpy of the system.

The first law of thermodynamics puts an upper limit on the amount of thermal energy that can be expected from any conversion process. No matter what type of process is employed, whether it be mechanical to thermal or electrical to thermal, it is never possible to obtain at the end of the process more thermal energy than was present to begin with. Because of this fact the first law is often restated in the following terms: "A perpetual motion machine of the first kind is impossible." A perpetual motion machine of the first kind is one which, in any one complete cycle of operation, would produce more energy that it began with. Thus the machine would run itself continuously and always produce "free" energy to run other machinery. The first law says that the existence of such a machine is impossible.

19-3 THE SECOND LAW

Imagine a rotating flywheel that is brought to rest by the friction in the bearings. The kinetic energy of the flywheel becomes thermal energy, and the temperature of the flywheel and bearings increases. On the other hand, the flywheel will never, by itself, reabsorb this thermal energy from the bearings and begin rotating again.

The gas in a flask when released in a large evacuated chamber will expand and occupy the entire volume of the chamber. But the gas in the chamber, by itself, will not reassemble completely in the flask. There would be no violation of the

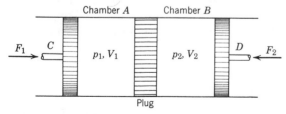

Figure 19-4 As piston C moves to the right, the gas in A diffuses through the porous plug into B. Piston C does an amount of work $p_1 v_1$ on the gas in A, while the gas in B does an amount of work $p_2 V_2$ on the piston D. Therefore, the net work done by the gas is $p_2 V_2 - p_1 V_1$.

first law of thermodynamics if the gas did reassemble in the flask or if the flywheel did reabsorb the thermal energy and begin rotating. Energy would still be conserved. Yet we see from experience that there is a preferred direction in which thermodynamic events proceed. There is evidently some thermodynamic property which, in any process, can change in only one direction. This property is called the *entropy* of the system. *Entropy is a measure of the disorder or lack of information in a system.* In the case of the flywheel, the orderly rotational kinetic energy of the flywheel changes into random, disorderly thermal energy, but the disorderly thermal energy will not, by itself, change into kinetic energy. When the gas is confined in the flask, there is less randomness to its motion and more is known about the general location of the particles in the gas. In the larger chamber, however, the randomness of the motion of the particles is increased and the precision to which the location of a particle is known is decreased. The second law of thermodynamics states that events always proceed in the direction from order to disorder, from a state having a certain amount of information to a state having a lesser amount of information. Put in other words, during any process carried out on a confined, isolated system, the entropy of the system either increases or remains the same. Note that the second law specifies isolated systems. In an actual process, the entropy of a nonisolated system may decrease, but it will be found that other systems in contact with the first will have their entropy increased by at least an equal amount.

There are many different ways of formulating the second law, all of which can be shown to be equivalent. Consider the following two formulations:

No process is possible that has as its sole result the removal of a certain amount of heat from a reservoir and the performance of an equivalent amount of work.

No process is possible that has as its sole result the removal of a certain amount of heat from a reservoir at one temperature and the absorption of an equal quantity of heat by a reservoir at a higher temperature.

The first statement is known as the Kelvin-Planck statement of the second law, while the second is attributed to Clausius. The process defined in the Kelvin statement would clearly produce a decrease in entropy (random thermal energy to ordered work). The Clausius statement, while less evident, can also be shown to be equivalent to the entropy formulation of the second law. Note once again that neither of the impossible processes in these two statements would violate the first law of thermodynamics.

A device that converts thermal energy into useful mechanical energy (work) is called an engine. A typical engine might have a gas as its working substance. The gas is taken into the engine at a high temperature. As it expands, it does work on its surroundings (a movable piston, perhaps) and its temperature decreases. Not all of the thermal energy lost by the gas can be converted into work (Clausius statement), so some thermal energy must be rejected from the engine at the lower temperature. To be useful, the engine has to be capable of continuous operation. Thus the gas must now be compressed back to its original temperature so that the cycle can be repeated. Compressing the gas requires doing work on the system. The net work output of the engine is the difference between the work done by the gas while expanding and the work done on the gas during the compression. If W represents the net work done by the engine, Q_1 the thermal energy absorbed at the high temperature, and Q_2 the thermal energy rejected at the low temperature, then from the first law of thermodynamics,

$$Q_1 = W + Q_2$$

The efficiency of an engine is the ratio of the work output to the heat input. Thus

$$\text{efficiency} = E = \frac{W}{Q_1} = \frac{Q_1 - Q_2}{Q_1}$$

$$= 1 - \frac{Q_2}{Q_1} \qquad (19\text{-}3)$$

The efficiency can never be 100%, but it can be increased by decreasing the ratio Q_2/Q_1. Figure 19-5 shows schematically the characteristics of an engine.

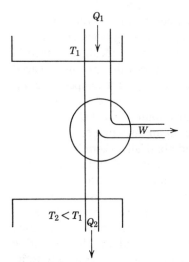

Figure 19-5 Schematic of an engine working between a high temperature reservoir T_1 and a low temperature reservoir T_2.

If an engine is operated in reverse, it is called a refrigerator. In a refrigerator, work is put into the system to draw an amount of thermal energy out of the cold reservoir and reject it to the warmer reservoir. The amount of work necessary to withdraw Q_2 calories from the cold reservoir and deposit Q_1 calories in the warm reservoir may be greater than the work output when the refrigerator was operating as an engine. Figure 19-6 is a schematic diagram of a refrigerator, with W' the input work. If $W = W'$, the engine is said to be reversible. The quality of a refrigerator is judged by its coefficient of performance, which is the amount of thermal energy removed from the cold reservoir divided by the work input. Thus

$$\text{coefficient of performance} = \eta = \frac{Q_2}{W}$$

$$= \frac{Q_2}{Q_1 - Q_2} \quad (19\text{-}4)$$

In Eq. 19-4 the refrigerator has been assumed reversible. *No engine can have a greater efficiency than a reversible engine when operating between the same two temperature reservoirs.* This can be proved by assuming that there is an engine with a greater efficiency than a reversible engine and showing that this leads to a violation of the second law.

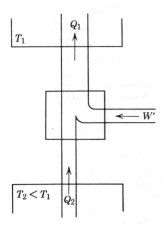

Figure 19-6 Schematic of a refrigerator.

Let the irreversible engine have the same work output as the reversible engine but let it take in an amount of heat $Q_1' < Q_1$ and reject $Q_2' < Q_2$. Further, let the reversible engine be operated as a refrigerator. The schematic for the system is shown in Fig. 19-7. Since the irreversible engine requires only Q_1' to operate, and the reversible engine delivers Q_1 to the hot reservoir, the system operates continuously with the sole result that an amount of thermal energy $Q_1 - Q_1'$ is transferred from the cold to the hot reservoir, in violation of the Clausius statement. Therefore, no irreversible engine can have a greater efficiency than a reversible engine.

By a similar argument it can be shown that all reversible engines operating between the same two temperature reservoirs must have the same efficiency.

Since all reversible engines operating between the same two temperature reservoirs have the same efficiency, and no engine operating between

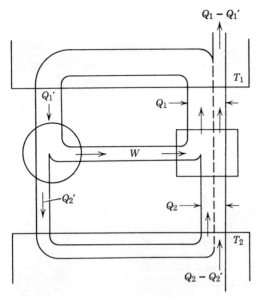

Figure 19-7 A system that violates the second law of thermodynamics. The refrigerator needs an input work W to supply Q_1 to the high temperature reservoir and the engine only needs Q_1' to supply an amount of work W to the refrigerator. Since $W = Q_1 - Q_2 = Q_1' - Q_2'$, we have $Q_1 - Q_1' = Q_2 - Q_2'$ equal to the amount of energy transferred from T_2 to T_1 in every cycle while the system operates continuously.

the same temperatures can have a greater efficiency than a reversible engine, the efficiency of any single reversible engine will serve to set a maximum on the possible efficiencies of engines. One such reversible engine was proposed by a French engineer, Sadi Carnot.

Carnot's engine operates on the following cycle (see Fig. 19-8). The working substance (solid, liquid, or gas, not necessarily ideal) contained in a cylinder fitted with a movable piston is placed in contact with a reservoir at a temperature T_1. The substance expands isothermally and reversibly, absorbing an amount of heat Q_1. The cylinder is next placed on an insulated stand where it expands further adiabatically, there being no net heat flow into the cylinder. Now the cylinder is placed in contact with a reservoir at temperature $T_2 < T_1$, where the working substance is compressed isothermally and reversibly, delivering an amount of heat Q_2 to the reservoir. Finally, the cylinder is again placed on the insulating stand, where it is compressed adiabatically to its original pressure, volume, and temperature. A p–V diagram showing the cycle is given in Fig. 19-9. The net work done by the engine is the

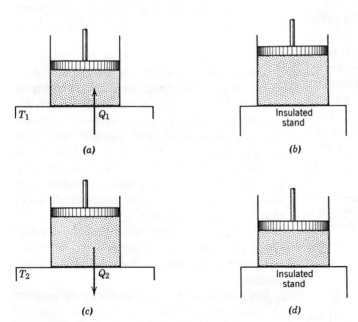

Figure 19-8 Successive steps in a Carnot cycle.

shaded area enclosed by the two isotherms and the two adiabats. Since no heat is gained or lost during the adiabatic processes, the efficiency of the Carnot engine is

$$E = \frac{W}{Q_1} = \frac{Q_1 - Q_2}{Q_1} = 1 - \frac{Q_2}{Q_1}$$

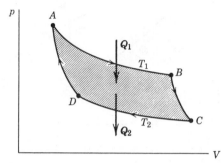

Figure 19-9 The adiabats (*BC* and *AD*) and isotherms (*AB* and *CD*) of a Carnot cycle.

It was suggested by Kelvin that since the Carnot cycles are independent of the working substance, a temperature scale based on the heats absorbed and rejected in Carnot cycles would depend only on heat transfers and not on the changes in pressures, volumes, lengths, or resistances of a substance. Such a scale would truly be an absolute scale. The scale is defined by setting the ratio

$$\frac{Q_2}{Q_1} = \frac{T_2}{T_1} \qquad (19\text{-}5)$$

and making the difference between the steam point and ice point equal to a specified number of degrees,

$$T_s - T_i = 100 \text{ (Kelvin)} \qquad (19\text{-}6a)$$
$$T_s - T_i = 180 \text{ (Rankine)} \qquad (19\text{-}6b)$$

Equations 19-5 and 19-6 define the absolute (Kelvin or Rankine) temperature scales, and they coincide exactly with the absolute temperatures discussed in Chapter 17.

In terms of absolute temperatures, the efficiency of a reversible engine and the coefficient of per-

formance of a refrigerator become

$$E = 1 - \frac{T_2}{T_1} \qquad (19\text{-}7)$$

$$\eta = \frac{T_2}{T_1 - T_2}$$

Example 19-3   A Carnot engine operating between the temperatures 400°K and 200°K absorbs 500 cal from the hot reservoir. (a) What is the efficiency of the engine? (b) How many calories are delivered to the cold reservoir? (c) How many joules of work does the engine do in one cycle?

Solution
(a) Using Eq. 19-7, we obtain

$$E = 1 - \tfrac{200}{400} = 1 - 0.5 = 0.5 \quad \text{or} \quad 50\%$$

(b) From Eq. 19-5

$$Q_2 = Q_1 \left(\frac{T_2}{T_1}\right) = 500 \left(\frac{200}{400}\right) = 250 \text{ cal}$$

(c) From the first law of thermodynamics

$$W = Q_1 - Q_2 = 500 - 250 = 250 \text{ cal}$$

Since 1 cal = 4.186 joules, $W = 250(4.186) =$ 1047 joules. ◀

Example 19-4   What is the maximum efficiency of an engine operating between temperatures of 27 and 327°C?

Solution.   The maximum efficiency of any heat engine is the efficiency of a Carnot engine operating between the same highest and lowest temperatures. Therefore, using Eq. 19-7, we find that

$$E_{\text{max}} = 1 - \frac{T_2}{T_1}$$

$$= 1 - \frac{300°\text{K}}{600°\text{K}} = 1 - 0.5$$

$$E_{\text{max}} = 0.5 \text{ or } 50\% \qquad \blacktriangleleft$$

Example 19-5   A Carnot engine has an efficiency of 20%. If the high temperature reservoir is 500°K, find the temperature at which energy is exhausted during the compression. What is the ratio of heat output to heat input?

Solution. We can use Eq. 19-7 to find the low temperature.

$$E = 1 - \frac{T_2}{T_1}$$

$$0.20 = 1 - \frac{T_2}{500°K}$$

$$\frac{T_2}{500°K} = 0.8$$

$$T_2 = 400°K = 127°C$$

The ratio of the heat output to heat input is determined from Eq. 19-5.

$$\frac{Q_2}{Q_1} = \frac{T_2}{T_1} = \frac{400°K}{500°K} = 0.8$$

This means that 80% of all the input energy is wasted in the exhaust; therefore, the engine has an efficiency of only 20%. ◀

In actual practice there are no reversible engines, because there are no real reversible processes. Real engines have their efficiencies increased by having the p–V diagram of the engine's cycle conform as closely as possible with that of an ideal cycle. Let us examine a few of the engines that are most often used.

19-4 THE FOUR-STROKE GASOLINE ENGINE

The ideal cycle used by the engineer for perfecting the four-stroke gasoline engine is the Otto cycle, named after a German engineer, Nikolaus Otto. Figure 19-10 illustrates the p–V diagram for the Otto cycle, which proceeds as follows. During the intake stroke a mixture of air and gasoline enters the cylinder through the open-intake valve. The valve then closes and the mixture undergoes compression. Just before the end of the compression stroke, the mixture is ignited by an electric

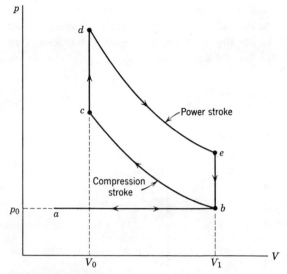

Figure 19-10 The otto cycle: *bc* and *de* represent adiabatic reversible processes.

spark. Combustion takes place, causing a sharp rise in temperature and pressure with little change in volume. This is followed by the expansion or power stroke. Just before the end of the power stroke, the exhaust valve is opened, the pressure in the cylinder reduces to atmospheric pressure, and the burned mixture is expelled during the exhaust stroke. Because the combustion takes place within the cylinder, the four-stroke gasoline engine is an example of an internal combustion engine.

The Otto cycle is more precisely termed the air-standard Otto cycle, because we imagine dry air to be the working substance rather than the actual gasoline–air mixture. In intake and exhaust, strokes (*ab* and *ba*) take place at the same pressure (atmospheric). The compression and power strokes (*bc* and *cd*) are assumed to be the result of adiabatic and reversible processes, and the combustion and depressurization processes are assumed to take place at the constant volumes V_0 and V_1. The efficiency of the ideal Otto cycle can be shown to be

$$E = 1 - \frac{1}{r^{\gamma-1}} \qquad (19-8)$$

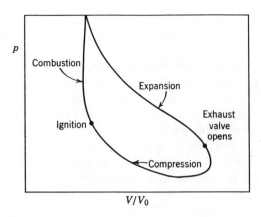

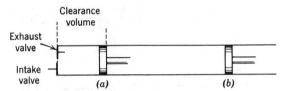

Figure 19-11 Indicator diagram for a four-stroke Otto cycle, where V_0 represents the unswept volume.

19-5 THE DIESEL ENGINE

The diesel engine has the following advantages over other internal combustion engines: (a) uses cheaper fuel, (b) does not require a special ignition system, and (c) can have a very high compression ratio. Only about one-third of the energy supplied by the fuel after combustion is used to perform external work. The efficiency of most diesels is greater than that of gasoline engines, however.

In the diesel engine, air is taken in and compressed during the intake and compression strokes, as in the gasoline engine. When the air is hot after the compression stroke, the fuel is injected as a fine spray and is ignited by the hot air. This process takes place at constant pressure whereas combustion in the gasoline engine occurs at constant volume. The expansion or power stroke follows. Near the end of this stroke, the exhaust valve opens and the pressure drops to atmospheric. The burnt gases are expelled during the exhaust stroke. The p–V diagram for an ideal diesel cycle is shown in Fig. 19-12. Here V_2/V_0 is

where $r = V_1/V_0$ = compression ratio and γ is the adiabatic constant for the working substance (air in this case). The adiabatic constant is the ratio of the specific heat at constant pressure to the specific heat at constant volume. These quantities will be discussed in Chapter 20. Since γ is greater than one, it would appear that the greater the compression ratio, the greater the efficiency. Actually, however, high compression ratios result in excessive heating of the gasoline and oil and result in preignition and power losses. For $r = 5.2$ and $\gamma = 1.4$, the efficiency of the ideal Otto cycle is about 50%. Figure 19-11 is an indicator diagram for an actual four-stroke gasoline engine with a schematic of a cylinder beneath it. The cylinder volume between the head of the cylinder and the piston in position a is called the clearance or unswept volume of the cylinder. The ratio of the cylinder volume when the cylinder is in position b to the clearance volume is the compression ratio. Notice how the indicator curve crudely approximates the Otto cycle. The actual engine whose cycle is shown by this curve has an efficiency of about 30%.

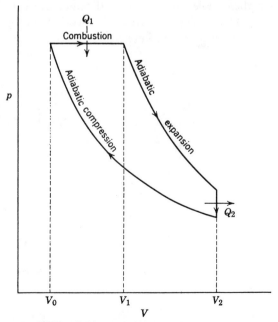

Figure 19-12 The diesel cycle.

the compression ratio r_c and V_2/V_1 is the expansion ratio r_e. The efficiency of this diesel engine can be shown to be

$$E = 1 - \frac{(r_e^{-\gamma} - r_c^{-\gamma})}{\gamma(r_e^{-1} - r_c^{-1})}$$

where $\gamma = c_p/c_v$ is the adiabatic constant of the working substance (Chapter 20).

In either the gasoline or diesel engines, the exhaust gases may be expelled and fresh fuel introduced by auxiliary equipment. In this way the engines may be converted to two-stroke cycles.

19-6 THE GAS TURBINE, JET ENGINE, AND ROCKETS

Schematic diagrams of the gas turbine and jet engines are shown in Fig. 19-13. In the gas turbine, air is compressed in the compressor from a pressure of p_0 to p_1. The fuel is then injected and burned in the combustion chamber. The products of the combustion begin to expand at constant pressure in the compressor. These products then enter the gas turbine where they expand further until the pressure is again p_0. The exhaust gases are then cooled to the original temperature in order to complete the cycle. The work done by the turbine over and above that needed to op-

erate the compressor is utilized externally. In Fig. 19-13(a) this work is used to operate an electric generator.

In the jet engine, the combustion products expand in the turbine to an intermediate pressure p'. Then they pass through the discharge nozzle, where the pressure falls to p_0 as external work is done by the jet. The pressure p' is chosen so that the work done by the turbine is just enough to operate the compressor. The ideal cycle for both the gas turbine and jet engine is called the Joule cycle and is shown in Fig. 19-14. The efficiency of these engines is given by

$$E = 1 - \left(\frac{p_0}{p'}\right)^{(\gamma-1)/\gamma} \qquad (19\text{-}10)$$

where γ is the adiabatic constant of the working substance.

Jet propulsion devices are classified as of either the self-contained rocket type or the air-stream type. The former type carries the liquid or dry fuel and the oxygen or other oxidizing agent needed for combustion in the rocket proper. In the air-stream type, oxygen from the air is used to burn the fuel. The jet engine and ramjet are examples of the air-stream type. In place of horsepower (hp), the output of jets of all types is measured by thrust in pounds.

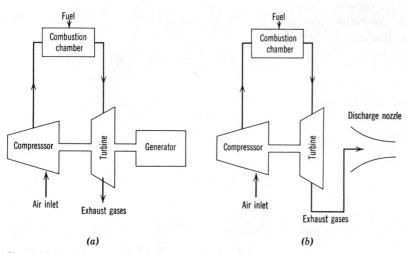

(a)   *(b)*

Figure 19-13 *(a)* The gas turbine. *(b)* The jet engine.

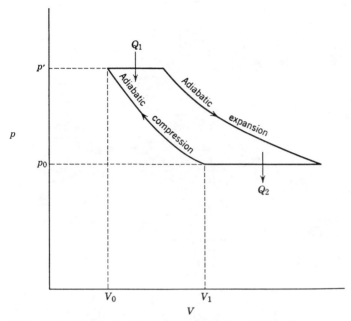

Figure 19-14   The Joule cycle for the gas turbine and jet engine.

The velocity a propellant can give to a rocket depends on two factors: (1) the length of time the powder can burn, (2) and the velocity of the escaping exhaust gases rushing out the rear of the rocket. The speed of the gases roaring out of the rear of the rocket is called the exhaust velocity. The greater the exhaust velocity, the faster the propellant can drive the rocket. A common assumption is that dry fuels must have a high exhaust velocity because they burn so rapidly. This assumption is not correct. The exhaust velocity of a fuel depends on its heat value, and the heat value of powder mixtures is relatively low.

The gas turbine is superior to the gasoline and diesel engines because it has only one moving part, other than accessory parts such as oil pumps and generators. It occupies less space for a given power output. On the other hand, the efficiency of the gas turbine is generally less than that of either the gasoline or diesel engine for a small engine. Two-thirds of the horsepower developed by a turbine is used in turning the compressor (see Fig. 19-13b).

19-7 THE ROTARY ENGINE

The unique rotary engine, often referred to as the Wankel engine, has received much attention in recent years. The engine is named after Felix Wankel, who devised the first rotary combustion engine in 1954. The first working model was built three years later.

The rotary engine is smaller than conventional engines, weighs less, and has fewer parts. There is less wear, less friction, and generally greater reliability.

The fundamental principle is that of replacing the conventional piston, cylinder, and crank assemblies with rotating disks. The disks have sections that are removed to form firing chambers (Fig. 19-15). Each of the three firing chambers forms a separate working chamber. The three working chambers undergo respective changes independently of each other. We can follow one chamber as it goes through a complete cycle by referring to Fig. 19-15.

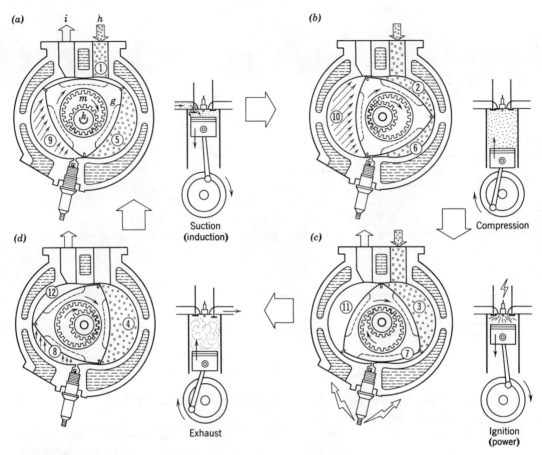

Figure 19-15   A Wankel rotary engine. One working chamber is followed through a complete cycle, *a–b–c–d–a*. Numbers 1 to 4 represent the intake stroke; numbers 5–7 represent the compression stroke; numbers 8–10 represent the ignition or power stroke; numbers 11 to 12 represent the exhaust stroke; *i* is the exhaust manifold; *h* is the intake manifold; and *m, k* are the output drive gears.

Phases 1 to 4 correspond to the intake stroke. The volume of the chamber, with the drawn-in mixture, reaches a maximum in phase 5. Compression takes place through phases 6 and 7, followed by ignition and power in phases 8 to 10. The burnt air–fuel mixture is exhausted during phases 11 and 12.

The rotary engine cycle is similar to the 4-stroke cycle in terms of gas replacement. It provides one power stroke for each rotation of the output shaft as does a 2-stroke engine. Therefore, it has the merits of both the 4-stroke engine and the 2-stroke engine.

19-8 REFRIGERATORS

Common refrigerators are not heat engines operated in reverse. All refrigerators use at least one irreversible process, usually a throttling process. The conventional vapor-compression refrigerator is shown in Fig. 19-16. The working substance is a

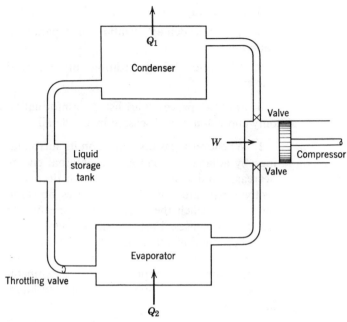

Figure 19-16   A schematic for a conventional vapor-compression refrigerator.

liquid which vaporizes readily through the throtting valve near room temperature. Among the most common refrigerants are ammonia, sulfur dioxide, and compounds of methane and ethane. Ammonia is widely used in industrial refrigerators, whereas Freon-12 (dichlorodifluoromethane) is usually used for household refrigeration.

The refrigerant partially vaporizes and cools in passing through the throttling valve. The mixed liquid and vapor absorbs heat from the interior of the refrigerator box and completely vaporizes. The vapor is then compressed by the compressor with its pressure and temperature increasing. In the condenser the refrigerant is cooled down, giving off heat to the cooler surroundings, usually air blown across the cooling coils of the condenser. After cooling down, the refrigerant is again a liquid and it returns through the liquid storage tank to the throttling valve to begin the cycle again.

A household refrigerator using Freon-12 and operating between the temperatures of -15 and

$30°C$ would have an ideal coefficient of performance of 4.8. Its actual coefficient of performance is about 2.8. A Carnot refrigerator operating between the same temperatures would have a coefficient of performance of 5.7.

The vapor-absorption-type refrigerating system is used in refrigerators such as the Servel Electrolux refrigerator. This system has no moving parts and uses an electric heater, gas flame, or kerosene burner to supply the external energy. Refrigerators of the vapor-absorption type are usually found in rural areas not supplied by electric power. Since it has no motor or mechanical pump, this refrigerator is practically noise-free.

19-9 CRYOGENICS

During the past three decades research scientists and engineers have been concentrating their efforts in the study of low temperature physics. By low temperature, we mean a temperature within,

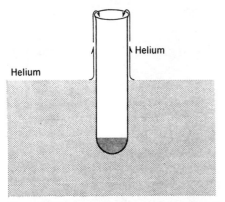

Figure 19-17   Liquid helium climbing the walls of a test tube.

say, 5 or 10° of absolute zero. Low temperature physics is generally referred to as *cryogenics*. Of the many interesting events that happen near absolute zero, two occurrences are of particular interest.

First, near absolute zero most, if not all, electrical conductors become superconducting, i.e., metals no longer offer any resistance to the passage of electrical charge. Second, below 2.18°K, liquid helium acquires strange characteristics: It has zero viscosity and literally climbs walls. If an empty test tube is inserted in liquid helium, as shown in Fig. 19-17, a thin film of helium will climb over the outside wall of the tube until the level of helium in the tube is the same as the outside level. If the tube is now slowly removed, the helium will climb back out, always leaving the level of helium in the tube the same as that outside. Even if the tube is withdrawn completely from the helium, the helium remaining in the tube will be seen to drip off the bottom of the tube back into the original beaker of helium. This behavior of liquid helium is called *superfluidity*.

The question of how objects are cooled to the neighborhood of absolute zero now arises. There are five common processes used to produce very low temperatures:

1. Adiabatic expansion of a gas as it does external work.

2. Expansion of a gas through a throttling valve.

3. Rapid evaporation of a liquid.

4. Adiabatic demagnetization of a paramagnetic salt.

5. Adiabatic demagnetization of polarized nuclei.

The last two processes are listed for information only and will not be discussed in any detail.

1. Whenever a gas does work on its surroundings by adiabatic expansion, its internal energy decreases and hence its temperature decreases. As the temperature of the gas decreases, however, the rate at which the temperature continues to fall also decreases. Further, the mechanical devices become difficult to operate at low temperatures. It is for these reasons that this method becomes somewhat impractical at low temperatures. Nevertheless, some gases have been liquefied by this method.

2. In this process a gas is cooled by forcing it through small openings, as through a porous wall. It turns out that this process is more efficient, the lower the initial temperature of the gas. The Collins helium-liquefying system utilizes process 1 to get helium to 15°K and process 2 to liquefy the helium.

3. To lower the temperature of the liquid helium, it is evaporated rapidly under reduced pressure. Temperatures as low as 0.71°K have been reached using this method.

4. This process can reduce the temperature of certain polarized atoms to 0.2°K.

5. By this method temperatures as low as 0.00002°K have been reached.

It should be noted that a temperature of absolute zero or less is unattainable. Consider a Carnot cycle operating between the temperatures T_1 and T_2, $T_1 > T_2$.

From Eq. 19-7, $E = 1 - T_2/T_1$. Since $E = W/Q_1$, we have

$$\frac{W}{Q_1} = 1 - \frac{T_2}{T_1}$$

or

$$\frac{T_2}{T_1} = 1 - \frac{W}{Q_1}$$

Multiplying by T_1 yields

$$T_2 = T_1\left(1 - \frac{W}{Q_1}\right)$$

Now from the second law, we know that $W < Q_1$. It follows then that T_2 can never be zero. The principle of the unattainability of absolute zero has been referred to as the third law of thermodynamics.

SUMMARY

The *first law of thermodynamics* is a restatement of the law of conservation of energy. It states, in symbols,

$$Q = W + \Delta U$$

where Q = the net heat flow into a system, W = the thermal equivalent of the work done by the system on its surroundings and ΔU = the change in the internal energy of the system.

The work done by a gas at a pressure, p, when it expands by an amount, ΔV, is given by $W = p \Delta V$. There is no work done when the volume does not change.

An *adiabatic* process is one for which $Q = 0$. *Entropy* is a measure of the disorder of a system.

The *second law of thermodynamics* states that for all real processes the entropy of the universe must increase or remain the same. The entropy of the universe can never decrease.

The *Carnot engine* has the greatest efficiency of all engines operating between the same highest and lowest temperatures.

The efficiency of an engine is

$$E = \frac{W}{Q(\text{input})} = 1 - \frac{Q(\text{output})}{Q(\text{input})}$$

The efficiency of a Carnot engine is

$$E = 1 - \frac{T(\text{low})}{T(\text{high})}$$

Cryogenics is the study of low temperature physics. The temperature of *absolute zero* is believed to be unattainable.

QUESTIONS

1. State the first law of thermodynamics in your own words.

2. How is the first law of thermodynamics related to the law of conservation of energy?

3. Give examples of mechanical transfers of thermal energy and nonmechanical transfers of thermal energy.

4. The work done by an expanding gas on its surroundings depends upon the way in which the pressure and volume of the gas change during the expansion. Explain the meaning of the statement with the use of several p–V diagrams.

5. What does the first law of thermodynamics say about perpetual motion machines?

6. Distinguish clearly between adiabatic and isothermal processes.

7. Explain the meaning of entropy. Give an example to illustrate the distinction between orderly and disorderly energy.

8. State the second law of thermodynamics in your own words.

9. What is a Carnot engine? Of what importance is it? Can you build a Carnot engine?

10. What is meant by the phrase "reversible process"? Are there such processes in practice?

11. Describe the steps in a 4-stroke gasoline engine.

12. How could a 4-stroke cycle be converted to a 2-stroke cycle?

13. Discuss the meaning of the terms *cryogenics* and *superfluidity*.

14. How does the Wankel rotary engine differ from the conventional internal combustion engines? In what way is the Wankel engine similar to the 4-stroke combustion engine? In what way is the Wankel engine similar to the 2-stroke engine?

PROBLEMS

(A)

1. How much work is done by a gas that expands by 400 m³ at a pressure of 3×10^5 N/m²?

2. How much work is done on a gas when its volume is changed from 40 to 18 m³ by a constant pressure of 2.1×10^5 N/m²?

3. How much work (measured in units of liters-atm) is done by a gas when it expands from 2 to 10 liters at a constant pressure of 2.5 atm?

4. By how much has the internal energy of a system changed if the system does 400 joules of work while absorbing 100 cal of thermal energy?

5. How much heat input is required for a system that is to do 500 cal of work while retaining 200 cal of internal energy?

6. One thousand calories of heat input is converted by an engine into some useful work. How much work is done if the internal energy of the working substance increases by 937 joules?

7. What is the efficiency of an engine that takes in 1800 cal of heat energy and exhausts 600 cal?

8. What is the efficiency of an engine that does 778 ft-lb of work for each 5 Btu of energy absorbed?

9. Find the heat exhausted by an engine that absorbs 2000 cal of energy and operates with an efficiency of 40%?

10. An engine has an efficiency of 60%. It does 2400 joules of work each cycle. Find:
    (a) The heat input.
    (b) The heat output.

11. Suppose that the engine in Problem 10 were a Carnot engine and the temperature during the isothermal expansion were 800°K. Find the temperature during the isothermal compression.

12. What is the efficiency of a Carnot engine operating between temperatures whose ratio is 2:5?

13. What is the coefficient of performance of the Carnot engine in Problem 12 when it operates as a refrigerator?

14. A Carnot refrigerator has a coefficient of performance of 1.5. What is the ratio of the high temperature reservoir to the low temperature reservoir?

(B)

1. A piston does 4000 ft-lb of work on a gas which then expands and does 3200 ft-lb of work on its surroundings. Find the change in the internal energy of the gas in Btu.

2. During an isothermal process a gas confined in a cylinder with a movable piston absorbs 2 Btu. If the piston weighs 1 ton, how high does the piston rise?

3. A 4000-lb automobile that is traveling at 50 ft/sec is brought to rest by applying the brakes. How many Btu of heat are produced in the brake linings?

4. In order to operate a clockwork mechanism, a 600-g bob is lifted up a distance of 150 cm. What is the maximum amount of work in ergs that can be delivered to the clock mechanism as the bob descends to its original level?

5. A 100-g block falls freely for a distance of 10 m and then strikes the ground. How many calories of thermal energy are produced during the impact?

6. The heat of combustion of gasoline is given as 30×10^6 cal/gal. A gasoline engine converts about 25% of the energy released by the combustion of gasoline into mechanical work. How much work in foot-pounds can be done by the engine for each gallon of gasoline that is consumed?

7. An inventor claims to have built an engine that absorbs 50 million cal from a fuel supply, expels 12.5 million cal in its exhaust, and does 12×10^7 ft-lb of mechanical work. Should you invest money to put this engine on the market? Explain in terms of the feasibility of this engine.

8. In a certain isobaric process 100 cal of thermal energy are added to a gas enclosed in a chamber equipped with a movable piston. The gas expands by 1 liter at a constant pressure of 2 atm. What was the change in the internal energy of the gas? (*Note:* 1 cal = 0.04 liters-atm.)

9. During an isothermal process, a system expanded $0.001 \ m^3$ while absorbing 200 cal. What was the average pressure of the system during the expansion?

10. A force pump pumps 20 gal of fuel oil per minute at an average pressure of 40 lb/in.² (*Note:* 1 gal = 231 in.³)
    (a) How much work in foot-pounds is done by the pump in 1 min?
    (b) What is the power output of the pump?

11. An electric motor is supplied with 600 watts of electric power from a power line. This power is converted into mechanical power at the rate of 60% by the motor, and the rest is dissipated into heat.
    (a) What is the useful power output of this motor?
    (b) How many calories of heat are produced during each minute the motor is operating?

12. What is the maximum efficiency of an engine that operates between the fixed temperatures of 500 and 400°K?

13. Steam flows into a turbine at a temperature of 300°C and flows out the exhaust at 75°C. What is the maximum possible efficiency of the turbine?

14. What is the efficiency of a heat engine that rejects 3000 joules of thermal energy to the cold reservoir while performing 2000 joules of useful work?

15. A Carnot heat engine has an efficiency of 40%. If the gas enters the engine at 450°C, what is the temperature of the exhaust? What is the ratio of heat input to heat output?

16. An engine operating at maximum efficiency between the temperatures 300 and 600°K rejects 1200 joules of thermal energy during each cycle. How much useful work does the engine do during each cycle?

17. The compression ratio of a particular ideal Otto cycle is 4.2. Assuming that the adiabatic constant of the working substance is 1.67, find the efficiency of the engine.

18. What is the efficiency of an ideal Diesel engine whose compression ratio is twice the expansion ratio when the working substance is air ($\gamma = 1.4$)? Give your answer in terms of r_e and evaluate for $r_e = 32$.

19. How much work must be done by a refrigerator pump to change 1 kg of water at 20°C to ice at −10°C in a refrigerator whose coefficient of performance is 4?

20. It was stated in this chapter that the Carnot cycle can be used to define a temperature scale, the Kelvin temperature. This means that the Carnot cycle can serve as a thermometer. Describe how one might measure the temperature of a reservoir by using only Carnot cycles as thermometers. Write down three equations from which the temperatures of the steam point, ice point, and reservoir can be determined. Which quantities are actually measured during the experiment?

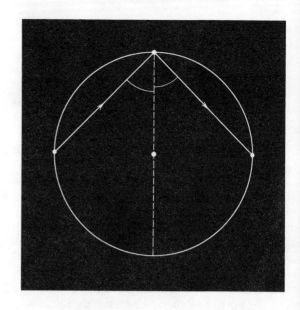

20.
Thermal Properties of Gases

In this chapter we shall be concerned with the relationships describing the pressure, volume, and temperature of an ideal gas. By an "ideal gas" we mean one that obeys Boyle's law at all temperatures. The relationships we develop for the ideal gas will be approximately valid for real gases over those ranges of pressure and temperature for which the forces between the·molecules of a gas can be considered negligible. When the temperature of a gas is about room temperature or higher, and the pressure is only a few atmospheres or lower, the ideal gas approximation may be used with confidence.

LEARNING GOALS

While studying this chapter, the reader should be alerted to the following major learning objectives:

1. *To learn Boyle's law, Charles' law, and the equation of state of an ideal gas.*
2. *To understand the properties of an ideal gas.*
3. *To understand Dalton's law of partial pressures.*
4. *To learn how to calculate the thermal energy changes of a gas for different processes.*
5. *To understand the distinction between adiabatic and isothermal processes.*

20-1 BOYLE'S LAW

Consider an ideal gas enclosed in a volume V, under a pressure P (Fig. 20-1). Let the pressure and volume be changed by varying the height of the column of mercury A. After allowing enough time for the gas to resume its original temperature, measure the new pressure and volume. Repeat the process a number of times and plot a graph of pressure versus volume (Fig. 20-2). The graph indicates that at *constant temperature* the pressure of a gas varies inversely as the volume, or

$$P = \frac{\kappa}{V} \qquad (20\text{-}1)$$

where κ is a proportionality constant. If we are interested in the relationship between the pressure and volume at two different points, we have

$$P_1 V_1 = \kappa \qquad (20\text{-}2a)$$

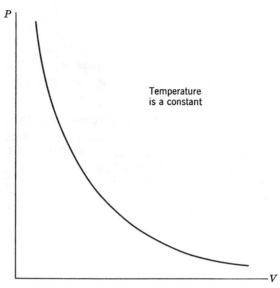

Figure 20-2 At constant temperature the pressure of an ideal gas varies inversely as the volume.

and $\qquad\qquad P_2 V_2 = \kappa \qquad (20\text{-}2b)$

where the subscripts refer to the two points under consideration. By eliminating the constant κ from Eq. 20-2, we obtain

$$P_1 V_1 = P_2 V_2 \quad \text{(for constant temperature)} \quad (20\text{-}3)$$

Equation 20-3 (or 20-1) is known as Boyle's law. It is to be accepted as an experimental observation of the thermodynamic behavior of gases. Later in this chapter we shall propose a theory about the dynamical behavior of gases. Boyle's law will result as a logical consequence from the theory.

Example 20-1 Ten liters of hydrogen gas at a pressure of 1 atm are allowed to expand at a constant temperature to a new volume of 18 liters. Find the new pressure of the gas.

Solution. From Eq. 20-3

$$P_1 V_1 = P_2 V_2$$
$$(1 \text{ atm})(10 \text{ liters}) = P_2 (18 \text{ liters})$$
$$P_2 = 1 \text{ atm} \times \frac{10 \text{ liters}}{18 \text{ liters}}$$
$$P_2 = \tfrac{5}{9} \text{ atm} \qquad \blacktriangleleft$$

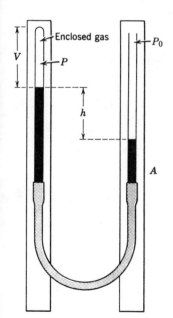

Figure 20-1 Boyle's law apparatus. The pressure of the enclosed gas is equal to the atmospheric pressure minus the gauge pressure caused by the column of mercury of height h. Thus $P = P_0 - (\rho g)_{Hg} h$. When the level of mercury in the left-hand column is lower than that in the right-hand column, h is considered negative and P is greater than P_0.

Example 20-2 A balloon whose volume is 10,000 ft^3 is to be filled with a gas at atmospheric pressure and constant temperature. If the gas is stored in containers with a volume of 4 ft^3 at a pressure of 100 lb/in.^{2}, how many containers are required?

Solution. Suppose we need n containers. Then the volume of gas in the n containers is n times the volume of each container, or $n(4 \text{ ft}^3) = 4n \text{ ft}^3$. Hence, using Eq. 20-3, we obtain

$$P_1 V_1 = P_2 V_2$$
$$(100 \text{ lb/in.}^2)(4n \text{ ft}^3) = (14.7 \text{ lb/in.}^2)(10,000 \text{ ft}^3)$$
$$400n = 14.7(10,000)$$
$$n = 367.5 \text{ containers} \qquad \blacktriangleleft$$

20-2 CHARLES' LAW

We now consider the expansion of a gas at constant pressure. According to Eq. 17-8, the new volume of the gas will be

$$V = V_0(1 + \beta \, \Delta t), \qquad (20\text{-}4)$$

where V_0 is the original volume, β is the coefficient of cubical expansion of the gas, and Δt represents the temperature change during the expansion. If the original volume is measured at 0°C, Eq. 20-4 can be rewritten

$$V = V_0(1 + \beta t) \qquad (20\text{-}5)$$

Equation 20-5 is that of a straight line when V is plotted as a function of t. If this equation is plotted for a few different gases, it is found that all the straight lines extrapolate to the same temperature when $V = 0$ (see Fig. (20-3). i.e., the intercept on the t-axis is identical for all gases, assumed ideal. This does not mean that if the volume were continuously decreased, the corresponding temperature would follow the extrapolated line, since the gas ceases to be ideal at low temperatures and high densities. Nevertheless, the graphs indicate that for the range of temperatures in which the gas may be considered ideal, the quantity $-1/\beta$, which represents the t-intercept in Eq. 20-5, is a constant for all gases.

The value of $-1/\beta$, as measured in the extrapolation process, is approximately -273°C, the absolute zero of Celsius temperature. Thus Eq. 20-5 becomes

$$V = V_0 \left[1 + \left(\frac{1}{273} \right) t \right] = \frac{V_0}{273}(273 + t)$$

or

$$V = \frac{V_0}{273} T = AT \qquad (20\text{-}6)$$

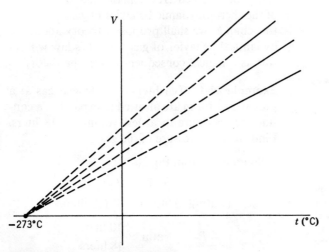

Figure 20-3 At constant pressure, the volume of an ideal gas varies directly as the absolute temperature.

where T is the absolute temperature and $A = V_0/273$ is constant for each gas. Equation 20-6 states that the volume of an ideal gas is directly proportional to the *absolute* temperature (Kelvin or Rankine) if the pressure is kept constant. Equation 20-6 is referred to as Charles' law.

Example 20-3   A 10-liter volume of hydrogen at standard temperature is expanded at constant pressure until the temperature increases to 300°C. Find the new volume of the gas.

Solution.   We must first convert the initial and final temperatures to absolute values. Using the Kelvin scale, we find that the standard temperature is 273°K and 300°C is equivalent to 573°K. Hence, from Eq. 20-6, we obtain

$$\frac{V_1}{V_2} = \frac{T_1}{T_2}$$

$$\frac{10}{V_2} = \frac{273}{573}$$

or $$V_2 = \frac{573}{273} \times 10 = 21 \text{ liters} \qquad \blacktriangleleft$$

Example 20-4   Twenty cubic feet of a gas at STP (0°C, and 1 atm of pressure) are first compressed at constant temperature to a pressure of 10 atm and then compressed at constant pressure until the final temperature is −20°C. Find the final volume occupied by the gas.

Solution.   We first represent the two processes on a P–V diagram as in Fig. 20-4. *AB* represents the compression at constant temperature and *BC* the compression at constant pressure. The pressure, volume, and temperature of the gas at *A* are given. The pressure and temperature at *B* and *C* are known. We require the volume at *B* from which we shall be able to evaluate the final volume.

Equation 20-3 is applicable between points *A* and *B*. Thus

$$P_A V_A = P_B V_B$$
$$1 \text{ atm} \times 20 \text{ ft}^3 = 10 \text{ atm} \times V_B$$
$$V_B = 2 \text{ ft}^3$$

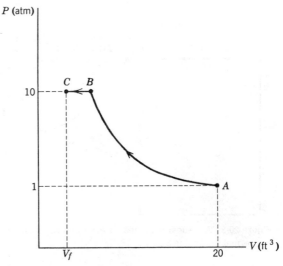

Figure 20-4   An isothermal compression *AB* followed by an isobaric compression *BC*.

Equation 20-6 applies in the interval from *B* to *C*. Thus

$$\frac{V_B}{V_C} = \frac{T_B}{T_C}$$

$$\frac{2}{V_C} = \frac{273}{253}$$

$$V_C = 2 \times \frac{253}{273} = 1.9 \text{ ft}^3 \qquad \blacktriangleleft$$

We now wish to combine Eqs. 20-3 and 20-6 to obtain a general result for all ideal gases, relating the pressure, volume, and temperature at two different points.

20-3 IDEAL GAS LAW

Consider an ideal gas which at a given instant is specified by the thermodynamic quantities (P_0, V_0, T_0). We first expand the gas at constant pressure to a new set of coordinates (P_1, V_1, T_1) and then expand at constant temperature so that the final coordinates of the gas are (P_2, V_2, T_2). In Fig. 20-5 the expansions have been plotted on a P–V diagram.

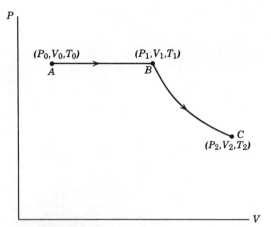

Figure 20-5   An isobaric expansion *AB* followed by an isothermal expansion *BC*.

Between points *A* and *B*, the gas obeys Charles' law; hence

$$P_0 = P_1 \qquad \text{(20-7a)}$$

and

$$\frac{V_0}{T_0} = \frac{V_1}{T_1} \qquad \text{(20-7b)}$$

From point *B* to point *C*, the gas obeys Boyle's law, giving

$$T_1 = T_2 \qquad \text{(20-8a)}$$

$$P_1 V_1 = P_2 V_2 \qquad \text{(20-8b)}$$

Substituting the value of $V_1 = V_0(T_1/T_0)$ from Eq. 20-7b into Eq. 20-8b, we obtain

$$P_1 \left(\frac{V_0}{T_0} \times T_1 \right) = P_2 V_2$$

or

$$\frac{P_1 V_0}{T_0} = \frac{P_2 V_2}{T_1} \qquad \text{(20-9)}$$

But, making use of Eqs. 20-7a and 20-8a, we can rewrite Eq. 20-9 as

$$\frac{P_0 V_0}{T_0} = \frac{P_2 V_2}{T_2} \qquad \text{(20-10)}$$

Equation 20-10 states that the specific combination of thermodynamic variables PV/T is a constant for any ideal gas. Equations 20-3 and 20-6 are special cases of the more general relationship in Eq. 20-10.

Example 20-5   A gas occupying a volume of 10 liters at standard temperature and pressure is expanded to a new volume of 25 liters. The pressure is then found to be one-half of an atmosphere. What is the final temperature of the gas?

Solution.   Using Eq. 20-10, we obtain

$$\frac{P_0 V_0}{T_0} = \frac{P_1 V_1}{T_1}$$

$$\frac{(1 \text{ atm})(10 \text{ liters})}{273°\text{K}} = \frac{(0.5 \text{ atm})(25 \text{ liters})}{T_1}$$

$$T_1 = (273°\text{K}) \frac{(0.5 \text{ atm})(25 \text{ liters})}{(1 \text{ atm})(10 \text{ liters})} = 341.3°\text{K} \quad ◀$$

To find the constant value of PV/T, we evaluate this relationship at standard temperature and pressure. *A mole is the mass in grams equal numerically to the molecular weight of a substance.* As an example consider carbon dioxide, CO_2. The atomic weight of carbon is 12 and the atomic weight of oxygen is 16. Therefore, the total molecular weight of the atoms in carbon dioxide is $12 + 16 + 16 = 44$. One mole of $CO_2 = 44$ g. The mole is also called the *gram-molecular weight*.

The ratio of the number of molecules, N, to the number of moles, n, is the number of molecules per mole or Avogadro's number, N_0. Avogadro's number is a constant for all substances and is equal to 6.02×10^{23} molecules/mole.

One mole of any gas at STP occupies a volume of 22.4 liters. The volume of the gas at STP is 22.4 liters/mole. Thus

$$\frac{PV}{T} = \frac{1 \text{ atm} \times 22.4 \text{ liters/mole}}{273°\text{K}}$$

$$= 0.0821 \frac{\text{liter-atm}}{\text{mole-°K}} = R \qquad \text{(20-11)}$$

Here R is known as the *universal gas constant*, since it has the same value for all gases. If we express pressure in newton/m², and volume in m³,

$$R = 8.31 \frac{\text{joules}}{\text{mole-°K}} = 2.0 \frac{\text{cal}}{\text{mole-°K}}$$

If the gas contains n moles, Eq. 20-11 becomes

$$\frac{PV}{T} = nR \qquad (20\text{-}12)$$

Equation 20-12 is called the *equation of state of an ideal gas*. The equation of state enables one to find the pressure, volume, or temperature at one point without knowing all three variables at any other point.

Example 20-6 Two moles of a gas occupy a volume of 5000 cm^3 at a temperature of 27°C. Find the pressure of the gas.

  Solution. In order to use Eq. 20-12, convert the volume to m^3 and the temperature to °K. Then,

$$\frac{PV}{T} = nR$$

$$\frac{P(5000 \text{ cm}^3 \times 1 \text{ m}^3/10^6 \text{ cm}^3)}{300°K}$$

$$= 2 \text{ moles} \times 8.31 \frac{\text{joules}}{\text{mole-}°K}$$

$$P = \frac{2 \text{ moles} \times 8.31 \text{ joules} \times 300°K}{0.005 \text{ m}^3 \times \text{mole-}°K}$$

$$P = 99.7 \times 10^4 \text{ joules/m}^3 \cong 10^6 \text{ newton/m}^2 \blacktriangleleft$$

Example 20-7 A certain gas occupies a volume of 1200 liters at a pressure of 1 atm and a temperature of 127°C. How many moles of this gas are present in the volume?

  Solution. Using Eq. 20-12,

$$\frac{PV}{T} = nR$$

$$n = \frac{PV}{RT} = \frac{1 \text{ atm} \times 1200 \text{ liters}}{0.0821 \text{ liter-atm/mole-}°K \times 400°K}$$

$$= \frac{1200}{0.0821 \times 400} \text{ moles}$$

$$n = 36.5 \text{ moles} \blacktriangleleft$$

It is often advantageous to be able to express the constant value of PV/T in terms of energy per *molecule* per Kelvin degree rather than energy per *mole* per Kelvin degree. To this end we note that there are 6.02×10^{23} molecules in every mole of a gas (Avogadro's number, N_0). Thus

$$\frac{R}{N_0} = \frac{8.31 \text{ joules/mole-}°K}{6.02 \times 10^{23} \text{ molecules/mole}}$$

$$\frac{R}{N_0} = 1.38 \times 10^{-23} \frac{\text{joules}}{\text{molecule-}°K}$$

$$= 1.38 \times 10^{-16} \frac{\text{ergs}}{\text{molecule-}°K} = k$$

Since R and N_0 are universal constants, k is also a universal constant and is known as *Boltzmann's constant*. Equation 20-12 can now be written

$$\frac{PV}{T} = nN_0\left(\frac{R}{N_0}\right) = nN_0k = Nk \quad (20\text{-}13)$$

where $N = nN_0$ represents the total number of molecules.

The equation of state of an ideal gas (Eq. 20-12) was derived on the basis of experimental observations (Boyle's law and Charles' law). We can, however, make some assumptions about the dynamical behavior and physical structure of ideal gases that will lead to a derivation of the equation of state on theoretical grounds. The following properties of ideal gases are assumed:

1. The molecules of the gas are all identical and spherical.

2. The size of a molecule is very small compared to the distance between molecules.

3. There are very many molecules in the smallest volumes without violating assumption 2.

4. The molecules of a gas exert no forces on each other except during collisions.

5. All collisions between molecules of the gas or with walls of a container are perfectly elastic, i.e., momentum and energy are conserved.

6. As a result of assumption 4, molecules move in straight lines between collisions.

Consider a single molecule confined in a spherical box of radius r (see Fig. 20-6). When the molecule, proceeding from A to B, collides at B, the molecule rebounds according to the laws of regular reflection so that angle

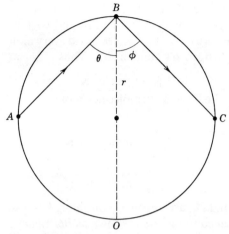

Figure 20-6   A molecule of an ideal gas making an elastic collision with the wall B of a spherical box. Why is triangle ABO a right triangle?

θ equals angle ϕ. Further, AB, BC, and BO (the normal) are all in the same plane. Clearly, $AB = BC$, and the time it takes the molecule to go from A to B, or from B to C, is given by

$$t = \frac{AB}{v}$$

where v is the molecular speed, which remains unchanged by the collision. Since triangle ABO is a right triangle,

$$AB = BO \cos \theta = 2r \cos \theta$$

Thus
$$t = \frac{2r \cos \theta}{v} \qquad (20\text{-}14)$$

There is a change in the normal component of the molecule's momentum due to the change in direction of its motion. Thus

$$
\begin{aligned}
\text{change in momentum} &= \Delta(mv)_{\text{normal}} \\
&= mv \cos \theta - (-mv \cos \phi) \\
&= mv \cos \theta - (-mv \cos \theta) \\
&= 2mv \cos \theta \qquad (20\text{-}15)
\end{aligned}
$$

The average impulsive force exerted during the collision with the wall is equal to the change in momentum due to the collision divided by the time between collisions. Thus

$$\bar{F} = \frac{\Delta(mv)}{t} = \frac{2\,mv \cos \theta}{(2r \cos \theta)/v} = \frac{mv^2}{r}$$

which is independent of θ. If there were N molecules, the total impulsive force exerted by the gas on the wall, would be

$$N\bar{F} = \frac{Nmv^2}{r}$$

The pressure exerted on the wall is the total force divided by the surface area of the spherical box. Thus

$$P = \frac{N\bar{F}}{A} = \frac{(Nmv^2)/r}{4\pi r^2} = \frac{Nmv^2}{4\pi r^3}$$

But the volume of a sphere is $V = \frac{4}{3}\pi r^3$. Hence

$$P = \frac{Nmv^2}{3V}$$

or
$$PV = \tfrac{1}{3}Nmv^2 \qquad (20\text{-}16)$$

Equation 20-16 relates the pressure and volume of the gas to the square of the molecular velocity. It was tacitly assumed here that all N molecules have the same speed. Actually, an averaging process is required; the average of the square of the velocity should be used. Thus

$$PV = \tfrac{1}{3}Nm(v^2)_{\text{av}} \qquad (20\text{-}17)$$

Now, $m(v^2)_{\text{av}} = 2$ (average kinetic energy of a molecule), or

$$
\begin{aligned}
PV &= \tfrac{1}{3}N(2\text{KE})_{\text{av}} \\
PV &= \tfrac{2}{3}N(\text{KE})_{\text{av}} \qquad (20\text{-}18)
\end{aligned}
$$

We can relate the average kinetic energy to the temperature of the gas, however, since it is precisely the average kinetic energy that is measured by a thermometer. Let us define the relationship between average KE and absolute temperature in such a way as to make Eqs. 20-18 and 20-13 agree. Thus

$$(\text{KE})_{\text{av}} = \tfrac{3}{2}kT \qquad (20\text{-}19)$$

where k is Boltzmann's constant. Then Eq. 20-18 becomes

$$PV = NkT \qquad (20\text{-}20)$$

It is important to realize that Eq. 20-19 is a result that we did not intentionally seek to obtain. Our derivation carried us to Eq. 20-18, which begins to look somewhat like Eq. 20-13. Equation 20-19 is a necessary consequence of the theory if Eq. 20-18 is to agree with Eq. 20-13. Thus we have a relationship between the average kinetic energy of a molecule and the absolute temperature. In Chapter 17 we intimated that such a relation must exist, since temperature is the measure of average kinetic energy. Now we have found the precise form of this relationship.

If there are N molecules in the gas, the total internal energy of the gas (there being no potential energy between molecules) is

$$U = N(KE)_{av} = \tfrac{3}{2}NkT$$

$$= \frac{3}{2}nN_0kT = \frac{3}{2}nN_0\left(\frac{R}{N_0}\right)T$$

$$U = \tfrac{3}{2}nRT \qquad (20\text{-}21)$$

where n is the number of moles of gas.

• **Example 20-8** Find the internal energy of 2 moles of oxygen at a temperature of 27°C.

Solution. From Eq. 20-21

$$U = \tfrac{3}{2}nRT$$

$$= \frac{3}{2}(2 \text{ moles}) \times 8.31 \, \frac{\text{joules}}{\text{mole-°K}} \times 300°\text{K}$$

$$= 3 \times 8.31 \times 300 \text{ joules} = 7479 \text{ joules}$$

$$= 7.48 \times 10^3 \text{ joules} \qquad \blacktriangleleft$$

• **Example 20-9** Using the data of the previous problem, find (a) the average kinetic energy per molecule, (b) the average of the square of the velocity, (c) the square root of part (b), which represents the root-mean-square velocity denoted by v_{rms}.

Solution. Two moles of oxygen contain $2 \times 6.02 \times 10^{23}$ molecules. Therefore,

(a) average $KE = \dfrac{U}{N} = \dfrac{7.48 \times 10^3 \text{ joules}}{12 \times 10^{23} \text{ molecules}}$

$$= 6.23 \times 10^{-21} \text{ joules/molecule}$$

(b) $(KE)_{av} = \tfrac{1}{2}m(v^2)_{av}$

or $(v^2)_{av} = \dfrac{2(KE)_{av}}{m}$

We note that the mass of an oxygen molecule is equal to the molecular weight divided by Avogadro's number.

Hence $m = \dfrac{32 \text{ grams/mole}}{6.02 \times 10^{23} \text{ molecules/mole}}$

$$m = \dfrac{0.032 \text{ kg/mole}}{6.02 \times 10^{23} \text{ molecules/mole}}$$

Thus

$$(v^2)_{av} = \frac{2 \times 6.23 \times 10^{-21} \text{ joules/molecule}}{\dfrac{0.032 \text{ kg/mole}}{6.02 \times 10^{23} \text{ molecules/mole}}}$$

$$(v^2)_{av} = \frac{2 \times 6.23 \times 10^{-21} \times 6.02 \times 10^{23}}{0.032} \text{ joules/kg}$$

$$(v^2)_{av} = 2.3 \times 10^5 \text{ m}^2/\text{sec}^2$$

(c) $v_{rms} = \sqrt{(v^2)_{av}} = 4.8 \times 10^2 \text{ m/sec} = 480 \text{ m/sec}$

At this point the reader should convince himself that v_{rms} is not the same as v_{av}. Consider the five velocities of magnitudes 1, 2, 3, 4, 5 in appropriate units. The average velocity is

$$v_{av} = \frac{1 + 2 + 3 + 4 + 5}{5} = 3$$

The v_{rms} is

$$v_{rms} = \left[\frac{(1^2 + 2^2 + 3^2 + 4^2 + 5^2)}{(5)}\right]^{1/2}$$

$$v_{rms} = \sqrt{11} = 3.31 \qquad \blacktriangleleft$$

The failure of the equation of state of an ideal gas to agree with experimental results for real gases is primarily due to the finite size of the molecules (see Table 20-1) and the interaction forces that are ever present between the molecules. An exact determination of the mathematical nature of the interaction forces would enhance the possibility of advancing a theory, similar to that developed for ideal gases, which would yield an equation of state for real gases. As yet no such determination has been made. There are various approximate equations of state depending on the approximation used to represent the potential energy of interaction. These equations of state are valid only over limited ranges of temperature and pressure.

One of the mildly successful equations of state is that due to van der Waals. In this approximation a constant correction term is subtracted from the volume to account for the size of the molecules and the pressure is increased by a term that is proportional to the square of the density to account

Table 20-1   Values of PV/RT at various pressures and temperatures for one mole of some real gases. (For an ideal gas, PV/RT is always one.)

Substance	Temperature, °K	1 atm	10 atm	100 atm
Air	300	0.99970	0.99717	0.9933
	500	1.00034	1.00348	1.0393
	1000	1.00033	1.00331	1.0333
Carbon dioxide	300	0.99501	0.9486	
	500	0.99927	0.99281	0.9365
	1000	1.0002	1.0022	1.0248
Oxygen	300	0.99939	0.99402	0.9541
	500	1.00022	1.00222	1.0256
	1000	1.00029	1.00288	1.0296
Hydrogen	200	1.0007	1.0068	1.0760
	300	1.0006	1.0059	1.0607
	500	1.0004	1.0040	1.0400

for the interaction forces. Thus van der Waals' equation of state can be written

$$\left(P + \frac{a}{V^2}\right)(V - b) = nRT \qquad (20\text{-}22)$$

The constants a and b vary for different gases and have been tabulated for many gases. We shall return for a closer look at Eq. 20-22 in Chapter 21, when we discuss the liquefaction of gases.

Suppose that we have a mixture of gases which do not react chemically with one another. We assume that the temperature and density of the mixture are such that the ideal gas approximation is still valid. It has been shown experimentally that the pressure of the mixture is equal to the sum of the partial pressure which each gas would exert if it alone occupied the entire volume of the mixture. For example, let the mixture consist of N_1 molecules of gas 1, and N_2 molecules of gas 2, and N_3 molecules of gas 3. Let the volume of the mixture be denoted by V. Then the partial pressure of each gas is

$$p_1 = \frac{N_1 kT}{V}, \quad p_2 = \frac{N_2 kT}{V}, \quad p_3 = \frac{N_3 kT}{V} \qquad (20\text{-}23)$$

where T is the equilibrium temperature common

to all three gases. The pressure of the mixture is then given by

$$P = p_1 + p_2 + p_3$$

This experimental law is known as *Dalton's law of partial pressures*.

Example 20-10   A 2 ft^3 vessel contains a partition. On one side of the partition there is $\frac{1}{2}$ ft^3 of nitrogen gas at a pressure of 70 cm of mercury. The other side of the partition contains $\frac{3}{2}$ ft^3 of oxygen at a pressure of 65 cm of mercury. What is the pressure of the mixture when the partition is removed?

Solution.   We first find the pressure of each gas if it alone occupied the entire 2 ft^3. Applying Boyle's law to each gas separately, we have for the nitrogen

$$P_1 V_1 = P_2 V_2$$
$$70 \times \tfrac{1}{2} = P_2 \times 2$$

Then $P_2 = \frac{70}{4} = 17.5$ cm of Hg for the nitrogen. For the oxygen

$$P_1 V_1 = P_2 V_2$$
$$65 \times \tfrac{3}{2} = P_2 \times 2$$

Then $P_2 = 65 \times \frac{3}{4} = 48.75$ cm of Hg for the oxygen. Finally, by Dalton's law

$$P = P_{\text{nitrogen}} + P_{\text{oxygen}}$$
$$P = 17.5 + 48.75 = 66.25 \text{ cm of Hg} \quad \blacktriangleleft$$

20-4 SPECIFIC HEATS OF A GAS

In Chapter 18, where the specific heat capacity of liquids and solids were discussed, it was assumed that there was no difference between processes which take place at constant pressure and those which take place at constant volume. When dealing with gases, this assumption cannot be made. We shall, in fact, encounter two different specific heat capacities for gases. One will apply whenever the gas undergoes a temperature change at constant pressure, and the other when the temperature change is at constant volume. The two specific heats will be related by a very simple relationship.

Consider an ideal gas that undergoes a temperature change ΔT at constant volume. From the first law of thermodynamics,

$$Q = \Delta U + W$$

Since there is no change in volume, $W = 0$. Therefore,

$$Q = \Delta U \qquad (20\text{-}24)$$

If C_v represents the specific heat capacity per mole at constant volume and there are n moles of gas, we can represent Q as $nC_v \Delta T$ (C_v is often referred to as the molar specific heat capacity at constant volume). Equation 20-24 says that

$$Q = \Delta U = nC_v \Delta T \qquad (20\text{-}25)$$

On the other hand, Eq. 20-21 indicates that the internal energy depends only on the temperature. Thus the change in the internal energy does not depend on *how* the change was made, but only on the temperature difference between the initial and final states of the gas. Therefore, Eq. 20-25 must be used to express the change in internal energy of a gas when its temperature changes by ΔT *even when the change does not take place at constant volume.*

Now let the gas be heated through the same temperature interval ΔT but at constant pressure. The heat supplied to the gas must now be represented by $Q = nC_p \Delta T$ with the subscript on C indicating that the pressure is held constant. Again, from the first law of thermodynamics,

$$Q = \Delta U + W$$
$$nC_p \Delta T = nC_v \Delta T + P \Delta V \qquad (20\text{-}26)$$

In Eq. 20-26 we have made use of the fact that the work done by the gas is equal to the pressure multiplied by the volume change. From Eq. 20-12

$$PV = nRT$$

or
$$P \Delta V = nR \Delta T \qquad (20\text{-}27)$$

since P, n, and R are constants. Thus Eq. 20-26 can be written as

$$nC_p \Delta T = nC_v \Delta T + nR \Delta T$$

or
$$C_p - C_v = R \qquad (20\text{-}28)$$

Equation 20-28 tells us that the molar specific heat at constant pressure is always larger than the molar specific heat at constant volume, and the difference is precisely the universal gas constant R. It should be noted, for emphasis, that this equation applies only to ideal gases. The ratio $(C_p - C_v)/R$, which is unity for ideal gases, is tabulated for a number of real gases near room temperature in Table 20-2. It is evident that the

Table 20-2 Molar specific heat capacities, at temperatures near room temperature

Gas	$\gamma = \dfrac{C_p}{C_v}$	C_p	C_v	R
Helium	1.66	5.04	3.04	2.00
Argon	1.67	5.04	3.02	2.02
Hydrogen	1.40	6.84	4.88	1.96
Oxygen	1.41	7.04	5.00	2.04
Nitrogen	1.40	7.00	5.00	2.00
Carbon monoxide	1.40	7.18	5.12	2.06
Carbon dioxide	1.30	8.80	6.77	2.03
Ammonia	1.31	8.96	6.84	2.12

ideal gas approximation is applicable for these gases near room temperature.

Example 20-11   Consider the following processes carried out in succession on 0.45 moles of an ideal gas for which $C_p = \frac{5}{2}R$. The gas originally has a volume of 10 liters at STP. (1) Double the pressure, keeping the volume constant. (2) Double the volume at constant pressure. (3) Halve the pressure at constant volume. (4) Compress to the original volume at constant pressure. The processes are shown on a P–V diagram in Fig. 20-7. (a) Find the coordinates at the corners of the rectangle. (b) Compute the heat absorbed by the gas during processes 1 and 2 in calories. (c) Compute the heat lost by the gas during processes 3 and 4. (d) Why are the answers to parts (b) and (c) not identical?

Solution.   The pressure at B is given as 2 atm and the volume is 10 liters. Thus from Eq. 20-10

$$\frac{P_A V_A}{T_A} = \frac{P_B V_B}{T_B}$$

$$\frac{1 \text{ atm} \times 10 \text{ liters}}{273°\text{K}} = \frac{2 \text{ atm} \times 10 \text{ liters}}{T_B}$$

$$T_B = 546°\text{K}$$

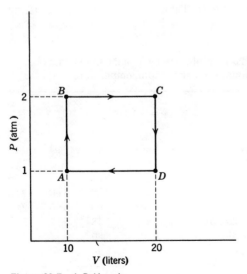

Figure 20-7   A P–V cycle.

From B to C the gas obeys Charles' law. Hence

$$\frac{V_B}{T_B} = \frac{V_C}{T_C}$$

$$\frac{10 \text{ liters}}{546°\text{K}} = \frac{20 \text{ liters}}{T_C}$$

$$T_C = 1092°\text{K}$$

To find the temperature at D, we can again use Eq. 20-10,

$$\frac{P_A V_A}{T_A} = \frac{P_D V_D}{T_D}$$

$$\frac{1 \text{ atm} \times 10 \text{ liters}}{273°\text{K}} = \frac{1 \text{ atm} \times 20 \text{ liters}}{T_D}$$

$$T_D = 546°\text{K}$$

The coordinates of A, B, C, and D are

$$A \ (1 \text{ atm}, 10 \text{ liters}, 273°\text{K})$$
$$B \ (2 \text{ atm}, 10 \text{ liters}, 546°\text{K})$$
$$C \ (2 \text{ atm}, 20 \text{ liters}, 1092°\text{K})$$
$$D \ (1 \text{ atm}, 20 \text{ liters}, 546°\text{K})$$

During process 1 there was no change in volume. Thus the heat absorbed by the gas is used to increase the internal energy of the gas. From Eq. 20-28

$$C_v = C_p - R = \tfrac{5}{2}R - R = \tfrac{3}{2}R = 3 \text{ cal/mole-}°\text{K}$$

Then

$$Q_{AB} = nC_v(T_B - T_A)$$
$$= 0.45 \text{ moles} \times 3 \text{ cal/mole-}°\text{K} \times 273\text{K}°$$
$$Q_{AB} = 367 \text{ cal}$$

Process 2 takes place at constant pressure so that

$$Q_{BC} = nC_p(T_C - T_B)$$
$$= 0.45 \text{ mole} \times 5 \text{ cal/mole-}°\text{K} \times (1092 - 546)\text{K}°$$
$$Q_{BC} = 1229 \text{ cal}$$

The third process is one for which the volume is constant so that

$$Q_{CD} = nC_v(T_D - T_C)$$
$$= 0.45 \text{ mole} \times 3 \text{ cal/mole-}°\text{K} \times (546 - 1092)\text{K}°$$
$$Q_{CD} = -737 \text{ cal}$$

During process 4, there is a change in volume and work is done on the gas. The process takes place at constant pressure so that the heat gained by the gas is

$$Q_{DA} = nC_p(T_A - T_D)$$
$$Q_{DA} = 0.45 \text{ moles} \times \tfrac{5}{2}R \text{ cal/mole-}°K$$
$$\times (273 - 546)K°$$
$$Q_{DA} = -614 \text{ cal}$$

Thus the heat absorbed by the gas during the first two processes is 1596 cal and the heat lost by the gas during the last two processes is 1351 cal. The difference between these two values was shown in Chapter 19 to be equal to the net work done by the gas on its surroundings, or the area enclosed by the rectangle. As a check, the area enclosed by the rectangle is equal to

$$(P_B - P_A)(V_D - V_A) = (2P_A - P_A)(2V_A - V_A)$$
$$= P_A V_A$$

From the equation of state,

$$P_A V_A = nRT_A$$

Expressing the right side of the last expression in calories, we have

$$nRT_A = 0.45 \text{ moles} \times 2 \text{ cal/mole-}°K \times 273°K$$
$$= 245 \text{ cal}$$

This is precisely the difference between the heat absorbed by the gas and the heat liberated. ◀

20-5 ADIABATIC PROCESSES

When a gas enclosed within walls that are impermeable to heat transfer is suddenly compressed or expanded, the process is said to be *adiabatic*. This means that $Q = 0$ and no heat can enter or leave the system. The equations that relate the pressure, volume, and temperature during adiabatic processes are

$$P_1 V_1{}^\gamma = P_2 V_2{}^\gamma \qquad \text{(20-29a)}$$
or
$$T_1 V_1{}^{\gamma-1} = T_2 V_2{}^{\gamma-1} \qquad \text{(20-29b)}$$
or
$$P_1^{1-\gamma} T_1{}^\gamma = P_2^{1-\gamma} T_2{}^\gamma \qquad \text{(20-29c)}$$

where $\gamma = C_p/C_v$ is known as the *adiabatic constant*. For monatomic gases $\gamma = 1.67$ and for diatomic gases $\gamma = 1.4$.

Example 20-12   A gas that occupies a volume of 10 liters at STP is compressed to a volume of 5 liters. The gas is diatomic. Find the final pressure and temperature if (a) the expansion is adiabatic and (b) the expansion is isothermal.

Solution.   Using Eq. 20-29a,

$$P_1 V_1{}^\gamma = P_2 V_2{}^\gamma$$
$$1 \text{ atm } (10 \text{ liters})^{1.4} = P_2 \, (5 \text{ liters})^{1.4}$$
$$P_2 = (\tfrac{10}{5})^{1.4} = 2^{1.4}$$
$$\log P_2 = 1.4 \log 2$$
$$\log P_2 = 1.4(0.3010)$$
$$\log P_2 = 0.4214$$
$$P_2 = 2.6 \text{ atm}$$

The temperature can be found from Eq. 20-10.

$$\frac{P_1 V_1}{T_1} = \frac{P_2 V_2}{T_2}$$

$$\frac{1 \text{ atm} \times 10 \text{ liters}}{273°K} = \frac{2.6 \text{ atm} \times 5 \text{ liters}}{T_2}$$

$$T_2 = 355°K$$

For the isothermal case, the final temperature is clearly 273°K. The pressure can be found from Boyle's law. Thus

$$P_1 V_1 = P_2 V_2$$
$$1 \text{ atm} \times 10 \text{ liters} = P_2 \times 5 \text{ liters}$$
$$P_2 = 2 \text{ atm}$$

It should be evident from this example that a familiarity with logarithms is a prerequisite for solving problems involving adiabatic processes. ◀

SUMMARY

There are several important laws related to the behavior of ideal gases:

$PV = \text{constant}$ if the temperature is kept fixed

$\dfrac{V}{T} = \text{constant}$ if the pressure is kept fixed

$\dfrac{P}{T} = \text{constant}$ if the volume is kept fixed

The *ideal gas law* relates the pressure, volume, and absolute temperature of an ideal gas

$$\frac{PV}{T} = nR$$

where n is the number of moles of gas and R is the universal gas constant.

A *mole* is the mass of a substance (in grams) equal numerically to the molecular weight of the substance. In other words, a mole is a *gram-molecular weight*.

The average kinetic energy of a molecule of a monatomic gas

$$(KE)_{av} = \tfrac{3}{2}kT$$

where k is Boltzmann's constant.

The total internal energy of an ideal gas

$$U = \tfrac{3}{2}nRT$$

Dalton's law of partial pressures states that the pressure exerted by a mixture of gases is equal to the sum of the separate pressures each gas would exert if it alone occupied the entire volume of the mixture.

For ideal gases and many real gases, the molar specific heat capacities at constant pressure and volume are simply related to the universal gas constant

$$C_p - C_v = R$$

For *adiabatic processes* there are relations among the pressure, volume, and absolute temperature, in addition to the ideal gas law:

$$PV^\gamma = \text{constant}$$

or

$$TV^{\gamma-1} = \text{constant}$$

or

$$P^{1-\gamma}T^\gamma = \text{constant}$$

QUESTIONS

1. Describe Boyle's law in your own words.
2. What is meant by an *ideal gas*?
3. Describe Charles' law in your own words.
4. What do the letters STP refer to when discussing the thermal behavior of gases?
5. What is the *ideal gas law*?
6. How is the term *mole* defined? What does *one mole of oxygen* mean?
7. How do the number of molecules in a mole of hydrogen compare with the number of molecules in a mole of iron?
8. How is the *average* kinetic energy of an ideal gas related to its absolute temperature? How is the *total* internal energy of such a gas related to its absolute temperature?
9. Why is it significant that Boyle's law can be derived theoretically, assuming that the gas molecules obey Newton's laws and are elastic?
10. What is van der Waal's equation? What does it attempt to do?
11. What is Dalton's law of partial pressures?
12. How are the two specific heat capacities of an ideal gas related to the universal gas constant? Is this relationship experimentally verified? Explain.

PROBLEMS

(A)

1. An ideal gas is originally at STP. Find the new pressure if the temperature is raised to 127°C at constant volume.
2. Twenty liters of an ideal gas at STP are compressed at constant pressure until the volume is 4 liters. Find the new temperature.
3. Ten liters of an ideal gas are at STP. How many moles of gas are present?
4. Refer to Problem 3. What is the new volume of the gas if the pressure is increased to 3.5 atm isothermally?
5. How many grams of oxygen occupy a volume of 4 liters at STP?
6. How many grams of hydrogen occupy a volume of 6 liters at a pressure of 5 atm and a temperature of 27°C?
7. An ideal gas occupies a volume of 3 liters at STP. Find the new pressure if the volume is doubled and the temperature is increased to 300°C?
8. What is the internal energy per mole of gas at a temperature of 127°C?

9. What is the internal energy per molecule of gas at a temperature of 127°C?

10. What is the average velocity of a hydrogen molecule at 127°C?

11. How many calories of energy are needed to double the temperature of 8 moles of diatomic nitrogen gas at constant pressure? At constant volume? (Assume $t_0 = 100°C$.)

12. Five moles of a nonatomic ideal gas are taken through the following processes:

   1. Start at a pressure of 10 atm and volume of 6 liters.

   2. Double the pressure at constant volume.

   3. Double the volume at constant pressure.

   Find the initial temperature and the temperature at the end of each process.

13. Refer to Problem 12. How many calories are required to complete each of the processes?

14. An ideal gas ($\gamma = \frac{5}{3}$) occupies a volume of 16 liters at a pressure of 0.25 atm. Find the new pressure if the gas is compressed isothermally to a volume of 2 liters.

15. Refer to Problem 14. What would be the final pressure be if the gas were compressed from 16 to 2 liters adiabatically?

(B)

1. A 4-ft^3 tank contains a gas at a temperature of 27°C and atmospheric pressure. What is the new pressure when the temperature is raised to 127°C?

2. How many moles are there in a gas that occupies a volume of 3 ft^3 at STP?

3. A cylinder with a movable piston contains 10 liters of gas at STP. What is the new volume of the gas at 2 atm of pressure and a temperature of 72°C? Determine the number of moles of gas within the cylinder.

4. A 4-liter tank contains gas at STP. The stopcock is opened and some gas is released. When the temperature is again 0°C, the pressure is found to be one-quarter of an atmosphere. How many moles of the gas escaped?

5. What is the mass of air in a 500 m^3 volume when the temperature is 27°C and the pressure is 76 cm of Hg? Take the average molecular mass of air as 28 g.

6. How many grams of helium are in a 5000-cm^3 tank when the pressure is 5×10^7 dynes/cm^2 and the temperature is 27°C?

7. A tank of volume 4000 cm^3 contains oxygen at STP. How many grams of oxygen must be pumped into the tank to raise the pressure to 80 atm with no change in temperature? Take the molecular weight of oxygen as 32.

8. What is the internal energy of 2 moles of an ideal gas at 27°C? What would be the rms velocity of a molecule if the gas were oxygen? Hydrogen?

9. The ideal gas law can be written as

$$pV = MR'T$$

where M = mass of gas (or weight) and R' = energy per unit mass (weight) per unit temperature. In this form R' is a different constant for each gas. Consider the following. A bottle of gas holds 3 ft^3 of gas that weighs 12 lb. The gas exerts a pressure of 100,000 lb/ft^2 absolute when the temperature is 100°F. Evaluate R' for this gas.

10. An inflated tire has a volume of 1200 in.^{3} When the air temperature is 60°F the gauge pressure in the tire is 300 lb/ft^2.
   (a) What will be the pressure in the tire when the air temperature is doubled?
   (b) Take R' to be 53 ft-lb/lb-F° and find the weight of air that the tire can hold.

11. When a manometer is connected to a tankful of gas, the level of mercury in the open side of the manometer is found to be 4 cm lower than the level in the connected leg when the barometric pressure is 76 cm Hg. Give the gauge pressure and absolute pressure of the gas in dynes/cm^2.

12. A volume of 200 cm^3 of air at 76 cm of Hg pressure is compressed to 50 cm^3. What is the new pressure if the process is carried out
   (a) isothermally, and
   (b) adiabatically?

13. A diatomic gas contained in a cylinder equipped with a movable piston occupies a 2-liter volume at STP. If the gas expands isothermally to a volume of 5 liters, what will be the new temperature? What would the temperature be if the expansion were performed adiabatically?

14. Two chambers are separated by a thin partition. Chamber 1 has a volume of 2 ft^3 and contains nitrogen at a pressure of 2 atm. Chamber 2 has a

volume of 3 ft^3 and contains oxygen at a pressure of 1 atm. Both chambers are at the same temperature. What would be the pressure of the mixed gases if the partition were removed?

15. Hydrogen has a specific heat of 3.4 cal/g-C°. 2000 calories must be removed from an electric generator in order to keep it cool. How many grams of hydrogen must circulate through the generator if the hydrogen enters at 15°C and leaves at 30°C?

16. Figure 20-8 represents a child's pop toy. Air enters the barrel at A. If 2.5 atm of pressure are required to pop the cork, how far into the barrel does the plunger get before the cork pops, assuming no temperature changes?

17. One mole of an ideal gas whose $C_v = 3$ cal/mole-K° are taken through the following consecutive processes:

1. Start at a pressure of 1 atm and a volume of 20 liters.

2. Double the pressure at constant volume.

3. Double the volume at constant pressure.

4. Halve the pressure at constant volume.

5. Halve the volume at constant pressure.

6. Show this cycle of processes on a P–V diagram:

(a) What was the original temperature of the system?

(b) What were the temperatures at the end of the second, third, and fourth processes?

(c) How much thermal energy was gained by the gas during process 2?

(d) How many calories of thermal energy were gained by the gas during process 3?

(e) How much net work was done by the gas during the complete cycle?

(f) What was the total change in internal energy of the gas at the conclusion of the cycle?

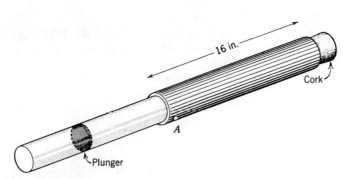

Figure 20-8

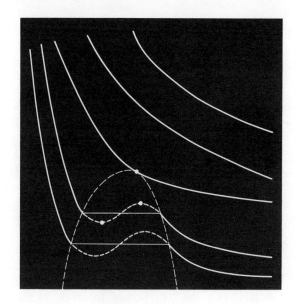

21.
Changes of Phase

In this chapter we shall study some processes that are exhibited by real gases. *Liquefaction, boiling, evaporation,* and *freezing* are discussed and analyzed with the help of diagrams that show the relationship among the thermodynamic variables during phase changes. The chapter concludes with a discussion of *humidity* and *dew point.*

LEARNING GOALS

This chapter has three major learning objectives:

1. *To understand the real processes of liquefaction, boiling, evaporation, and freezing.*

2. *To learn the p–V and p–T diagrams for the different processes mentioned in the first objective.*

3. *To understand the meaning of relative humidity, absolute humidity, and dew point.*

21-1 LIQUEFACTION

It has been mentioned in the previous chapter that there are no forces between the molecules of an ideal gas. Consequently, an ideal gas cannot be liquefied, since there are no intermolecular binding forces that are needed to form the liquid molecular structure. To understand this concept of binding, consider the analogy of a rocket ship hurtling through space toward a planet. While the two are far apart, their separate motions may be considered independent of one another. As the rocket nears the planet, the gravitational force which the planet exerts on the rocket, and which, by Newton's third law, the rocket exerts on the planet, draws the two bodies together. If the rocket has the right amount of initial kinetic energy, it may be "caught" by the planet and revolve about the planet as a satellite. The motion of the two individual bodies is no longer independent; rather, they have combined to form a new system. To revert to the original system of two independent bodies, energy must be supplied to separate the bodies to the extent that the gravitational force again becomes negligible. If there were no gravitational force between these two bodies, the motions would always be independent of each other, no matter how close together they were, barring collisions. Thus ideal gases cannot be liquefied. A plot of pressure versus volume for different values of temperature for an ideal gas will always yield hyperbolas characteristic of Boyle's law.

Real gases, unlike ideal gases, exert intermolecular forces. At pressures and temperatures where the molecules of the gas are brought close to each other, there will be a departure from Boyle's law. The molecules will begin to form new structures, which will have properties very different from those of a gas. When these new structures begin to form, the gas is said to be liquefying. Figure 21-1 shows a plot of pressure versus volume for different temperatures of a real gas, for example, carbon dioxide. Curves A and B, for which the temperatures are still high, are hyperbolas conforming to Boyle's law. In curve C we

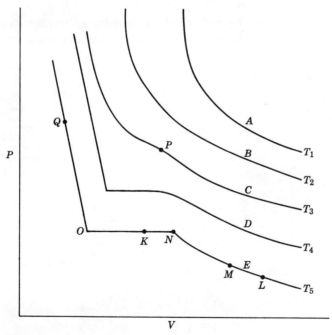

Figure 21-1 Variations of pressure with volume in a real gas at different temperatures.

see a departure from Boyle's law. This curve is no longer completely hyperbolic. A small bump has formed at point P. At still lower temperatures, curves D and E show complete departures from Boyle's law. A flat plateau appears connecting the low-pressure, large-volume portion of the curves with the high-pressure, small-volume portion. To understand the significance of these curves, let us consider curve E. All points on this curve are at the same temperature, T_5. When the system has the pressure and volume associated with points L or M, the carbon dioxide is in the gaseous state. At point N the carbon dioxide is still a gas, but all further decreases in the volume do not affect the pressure until point O is reached. Beyond point O the system is extremely difficult to compress. Along the plateau N–O the temperature and pressure of the system are both constant while the volume changes. This is characteristic of a change of phase. The relative incompressibility of the system beyond O indicates that the state of the system at O is liquid. The fraction of the system that has already turned to liquid at point K is found by taking the ratio of the lengths N–K to N–O.

It is important to note that each curve has only one plateau, i.e., there is only one pressure, for a given temperature, at which the gas will liquefy. At temperatures higher than that of curve C, there is no pressure at which the gas can be liquefied. Point P on curve C is called the *critical point*. The temperature and pressure at P represent the highest temperature and pressure at which the gas can be liquefied. The values of the

pressure, density, and temperature at the critical point are called the *critical values* (see Table 21-1).

Many attempts have been made to find the form of the intermolecular forces so as to obtain an equation of state that looks like Fig. 21-1. One such attempt leads to van der Waals' equation of state

$$\left(p + \frac{a}{V^2}\right)(V - b) = nRT \qquad (21-1)$$

which has been mentioned in Chapter 20. The plot of Eq. 21-1 appears in Fig. 21-2. Exclusive of the region inside the dashed curves Fig. 21-2 agrees with Fig. 21-1. Point P corresponds to the critical point of Fig. 21-1. The dashed portions of the curves show the pressure and volume both decreasing simultaneously, such as RS, are physically impossible, since the pressure and volume cannot both decrease or increase simultaneously.

However, in the region where van der Waals' equation fails to agree with experimental results, the plateau can be inserted so that the areas of the two lobes I and II are equal (see Fig. 21-2). With this understanding, van der Waals' equation can be used as a fair approximation of the behavior of real gases. The values a and b in Eq. 21-1 are constant for specific gases. Attempts have been made to explain the portions ST and UR of the curves, which are not physically impossible (the slopes of these portions are negative), by asserting that they refer to the states of supercooled gases and superheated liquids.

From Fig. 21-2 it is clear that at higher pressures the liquefaction process takes place at

Table 21-1 Critical values for some real gases

Substance	Temperature, °C	Pressure, atm	Density, g/cm³
Air	−140.7	37.2	0.35
Ammonia	132.4	111.5	0.235
Carbon dioxide	31.1	73.0	0.460
Chlorine	144.0	76.1	0.573
Helium	−267.9	2.26	0.0693
Hydrogen	−239.9	12.8	0.031
Sulfur dioxide	157.2	77.7	0.52
Water	374.0	217.7	0.4

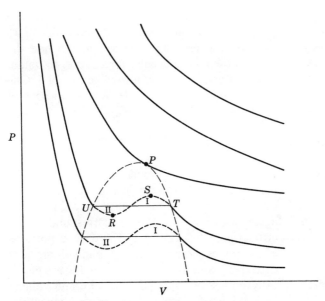

Figure 21-2 Isotherms corresponding to van der Waals' equation of state.

higher temperatures. The temperature at which a gas liquefies or a liquid vaporizes is called the *boiling point* and depends on the pressure of the system. Figure 21-3 is a plot of pressure versus temperature for those pairs of values at which water boils. This curve will be referred to as the boiling point curve. (For values of vapor pressure at various temperatures for other liquids, see Table 21-2.) The boiling point curves for all other substances are similar in shape to that for water, although the corresponding pressure–temperature values at which boiling takes place

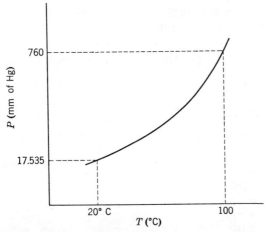

Figure 21-3 The boiling point curve for water is also known as the vapor-pressure curve.

Table 21-2 Vapor pressure of liquids

Liquid	Temperature, °C	Pressure, cm Hg
Alcohol (ethyl)	20	4.4
Alcohol (ethyl)	50	22.2
Alcohol (ethyl)	78.5	76.00
Mercury	20	0.00012
Mercury	100	0.0273
Mercury	356.7	76.00
Mercury	400	157.4
Water	0	0.458
Water	20	1.75
Water	50	9.25
Water	100	76.00
Water	150	357.0

may be quite different. Thus, to boil at 20°C, water requires a pressure of 1.75 cm of Hg, mercury requires 0.00012 cm of Hg, and alcohol requires 4.4 cm of Hg (Table 21-2).

21-2 EVAPORATION, BOILING, AND FREEZING

It is important to understand the distinction between boiling and evaporation. *Evaporation* is a process whereby molecules of a liquid acquire enough kinetic energy to break out of the liquid and enter the surrounding atmosphere as vapor molecules. Evaporation is a surface phenomenon; i.e., only the molecules on or very close to the surface of the liquid take part in the evaporation process. Although the average kinetic energy of all the molecules of the liquid is fixed for a given temperature, at any instant there are many molecules with kinetic energies greater than the average. These molecules will make many collisions with each other, and energy transfers will take place during each collision. If a fast-moving molecule is at the surface of the liquid, it may be able to use its energy to escape, since there are fewer molecules in its path with which to make collisions.

If there are many air and vapor molecules in the atmosphere directly above the liquid, an escaping vapor molecule may be reflected after collision back into the liquid. The rate of evaporation can, therefore, be increased by reducing the pressure above the liquid. The rate of evaporation can also be increased by providing a means for blowing away or carrying away vapor molecules before they are reflected back into the liquid. This is the reason that clothes dry faster on a warm, windy day than on a warm, calm day. The rate of evaporation can also be increased by heating the liquid (increasing the kinetic energies of the molecules) and by increasing the surface area of the liquid, giving more of the molecules an opportunity to escape.

The pressure exerted by the vapor molecules on the surrounding atmosphere and liquid is called the *vapor pressure*. The vapor pressure depends on the temperature of the liquid. The higher the temperature, the greater the vapor pressure. When a bubble forms beneath the surface of a liquid, the pressure inside the bubble is the vapor pressure at that temperature. If this vapor pressure is equal to or greater than the atmospheric pressure, the bubble will be able to rise and escape the liquid. We then say that evaporation is taking place throughout the liquid. The latter state of evaporation is called *boiling*. The *boiling point* of a liquid is that temperature at which the vapor pressure of the liquid is equal to the atmospheric pressure. Thus, water boils at 100°C under 14.7 lb/in.^{2} of pressure because, at that temperature, the vapor pressure of water is precisely 14.7 lb/in.^{2} The boiling point curve (Fig. 21-3) is then a plot of vapor pressure versus temperature. At higher altitudes where the atmospheric pressure is less than 14.7 lb/in.^{2} (e.g., Denver, Colorado) water will boil at temperatures below 100°C. If the pressure above the liquid is increased, as in a pressure cooker, water will not boil until it reaches temperatures above 100°C.

If, as shown in Fig. 21-1, we continue to compress the liquid beyond point Q, we encounter a second plateau. It is here that the liquid undergoes a change of phase into a solid (see Fig. 21-4).

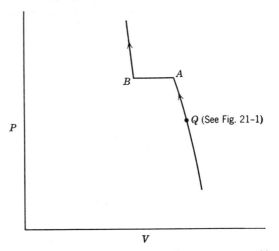

Figure 21-4 Liquid-to-solid phase change represented by the plateau *AB*.

The temperature at which a substance freezes at a particular pressure is called the *freezing point*. Figure 21-5 shows two freezing point curves. Figure 21-5*a* is typical of substances that expand on freezing, such as water, while Fig. 21-5*b* is characteristic of substances that contract on freezing. As the pressure on a solid is increased, the freezing point is lowered for those substances which behave like water and raised for those substances which contract on freezing. That the freezing point of ice is lowered when the pressure on the ice is increased can readily be seen from the following experiment.

A thin wire with two large weights suspended from each end is draped across a large cake of ice, as in Fig. 21-6. Gradually, the weights and wire cut through the cake of ice while the ice remains in one piece. This process is known as *regelation*. The increased pressure due to the wire and weights lowers the freezing point of the top layer of ice. This layer melts allowing the wire to pass through, and then immediately refreezes since the pressure has been removed. This process continues until the wire has passed through the entire cake of ice.

At all points on the boiling point curve, the liquid and vapor can coexist in equilibrium and at all points on the freezing point curve, the solid and liquid can coexist. At the intersection of these two curves is the *triple point*. At this point

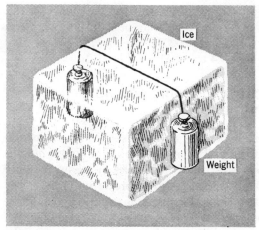

Figure 21-6   Apparatus for performing a regelation experiment.

all three phases can coexist in equilibrium. Some substances (e.g., iodine and dry ice) can change from the solid to vapor without passing through the liquid phase. This process is known as *sublimation*. The loci of the points at which a solid and its vapor can coexist define the sublimation curve. Figure 21-7 shows the boiling, freezing, and sublimation curves with their common intersection, the triple point, for substances that contract on freezing and for those that expand on freezing. Note the difference in the slopes of the freezing point curves in the two diagrams.

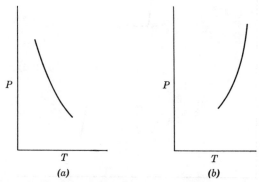

Figure 21-5   Freezing point curves. (*a*) Substances that expand upon freezing such as water. (*b*) Substances that contract upon freezing, such as mercury.

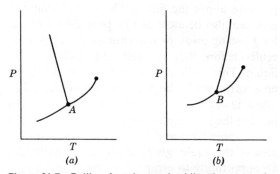

Figure 21-7   Boiling, freezing, and sublimation curves for (*a*) substances which expand upon freezing and (*b*) substances which contract upon freezing. Points *A* and *B* refer to the respective triple points.

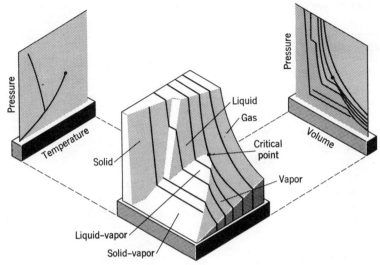

Figure 21-8   The *P–V–T* surface for water and its projections on the *P–V* and
P–T planes.

An equation of state that describes the changes in a thermodynamic system is generally a function of the three variables: pressure, volume, and temperature. An equation such as

$$f(p, V, T) = 0 \qquad (21\text{-}2)$$

where f is a function of p, V, and T describes a three-dimensional P–V–T surface. When this surface is projected on the P–V plane, we obtain the variations in P–V for constant temperature. When the surface is projected on the P–T plane, we get the variations in P–T for fixed volumes, and on the V–T plane we see the variations for constant pressures. In Fig. 21-8, the P–V–T surface of water is shown with its projections on the two perpendicular planes. Note that the projection on the P–T plane is precisely Fig. 21-7a.

21-3 HUMIDITY AND DEW POINT

The absolute humidity is the mass of water vapor present per unit volume of the atmosphere. Absolute humidity is usually expressed in g/m^3, grains/ft^3, or lb/ft^3. The amount of water vapor which the atmosphere can hold depends on the

atmospheric temperature (see Tables 21-3, 21-4, and 21-5). When the atmosphere is holding the

Table 21-3   Absolute humidity of saturated air

Temperature, °F	Absolute Humidity, g/m³	Temperature, °F	Absolute Humidity, g/m³
26	3.7	62	14.0
28	4.0	64	15.0
30	4.4	66	16.0
32	4.8	68	17.1
34	5.2	70	18.2
36	5.6	72	19.4
38	6.0	74	20.6
40	6.4	76	22.0
42	6.9	78	23.5
44	7.5	80	25.0
46	8.1	82	26.5
48	8.7	84	28.2
50	9.3	86	30.0
52	10.3	88	32.0
54	10.7	90	34.0
56	11.4	92	36.0
58	12.2		
60	13.0		

Table 21-4 Vapor pressure of water vapor for various temperatures, °F

Temperature, °F	Pressure in mm of Hg	Temperature, °F	Pressure in mm of Hg
32	4.58	68	17.54
36	5.37	72	20.07
40	6.31	76	22.92
44	6.31	80	26.12
48	8.55	84	30.00
52	9.90	88	33.85
56	11.48	92	38.58
60	13.29	96	43.60
64	15.28	100	49.16

Table 21-5 Saturated vapor pressure of water at various temperatures, °C

Temperature, °C	Pressure in cm of Hg	Temperature, °C	Pressure in cm of Hg
0	0.458	70	23.37
5	0.654	80	35.51
10	0.921	90	52.58
20	1.75	95	63.39
30	3.17	100	76.00
40	5.53	110	107.4
50	9.25	120	148.4
60	14.94	150	357.0

maximum amount of water vapor at its particular temperature, we say that the atmosphere is saturated. The relative humidity is the ratio of the mass of water vapor per unit volume present in the air to the mass of water vapor per unit volume in saturated air at the same temperature. Since the pressure exerted by water vapor is approximately proportional to the mass per unit volume, we can replace the mass of vapor per unit volume by the pressure of the water vapor in the last ratio.

Example 21-1 What is the pressure of water vapor in the air on a day when the temperature is 86°F and the relative humidity is 60%? The pressure of saturated water vapor at 86°F is 31.7 mm.

Solution. Using our ratio, we find that relative humidity

$$= \frac{\text{pressure of water vapor}}{\text{pressure of water vapor in saturated air}}$$

$$0.60 = \frac{x}{31.7 \text{ mm}}$$

$$x = 31.7(0.6) \text{ mm}$$

$$= 19.0 \text{ mm} \quad \blacktriangleleft$$

In climates where the relative humidity is high, evaporation takes place very slowly. Such climates are usually uncomfortable in warm weather because perspiration does not evaporate from the body quickly.

In discussing relative humidity, the meteorologist often speaks of *dew point*. Dew point is a temperature. By definition it is that temperature at which the air during cooling becomes saturated and water droplets form as a result of condensation. The formation of dew is common on cool surfaces such as grass and plants after they have lost much of their heat through radiation. We are all aware that dew forms during the night and early morning when the temperature of the earth has fallen. The moist air surrounding the surface of the earth is cooled by contact with objects on the earth. Its vapor capacity is thereby reduced, the air becomes saturated, and, as the cooling goes on, some of its moisture condenses out as dew.

The scientist would find it inconvenient to wait for dew to form on grass and plants in order to measure the dew point. Being resourceful, he finds it experimentally, employing a method that is easily duplicated. Take a container partly filled with water and add to it small pieces of ice. When the water and container reach the dew point, small droplets of water collect on the outside of the container. The temperature at which these drops of moisture first form gives a very close approximation to the dew point. If there is much water in the air, the moisture will form quickly and the dew point will be relatively high; if there is little water in the air, it will take longer to reach the dew point, and consequently the dew point will be lower. It should be clear that the dew point is an indication of the amount of water that is in the air as vapor. This information, coupled with the fact that evaporation is a cooling process, provides the necessary data for constructing an instrument for measuring relative humidity.

One instrument for measuring relative humidity is the *hygrometer*. The hygrometer records both relative humidity and temperature. The relative humidity measurement depends on the expansion and contraction of human hair with fluctuations in the water content of the air. The hair (blond hair is used) expands in length with an increase in humidity and contracts with a decrease in humidity. In a recording hygrometer the fluctuations are transmitted to a revolving drum by means of a stylus, and a graph is traced on a scale attached to the drum.

SUMMARY

Real gases do not obey Boyle's law at all temperatures but begin to liquefy at sufficiently low temperatures. The greatest pressure and temperature at which a gas will liquefy determines its *critical point.*

Vapor pressure is the pressure exerted by the vapor molecules on the liquid and surroundings.

A liquid begins to liquefy or vaporize throughout its volume when its vapor pressure is equal to the atmospheric pressure.

The solid, liquid, and vapor phases all coexist at the *triple point.*

Sublimation is the process of changing from solid to vapor directly without passing through the liquid phase.

Absolute humidity is the mass of water vapor per unit volume of atmosphere. The atmosphere is *saturated* when it is holding the maximum amount of water vapor at a particular temperature.

Relative humidity is the ratio of the absolute humidity at the present temperature to the absolute humidity of saturated air at the same temperature. Relative humidity can also be measured by the ratio of the pressure of water vapor in the air to the pressure of water vapor in saturated air at the same temperature.

Dew point is the temperature at which the air, during cooling, becomes saturated and water droplets form as a result of condensation.

QUESTIONS

1. Explain how a real gas condenses into a liquid when the temperature decreases.

2. What is meant by the critical temperature?

3. How successful is van der Waal's equation of state as a description of the pressure–volume–temperature relationship of a real gas?

4. What is the boiling point of a liquid?

5. Why are the boiling point curve and vapor pressure curve of a liquid essentially the same?

6. What is the difference between evaporation and boiling?

7. Why would clothes hung on a line dry more quickly on a warm, windy day than on a warm, calm day?

8. What is regelation? Sublimation?

9. What is the triple point of a substance?

10. How is relative humidity defined?

11. What is the dew point?

6. Without referring to the text, explain why ideal gases cannot undergo condensation.

7. Without referring to the text, draw the P–T diagrams showing the boiling–freezing–sublimation curve for a substance that contracts upon freezing.

8. Repeat Problem 7 for a substance that expands upon freezing. How does the freezing point curve of Problem 7 differ from the one in this question?

9. If the relative humidity is 53% when the temperature is 10°C, what is the dew point?

10. It is noted that droplets of water begin to collect on a pitcher of cold water when its temperature is 20°C and the room temperature is 30°C. What is the relative humidity?

PROBLEMS

(A)

(*Note*: Refer to Tables 21-4 and 21-5 as needed.)

1. Explain why water often appears under an ice-skater's skates even when the temperature is 0°C.

2. Under what circumstances will water boil at temperatures greater than 100°C? at temperatures less than 100°C?

3. On a clear day condensation started at a temperature of 10°C when the air temperature was 30°C. What was the relative humidity? (The pressure of saturated vapor at 10°C is 9.21 mm of Hg; at 30°C it is 31.7 mm of Hg.)

4. If the air temperature is 86°F and the dew point is 59°F, what is the relative humidity? (The pressure of saturated vapor at 59°F is 12.8 mm of Hg; at 86°F it is 31.7 mm.)

5. What is the pressure of water vapor in the air on a warm day when the temperature is 30°C and the relative humidity is 80%? (See the information given in Problem 3.)

(B)

1. What must be the temperature of a piece of metal if it is to show moisture in a room where the temperature is 68°F and the relative humidity is 50%?

2. If the relative humidity is 30% when the air temperature is 30°C, what is the absolute humidity?

3. If the absolute humidity is 12 g/m³ when the temperature is 76°F, what is the relative humidity?

4. When the relative humidity is 80% and the air temperature is 92°F, what is the absolute humidity?

5. A test at 64°F indicated that there was 7.00×10^{-4} lb/ft³ of water vapor per unit volume in the atmosphere. At this temperature the amount of water vapor needed for saturation is 9.60×10^{-4} lb/ft³. What was the relative humidity?

6. Air, saturated with water vapor, is confined in a chamber at 50°F. The air in the chamber is then heated to 88°F. What is the relative humidity at this higher temperature?

7. What pressure in mm of Hg was exerted by the water vapor present in Problem 5?

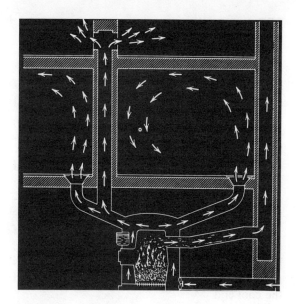

22.
Heat Transfer

In the preceding chapters we have studied the nature of thermal energy and the interchanges of heat that take place when two or more substances at different initial temperatures are mixed together. The method by which the thermal energy was transferred from one substance to a second was not of paramount importance. In this chapter we concern ourselves directly with the ways by which thermal energy is transferred in solids, liquids, and gases. We shall be particularly interested in discovering how heat is circulated in a thick slab or rod whose ends are maintained at two different temperatures, how thermal energy distributes itself through a large tank filled with a liquid or through the air in a room, and how thermal energy from the sun is transferred to us through empty space.

Heat can be transferred in three ways: *conduction*, *convection*, and *radiation*. Conduction, the principal mode of transfer in solids, involves transfer through atomic and molecular collisions. In convection, the principal method of transfer in liquids and gases, warm currents of the liquid or gas are established through the difference in density between the warmer fluids. Radiation is the means by which thermal energy is transferred by electromagnetic waves.

LEARNING GOALS

The major learning objectives of this chapter are

1. *To understand the processes of conduction, convection, and radiation.*

2. *To learn to use the equation that relates the rate at which energy is conducted in a rod to the temperature gradient along the rod.*

3. *To use the Stefan–Boltzmann law to calculate the energy radiated or absorbed by an object.*

22-1 CONDUCTION

Consider a metal rod with one end in contact with a hot reservoir. The molecules of the metal in contact with the reservoir are heated by collisions with the molecules of the reservoir. The layer of molecules next removed from the reservoir collides with the first layer of molecules and gains kinetic energy during these collisions. By continuing this process, thermal energy is transferred throughout the rod (Fig. 22-1).

When a steady state has been reached, each cross section of rod will have a fixed temperature. The temperature at each cross section decreases uniformly along the rod from the hot end toward the cool end.

The rate at which thermal energy is transferred through the rod at steady state is directly proportional to both the cross-sectional area of the rod and the temperature gradient of the rod. The temperature gradient is the difference in temperature Δt between the two ends of the slab divided by the length of the slab. Thus

$$H = kA \frac{\Delta t}{L} \qquad (22\text{-}1)$$

where H is the number of cal/sec or Btu/hr transferred through the slab; A is the cross-section area of the slab; $\Delta t/L$ is the temperature gradient; and k is the constant of proportionality, called the

thermal conductivity. The thermal conductivity, k, depends on the material of which the slab or rod is made. The dimensions of k depend on the system of units used for H, A, and $\Delta t/L$. In the cgs or scientific system, H is in cal/sec, A in cm^2, and $\Delta t/L$ in C°/cm. Then,

$$(k) = \left(\frac{H}{A} \times \frac{L}{\Delta t} \right) = \text{cal/sec-cm-C}°$$

In the engineering system, H is in Btu/hr, A in ft^2, and $\Delta t/L$ in F°/in. Then,

$$(k) = \left(\frac{H}{A} \times \frac{L}{\Delta t} \right) = \text{Btu-in./hr-ft}^2\text{-F}°$$

Note that in the engineering system two different units of length are used. The length of the slab or rod must be expressed in inches whereas the cross-section area is given in square feet. The thermal conductivity of various materials is given in Table 22-1.

If two rods of different thermal conductivities but of the same cross section are connected, the number of cal/sec or Btu/hr transferred through each rod will be the same if there are no heat sources or sinks in the rods. If the temperatures of the free ends of the rods are known we can calculate the temperature at the junction by equating the rate of transfer in the first rod to that of the second rod.

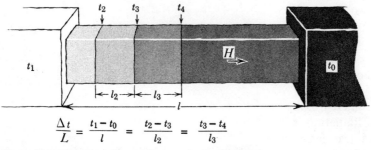

$$\frac{\Delta t}{L} = \frac{t_1 - t_0}{l} = \frac{t_2 - t_3}{l_2} = \frac{t_3 - t_4}{l_3}$$

Figure 22-1   Transfer of thermal energy in a rod or bar by conduction. At equilibrium, each cross section of the conductor is at a different constant temperature, denoted here by t_2, t_3, and t_4.

Table 22-1 Thermal conductivity of different substances

Substance	cgs Units (cal/sec-cm-C°)	Engineering Units (Btu-in./hr-ft²-F°)
Aluminum	0.50	1400.0
Brass	0.26	750.0
Copper	0.92	2600.0
Silver	0.97	2900.0
Iron	0.16	430.0
Steel	0.11	300.0
Lead	0.08	
Mercury	0.02	
Tile	0.002	
Glass	0.0025	6.0 .
Paper	0.0003	
Felt	0.00004	0.30
Wood		0.80
Abestors		4.0

Example 22-1 An aluminum pot is held over a flame. The bottom of the pot is 0.5 cm thick and has an area of 500 cm². If the lower surface of the bottom is kept at 130°C and the upper surface of the bottom is kept at 105°C, how many calories are transferred through the bottom of the pot in 2 min?

Solution. From the table, the thermal conductivity of aluminum in cgs units is 0.50. Then,

$$H = \frac{kA}{L}(t_2 - t_1)$$

$$= \frac{0.50(500)}{0.5}(130 - 105)$$

$$= 12,500 \text{ cal/sec}$$

For 120 sec,

$$Q = 12,500 \text{ cal/sec (120 sec)}$$
$$= 1,500,000 \text{ cal} \qquad \blacktriangleleft$$

Example 22-2 A steel rod, 20 cm long, is rigidly attached at one end to a silver rod, 50 cm long. Both rods have cross sections of 5 cm². The free end of the steel rod is kept at 10°C, and the free end of the silver rod is kept at 90°C. Find the temperature at the junction and the rate at which calories are conducted along each rod.

Solution. Since there are no sinks for losing heat and no sources of heat in the rods, the rate of transfer in each rod is the same. Thus, letting t equal the temperature at the junction, we obtain

$$H_{steel} = H_{silver}$$

$$k_{steel}\frac{A(t - 10°)}{L_{steel}} = k_{silver}\frac{A(90° - t)}{L_{silver}}$$

Substituting the given values for the lengths of steel and silver and the values of k from Table 22-1, we have

$$0.11(5)\left(\frac{t - 10}{20}\right) = 0.97(5)\left(\frac{90 - t}{50}\right)$$

$$50(11)(t - 10) = 20(97)(90 - t)$$
$$55t - 550 = 17,460 - 194t$$
$$249t = 18,010$$
$$t = 72°C$$

Then, $$H = 0.11(5)\left(\frac{72 - 10}{20}\right)$$

$$= 1.71 \text{ cal/sec} \qquad \blacktriangleleft$$

22-2 CONVECTION

Consider a rectangular tube filled with a liquid, as shown in Fig. 22-2, with a Bunsen burner under point A. As the fluid immediately above point A is heated it expands, becomes less dense than the cooler fluid, and diffuses upward in arm B of the tube. The cooler fluid in arm C moves into the space formerly occupied by the warmer fluid and is in turn heated. This process continues until a current of warm fluid is circulating through the liquid in the tube. Such a current is called a *convection current*, and this process of transferring heat is called convection. If a drop of ink is placed into the tube at D, the ink will be dragged along

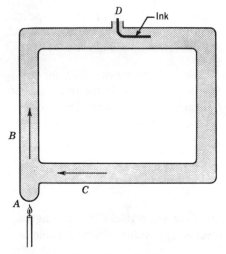

Figure 22-2   Convection in liquids.

by the convection current. The fluid to the left of
D will not become colored until the ink has moved
around the entire tube.

At one time some automobile engine blocks
were cooled entirely by convection. However, all
standard water-cooled engines today use a water
pump to circulate the water by pumping the
cooled water from the bottom of the radiator
(see Fig. 22-3) to the lower section of the water
jacket of the engine's cylinders. This is one example of forced convection.

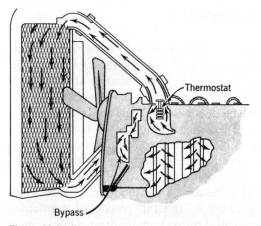

Figure 22-3   Forced convection in an automobile cooling
system.

In addition, circulating-hot-water heating systems use "forced convection" by using a pump in
the same manner. Hot air household heating systems were once entirely dependent on convection.
Today the more modern systems use a fan to blow
cold air into the bottom of the convection chamber inside the hot air furnace, as in Fig. 22-4. This
is another example of forced convection.

Liquids and gases are poorer conductors of
heat than solids because of the larger separation
between the molecules of a fluid. Convection, then,
is the principal method of heat transfer in fluids.

Breezes that blow in from ocean to land in the
morning and from land to ocean in the evening
are caused by convection currents. The ocean
takes longer than the land to gain or lose heat
because of its high specific heat capacity. Thus,
when the air above the ocean is still cool from
the previous evening, the air above the land has
already been warmed by the early sun. This leads
to a convection current from the sea toward the
land (see Fig. 22-5). In the evening the air above
the sea, which has been warmed through the day,
has retained some of its warmth whereas the air

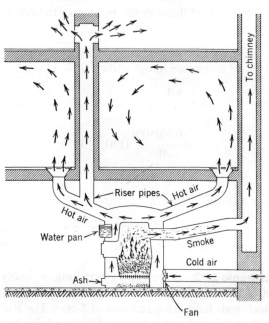

Figure 22-4   Convection currents in gases.

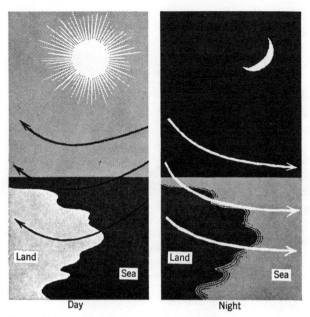

Figure 22-5 Land and sea breezes.

above the land has already been cooled. We then have a convection current from land to sea.

22-3 RADIATION

Perhaps the most important supplier of energy to the earth is the sun. Without the thermal energy supplied by the electromagnetic rays (visible light, ultraviolet, and infrared) from the sun, life on earth would be impossible. How is this vital thermal energy transferred to us from the sun, which is approximately 93,000,000 miles away? A little careful reflection leads us to the conclusion that neither of the two processes discussed in the previous sections is applicable. Since the space between the earth and sun is essentially a vacuum, neither conductive nor convective processes are possible. The method by which thermal energy is transferred by electromagnetic rays is called *radiation.*

According to the modern (quantum) theory of radiation, light is composed of bundles (quanta) of energy. The energy of each quantum is directly proportional to the frequency of the light. When light strikes a surface, these quanta deliver their energy to the surface. This energy increases the average kinetic energy of the molecules of the surface.

The rate at which an object radiates energy is a function of the absolute temperature of the object. The best radiators of thermal energy are also the best absorbers. An object that absorbs all of the radiation impinging upon its surface is called an *ideal absorber.* Such an object will also be an ideal radiator. Practically, there are no ideal radiators, but the blacker an object is, the greater is its absorbing and radiating power. For this reason, those objects which are approximately ideal radiators are called *blackbodies.* The rate at which a blackbody radiates per unit surface area is given by

$$R = \sigma T^4 \qquad (22\text{-}2)$$

where R = the energy radiated per unit time per unit area, T = the absolute temperature of the radiator, and σ = a universal constant whose value is independent of the nature of the radiation. Equation 22-2 is called the Stefan–Boltzmann law and σ is known as Stefan's constant. The value of

σ is 5.67×10^{-5} erg/cm²-sec-°K⁴. If energy and area are expressed in different units, σ must be adjusted accordingly. For objects that are not ideal, the right-hand member of Eq. 22-2 must be multiplied by a constant ε, called the *emissivity*. Its range of values is limited between nearly 0 and 1, depending on the nature of the material and surface of the radiating object. For blackbodies the emissivity is unity.

Example 22-3   Assuming the sun to be an ideal radiator, at what rate is energy radiated per square centimeter of the sun's surface? Take the temperature of the sun to be 6000°K.

Solution.   Using Eq. 22-2

$$R = \sigma T^4$$
$$= 5.67 \times 10^{-5} \times (6000)^4$$
$$= 5.67(6)^4 \times 10^{-5} \times 10^{12}$$
$$= 7.35 \times 10^{10} \text{ erg/sec-cm}^2$$
$$= 7.35 \times 10^3 \text{ joules/sec-cm}^2$$
$$= 7.35 \text{ kilowatts/cm}^2 \qquad \blacktriangleleft$$

Consider an object at a temperature T_1 surrounded by walls at a temperature T_2, where $T_1 > T_2$. Both of these objects are radiating, according to Eq. 22-2. The net rate of energy emitted by the hotter object per unit area is the difference between the rate at which it radiates per unit area and the rate at which it absorbs energy from its surroundings per unit area. Assuming only radiative processes, we have

$$R = \varepsilon\sigma(T_1{}^4 - T_2{}^4) \qquad (22\text{-}3)$$

for the net rate of energy emitted by a radiator of temperature T_1 in the presence of surroundings at a temperature T_2.

22-4 HEAT INSULATION

In order to prevent the loss of thermal energy from an object, an insulation system must be devised that takes account of all three methods of heat transfer. The poorest conductors (those with

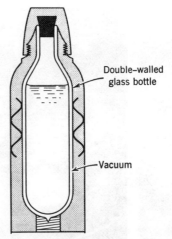

Figure 22-6   Dewar, or thermos flask. (From O. H. Blackwood, W. C. Kelly, and R. M. Bell, *General Physics*, 2nd ed., John Wiley & Sons., Inc., New York, 1962, p. 193.)

the lowest thermal conductivities) are generally porous and entrap much air. Furs and knitted woolens are used to make warm clothes because of their poor conductivity. Convection losses are prevented by keeping the entrapped air from circulating. A fine metallic mist, which adheres to fibers and reflects light, will prevent radiation losses. In building insulation, rock wool, glass wool, or cellulose are used. In addition aluminum foil is sometimes attached to one side of the insulating sheets to prevent radiation losses.

The thermos or vacuum bottle exhibits the methods of heat insulation (see Fig. 22-6). The inner and outer walls are made of glass, which is a poor conductor. The vacuum separating the inner and outer vessels prevents convection losses, and the walls forming the boundary of the vacuum are silvered to prevent radiation losses.

SUMMARY

Heat is transferred in three ways: *conduction*, *convection*, and *radiation*.

In steady state, the temperature gradient along a rod is constant.

In steady state, the rate at which heat flows down a rod is given by

$$H = k \frac{A \, \Delta t}{L}$$

Conduction is the principal mode of energy transfer in solids.

Convection is the principal mode of energy transfer in liquids and gases.

Radiation is the means by which energy is transferred by electromagnetic waves.

Ideal radiators and absorbers are called *blackbodies*.

The rate per unit area at which an object radiates energy depends upon the fourth power of the absolute temperature

$$R = \sigma T^4$$

QUESTIONS

1. In which three ways is heat transferred?

2. Describe in your own words each of the three methods of heat transfer.

3. Why are liquids and gases poor conductors of heat?

4. A lit match is held above the hand. The hand feels warm. Which method of heat transfer is being demonstrated? Explain.

5. Design an experiment to illustrate that different metals have different thermal conductivities.

6. How does the thermos bottle keep liquids hot or cold?

7. How are land and sea breezes produced?

8. By which method do the "radiators" in your home actually warm a room? Can you think of a more appropriate name for a home radiator?

9. Which do you think is a faster process, conduction or convection?

PROBLEMS

(A)

1. The ends of a 2-m bar of steel are held at temperatures of 20 and 100°C. What is the temperature gradient in C°/m? In C°/cm?

2. How many calories/sec flow through the steel bar in Problem 1? Assume that the cross-section area is 40 cm^2.

3. One end of a copper rod is immersed in a large ice bath at 0°C and the other end is immersed in a steam vat at 120°C. The rod is 50 cm long and has a cross-section area of 4.8 cm^2.
   (a) At what rate do calories flow through the rod?
   (b) How many calories pass a given point in 1 min?

4. What thickness of copper would have the same insulating effect as 4 cm of glass?

5. What thickness of aluminum would have the same insulating effect as 2 cm of lead?

6. An iron rod is 30 cm long and has a cross section of 5 cm^2. How many calories flow through the rod in 6 min if one end is kept at 10°C and the other at 85°C?

7. A glass window (2 ft × 3 ft) is $\frac{3}{8}$ in. thick. How many Btu are conducted through the glass each hour if the room temperature is 74°F and the outside temperature is 34°F?

8. At what rate does a star radiate energy if its surface temperature is 10,000°K?

9. At what rate does a wire radiate energy when its temperature is 727°C? Assume the wire to be an ideal radiator.

10. A spherical blackbody, with radius 2 cm, is maintained at a temperature of 500°C. How many joules of energy are radiated from its surface each second?

11. At what rate (cal/sec) does each square centimeter of a star radiate if its surface temperature is 8000°K?

12. When paper is wrapped tightly around a copper rod and held over a flame, the paper does not burn. Explain.

13. It requires less air conditioning to keep a white Cadillac cool in the summer than a black Cadillac of the same size. Explain.

14. The difference in the temperature on two sides of a metal plate is 40C°. The plate is 0.5 cm thick and is found to transmit 0.2 kilocalories per hour through a cross section of 5 cm^2. Calculate the thermal conductivity for this metal.

(B)

1. The diameter of a copper rod is 6 cm. The temperature gradient maintained along the rod is 2.5C°/cm.

How many calories are conducted along the rod in 30 min?

2. What difference in temperature exists between the two ends of an aluminum plate, 10 cm thick, in order to transmit 10 cal/sec for each square centimeter of cross section?

3. A steel plate 4 cm thick has a cross section of 2000 cm^2. One side of the plate is kept at 30°C while the other side is kept at the temperature of boiling water. How many calories are transmitted through the plate in 1 min?

4. A wire filament in a lamp radiates at the rate of 100 W, at an operating temperature of 927°C. The envelope surrounding the wire is kept at a temperature of 327°C. If the filament has an emissivity coefficient of 0.3, find the surface area of the filament.

5. A 2-ft brass slab is rigidly fastened to a 1-ft cast-iron slab, both having a cross section of 2 ft^2. The cast-iron end is in contact with a mixture of ice and water in equilibrium; the brass end is in contact with a mixture of steam and water in equilibrium.

(a) What is the temperature at the junction?
(b) How many Btu are transmitted through the combination slab in 5 hr?

6. An aluminum rod 36 cm long, having a cross section of 10 cm^2 is rigidly fastened to a 40-cm copper rod of the same cross section. The aluminum end is held at a fixed temperature of 40°C; the free copper end is held at a temperature of 100°C.
(a) What is the temperature of the junction?
(b) How many calories flow through the combination rod each minute?

7. A sphere of radius 5 cm is surrounded concentrically by a second sphere of radius 10 cm. The inner sphere is at a temperature of 227°C and the outer sphere is at a temperature of 127°C. How many calories are radiated from the inner sphere each minute?

8. A cube of ice fills an aluminum container 4 cm on a side. The aluminum walls are 1-mm thick. If the temperature surrounding the aluminum is kept at 25°C, how many grams of ice will melt each second?

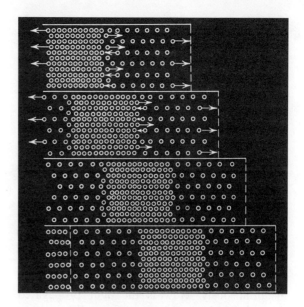

23.
Wave Motion

The basic features of *wave motion* are analyzed in this chapter. The general properties of this motion are studied by considering waves produced on a rope. The effects of the superposition of waves are discussed, leading to the concepts of *standing waves, interference, resonance,* and *beats.*

LEARNING GOALS

The major learning objectives of this chapter are

1. *To distinguish between transverse and longitudinal wave motion.*

2. *To understand the process by which energy is transmitted through a medium by wave motion.*

3. *To learn to use the appropriate equations to find the velocity with which a wave moves through a medium.*

4. *To distinguish between traveling waves and standing waves.*

5. *To understand the conditions that produce resonance of waves on fixed strings, in air columns, and in rods.*

6. *To learn the meaning of fundamental, harmonic, overtone, node, antinode, interference, beat, and frequency.*

23-1 TRANSVERSE AND LONGITUDINAL WAVES

In earlier chapters we have seen how the physicist describes various types of motion. First there was rectilinear motion; then, projectile, circular, and simple harmonic motion. In this chapter we shall learn how the physicist describes still another form of motion: *wave motion*. A piano is sounded on the stage of a great auditorium. This is the point of origin of the sound (disturbance). In a very short time this sound may be heard throughout the entire hall. How does the sound spread? Wave motion. A match is struck in a large theatre. Everyone can see the match almost instantaneously. How is the light propagated through the theatre? Wave motion. Whenever a medium is disturbed, and this disturbance is heard, seen, or otherwise observed, in another part of the medium, we say that the disturbance has been *propagated* through the medium.

Wave motion is the means by which energy is propagated from place to place without the transfer of mass.

Before discussing the more abstract case of sound waves, let us look at a form of wave motion with which we are all familiar: waves on a rope or tight spring.

Consider a long rope with one end fastened to a post and the other end held in the hand. If the free end of the rope is moved quickly up and down, a single pulse will be sent down the length of the rope and reflected from the fixed end (see Fig. 23-1). If the free end is driven up and down rhythmically by a simple harmonic oscillator, a train of waves will travel along the rope and be reflected at the far end. Each particle of rope executes its own simple harmonic motion in the vertical direction. Since the waves travel with finite velocities, they reach each particle of rope at a different time. Hence the vibrations of neighboring portions of the rope are out of phase. Two particles of rope always have their vertical displacements in phase when the distance between them is exactly one wavelength (λ) (see Fig. 23-2).

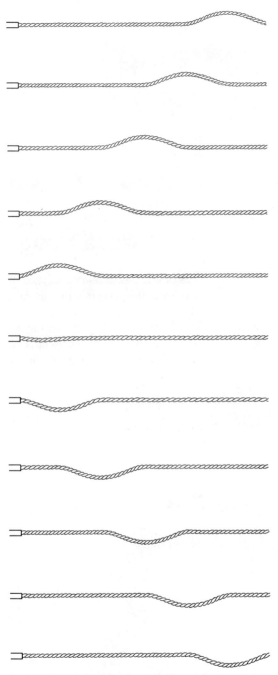

Figure 23-1   Reflection of a pulse from a fixed end. The reflected pulse is upside down.

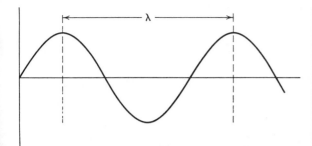

Figure 23-2 Definition of the wavelength, λ. Two sections of a wave that are moving in the same sense and have the same displacement at the same time are said to be in phase.

A particle completes one cycle of its harmonic motion when one wave has passed it by (Fig. 23-3). Thus the velocity of propagation of the waves is the wavelength divided by the time it takes a particle to complete one vibration. Hence

$$v = \frac{\lambda}{T} \qquad (23\text{-}1)$$

where T is the period of the harmonic motion. Since the frequency is the reciprocal of the period,

$$v = \lambda f \qquad (23\text{-}2)$$

The waves on a stretched rope are called *transverse* waves, because the rope particles vibrate in a direction perpendicular to the direction of propagation of the waves. When the particles execute the simple harmonic motion in the same direction as the direction of propagation, the waves are called *longitudinal*. Such is the case of sound waves.

When a medium such as air is disturbed by the snapping of a finger, or the striking of a tuning fork, the molecules of the air are set to vibrating to and fro along the direction in which the sound is traveling. A picture of the distribution of molecules at any instant shows regions of comparatively many molecules and other regions of few molecules. The dense regions where many molecules are packed together are called *compressions*, while the regions with few molecules are

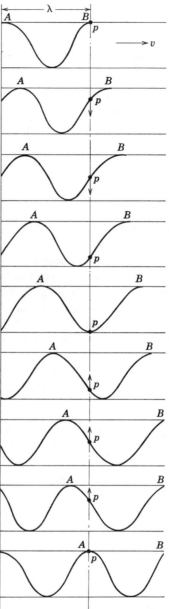

Figure 23-3 In the time it takes the wave to move to the right a distance λ (section *AB*), a particle (*p*) completes one cycle of its harmonic motion (down and up).

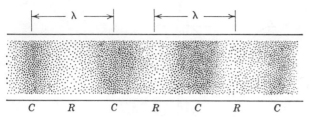

Figure 23-4   Compressions and rarefactions in a sound wave in air.

called *rarefactions* (see Fig. 23-4). These compressions and rarefactions alternate throughout the medium. As time proceeds, the molecules that previously formed a compression spring back, half to the left and half to the right, leaving a rarefaction in place of the previous compression, and forming compressions on either side where there previously were rarefactions (see Fig. 23-5). The distance between two successive compressions or rarefactions is one wavelength.

Equations 23-1 and 23-2 express the velocity of propagation of waves in terms of the wavelength and frequency. The velocity with which waves travel through a medium is a constant that de-

pends only on the physical characteristics of the medium. The following formulas illustrate this idea.

For *transverse* vibrations on a string of mass per unit length μ, under a tension T

$$v = \sqrt{\frac{T}{\mu}} \qquad (23\text{-}3)$$

For *longitudinal* vibrations in a fluid of bulk modulus B and mass per unit volume ρ

$$v = \sqrt{\frac{B}{\rho}} \qquad (23\text{-}4)$$

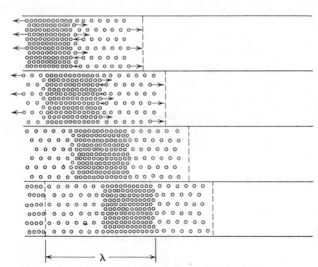

Figure 23-5   This series of diagrams depicts the motion of the compressions and rarefactions to the right. One wavelength contains one rarefaction and one compression.

For *longitudinal* vibrations in a solid of Young's modulus Y and mass per unit volume ρ

$$v = \sqrt{\frac{Y}{\rho}} \qquad (23\text{-}5)$$

For *longitudinal* vibrations of a gas (sound waves)

$$v = \sqrt{\frac{\gamma p}{\rho}} = \sqrt{\frac{\gamma RT}{M}} \qquad (23\text{-}6)$$

where γ = adiabatic constant (see Chapter 20), p = pressure, ρ = mass per unit volume, R = universal gas constant, T = absolute temperature, and M = mean molecular weight of the gas.

From Eq. 23-6 we see that the velocity of sound in air is a function of temperature. At STP the velocity of sound is 1090 ft/sec or 331 m/sec. This velocity increases by approximately 2 ft/sec or 0.6 m/sec for each Celsius degree rise in temperature. The approximation is only valid in the region between 0°C and room temperature.

Example 23-1   What is the velocity of transverse waves on a string that has a mass per unit length 0.01 g/cm and is under a tension of 1.6×10^7 dynes.

Solution.   From Eq. 23-3,

$$v = \sqrt{\frac{T}{\mu}}$$

$$= \sqrt{\frac{1.6 \times 10^7 \text{ dynes}}{0.01 \text{ g/cm}}} = \sqrt{16 \times 10^8 \text{ cm}^2/\text{sec}^2}$$

$$v = 4 \times 10^4 \text{ cm/sec} \qquad \blacktriangleleft$$

Example 23-2   It is required that in a certain solid, having a mass density of 2 g/cm³, longitudinal waves should have a velocity of 5×10^4 cm/sec. What should Young's modulus be for this solid?

Solution.   From Eq. 23-5,

$$v = \sqrt{\frac{Y}{\rho}} \qquad \text{and} \qquad v^2 = \frac{Y}{\rho}$$

$$Y = \rho v^2$$
$$= 2 \text{ g/cm}^3 \, (5 \times 10^4 \text{ cm/sec})^2 = 2(25 \times 10^8)$$
$$= 5 \times 10^9 \text{ dynes/cm}^2 \qquad \blacktriangleleft$$

Example 23-3   Calculate the speed of sound in air (mean molecular weight = 28.8 g/mole) at a temperature of 27°C.

Solution.   From Eq. 23-6, we have

$$v = \sqrt{\frac{\gamma RT}{M}}, \qquad \text{where } R = 8.31 \text{ joule/mole-K}°$$

In this example, $\gamma = 1.4$ because air is essentially a mixture of diatomic molecules, $T = 300°K$ and $M = 28.8 \times 10^{-3}$ kg/mole. Therefore,

$$v = \left(\frac{1.4 \times 8.31 \times 300}{28.8 \times 10^{-3}} \right)^{\frac{1}{2}}$$

$$= (12.1 \times 10^4)^{\frac{1}{2}}$$

$$= 348 \text{ m/sec} \qquad \blacktriangleleft$$

23-2 TRAVELING WAVES AND STANDING WAVES

What is the mathematical representation of a traveling wave? The derivation of such an equation is beyond the scope of this text, but it can be shown that

$$y = A \cos \frac{2\pi}{\lambda} (x - vt) \qquad (23\text{-}7)$$

is a representation of a traveling wave. In Eq. 23-7, A is the amplitude, or maximum displacement of any point on the wave; λ is the wavelength; v is the velocity of propagation; and y measures the displacement of the particle at position x at a time t. Since $v = \lambda/T$, Eq. 23-7 can also be written

$$y = A \cos 2\pi \left(\frac{x}{\lambda} - \frac{t}{T} \right) \qquad (23\text{-}8)$$

Figure 23-6 plots Eq. 23-8 for $t = 0, T/8, 2T/8 \ldots$, $4T/8$. It is easy to see that Eq. 23-7 or 23-8 represents a wave traveling to the right. If at a time

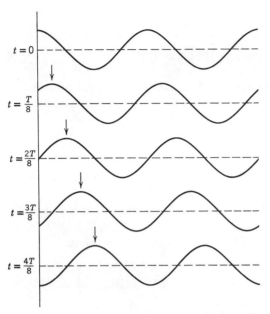

$t = 0$

$t = \dfrac{T}{8}$

$t = \dfrac{2T}{8}$

$t = \dfrac{3T}{8}$

$t = \dfrac{4T}{8}$

Figure 23-6   A traveling wave. Arrows depict the motion of the maximum toward the right.

$t + \Delta t$ (Δt is a small increase in time) we replace x by $x + \Delta x$, the displacement is

$$y = A \cos \frac{2\pi}{\lambda} [x + \Delta x - v(t + \Delta t)]$$

$$y = A \cos \frac{2\pi}{\lambda} [x - vt + \Delta x - v \, \Delta t]$$

Since

$$v = \frac{\Delta x}{\Delta t}$$

$$\Delta x = v \, \Delta t$$

Therefore,

$$y = A \cos \frac{2\pi}{\lambda} (x - vt)$$

Thus the waveform is exactly the same at $t + \Delta t$ when the wave has moved a distance Δx to the right ($+\Delta x$), as it was at time t.

If we had chosen as the representation

$$y = A \cos \frac{2\pi}{\lambda} (x + vt) \tag{23-9}$$

instead of Eq. 23-7, we would have had a wave that moves to the left. In general, the displacement of a particle in a disturbed medium (rope, spring, gas, solid, etc.) may be represented by a combination of terms such as appear in Eqs. 23-7 and 23-9 as well as similar expressions using a sine function instead of the cosine.

Let us now return to the motion of waves on a rope. Let the rope be fixed at both ends. A wave train initiated by plucking the rope will travel in both directions and be reflected from the fixed ends (Fig. 23-7). The resulting shape of the rope will be determined by a superposition of both the original wave train and the reflected waves

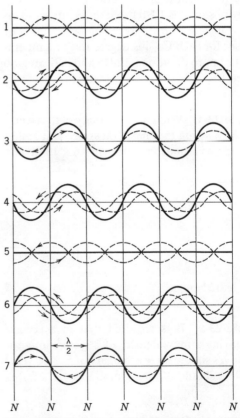

Figure 23-7   Method of adding two waves of equal lengths and equal amplitudes traveling in opposite directions to produce a stationary wave. Nodal points are labeled N.

and is given by the expression

$$y = 2A \cos 2\pi ft \sin \frac{2\pi x}{\lambda} \qquad (23\text{-}10)$$

Equation 23-10 sets a condition on the possible frequencies that can be maintained on the rope. Since $y = 0$ when $x = l$, we must have

$$\sin \frac{2\pi l}{\lambda} = 0$$

or $\qquad \dfrac{2\pi l}{\lambda} = n\pi \qquad n = 1, 2, 3, \ldots$

and $\qquad\qquad \lambda = \dfrac{2l}{n} \qquad\qquad (23\text{-}11)$

Thus, since the velocity of the waves is a constant for given density and tension, only those frequencies are possible for which

$$f = \frac{v}{\lambda} = \frac{n}{2l} v = \frac{n}{2l} \sqrt{\frac{T}{\mu}} \qquad (23\text{-}12)$$

The smallest possible frequency corresponds to $n = 1$, and is called the *fundamental*. All other possible frequencies are called *overtones*. If the ratio of the frequency of an overtone to that of the fundamental is an integer, the overtone is called a *harmonic*: an even harmonic if the integer is even and an odd harmonic if the integer is odd. The *quality* of sound that distinguishes the same frequency played on two different instruments is determined by the number and prominence of the overtones sounded with the fundamental. We note that a string fixed at both ends has an infinite number of *natural frequencies* of vibration in contrast to the examples of simple harmonic motion studied in Chapter 15. The systems that exhibited simple harmonic motion had only a single natural frequency of oscillation.

Returning to Eq. 23-10, we note that there are points on the string which never vibrate; i.e., $y = 0$ at all times for these points. The positions of these points are given by

$$\sin \frac{2\pi}{\lambda} x = \sin \frac{\pi n}{l} x = 0$$

or $\qquad \pi \dfrac{nx}{l} = m\pi \qquad m = 0, 1, 2, \ldots, n$

or $\qquad\qquad x = \dfrac{ml}{n} \qquad\qquad (23\text{-}13)$

For $m = 0$ and $m = n$ we get the points $x = 0$ and $x = l$, which are both fixed. There are also $(n - 1)$ additional points on the string that do not vibrate. For example, for $n = 3$ (second overtone, third harmonic), the points that do not vibrate are located at

$$x = 0, \frac{l}{3}, \frac{2l}{3}, l$$

Because there are points that never vibrate, Eq. 23-10 is said to represent *standing* or *stationary* waves. The points that have zero displacement are called *nodes* or *nodal points*. Between every two adjacent nodes there is a point that reaches the maximum displacement $2A$. These points are called *antinodes* (see Fig. 23-8).

Stationary waves can also be produced in tubes. Figure 23-9 shows a Kundt's tube, which is closed at one end and fitted at the other end with an adjustable piston. This piston is clamped at its midpoint so that the wavelength of the vibration in the rod will be easily calculable, as will be shown later. Cork dust has been spread along the inside of the tube. If the rod is pulled along its free end B, with a heavy cloth that has been rubbed with some resin, longitudinal waves will be set up in the rod. These waves will travel down the rod, causing the piston to vibrate, and it in turn will cause the air in the tube to vibrate. The motion of the air will cause the cork dust to collect in heaps, which are evenly spaced down the length of the tube. The dust is swept away by the air from the points of maximum vibration and comes to rest at the points of least activity. Thus, the distance between two dust heaps is one-half wavelength. If the velocity of sound in air at the temperature of the air in the tube is known, the frequency of the vibrations in the tube can be determined.

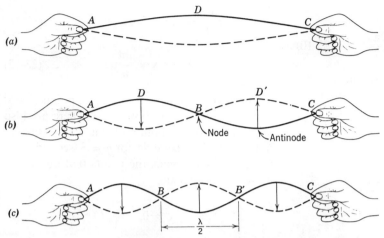

Figure 23-8   Examples of standing waves produced on a string fixed at both ends. (a) Fundamental mode with antinode at D and nodes at A and C. (b) First overtone with antinodes at D and D' and nodes at A, B, C. (c) Second overtone with nodes at A, B, B', C. (From O. H. Blackwood, W. C. Kelly, and R. H. Bell, *General Physics*, 2nd ed., John Wiley & Sons, Inc., New York, 1962, p. 200.)

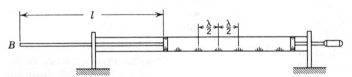

Figure 23-9   Kundt's tube apparatus.

Example 23-4   A 40-cm-long violin string is fixed at both ends and vibrates in the fifth harmonic. At what positions along the string will there be nodal points?

Solution.   We find the nodal positions by using Eq. 23-13. Because the string is vibrating in the fifth harmonic, $n = 5$ and $m = 0, 1, 2, 3, 4, 5$. Then

$$x = 0, \frac{l}{5}, \frac{2l}{5}, \frac{3l}{5}, \frac{4l}{5}, l$$

and the nodes are at

$$x = 0, 8 \text{ cm}, 16 \text{ cm}, 24 \text{ cm}, 32 \text{ cm}, 40 \text{ cm} \blacktriangleleft$$

Example 23-5   A standing wave is represented by the equation, $y = 2 \cos 160 \pi t \sin 0.02\pi x$, where t is measured in seconds and x and y are

measured in meters. Find the frequency, wavelength, and velocity of propagation of the standing wave.

Solution.   We compare the representation given for the wave with the general expression given in Eq. 23-10.
If we compare

$$y = 2A \cos 2\pi f t \sin \frac{2\pi}{\lambda} x$$

with $y = 2 \cos 160\pi t \sin 0.02\pi x$

we see immediately that $f = 80$ Hz. We note also that $2/\lambda = 0.02$ so that

$$\lambda = \frac{2}{0.02} = 100 \text{ m}$$

Therefore, $v = f \cdot \lambda = 80(100)$

$\qquad\qquad = 8000 \text{ m/sec}$ ◀

23-3 VIBRATIONS IN AIR COLUMNS

Consider the standing waves produced in a closed tube filled with air. Since the fixed end of the tube cannot vibrate, this point must be a displacement node. The air at the open end of the pipe has the greatest freedom of motion, so the open end is always an antinode. Thus the first possible frequency of vibration is the one for which the length of the tube corresponds to one-quarter of a wavelength (Fig. 23-10a). The next possible frequency occurs when we introduce one more node inside the pipe (Fig. 23-10b). The length of the tube now contains three-quarters of a wavelength. If we introduce one more node, the length of the tube will correspond to five-fourths of a wavelength. See Table 23-1. Thus the overtones in a closed tube or pipe are odd harmonics of the fundamental. There are no even harmonics as there were on the string fixed at both ends.

Figure 23-11 shows the fundamental vibration and first two overtones of an open pipe (organ, flute, trumpet, etc.). From Fig. 23-11a we can see that the fundamental wavelength is twice the length of the tube. By introducing nodes, one at a time, it becomes evident that the overtones consist of all the even and odd harmonics just as in the case of the fixed string. See Table 23-2.

Example 23-6   A closed pipe is 6 in. long. What is the frequency of the fundamental and first two overtones if the air temperature is 25°C? At 25°C

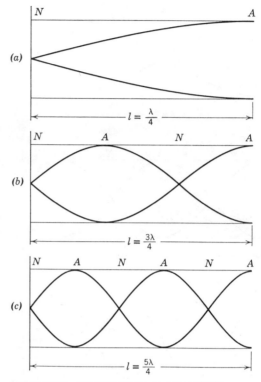

Figure 23-10   Standing waves in a closed tube.

the velocity of sound in air can be taken to be $v = 1090 \text{ ft/sec} + 25(2 \text{ ft/sec}) = 1140 \text{ ft/sec}.$

Solution.   The fundamental wavelength $= 4l = 4(\frac{1}{2} \text{ ft}) = 2 \text{ ft}.$ The wavelength of the first two overtones is $\frac{2}{3}$ ft and $\frac{2}{5}$ ft.

Thus

$$\text{fundamental frequency} = \frac{v}{\lambda_0} = \frac{1140}{2}$$

$$= 570 \text{ cps} = 570 \text{ Hz}$$

Table 23-1   Closed tube

Fundamental wavelength	$\lambda_0 = 4l$
First overtone	$\lambda_1 = \frac{4}{3}l$
Second overtone	$\lambda_2 = \frac{4}{5}l$
In general,	$\lambda_n = \dfrac{4l}{2n + 1} \qquad n = 0, 1, 2, 3 \ldots$

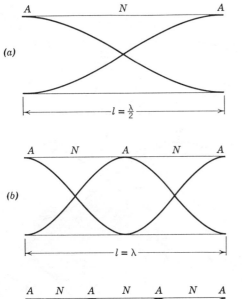

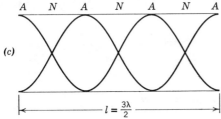

Figure 23-11   Standing waves in an open tube.

Table 23-2   Open tube

Fundamental wavelength	$\lambda_1 = 2l$
First overtone	$\lambda_2 = \frac{2}{2}l$
Second overtone	$\lambda_3 = \frac{2}{3}l$
In general,	$\lambda_n = \dfrac{2l}{n}$ $n = 1, 2, 3 \ldots$

$$\text{first overtone frequency} = \frac{1140}{\frac{2}{3}} = 1710 \text{ Hz}$$

$$\text{second overtone frequency} = \frac{1140}{\frac{2}{5}}$$

$$= 2850 \text{ Hz} \quad \blacktriangleleft$$

Example 23-7   Repeat the previous example for an open pipe 6 in. long.

Solution.   Here the first possible wavelength is

$$\lambda_1 = 2l = 2(\tfrac{1}{2} \text{ ft}) = 1 \text{ ft}$$

The wavelength of the next two overtones is

$$\lambda_2 = \tfrac{2}{2}l = \tfrac{1}{2} \text{ ft}$$
$$\lambda_3 = \tfrac{2}{3}l = \tfrac{1}{3} \text{ ft}$$

The frequencies are

$$f_1 = \frac{v}{\lambda_1} = \frac{1140 \text{ ft/sec}}{1 \text{ ft}} = 1140 \text{ Hz}$$

$$f_2 = \frac{v}{\lambda_2} = \frac{1140}{\frac{1}{2}} = 2280 \text{ Hz}$$

$$f_3 = \frac{v}{\lambda_3} = \frac{1140}{\frac{1}{3}} = 3420 \text{ Hz} \quad \blacktriangleleft$$

Example 23-8   What length of open pipe would give the same fundamental frequency as the closed pipe in Example 23-6 at the same temperature?

Solution.   Since the velocity is 1140 ft/sec and the frequency is to be 570 Hz, the required wavelength

$$\lambda = \frac{1140 \text{ ft/sec}}{570 \text{ Hz}} = 2 \text{ ft}$$

In an open pipe the fundamental wavelength is twice the length of the pipe. Hence

$$\lambda = 2l = 2 \text{ ft}$$
$$l = 1 \text{ ft}$$

Thus an open pipe must be twice as long as a closed pipe in order to give the same frequencies. On most musical instruments the length of the pipe can be changed by covering or uncovering holes along the pipe. This changes the effective length of the pipe and gives rise to different frequencies. In a trombone the length of the pipe is changed directly by sliding the movable metal tube in and out. ◀

23-4 FORCED VIBRATIONS

Suppose we strike a tuning fork with a rubber hammer and place the stem in contact with a table top. The intensity of the sound will be suddenly increased. The vibrations of the tuning fork are transmitted to the tabletop, forcing the molecules of the tabletop to vibrate. The vibrations, now taking place over a much larger surface area, are intensified. When the tuning fork is removed from the tabletop, the intensity quickly returns to its original level. The vibrations of the tabletop are said to damp out very quickly. The vibrations of the tabletop while in contact with the tuning fork are called *forced vibrations*.

23-5 SYMPATHETIC VIBRATIONS (RESONANCE)

Now consider identical boxes, each open at one end, arranged with open ends facing each other. Each box has a tuning fork mounted vertically from its top face, both forks having identical frequencies (see Fig. 23-12). If we strike one tuning fork with a rubber rod and then stop the vibrations by grasping the fork, the sound will persist. Careful observation will indicate that the second fork is now vibrating. The vibrations from the first fork set the column of air in the first cavity box to vibrating at the same frequency. These vibrations entered the second cavity and set the second

tuning fork to vibrating, again with the same frequency. Here is an example of forced vibrations that persist after the original disturbance has been removed. These vibrations are called *sympathetic vibrations*, and the second fork is said to resonate with the first. A system resonates to an incoming or incident frequency of vibration when the natural vibration frequency of the system is identical to that of the incident frequency.

Another easy demonstration of resonance can be performed with the apparatus of Fig. 23-13. After the tuning fork is struck with a hammer, the water level in the tube is quickly lowered by lowering tank T. When the level reaches point C, a sudden reinforcement of the sound takes place. To understand this, consider Fig. 23-14, which shows three positions of one of the prongs of the tuning fork. A cycle begins when the prong starts at A and moves down toward B. At this instant a compression is formed below the fork and sent down into the tube. The second quarter of the cycle has the prong moving from B back toward A. Here the halfway point is reached. If the compression sent out at the beginning of the cycle is reflected from the bottom of the air column and returns just as the prong is at A, at the halfway point, the motion of the prong through A will be aided and strengthened by this reflected compression. Thus the amplitude of this next wave will be increased and a louder sound, since the loudness depends on the amplitude, will be produced. Because the sound has to travel down the tube

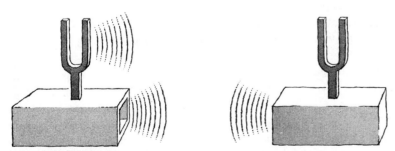

Figure 23-12   Demonstration of resonance by using two resonance boxes with identical tuning forks.

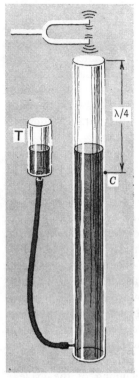

Figure 23-13 Resonances in a closed tube with variable length.

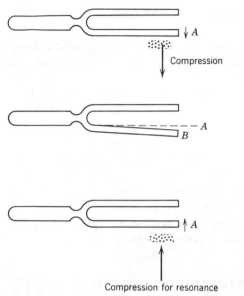

Figure 23-14 Motion of a tuning-fork prong for one-half cycle.

and back during one-half of a cycle, the length of the air column for resonance must be one-quarter of the wavelength being emitted by the tuning fork. If the tube is lengthened by one-half a wavelength, resonance will occur again. Now the reflected wave will return to A when the prong is moving upward the second time after producing the original compression. Clearly, we can continue to increase the length of the tube by half-wavelengths and find resonance points.

The term resonance will be encountered again in the study of electrical resonance in an ac circuit.

Example 23-9   A tuning fork is sounded above a tube filled with water. As the water is let out of the tube, resonances are heard when the water level is at two different points, measured to be 13.5 in. apart. The temperature of the air in the tube is measured at 17.5°C. What is the experimental value of the frequency of the tuning fork?

Solution.   At 17.5°C the velocity of sound in air is approximately 1125 ft/sec. The distance between the two resonances is a half-wavelength. Hence

$$\lambda = 2(13.5 \text{ in.}) = 27 \text{ in.} = 2.25 \text{ ft}$$

Therefore, $f = \dfrac{v}{\lambda} = \dfrac{1125}{2.25} = 500 \text{ Hz}$ ◀

23-6 INTERFERENCE AND BEATS

The result of the interaction of two or more tones of the same or slightly different frequencies that are heard simultaneously is called *interference*. In Fig. 23-15 point B is at a distance x from source C and at a distance $x + \Delta x$ from source D. Both sources C and D produce signals of the same amplitude and frequency. What is the resultant vibration at B; i.e., what would an observer at

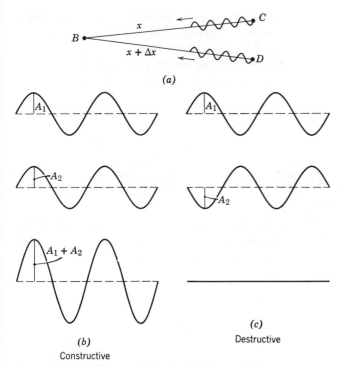

(a)

(b)
Constructive

(c)
Destructive

Figure 23-15 (a) The resultant effect at B is due to the superposition of the signals produced by sources C and D. (b) $\Delta x = n\lambda$. (c) $\Delta x = (n + \frac{1}{2})\lambda$.

B hear? Using Eq. 23-7, the resultant vibration at B is

$$y = y_C + y_D$$

$$y = A \cos \frac{2\pi}{\lambda}(x - vt)$$

$$+ A \cos \frac{2\pi}{\lambda}(x + \Delta x - vt) \quad (23\text{-}14)$$

The term Δx in the second member on the right-hand side of Eq. 23-14 is called the *path difference*. It measures how much farther one source is from the observer than the other. The phase difference between the vibrations from C and D is related to the path difference by

$$\delta(\text{phase difference})$$

$$= \frac{2\pi}{\lambda} \Delta x \text{ (path difference)} \quad (23\text{-}15)$$

There are two cases of special interest.

1. Suppose $\Delta x = n\lambda$, an integral multiple number of wavelengths. Then $\delta = 2n\pi$ and

$$A \cos \frac{2\pi}{\lambda}(x + \Delta x - vt)$$

$$= A \cos \left[\frac{2\pi}{\lambda}(x - vt) + 2n\pi\right]$$

$$= A \cos \frac{2\pi}{\lambda}(x - vt)$$

and $\qquad y = 2A \cos \frac{2\pi}{\lambda}(x - vt)$

This is known as *constructive interference*. The two sounds reinforce each other, producing a resultant sound of twice the amplitude of either individual tone.

2. Suppose $\Delta x = (n + \frac{1}{2})\lambda$, a half-integral multiple number of wavelengths. Then

$$\delta = (2n + 1)\pi$$

and

$$A \cos \frac{2\pi}{\lambda}(x + \Delta x - vt)$$

$$= A \cos \left[\frac{2\pi}{\lambda}(x - vt) + (2n + 1)\pi \right]$$

$$= -A \cos \frac{2\pi}{\lambda}(x - vt)$$

and, from Eq. 23-14 $y = 0$

This is called *destructive interference*. The two sounds cancel each other out, producing a tone of zero amplitude or no tone. Constructive and destructive interference are more readily observable with optical sources, and interference effects in light will be discussed in Chapter 38.

Figure 23-16 illustrates a second type of interference. Here we have two sources of slightly different frequency situated at equal distances from the observer. The resultant amplitude is found by adding the effect of each source. If the difference in the two frequencies is Δf, the resultant displacement at a point x will be

$$y = y_1 + y_2$$

$$= A \cos \left(\frac{2\pi x}{\lambda} - 2\pi ft \right)$$

$$+ A \cos \left[\frac{2\pi x}{\lambda} - 2\pi(f + \Delta f)t \right]$$

It has been assumed that the amplitude of the disturbance from each source is A. If Δf is small compared to f, it can be shown that

$$y = A(1 + \cos 2\pi \, \Delta ft) \cos \left(\frac{2\pi x}{\lambda} - 2\pi ft \right)$$

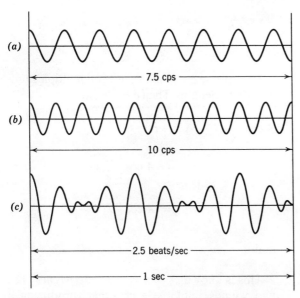

Figure 23-16   When the waves from (a) and (b) interfere constructively, we get a loud tone, represented by the larger amplitudes in (c). When the waves from (a) and (b) interfere destructively, we get the quiet periods represented by the smaller amplitudes in (c). These regular pulsations occur at a frequency equal to the difference between the frequencies of (a) and (b) and are called beats.

which represents a wave having a frequency f and an amplitude $A(1 + \cos 2\pi \Delta f t)$.

The amplitude is a maximum of $2A$ when $\cos 2\pi \Delta f t = 1$, and zero when $\cos 2\pi \Delta f t = -1$. The specific frequency with which the amplitude waxes and wanes between these two limiting values is clearly $\Delta f = f_2 - f_1$; i.e., there are $f_2 - f_1$ maxima and minima for the amplitude every second. This variation in amplitude which occurs when two frequencies that are nearly equal are sounded together is called *beating*. The number of beats heard per second is the difference between the two frequencies of the two original tones. Thus, if a 256 Hz tuning fork were sounded with a 250 Hz tuning fork, 6 beats/sec would be heard.

If the difference in frequencies is greater than 10 Hz, the beating is not very pronounced and the ear will not readily detect the effect. Beats are generally rhythmic pulses, but if there is a large difference between the two original frequencies, the resultant sound has a rough or discordant character. (Actually, when the frequency difference is greater than 200 Hz, the sound again becomes melodious with a combination tone being heard, the frequency of which is equal to the difference between the two frequencies.) If two tuning forks with a frequency difference of less than 10 Hz are not available, beats may be produced with two identical forks by placing a rubber band on one prong of one of the forks. This will slow up the fork just enough to produce audible beats.

23-7 VIBRATIONS IN RODS

When a rigid bar is stroked along its axis, longitudinal vibrations are set up in the bar. Figure 23-17 illustrates the first two possible modes of vibration for a rod clamped at its center. Note that when the rod is clamped at its center and free at both ends, the fundamental wavelength in the rod is twice the length of the rod. In the Kundt's tube experiment discussed earlier, the frequency of the vibration, transferred from the clamped rod to the

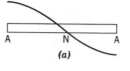

(a)

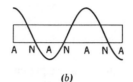

(b)

Figure 23-17 Vibrations in a rigid rod free at both ends and clamped at the center. The nodes and antinodes are indicated for the fundamental vibrations (*a*) and the first overtone (*b*).

air in the tube, is calculated. From a knowledge of the wavelength of the sound in the metal rod, the velocity of sound in the rod can be calculated. The velocity of sound in metals is from 10 to 15 times greater than the velocity of sound in air.

Example 23-10 In a Kundt's tube experiment, a 100-cm metal rod is used. The velocity of sound in air is known to be 340 m/sec and the average distance between the heaps of cork dust is 8.5 cm. Find the velocity of sound in the metal rod.

Solution. The wavelength of the sound in the tube is $2(8.5 \text{ cm}) = 17 \text{ cm}$ and in the rod it is equal to 200 cm. The frequency is the same in the rod as it is in the air. Thus

$$f = v_{air}/\lambda_{air} = v_{rod}/\lambda_{rod}$$
$$v_{rod} = (\lambda_{rod}/\lambda_{air})v_{air}$$
$$= (200 \text{ cm}/17 \text{ cm})\ 340 \text{ m/sec}$$
$$v_{rod} = 4000 \text{ m/sec}$$

SUMMARY

Wave motion is the method by which energy is transferred from one place to another without the transfer of mass.

The *wavelength* is the shortest distance between two points on a wave moving exactly in phase.

The *frequency* of a wave represents the number of waves passing a particular point per unit time.

The *velocity* of a wave is always related to the wavelength and frequency

$$v = \lambda f$$

In a *transverse wave*, the direction of propagation of the wave is perpendicular to the direction of vibration of the medium.

In a *longitudinal wave*, the direction of propagation is parallel to the direction of vibration of the medium.

A *sound wave* is composed of *compressions* (regions where molecules are packed together) and *rarefactions* (regions with relatively few molecules).

The velocity of waves in a medium depends on the characteristics of the medium:

$$v = \sqrt{\frac{T}{\mu}} \qquad \text{transverse vibrations on a string}$$

$$v = \sqrt{\frac{B}{\rho}} \qquad \text{longitudinal vibrations in a fluid}$$

$$v = \sqrt{\frac{Y}{\rho}} \qquad \text{longitudinal vibrations in a solid}$$

$$v = \sqrt{\frac{\gamma RT}{M}} \qquad \begin{array}{l}\text{longitudinal vibrations in a gas}\\ \text{(sound waves)}\end{array}$$

Expressions for Special Waves

Traveling wave moving toward right:

$$y = A \cos \frac{2\pi}{\lambda}(x - vt)$$

Traveling wave moving toward left:

$$y = A \cos \frac{2\pi}{\lambda}(x + vt)$$

Standing wave: $y = 2A \cos 2\pi f t \sin \dfrac{2\pi x}{\lambda}$

Standing waves are possible in a given structure for many different frequencies, called *natural frequencies*:

Fixed string of length l:

$$f = n\left(\frac{v}{2l}\right), n = 1, 2, 3 \ldots$$

Open pipe of length l:

$$f = n\left(\frac{v}{2l}\right), n = 1, 2, 3 \ldots$$

Closed pipe of length l:

$$f = n\left(\frac{v}{4l}\right), n = 1, 3, 5, 7 \ldots$$

Rigid rod of length l, fixed at center:

$$f = n\left(\frac{v}{2l}\right), n = 1, 3, 5, 7 \ldots$$

The lowest frequency at which a structure can maintain a standing wave is called the *fundamental* frequency or tone. Multiples of the fundamental are called *odd or even harmonics*, depending upon whether they are odd or even multiples. All frequencies greater than the fundamental tone are called *overtones*.

Forced vibrations occur when a system is caused to vibrate at a frequency other than one of its natural frequencies. When the cause of vibrations is removed, the forced vibrations stop. When the system continues to vibrate after the cause of vibrations has been removed, we say that the system is *resonating* with the initial vibration. This occurs only when the initial vibration is one of the natural frequencies of the system.

Beats are a variation in the amplitude of the sound which results when two frequencies that are nearly the same are sounded together.

QUESTIONS

1. How is wave motion different from the types of motion previously studied?

2. In what way are longitudinal waves different from transverse waves?

3. What are compressions? Rarefactions?

4. From your knowledge of how sound waves carry energy in a gas, explain why the speed of sound should be greater in solids than liquids and greater in liquids than gases.

5. How are traveling waves different from standing waves?

6. What is meant by the *quality* of a sound?

7. Distinguish clearly between *harmonic* and *overtone*.

8. What are beats and how are they produced?

9. What are sympathetic vibrations?

10. Describe an experiment that can be used to demonstrate resonance.

11. How is Kundt's tube used to measure the speed of sound in metals?

PROBLEMS

(A)

1. What is the velocity of sound in air at a temperature of 12°C?

2. The wavelength of a transverse wave on a string is 80 cm. The frequency of vibration is 1000 Hz. The mass of the string is 0.02 g/cm. Find the velocity of the wave motion and the tension in the string.

3. A rope is under a tension of 490 newtons and has a mass of 0.001 kg/m. Find the velocity of transverse waves on the rope.

4. An open pipe is 16 in. long. What are the frequencies of the fundamental and first three overtones at a temperature of 15°C?

5. The density of a certain solid is 12 g/cm^3 and Young's modulus is 3×10^{12} dynes/cm^2. Find the velocity of sound in this material.

6. What is the speed of sound in oxygen (molecular weight 32) at a temperature of 600°C? Assume that $\gamma = 1.4$.

7. The lowest note on a pipe organ is 16.35 vib/sec. What length of closed pipe gives this note at 20°C?

8. A closed pipe is 4 ft long. What are the frequencies of the first three overtones at a temperature of 20°C?

9. The density of copper is 8.8 g/cm^3. Determine the speed of sound in copper. Young's modulus for copper is 1.1×10^{12} dynes/cm^2.

10. In a resonance experiment a closed tube of movable length is found to resonate with a tuning fork when the tube is first 6 in. and then 18 in. long. If the temperature of the air in the tube is 5°C, find the frequency of the tuning fork.

11. What is the frequency of a sound wave moving in air of temperature 15°C when the wavelength is 2 ft?

12. Compute the wavelength of the sound produced by a 288-Hz tuning fork vibrating in air at a temperature of 25°C.

13. Find the velocity of transverse waves in a wire that is under a tension of 800 newtons. The wire has a mass of 0.02 g/cm.

14. The wavelength of a transverse wave on a string is 60 cm. The frequency of the vibration is 500 Hz. What is the velocity of the wave motion? What is the tension in the string if the mass of the string is 0.04 g/cm.?

15. A string is 60 cm long and has a mass of 3 g. What must be the tension in the string if the frequency of the first overtone for transverse waves is 200 vib/sec.?

16. A string of mass 0.5 g and length 100 cm is under a tension of 44.1 newtons.
    (a) Find the velocity of transverse waves in the string.
    (b) What are the frequencies of the fundamental and first three overtones?

(B)

1. Find the velocity of longitudinal compression waves in water. ($B = 2 \times 10^{10}$ dynes/cm^2.)

2. What is the speed of sound in helium gas at 900°C? The molecular weight of helium is 4.0 and $\gamma = 1.66$.

3. What is the velocity of sound in carbon dioxide at STP? ($\gamma = 1.30$.)

4. In a Kundt's tube experiment the average distance between dust heaps is found to be 7 cm. The rod is 1 m long and clamped at its midpoint. If the speed of sound in air is 332 m/sec, determine the frequency and velocity of sound in the metal rod.

5. A bar 9 m long is clamped at its center. When vibrating in its first overtone, the bar resonates with a tuning fork marked 800 Hz. Compute the speed of sound in the bar.

6. The equation of a traveling transverse wave on a string is known to be

$$y = 4 \cos 0.01\pi(x - 400t)$$

where y is in cm, x is in cm, and t is in seconds. By placing the equation in the form

$$y = A \cos \frac{2\pi}{\lambda}(x - vt)$$

determine:
(a) The amplitude.
(b) The velocity.
(c) The wavelength.
(d) The period.

7. If the string in Problem 6 has a mass of 0.02 g/cm, what is the tension in the string?

8. A steel wire has a fundamental frequency of vibration of 400 Hz. By how much will the frequency change if the tension in the wire is increased by 5%?

9. A long aluminum rod is struck with a hammer at one end. The sound is heard, by an observer 2 km away, through the air and through the rod. What is the time interval between sounds?

10. Observers on board a ship witness an underwater explosion. The sound reaches them through the water 3 sec before the sound arrives through the air. How far away is the sight of the explosion?

11. An open pipe is sounded at the same time as a closed pipe. The open pipe is 85 cm long and the closed pipe is 45 cm long. How many beats per second will be heard if the speed of sound is 340 m/sec?

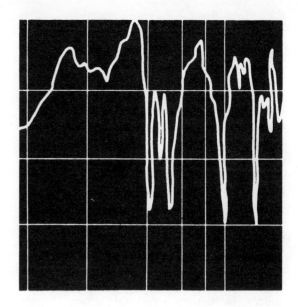

24.
Intensity and
Quality of
Sound Waves

In this chapter we discuss the effect on relative motion between source and observer on the apparent frequency heard by the observer. This phenomenon is known as the *Doppler effect*. The intensity of sound and the measurement of intensity levels are also analyzed. The chapter concludes with a description of high fidelity systems and ultrasonics.

LEARNING GOALS

This chapter has four main learning objectives:

1. *To study the Doppler effect and learn to use the equations to calculate apparent frequencies.*
2. *To learn the meaning of intensity and quality of sound.*
3. *To study how intensity levels are measured.*
4. *To learn some of the features of the hearing process and high fidelity systems.*

24-1 DOPPLER EFFECT

One important aspect of wave motion is the change in the frequency and wavelength if the source has a velocity relative to the observer. Perhaps you have noticed that a train whistle or siren seems to increase in pitch as it approaches, and decrease in pitch as it passes by. When the pitch is increasing, the frequency increases; when the pitch is decreasing, the frequency decreases. Thus a shift toward a longer wavelength is observed with a spectroscope when pure red light is moving away from the observer. When the light is approaching, the red shifts toward the orange, which has a shorter wavelength or higher frequency. In modern aerial navigation, Doppler radar uses radio waves to determine the precise position of an airplane by the Doppler shift.

The simplest way to see a Doppler effect change the wavelength and frequency of a wave is in a transparent tray of water. In Fig. 24-1 a vibrator sets up periodic circular waves. Now, if the vibrator is moved, the waves on the leading side (in the direction of motion) are shorter and the frequency is higher. Also note that on the trailing side the wavelength is greater and the frequency is less.

When a source of sound approaches an observer, the pitch increases; when a source of sound moves away from an observer, the pitch decreases. Let us call the frequency of the sound emitted from the source, f. As the sound approaches, the number of waves per second reaching the ear increases. Let f' represent this increased frequency. In t seconds, the source emits $(f)(t)$ waves. During this time the first of these waves has traveled a distance vt, where v is the speed of sound, and the source has traveled $v_s t$, where v_s is the speed of the source. Therefore, all of the $(f)(t)$ waves are crowded into a distance $vt - v_s t$. This squeezing effect results in a decrease of the wavelength. The new wavelength is

$$\lambda' = \frac{\text{length of wave train}}{\text{number of waves}}$$

$$= \frac{(v - v_s)t}{(f)(t)} = \frac{v - v_s}{f}$$

Therefore, the increased frequency is

$$f' = \frac{v}{\lambda'} = \left(\frac{v}{v - v_s}\right)f \qquad (24\text{-}1)$$

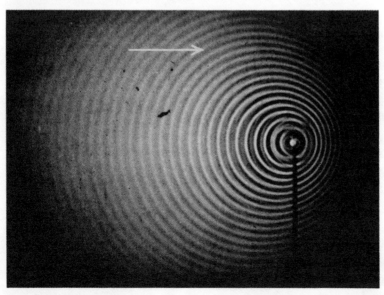

Figure 24-1 Doppler effect.

Example 24-1 A tuning fork with a frequency of 256 Hz is moving toward an observer at a velocity of 4 m/sec. What frequency is heard by the stationary observer? (The speed of sound in air at 0°C is 331 m/sec, or 1090 ft/sec).

Solution. To find the apparent frequency, we use Eq. 24-1,

$$f' = \frac{v}{v - v_s} f$$

$$= \left(\frac{331}{331 - 4}\right) 256 \text{ Hz}$$

$$= 259 \text{ Hz} \qquad \blacktriangleleft$$

If the source were moving away from the observer, the length of the wave train after a time t would be $(v + v_s)t$ and the wavelength would have been increased so that

$$\lambda' = \frac{(v + v_s)t}{(f)(t)} = \frac{v + v_s}{f}$$

Then, the apparent frequency heard by the observer would be

$$f' = \frac{v}{\lambda'} = \left(\frac{v}{v + v_s}\right) f \qquad (24\text{-}2)$$

Equation 24-2 shows that the apparent frequency heard by the observer when the source is moving away from him is less than the frequency emitted by the source.

Equations 24-1 and 24-2 can be written as one equation

$$f' = \frac{v}{v \pm v_s} f \qquad (24\text{-}3)$$

where it is to be remembered that the positive sign in the denominator is used when the source recedes from the observer and the negative sign is used when the source approaches the observer.

Example 24-2 What is heard by the observer in the previous example when the source moves away from the observer?

Solution. Using Eq. 24-2, we find that

$$f' = \frac{v}{v + v_s} f$$

$$= \left(\frac{331}{331 + 4}\right) 256 \text{ Hz}$$

$$= 253 \text{ Hz} \qquad \blacktriangleleft$$

Consider next what occurs when the observer moves with a speed v_0 and the source remains stationary. If the observer moves toward the source, more waves reach the observer each second. Less waves reach the observer each second if the observer moves away from the source. The speed of sound waves as seen by the observer is not v. The speed of sound relative to the moving observer is $(v + v_0)$ if he moves toward the source, and $(v - v_0)$ if he moves away from the source. Therefore,

$$f' = \frac{v + v_0}{\lambda} = \frac{v + v_0}{v/f} = \left(\frac{v + v_0}{v}\right) f \qquad (24\text{-}4)$$

when the observer approaches the source, and

$$f' = \frac{v - v_0}{\lambda} = \frac{v - v_0}{v} f \qquad (24\text{-}5)$$

when the observer moves away from the source.

Example 24-3 A man is approaching a stationary fire siren that is producing waves with a frequency of 512 Hz. He is moving at a speed of 20 m/sec. What is the observed frequency? Assume that the velocity of sound is 331 m/sec.

Solution. Using Eq. 24-4, we obtain

$$f' = \frac{v + v_0}{v} f$$

$$f' = \left(\frac{331 + 20}{331}\right) 512 \text{ Hz}$$

$$f' = 543 \text{ Hz}$$

If the observer were moving away from the siren at the same speed, we would find the apparent frequency from Eq. 24-5.

$$f' = \left(\frac{v - v_0}{v}\right) f$$

$$f' = \left(\frac{331 - 20}{331}\right) 512 \text{ Hz}$$

$$f' = 481 \text{ Hz} \qquad \blacktriangleleft$$

Equations 24-4 and 24-5 can be written as a single equation

$$f' = \frac{v \pm v_0}{v} f \qquad (24\text{-}6)$$

where the positive sign in the numerator is used when the observer approaches the source, and the negative sign is used when the observer moves away from the source.

When both source and observer are moving, the appropriate expression for the apparent frequency heard by the observer is a combination of Eqs. 24-3 and 24-6,

$$f' = \left(\frac{v \pm v_0}{v \pm v_s}\right) f \qquad (24\text{-}7)$$

where the correct choice of signs is determined by the relative motions (Table 24-1).

Table 24-1   The four common relative motions of observer and source and their apparent frequencies.

Source	Observer	Equation
(a) $\bullet\!\!\rightarrow v_s$	$\circ\!\!\rightarrow v_0$	$f' = \left(\dfrac{v - v_0}{v - v_s}\right) f$
(b) $\bullet\!\!\rightarrow v_s$	$v_0 \leftarrow\!\!\circ$	$f' = \left(\dfrac{v + v_0}{v - v_s}\right) f$
(c) $v_s \leftarrow\!\!\bullet$	$v_0 \leftarrow\!\!\circ$	$f' = \left(\dfrac{v + v_0}{v + v_s}\right) f$
(d) $v_s \leftarrow\!\!\bullet$	$\circ\!\!\rightarrow v_0$	$f' = \left(\dfrac{v - v_0}{v + v_s}\right) f$

Example 24-4   A tuning fork has a frequency of 400 Hz when the fork is stationary. The fork is moved toward an observer with a velocity of 6 m/sec. The observer moves toward the fork with a speed of 19 m/sec. What is the observed frequency of the sound if the velocity of sound is 331 m/sec?

Solution.   The problem is solved directly by using Eq. 24-7. Since the source and the observer are moving toward each other, we use the form of the equation indicated in Table 24-1(b).

$$f' = \frac{v + v_0}{v - v_s} f$$

$$= \left(\frac{331 + 19}{331 - 6}\right) 400 \text{ Hz}$$

$$= \left(\frac{350}{325}\right) 400 \text{ Hz}$$

$$f' = 431 \text{ Hz} \qquad \blacktriangleleft$$

When objects are moving at supersonic speeds, the source is moving faster than the speed of sound. When this occurs, we get a wave form similar to that in Fig. 24-2b. If v' is the velocity of the supersonic source and v is the velocity of sound, the half-angle of the cone (θ) in Fig. 24-2a can be found by noting that

$$\sin \theta = \frac{vt}{v't} = \frac{v}{v'} \qquad (24\text{-}8)$$

The ratio (v/v') is the reciprocal of the *Mach number*.

Example 24-5   A jet plane moving at supersonic speed creates a bow wave similar to Fig. 24-2, with $\theta = 37°$. If the speed of sound in air is 350 m/sec, how fast is the jet moving?

Solution.   We can solve for the velocity of the plane by using Eq. 24-8.

$$\sin \theta = \frac{v}{v'}$$

$$\sin 37° = \frac{350 \text{ m/sec}}{v'}$$

$$v' = \frac{350 \text{ m/sec}}{\sin 37°}$$

$$= \frac{350}{0.6} \text{ m/sec}$$

$$v' = 583 \text{ m/sec} \qquad \blacktriangleleft$$

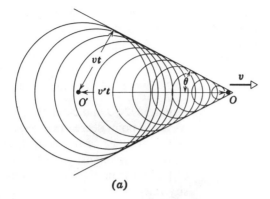

(a)

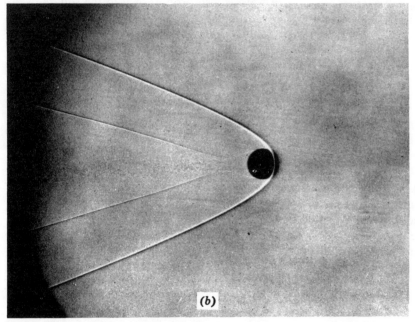

(b)

Figure 24-2 (a) A group of wave fronts associated with a projectile moving with supersonic speed. The wave fronts are spherical and their envelope is a cone. (b) A spark photograph of a projectile undergoing this motion. (U.S. Navy photograph. From R. Resnick and D. Halliday, *Physics for Students of Science and Engineering*, John Wiley & Sons, Inc., New York, 1962, p. 443.)

Example 24-6 A supersonic jet is flying at a speed of Mach 1.5 (i.e., 1.5 times the speed of sound). Find the half-angle of the cone-shaped bow wave that it leaves behind.

Solution. The Mach number is equal to the ratio v'/v which, from Eq. 24-8, is the reciprocal of $\sin\theta$. Therefore,

$$\frac{v'}{v} = 1.5 = \frac{1}{\sin\theta}$$

$$\sin\theta = \frac{1}{1.5} = \frac{2}{3}$$

$$\theta = \sin^{-1}\tfrac{2}{3} = \sin^{-1}(0.667)$$

From the table of trigonometric values in the appendix, we find $\theta \approx 42°$. ◀

24-2 INTENSITY OF SOUND

Waves transmit energy as they move from one point in space to another. The rate at which energy is transported is called the *power*. The *intensity* of a wave is the power transmitted per unit cross-section area (Fig. 24-3).

Intensity is measured in units of watts per m². The intensity of a wave decreases as the wave moves further away from the source. At any particular distance r from the source, the total power emitted by the source is spread over a sphere of radius r surrounding the source. Therefore, the intensity at this distance is the power divided by the area of the sphere,

$$I = \frac{\text{power}}{\text{area}} = \frac{P}{4\pi r^2} \qquad (24\text{-}9)$$

Equation 24-9 indicates that the intensity of a wave at a point decreases inversely with the square of the distance of the point from the source. A wave with an intensity of 10^{-8} W/m² at a distance of 1 meter from the source, would have an intensity one-hundredth as large (10^{-10} W/m²) at a distance of 10 m from the source. The intensity of a wave is also proportional to the square of the wave amplitude. A wave with twice the amplitude would have an intensity four times as great.

Acoustic experts have defined the threshold intensity of sound to be 10^{-12} W/m². This intensity is denoted by the symbol I_0. To the human ear, a sound 10 times as intense as I_0 does not seem 10 times louder. We therefore measure sound intensity on a logarithm scale. The unit for measuring intensity levels is called the *bel* (B), in honor of Alexander Graham Bell.

$$\text{intensity level (in bels)} = \log_{10}\frac{I}{I_0} \qquad (24\text{-}10)$$

In Eq. 24-10 the intensity level is defined as the logarithm (base 10) of the relative intensity of the sound compared to the threshold intensity. Because the bel turns out to be a very large unit, we introduce the *decibel*, which is one-tenth of a bel. There are 10 decibels to the bel. Therefore,

intensity level (in decibels)

$$= 10 \text{ intensity level (bels)}$$

$$= 10 \log_{10}\frac{I}{I_0} \qquad (24\text{-}11)$$

The intensity level in decibels is indicated by the symbol dB.

Example 24-7 Two sound waves have intensities of $I_1 = 5 \times 10^{-10}$ W/m² and $I_2 = 5 \times 10^{-12}$ W/m². How many decibels louder is one sound than the other?

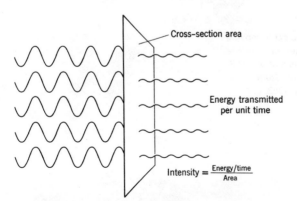

Cross–section area

Energy transmitted per unit time

$$\text{Intensity} = \frac{\text{Energy/time}}{\text{Area}}$$

Figure 24-3 The intensity of a wave is the rate at which its energy is transported across unit cross section area.

Solution. We can get the dB level for each intensity and then take the difference. By the use of Eq. 24-11,

$$dB_1 = 10 \log \frac{I_1}{I_0} = 10 \log \frac{5 \times 10^{-10}}{10^{-12}} = 10 \log 500$$

$$dB_2 = 10 \log \frac{I_2}{I_0} = 10 \log \frac{5 \times 10^{-12}}{10^{-12}} = 10 \log 5$$

$$dB_1 - dB_2 = 10(\log 500 - \log 5)$$
$$= 10 \log \tfrac{500}{5} = 10 \log 100$$
$$= 20$$

The same result could have been found by comparing the two intensities directly. Thus

$$dB_1 - dB_2 = 10 \log \frac{I_1}{I_2}$$

$$= 10 \log \frac{5 \times 10^{-10}}{5 \times 10^{-12}}$$

$$= 10 \log 100$$
$$= 20$$

Example 24-8 Two sound waves have intensities of 50 and 1000 microwatts/cm^2 each. How many decibels louder is one sound than the other?

Solution. Comparing the two sounds directly, we obtain

$$dB_1 - dB_2 = 10 \log \frac{I_1}{I_2}$$

$$dB_1 - dB_2 = 10 \log \frac{(1000 \text{ microwatts/cm}^2)}{(50 \text{ microwatts/cm}^2)}$$

$$= 10 \log 20$$

$$dB_1 - dB_2 = 10(1.3010) = 13.01 \text{ dB} \quad \blacktriangleleft$$

One ten-millionth of a watt, or 10^{-7} watts, of power in sound is just above the bare threshold to which a normal ear will respond. Thus at close to 0 dB we can just barely hear the sound. In a quiet room the noise level is on the order of 10 dB, in a quiet garden the level is 20 dB. In a quiet street with no traffic, we hear about 30 dB, however, in the so-called quiet street of a city at night, the level is 40. Now, 40 dB is not 40 times the standard but $10 \times 10 \times 10 \times 10$ or 10^4 times. A moving car about 10 to 20 ft away is heard at 50 dB. A large, busy store averages 60 dB, and heavy traffic on the street about 70 dB. Talking to a friend in a normal conservational tone takes place at a level somewhere between 60 and 70 dB. An elevated railroad with a train passing is about 80 dB. The roar of Niagara Falls varies from 80 to 90 dB. A compressor-driven air drill breaking up pavement is at about 90 dB heard from only a few feet away. The sound level of a riveting hammer in a shipyard is 115 dB. The roar of an airplane engine idling at 1000 rev/min and just under 20 ft away is 115 dB. Decibels approaching 100 are painful to the ear. Audible noise that is distracting can affect productivity and safety in industrial and research plants. Acoustical engineers and technicians design rooms to reduce or eliminate noise. Special sound-absorbing materials can be used on walls, ceilings, and floors or can be built into buildings under construction.

24-3 FREQUENCY RESPONSE OF THE EAR

The average human being with normal hearing can hear frequencies from 20 Hz to 16,000 Hz. Frequencies above 16,000 Hz can rarely be heard by humans. Increasing age normally causes a decrease in the ability to hear higher frequencies. The sounds that are generated by speech are made of many different frequencies. In addition, the decibel level of each of the frequencies is different. Fourier, a leading French mathematician of the early nineteenth century, developed a theory that enables us to separate out and determine the frequencies and amplitudes present in any complicated wave. The theory states that the sound was simply the sum of many sine waves. No matter how complicated the sound wave, these sine waves can be isolated by the method we today call Fourier analysis.

We have all heard many different kinds of instruments playing the same note. The same note, middle C (256 vibrations per second), played on different instruments is readily distinguished. Even without being able to recognize the frequency, it is definitely possible to tell a violin from a piano, a clarinet, or a triangle all playing the same note. This quality that distinguishes the sounds of one instrument from another is called *timbre*. Timbre depends on the number and prominence of the overtones produced by a particular instrument.

In the preceding chapter we studied the interference of waves. Sound waves can produce both destructive and constructive interference. Cancellation or destructive interference occurs at positions that are one-half wavelength apart. In some auditoriums there are dead spots caused by destructive interference.

24-4 HIGH FIDELITY

What do people mean when they use the term high fidelity or hi-fi when they refer to a radio receiver, record player, or tape recorder? From a layman's point of view, what the customer wants is a system that reproduces the music faithfully and as close to the original live playing or singing as possible. Unfortunately, the average human

ear is not a good judge of these qualities. With human voices, especially in speech, we want to be able to recognize the voice.

In Fig. 24-4, the incoming signals may be supplied by a tuner, a microphone tape recorder head, or a magnetic cartridge of the pickup head. Where does the distortion take place that reduces the faithfulness of reproduction? First, very few loudspeakers in use are truly of high fidelity. In Fig. 24-5 the response curve of two speakers is shown. The second one is called a flat response, but it is flat only by comparison with the other curve. In frequencies where the poor speaker reproduces only the abrupt variations, the deeper peaks and valleys are seen. One factor that is often overlooked by amateurs is that records have a mechanical problem in the grooves that limits the fidelity of reproduction. This limit does not exist for live radio or a tape recorder. We are all well acquainted with the record noise known as "scratch." In tape this is called "frying."

In stereophonic systems two signals are sent out from the original recorded two sets of signals of the same sound with different locations. In theory, if you place yourself in the correct position between the two speakers of the stereo set, a sym-

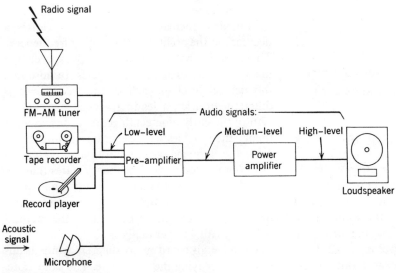

Figure 24-4 High fidelity system components.

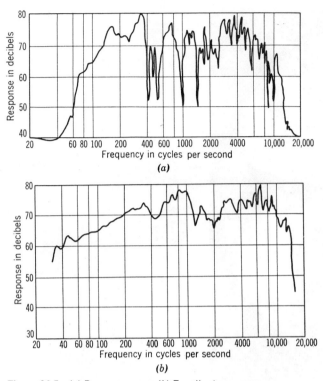

Figure 24-5 (*a*) Poor response. (*b*) Excellent response.

phony will sound like it did in the concert hall. With high quality stereo, this is true. But in a concert hall if you turn your head, you notice no change in the music. Sometimes, with stereo, if you turn your head, one signal will be heard with more amplitude than the other, and you will get the effect of a single speaker. However, in a properly designed room in which the speakers are correctly placed you can get the total effect of music in a concert hall.

24-5 ULTRASONICS

Let us start by defining the terms *supersonic* and *ultrasonic*. We use the term *ultrasonic* for sounds with frequencies above the normal hearing range. The term *supersonic* is applied only to the motion of objects, such as that of an airplane traveling faster than the speed of sound (Fig. 24-6).

Although ultrasonic sounds can be made by designing a whistle with a very short wavelength (high frequency), or by mechanical means, today almost all ultrasonic generators are crystals. There are many crystals that can be used. The crystal has its atoms set into vibration by applying a voltage. In some cases a high frequency radio wave puts the crystal structure in vibration.

Among the more common everyday uses of ultrasonics is the cleaning of parts of equipment. An ultrasonic drill can cut holes of any shape. The shape of the hole is determined by the shape of the vibrating part although the cutting tool itself does not touch the material being cut. A cutting powder is set into vibration between the tool bit and the stock, and the cutting powder does the work. Ultrasonics can also be used to quickly test the quality of forgings or castings.

Sonar, the detection of ships, submarines, and of schools of fish by ultrasonic waves, is used

Figure 24-6   Supersonic shock wave produced by air streaming past model plane at 1.5 times the speed of sound in a wind tunnel. (NASA photo)

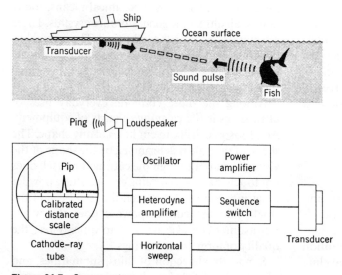

Figure 24-7   Sonar system.

underwater as is radar in space. The time it takes the signal to be emitted and then reflected (as in Fig. 24-7) is converted into distance. This is possible because the speed of sound in water is known for that depth and temperature of water.

SUMMARY

When a source of sound is moving relative to an observer, the frequency of the sound heard by the observer is different from the true frequency of the source. This phenomenon is referred to as the *Doppler effect*. The apparent frequency heard by the observer will depend upon the relative motion of the observer and the source according to Table 24-2:

Table 24-2

Observer Stationary

(a) Source moving toward observer: $f' = \dfrac{v}{v - v_s} f$

(b) Source moving away from observer: $f' = \dfrac{v}{v + v_s} f$

Source Stationary

(a) Observer moving toward source: $f' = \dfrac{v + v_0}{v} f$

(b) Observer moving away from source: $f' = \dfrac{v - v_0}{v} f$

If the observer and the source are both moving, the apparent frequency is

$$f' = \frac{v \pm v_0}{v \pm v_s} f$$

where the choice of signs in the numerator and denominator depends on the relative locations of the observer and source (Table 24-1).

The ratio of an object's speed through a medium to that of sound in that medium is called the *Mach number*. An object moving faster than the speed of sound in the medium creates a bow wave. The half-angle of the conelike wave depends on the ratio of the velocity of sound, v, to that of the object, v':

$$\sin \theta = \frac{v}{v'}$$

The *intensity* of a sound wave at a point in space is the power transmitted through the unit area at that point. Intensity levels are measured logarithmically, relative to a standard intensity called the threshold. The threshold intensity is 10^{-12} watts/m^2, generally denoted by the symbol I_0. Relative to I_0, other intensity levels are measured in decibels according to the equation:

$$dB = 10 \log_{10} \frac{I}{I_0}$$

The average human being with good hearing can detect frequencies from approximately 20 to 16,000 Hz.

Ultrasonic refers to sounds whose frequencies are above the normal hearing range.

QUESTIONS

1. Describe the Doppler effect.
2. What are the metric units for power and intensity?
3. How is the bel defined? The decibel?
4. Why do we use a logarithmic scale to measure intensity levels?
5. What is the intensity at the zero level of human hearing adopted by acoustic experts?
6. What is the range of frequencies that can be sensed by the average human ear?
7. Describe the bow waves created by a jet flying at supersonic speeds.
8. What do we call the quality that distinguishes the same sounds played on different instruments?
9. Describe some factors that can limit the quality of high fidelity systems.
10. What are some of the practical industrial uses being made of ultrasonic systems today?

PROBLEMS

(A)

1. A tuning fork vibrates with a frequency of 280 Hz.
   (a) How many wavelengths are produced in 5 sec?
   (b) If the sound travels at 330 m/sec, how far does the sound travel in the 5 sec?

2. The source in Problem 1 is moving with a velocity of 30 m/sec.
   (a) How far has the source moved in 5 sec?
   (b) What is the apparent wavelength of the emitted sound in front of the moving source?
   (c) What is the apparent wavelength of the emitted sound to the rear of the moving source?

3. Refer to Problem 2.
   (a) What is the apparent frequency heard by a stationary observer in front of the moving source?
   (b) What is the apparent frequency heard by a stationary observer located to the rear of the moving source?

4. A glider is moving at 100 km/hr directly toward a stationary siren whose frequency is 600 Hz. If the speed of sound is 340 m/sec, what is the frequency heard by the pilot of the glider?

5. A car approaches a stationary source that is emitting a signal of frequency 256 Hz. If the car is moving at 50 miles/hr, what apparent frequency is heard by the car passengers? (Assume that the speed of sound is 750 miles/hr.)

6. A train whistle is emitting a signal with a frequency of 300 Hz. The train is moving toward a platform with a speed of 80 ft/sec. The air temperature is 15°C. What frequencies are heard by a person on the station platform as the train rushes past the station?

7. How fast must an observer be running toward a source vibrating at 250 Hz if the frequency appears to be 1 Hz higher? Assume that the speed of sound is 330 m/sec.

8. A source is vibrating at 250 Hz. How fast must the source move toward a stationary observer for the observer to detect an increase of 1 Hz in the frequency? Assume the speed of sound to be 331 m/sec.

9. A supersonic jet creates a bow wave of half-angle 30°. How fast is the jet moving if the speed of sound is 750 miles/hr?

10. A bow wave is established by a jet plane moving at 720 m/sec. If the speed of sound is 340 m/sec, find the half-angle of the bow wave.

11. What is the ratio of intensities of two sounds if one is 12 dB louder than the other?

12. How much more intense is a sound that is 100 dB louder than the threshold intensity?

13. Two sounds have intensities of 10^{-9} and 10^{-3} W/m². Calculate the difference in intensity level in decibels.

(B)

1. Train A is moving west at a speed v_A and train B is moving east at a speed v_B. Train B is emitting a continuous sound of frequency 250 Hz. A passenger on the train moving west measures the sound from train B at a frequency of 300 Hz as it approaches, and 210 Hz after it passes. If the speed of sound is 340 m/sec, how fast is each train moving relative to the ground?

2. A student, carrying a 300-Hz tuning fork, runs toward a smooth, stationary reflecting surface, at a speed of 20 ft/sec. How many beats per second will the student hear?

3. Starting with Eq. 24-2, show that $v_s/v = \Delta f/f'$, where $\Delta f = f - f'$. What is the velocity of the source if a signal with a true frequency of 256 Hz appears to have a frequency of 250 Hz to a stationary observer? Assume the speed of sound to be 330 m/sec.

4. Starting with Eq. 24-4, show that $v_0/v = \Delta f/f$, where $\Delta f = f' - f$. What is the velocity of the observer if he notices an increase in frequency of 2%?

5. What is the half-angle of the bow wave produced by a supersonic jet moving at Mach 2.5?

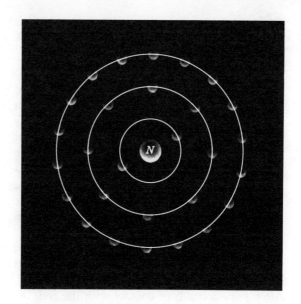

25.
Coulomb's Law

This chapter begins the study of electromagnetism, which includes electricity and magnetism and their interrelationship. The study begins with the nature of *electrification* and the law of force between stationary charges. In this chapter we shall also study the Rutherford model of the atom and the difference between *conductors* and *insulators*.

LEARNING GOALS

The learning objectives for this chapter are

1. *To learn the definitions of positive and negative charge.*

2. *To understand the basic features of the Rutherford model of the atom.*

3. *To learn the function of an electroscope.*

4. *To recognize the difference between conductors and insulators.*

5. *To understand the process of charging an object by induction.*

6. *To learn to use Coulomb's law to calculate the electrical force between pairs of charged objects.*

25-1 CHARGE

Undoubtedly, you have witnessed how a hard rubber comb, after having been brushed through the hair, attracts tiny pieces of paper. You may have speculated about the "magical powers" that the comb acquires after being in contact with the hair. As a matter of fact, a glass rod rubbed in a silk cloth will have exactly the same effect on the pieces of paper. Yet, it can be shown that the behavior of the glass rod is quite different from that of the rubber comb.

When a rubber rod is rubbed in wool or cat's fur and touched to two small, metallically coated pith balls, the pith balls are attracted to the rod. After a few seconds of contact with the rod, the balls are repelled away from the rod and are seen to repel each other (see Fig. 25-1). The same sequence of events takes place when a glass rod is rubbed in silk and brought into contact with two pith balls. However, a pith ball that has been in contact with a rubber rod will be attracted to one that has been in contact with a glass rod (see Fig. 25-2). Thus, although the two phenomena appear to be identical, there is a fundamental difference between the nature of the rubber rod after contact with the cat's fur and the glass rod after contact with the silk. We say that the rubber and glass rods are *charged* and define two kinds of charge. The charge possessed by a rubber rod after being rubbed in cat's fur is defined to be

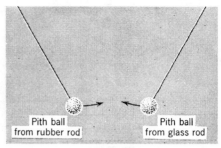

Figure 25-2 Attraction between oppositely charged pith balls.

negative charge. The charge that a glass rod has attained after being rubbed in silk is defined to be *positive* charge. Benjamin Franklin defined objects charged opposite to glass as negative and those charged the same as glass as positive. What are these charges and where do they come from?

Scientists have learned from experiments, such as the Rutherford experiments to be discussed in the next section, that atoms are composed of a positively charged central core, called the *nucleus*, surrounded by negatively charged particles. The total negative charge of the surrounding particles, which move about the nucleus like earth satellites move about the earth, is equal to the positive core. This is why the atom is said to be *neutral*. The negatively charged particles, called *electrons*, do not move about the nucleus in haphazard fashion. Electrons may only revolve about the nucleus in orbits (here assumed circular) with specific radii. The number of electrons that may occupy any one orbit is limited. For example, only two electrons may occupy the first orbit, eight the second, etc. We say that the electrons occupy *shells* about the nucleus, the first shell holding two electrons, the second eight, and so on (see Fig. 25-3).

The simplest element in nature is hydrogen, which has one positive charge, called a proton, in its nucleus and one electron in the first shell. The next element, helium, has two protons in the nucleus and two electrons in the first shell, completing that shell. Lithium has three protons in the nucleus, two electrons in the first shell, and the third electron in the second shell. (There are

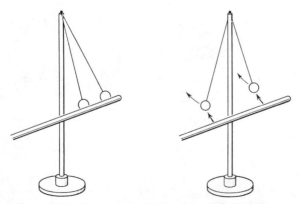

Figure 25-1 After contact with the charged rod the pith balls repel the rod and each other.

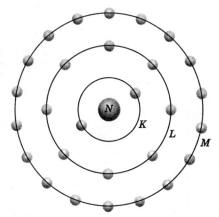

Figure 25-3 First three electron shells, *K*, *L*, and *M*. At the center is the nucleus, *N*.

charged. The reverse is true for the glass rod and silk. The glass rod gives up electrons and becomes positively charged, whereas the silk gains electrons and becomes negatively charged.

When the pith balls are in contact with the rubber rod, some of the excess electrons move onto the pith balls. The pith balls then become negatively charged and are repelled by the negatively charged rod and by each other. When the pith balls are in contact with the glass rod, some of the electrons from the pith balls are drawn to the glass rod to help neutralize the positive charge. The pith balls are thus left positively charged and are repelled by the rod. However, a negatively charged pith ball is attracted to a positively charged ball. Thus, experimentally, we see that like charges repel each other while unlike charges attract each other.

other particles in a nucleus with which we do not presently concern outselves.) In such a manner, with a complete knowledge of the number of electrons permissible in each shell, we can build up a table showing the distribution of electrons in the various shells for each element.

When the rubber rod is rubbed in the wool, electrons are removed from the wool, which tends to release those electrons that are not in complete shells, and are deposited on the rubber rod. The wool now has more positive charges in all of its nuclei than negative charges. The wool is then positively charged. The rubber, on the other hand, has more electrons than protons and is negatively

25-2 RUTHERFORD ATOM

The picture of the atom as having a strong positive core surrounded by external electrons was first offered by Rutherford in 1911. He was led to this theory after studying the experimental results obtained by Marsden and Geiger on the scattering of alpha particles (helium nuclei) by thin foils of gold and silver. These two experimentalists found that some alpha particles were scattered more than 90° (see Fig. 25-4). In order for such a large

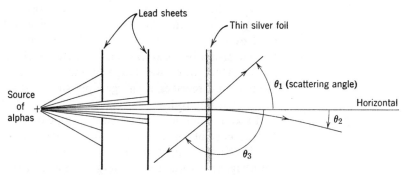

Figure 25-4 The beam of alpha particles coming from a radioactive source is narrowed (collimated) by several sheets of lead. Some alphas are scattered through angles larger than 90° (θ_3).

scattering angle to be possible, Rutherford reasoned that there must be a strong central core of positive charge in the atoms of the foil and that the electrons of an atom are situated outside the nucleus. There had been other proposals regarding the model of the nucleus. One model suggested that the electrons move among the protons in the nucleus proper, occupying instantaneous positions much like the raisins in a rice pudding. For such a model, however, the intensity of the strong positive charge would be offset by the electrons, and 90° scattering angles would not be possible. More scattering experiments were carried out in 1913 and 1920, and the results substantiated Rutherford's theory.

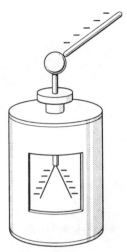

Figure 25-6   Charging the electroscope by contact.

25-3 THE ELECTROSCOPE

A device used to detect the presence of charge is called an *electroscope*. A simple electroscope is pictured in Fig. 25-5. It consists of a metal rod *A* which is round at one end and connected to two flat gold leaves *B* at the other end. This assembly is suspended in a cylinder that has two windows for viewing the leaves. The knob at the end of the metal rod is insulated from the cylinder by a plastic stopper. When a charge is placed on the knob, it spreads to the leaves which then diverge, since like charges repel (see Fig. 25-6). This is called charging by direct contact.

Figure 25-5   A simple uncharged electroscope.

25-4 INDUCTION

The electroscope may also be charged indirectly by induction. Suppose that a negatively charged rod is brought near the knob, but not in contact with it. The electrons in the metal rod will be repelled to the leaves of the electroscope (see Fig. 25-7a). Now suppose that we draw off these electrons by providing a path to the ground (known as *grounding*). This may be done by touching the knob with a finger (see Fig. 25-7b). We now remove the path to ground, leaving the electroscope with a net positive charge. This positive charge is only on the knob, because the negatively charged rod has not yet been removed (see Fig. 25-7c). When we remove the rod, the positive charge on the knob becomes generally distributed over the metal rod and the gold leaves of the electroscope, and the leaves diverge (Fig. 25-7d). Thus, when we charge by induction, the electroscope receives a charge opposite to that of the charging rod.

It should be noted that an electroscope detects the presence of charge but does not measure how much charge is present. (We have, in fact, not yet assigned a value to the charge on an electron or proton.) However the leaves of an electroscope continue to diverge as the charge on the electro-

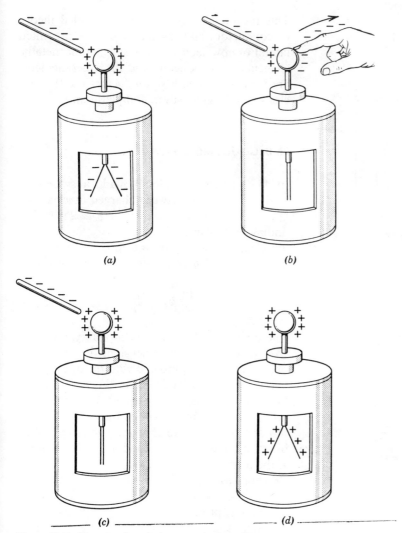

Figure 25-7   Charging the electroscope by induction.

scope is increased. Thus, the force of repulsion between charges appears to be proportional to the amount of charge present on each leaf.

25-5 CONDUCTORS AND INSULATORS

Suppose that we connect the knob of an electroscope to an insulated metal rod by a piece of silk thread (see Fig. 25-8). Touch the metal rod *A* with a charged rod. Nothing unusual appears to happen at the electroscope. Now replace the silk thread by a copper wire (see Fig. 25-9). If we again touch the metal stand with a charged rod, the leaves on the electroscope diverge. Evidently, the copper wire conducts charge from the stand to the electroscope and the silk thread does not. Materials through which charges are readily conducted are called *conductors*; materials that do not conduct charge are called *insulators*.

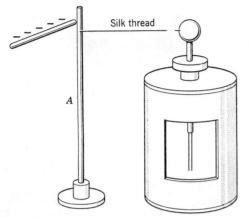

Figure 25-8 Demonstration of an insulator.

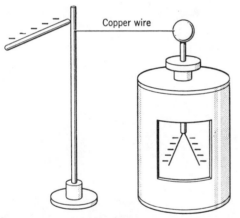

Figure 25-9 Flow of charge through a conducting wire.

When a charge is placed on an insulator, it remains localized. The charges do not distribute themselves throughout the material. On a conductor, however, the charges move about rather freely and quickly distribute themselves throughout the conductor. Actually, the designation of certain substances as conductors and others as insulators is relative. Many insulators become conductors under special conditions. For example, you have seen sparks jump across air gaps. Air is an insulator, but under certain conditions its resistance to the passage of charge is broken.

This results in a spark. A more detailed theory of conductors and insulators will be presented later. For now, let it suffice to say that the metallic elements are conductors and the other elements are insulators. The best conductors are the precious metals: gold, silver, and platinum.

25-6 COULOMB'S LAW

In 1789 Charles Coulomb made a systematic study of the force between charged objects. The apparatus he used is shown in Fig. 25-10. His findings result in an inverse square law of force for electrostatics. Thus the force between two charges, q and q', separated by a distance r, is given by

$$F = \frac{kqq'}{r^2} \qquad (25\text{-}1)$$

where k is a constant of proportionality. Equation 25-1 assumes that the dimensions of the charged objects are small compared with the distance between the objects. The force is directed along the line joining the centers of the charges. The force on q, produced by q', will be directed toward q' if q and q' are unlike charges. The force will be directed away from q' if q and q' are like charges. If the direction of the force is known, the sign of the charges need not be included in the application of Eq. 25-1. Because of Newton's third law, the force on q', produced by q, is equal in magnitude and oppositely directed to the force on q, produced by q'.

There are two systems of units for defining charge and the constant k. In the cgs system, k is defined to be unity in a vacuum, and one electrostatic unit of charge (esu) is one which, when placed in a vacuum a distance of one centimeter away from a second similar charge, experiences a force of repulsion of one dyne. The esu is often called a *statcoulomb*.

In the mks system the unit of charge, the *coulomb*, is derived from experiments on moving charges that define the ampere. Here we relate the coulomb to the statcoulomb. One coulomb

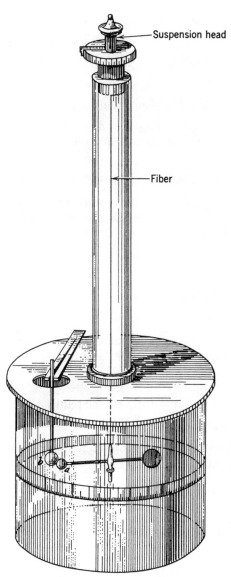

Figure 25-10 Coulomb's torsion balance from his 1785 memoir to the French Academy of Sciences. (From R. Resnick and D. Halliday, *Physics for Students of Science and Engineering*, John Wiley & Sons, Inc., New York, 1962.)

of charge is equivalent to 3×10^9 statcoulomb. The constant k for the mks system is determined by experiment and found to be equal numerically to 9×10^9. The units of k are force $\times$ distance^2/

charge^2. Thus

$$k = 9 \times 10^9 \, \frac{\text{newton-meter}^2}{\text{coulomb}^2} \qquad (25\text{-}2)$$

Equation 25-1 represents the force on a charged particle and is thus a vector equation. If the force on a charged particle is caused by two or more other charges, then contributions must be added vectorially to obtain the resultant force.

Example 25-1 Two charges, each of a magnitude of 9×10^{-5} coulomb, are separated by the distance of 30 cm. Find the force of repulsion between them.

Solution. From Eq. 25-1

$$F = k\frac{qq'}{r^2}$$

$$= \frac{9 \times 10^9 (9 \times 10^{-5})(9 \times 10^{-5})}{(0.3)^2}$$

$$= \frac{729 \times 10^{-1}}{(0.09)} = 810 \text{ newtons} \qquad \blacktriangleleft$$

Example 25-2 The force of repulsion between two identical positive charges is 0.9 newton when the charges are 1 m apart. Find the value of each charge.

Solution. We can use Coulomb's law, Eq. 25-1, to find the charge.

$$F = \frac{kqq'}{r^2} = \frac{kq^2}{r^2}$$

$$0.9 \text{ N} = 9 \times 10^9 \, \frac{\text{N} - \text{m}^2}{\text{C}^2} \cdot \frac{q^2}{(1 \text{ m})^2}$$

$$0.9 = 9 \times 10^9 \, q^2$$

$$q^2 = \frac{0.9}{9 \times 10^9} = 10^{-10} \text{ C}^2$$

$$q = 10^{-5} \text{ C} \qquad \blacktriangleleft$$

Example 25-3 A charge of -2×10^{-6} C exerts a force of 720 newtons on a charge of 6×10^{-6} C. How far apart are the charges?

Solution. We use Eq. 25-1 and solve for the distance of separation.

$$F = \frac{kqq'}{r^2}$$

$$-720 \text{ N} = 9 \times 10^9 \frac{\text{N} - \text{m}^2}{\text{C}^2}$$

$$\cdot \frac{(-2 \times 10^{-6})(6 \times 10^{-6} \text{ C})}{r^2}$$

$$720 = \frac{(9)(12) \cdot 10^{-3}}{r^2}$$

$$r^2 = \frac{108 \times 10^{-3}}{720} = 1.5 \times 10^{-4} \text{ m}^2$$

$$r = 1.22 \times 10^{-2} \text{ m} \qquad \blacktriangleleft$$

Example 25-4 A charge of 8×10^{-6} coulomb (8 microcoulomb = 8 μC) is placed at one corner of an equilateral triangle of side 1 meter. A second charge of magnitude -8 μC is placed at one of the other corners. What would be the resultant force on a charge of 6 μC when placed at the vacant corner (see Fig. 25-11)?

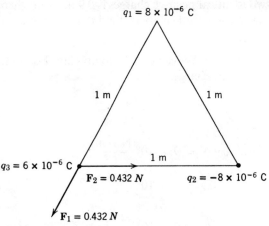

$q_1 = 8 \times 10^{-6}$ C

1 m   1 m

$q_3 = 6 \times 10^{-6}$ C

1 m

$F_2 = 0.432 \text{ } N$ $q_2 = -8 \times 10^{-6}$ C

$F_1 = 0.432 \text{ } N$

Figure 25-11

Solution. Charge $q_1 = 8 \times 10^{-6}$ C exerts a force on $q_3 = 6 \times 10^{-6}$ C which is directed away from q_1. The magnitude of the force is given by Eq. 25-1.

$$F_1 = \frac{kqq'}{r^2}$$

$$= \frac{9 \times 10^9(8 \times 10^{-6})(6 \times 10^{-6})}{1^2}$$

$$= 0.432 \text{ newton (repulsion)}$$

The force exerted by $q_2 = -8 \times 10^{-6}$ C on q_3 will have the same magnitude and will be directed toward q_2.

$$F_2 = 0.432 \text{ newton (attraction)}$$

To find the resultant force, F_1 must be resolved into its horizontal and vertical components (Fig. 25-12a). Thus

$$F_{1x} = F_1 \sin 30° \text{ (to the left)}$$

$$= -(0.432)(0.5) = -0.216$$

$$F_{1y} = F_1 \cos 30 \text{ (down)}$$

$$= -(0.432)(0.866) = -0.374$$

The components of the resultant force are found by adding the components of F_1 and F_2.

$$R_x = F_{1x} + F_2 = -0.216 + 0.432$$

$$= 0.216 \text{ (to the right)}$$

$$R_y = F_{1y} \qquad = -0.374 = 0.374 \text{ (down)}$$

The resultant magnitude is now determined from the right triangle in Fig. 25-12b.

$$R = \sqrt{R_x{}^2 + R_y{}^2}$$

$$R = \sqrt{(0.216)^2 + (-0.374)^2}$$

$$R = 0.432 \text{ newton}$$

at an angle θ such that

$$\tan \theta = \frac{R_y}{R_x}$$

$$\tan \theta = \frac{-0.374}{0.216} = -1.73$$

and $\theta = 60°$ below the x axis.

How does the charge on a single electron or proton compare with the coulomb? The Millikan experiment, which will be discussed in Chapter 26, demonstrated that the charge on one electron is

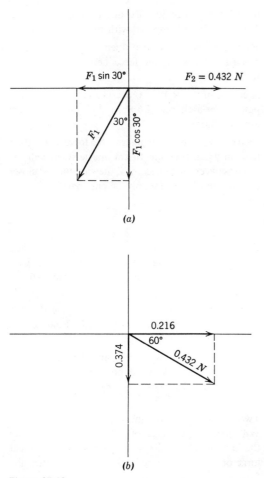

(a)

(b)

Figure 25-12

showed that a strong positive nuclear core was responsible for the large scattering angles, had as its basic assumption the Coulomb law of force. It is known now that the inverse-square law of force for electrostatics is correct within an error of one part in 10^9.

SUMMARY

There are two types of charge: *positive* and *negative*. Matter is usually in an electrically neutral state. If an object has more of one charge than the other, we saw the object is *charged*. The force between charges is calculated with the aid of *Coulomb's law*:

$$F = \frac{kqq'}{r^2}$$

Like charges repel each other and unlike charges attract each other.

An object can become charged by direct contact or through the process of *induction*. An *electroscope* is a device that can be used to detect the presence of charge. Materials that readily allow the movement of charge through them are called *conductors*. Nonconducting materials are called *insulators*.

The Rutherford picture of an atom consists of a positively charged nucleus surrounded by negative charges (electrons) moving around the nucleus much like the planets move around the sun.

The elementary unit of charge is the charge on the electron or proton. This charge is 1.6×10^{-19} coulomb.

equivalent to 1.6×10^{-19} coulomb. This charge, which is the same for one proton, is denoted by the symbol e. Thus

$$e = 1.6 \times 10^{-19} \text{ coulomb}$$
$$1 \text{ coulomb} = 6.25 \times 10^{18} \text{ electrons}$$

One coulomb of charge is the aggregate charge of 6.25×10^{18} electrons.

Before closing this chapter we should state that Rutherford's success in explaining alpha-scattering experiments constitutes one proof of Coulomb's law (Eq. 25-1). Rutherford's theoretical derivation of the scattering angle, in which he

QUESTIONS

1. Describe an experiment which shows that a rubber rod can pick up pieces of paper.

2. Draw a picture showing how an atom of helium might look.

3. How many positive charges would be in the nucleus of the neutral atom depicted in Fig. 25-3?

4. Describe how a rubber rod and wool can be used to charge a pith ball positively.

5. What experimental fact led Rutherford to the conclusion that the center of an atom must be strongly positive?

6. Two bodies attract each other. Must both bodies be charged? Must they both be charged if they repel each other?

7. Which materials are the best electrical conductors? Which are the best conductors of heat? Is there a general relationship between the two processes?

8. Draw a series of diagrams showing how a neutral object would be attracted to a nearby charged object.

PROBLEMS

(A)

1. Two charges of -2×10^{-8} coulomb and 1.8×10^{-8} coulomb are 0.2 m apart in a vacuum. What is the electrostatic force between the charges? Is it repulsive or attractive?

2. When a 10 μC (1 $\mu C = 10^{-6}$) charge is placed 0.10 cm from a charge q in a vacuum, the force between the charges is 0.9 newtons. What is the charge q?

3. Two charges of $+4$ μC and -4 μC are placed 1 m apart. What is the electrostatic force between them?

4. A positron is a positively charged electron. What is the attractive force between an electron and a positron separated by 10^{-9} m?

5. An alpha particle consists of two protons and two neutrons (no charge). What is the repulsive force between two alpha particles that are separated by 3 millimicrons (1 millimicron = 10^{-9} m)?

6. What is the coulomb force of repulsion between a sodium nucleus and a chlorine nucleus separated by 10^{-8} cm? (See the periodic table in the appendix at the end of the book.)

7. A sphere has a charge of 16 μC. How many excess electrons are on the sphere?

8. How many excess electrons must be placed on a body to give it a charge of 2×10^{-8} C?

9. An electron and proton are separated by 5×10^{11} m.
   (a) What is the gravitational force between them?
   (b) What is the electrical force between the charges?

(c) Would you be justified in ignoring the gravitational force compared with the electrical force?

10. How far apart are two identical charges ($q = 5 \times 10^{-6}$ C) when the force between them is 45 newtons?

11. A charge of -3×10^{-9} C exerts a force of 2×10^{-3} newton on a charge of 6×10^{-8} C. How far apart are the charges?

12. Two charges repel each other with a force of 4×10^{-3} newton when they are 10 cm apart. What will the force between them become if they are moved closer together until they are only 2.5 cm apart?

(B)

1. Charges of -2, 3, and -6 μC are placed at the vertices of an equilateral triangle of side 8 cm. Calculate the resultant force acting on the -6 μC charge.

2. Four equal charges of 2×10^{-8} C are placed at the corners of a square 20 cm on a side. Find the resultant force on each charge.

3. Two charges of 16×10^{-8} C and 4×10^{-8} C are held 8 cm apart. At what point between the two charges will a charge of -4×10^{-8} C experience no net force?

4. Two charges of 5 μC and 6 μC are 10 cm apart. Where should a charge of -4 μC be placed along the line joining the first two charges so that the net force on the 6 μC charge is zero? What is then the force on the -4 μC charge?

5. Two equally charged spheres are suspended from a common point by strings 10 cm long. When the system is in equilibrium, the spheres are 16 cm apart. Each sphere has a mass of 0.1 g. Determine the charge on each sphere.

6. Three equally charged spheres are suspended from a common point by strings 10 cm long. When the system is in equilibrium, the spheres are 16 cm apart. Each sphere has a mass of 0.1 g. Determine the charge on each sphere.

7. What is the force between an electron and a proton in a hydrogen atom when they are separated by a distance of 5×10^{-9} cm? If this force produces the centripetal acceleration of the electron as it circles the proton, how many times per second does the electron orbit the proton? (Electron mass = 9.1×10^{-31} kg.)

8. A movable 0.02 g pith ball has a charge of 10 μC. The ball is held 50 cm from a charge of 40 μC.
   (a) What is the coulomb force acting on the pith ball?
   (b) If the ball is released, with what initial acceleration will it begin to move?
   (c) Does this acceleration remain constant? Explain.

9. Two equal charges are separated by a distance of x meters. If the charges are brought 2 m closer together, the force between them is doubled. Find the original distance of separation.

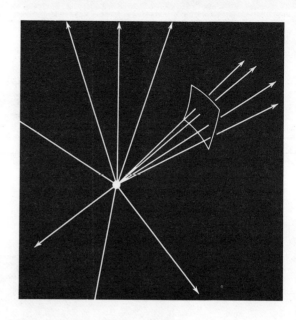

26.
Electric Fields and
Electric Lines of Force

In this chapter we develop the concept of an *electric field* and introduce basic techniques for calculating the intensity of the electric field produced by a set of charges. The representation of the electric field by drawings of the *lines of force* is explained and the characteristics of the electric field intensity within and near conductors are discussed. The chapter concludes with an analysis of *Millikan's experiment*, which led to the discovery of the magnitude of the elementary charge on an electron.

LEARNING GOALS

While studying this chapter, the reader should concentrate on the following major objectives:

1. *To understand the definition of electric field intensity.*

2. *To learn to calculate the electric field intensity produced by one or more point charges.*

3. *To understand the concept of lines of force and the principle of Gauss' law.*

4. *To learn how to apply Gauss' law to several examples featuring symmetric charge distributions.*

5. *To learn the essential properties related to charged conductors and the electric field in and on the surface of conductors.*

6. *To understand Millikan's experiment to determine the elementary charge.*

26-1 ELECTRIC FIELD INTENSITY

The electrostatic force discussed in the previous chapter is an example of an "action-at-a-distance" force. There is no direct contact between the two charges, as there is between a hammer and the nail it is striking.

It is generally advisable to revise physical theories that rely upon "action-at-a-distance" forces by introducing a mathematical concept which allows us to reinterpret such forces as contact forces. Such mathematical concepts are called *fields*.

To help understand the concept of a field, consider the following. A very large canvas sheet is stretched taut in a horizontal plane. A very large mass is deposited at the center of the sheet, causing a sag at the center and a general downhill sloping throughout the sheet (see Fig. 26-1). A second, smaller mass is placed somewhere on the sheet. It is observed that this smaller mass rolls down the incline toward the larger mass at the center. There are now two possible ways of explaining the behavior of the smaller mass.

1. The smaller mass was attracted by the larger mass and was thus accelerated toward the center.

2. The smaller mass was caused to move by the deformation in the sheet (the incline), the nature of the force exerted by the deformation

being somehow related to the existence of the larger mass.

The first explanation views the cause of the motion as an "action-at-a-distance" force. The second, however, *makes that part of the canvas directly in contact with the smaller mass* the immediate cause of the motion. The deformations in the canvas play the role of the field. The first mass, by its presence, establishes the field which in turn acts on the second mass.

In the case of electrostatics we introduce the electric field by defining its intensity. The electric field intensity, $\mathbf{E}$, is a vector quantity, whose direction at any point in the field is the direction in which a *positive* charge would move if placed at that point, and whose magnitude is given by

$$E = \frac{F}{q'} \qquad (26\text{-}1)$$

where F is the magnitude of the electrostatic force acting on the positive charge q'. The units of E are dynes/statcoulomb (in the cgs system) or newtons/coulomb (in the mks system).

Since the positive charge q' would move in the direction of $\mathbf{F}$, the electric field intensity is in the same direction as the electrostatic force and Eq. 26-1 can be rewritten in vector form,

$$\mathbf{E} = \frac{\mathbf{F}}{q'} \qquad (26\text{-}2)$$

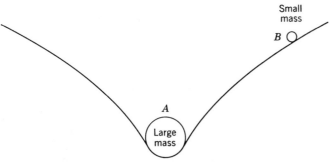

Small mass

B

A

Large mass

Figure 26-1 Cross section of stretched sheet depressed at center by a large mass (*A*). It is now assumed that the second mass (*B*) is much closed to *A* than to any other mass and is therefore affected only by *A*.

From Coulomb's law, the electrostatic force between a pair of point charges q and q' is given by

$$F = \frac{kqq'}{r^2}$$

Then Eq. 26-1 becomes

$$E = \frac{F}{q'} = \frac{kqq'}{r^2q'}$$

or $\qquad E = \frac{kq}{r^2}$ $\qquad$ (26-3)

Equation 26-3 gives the *magnitude of the electric field intensity produced by a point charge q.*

If q is a positive charge, any positive charges near it will be repelled. Hence the electric field produced by q points in the direction away from q if q is positive. If q is negative, any positive charges near q will be attracted toward q. Hence the electric field produced by q points in the direction toward q if q is a negative charge. Figure 26-2a shows the direction of E produced by a positive charge, and Fig. 26-2b indicates the direction of E produced by a negative charge.

From Eq. 26-2 it is clear that the force exerted by an electric field on a charge q' is

$$\mathbf{F} = q'\mathbf{E} \qquad (26\text{-}4)$$

If q' is positive, the force on q' is in the direction of the electric field $\mathbf{E}$. If q' is negative, Eq. 26-4 indicates that the force is in the direction opposite to the direction of $\mathbf{E}$.

Example 26-1   What is the electric field intensity 3 m away from a charge of 9×10^{-8} coulomb?

Solution.   From Eq. 26-3

$$E = \frac{kq}{r^2} = \frac{9 \times 10^9 \times 9 \times 10^{-8}}{3^2}$$

$$= 90 \text{ newtons/coulomb}$$

in the direction away from the charge, since q is positive. ◄

Example 26-2   Using the data of Example 1, find the force acting on a charge of -9×10^{-8} coulomb placed 3 m away from the first charge.

Solution.   From Eq. 26-4

$$\mathbf{F} = q'\mathbf{E} = -9 \times 10^{-8}(90)$$

$$= -810 \times 10^{-8} \text{ newton}$$

$$= -8.1 \times 10^{-6} \text{ newton}$$

directed toward the positive charge. ◄

Example 26-3   Compute the force acting on the charge of Example 2, directly from Coulomb's law.

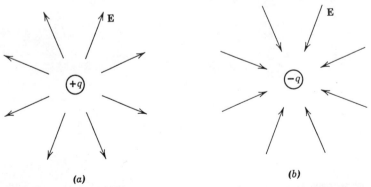

(a)

(b)

Figure 26-2   Electric field (a) near a positive charge and (b) near a negative charge.

Solution. We have from Coulomb's law,

$$F = \frac{kqq'}{r^2}$$

$$= \frac{9 \times 10^9(9 \times 10^{-8})(-9 \times 10^{-8})}{3^2}$$

$$= -81 \times 10^{-7} \text{ newton}$$

$$= -8.1 \times 10^{-6} \text{ newton}$$

directed toward the positive charge. ◀

If we have more than one charge producing a field, the resultant field intensity at any point is the *vector sum* of the separate field intensities produced at that point by each charge. Thus, for a charge distribution of n charges,

$$\mathbf{E} = \sum \frac{kq_i}{r_i^2} \text{ (vector sum)} \qquad (26\text{-}5)$$

The x and y components of E are given by

$$E_x = E_{1x} + E_{2x} + E_{3x} + \cdots,$$
$$E_y = E_{1y} + E_{2y} + E_{3y} + \cdots \qquad (26\text{-}6)$$

Example 26-4 Two charges each of magnitude 9×10^{-7} coulombs are situated at the points $(3, 0)$, $(-3, 0)$ on the x–y plane. Find the electric field intensities at the following points: (a) $(0, 4)$, (b) $(2, 0)$, and (c) $(-3, -8)$. Refer to Fig. 26-3. All distances are in meters.

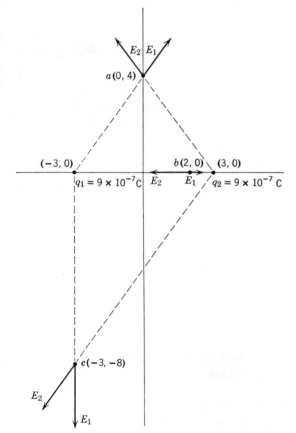

Figure 26-3

Solution. (a) Both $\mathbf{E}_1$ and $\mathbf{E}_2$ will be resolved into their x and y components. The magnitude of both E_1 and E_2 is

$$E_1 = E_2 = \frac{kq}{r^2} = \frac{9 \times 10^9 (9 \times 10^{-7})}{25} = 324 \text{ N/C}$$

Resolving, we have,

$$E_{1x} = 324 \cos 53° = 324(0.6)$$
$$E_{1y} = 324 \sin 53° = 324(0.8)$$
$$E_{2x} = -324 \cos 53° = -324(0.6)$$
$$E_{2y} = 324 \sin 53° = 324(0.8)$$

Therefore, from Eq. 26-6

$$E_x = E_{1x} + E_{2x} = 0$$
and $$E_y = E_{1y} + E_{2y} = 2 \times 324(0.8)$$
$$= 518.4 \text{ N/C}$$
Hence $\quad \mathbf{E} = \mathbf{E}_y = 518.4 \text{ N/C}$

in the positive y direction.

(b) The magnitudes of the two field intensities $\mathbf{E}_1$ and $\mathbf{E}_2$ are different, but they both lie along the x axis. Thus

$$E_{1x} = \frac{kq}{r^2} = \frac{9 \times 10^9 (9 \times 10^{-7})}{5^2} = 324 \text{ N/C} \rightarrow$$

$$E_{2x} = \frac{kq}{r^2} = \frac{9 \times 10^9 (9 \times 10^{-7})}{1^2} = 8100 \text{ N/C} \leftarrow$$

$$E_{1y} = 0 \quad \text{and} \quad E_{2y} = 0$$

Therefore, by Eq. 26-6,

$$E = E_{1x} + E_{2x} = 324 - 8100$$
$$= -7776 \text{ N/C} = 7776 \text{ N/C} \leftarrow$$

(c) The intensity due to the charge at (3, 0) has both x and y components, whereas the intensity due to the charge at $(-3, 0)$ has only a y component. Thus

$$E_{2x} = -\left(\frac{kq}{r^2}\right) \cos 53°$$

$$= -\frac{9 \times 10^9 (9 \times 10^{-7})(0.6)}{100} = -48.6 \text{ N/C}$$

$$E_{2y} = -\left(\frac{kq}{r^2}\right) \sin 53°$$

$$= -\frac{9 \times 10^9 (9 \times 10^{-7})(0.8)}{100} = -64.8 \text{ N/C}$$

$$E_{1x} = 0$$

$$E_{1y} = -\frac{kq}{r^2} = -\frac{9 \times 10^9 (9 \times 10^{-7})}{64}$$

$$= -126.6 \text{ N/C}$$

Then, by Eq. 26-6

$$E_x = E_{1x} + E_{2x} = -48.6 \text{ N/C}$$
$$E_y = E_{1y} + E_{2y} = -191.4 \text{ N/C}$$

and from Fig. 26-4

$$E = \sqrt{(E_x)^2 + (E_y)^2} = \sqrt{(-48.6)^2 + (-191.4)^2}$$
$$= 197.5 \text{ N/C}$$

at an angle $\theta = \tan^{-1}(-191.4/-48.6)$. ◀

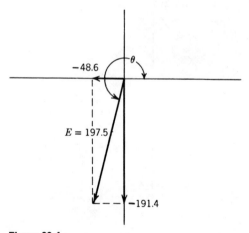

Figure 26-4

26-2 LINES OF FORCE AND GAUSS' LAW

In Fig. 26-5 the lines of force drawn from the charge q indicate the direction of the electric field intensity. It would be helpful if these lines also served as a measure of the field strength.

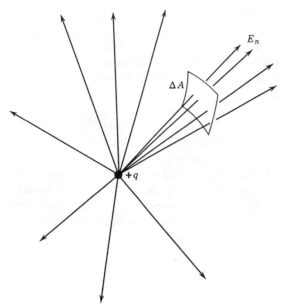

Figure 26-5 Lines of force drawn from a point charge. The number of lines per unit area crossing ΔA is proportional to E_n; E_n is the component of E perpendicular to ΔA and directed outward from the surface. The product $E_n \Delta A$ will be considered negative when E_n is directed inward.

Consider an element of area ΔA, situated at a point in the field where the intensity is **E**. Let E_n be the outward component of E perpendicular to ΔA, and let ΔN represent the number of lines passing through ΔA. We choose ΔN so that the number of lines per unit area passing through ΔA is proportional to E_n. Thus

$$\frac{\Delta N}{\Delta A} \sim E_n \qquad (26\text{-}7)$$

In a region of large E, the lines of force will be crowded together, whereas in a region of small E, the lines will be greatly separated.

It has been found that the most advantageous choice for the constant of proportionality in Eq. 26-7 is ε_0, the *permittivity of free space*. Thus

$$\frac{\Delta N}{\Delta A} = \varepsilon_0 E_n \qquad \text{or} \qquad \Delta N = \varepsilon_0 E_n \Delta A \quad (26\text{-}8)$$

The total number of lines of force passing perpendicularly through a surface, when E_n is constant over the entire surface, is

$$N = \varepsilon_0 E_n A \qquad (26\text{-}9)$$

Consider a sphere of radius r surrounding a point charge q, situated at $r = 0$. How many lines of force pass through the surface of the sphere? From Eqs. 26-9 and 26-3

$$N = \varepsilon_0 E_n A, \qquad E_n = \frac{kq}{r^2}, \qquad A = 4\pi r^2$$

$$N = \varepsilon_0 \left(\frac{kq}{r^2}\right)(4\pi r^2)$$

$$= \varepsilon_0 4\pi k q = q$$

if we define ε_0 so that

$$\varepsilon_0 = \frac{1}{4\pi k} = 8.85 \times 10^{-12} \text{ C}^2/\text{N-m}^2$$

Because of the way in which we chose our constant of proportionality ε_0, the total number of lines of force passing perpendicularly through a surface is numerically equal to the charge contained within the surface. Although this assertion has only been shown to be true for a sphere with the charge at the center, it can be demonstrated to apply to any shaped surface, with E_n not necessarily constant over the surface. The more general version of Eq. 26-9 is

$$N = \sum_{\text{surface}} \varepsilon_0 E_n \Delta A = q \qquad (26\text{-}10)$$

where E_n varies over the surface and the sum is taken over all elements of area, ΔA, on the surface. This equation is known as Gauss' law. Equation 26-10 reduces to Eq. 26-9 whenever E_n is constant over the surface. Some examples of this will be discussed presently.

26-3 APPLICATIONS OF GAUSS' LAW

The fundamental advantage of Gauss' law is to enable us to determine the field intensity due to uniform symmetric distributions of charge on a line or surface, or in a volume bounded by some surface. Since the number of charges present may be very large, and the exact location of each charge may not be known at all, the calculation of E, in these cases, by use of Eq. 26-3 or an equivalent is complicated. We illustrate Gauss' law by applying it to several problems.

Example 26-5   Find the electric field intensity at a distance r from an infinite line, along which there is a charge of λ per unit length. Such a charge distribution is known as a line charge of charge density λ.

Solution.   We first show that the field is everywhere perpendicular to the line. Figure 26-6 shows the line charge and a point P at which we investigate the direction of the field intensity. Clearly, for each small element of charge, Δq, situated to the left of OP that produces a horizontal component of E in the positive x direction, there will be an equal element of charge Δq situated to the right of OP that will produce a horizontal component of E, of equal magnitude, in the negative x direction. *This is true only because the line is infinitely long.* Thus the intensity is everywhere directed along the radii r. Further, from symmetry, the magnitude of E is the same in all directions, at the same distance r from the line. In order to use Gauss' law, we must now construct an imag-

inary surface about the line and determine the total number of lines N that pass through this surface. We choose as our Gaussian surface (as the imaginary surface is called) a cylinder of radius r, length l, and axis coincident with the line charge, as in Fig. 26-7.

This choice fits the symmetry of the problem. The electric field on the curved surface will be the same at each point and the direction at each point will be along the normal to the surface. There are no lines of force passing perpendicularly through the circular ends of the cylinder because the horizontal component of E is everywhere zero. Since the charge enclosed within the surface is $q = \lambda l$, we have, using Eq. 26-10

$$q = \lambda l = \varepsilon_0 EA = \varepsilon_0 E 2\pi r l$$

or
$$E = \frac{\lambda}{2\pi\varepsilon_0 r} \qquad (26\text{-}11)$$

for the electric field intensity at a distance r from a line charge of charge density λ.   ◀

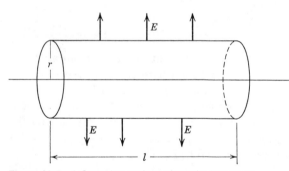

Figure 26-7   A Gaussian surface about the line charge. E is perpendicular to the curved surface and directed outward. The area of the curved surface is $2\pi r l$.

Example 26-6   Find the electric field intensity at a distance r from a charged infinite sheet of charge σ per unit area.

Solution.   Once again using symmetry arguments, similar to those used in the previous example, we see that the intensity must everywhere be perpendicular to the sheet. We choose as our

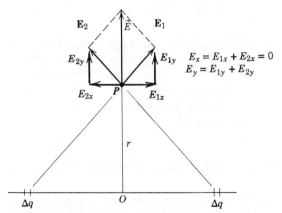

$E_x = E_{1x} + E_{2x} = 0$
$E_y = E_{1y} + E_{2y}$

Figure 26-6   The resultant field at P is directed radially outward from the line.

Gaussian surface a cylinder of length $2r$ and end area S, intersecting the sheet as shown in Fig. 26-8. Here E has the same value on both ends, S, and the charge within the cylinder is $q = \sigma S$. Thus, using Eq. 26-10,

$$q = \sigma S = 2\varepsilon_0 ES$$

taking into account both ends. We then have

$$E = \frac{\sigma}{2\varepsilon_0} \qquad (26\text{-}12)$$

on both sides of an infinite sheet. Note that E is *independent of any distance*. Therefore, the field has the same intensity everywhere in space. ◀

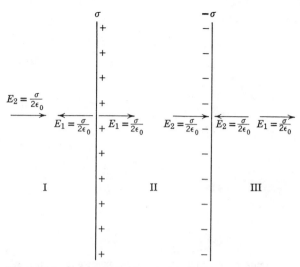

Figure 26-9 Calculation of the electric field between and outside a pair of parallel plates.

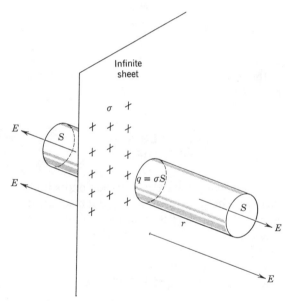

Figure 26-8 The Gaussian surface through a charged infinite sheet. There are no lines of the electric field through the curved surface.

Example 26-7 What is the electric field intensity in all space due to two infinite plates of opposite charge separated by a distance d? The charge distribution on both sheets is σ.

Solution. A pair of parallel plates, oppositely charged, is called a condenser or capacitor. Figure 26-9 shows the system and the three regions of space, for which a value of E must be calculated. From Example 7 it is known that the magnitude of E, due to each plane surface, is always $\sigma/2\varepsilon_0$. In region I, the field due to the positive plate is to the left, and that due to the negative plate is to the right. Thus E in region I is zero. A similar argument shows that E in region III is zero. In region II both fields are in the same direction, and

$$E = \frac{\sigma}{2\varepsilon_0} + \frac{\sigma}{2\varepsilon_0} = \frac{\sigma}{\varepsilon_0} \qquad (26\text{-}13)$$

directed from the positive to the negative plate. Note that the distance between the plates does not enter into the calculation of E. ◀

Example 26-8 Find the electric field intensity at a distance r from a positive point charge q by using Gauss' law.

Solution. We construct a Gaussian surface of radius r about the point charge (Fig. 26-10). From symmetry we know that E will have the same magnitude at equal distances from q. Thus E is constant on the surface, S. Further, E is directed outward at S and is perpendicular to

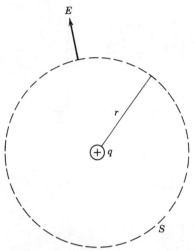

Figure 26-10 *E has the same magnitude at all points on the surface S and is directed outward from S.*

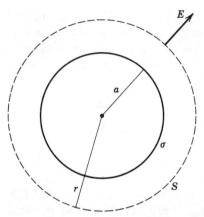

Figure 26-11 A thin-shelled sphere of radius *a* has a uniformly charged surface.

S. Hence, by using Eq. 26-10,

$$q = \varepsilon_0 E A$$
$$= \varepsilon_0 E (4\pi r^2)$$

$$E = \frac{q}{4\pi\varepsilon_0 r^2}$$

which agrees with Eq. 26-3. ◀

Example 26-9 Find the electric field intensity at a distance *r* from a hollow sphere of radius *a* with a uniform charge distribution on its surface. Assume that *r* > *a*.

Solution. We choose the Gaussian surface to be a sphere of radius *r* concentric with the given sphere (Fig. 26-11). Let the charge on the surface of the given sphere be *Q*. For *r* > *a*, the electric field intensity on the Gaussian surface *S* is constant in magnitude and directed outward from *S*. Thus

$$\sum \varepsilon_0 E_n \, \Delta A = Q$$
$$\varepsilon_0 \cdot E \cdot 4\pi r^2 = Q$$

and

$$E = \frac{Q}{4\pi\varepsilon_0 r^2}$$

We see that the electric field outside the charged surface behaves as if all of the charge were concentrated at the center of the hollow sphere. What is the electric field intensity at a point *r*, where *r* < *a*? ◀

26-4 ELECTRIC FIELDS AND CHARGED CONDUCTORS

Suppose that we have a charged conductor and all of the charges are known to be at rest. Consider Gauss' law applied to a surface within the conductor, as indicated by the dashed line in Fig. 26-12. Since the charges are assumed to be at rest, the electric field intensity within the conductor must be zero. For if *E* were not zero,

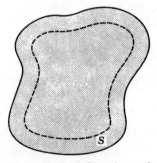

Figure 26-12 A Gaussian surface *S* inside a conductor.

charges would be forced to move in the direction of E according to Eq. 26-4. Since $E = 0$ within the conductor, there can be no charge within the dashed curve of Fig. 26-12. We conclude that all the charge on the conductor must reside on its outermost surface. The electric field intensity on the surface of the conductor must be perpendicular to the surface, for if there were a component of E parallel to the surface, the charges could not be at rest.

It was Michael Faraday with his famous ice pail experiment who demonstrated that the charge on a conductor resides on the outermost layer of the surface. In Fig. 26-13, A is a conducting pail, electrically neutral, and B is a charged sphere held on an insulated piece of string. The charged sphere is placed in contact with the inner surface of B. When the inside of the pail is now probed by a charge-sensing device, no charge is detected. All of the charge now appears on the outermost surface of the pail.

Another very important characteristic of electric field intensities on conductors concerns the curvature of the conductor. Figure 26-14 shows a conductor with the greatest outward curvature at point A and negative (inward) curvature at point B. It can be shown that E on the conductor is greatest at A and least at B. A sharply pointed conductor can be used to render the surrounding atmosphere conducting. Charge can then be sprayed off the points of the conductor. This prin-

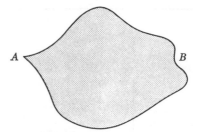

Figure 26-14   The electric field intensity is greatest at point A.

ciple is used in lightning rods and in electrostatic generators such as the Van de Graaff. It is important to remove all sharp edges from electrical equipment in order to minimize charge leakage.

26-5 MILLIKAN'S EXPERIMENT

We now turn to the question of whether or not there is a smallest charge. The original experiment designed to answer this question was performed by Robert Millikan in 1909, and is known as the oil drop experiment. We shall describe a similar experiment using very small latex spheres (with a diameter on the order of 10^{-6} m) instead of oil drops. The theory of the experiment is rather simple. When an object falls in a viscous medium, it reaches a terminal velocity, after which the acceleration is zero. The magnitude of the terminal velocity is proportional to the original unbalanced force acting on the object. The schematic for the experiment is shown in Fig. 26-15. Initially, the latex sphere is uncharged. The sphere accelerates until the upward viscous drag equals the gravitational pull on the sphere. After this point has been reached, the sphere moves with a terminal velocity, e.g., v_0. If the sphere is now charged by using an X-ray beam, there will be an extra driving force acting on the sphere, namely, the electrostatic force. The new terminal velocity will be v. For different charges on the sphere, different terminal velocities are reached. The difference between the terminal velocities when the sphere is charged and uncharged is

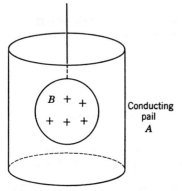

Figure 26-13   Faraday ice pail experiment. B is a charged sphere.

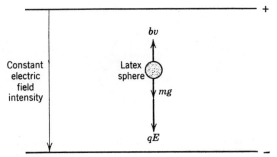

Figure 26-15 Millikan's experiment using latex spheres instead of oil drops. After the sphere has been charged by an X-ray beam, the sphere will accelerate until a terminal velocity v has been reached such that $bv = mg + qE = bv_0 + qE$, where v_0 is the terminal velocity of an uncharged sphere and b is a constant that depends upon the viscosity of the medium.

proportional to the electrostatic force. The terminal velocities are measured and it is found that these differences are always an integral multiple of some smallest value. Since the electric field between the plates is constant throughout the experiment, the charges on the sphere during the different parts of the experiment must be an integral multiple of some smallest value. The most recent experiments indicate that the best value of the elementary charge (that of a single electron or proton) is

$$e = (1.601864 \pm 0.000025) \times 10^{-19} \text{ Coulomb}$$

SUMMARY

The *electric field intensity* at a point is defined as the force per unit charge at that point:

$$\mathbf{E} = \frac{\mathbf{F}}{q'}$$

The electric field intensity is a vector quantity.

The force on a charge, Q, which is placed in an electric field whose intensity is E, will be

$$\mathbf{F} = Q\mathbf{E}$$

Gauss' law relates the normal component of the electric field intensity on a surface to the total charge within the surface:

$$\sum \varepsilon_0 E_n \, \Delta A = Q$$

Gauss' law is particularly useful when E_n is constant over the entire surface. When this is the case,

$$E = E_n = \frac{Q}{\varepsilon_0 \sum \Delta A}$$

The electric field is pictured by using *lines of force*. The number of lines of force per unit area at a point is proportional to the electric field intensity at that point in space. The total number of lines of force leaving a surface enclosing a net charge, q, is equal to the charge.

The electric field intensity outside a spherically symmetric charge distribution is the same as if all of the charges were concentrated at the center of the charge distribution.

The electric field intensity inside a conductor, whose charges are not moving, is zero. For such a conductor, all of the excess charge must reside on the outermost surface of the conductor.

The electric field intensity on a conductor is greatest at the sharply pointed regions where the charge per unit area is greatest.

The Millikan oil drop experiment indicates that charges always appear on objects in multiples of a smallest number, called the elementary charge. This charge is 1.6×10^{-19} C.

QUESTIONS

1. In what way is an electric field similar to a gravitational field? How can these fields be distinguished from each other?

2. Why can lines of force never cross (intersect) in space?

3. What can be learned about the force that would be exerted on a charge in a field, by looking at a sketch of the lines of force of the field?

4. Sketch the lines of force between two positive charges of the same magnitude. Do the same for two exactly opposite charges.

5. Can there be an electric field at a point in space where there is no charge? Explain.

6. A charge is placed somewhere in space. It experiences no force. Is there an electric field at this point in space? Explain.

7. The electric field intensity at the surface of a conductor is always perpendicular to the surface if the surface charges are at rest. Why must this be the case?

8. Describe Millikan's oil drop experiment. Interpret the results of the experiment.

9. Discuss Faraday's ice pail experiment. What did it prove?

PROBLEMS

(A)

1. A charge of 10 μC experiences a force of three newtons in an electric field. What is the magnitude of the electric field intensity?

2. An electric field intensity of 100 N/C is measured at a distance of 10 cm from a charge q. Determine the charge q.

3. What is the electric field intensity 30 cm from a charge of 9×10^{-8} C? What force would an electron experience if placed at this point?

4. What is the electric field intensity 2 millimicrons away from an alpha particle?

5. A charge of 3 μC experiences a force of 12×10^{-4} newtons. What is the electric field intensity?

6. What force will act upon a charge of 5 μC when placed in an electric field intensity of 600 N/C?

7. The electric field intensity at a point is 300 N/C. How far is this point from the charge ($q = 2 \times 10^{-7}$ C) that is producing the field?

8. Refer to Fig. 26-16. Find the electric field intensity at points a and b. Take $q_1 = 8 \times 10^{-9}$ C and $q_2 = 8 \times 10^{-9}$ C.

9. Refer to Fig. 26-16. At what point along the line passing through q_1 and q_2 would the electric field intensity be zero? Take $q_1 = 8 \times 10^{-9}$ C and $q_2 = -12 \times 10^{-9}$ C.

10. The radii of the permissible electronic orbits in the hydrogen atom are given by $r = n^2 r_0$, where n is an integer and $r_0 = 0.53 \times 10^{-8}$ cm. Find the electric field intensity produced by the proton at the first four radii.

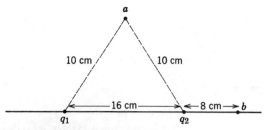

Figure 26-16   (Problems 8 and 9)

11. Two parallel plates, 40 cm × 50 cm each, carry equal but opposite charges. The electric field intensity between the plates is 100 N/C. Find the magnitude of the charge on each plate.

(B)

1. Two charges of 4×10^{-8} C and 7×10^{-8} C are 12 cm apart. At what point between them is the electric field intensity zero?

2. Equal charges of -4×10^{-7} C are placed at two of the vertices of an equilateral triangle of side 12 cm. Find the electric field intensity at the unoccupied vertex.

3. Two parallel plates are 2 cm apart. The electric field intensity between the plates is 2000 N/C. An electron is released from the negative plate. With what kinetic energy does it strike the positive plate? With what velocity?

4. The electric field intensity between a pair of parallel plates is 1000 N/C downward. What is the acceleration of an electron placed between the plates? Neglect the weight of the electron.

5. A 4-g mass is in equilibrium when placed between two horizontal parallel-plate conductors. If the electric field intensity between the plates is 4×10^5 N/C, find the charge on the 4-g mass.

6. Figure 26-17 depicts a pair of parallel plates. An electron enters at the middle of the plates with a horizontal velocity of 2×10^7 m/sec. The electric field intensity is 1000 newtons/coulomb. Does the electron escape from between the plates?

7. Compare the acceleration of a proton in an electric field of intensity 600 N/C to the acceleration due to gravity. The mass of a proton is 1.67×10^{-27} kg.

8. A sphere of radius R is uniformly charged throughout its volume with a charge density of ρ C/m^3.

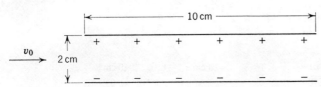

Figure 26-17

Show that for $r > R$, the electric field intensity is the same as if all of the charge were concentrated at the center of the sphere. (Use Gauss' theorem choosing a sphere of radius $r > R$ as the Gaussian surface.)

9. Using the information of Problem 8, show that the electric field intensity for $r < R$ is

$$E = \frac{\rho r}{3\varepsilon_0}$$

10. A sphere has a surface charge density of 8.85×10^{-9} C/m^2. The electric field intensity 2 m from the surface of the sphere is 160 N/C. Find the radius of the sphere. Refer to Fig. 26-11.

11. Two charges, A and B, of 8×10^{-5} C and 10×10^{-5} C, respectively, are 12 cm apart. Find the electric field intensity at a point 9 cm distant from A and 15 cm distant from B.

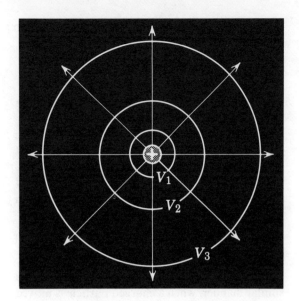

27.
Electric Potential Energy and Electric Potential

This chapter develops one of the most important quantities in electricity, the volt.

LEARNING GOALS

1. *An understanding of the meaning of electric potential, which is measured in volts.*

2. *An understanding of the difference between the concepts of electric potential and electric potential energy.*

3. *To be able to calculate the potential at a point in space.*

4. *To be able to calculate the speed of an electron as it travels between points of different potentials.*

5. *To learn the meaning of the electron-volt.*

27-1 DIFFERENCE OF POTENTIAL ENERGY

It should be recalled from Section 12-4 that if a book of weight w is slowly lifted (without acceleration) to a table surface at a height h above the ground, an amount of work, $W_{1\to2} = wh$, must be done on the book against the gravitational force. This work is not lost but is transformed into gravitational potential energy so that (in Fig. 27-1a)

$$W_{1\to2} = wh = \Delta U_g = U_{g2} - U_{g1} \quad (27\text{-}1)$$

where ΔU_g is the difference of potential energy possessed by the book at the two positions. In a similar manner, when a charged particle is moved slowly in an electric field, as in Fig. 27-1b, work has to be done against the coulomb force, F_C acting on the particle. This work changes the *electric potential energy* of the particle by an amount ΔU_E, or

$$W_{1\to2} = \Delta U_E = U_2 - U_1 \quad (27\text{-}2)$$

These considerations arise from the work–energy principle, which can be stated in the following manner: In the absence of frictional forces, the external work done on a body is equal to the change in its kinetic energy (ΔK) plus the change in its potential energy (ΔU). This can be expressed as

$$W_{1\to2} = \Delta K + \Delta U \quad (27\text{-}3)$$

If a body is moved at constant speed (as it is in Fig. 27-1) the change in speed is zero and the change in kinetic energy is also zero. Under these circumstances, Eq. 27-3 reduces to Eqs. 27-1 and 27-2.

Example 27-1   The sphere in Fig. 27-1b carries a uniformly distributed positive charge of 6×10^{-8} C. How much work is required to slowly move a particle carrying a charge of $q = -10^{-12}$ C from position 1, which is 1 m from the center of the sphere, to position 2, a distance $\Delta s = 0.01$ m? What is the difference in its electric PE at these positions?

Solution.   While at position 1, the particle experiences a force, F_C toward the left which, from Coulomb's law, (Eq. 25-1) is

$$F_C = \frac{kQq}{s_1{}^2} = 9 \times 10^9 \times 6 \times 10^{-8}(-10^{-12})$$

$$F_C = -5.4 \times 10^{-10} \text{ newton}$$

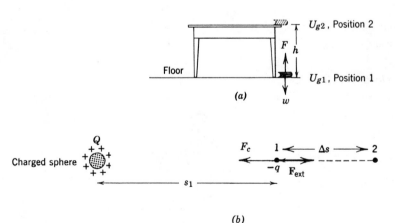

(a)

(b)

Figure 27-1   (a) An outside agent has to perform work on the book in order to slowly lift it from the floor to table top. (b) Work has to be done on the charge $-q$ in order to move it from positions 1 to 2.

In order to move the particle at constant slow speed to position 2, an external agent must provide an equal force, F_{ext} toward the right. $F_{ext} = -F_C = 5.4 \times 10^{-10}$ N. If Δs is small compared with s_1, the applied force can be considered constant and

$$W_{1\to2} = F_{ext} \times \Delta s$$
$$= 5.4 \times 10^{-10} \text{ newton} \times 10^{-2} \text{ m}$$
$$= 5.4 \times 10^{-12} \text{ joule}$$

Since there are no frictional influences, this contributes toward the increase in electric PE of the particle. By Eq. 27-2, $\Delta U_E = 5.4 \times 10^{-12}$ joule $= U_2 - U_1$, where U_2 and U_1 are, respectively, the electric potential energies of the small charge at points 2 and 1. Because the work done is a positive number, ΔU_E is also positive and $U_2 > U_1$. Notice that the potential energy is greater at a greater separation from Q. This is so when attractive forces are involved. (Compare with the gravitational situation.)   ◀

Example 27-2   How much external work is required to slowly move (without acceleration) a particle carrying a small charge, $q = +10^{-12}$ C from point 1 to point 2 (in Fig. 27-1c), a small distance, $\Delta s = 0.01$ m? What is the difference in potential energy possessed by the particle at these positions and where is it greater?

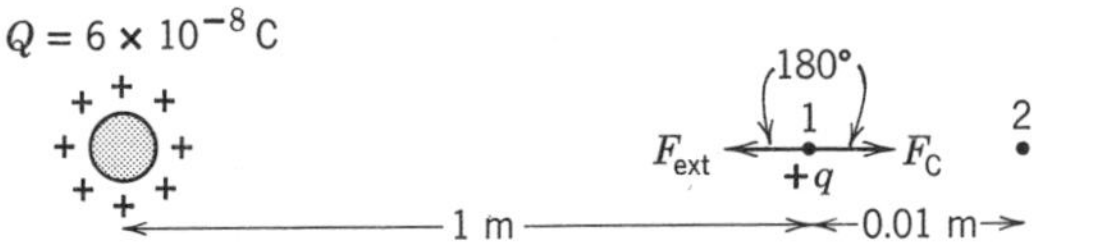

$Q = 6 \times 10^{-8}$ C

Figure 27-1   (c)

Solution.   Notice that here we are dealing with a repulsive coulomb force. This force, F_C would accelerate the particle toward point 2. To solve this problem, we can imagine applying on the particle an external force, F_{ext}, which is equal and opposite to F_C. The vector sum of these forces would thus be zero, and the particle would move at *constant velocity* as this example stipulates.

From the definition of work, (Eq. 12-1) $W_{1\to2} = F_{ext} \Delta s \cos \theta$, where F_{ext} is the magnitude of the external force, Δs is the magnitude of the displacement, and θ is the angle between the directions of the force and displacement. In this example, $\theta = 180°$ and $\cos \theta = -1$.

$$F_C = \frac{kQq}{s^2} = \frac{9 \times 10^9 \times 6 \times 10^{-8} \times 10^{-12}}{(1)^2}$$
$$= 5.4 \times 10^{-10} \text{ N}$$

Since the magnitudes, $F_{ext} = F_C = 5.4 \times 10^{-10}$ N, $W_{1\to2} = 5.4 \times 10^{-10} \times 10^{-2}(-1) = -5.4 \times 10^{-12}$ joule. But, by Eq. 27-2, the work done is equal to the change in PE. Therefore, -5.4×10^{-12} joule $= U_2 - U_1$. Since $W_{1\to2}$ is negative, $U_1 > U_2$.

When repulsive forces are involved, the mutual potential energy *decreases* with greater separations.   ◀

27-2 POTENTIAL ENERGY AT A POINT IN SPACE

It is sometimes useful to consider the potential energy of a body at a given point in space rather than the difference in potential energy between two points. What is the potential energy of the book while resting on the table top (see Fig. 27-1a)? Equation 27-1 gives a value for the difference in potential energy of the book but not the potential energy, U_{g2}, of the book while at a definite location like the table top. Suppose that we arbitrarily assume that the potential energy of the book on the floor, U_{g1} is zero. We can now solve for U_{g2}:

$$W_{1\to2} = U_{g2} - U_{g1} = U_{g2} - 0$$

By knowing $W_{1\to2}$, we also know U_{g2}. This brings out an important point regarding potential energy. The potential energy of a body at a point in space is calculated with respect to an arbitrarily chosen position where the PE is considered zero.

What is the electric potential energy of the charged particle q' in Fig. 27-2? This question

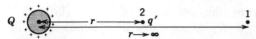

Figure 27-2 Point 1 is at an infinite distance from Q where the potential energy is zero.

really seeks the potential energy of q' with respect to a position where its potential energy is zero. In electricity, the position for zero potential energy is often taken to be at infinity. Applying Eq. 27-2, we get

$$W_{1 \to 2} = W_{\infty \to 2} = U_2 - U_\infty = U_2 - 0$$

$$(27\text{-}4)$$

In order to find U_2 we must calculate $W_{\infty \to 2}$ or the amount of work required to bring a charged particle from infinity to point 2. This calculation is performed in Section 27-7. There it is shown that the work done on a small charge, q' in bringing it from infinity to a point outside a uniformly charged sphere of net charge Q and at a distance r from its center is

$$W = \frac{kQq'}{r} \qquad (27\text{-}5)$$

Substituting this expression into Eq. 27-4 and dropping the subscripts for generality, we get

$$U = \frac{kQq'}{r} \qquad (27\text{-}6)$$

A dimensional analysis reveals that the electric potential energy is in units of joules. Thus from Eq. 27-6

$$U = \frac{\text{N-m}^{\cancel{2}}}{\cancel{C^2}} \times \frac{\cancel{C} \times \cancel{C}}{\cancel{m}} = \text{N-m} = \text{joule}$$

Electric potential energy, like any other type of energy, is a scalar quantity. Potential energy, in general, involves two bodies. In Eq. 27-6, Q represents the body that causes the electric field and q' represents the small charge carried by another body. It can be shown that a uniformly charged sphere behaves as if its total charge is concentrated at its center point. Equations 27-5, 27-6, therefore, also hold for a charge, Q, that is concentrated in a small volume approaching a point.

Example 27-3 Calculate the electric potential energy of the point charge $q = 2 \times 10^{-12}$ C located at point P in Fig. 27-3.

Solution. The electric potential energy of the 2×10^{-12} C point charge due to the 4×10^{-8} C charged sphere is, from Eq. 27-6,

$$U_1 = \frac{9 \times 10^9 \times 2 \times 10^{-12} \times 4 \times 10^{-8}}{2\sqrt{2}}$$

$$= 2.55 \times 10^{-10} \text{ joule}$$

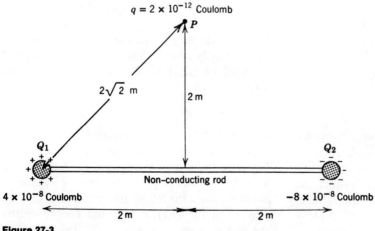

Figure 27-3

The potential energy due to Q_2 is

$$U_2 = \frac{9 \times 10^9 \times 2 \times 10^{-12}(-8 \times 10^{-8})}{2\sqrt{2}}$$

$$= -5.09 \times 10^{-10}\, joule$$

The total potential energy of the point charge is

$$2.55 \times 10^{-10}\, joule + (-5.09 \times 10^{-10})\, joule$$
$$= -2.54 \times 10^{-10}\, joule$$

The minus answer signifies that $+2.54 \times 10^{-10}$ joule of external work is required to transport q from point P to an infinite separation from the rod.   ◀

27-3 ELECTRIC POTENTIAL; THE VOLT

The electric *potential*, V, at a point in an electric field is defined as the electric potential energy of a small positive test charge q' located at that point, divided by the magnitude of the test charge. Thus

$$V = \frac{U}{q'} \qquad (27\text{-}7)$$

The potential at a point can also be expressed in terms of work. Since the *potential energy*, U in Eq. 27-7 depends on the work, W done on it (by Eq. 27-4) the potential at a point in space can also be expressed as

$$V = \frac{W}{q'} \qquad (27\text{-}8)$$

If one joule of work is required to transport one coulomb from infinity to a point in an electro-static field, then the potential at this point is one volt.

The difference of potential, $(V_2 - V_1)$, between two points in an electric field is defined as

$$V_2 - V_1 = \frac{W_{1 \to 2}}{q'} \qquad (27\text{-}9)$$

where $W_{1 \to 2}$ is the external work required to slowly transport a small positive charge from points 1 to 2.

When U or W in Eqs. 27-7 or 27-8 are ex-pressed in joules and q' in coulombs, V is ex-pressed in joule/coulomb or the volt. Since both energy and charge are scalar quantities, dividing one into the other also yields a scalar. The volt, therefore, is a scalar.

The electric potential, like the electric field intensity (see Chapter 26), is a quantity that de-scribes the electric field. Both are very useful in describing the electrical properties of the space in the vicinities of charged particles, spheres, wires, and plates. There is, of course, an associa-tion between potential and field intensity. One such association is described by Eq. 27-17.

The potential (not potential energy) at a point outside a uniformly charged sphere, of net charge Q, at a distance r from its center, can be found from Eqs. 27-6 and 27-7.

$$V = \frac{U}{q'} = \frac{kQq'}{rq'} = \frac{kQ}{r}$$

$$V = \frac{kQ}{r} \qquad (27\text{-}10)$$

Obviously, V can be either positive or negative, depending on Q. This equation is also valid when Q is concentrated in a very small volume that is nearly a point.

Example 27-4   In Fig. 27-4 a charge,

$$Q = 6 \times 10^{-10}\, C$$

is distributed uniformly on a sphere. Calculate the potential at (a) point 1, and (b) point 2. What is the potential difference (c) $(V_1 - V_2)$, and (d) $(V_2 - V_1)$?

Solution
(a) By Eq. 27-10, $V_1 = kQ/r_1$

$$V_1 = \frac{9 \times 10^9 \times 6 \times 10^{-10}}{0.5} = 10.8\ volts$$

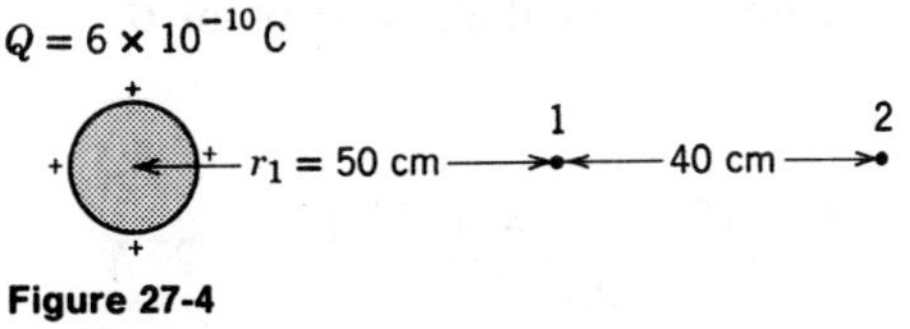

Figure 27-4

(b) $V_2 = \dfrac{kQ}{r_2} = \dfrac{9 \times 10^9 \times 6 \times 10^{-10}}{0.9} = 6$ V

(c) $V_1 - V_2 = 10.8$ V $- 6$ V $= 4.8$ V

(d) $V_2 - V_1 = 6$ V $- 10.8$ V $= -4.8$ V

The answers to (c) and (d) indicate that point 1 is at a higher potential than point 2. ◀

Example 27-5 In Fig. 27-5, q_1 and q_2 are point charges located at points A and B of the rectangle. (a) Calculate the potential, V_C at point C.

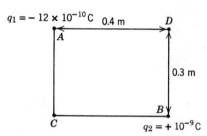

$q_1 = -12 \times 10^{-10}$ C   0.4 m

0.3 m

$q_2 = +10^{-9}$ C

Figure 27-5

Solution. The potential at any point is the algebraic sum of the potentials at that point produced by the various charges.

If V_{C1} and V_{C2} represent, respectively, the potentials at point C produced by q_1 and q_2, then by Eq. 27-10

$$V_{C1} = 9 \times 10^9 \, \dfrac{\text{N-m}^2}{\text{coulomb}^2} \left(\dfrac{-12 \times 10^{-10} \, \text{coulomb}}{0.3 \text{ m}} \right)$$

$$= -36 \text{ volts}$$

$$V_{C2} = \dfrac{9 \times 10^9 \times 10^{-9}}{0.4} = +22.5 \text{ volts}$$

$$V_C = V_{C1} + V_{C2} = -36 \text{ V} + 22.5 \text{ V} = -13.5 \text{ V}$$

(b) Calculate the potential, V_D at point D.

Solution

$$V_{D1} = \dfrac{9 \times 10^9(-12 \times 10^{-10})}{0.4} = -27 \text{ V}$$

$$V_{D2} = \dfrac{9 \times 10^9 \times 10^{-9}}{0.3} = +30 \text{ V}$$

$$V_D = 30 \text{ V} + (-27 \text{ V}) = +3 \text{ V}$$

Point D is at a higher potential than point C.

(c) What is the potential difference, $(V_C - V_D)$?

$$(V_C - V_D) = -13.5 \text{ V} - 3 \text{ V} = -16.5 \text{ V}$$

(d) What is the potential difference, $(V_D - V_C)$?

$$(V_D - V_C) = +3 \text{ V} - (-13.5 \text{ V}) = +16.5 \text{ V} ◀$$

Equation 27-10 indicates that the potentials are the same at all points that are equidistant from a point charge, Q, or equidistant from the center of a uniformly charged sphere of charge Q. The equidistant points form lines that are circles (really spheres) in Fig. 27-6. Such lines are called *equipotential lines* because all points on these lines are at the same potential. Notice that the lines representing the electric field intensity E are perpendicular to the equipotential lines.

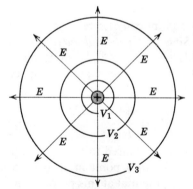

Figure 27-6 Each circle represents an equipotential line. V_1 is larger than V_2; V_2 is larger than V_3.

27-4 KINETIC ENERGY OF A CHARGED PARTICLE

A charged particle will accelerate when placed in an electric field. If the particle carries a net positive charge, it will move and speed up toward points of *lower* potential; if it carries a net negative charge, it will move and speed up toward points of *higher* potential. Since these particles also have a mass, they will experience a change in kinetic energy while moving in an electric field.

In Fig. 27-7, V_a represents the potential at point a and V_b, the potential at point b. Thus $V_a > V_b$ (why?). At point a, the positive particle experiences a force that is directed toward point b (the similar charges Q and q' repel). Therefore, q' will accelerate toward point b *without external work.* It will experience an increase in kinetic energy,

$$\Delta K = \tfrac{1}{2}mv_b{}^2 - \tfrac{1}{2}mv_a{}^2 \qquad (27\text{-}11)$$

where m is the mass of the charged particle, and v_a and v_b are, respectively, its speed at points a and b. The small charge q' will also experience a change in potential energy, which by Eq. 27-7 is

$$\Delta U = U_b - U_a = q'(V_b - V_a) \quad (27\text{-}12)$$

We can calculate the increase in the particle's kinetic energy (ΔK) in its motion from points a to b in Fig. 27-7 by Eq. 27-3, $W_{\text{ext}} = \Delta K + \Delta U$. Noting that the external work is zero and that there are no frictional forces involved, we get

$$-\Delta U = \Delta K \qquad (27\text{-}13)$$

Combining this equation with Eqs. 27-11 and 27-12 gives

$$q'(V_a - V_b) = \tfrac{1}{2}mv_b{}^2 - \tfrac{1}{2}mv_a{}^2 \quad (27\text{-}14)$$

The increase of the particle's kinetic energy is given by the product of its charge and potential difference. If the particle's mass and speed at point a are known, its speed at point b can be determined from a knowledge of the potential difference. Although the considerations in this section were based on the field produced by a charged sphere, the equations are valid for any field. If a charged particle of charge q and mass m

is accelerated from rest between two points at different potentials, the particle's speed, v at arrival can be determined from

$$qV = \tfrac{1}{2}mv^2 \qquad (27\text{-}15)$$

where V is the potential difference between the two points. This equation is useful in determining the speed with which an electron arrives at the positive plate in electron tubes.

Example 27-6 An X-ray tube consists essentially of a cathode, which emits electrons, and a tungsten anode both in an evacuated glass enclosure. The anode is at a high potential with respect to the cathode. This causes the electrons to accelerate toward the anode and strike it at very high speeds. The electron's energy is transferred to the atoms of the tungsten anode, causing the latter to emit X rays (Fig. 27-8).

If the potential difference is 50,000 volts (a value usually used in medical X rays), calculate the speed of an electron at the moment before it strikes the anode. Assume that the electrons leave the cathode at zero speed. The mass of an electron is 9.1×10^{-31} kg and the magnitude of its charge is 1.6×10^{-19} C.

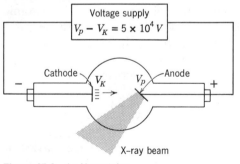

Figure 27-8   An X-ray tube.

Solution.   By Eq. 27-15, $v = \sqrt{2qV/m}$ where q is the magnitude of the electron's charge,

$$v = \sqrt{\frac{2 \times 1.6 \times 10^{-19} \times 5 \times 10^4}{9.1 \times 10^{-31}}}$$

$$= 1.33 \times 10^8 \text{ m/sec}$$

Q

Figure 27-7   A positively charged particle, q' will accelerate toward points of lower potentials.

We next give a more lengthy solution to this question that brings out some of the important principles. We apply Eq. 27-13, $-\Delta U = \Delta K$. The electrons will accelerate toward the anode. Their increase in kinetic energy is $\Delta K = \frac{1}{2}mv_p{}^2 - \frac{1}{2}mv_k{}^2$, where v_p is their speed at the anode and v_k their speed at the cathode, which is zero in this example.

In moving from cathode to anode, the electrons undergo a change in potential energy, $\Delta U = U_p - U_k$. By Eq. 27-12, $\Delta U = e(V_p - V_k)$, where e is the charge carried by an electron. Therefore,

$$-\Delta U = \Delta K$$
$$e(V_k - V_p) = \frac{1}{2}mv_p{}^2$$

Since the potential difference $(V_p - V_k) = 50,000\,\text{V}$, $(V_k - V_p) = -50,000\,\text{V}$.

$$-1.6 \times 10^{-19}(-50,000) = \frac{1}{2}mv_p{}^2$$

$$v_p = \sqrt{\frac{16 \times 10^{-15}}{9.1 \times 10^{-31}}} = 1.33 \times 10^8 \text{ m/sec} \blacktriangleleft$$

27-5 THE ELECTRON VOLT

The joule is a convenient and practical unit of energy in studies of macroscopic events. The *electron volt* (eV) is a more convenient unit of *energy* in studies involving atomic and subatomic particles. The joule is associated with electrical quantities by Eq. 27-8, 1 volt = 1 joule/1 coulomb. Since 1 coulomb represents the combined charge carried by 6.25×10^{18} electrons (or protons),

$$1 \text{ volt} = \frac{1 \text{ joule}}{6.25 \times 10^{18}e}$$

where e is the magnitude of the charge carried by the electron or proton.

The significance of the last equation can be brought out by an electron tube similar to the X-ray tube described in Example 27-6. If the potential difference between the anode and cathode were 1 volt, then 6.25×10^{18} electrons ($= 1$ C) would gain 1 joule in energy upon arrival at the anode. The electron-volt represents the energy gained by a *single* electron (instead of 6.25×10^{18} electrons) when it moves through a potential difference of 1 volt. To determine the relation between the electron-volt and joule, we divide the numerator and the denominator in 1 volt = 1 joule/$6.25 \times 10^{18}e$ by the number 6.25×10^{18}. Thus

$$1 \text{ volt} = \frac{1}{6.25 \times 10^{18}} \text{ joule} \div e$$
$$= \frac{1.6 \times 10^{-19} \text{ joule}}{e}$$

We call 1.6×10^{-19} joule, 1 electron volt.

$$1 \text{ electron volt} = 1\,\text{eV} = 1.6 \times 10^{-19}\,\text{joule} \quad (27\text{-}16)$$

$$\text{one million or } 10^6 \text{ electron volt} = 1\,\text{MeV}$$
$$= 1.6 \times 10^{-13}\,\text{joule}$$
$$\text{one billion or } 10^9 \text{ electron volt} = 1\,\text{BeV}$$
$$= 1.6 \times 10^{-10}\,\text{joule}$$

27-6 POTENTIAL DIFFERENCE BETWEEN CHARGED PLATES

Figure 27-9 depicts two closely spaced parallel plates carrying equal and opposite charges that are uniformly distributed. It is useful to express the potential difference between the plates, $(V_1 - V_2)$, in terms of the electric field intensity E. It should be recalled from Example 26-6 that E is constant between the plates. This means that when a small positive charge, q' is placed at different points within the space between the plates, it will experience the same force, F (in magnitude and direction). Since $E = F/q'$ (Eq. 26-2), $F = Eq'$ and F will point in the same direction as E, which is toward plate 2 in Fig. 27-9.

To move q' at constant speed toward plate 2, we can imagine an external force, F_{ext} which is equal and opposite to F has to be applied on q'. The work, $W_{1 \to 2}$ done by F_{ext} in bringing q' from plates 1 to 2 through the displacement, d, is given by Eq. 12-1 as $W_{1 \to 2} = F_{\text{ext}}\,d \cos\theta$, where $\theta = 180°$, (Example 27-2). But $F_{\text{ext}} = F = Eq'$ (in magnitude). Therefore,

$$W_{1 \to 2} = Eq'd(-1) = -Eq'd$$

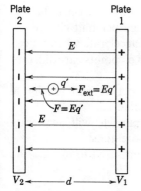

Figure 27-9 The potential difference between the oppositely charged parallel plates is $V_1 - V_2 = Ed$

27-7 ELECTRIC POTENTIAL ENERGY DUE TO A CHARGED SPHERE

We wish to derive an expression for the electric potential energy of a charge q' at a point P_4 in the vicinity and outside of a uniformly charged sphere. (See Fig. 27-10). This will be accomplished by calculating the amount of work, $W_{\infty \to p}$ required to bring q' from infinity to P_4. The work $W_{1 \to 2}$ required to bring the small charge from point P_1 to P_2, a small distance Δr is

$$W_{1 \to 2} = F_{1 \to 2} \times \Delta r \qquad (27\text{-}18)$$

where $F_{1 \to 2}$ is the force that has to be exerted by an outside source on q' to counteract the coulomb force exerted on it by Q.

But by Eq. 27-9, $W_{1 \to 2} = (V_2 - V_1)q'$. Therefore,

$$-Eq'd = (V_2 - V_1)q'$$

or
$$V_1 - V_2 = Ed \qquad (27\text{-}17)$$

The units for the electric field intensity E in Eq. 27-17 are volt/m. These units can be reconciled with newton/coulomb for E (as given in Section 26-1) by noting that

$$\text{volt} = \frac{\text{joule}}{\text{coulomb}} = \frac{\text{newton} \times \text{meter}}{\text{coulomb}}$$

$$\frac{\text{volt}}{\text{meter}} = \frac{\text{newton} \times \text{meter}}{\text{coulomb} \times \text{meter}} = \frac{\text{newton}}{\text{coulomb}}$$

Equation 27-17 is useful because it renders a method of computing the electric field intensity from the easily measured quantities, voltage difference and distance.

Example 27-7 A capacitor (a device commonly used in radios) consists of two parallel plates having an equal and opposite charge impressed on them. If the potential difference between the plates is 1 volt, calculate the electric field intensity if the separation is 2 cm.

Solution. From Eq. 27-17, $E = (V_1 - V_2)/d$ where $V_1 - V_2 = 1$ V.

$$E = \frac{1 \text{ volt}}{0.02 \text{ m}} = 50 \frac{\text{volt}}{\text{m}} = 50 \frac{\text{newton}}{\text{coulomb}} \blacktriangleleft$$

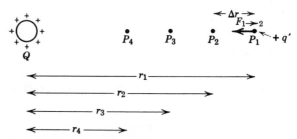

Figure 27-10 The work required to move the point charge $+q'$ from P_1 to P_2 is $F_{1 \to 2} \Delta r$, where $F_{1 \to 2}$ represents the average coulomb force on $+q'$ while in the region between P_1 and P_2.

From Coulomb's law, $F = kQq'/r^2$, where r is the separation between Q and q'. Since r varies with the movement of the charge from position 1 to 2, we can use an average distance, $\bar{r} = (r_1 + r_2)/2$. The expression for the work then becomes

$$W_{1 \to 2} = \frac{kQq'}{\bar{r}^2} \Delta r = kQq' \frac{\Delta r}{[(r_1 + r_2)/2]^2}$$

From Fig. 27-10, $r_1 = r_2 + \Delta r$. Substituting this value for r_1 in the last equation, we get

$$W_{1 \to 2} = kQq' \left[\frac{4 \, \Delta r}{(r_2 + \Delta r + r_2)^2} \right]$$

$$= kQq' \left[\frac{4 \, \Delta r}{4r_2{}^2 + 4r_2 \, \Delta r + (\Delta r)^2} \right]$$

Since Δr is small, $(\Delta r)^2$ is much smaller and can be neglected. The last equation can now be written as

$$W_{1 \to 2} = kQq' \left[\frac{\Delta r}{r_2(r_2 + \Delta r)} \right]$$

Noting that $\Delta r = r_1 - r_2$, the last expression becomes

$$W_{1 \to 2} = kQq' \left(\frac{r_1 - r_2}{r_1 r_2} \right) = kQq' \left(\frac{r_1}{r_1 r_2} - \frac{r_2}{r_1 r_2} \right)$$

$$W_{1 \to 2} = kQq' \left(\frac{1}{r_2} - \frac{1}{r_1} \right) \tag{27-19}$$

Equation 27-18 expresses the work required for a charge q' to be taken from a point that is at a distance r_1 to a point that is at a distance r_2 from the center of a charged sphere. The work necessary to take q' from point 2 to point 3 is deduced in the same manner:

$$W_{2 \to 3} = kQq' \left(\frac{1}{r_3} - \frac{1}{r_2} \right)$$

Similarly, $$W_{3 \to 4} = kQq' \left(\frac{1}{r_4} - \frac{1}{r_3} \right)$$

The total work, W_T, required for getting q from positions 1 to 4 is simply the sum

$$W_T = W_{1 \to 2} + W_{2 \to 3} + W_{3 \to 4}$$

or $$W_T = kQq' \left(\frac{1}{r_2} - \frac{1}{r_1} + \frac{1}{r_3} - \frac{1}{r_2} + \frac{1}{r_4} - \frac{1}{r_3} \right)$$

$$= kQq' \left(\frac{1}{r_4} - \frac{1}{r_1} \right)$$

Finally, we let r_1 approach ∞; then $1/r_1$ approaches 0. Omitting the subscripts for generality, we get

$$W = \frac{kQq'}{r} \tag{27-5}$$

This is the expression for the electric potential energy possessed by a charged particle of charge q', which is located at a point outside a uniformly charged sphere of charge Q and at a distance r from the sphere's center.

SUMMARY

The electric *potential energy* of a small charge in an electric field is the external work required to transport the charge from infinity to that point.

Electric *potential* is measured in volts.

The potential at a point in an electric field is the electric potential energy per unit charge.

A charged particle will accelerate when placed in a region between two points at different potentials.

A positively charged particle will accelerate toward points of lower potential.

A negatively charged particle will accelerate toward points of higher potentials.

The electron-volt is a unit used to express the energy of small particles.

QUESTIONS

1. Both electric and gravitational potential energy are similarly defined. What are the similarities? State the units for each.

2. Differentiate between electric potential energy and electric potential by defining and giving the units for each.

3. How is the principle of conservation of energy involved with potential energy?

4. Does a charge that is located very far away from any other charge possess electric potential energy? Explain.

5. Compare the changes in energies of a falling book with that of an electron moving toward a positively charged sphere.

6. The volt is a unit of potential. Is an electron volt a unit of potential? Explain.

7. Why does the charge q' in Eq. 27-8 have to be small?

8. What is meant by equipotential lines?

9. The potential at point A is -5000 volts and the potential at point B is -1000 volts. Which point is at a higher potential.

10. If an electron is released somewhere in the region between points A and B in the previous question, toward which point would the electron accelerate?

PROBLEMS

(A)

1. A sphere is given a net charge of $+6 \times 10^{-10}$ C, which is uniformly distributed.

(a) How much work is required to slowly transport a charge, $q = +6 \times 10^{-14}$ C, from a very large distance (from infinity) to a point that is 150 cm from the center of the sphere?

(b) From the conservation of energy principle point of view, what has happened to the work?

2. Calculate the electric potential energy of the small charge (in Problem 1) at the 150-cm point.

3. What is the electric potential V_A at a point 75 cm from the center of a sphere having a net charge of 2 μC? (One μcoulomb = 1 microcoulomb = 10^{-6} coulomb.)

4. What is the electric potential V_B at a point 150 cm from the center of a 2 μC sphere?

5. Calculate $(V_A - V_B)$ and $(V_B - V_A)$ in the preceding two problems.

6. How much work is required to slowly transport a charged particle, $q' = +100$ $\mu\mu$C, from a point that is 1.5 m to a point 0.5 m away from the center of a sphere with a uniformly distributed charge of 100×10^{-9} C? One $\mu\mu$coulomb = 1 micromicro-coulomb = 10^{-12} C.

7. Determine the potential difference $(V_A - V_B)$ between point A, which is 0.25 m from the center of a sphere that has a uniformly distributed charge of $+6 \times 10^{-10}$ C and point B which is 1.25 m from the center of the sphere.

8. Calculate the minimum amount of work that is required to transport an electron from points A to B in Problem 7. The charge on an electron is -1.6×10^{-19} C.

9. The potential difference between two large parallel plates is 300 V, and the plates are separated by 1 cm of air. Calculate the electric field intensity between the plates.

10. The electric field intensity between two large parallel plates that are separated by 0.5 cm of air is $E = 5000$ V/m. Determine the potential difference between the plates.

(B)

1. An electron is accelerated from rest at the cathode to the anode of a radio tube. The difference of potential between the anode and cathode in this tube is 90 V. How much kinetic energy does the electron acquire at the moment before striking the anode? State the answer in electron volts and in joules.

2. Calculate the speed of the electron in Problem B-1 at the moment before striking the anode. The mass of the electron is 9.1×10^{-31} kg.

3. The cathode and anode of an X-ray tube are at a potential difference of 10^5 V. Calculate the energy of an electron when it arrives at the anode. State your answer in electron volts and joules.

4. Consider the three fixed charges in Fig. 27-3 to be far from any other charges.

(a) How much work was required to bring q to its position in the figure, assuming that the other two charges were not present?

(b) How much work was then required to bring Q_1 from infinity to its position in the figure, assuming that q was already at point P?

(c) How much work was then required to bring Q_2 (wc) to its position in Fig. 27-3?

5. Calculate the total work required to assemble all three charges in Fig. 27-3. Does the order of delivery matter?

6. Assume that the center of a sphere which has a uniformly distributed charge of $+4 \times 10^{-10}$ C on it is at the origin of an x, y coordinate system. Calculate the electric potential at the following points:

(a) $x = 0$, $y = 1.5$ m.

(b) $x = 0$, $y = -1.5$ m.

(c) $x = -3$ m, $y = -4$ m.

7. Calculate the electric potential energy of an electron that is located at the following coordinates due to the charged sphere in Problem B-6:

(a) $x = 1.5$ m, $y = 0$.

(b) $x = -1.5$ m, $y = 0$.

(c) $x = 3$ m, $y = 4$ m.

8. A point in space has a potential of 100 V. How much work is required to bring a charged particle, $q = +10^{-9}$ C, from a large distance (infinity) to that point?

9. How much work is required to bring a $+10^{-9}$ C particle from a point that is at 100 V to a point that is 200 V?

10. How much work is required to bring a charged particle from a point in space that is at 200 V to another point that is also at 200 V?

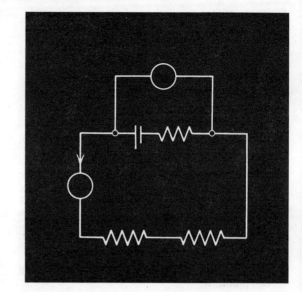

28.
The Electric Circuit

In Chapter 27 potentials and potential differences between points in space are discussed. This chapter considers potentials and potential differences between points in a wire. Potential differences between the ends of a wire cause an electric current to be established in the wire.

LEARNING GOALS

1. *To understand the meaning of and learn the equation for a steady electric current.*

2. *To understand the meaning of electromotive force.*

3. *To learn the equation for Ohm's law and how to apply it to elements of an electric circuit.*

4 *To understand the difference between series and parallel circuits.*

5. *To learn how to use Kirchhoff's rules for the solution of a circuit.*

6. *To learn how to calculate the energy and power dissipated in a circuit.*

28-1 ELECTRIC CURRENT IN A WIRE

If a metallic wire is connected to the terminals of a good battery (Fig. 28-1), a steady stream of electrons will move in the wire toward the positive terminal of the battery. When one coulomb ($= 6.25 \times 10^{18}$ electrons) passes through any cross-sectional area of the wire in a time of one second, the current in the wire is said to be one *ampere*. In equation form, a *steady current, i,* is expressed by

$$i = \frac{q}{t} \qquad (28\text{-}1)$$

where i is measured in amperes when the constant charge q is in coulombs and the time t is in seconds.

The current is the same through every part of the wire even if some parts are thicker than others. Thus, if a current of 2 amp is established in the wire of Fig. 28-1, 2 C/sec will pass through the cross-sectional area A_1 and 2 C/sec will pass through area A_2 and every other cross section.

Example 28-1 A steady current of 5 amp is maintained in a copper wire by means of a 12-volt battery. How many electrons pass through any cross section of the wire in 5 sec?

Solution. From Eq. 28-1, $q = it = 5$ amp $\times$ 5 sec $= 25$ C. Since there are 6.25×10^{18} electrons in a coulomb,

$$25 \text{ C} = 25 \text{ C} \times 6.25 \times 10^{18} \text{ electrons/C}$$
$$= 156.25 \times 10^{18} \text{ electrons} \qquad \blacktriangleleft$$

A given electron travels but a short distance in the wire before it is captured by a neighboring atom that has lost a conduction electron to its neighboring atom. The electrons constituting the current move at a constant speed called the *drift speed* in spite of the fact that an electric force is exerted on them. There is no net acceleration because of their collisions with the atomic particles of the wire.

The magnitude of the drift speed is low and depends on the magnitude of the current, the chemical composition of the wire, and the diameter of the wire. For example, the drift speed in a copper wire, a few millimeters in diameter and carrying a current of a few amperes is of the order of 10^{-2} cm/sec. In addition to the directed drift motion, the electrons also move at very high speeds in random directions even when the wire is not connected to the battery. On the average, as many move in one direction as in another.

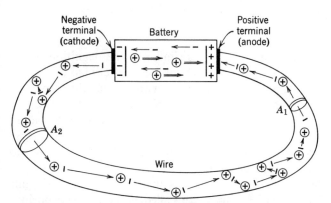

Figure 28-1   Illustration of the motion of an outer electron of the wire's atoms toward an adjacent atom which has lost an electron. The electric field within the wire causes a net drift of these electrons toward the positive terminal of the battery.

This random motion is not associated with the electric current.

28-2 DIFFERENCE OF POTENTIAL AND ELECTRIC CURRENT

There exists a potential difference between the terminals of a battery. Thus the potential at the positive terminal of a 12-volt auto battery is 12 volts above the potential at the negative terminal. When the ends of a metallic wire are connected to the battery terminals, a potential difference is also established between the ends of the wire. In Section 27-3 it was pointed out that an electric field is associated with a potential difference. Therefore, an electric field is established within the wire when it is connected to a battery. The significance of an electric field involves a force on a charged particle located in the field. Hence the atomic electrons of the wire experience a force. But only those electrons called *conduction electrons*, which are in the outer ring of the atom, will respond to this force by a net directed motion toward the positive terminal of the battery. The electrons in the inner rings are too tightly bound to the nucleus.

In order for a current to last in a wire, the battery must transport electrons and positive charges within the battery, to the negative and positive terminals, respectively. A continuous current in a wire must be accompanied by a continuous current in the *electrolyte*, which is the fluid in a storage battery.

28-3 ELECTROMOTIVE FORCE

A battery is a source of electrical energy. This energy is converted to heat and light energy when, for instance, a flashlight bulb is connected to it; the electrical energy furnished by a battery that is connected to a toy train is transformed, by means of a motor, to mechanical energy of motion and to heat energy in the connecting wires.

The principle of conservation of energy is upheld in these transformations. A battery is referred to as a *seat of electromotive force* (emf). A seat of emf is a device that changes any form of energy into electrical energy. An electric generator is also a seat of emf because it changes mechanical energy into electrical energy. A seat of emf also maintains a constant potential difference that is independent of the current drawn.

The *electromotive force* of a battery, abbreviated as emf and symbolized by $\mathscr{E}$, is defined as

$$\mathscr{E} = \frac{\Delta W}{\Delta q'} \qquad (28\text{-}2)$$

where ΔW is the work done in transporting a net charge, $\Delta q'$, through the battery. The unit for emf is the joule/coulomb, which is the same as the volt. Thus a battery having an emf of 12 V does 12 joules of work in transporting one coulomb of charge. It should be noticed that emf is measured in volts and not in units of force.

We interpret ΔW in the above equation as the work done *on* the charges *by* the chemical action of the battery. From the conservation of energy principle, this is equal to the loss in chemical energy of the electrolyte and also equal to the *gain* in electrical potential energy, ΔU, of the charges near or at the plates. We can therefore rewrite Eq. 28-2 as

$$\mathscr{E} = \frac{\Delta U}{\Delta q'} \qquad (28\text{-}3)$$

This equation is similar to the definition of potential. Each terminal of a battery has an electric potential associated with it, and there exists a potential difference between the terminals. The terminal connected to the side of the battery having an excess of positive charges is at a higher potential than the terminal that is connected to the side having an excess of electrons.

The battery cells that are used in flashlights and for toys are called *dry cells*. The positive terminal, which is usually located at the center of the top surface, is made of carbon; the negative terminal is made of zinc (which is also the container of the cell). Dry cells are good for short, intermittent uses and are, in general, not recharged.

The type of battery that is used in automobiles is called the *lead storage battery*. It consists of a few connected cells, each cell consisting essentially of a plate coated with lead and another plate coated with lead dioxide. The two plates are separated by dilute sulfuric acid. In 1800, Alessandro Volta discovered that when the surfaces of two dissimilar metal plates are briefly touched and then separated, one plate takes on an excess of electrons, leaving the other plate with an excess of protons. He found this result to be intensified by placing layers of electrolytes between the plates. An *electrolyte* is a liquid conductor of electricity, such as dilute water solutions of salts, acids, or bases. Modern storage batteries are similar to Volta's model of almost 200 years ago.

In the lead-acid storage battery (Fig. 28-2) the chemical reactions within each cell cause the lead dioxide plate to become richer in positive charges, and the lead plate richer in negative charges. The positive plates terminate at a post on the outside of the battery called the *positive terminal*; the negative plates terminate at a post called the *negative terminal*.

The reasons for the migration of charges within the cell are complicated and treated in studies of electrochemistry. For our purposes, however, we can use the law of conservation of energy to enhance the understanding of an electric circuit. The molecules of lead, lead dioxide, and sulfuric acid possess a certain amount of chemical energy before their chemical interactions. This chemical energy stems from the short-range electrical forces, i.e., forces that are exhibited by charged particles when they are separated at short distances, as in atoms and molecules. After these molecules combine within the battery, the chemical reactions give rise to molecules that have a smaller amount of chemical energy. The loss in chemical energy results in a gain in electric po-

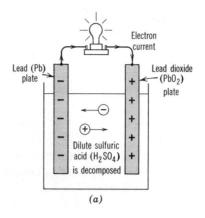

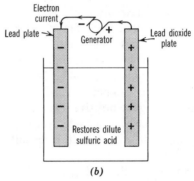

Figure 28-2 A lead storage battery-cell. (*a*) Chemical reactions leave the lead dioxide plate with an excess of positive and the lead plate with an excess of negative charges. In time, both plates become coated with lead sulfate and the sulfuric acid decomposes, causing the excess charges at the plates to diminish. (*b*) The battery is recharged by a dc generator which drives electrons through the lead plate. This causes chemical reactions which will rid the plates of lead sulfate and restore the sulfuric acid.

tential energy of the charges that are assembled near or at the lead and lead dioxide plates in the battery.

28-4 OHM'S LAW

In Section 28-2 it was stated that a potential difference established between the ends of a metallic wire will cause a current in the wire. The magnitude of the current i depends on the *potential difference V* and on a property of the wire called *resistance*, which is abbreviated as R. This dependence can be expressed mathematically as $i = V/R$ or

$$\frac{V}{i} = R \qquad (28\text{-}4)$$

where V is measured in volts, i in amperes, and R in *ohms*. Equation 28-4 is referred to as Ohm's law. Ohm's law holds only for metallic conductors and does not apply for currents in liquids and gases.

Applications of this law involve *circuit diagrams*, which are a schematic way of representing the important elements and their connections in a given electric circuit. The resistance of a wire is sketched as a zig-zag line, ⚡ (see Fig. 28-3). A battery cell is sketched as two parallel lines, ⊣|⊢, the longer line indicating the positive side and the shorter one indicating the negative side. A storage battery having more than one cell is usually sketched as ⊣|||⊢.

The arrow associated with i in Fig. 28-3a indicates the *direction* of the *electron current*. However, *by convention the current direction is taken as opposite to the direction of the electron current*, i.e., *from the positive toward the negative battery terminal in the wire*, as indicated in Fig. 28-3b. We shall henceforth follow this convention. Thus, whenever a current is discussed, the *conventional direction* will be implied. This will in no way alter the principles thus far developed.

Example 28-2 A current of 5 amp is established in the circuit of Fig. 28-4. Determine the potential difference across (a) R_1 and (b) R_2.

Solution. The current in this loop is everywhere the same and equal to 5 amp.

(a) The potential difference V_1 across R_1 is determined by Ohm's law: $V_1 = iR_1$; $V_1 = 5$ amp $\times$ 15 ohms $= 75$ V.

(b) By Ohm's law, the voltage difference, V_2, across R_2 is $V_2 = iR_2 = 5$ amp $\times$ 5 ohms $= 25$ V. Notice that the sum of the potential differences across both resistors is equal to the battery emf. ◀

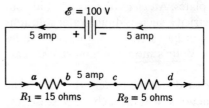

Figure 28-4

28-5 RESISTANCE, RESISTIVITY

Different materials offer varying resistances to electric currents. The resistance, R, of a wire, which is made of a given metal, say copper, depends on the wire's length and cross-section area. The resistance is greater for a longer than

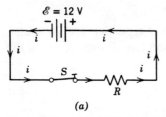

(a)

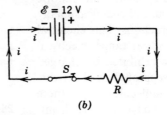

(b)

Figure 28-3 (a) A circuit diagram depicting a switch S and a resistor R connected to a battery having an emf of 12 volts. When the switch is pushed down until it makes contact with the lower wire, an electron current is established in the entire circuit, in the direction shown. (b) Conventional current direction.

for a shorter wire, and greater for a thinner than for a thicker wire. The geometric dependence is summarized by

$$R \propto \frac{l}{A} \qquad (28\text{-}5)$$

where l is the wire's length and A, its cross-sectional area.

Resistance also depends on the temperature of the wire. For most metallic conductors, the resistance increases as the temperature increases. For liquid conductors, or electrolytes, the resistance decreases as the temperature is increased. The dependence of resistance on chemical composition and temperature is summarized by the quantity called *resistivity*, which is symbolized by the Greek letter, ρ (rho).

$$R = \rho \frac{l}{A} \qquad (28\text{-}6)$$

Solving for the units for ρ yields

$$\rho = \frac{\text{ohm-m}^2}{\text{m}} = \text{ohm-meter}$$

The dependence of resistivity on temperature is given by

$$\rho = \rho_0(1 + \alpha t) \qquad (28\text{-}7)$$

where ρ_0 is the resistivity at $0°C$, t the wire's temperature in degrees Celsius and α (alpha) the *temperature coefficient of resistivity*. Since resistivity is temperature-dependent, resistivities are given at definite temperatures; for example, the resistivity of a copper wire at $20°C$ is 1.7×10^{-8} ohm-meter. Equations 28-6 and 28-7 can be combined to yield

$$R = R_0(1 + \alpha t) \qquad (28\text{-}8)$$

where R_0 is the resistance at $0°C$. See Table 28-1.

No material has a zero resistance nor does any material have an infinite resistance. Materials having very low resistance are called *conductors*; those having high resistance are called *insulators*. Some alloys, called *superconductors*, exhibit almost zero resistance at very low temperatures in the neighborhood of a few degrees Kelvin. This means that once a current is started in a superconductor it will continue for hours without an outside emf.

Table 28-1 Resistivities (ρ) and temperature coefficients of resistivity (α)

Material		ρ in ohm-m (at 20°C)	α, per C°
Aluminum	↑	2.7×10^{-8}	3.9×10^{-3}
Carbon		3.5×10^{-8}	-5×10^{-4} (resistance decreases with increasing temperature)
Copper	Good Conductors	1.7×10^{-8}	4×10^{-4}
Manganin (alloy of Cu, Mn, and Ni)		4.4×10^{-7}	0.000 (resistance does not change with temperature)
Steel	↓	1.8×10^{-7}	3×10^{-3}
Amber	↑	5×10^{14}	
Glass	Good Insulators	$10^{10} - 10^{14}$	
Quartz		10^{16}	
Rubber		$10^{13} - 10^{16}$	
Wood	↓	$10^8 - 10^{11}$	

Example 28-3 Compare the resistance of 3 m of steel wire having a cross section of 0.03 cm² with 3 m of steel wire having a cross-section area of 0.06 cm². Assume that both wires are at the same temperature.

Solution. Equation 28-6 indicates that R is inversely proportional to A. If A is doubled, R is halved, provided that ρ and l remain the same. Therefore, the 0.03-cm² wire has twice the resistance as the 0.06-cm² wire. For this reason larger currents are established in a thicker than in a thinner wire when each is connected to the same battery (provided that both are of the same length and material). ◀

28-6 KIRCHHOFF'S LOOP RULE; POTENTIAL AT A POINT

Example 28-2a seeks the potential *difference* across R_1 in Fig. 28-4. The phrase "potential difference across R_1" implies that one end of R_1 is at a higher potential than the other end. A potential (or voltage) is associated with a given point in a circuit. For example, the voltage, V_a at point a in Fig. 28-4 is greater than the voltage, V_b at point b. Hence there is a voltage *drop* of 75 V when R_1 is traversed from a to b. There is a voltage *gain* of 75 V when R_1 is traversed from b to a.

Let us imagine that a small *positive* charge, q' moves completely around the circuit of Fig. 28-5. Let the charge move from the negative terminal of the battery, through the battery, to the positive terminal, then through R_1 and R_2 and back to point a. The chemical action of the battery has to do work on the positive charge to bring it from the negative terminal to the positive terminal. As a result, the positive charge possesses a maximum amount of potential energy while at point b. Some of this energy is lost in moving through R_1. Therefore, its energy at point c is less than at point b. The charge then reaches R_2. (It has the same energy at points c and d because there are no intervening resistors.) Since it loses some energy in traversing R_2, its energy at point e is less than at point d. The imaginary positive charge has now completely traversed the loop in the *same direction as the conventional current*. Notice that at point a the charge is neutralized and has lost all the energy that it has received from the battery. To bring it to point b again, the battery will have to supply more energy.

$$\text{chemical work} = \text{energy lost in } R_1$$
$$+ \text{ energy lost in } R_2$$

or

$$\text{chemical work} - \text{energy lost in } R_1$$
$$- \text{ energy lost in } R_2 = 0 \quad (28\text{-}9)$$

Dividing each term in the last equation by q' we get:

$$\frac{\text{Chemical work}}{q'} = \mathscr{E}, \text{ the battery emf}$$
$$(\text{by Eq. 28-2});$$

$$\frac{\text{The energy lost in } R_1}{q'} = \frac{\text{the potential difference}}{\text{across } R_1 \text{ (by Eq. 27-12)}};$$

$$\frac{\text{The energy lost in } R_2}{q'} = \frac{\text{the potential difference}}{\text{across } R_2 \text{ (by Eq. 27-12)}}$$

Therefore, Eq. 28-9 can be rewritten as

$$\mathscr{E} - \Delta V_1 - \Delta V_2 = 0 \quad (28\text{-}10)$$

where $-\Delta V_1$ represents the *potential drop* across R_1, provided that R_1 is traversed in the *same* direction as the conventional current. This means that the potential, V_b, at point b in Fig. 28-5 is greater than the potential, V_c, at point c. Similarly, $-\Delta V_2$ represents the *drop in potential* across R_2, provided that R_2 is traversed in the same direction as the current.

By Ohm's law, $\Delta V_1 = iR_1$ and $\Delta V_2 = iR_2$. Therefore, Eq. 28-10 can be written as

$$\mathscr{E} - iR_1 - iR_2 = 0 \quad (28\text{-}11)$$

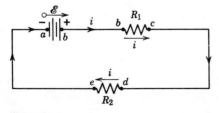

Figure 28-5

In general, a single-loop circuit may have more than one battery. Equation 28-11 can be put in the more general form,

$$\sum \mathcal{E} - \sum iR = 0 \qquad (28\text{-}12)$$

The last equation is referred to as *Kirchhoff's loop rule*. It is to be noted that this equation expresses the law of conservation of energy as applied to potential changes in an electric circuit. Equation 28-12 is very helpful in analyzing circuits that are more complex than the one depicted in Fig. 28-5. To help solve for the unknowns in a circuit by the loop rule, (1) the emfs are assigned a sense of direction (indicated by an arrow with a little circle at the end $\overset{\mathcal{E}}{\circ\!\!\rightarrow}$) that points from the negative to the positive terminal, within the battery and (2) changes in potential across a resistor are taken as negative when the resistor is traversed in the direction of the conventional current and as positive when the resistor is traversed in a direction opposite to the current.

Example 28-4 The current in the single-loop circuit of Fig. 28-6 is 3 amp. Determine the value of the unknown resistor, R_x. The Greek letter Ω (omega) is used as a symbol for the ohm.

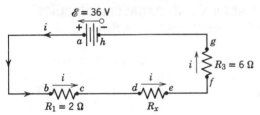

Figure 28-6

Solution

Notice that the direction of the conventional current is counterclockwise. We therefore expect voltage drops across the resistors as we traverse each in the direction of the current. Let us start (arbitrarily) at point a and traverse the complete loop, in a counterclockwise direction, back to point a. Points a and b are at the same potential because there is no intervening resistor between them. There is a voltage *drop* of $iR_1 = 3R_1$ across R_1. The next voltage drop of $3R_x$ is encountered in traversing R_x from points d to e. There is another drop of $3R_3$ from points f to g. This brings us to point h, which is the negative terminal of the battery. In traversing the battery back to point a, we encounter a *gain* in potential, which is the emf, $\mathcal{E}$. By Eq. 28-11,

$$-3 \times 2 - 3R_x - 3 \times 6 + 36 = 0$$
$$R_x = 4 \text{ ohms} = 4\,\Omega$$

We would achieve the same result if we started from point h and traversed the full loop in a *clockwise* direction. Notice that we are now traversing each resistor in a direction that is opposite to the current in that resistor. We shall now encounter a *gain* in potential across each resistor. In traversing the battery from the positive terminal to the negative terminal (against the emf arrow), we encounter a *drop* in potential.

$$+3 \times 2 + 3 \times R_x + 3 \times 6 - 36 = 0$$
$$R_x = 4\,\Omega \qquad \blacktriangleleft$$

Example 28-5 The two circuits in Fig. 28-7 are identical except that each is grounded at different

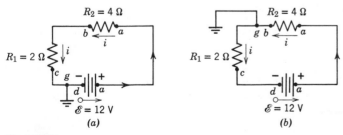

(a)   (b)

Figure 28-7

points. For each circuit, calculate (a) the current, (b) the potential difference $(V_a - V_b)$ and $(V_c - V_b)$, and (c) the potentials V_d at point d and V_b at point b.

Solution. Point g in both circuits is said to be *grounded.* Grounding a point in a circuit can be achieved by connecting a wire from it to a conductor such as a pipe, which leads into the earth. The symbol for such a connection is $\underline{\underline{\perp}}$. The earth is arbitrarily assumed as having a zero potential; potentials of other points in the circuit are figured with respect to the zero potential of the ground.

(a) The direction of the current in both circuits is counterclockwise. Applying the loop rule and traversing the complete loop from point c in a counterclockwise direction, we get (for each circuit)

$$+12\text{ V} - 4i - 2i = 0$$
$$i = 2\text{ amp}$$

The current is the same in the entire loop.

(b) *For both circuits*: Notice that if we move from points a to b in a counterclockwise direction, we encounter a *drop* in potential (an iR drop) because we are moving in the direction of the current. Thus

$$V_a - iR_2 = V_b$$
$$V_a - V_b = iR_2 = 2\text{ amp} \times 4\,\Omega = 8\text{ V}$$

If we had chosen to move from points b to a in a clockwise direction, we would get the same answer. But now we would move in a direction that is opposite to the current.

$$V_b + iR_2 = V_a$$
$$(V_a - V_b) = iR_2 = 8\ V$$

To determine $(V_c - V_b)$, we can move from points b to c in a counterclockwise direction:

$$V_b - iR_1 = V_c$$
$$(V_c - V_b) = -iR_1 = -2\text{ amp} \times 2\,\Omega = -4\text{ V}$$

(c) *For the circuit of Fig. 28-7a*: $V_d = 0$ because it is connected to the ground without any intervening resistors. Therefore, $V_a = 12$ V.

$$V_a - iR_2 = V_b$$
$$12\text{ V} - 2\text{ amp} \times 4\,\Omega = 4\text{ V} = V_b$$

For the circuit of Fig. 28-7b: $V_b = 0$ because point b is directly connected to the ground. In moving from points b to c in a counterclockwise direction, we get

$$V_b - iR_1 = V_c$$
$$0 - 2\text{ amp} \times 2\,\Omega = -4\text{ V} = V_c$$

Notice that $V_b = 0$ is larger than $V_c = -4$ V. Since $V_c = V_d$ because there are no intervening resistors, $V_d = -4$ V. In this circuit the positive side of the battery is at 8 V above-ground voltage and the negative side is at 4 V below-ground voltage. ◀

28-7 SERIES AND PARALLEL CIRCUITS, KIRCHHOFF'S JUNCTION RULE

In the circuit of Fig. 28-8a, there is only one possible path that the current can take, and that is through R_1, R_2 and R_3. The resistors in this

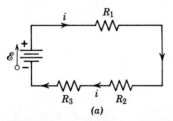

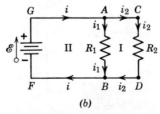

(a)   (b)

Figure 28-8 (a) Resistors in series—the current is the same in each resistor. (b) Resistors in parallel—the voltage is the same across each resistor.

type of a circuit are said to be *connected in series* and the circuit is called a *series circuit*. In the circuit of Fig. 28-8*b*, the current at junction *A* will split up into two branches, part of it going through R_1 and the remainder through R_2. The branches R_1 and R_2 are said to be *connected in parallel* and the circuit is called a *parallel circuit*.

Point *A* in Fig. 28-8*b* is called a *junction* because currents from more than two branches of the circuit either arrive or leave it. Thus *i* arrives at point *A* and i_1 and i_2 leave it. Point *B* in this figure is also a junction because i_1 and i_2 heads toward this point and *i* leaves it. If we assign a positive sense to the currents arriving at a junction and a negative sense to the current leaving the same junction, then at a given junction,

$$\sum i = 0 \quad \text{(junction rule)} \quad (28\text{-}13)$$

Applying this equation to junction *B*, we get $i_1 + i_2 - i = 0$. The junction rule signifies that charges in an electric circuit cannot "pile up" at a point. The number of charges per unit time that arrive at a point is equal to the number of charges per unit time that leave that point. Equation 28-13 is referred to as Kirchhoff's junction rule.

We seek a single equivalent resistance, R_{equiv}, which can replace the three series resistors of Fig. 28-8*a*. What this means is that we seek the value of a *single* resistor which, if attached to the same battery, will draw the same current, *i*, as is drawn by R_1, R_2, and R_3. This is demonstrated by Fig. 28-7*c*.

Application of the loop rule to the circuit of Fig. 28-8*a*, leads to

$$\mathscr{E} = iR_1 + iR_2 + iR_3$$

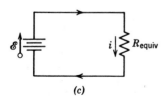

(c)

Figure 28-8   *(c)* Equivalent circuit for the one in Figure 28-8*a*.

also, from Fig. 28-7*c*, $\mathscr{E} = iR_{\text{equiv}}$. Therefore,

$$iR_{\text{equiv}} = iR_1 + iR_2 + iR_3$$

upon canceling the *i*'s, we finally get

$$R_{\text{equiv}} = R_1 + R_2 + R_3 \quad (28\text{-}14)$$

The equivalent resistance for resistors connected in series is equal to the sum of the individual resistances.

We now seek the equivalent resistor for the parallel resistors R_1 and R_2 in Fig. 28-8*b*. Again, we seek a single resistance, R_{equiv}, which will draw the same current as is drawn by both R_1 and R_2.

Application of Kirchhoff's loop rule to the closed loop II in Fig. 28-8*b* leads to

$$\mathscr{E} = i_1 R_1 \quad (28\text{-}15)$$

Applying the loop rule to the closed loop *GCDFG* leads to

$$\mathscr{E} = i_2 R_2 \quad (28\text{-}16)$$

Therefore, $$i_1 R_1 = i_2 R_2$$

The voltage (difference) *is the same across resistors that are connected in parallel with each other.*

The junction rule applied to junction *B* in Fig. 28-8*b* lead to

$$i = i_1 + i_2 \quad (28\text{-}17)$$

Since R_{equiv} is the equivalent resistance that would draw the same total current as is drawn by the two parallel resistors, R_1 and R_2,

$$\mathscr{E} = iR_{\text{equiv}} \quad (28\text{-}18)$$

It follows from Eqs. 28-15, 28-16, and 28-18 that $i_1 = \mathscr{E}/R_1$, $i_2 = \mathscr{E}/R_2$, and $i = \mathscr{E}/R_{\text{equiv}}$. By substituting these values in Eq. 28-17, we obtain

$$\frac{\mathscr{E}}{R_{\text{equiv}}} = \frac{\mathscr{E}}{R_1} + \frac{\mathscr{E}}{R_2}$$

or $$\frac{1}{R_{\text{equiv}}} = \frac{1}{R_1} + \frac{1}{R_2} \quad (28\text{-}19)$$

The reciprocal of the equivalent resistance is equal to the sum of the reciprocals of the individual parallel resistances.

Example 28-6   The circuit of Fig. 28-9 is a combination of resistors in series and in parallel. The numbers indicate the resistance in ohms. Calculate (a) the equivalent resistance of the 10- and 25-ohm resistors; (b) the equivalent resistance of the 40-, 60-, and 40-ohm resistors; and (c) the equivalent resistance of the entire circuit, i.e., what single resistor will carry the same total current as the five resistors combined in the figure? (d) Simplify the circuit diagram of Fig. 28-9 by drawing an equivalent series circuit. (e) Calculate the total current, i.e., the current leaving the battery; (f) the current in each resistor; and (g) the voltage drop across each resistor.

Solution
(a) The 10- and 25-ohm resistors are in series. From Eq. 28-14

$$R_{equiv} = R_1 + R_2 = 25 \text{ ohms} + 10 \text{ ohms}$$
$$= 35 \text{ ohms}$$

(b) The three other resistors are in parallel. From Eq. 28-19

$$\frac{1}{R_{equiv}} = \frac{1}{R_1} + \frac{1}{R_2} + \frac{1}{R_3} = \frac{1}{40} + \frac{1}{60} + \frac{1}{40}$$

Solving for R_{equiv}, we get $1/R_{equiv} = \frac{8}{120}$, or $R_{equiv} = 15\ \Omega$. Note that the equivalent resistance in a parallel circuit is less than any of the individual resistors.

(c) and (d) Since the parallel resistors can be replaced by one 15-ohm resistor, Fig. 28-9 can be redrawn as in Fig. 28-10. This results in a series circuit. The equivalent resistance (from Eq. 28-14) is

$$R_{equiv} = 25 \text{ ohms} + 10 \text{ ohms} + 15 \text{ ohms}$$
$$= 50 \text{ ohms}$$

(e) Since the equivalent resistance is 50 ohms, the total current, by Ohm's law, is

$$i_{total} = \frac{\mathcal{E}}{R_{total}} = \frac{100 \text{ volts}}{50 \text{ ohms}} = 2 \text{ amp}$$

Since the equivalent resistance of the three parallel branches is 15 ohms [part (b) above], and the total current through them is 2 amp, we get for the potential difference between the ends of each by Ohm's law:

$$V_{DC} = 2 \text{ amp} \times 15 \text{ ohms} = 30 \text{ volts}$$

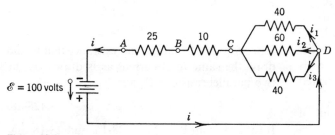

Figure 28-9

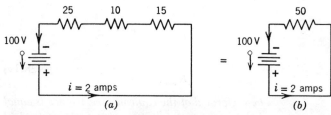

Figure 28-10   Simplified equivalent circuits for Figure 28–9.

(f) The current i_1, through *each* 40-ohm resistor (by Ohm's law) is

$$i_1 = \frac{30 \text{ volts}}{40 \text{ ohms}} = 0.75 \text{ amp} = i_3$$

The current i_2 through the 60-ohm resistor is, by Ohm's law

$$i_2 = \frac{30 \text{ volts}}{60 \text{ ohms}} = 0.5 \text{ amp}$$

Thus $i_1 + i_2 + i_3 = 2$ amp.

(g) $V_B - V_A \equiv V_{BA} = iR_{BA} = 2 \text{ amp} \times 25 \text{ ohms}$
$$= 50 \text{ volts}$$

$V_C - V_B \equiv V_{CB} = iR_{CB} = 2 \text{ amp} \times 10 \text{ ohms}$
$$= 20 \text{ volts}$$

$V_D - V_C \equiv V_{DC} = 30 \text{ volts as computed in}$
part e.

$V_{BA} + V_{CB} + V_{DC}$ should equal the battery emf. ◀

Example 28-7   In Fig. 28-11, let $R_1 = 2$ ohms, $R_2 = 4$ ohms, and $R_3 = 8$ ohms, $\mathscr{E}_1 = 24$ volts, and $\mathscr{E}_2 = 96$ volts. Determine the current in each resistor by Kirchhoff's rules.

Solution.   We assign current directions in each resistor as shown in Fig. 28-11. At junction D, Kirchhoff's junction rule leads to

$$i_2 = i_1 + i_3 \qquad (28\text{-}20)$$

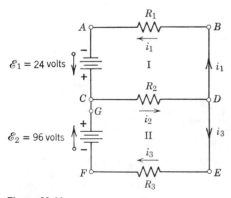

Figure 28-11

Kirchhoff's loop rule, applied from point A around loop I, in a clockwise direction, and back to A leads to

$$+i_1R_1 + i_2R_2 - \mathscr{E}_1 = 0 \qquad (28\text{-}21)$$

For loop II

$$+i_3R_3 + i_2R_2 - \mathscr{E}_2 = 0 \qquad (28\text{-}22)$$

Substituting for i_1, $i_2 - i_3$ in Eq. 28-21, we get

$$i_2(R_1 + R_2) - i_3R_1 = \mathscr{E}_1$$

Solving this equation simultaneously with Eq. 28-22, and filling in the known values, leads to

$$6i_2 - 2i_3 = 24$$
$$4i_2 + 8i_3 = 96$$
$$i_2 = 6.9 \text{ amp} \qquad \text{and} \qquad i_3 = 8.6 \text{ amp}$$

In order to solve for i_1, we substitute the known values for i_2 and i_3 in Eq. 28-20.

$$i_1 = 6.9 \text{ amp} - 8.6 \text{ amp} = -1.7 \text{ amp}$$

with the negative sign of the current for i_1 showing that its direction was incorrectly chosen. The direction of i_1 is from left to right in R_1. ◀

28-8 BATTERY ELECTROMOTIVE FORCE AND TERMINAL VOLTAGE—INTERNAL RESISTANCE

In the previous sections it was assumed that the resistance to current stemmed from parts of the circuit that are external to the battery. Actually, however, all voltage sources possess an inherent resistance, called *internal resistance*. In Fig. 28-12 the internal resistance, r, which really resides within the battery, is drawn outside of it for clarity, and points a and b indicate the battery terminals. The voltage difference, V_{ab}, between these points is called the *terminal voltage*. To compare the emf of a battery with its terminal voltage, we apply Kirchhoff's loop rule to this circuit. Starting at point a and traversing the loop in a counterclockwise direction, we get

$$-iR - ir + \mathscr{E} = 0$$

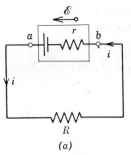

Figure 28-12 The terminal voltage $(V_a - V_b)$ is lower than the battery emf because of the battery's internal resistance. The box around the battery indicates that the internal resistance is within the battery.

or
$$\mathscr{E} - ir = iR$$

The quantity $(\mathscr{E} - ir)$ represents the potential difference between the battery terminals or terminal voltage. It can be seen that $\mathscr{E}$ is larger than the quantity $(\mathscr{E} - ir)$, or the voltage across the terminals of a battery is smaller than the emf of a battery. When a battery is new, r is approximately equal to zero and the terminal voltage approximately equals $\mathscr{E}$.

Example 28-8   What would be the reading on a good voltmeter when connected to the terminals of the battery in Fig. 28-13? What would an ammeter reading be? The battery emf is 12 volts, its internal resistance is 1 ohm, $R_1 = 12$ ohm, and $R_2 = 35$ ohm.

Solution. An ammeter is an instrument that indicates current and is connected in series with the circuit whose current is to be determined. Since such a connection establishes a series circuit, the current through all resistors and ammeters is the same. A voltmeter is an instrument that indicates potential difference and is connected in parallel with the part of the circuit whose potential difference is sought. By Kirchhoff's loop rule,

$$iR_1 + iR_2 + ir = \mathscr{E}$$

$$i = \frac{\mathscr{E}}{R_1 + R_2 + r} = \frac{12 \text{ volts}}{48 \text{ ohms}} = \frac{1}{4} \text{ amp}$$

$$V_A - iR_2 - iR_1 = V_B$$

The terminal voltage $V_A - V_B = iR_2 + iR_1 = 11.75$ V. The terminal voltage is lower than the emf.   ◀

Example 28-9   The circuit of Fig. 28-14 depicts the storage battery 1, having an emf of $\mathscr{E}_1 = 6$ V being "charged up" by the storage battery 2 having an emf of $\mathscr{E}_2 = 12$ V. The internal resistance $r_1 = r_2 = R_2 = 2$ Ω. Calculate (a) the terminal voltage V_{bc} of battery 2, and (b) the terminal voltage V_{da} of battery 1.

Solution. The current direction is determined by the battery having the larger emf. Thus its

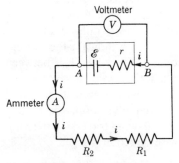

Figure 28-13

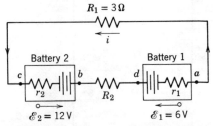

Figure 28-14 Battery 1 is being "charged up" by battery 2. The connecting cable R_1 connects the negative terminals and connecting cable R_2 connects the positive terminals.

direction is counterclockwise as indicated in Fig. 28-14.

(a) We first determine the current by Kirchhoff's loop rule.

$$-2i + 12 - 2i - 6 - 2i - 3i = 0$$
$$i = \tfrac{2}{3} \text{ amp}$$

To determine the terminal voltage $V_{da} = (V_d - V_a)$, we start at point d and add all the changes in potential to point a. Thus

$$V_d - \mathscr{E}_1 - ir_2 = V_a$$
or $V_d - V_a = \mathscr{E}_1 + ir_2 = 6 + \tfrac{2}{3} \times 2 = 7\tfrac{1}{3}$ V.

The terminal voltage of battery 1 is larger than its emf. Therefore, a charge is driven into this battery.

(b) To determine the terminal voltage $V_{bc} = V_b - V_c$, we start at point b and add all the changes in potential to point c.

$$V_b - 12 + ir_1 = V_c$$
$$V_b - V_c = 12 - \tfrac{2}{3} \times 2 = 10\tfrac{2}{3} \text{ V.}$$

The terminal voltage of battery 2 is less than its emf.   ◀

28-9 THE WHEATSTONE BRIDGE

The circuit in Fig. 28-15 illustrates the Wheatstone bridge, an instrument that is used to determine the values of unknown resistors. In this circuit the value of R_x is sought; R_1, R_2, and R_3 are variable but known resistors. In general, when the switch S is pressed down, the ammeter will indicate a current between points A and C through G. By correctly varying the values of any one or combination of the known resistors, the ammeter deflection will be zero, indicating zero current between points A and C through G. Under these circumstances, the loop rule, applied to the closed loop I ($ABCA$) yields, $-i_1R_1 + i_2R_x = 0$, or

$$i_1R_1 = i_2R_x \qquad (28\text{-}23)$$

By applying the loop rule to the closed loop II ($ACDA$), we get $-i_1R_2 + i_2R_3 = 0$, or

$$i_1R_2 = i_2R_3 \qquad (28\text{-}24)$$

Solving for R_x in the simultaneous Eqs. 28-23 and 28-24, we get

$$R_x = \frac{R_3R_1}{R_2} \qquad (28\text{-}25)$$

28-10 POTENTIOMETER

A voltmeter does not accurately measure the electromotive force of a battery for two reasons. First of all, when the voltmeter is connected to the terminals of a battery, it will measure the terminal voltage, which is smaller than the emf because of the internal battery resistance (see Example 28-8). Also, the principle of the operation of a voltmeter (see Section 30-5) is based on a small current's passing through the moving coil within a voltmeter. This also introduces an error.

An instrument that accurately renders the emf is called a *potentiometer*. It is based on an arrangement such that zero current is drawn from the battery whose emf is sought. Therefore, the internal resistance becomes unimportant. Figure 28-16 depicts the theory.

The double-throw switch, S, allows for a connection to the cell of unknown emf, $\mathscr{E}_x$, or to a standard cell whose emf, $\mathscr{E}_s$, is known. If the switch is thrown such that $\mathscr{E}_s$ is in the circuit, the ammeter A will indicate that a current is drawn from the cell. However, by moving the sliding

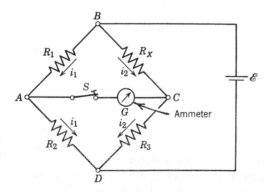

Figure 28-15

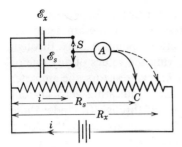

Figure 28-16

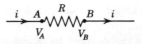

Figure 28-17

contact C to an appropriate position on the resistance wire, a position will be reached where the ammeter will indicate zero current drawn from the cell. Under these circumstances Kirchhoff's loop theorem for the closed loop containing the cell is

$$-iR_s + \mathscr{E}_s = 0 \quad \text{or} \quad \mathscr{E}_s = iR_s$$

The switch is then thrown so that the cell of unknown emf, $\mathscr{E}_x$, is connected to the circuit and the sliding contact moved to a position where A again reads zero. Kirchhoff's loop theorem for the loop containing $\mathscr{E}_x$ yields

$$\mathscr{E}_x = iR_x$$

Solving the last two equations simultaneously yields

$$\mathscr{E}_x = \frac{\mathscr{E}_s R_x}{R_s} \qquad (28\text{-}26)$$

The emf of the unknown cell can thus be computed from the known quantities, $\mathscr{E}_s$, R_x, and R_s.

28-11 ENERGY AND POWER—JOULE'S LAW

By touching an insulated wire that is connected to a battery, one often experiences an increase in the warmth of the wire. The increase in heat energy results from a loss in the electrical energy of the charges traversing the wire. In Fig. 28-17, the potential energy, U_A, at point A of a steady charge q which traverses R toward the right is greater than its potential energy U_B at point B.

But by Eq. 27-7, $U_A = V_A q$ and $U_B = V_B q$, where V_A and V_B are, respectively, the potentials at points A and B. Therefore, the loss in PE is $U_A - U_B = q(V_A - V_B)$. By the conservation of energy principle, this loss is equal to the heat gained by the resistor. The many collisions of the electrons constituting the current, with the atoms of the wire, causes the speed of the atoms to increase, thereby raising the temperature of the wire.

Calling $(V_A - V_B)$ V and noting that $q = it$ (from Eq. 28-1), we arrive at the expression for the heat (H) gained by a resistor: $H = q(V_A - V_B) = Vq = Vit$ so that

$$H = Vit \qquad (28\text{-}27)$$

The product of the current in a resistor, the duration (time) that the current is applied, and the potential difference across the resistor, yields the amount of heat gained by the resistor. A dimensional analysis of this equation shows that

$$H = \text{volt} \times \text{ampere} \times \text{sec}$$

$$= \frac{\text{joule}}{\text{coul}} \times \frac{\text{coul}}{\text{sec}} \times \text{sec} = \text{joule}$$

But heat energy is also measured in calories. Experimental results indicate that

$$4.186 \text{ joules} = 1 \text{ cal.}$$

The power, P that is dissipated is, by definition, the energy per unit time, and from Eq. 28-27

$$P = \frac{Vit}{t} = (Vi)\frac{\text{joules}}{\text{sec}}$$

$$P = Vi \text{ (watt)} \qquad (28\text{-}28)$$

Since 1 watt = 1 joule/1 sec, 1 joule = 1 watt × 1 sec, or watt-sec. Energy can therefore be expressed in units of power × time, e.g., watt-sec, watt-hr. Equation 28-28 can be put in two other useful ways. Note that V in this equation can be replaced by iR (from Ohm's law). Equation 28-28 then becomes

$$P = Vi = i^2R \qquad (28\text{-}29)$$

This equation is referred to as Joule's law.
Also, i in Eq. 28-28 can be replaced by V/R (from Ohm's law) so that a second variation of Eq. 28-28 is

$$P = Vi = \frac{V^2}{R} \qquad (28\text{-}30)$$

Example 28-10 (a) Calculate the heat energy gained by each resistor in the circuit of Example 28-3 in a time of 10 sec. (b) How much electrical power is dissipated in each resistor and in the entire circuit?

Solution. (a) The heat energy H_1 gained by R_1 and H_2 gained by R_2 is, from Eq. 28-27,

$H_1 = 75$ volts $\times$ 5 amp $\times$ 10 sec
$\quad = 3.75 \times 10^3$ joules
$H_2 = 25$ volts $\times$ 5 amp $\times$ 10 sec
$\quad = 1.25 \times 10^3$ joules

(b) $P_1 = 75$ volts $\times$ 5 amp
$\quad = 375$ watts (by Eq. 28-28)

or, by Eq. 28-29

$P_1 = i^2R_1 = (5 \text{ amp})^2 \times 15 \text{ ohms} = 375 \text{ watts}$
$P_2 = i^2R_2 = (5 \text{ amp})^2 \times 5 \text{ ohms} = 125 \text{ watts}$
$\sum P = P_1 + P_2 = 500 \text{ watt}$

or, by Eq. 28-30

$$\sum P = \frac{(\sum V)^2}{\sum R} = \frac{(100 \text{ volts})^2}{20\,\Omega} = 500 \text{ watts} \quad \blacktriangleleft$$

SUMMARY

The electromotive force of a battery is measured in volts and stems from the chemical energy in the battery.

The terminal voltage of a battery is equal to the electromotive force if the internal resistance

is zero and is lower than the electromotive force when the internal resistance is greater than zero.

A thicker wire of the same length, and at the same temperature, has a smaller resistance than a thinner wire.

$$R = \rho\frac{l}{A}$$

The resistivity ρ depends on the kind of metal and temperature.

Ohm's law leads to $V = iR$, where V is the potential difference.

The equivalent resistance, R_{equiv}, for resistors connected in

(a) Series: $R_{equiv} = R_1 + R_2 + R_3 + \text{———}.$

(b) Parallel: $\dfrac{1}{R_{equiv}} = \dfrac{1}{R_1} + \dfrac{1}{R_2} + \dfrac{1}{R_3} + \text{———}.$

Kirchhoff's rules are
(a) Loop-rule: The sum of the changes of potential around a closed loop is zero.
(b) Junction rule: $\sum i = 0$ (at a junction).
The heat energy gained by a resistor is $H = Vit$ joules.

The electric power dissipated in a resistor is

$$P = Vi = i^2R = \frac{V^2}{R} \text{ watts}$$

QUESTIONS

1. What is meant by the emf of a battery; its terminal voltage?
2. Does a sustained current in a wire connected to a battery involve the motion of electrons in the wire only? Explain.
3. Will a given conduction electron in a current-carrying wire necessarily reach the positive terminal of the battery? Explain.
4. Would it be better to use a thin or a thick jumper cable from a good battery to start your car? Explain.
5. A friend might say "shut off your headlights, you're using up the battery current." Is current really used up?

6. When recharging a weak battery by another battery, how should the jumper cables be connected? Why?

7. Which will exhibit a lower resistance, a copper wire at 50°C or the same wire at 0°C?

8. How many different resistors can be constructed from three wires of identical size and material, using all three?

9. Do you pay your electric power company for power or energy?

10. There is a current established in a wire. Does that mean that the total number of electrons at any one instant in the wire is greater than when there is no current in the wire? Explain.

11. What is meant by *superconductivity*?

12. Which of these quantities are of the same dimensions: megawatt, watt-hr, joules, joules/sec, and horsepower?

PROBLEMS

(A)

1. A steady charge of 10 C passes a cross section of a wire in a time of 3 sec. Calculate the current in the wire.

2. Does your answer to Problem 1 depend on the length or thickness of the wire? Explain.

3. How many electrons will move through the cross section of the wire in Problem 1 in 1 sec?

4. How much time is required to pass 50,000 C in a circuit having a steady current of 8 amp?

5. The resistance of a certain wire having a length of 2m and a cross-sectional area of 1 cm² is 5 ohms. The wire is cut into two 1-m lengths.
   (a) What is the resistance of each half?
   (b) The two 1-m lengths are twisted around each other so that an effective cross-sectional area of 2 cm² is achieved. What is the resistance of this combination?

6. A series circuit is constructed with a 20-ohm resistor, a 40-ohm resistor, and a battery having a terminal voltage of 100 V and negligible internal resistance.
   (a) Draw the circuit diagram. Calculate
   (b) the current, and
   (c) the voltage drop across each resistor.

7. Calculate the equivalent resistance of 3 resistors, each having a resistance of 50 ohms when these are connected
   (a) in a series, and
   (b) in parallel.

8. A 20-ohm resistor is connected in parallel with a 40-ohm resistor and a new 100-V battery.
   (a) Draw the circuit diagram.
   (b) Calculate the voltage drop across each resistor.
   (c) Calculate the current established in each resistor.

9. In the circuit of Problem (A)8, calculate
   (a) the heat energy gained by each resistor in a time of 30 sec, and
   (b) the power dissipated by each resistor.

10. An advertising sign consists of 10 identical electric lamps that are connected in series to a 110-V line. Calculate the potential difference across each lamp.

11. (a) What is the resistance at 20°C of a steel wire which is 2 m in length and 0.2 cm² in cross sectional area?
    (b) What would the resistance of such a wire be if its length is halved and its cross-sectional area doubled?

12. Calculate the voltage drop across a 250-ohm resistor carrying a 7-amp current.

(B)

1. When another resistor is connected in series to the one in Problem (A)12, the current drops to 3 amp. Calculate the value of this resistor and the voltage across each.

2. The current in a 100-V series circuit, consisting of a resistor in series with a 750-ohm resistor, is 0.1 amp. Calculate the value of the unknown resistor.

3. Three resistors, 50 ohms, 70 ohms, and 80 ohms, are connected in parallel. What resistor connected in parallel to the above resistors will yield an effective resistance of 10 ohms?

4. Assume that the headlamps of a car draw 2 amp from the storage battery having an emf of 12 V and an internal resistance of 2 Ω. Calculate the terminal voltage of the battery when the headlamps are turned on.

5. Assume that the starting motor of an auto draws 25 amp during the interval that the car is started. If the battery emf is 12 V and its internal resistance

is 0.2 Ω, calculate the battery's terminal voltage during the interval that the car is started.

6. (a) If the electric power company charges you $0.08 per kilowatt-hour, are you paying for power or energy?
   (b) Assume that you illuminate your bedroom by two 50-watt lamps and one 100-watt lamp for 4 hr every day during a 30-day month. How much will this cost for the month?

7. The maximum current that a certain circuit can carry is 5 amp. What should be the smallest resistance of this circuit when connected to a 110-V line?

8. Calculate the current in each resistor of the circuit in Fig. 28-18.

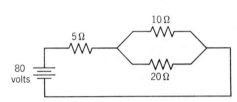

Figure 28-18

9. Calculate the power dissipated in each of the resistors in Problem (B)8.

10. In the circuit of Fig. 28-19, r_1 is 20 ohms, r_2 is 30 ohms, and r_3 is 10 ohms, with r_4 equal to 15 ohms. Calculate:
    (a) The current in each resistor.
    (b) The voltage drops in each resistor.
    (c) The power dissipated in the entire circuit.

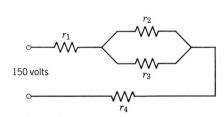

Figure 28-19

11. An electric lamp, having a 75-ohm filament, is connected to 110-V direct-current line.
    (a) How much current will flow through the lamp?
    (b) How many coulombs will flow through the circuit in 5 min.?
    (c) How much energy, in watt-sec, is consumed in this time?

12. Compute the current in each of the resistors of the circuit shown in Fig. 28-20.

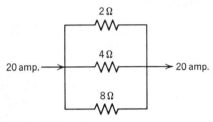

Figure 28-20

13. How much energy is consumed in 100 sec by each of the resistors in Problem (B)12?

14. Using Kirchhoff's rules, determine the current and its direction in each resistor of the circuit in Fig. 28-21.

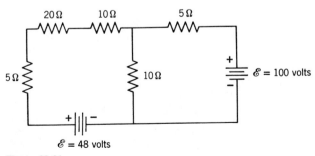

Figure 28-21

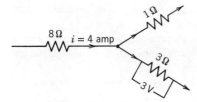

Figure 28-22

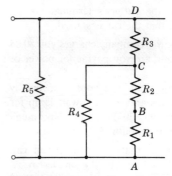

Figure 28-23

15. Calculate the currents in the (a) 1 Ω resistor (b) 3 Ω resistor and (c) the voltage drop across the 1 Ω resistor in Fig. 28-22.

16. $R_1 = R_2 = R_3 = 50 \, \Omega$, $R_4 = R_5 = 30 \, \Omega$ and the potential difference between points A and D is 200 V. Calculate (a) the equivalent resistance between points A and D and (b) the potential difference between B and C in Fig. 28-23.

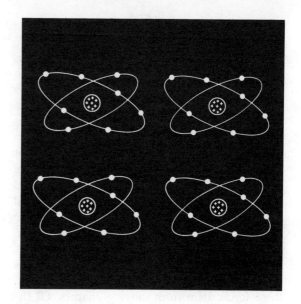

29.
Magnetism and
Electromagnetism

Electricity and magnetism are intimately associated. When an electric current is established in a wire, the space near the wire exhibits magnetic characteristics. In this chapter, some properties of magnets are discussed and some principles of electromagnetism are developed.

LEARNING GOALS

1. *To learn the equation (and how to apply it) for the forces between magnets.*
2. *To understand the concept of the magnetic field.*
3. *To learn about electric currents giving rise to magnetic fields.*
4. *To learn the right-hand rule for determining the direction of the field caused by a current in a wire.*
5. *To understand the concept of hysteresis.*

Certain materials, like iron ores, exhibit forces of attraction or repulsion when placed near each other. Such materials are called *magnetic materials* and these forces are referred to as *magnetic forces*. So far this text has treated two fundamental types of forces: forces between charged bodies, or electrical forces, and forces between any objects, or gravitational forces. In addition to these, there exists in nature another type of fundamental force, namely, the nuclear force. It is to be noted that gravitational forces are only of the attractive type, whereas electrical and magnetic forces can be either attractive or repulsive (see Fig. 29-1). Magnetic forces are not usually considered as fundamental because, as will be pointed out later, these forces stem from moving charges. Of the three fundamental forces, the nuclear force is the strongest, and the gravitational force the weakest. The electrical force is many orders of magnitude

larger than the gravitational force and only about one order of magnitude smaller than the nuclear force. The fundamental forces are noncontact forces, which makes their causes difficult to understand.

29-1 FORCES BETWEEN MAGNETS

Figure 29-1c demonstrates that the ends of two bar magnets will attract or repel one another, depending on how they are placed. If the ends marked N (for north pole) or the ends marked S (for south pole) are placed near each other, they will be repelled; if one of these magnets is reversed so that a north pole faces a south pole, the magnets will attract each other. Figure 29-2 demonstrates that the magnetic properties of

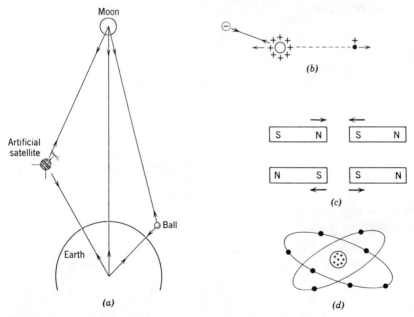

Figure 29-1   Four types of physical forces. (a) Gravitational forces are always attractive and occur between any two bodies. (b) Electrical forces occur between charged bodies. Like charges repel and opposite charges attract. Only two types of electrical charges are known. (c) Magnetic forces occur between magnets. Like poles repel and opposite poles attract. Only two types of magnetic poles are known. (d) The nucleus of an oxygen atom contains eight protons. Nuclear forces prevent the protons from flying apart.

Figure 29-2 Iron powder is concentrated at the ends of a bar magnet that has been dipped into iron powder. This indicates that the magnetic properties of a bar magnet are concentrated at its ends.

magnets are concentrated in two regions called *poles*, located near the ends. All magnets, no matter how small, have two poles (see Fig. 29-3). Because one end of a horizontally suspended bar magnet will point in a northerly direction, that end of the magnet is called the north-seeking pole. The opposite end of the magnet is called the south-seeking pole. The word "seeking" is usually omitted.

scientist, Charles Augustine de Coulomb (1736–1806), experimentally determined a quantitative expression for the force of interaction between magnets. He used a torsion balance which is similar to the one depicted in Fig. 29-4. This

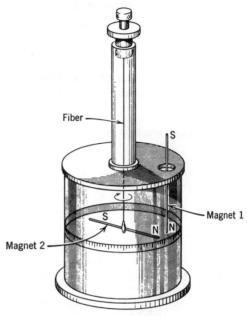

Figure 29-4 Torsion balance for determinations of magnetic forces.

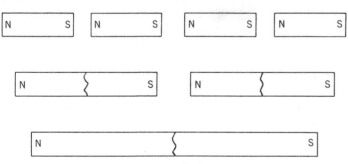

Figure 29-3 When a magnet is cut, each piece no matter how small, will exhibit a north and south pole.

29-2 COULOMB'S LAW FOR MAGNETISM

The qualitative characteristics of magnets were known in the times of ancient Greece, but it was not until the eighteenth century that the French

instrument is essentially an evacuated cylinder into which two bar magnets are inserted. Magnet 1 is fixed so that it cannot move, and magnet 2 is suspended by a fine fiber. The force of interaction (attraction or repulsion) between the magnets gives rise to a torque on the suspended magnet, causing it to twist. The force acting on the suspended magnet can be deduced from its angular displacement. After trying the same magnets at various distances apart, and multiples of identical magnets at the same distance, it can be determined that:

1. For the same pair of magnets, the force F of attraction or repulsion is inversely proportional to the square of the distance r between the poles (ends) of the two magnets, or

$$F \propto \frac{1}{r^2} \text{ (for the same magnets)}$$

2. For the same pole distances, the force of attraction or repulsion is directly proportional to the pole strength p of each magnet or

$$F \propto p_1 p_2 \text{ (constant distance)}$$

The last two mathematical expressions can be combined into one equation, known as Coulomb's law for magnetism:

$$F = k \frac{p_1 p_2}{r^2} \qquad (29\text{-}1)$$

(Does this recall Coulomb's law for electricity?) Here k is a proportionality constant, the value of which depends on the units used, and on the medium in which the magnets are immersed.

There exists no experimental evidence to indicate that the poles of a given magnet are located at geometrical points, or even, very small regions. Equation 29-1 is useful in describing the magnetic field caused by a given magnet (Sec. 29-3) and serves as a definition of a *unit magnetic pole*. In the cgs system, a unit pole can be defined by letting $k = 1$. Thus one cgs unit pole when placed 1 cm away from an identical unit pole (in vacuum) will repel it with a force of 1 dyne.

Example 29-1   A 10-g and a 5-g magnet are held down on a flat, frictionless surface with their south poles facing and 10 cm apart. See Fig. 29-5. Each pole has a strength of 200 cgs unit poles and each magnet is 10 cm long. (a) Calculate the acceleration of the 5-g magnet at the moment that it is released. (b) Calculate the acceleration of the 10-g magnet at the moment it is released, while the 5-g magnet is held at its original position. Assume the poles to be at the ends of each magnet.

Solution.   There are four interactions involved: N_1 with S_2, N_1 with N_2, S_1 with S_2, and S_1 with N_2. Applying Eq. 29-1 to each interaction and putting $k = 1$, we obtain for the forces on the 5-g magnet:

(a) The force on N_1 due to S_2 is

$$F_{N_1 \rightarrow S_2} = \frac{1 \times 200 \text{ unit poles} \times 200 \text{ unit poles}}{(20 \text{ cm})^2}$$

$$= 100 \text{ dynes (toward the left)}$$

$$F_{N_1 \rightarrow N_2} = \frac{1 \times 200 \text{ unit poles} \times 200 \text{ unit poles}}{(30 \text{ cm})^2}$$

$$= 44.4 \text{ dynes (toward the right)}$$

$$F_{S_1 \rightarrow S_2} = \frac{1 \times 200 \text{ unit poles} \times 200 \text{ unit poles}}{(10 \text{ cm})^2}$$

$$= 400 \text{ dynes (toward the right)}$$

$$F_{S_1 \rightarrow N_2} = \frac{1 \times 200 \text{ unit poles} \times 200 \text{ unit poles}}{(20 \text{ cm})^2}$$

$$= 100 \text{ dynes (toward the left)}$$

The resultant force on the 5-g magnet is the vector sum of the four forces just given. Assigning a negative value to the forces pointing to the left

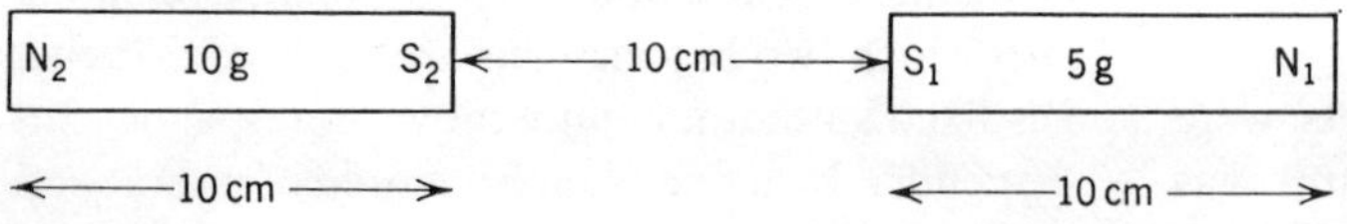

Figure 29-5

and a positive value to those pointing toward the right, we get

$$\sum \vec{F} = -100 \text{ dynes} + 44.4 \text{ dynes} + 400 \text{ dynes}$$
$$- 100 \text{ dynes}$$
$$= 244.4 \text{ dynes (to the right)}$$

$$\sum \vec{F} = m\vec{a}; \quad \vec{a} = \frac{\sum F}{m} = \frac{244.4 \text{ dynes}}{5\text{-g}}$$
$$= 48.9 \text{ cm/sec}^2$$

Thus the acceleration of the 5-g magnet is 48.9 cm/sec² toward the right, since the resultant force is to the right.

(b) A similar analysis of the magnetic forces acting on the 10-g magnet will show the resultant force to be also 244.4 dynes, but pointing toward the left. This can also be deduced from Newton's third law.

$$\vec{a} = \frac{\sum \vec{F}}{m} = \frac{-244.4 \text{ dyne}}{10\text{-g}}$$
$$= -24.4 \text{ cm/sec}^2 \text{ (to the left)} \quad \blacktriangleleft$$

29-3 MAGNETIC FIELD AND FIELD STRENGTH

A magnet modifies the space surrounding it. When any magnet is inserted into a region (see Fig. 29-6a), the magnetic characteristics of this region change in the sense that, now, when another magnet is brought into this region of space, it will experience magnetic forces. The space has thus acquired magnetic properties and is said to constitute a *magnetic field*. (Similar arguments prevail for gravitational and electrical fields.) A magnetic field is described qualitatively by means of a sketch showing *magnetic lines of force* (Fig. 29-6c, d). *Lines of force start at the north end of a magnet and end at the south end. The direction of the magnetic field at any point is the direction of the force exerted by the field on a unit north pole placed at that point.*

Points in a magnetic field are described quantitatively by the force per unit north pole at that point. The force per unit pole is called the *magnetic field strength, B,* and

$$B = \frac{F}{p} \quad (29\text{-}2)$$

In the cgs system of units, B is measured in *gauss* (after the German mathematician, Karl Friedrich Gauss) when F is in dynes and p in cgs unit poles. One gauss is therefore one dyne/unit pole. The field, B is a vector because it is found by dividing a vector, F by a scalar, p. The direction of B is the same as that for the force.

Example 29-2 Determine the magnitude and direction of the magnetic field strength at point P of Fig. 29-6a, which lies on the perpendicular bisector of the magnet and at a distance of 10 cm from each pole. The magnet has a pole strength of 300 cgs unit poles.

Solution. A unit magnetic north pole is imagined at point P, and the forces exerted on it by the north (F_N) and south (F_S) poles of the magnet are calculated from Eq. 29-2. The force exerted by the north pole is

$$F_N = \frac{1 \times 300 \text{ unit poles} \times 1 \text{ unit pole}}{(10 \text{ cm})^2} = 3 \text{ dynes}$$

This force acts on the unit north pole at P on a line joining P and N, and is directed away from N.

The force F_S that the south pole exerts on the unit pole is

$$F_S = \frac{1 \times 300 \text{ unit poles} \times 1 \text{ unit pole}}{(10 \text{ cm})^2} = 3 \text{ dynes}$$

This force is on the line connecting P and S and is directed toward S (since unlike poles attract).

The resultant force, F on the unit pole is

$$F = \sqrt{(3 \text{ dynes})^2 + (3 \text{ dynes})^2} = 3\sqrt{2} \text{ dyne}$$

and is directed toward the bottom of the page. From Eq. 29-2 the magnitude of the field strength, $B = 3\sqrt{2}$ dyne/pole $= 3\sqrt{2}$ gauss. The direction of B is the same as that for F. The field strength at any other point can be computed in the same manner. $\quad \blacktriangleleft$

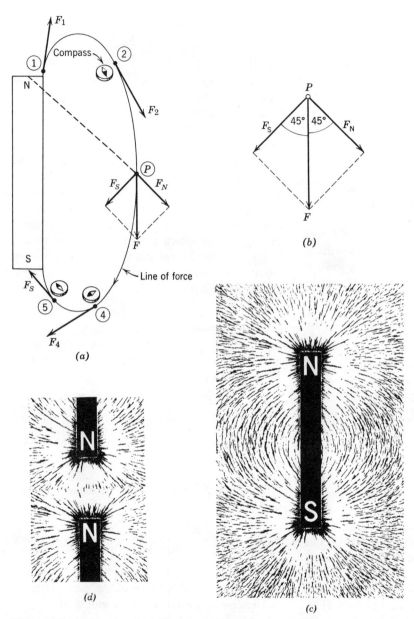

Figure 29-6 (a) The direction of the magnetic field at any point (as for instance 1 and 2) is the direction of the magnetic force on a unit north pole placed at that point. A compass will point along a line that is tangent to the line of force. (b) F is the vector sum of F_N and F_S (see example 29-2). (c) Patterns of iron powder near a bar magnet. The curved lines constitute the magnetic lines of force. (d) Patterns of iron powder in the vicinity of two north poles.

29-4 THE MAGNETISM OF THE EARTH

When a thin and narrow strip of magnetic material, called a *magnetic needle,* is mounted horizontally on a near frictionless bearing in a *magnetic compass,* the needle will oscillate and soon settle in an equilibrium position. At the equilibrium position, one end of the needle will point in approximately a northerly direction and the other end in an approximate southerly direction. Since this occurs everywhere on earth, we conclude that the earth is surrounded by a magnetic field. Figure 29-7 is a sketch of the earth's field. An *imaginary* giant bar magnet, but smaller than the earth's diameter, passing through the center of the earth, would give rise to similar field lines (compare with Fig. 29-6c). Notice that the north pole of the imaginary magnet is in the southern geographic hemisphere and the south pole in the northern geographic hemisphere.

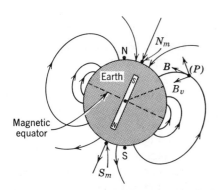

Figure 29-7   The magnetic field of the earth.

The earth's *magnetic north pole,* N_m does not coincide with the *geographic north pole.* Their separation is about 1150 miles. That is why a compass does not point in the true northerly direction. Let us choose an arbitrary point, point P in Fig. 29-7 on a line of force, say, in the New York region. The direction of the field, *B,* at this point is along a line that is tangent to the curved line. Since *B* is a vector, we can resolve it into a vertical component, B_v (the vertical is a line going

through point *P* and the center of the earth) and into a horizontal component (perpendicular to the vertical). A magnetic compass needle that is mounted in a horizontal plane will align itself with the horizontal component of the field. In New York City the horizontal component points in a direction that is approximately 12° west of true north. The angle between the horizontal component and the true north is called the *declination.* The magnitude of the horizontal component in the New York region is about 0.2 gauss. The vertical component has an approximate magnitude of 0.6 gauss. The earth's magnetic field at a given region on the earth does not remain constant and varies with time. The understanding of these variations awaits a clear explanation of the causes for the earth's magnetism.

If the compass needle is suspended about a horizontal axis so that the needle can swing in a vertical plane, the needle is called a *dip needle.* In the New York area the dip needle will dip approximately 72° below the horizontal. This is called the *angle of dip.* In the regions of the magnetic north pole the horizontal component of the field is zero; there is only a vertical component (see Fig. 29-7). The same is true at the magnetic south pole. The angle of dip in these regions is 90°, i.e., a dip needle would point vertically toward the center of the earth. In the regions of the magnetic equator, the vertical component is zero and the field has only a horizontal component. The causes for the earth's magnetism are not understood too well. One explanation involves the motion of charged particles within the earth.

29-5 A BAR MAGNET IN A UNIFORM FIELD

A bar magnet will align itself with an external magnetic field. A uniform horizontal field *B* is depicted in Fig. 29-8. The expression "uniform field" in a region of space indicates that the field in the region has the same direction and the same magnitude throughout the region. As is indicated by Fig. 29-6a, a bar magnet does not produce a

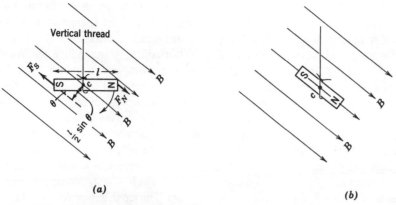

Figure 29-8   (a) The net torque will rotate the magnet. (b) The net torque is zero. The magnet is aligned with the external field.

uniform field. For instance, the directions of the fields at points 1, 2 and 4 of Fig. 29-6a are different. However, the region between the opposite poles of two closely spaced, thick bar magnets does exhibit an almost uniform field.

If a bar magnet is suspended horizontally by a fine thread as in Fig. 29-8a, it will rotate until it aligns itself with the field, Fig. 29-8b. The reason for this can be deduced by the following considerations. The force F_N that the field B exerts on the north pole of the magnet is, from Eq. 29-2, $F_N = pB$, where p is the pole strength of the suspended magnet. The field will also interact with the south pole, exerting a force on it, $F_S = pB$. In spite of the fact that the resultant of these two forces is zero, they will produce an unbalanced torque.

The clockwise torque produced by F_N around an axis going through the magnet's center, C, is $F_N \times (l/2) \sin \theta$, where $l/2 \sin \theta$ is the lever arm distance, l is the length of the magnet, and θ is the angle that the magnet makes with the direction of the field. The force F_S also produces a clockwise torque, $F_S \times (l/2) \sin \theta$. The total torque M is

$$M = F_N \times \frac{l}{2} \sin \theta + F_S \times \frac{l}{2} \sin \theta$$

But $F_N = F_S = pB$. The torque equation can

therefore be written as

$$M = lpB \sin \theta \qquad (29\text{-}3)$$

This torque will act on the magnet until it aligns itself with the field, Fig. 29-8b. At this orientation the vector sum of the forces and torques on the magnet is zero. The quantity lp in Eq. 29-3 depends on the characteristics of a given magnet and is called the *magnetic moment*.

As stated earlier, the phrase "north and south pole" describes in a general way the difference between the two ends of a magnet but does not imply that each pole is at a point. Experiments (such as depicted in Fig. 29-2) do indicate that the magnetism of a magnet is concentrated somewhere near the two ends but do not precisely locate the poles. The magnetic characteristics of a given magnet, i.e., the product of the pole strength and pole separation (lp) can be determined from an experiment suggested by Fig. 29-8 and Eq. 29-3 without an individual knowledge of the pole locations.

29-6 ELECTROMAGNETISM

In 1820, Hans Christian Oersted discovered a fundamental association between electricity and magnetism: *A moving charge* (electric current) *gives rise to a magnetic field.* Oersted's experimental set-up was similar to the one depicted in Fig. 29-9.

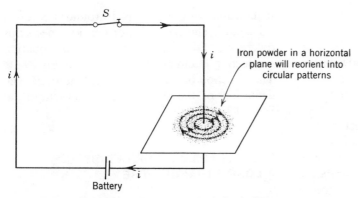

Figure 29-9 The arrows on the concentric circles indicate the direction of the magnetic lines of force produced by a downward current in a vertical wire.

With no current in the wire, a compass needle points toward the north. But when a current is established, the compass will turn away from its previous orientation, showing the presence of a magnetic field other than that of the earth. Experiments such as this indicate that an electric current gives rise to a magnetic field. In fact, modern theory of magnetism holds that the magnetism of permanent magnets is caused by the orbital motion of electrons (electric current) in the atoms of these magnets.

Assume that the wire in Fig. 29-10 is vertical. The magnetic field produced by the current in this wire is directed northward at points directly to the right of the wire and southward at points located directly to the left of the wire. These directions are indicated in Fig. 29-10 by crosses (xx), which signify a direction heading into the

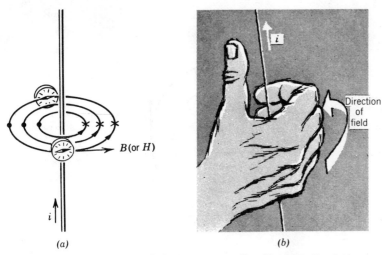

(a) (b)

Figure 29-10 (a) The north end of a compass needle will point to the right when placed directly in front of the vertical wire and toward the left when placed directly behind the wire. (b) Right-hand rule—a method for determining the direction of the magnetic field caused by a current-carrying wire.

page, and by dots (..), which signify a direction coming out of the page. The *right-hand rule* is helpful in determining the direction of the magnetic field. If the thumb of the right hand is placed in the direction of the conventional current, the tips of the remaining fingers will point in the direction of the field.

So far we have been using the vector *B* to describe the interactions between poles. In the sections that follow, we shall use the vector *B*, also called the *magnetic flux density*, to describe the magnetic field caused by a current. Another quantity, the magnetic field intensity, *H* which does not depend on the medium, is also used to describe the field. For instance, the magnetic field resulting from a current-carrying wire that is wrapped around a piece of iron is stronger than

if the same wire is wrapped around a piece of cardboard. The magnetic flux density reflects this difference. Of course, there is an equation connecting *B* and *H* which is given by Eq. 29-10. The direction of *B* is the same as that for *H*. The definition of the flux density is developed in Sec. 30-2.

29-7 THE FIELD PRODUCED BY A LONG STRAIGHT WIRE

In the last section, the direction of the magnetic field created by a current-carrying wire was discussed. But what about its magnitude? In general, the magnitude of the field depends on the shape of the wire.

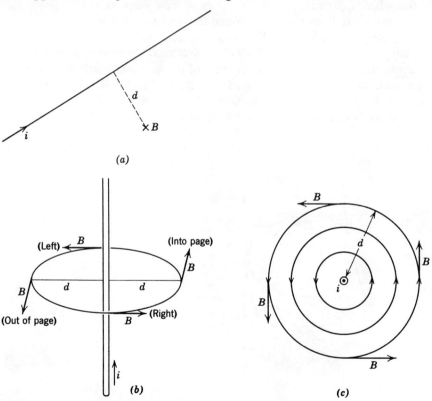

(a)

(b)

(c)

Figure 29-11   (a) The magnetic field *B* at a perpendicular distance *d* from a straight current-carrying wire. (b) The direction of the field caused by a vertical wire carrying a current upwards. (c) The wire in (b) viewed from above. The dot near *i* indicates the wire to be perpendicular to the page with the current heading out of the page.

The field B at a perpendicular distance d meters from a wire carrying a current of i amperes (Fig. 29-11) is given by

$$B = \frac{\mu_0 i}{2\pi d} \qquad (29\text{-}4)$$

where B is measured in units of weber/m². As is seen from this equation, the field does not depend on the length of the wire, provided that it is long compared with d. The quantity μ_0 is called the *permeability constant of free space*. Its value is

$$\mu_0 = 4\pi \times 10^{-7} \text{ weber/ampere-meter}$$

The direction of B is determined by the right-hand rule. Thus, in Fig. 29-11b, B points into the page at points located directly at the right of the wire (when looking in the direction of the current) and out of the page at points directly at the left of the wire.

Example 29-3   Two parallel long wires, 20 cm apart carry a current of 10 amp in the same direction. What is the flux density at a point halfway between the wires?

Solution.   From Eq. 29-4, the flux density B at a midpoint caused by each wire is

$$B = 4\pi \times 10^{-7} \frac{\text{weber}}{\text{amp-m}} \times \frac{10 \text{ amp}}{2\pi \times 0.1 \text{ m}}$$

$$= 20 \times 10^{-6} \text{ weber/m}^2$$

By the right-hand rule, the direction of B produced by one wire is opposite to that produced by the other wire. The two fields, being equal in magnitude and of opposite direction, cancel. The flux density at a midpoint is therefore zero. ◀

29-8 THE FIELD AT THE CENTER OF A CIRCULAR WIRE

The field B at the center of a circular loop (Fig. 29-12) having N turns of wire is given by

$$B = \frac{\mu_0 N i}{2r} \qquad \text{(center of circular loop)} \quad (29\text{-}5)$$

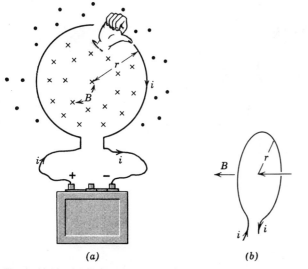

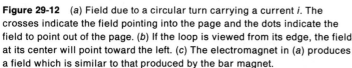

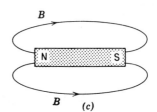

(a)   (b)   (c)

Figure 29-12   (a) Field due to a circular turn carrying a current *i*. The crosses indicate the field pointing into the page and the dots indicate the field to point out of the page. (b) If the loop is viewed from its edge, the field at its center will point toward the left. (c) The electromagnet in (a) produces a field which is similar to that produced by the bar magnet.

The current i is expressed in amperes, r, the radius of the loop, is expressed in meters, and B in weber/m^2. The direction of B is determined by the right-hand rule.

29-9 FIELD PRODUCED BY A SOLENOID

A solenoid consists of many circular turns of wire closely wound in the shape of a cylinder. The current in every turn is in the same direction. A solenoid produces a magnetic field within the cylinder formed by the wires. The field is very nearly constant near the axis (see Fig. 29-13). This feature makes it a useful instrument for many laboratory and industrial applications. The field near the axis of a long solenoid is given by

$$B = \frac{\mu_0 N i}{l} \quad \text{(solenoid)} \quad (29\text{-}6)$$

The flux density B is expressed in weber/m^2, when the current i is in amperes, the length l in meters, and μ_0, the permeability constant, is expressed in weber/amp-m. The number of turns is denoted by N. A solenoid is considered long if its length is a few times greater than its diameter.

Example 29-4 A solenoid (Fig. 29-14), 20 cm long, consists of 200 turns wound on a cardboard hollow cylinder. Calculate the magnitude of the magnetic field near its axis if the current is 5 amp. What is the direction of the field?

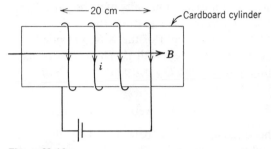

Figure 29-14

Solution. From Eq. 29-9,

$$B = \frac{4\pi \times 10^{-7} \times 200 \times 5}{0.2}$$

$$= 20\pi \times 10^{-4} \text{ weber/m}^2$$

The current in every wire on the side shown is directed down. The right-hand rule indicates that

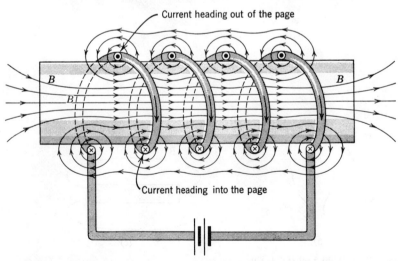

Figure 29-13 A solenoid. The field due to a few continuous circular loops carrying current in the same direction. Each turn contributes to the field causing the magnetic field to increase in the direction shown. B is constant in magnitude and direction in the central axis.

B points from left to right within the solenoid. This is equivalent to a bar magnet, oriented as in Fig. 29-14b with the north side on the right. ◀

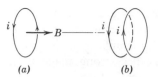

Figure 29-15   (a) A current in a circular loop causes a magnetic field. (b) The resultant field is zero when the direction of the current is opposite in two adjacent loops.

29-10 MAGNETIC THEORY

All magnets, no matter how small, have a north and a south end. A single pole has never been isolated. In Sections 29-8 it was stated that a circulating current in a wire gives rise to an electromagnet having a north and south end. It is believed that the magnetism in a permanent magnet also arises from circulating charges. The atomic electrons of all materials orbit around the nucleus. In addition, electrons also spin about an axis passing through them, in a manner similar to the earth's spin. We thus associate the magnetism of a permanent magnet with circulating charges. The smallest magnet is therefore a spinning electron.

All materials, including wood, aluminum, iron, gases, and liquids possess spinning and orbiting electrons, but only iron (and a few other elements) exhibit pronounced magnetic characteristics. An understanding of the reason for this is strengthened by a macroscopic analogy, the electromagnet of Fig. 29-12. Consider a circular loop of wire,

carrying a current i, Fig. 29-15a. If we place another identical loop carrying the same current but in the opposite direction (Fig. 29-15b) very near and parallel to the first loop, it will give rise to a magnetic field that is equal and opposite to that of the first loop. The resultant magnetic field due to both loops is zero. On the atomic level, the directions of spins of the electrons in some materials like wood or aluminum are such as to neutralize the magnetic properties of these atoms. In iron the magnetic effects are not neutralized.

The tiny fields produced by the atoms in iron do not point in one direction, as shown in Fig. 29-16a. However, there is evidence that in a given sample of iron there exist many very small regions, in which the field does effectively point in one direction. Such regions are called *domains* (see Fig. 29-16c). Notice that the directions of the

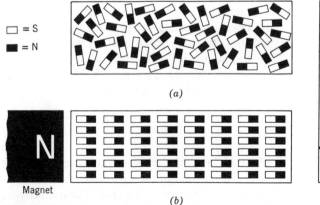

□ = S

■ = N

(a)

(b)

Magnet

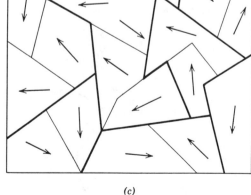

(c)

Figure 29-16   (a) Unmagnetized iron. (b) Totally magnetized iron. (c) Domains. The directions of the resultant fields in the different domains of unmagnetized iron are different.

fields in different domains are, in general, different. Such a chunk of iron is said to be *unmagnetized*. When such a sample of iron is brought in the vicinity of another strong magnet, the latter will exert a magnetic force causing the domains to line up in the direction of the external field (Fig. 29-16b). When the external field is then removed, the magnetic polarity of the chunk of iron persists and the iron retains a strong north and south pole. This is why such magnets are called "permanent" magnets. If the atoms of a permanent magnet are set into violent motion, as for instance by heating the magnet to a high temperature, or by strong impacts, the alignment of the atoms is disturbed and the magnetism is reduced. At temperatures higher than about 770°C, an iron permanent magnet will lose much of its magnetism.

A quantity that indicates the relative magnetic strength of a given material is the *relative permeability*. The material whose relative permeability is sought is shaped into a Rowland ring (named after the American physicist) and the flux density B_m is experimentally determined (see Fig. 29-17). The flux density is partly due to the electromagnet produced by the current in the wire around the ring and also to the magnetism of the material of the ring. Then the flux density, B_0, produced by the same current in the wire, wound in the same manner but this time in vacuum, without the material, is determined. The relative permeability of the material K_m is given by

$$K_m = \frac{B_m}{B_0} \qquad (29\text{-}7)$$

The maximum value for the relative permeabilities for different materials varies from a number that is slightly less than 1 to approximately 80,000. For instance, the maximum value for iron is about 8000; for an alloy called permalloy, which consists of about 80% nickel and 20% iron, $K_m = 80{,}000$. Materials whose K_m is much greater than one are called *ferromagnetic* materials; examples are the elements iron, nickel, and their alloys. Materials whose relative permeabilities are slightly larger than 1 are called *paramagnetic*. Platinum is an example, $K_m = 1.0002$. Substances having relative permeabilities that are slightly less than 1 are called *diamagnetic*. Bismouth is an example, $K_m = 0.09998$.

29-11 HYSTERESIS

The relative permeability of a ferromagnetic material is not constant. That is why the maximum values for K_m were cited in the preceding section. The relative permeability of a piece of iron can have values of a few hundred, indicating slight magnetization, or up to 8000 indicating a very high degree of alignment of the atomic poles. The degree of magnetization of a given piece of ferromagnetic material depends on its previous history, i.e., on its previous exposure to an external magnetic field.

Magnetization curves for a given sample are obtained by shaping the sample into a Rowland ring and obtaining experimentally the flux density B_m and the magnetic intensity H. The theoretical expression for the flux density for a material shaped into a Rowland ring is

$$B_m = \frac{\mu_m N i}{l} \qquad (29\text{-}8)$$

where N is the number of turns of wire around the ring, i the current in the wire, and l the mean circumferential length of the ring. The quantity μ_m is the permeability of the metal. The equation for the magnetic intensity H is the same as that for B_m except that it does not depend on μ_m.

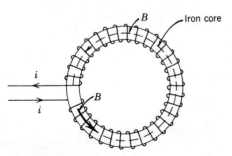

Figure 29-17   Rowland ring.

$$H = \frac{Ni}{l} \text{ ampere-turns/meter} \qquad (29\text{-}9)$$

Substituting H for Ni/l in Eq. 29-8 yields

$$B_m = \mu_m H \qquad (29\text{-}10)$$

The field intensity H given by Eq. 29-9 is in mks units and involves a current in a wire. The flux density as given by Eq. 29-2 is in cgs units and describes the field produced by a magnet.

A plot of B_m versus H, called a *hysteresis curve*, reveals the magnetic properties of a given ferromagnetic material. A typical hysteresis curve for iron is given in Fig. 29-18. As the current in the coil is increased, H, the magnetic intensity, which is associated with the current (not with the magnetic material), is also increased. So does the flux density, B_m, associated with the material, increase. In particular, in the excursion from point O to point A in Fig. 29-18, along the route marked ①, values H_A and B_A are reached at point A. If the current is then reduced to zero, H becomes zero, but B_m maintains a value that is higher than zero. This indicates that ferromagnetic materials have a *retentivity*. The magnetic properties of a given sample of a ferromagnetic material depend

on its past history. This characteristic is called *hysteresis*. Both quantities B_m and μ_m are not constants for a given sample of a magnetic material.

29-12 THE MAGNETIC CIRCUIT

The magnetism in a Rowland ring is almost entirely confined within the ring, Fig. 29-19. The magnetic lines of force are closed loops, which suggests similarities to the closed current loops in an electric circuit. The flux density in the ring is

$$B_m = \frac{\mu_m Ni}{l} \qquad (29\text{-}8)$$

The magnetic flux ϕ (phi) is $\phi = BA$, where A is the uniform cross-section area of the ring. Therefore,

$$\phi = \frac{\mu_m NiA}{l}$$

If we rearrange the last equation as

$$\phi = \frac{Ni}{l/\mu_m A}$$

we get an equation that resembles the form of Ohm's law, $i = \mathscr{E}/R$. Thus ϕ is analogous to i, Ni is analogous to $\mathscr{E}$, the electromotive force and $l/\mu_m A$ is analogous to resistance. In the electric circuit, $\mathscr{E}$ is the "cause" of the current; in the magnetic circuit, Ni is the cause of the magnetic

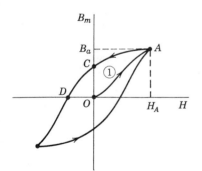

Figure 29-18   Hysteresis loop for a ferromagnetic material. (1) $0 \to A$ (on the curve) The nonmagnetized core becomes magnetized. The flux density B increases as the magnetizing field H is increased. (2) $A \to C$ (on the curve) The magnetizing field H, (Ni/L) is reduced to zero by reducing the current to zero. The flux density does not return to zero and the core retains some magnetization.

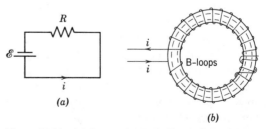

Figure 29-19   (a) Current loop. (b) Magnetic loops.

flux ϕ. In the electric circuit, R is the resistance to current; in the magnetic circuit, the quantity $l/\mu_m A$ represents the opposition to flux. The quantity Ni is called the *magnetomotive force*, mmf and $l/\mu_m A$ is called the *reluctance $\mathscr{R}$*. Thus

$$\phi = \frac{Ni}{l/\mu_m A} = \frac{\text{mmf}}{\mathscr{R}} \qquad (29\text{-}11)$$

where mmf is measured in ampere-turns $\mathscr{R}$ in ampere-turns/weber, and ϕ in webers. The last equation is true for magentic circuits of any shape (not necessarily a ring). The analysis of the magnetic circuit is important in instruments and machines utilizing magnets, e.g., electric generators, motors, and voltmeters.

Example 29-5 An iron core, having a relative permeability of 1000, has 100 turns of wire wound around it. The current in the wire is 7 amp, the cross-section area of the core is 7 cm², and the mean length of the core l is 70 cm. Compute (a) the magnetomotive force, (b) the reluctance, and (c) the flux in the core.

Solution

(a) mmf $= Ni = 100$ turns $\times$ 7 amp

$\qquad = 700$ amp-turns

(b) $\mathscr{R} = \dfrac{l}{\mu_m A}$; $l = 0.7$ m,

$$A = 7 \text{ cm}^2 \times \frac{1 \text{ m}^2}{10^4 \text{ cm}^2} = 7 \times 10^{-4} \text{ m}^2$$

We can deduce μ_m from the following considerations. The flux density, B_m in the core of a Rowland ring is given in Eq. 29-8, $B_m = \mu_m Ni/l$. If the turns of wire were wrapped around the ring, which is a vacuum, the flux density is $B_0 = \mu_0 Ni/l$. It can be seen from the last two equations that $(B_m/B_0) = (\mu_m/\mu_0)$, where μ_0 is the permeability of free space (vacuum). Therefore, from Eq. 29-7 $K_m = (B_m/B_0) = (\mu_m/\mu_0)$ and $\mu_m = K_m\mu_0$. In the problem K_m is given to be 1000, and μ_0 is $4\pi \times 10^{-7}$ weber/amp-m. Therefore, $\mu_m = K_m\mu_0 = 1000 \times 4\pi \times 10^{-7} = 4\pi \times 10^{-4}$ weber/amp-m.

$$\mathscr{R} = \frac{l}{\mu_m A} = \frac{0.7}{4\pi \times 10^{-4} \times 7 \times 10^{-4}}$$

$$= 8 \times 10^5 \text{ amp-turns/weber}$$

(c) $\phi = \dfrac{\text{mmf}}{\mathscr{R}} = \dfrac{700}{8 \times 10^5}$

$\qquad = 8.8 \times 10^{-4}$ weber. ◀

SUMMARY

All magnets, no matter how small, exhibit a north and south end.

The exact location of the two poles of a given magnet cannot be determined. Each pole is located approximately near each end of the magnet.

One end of a compass needle, called its north pole, will point toward the north when the needle is suspended horizontally by a fine thread. That end of the compass is called its north pole.

When the north pole of one magnet is placed near the south pole of another magnet, each magnet will experience a force of attraction; when two similar poles are placed near each other, each magnet will experience a force of repulsion.

A magnet modifies the magnetic characteristics of the space surrounding it. The magnetic field strength is a measure of the magnetic field at a point near a magnet.

Charges in motion (electric current) give rise to a magnetic field.

A current in a circular loop of wire gives rise to a magnetic north pole on one side of the loop and a south pole on the other side.

The magnetic characteristics of a given permanent magnet can be greatly diminished by strong impact or by heating it to a high temperature.

QUESTIONS

1. Does a horsehoe magnet also have a polarity (like a bar magnet)? Describe an experiment that can prove your answer.

2. In what sense is the magnetic force not a fundamental force in nature?

3. State Coulomb's law for the forces between magnets and for the forces between electrical charges. What are the similarities?

4. Imagine an iron bar magnet to be ground into iron dust. Would each particle of dust exhibit a north and south pole? Explain.

5. What is meant by the horizontal and vertical components of the magnetic field at a given region on the earth?

6. Would a horizontally mounted magnetic compass react when placed at either the magnetic north or south poles of the earth? Explain.

7. Explain the meaning of the angle of declination.

8. A narrow iron cylinder is pulled into a solenoid in in which a current is established. Why?

9. Can a permanent magnet be demagnetized? How?

10. What is meant by the retentivity of a permanent magnet?

11. What is meant by hysteresis?

12. How can a sensitive laboratory instrument be magnetically insulated from external magnetic effects?

PROBLEMS

(A)

1. Two long bar magnets lie on a straight line with their north ends facing and 20 cm apart. Calculate the force of repulsion (in dynes) if the pole strength of each pole is 600 unit poles.

2. If one of the magnets in Problem 1 has a mass of 80 g, what will be its acceleration at the moment when the north ends are 20 cm apart?

3. A long and straight vertical wire carries a current of 10 amp directed toward the center of the earth. Calculate the magnetic flux density at a point P that is located at a perpendicular distance of 30 cm directly east of the wire.

4. What is the direction of the field at point P in Problem 3?

5. A long and straight horizontal wire carries a current of 10 amp that is directed toward the west. Calculate the magnetic flux density at a point P that is 30 cm directly above the wire.

6. What is the direction of the field at point P in Problem 5?

7. A long wire carries a current that is directed vertically upward. What is the direction of the magnetic field at a point near the wire which is located directly
   (a) southward,
   (b) eastward,
   (c) northward, and
   (d) westward from the wire?

8. Calculate the magnetic flux density at the center of a circular loop of wire having one turn in which a current of 5 amp is established. The radius of the loop is 20 cm.

9. Assume that the direction of the current in the horizontal loop of Problem 8 is counterclockwise. Determine the direction of the field at the center.

10. Calculate the magnitude of the magnetic field at the center of the loop in Problem 8 if the loop has 100 turns.

(B)

1. A current of 3 amp is established in a long solenoid consisting of 500 turns. Calculate the flux density at points well within the solenoid that lie on its long central axis. The length of the solenoid is 30 cm.

2. What is the direction of the field within a solenoid whose axis is on a north-south line with the current directed in a clockwise sense when viewed from the north end?

3. Sketch a bar magnet with the appropriate polarity that would produce a field which is similar to the one produced by the solenoid in Fig. 29-13.

4. Two long vertical wires, 20 cm apart, carry a current as is indicated by Fig. 29-20. Calculate the magnitude of the resultant flux density at point A which is midway between the wires.

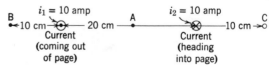

Figure 29-20

5. What is the magnitude and direction of the resultant flux density at point B in Fig. 29-20?

6. What is the magnitude and direction of the resultant field at point C of Fig. 29-20?

7. The two cylindrical magnets in Fig. 29-21 have the same dimensions and each has a pole strength of 1000 unit poles. The magnets are held on a frictionless table. Calculate the magnitude and direction of the force (in dynes) that will be experienced by magnet
(a) I and
(b) II

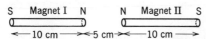

Figure 29-21

8. If the mass of magnet II in Fig. 29-21 is 80 g, what will be its acceleration at the moment that it is released?

9. Repeat Problem B7 with the magnets oriented as in Fig. 29-22.

10. A current of 5 amp is established in the 1000 turns of wire wrapped around the iron in Fig. 29-23. Calculate:

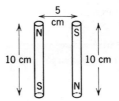

Figure 29-22

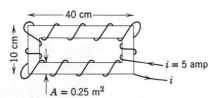

Figure 29-23

(a) The magnetic intensity H produced.
(b) The magnetomotive force.
(c) Reluctance in the magnetic circuit if the permeability of the iron is 10^{-4} weber/amp-m.
(d) The magnetic flux.

11. Calculate the relative permeability of the iron in Fig. 29-23.

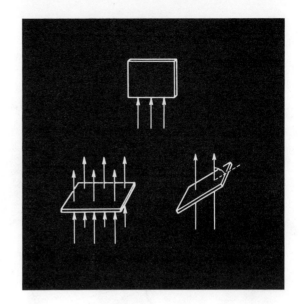

30.
Electromagnetic Theory

Electromagnetic Theory is an important branch of physics. Practical applications of this theory are responsible for the advanced technological level of modern society. Such important machines and instruments as electric motors, generators, voltmeters, ammeters, and cyclotrons, to mention a few, are based on the principles developed in this chapter.

LEARNING GOALS

1. *To learn about the electromagnetic force on a current-carrying wire that is located in a magnetic field.*
2. *To understand the left-hand rule, which is used to determine the direction of this force.*
3. *To learn that a current is induced in a closed loop of wire that moves in a magnetic field.*
4. *To understand Faraday's law by which the magnitude of the induced electric potential and current can be calculated.*
5. *To understand Lenz's law from which the direction of the induced current can be determined.*

Chapter 29 deals with one association between electricity and magnetism namely that an electric current in a wire produces magnetic effects (a magnetic field) outside the wire. In this chapter, two other associations are discussed. These are:

(1) A current carrying wire that is located in a magnetic field will experience a force, and

(2) relative motion between a wire and a magnet can produce an emf and a directed current in the wire.

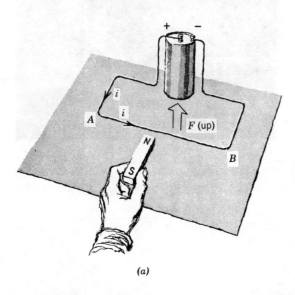

(a)

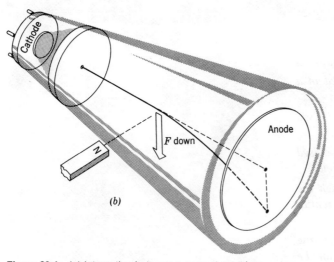

(b)

Figure 30-1 *(a)* Interaction between a current-carrying wire and a magnetic field. The wire will jump vertically up. *(b)* Cathode-ray tube showing electron deflection in the presence of a magnetic field. The visible stream of electrons travels in a straight line from the cathode to the anode where it produces a bright spot. With the magnet near the tube the electron beam will deflect in a downward direction.

30-1 INTERACTION BETWEEN A MAGNETIC FIELD AND ELECTRIC CURRENT

In Fig. 30-1a, an electric current is directed toward the right in the aluminum wire A–B. If a magnet with its north pole pointing toward it is brought near the wire, the wire will jump vertically up. If there is no electric current in the wire, the wire will remain stationary. In Fig. 30-1b a cathode ray tube produces an electron stream from the cathode to anode. When a north pole is brought in horizontally on the same level, the electron stream will bend vertically downward. From these and other demonstrations it can be concluded that

an external magnetic field exerts a force on charged particles if

(1) the charged particles are in motion, and

(2) the motion of the charged particles is *not* parallel to the magnetic field.

30-2 FORCE ON A CURRENT-CARRYING WIRE

It is found from experiments suggested by Fig. 30-1a that the force on a current-carrying wire is directly proportional to the strength of the magnetic field in which the wire is immersed. The equation expressing this relationship is

$$F = Bil \sin \theta \qquad (30\text{-}1)$$

or, solving for B,

$$B = \frac{F}{il \sin \theta} \qquad (30\text{-}2)$$

When the force F on the wire is expressed in newtons, the current i in amperes, and the length l of the wire in meters, B is expressed in weber/meter^2. The angle between the direction of B and the direction of the positive current is θ. Equation 30-2 suggests another definition for the strength of a magnetic field (one different from Eq. 29-2).

* The tesla, in honor of Nikola Tesla (1856–1943) who contributed to the development of electric power generation and

Thus, if a certain magnetic field exerts a force of one newton on a wire one meter long that is carrying a one-ampere current directed perpendicularly to the field, the strength of the field is then one weber/m^2. If the force is 0.5 newtons in the same circumstances, then the strength of the field is 0.5 weber/m^2. Actually, the phrase "magnetic flux density" or "magnetic induction" is used in connection with B,* rather than the phrase "strength of magnetic field" or "magnetic field strength."

There are three directed quantities in Eq. 30-1. These are B, F, and i. The direction of each of these can be deduced from a knowledge of the other two by means of the left-hand rule (Fig. 30-2). The index finger, middle finger, and thumb of the left hand are extended so that each forms a right angle with the others. When the middle finger is pointed along the *conventional* current and the index finger in the direction of the flux density, the thumb will indicate the direction of the force exerted on the wire. The force is always perpendicular to the directions of both B and i.

Example 30-1 A straight wire, 1 m in length, carrying a current of 10 amp is placed in a uniform magnetic field of flux density 1 weber/m^2 directed toward the east. Calculate the magnitude and direction of the force on the wire if the direction of the current is (a) westerly, (b) northerly, (c) southerly, and (d) 30° north of east.

Solution. $F = Bil \sin \theta$
(a) $F = 0$ because sine 180° is zero.
(b) $\theta = 90°$, sin 90° = 1,
    $F = 1 \times 10 \times 1 \times 1 = 10$ newtons.
The left-hand rule gives the direction as vertically down.
(c) $F = 1 \times 10 \times 1 \times 1 = 10$ newtons, vertically up.
(d) $F = 1 \times 10 \times 1 \times \frac{1}{2} = 5$ newtons, vertically down. ◀

Example 30-2 The two parallel wires in Fig. 30-3 carry a current in the same direction. Determine the magnitude and direction of the force on

distribution, is also used as a unit for magnetic flux density;

$$1 \text{ tesla} = \frac{1 \text{ weber}}{\text{m}^2} = 10^4 \text{ gauss}.$$

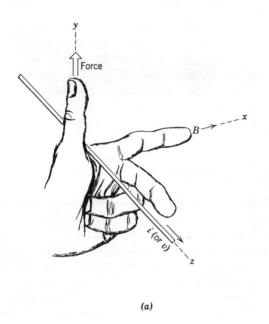

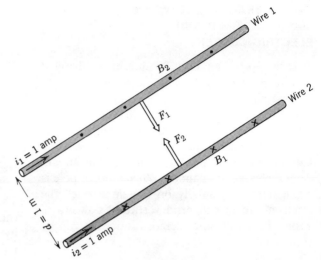

Figure 30-3 Two parallel wires carrying a current in the same direction will attract each other.

(a)

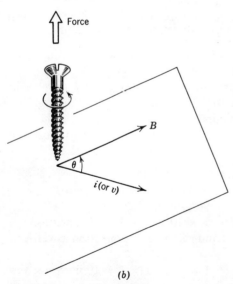

(b)

Figure 30-2 Two rules for finding the direction of the force. (*a*) The Left-Hand Rule is convenient when the field is perpendicular to the current. (*b*) The Screw-Rule. A "right-hand" screw is imagined to be placed at right angles to *both* the current and *B*. The direction of the force is given by the advance of the screw when turned *from* the wire *toward B*. This rule is more convenient when θ is not a right angle.

(a) wire 1, and (b) wire 2. The length of each wire is 1 m.

Solution. Wire 1 will produce a magnetic field in the vicinity of wire 2. The magnitude of this field (by Eq. 29-4) is

$$B_1 = \frac{\mu_0 i_1}{2\pi d} = \frac{4\pi \times 10^{-7} \times 1}{2\pi \times 1}$$
$$= 2 \times 10^{-7} \text{ weber/m}^2$$

The direction of this field at the vicinity of wire 2 is into the page (by the right-hand rule, see Section 29-6). This field exerts a force F_2 on wire 2. Then $F_2 = B_1 i_2 l \sin \theta = 2 \times 10^{-7}$ newton, which is directed toward wire 1 (by the left-hand rule). The magnitude of the flux density B_2 produced by wire 2 in the vicinity of wire 1 is the same as B_1. The direction of B_2 is out of the page. The resulting force on wire 1 is $F_1 = B_2 i_1 l \sin \theta = 2 \times 10^{-7}$ newton. This force is directed toward wire 2 (by the left-hand rule). If the current in the two wires is in the same direction, the wires attract; if the currents are in opposite directions, the wires repel.

This example demonstrates the definition of the ampere. The National Bureau of Standards maintains a "current balance" which can measure very accurately the force of attraction. If the force on each of two parallel wires, 1 m long, separated in vacuum by 1 m and carrying an identical current is 2×10^{-7} newton, then the current in each wire is defined as 1 ampere. ◀

30-3 FORCE ON A MOVING CHARGED PARTICLE

The equation for the force on a current-carrying wire (Eq. 30-1) can be modified to express the force on the moving charges that constitute the current. Figure 30-4 depicts a wire carrying a current that is perpendicular to the magnetic field. The field is directed out of the page. The steady current i can be expressed as q/t (Eq. 28-1).

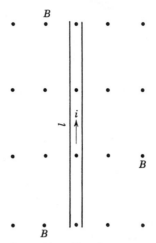

Figure 30-4 The wire carries a current that is perpendicular to a uniform magnetic field heading out of the page.

In other words the current will transfer a charge q coulombs in t seconds. Since the drift velocity, v of the charges constituting the current is constant, the charges will move a distance, $l = vt$ in t sec (If you run at a constant speed of 3 m/sec, you will cover a distance, $l = vt = 3$ m in 1 sec.) Applying this consideration to Eq. 30-1, we get

$$F = Bi\, l \sin \theta = B \frac{qvt}{t} \sin \theta$$
$$F = Bqv \sin \theta \tag{30-3}$$

The quantities, F, v, and B are vector quantities, the direction of each being determined by the same left-hand rule as in Fig. 30-2. The direction of the velocity vector v in Eq. 30-3 is that of the positive charges.

Example 30-3 Assume that an electron moves horizontally from east to west at a speed of 3×10^6 m/sec in a region on the earth where the horizontal component of the earth's field is 2×10^{-5} weber/m^2 pointing north. Calculate the magnitude and direction of the force exerted on the electron. The charge on an electron is 1.6×10^{-19} coulomb.

Solution

$$F = qvB \sin \theta, \quad \theta = 90°, \quad \sin \theta = 1$$
$$F = 1.6 \times 10^{-19} \times 3 \times 10^6 \times 2 \times 10^{-5} \times 1$$
$$= 9.6 \times 10^{-18} \text{ newton}$$

To determine the direction of the force on the electron, we can first assume it to be a positive charge and apply the left-hand rule. This would lead to vertical force that is straight up. But since the electron is a negative particle, we must reverse this direction. The direction of the force on the electron is vertically down.

It is interesting to compare this force with the force on the electron caused by its weight. The mass of an electron is 9.1×10^{-31} kg.

$$w = mg$$
$$9.1 \times 10^{-31} \times 9.8 = 89.5 \times 10^{-31} \text{ newton}$$

The force on a charged particle moving in magnetic field is many orders of magnitude greater than its weight. ◄

30-4 ORBITS OF CHARGED PARTICLES

A careful analysis of Eq. 30-3 will show that a stream of charged particles that are released at right angles to a uniform magnetic field will orbit in a circle. Consider Fig. 30-5, which depicts positively charged particles originating with a velocity v at point A. During a very short time interval the charges will undergo a small displacement, Δs and will also experience a force, F, which is normal to this displacement. Application of the definition of work (Eq. 12-1) $W = F$ $\Delta s \cos \theta$ to this situation will show that the electromagnetic force does zero work on the charged particle, because the angle between the force and displacement is 90° and $\cos 90° = 0$. This means that this force will not speed up the particle but will change the particle's direction. This is analogous to an orbiting satellite, where the gravitational force provides the centripetal force required for the satellite to move in a circular orbit.

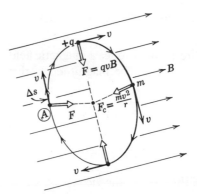

Figure 30-5 A charge will move in a circular orbit when projected at right angles to a uniform magnetic field. The plane of the circle is ⊥ to B.

By equating the electromagnetic force, $F = qvB$ with the centripetal force, $F_c = mv^2/r$ (see Section 10-3) we get

$$qvB = \frac{mv^2}{r} \qquad (30\text{-}4)$$

and

$$r = \frac{vm}{qB} \qquad (30\text{-}5)$$

The radius r of the orbit will remain the same as long as the particle's speed, mass, charge, and magnetic field do not change.

The *cyclotron*, which is one of the most important instruments used in nuclear physics, is based on these principles. The American physicist, Ernest Lawrence designed and assembled the first cyclotron at Berkeley in 1932. The purpose of this instrument is to accelerate charged particles to very high speeds so that they can serve as "bullets" in the bombarding and shattering of nuclei. A study of the nuclear fragments can provide information on the constituent particles within the nucleus.

The 2 D's of the cyclotron (Fig. 30-6) are evacuated and connected to an alternating emf which changes the polarity of the D's many thousands of times per second. This can easily by accomplished by electronic instruments called oscillators. If the charged particle is a proton and is released from point A, when the left D is negative, the proton will enter it and be bent into a circular orbit of radius r_1. If at the moment that

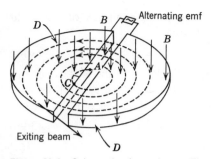

Figure 30-6 Schematic of a cyclotron. The speed of the circulating charged particle increases as the radius of the orbit increases.

it arrives at point C, the polarity of the D's change so that the right D is negative, the proton will be accelerated by the electric field between the D's and enter the right D at a higher speed. By Eq. 30-5 its radius will now increase. As the polarities keep changing at the proper instants, the proton continues its path and is accelerated every time it passes the gap between the D's. Finally, at the largest effective working radius, r_{max} of a particular cyclotron, the proton emerges with a very high speed which, from Eq. 30-4, is given as

$$v_{max} = \frac{qBr_{max}}{m} \qquad (30\text{-}6)$$

We can solve for the proper frequency f of the alternating emf across the D's by noting that $v_{max} = 2\pi r_{max}/T$, where T is the period and $f = 1/T$. Therefore, $v_{max} = 2\pi r_{max}f$. Inserting this value for v_{max} in Eq. 30-6 and solving for f, we get

$$f = \frac{qB}{2\pi m} \qquad (30\text{-}7)$$

The frequency does not depend on the particles speed or radius. This fact makes the cyclotron feasible.

Example 30-4 Assume that the magnets of a given cyclotron produce a constant field of 1 weber/m². (a) With what frequency should the polarity of the D's change when the charged particles are protons? The mass of a proton is 1.67×10^{-27} kgm and its charge is 1.6×10^{-19} C. (b) With what maximum speed will a proton

emerge from the cyclotron if its maximum working radius is 1 m? (c) What will be the kinetic energy of an emerging proton?

Solution

(a) $f = \dfrac{qB}{2\pi m}$ $\qquad (30\text{-}7)$

$$f = \frac{1.6 \times 10^{-19} \times 1}{2\pi \times 1.67 \times 10^{-27}} = 1.5 \times 10^7 \text{ cycles/sec}$$

This frequency can be achieved with electronic instruments.

(b) $v_{max} = \dfrac{qBr_{max}}{m}$ $\qquad (30\text{-}6)$

$$v_{max} = \frac{1.6 \times 10^{-19} \times 1 \times 1}{1.67 \times 10^{-27}} = 9.5 \times 10^7 \text{ m/sec}$$

This is approximately one-third of the speed of light. Some of the more recent accelerators have maximum radii that are approximately 100 m. The speed of the emerging particles is therefore increased, but a speed of a particle that is equal to, or greater than, the speed of light is impossible to attain.

(c) $K = \frac{1}{2}mv^2 = \frac{1}{2} \times 1.67 \times 10^{-27}(9.5 \times 10^7)^2$

$K = 75.36 \times 10^{-13}$ joule

We can express this energy in electron volts (see Section 27.5) as

$$K = 75.36 \times 10^{-13} \text{ joules} \times \frac{1 \text{ eV}}{1.6 \times 10^{-19} \text{ joules}}$$

$K = 47 \times 10^6 \text{ eV} = 47 \text{ MeV}$

To achieve the same energy without a cyclotron, the proton would have to be accelerated through a potential difference of 47 million volts. Such high voltages cannot ordinarily be attained. If we assume that the potential difference between the D's is 94,000 V, then the energy gain in one complete revolution is $2 \times 94 \times 10^3$ eV. The proton would therefore have to make

$$\frac{47 \times 10^6}{2 \times 94 \times 10^3} = 250$$

revolutions to achieve the energy of 47 MeV. ◀

30-5 FLUX AND FLUX DENSITY

A region in space is in a *uniform* magnetic field if the magnitude and direction of the field within that entire region does not change. Consider a small regional area A_1 in the field of a bar magnet (Fig. 30-7a). The lines of force going through that area are not parallel, that is, the direction of the force on a unit north pole is different for different points within the area. The field enclosed by this area is, therefore, not uniform. An almost uniform field can be attained within the area A_2 between large pole pieces placed near each other, as in Fig. 30-7b.

It is convenient to characterize the magnetic properties in a given region of a magnetic field, such as area A_2 and A_3, in terms of imaginary lines, called magnetic *lines of force* (see Sec. 29-3) piercing this area. The stronger the field encompassed by this area, the more lines pierce it at an angle of 90°. Area A_4 in Fig. 30-7d, for instance, has no lines threading it (piercing through it). The *magnetic flux*, denoted by the Greek letter ϕ and measured in units of webers in the mks system, expresses the magnitude of the field piercing a given area

$$\phi = BA \qquad (30\text{-}8)$$

where B is perpendicular to A, and is in weber/m^2, ϕ is in webers, and A in m^2. In general, when B makes any angle θ with the surface of the area

$$\phi = BA \sin \theta \qquad (30\text{-}9)$$

Flux is a scalar quantity.

Example 30-5   (a) Calculate the flux piercing area A_2 in Fig. 30-7b if the flux density B is 0.3 weber/m^2 and directed everywhere normal to area A_2. The area A_2 is 0.05 m^2. (b) Calculate the magnetic flux through the surface A_3 in Fig. 30-7c if a uniform flux density threads it, at an angle of $\theta = 30°$. The area of the surface is 0.05 m^2 and the magnitude of B is 0.3 weber/m^2.

Solution.   (a) B is normal to the surface; therefore, $\sin \theta = 1$ and from Eq. 30-8

$$\phi = 0.3 \, \frac{\text{weber}}{\text{m}^2} \times 0.05 \times 1$$

$$= 15 \times 10^{-3} \text{ weber}$$

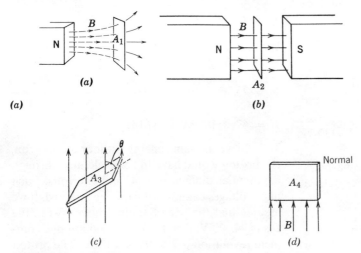

(a)

(b)

(c)

(d)

Figure 30-7   (a) The field in the region A_1 is not constant because the direction of B varies within this region. (b) The field is constant and perpendicular to A_2. The flux through A_2 is BA_2. (c) B is constant and forms an angle θ with the surface. The flux through A_3 is $BA_3 \sin \theta$. (d) B is constant but does not pierce A_4. The flux through A_4 is zero.

Since ϕ is a scalar quantity, it has no direction.

(b) The angle between B and the surface is $30°$ and $\sin 30° = \frac{1}{2}$. By Eq. 30-9

$$\phi = 0.3 \frac{\text{weber}}{m^2} \times 0.05 \, m^2 \times \frac{1}{2}$$

$$= 7.5 \times 10^{-3} \text{ weber}$$

30-6 ELECTROMAGNETIC INDUCTION

Another fundamental association between electricity and magnetism was discovered by Michael Faraday in England and Joseph Henry in the United States in about 1831. A loop consisting of several turns of wire (see Fig. 30-8a) is located

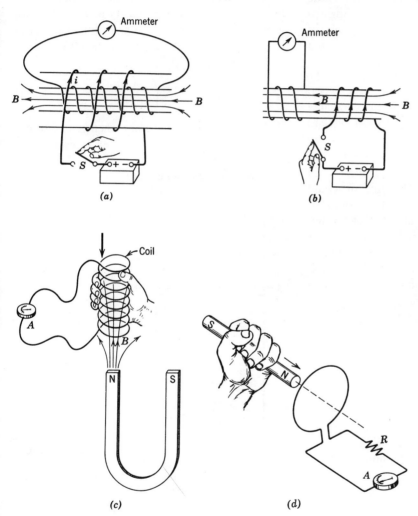

(a)

(b)

(c)

(d)

Figure 30-8   Induced emf. (a), (b) During the short moment when the switch, S contacts or loses contact with the wire of one coil, an emf is induced in the other coil. Both coils are electrically insulated from each other. The emf causes a momentary induced current which is indicated by the deflection of the ammeter. (c) The act of moving the coil towards a magnet or (d) moving a magnet towards a coil also induces an emf in the coil.

within another loop, *both insulated from each other*. The outer loop is connected to a battery and a switch. *While* the switch is being closed, there is a *momentary* deflection in the ammeter showing a momentary current. Immediately after the moment that the switch is completely closed, the current in the inner coil ceases. While the switch is opened, there is again a *momentary* ammeter deflection, but in the opposite direction, after which the current goes to zero. Since a current can only be established between points of different electrical potential, the act of closing and opening the switch in the outer coil *induced* (gave rise to) a difference of potential in the inner coil.

The induced voltage in the inner coil is as real as if there were a battery momentarily connected to this coil. This phenomenon is referred to as an *induced emf* (for the meaning of emf, refer to Section 28-3). If some agent were to continue to push the switch down and up, a current would continually exist in this coil. This current would be alternating because its direction would be changing. Figures 30-8*b*, *c*, and *d* depict other methods for inducing an emf. These methods have one thing in common: there is a change in magnetic flux through the coil in which an emf is induced. A necessary condition for voltage induction is a *changing* flux. The closing of the switch in Fig. 30-8*a* will produce a momentary increase in the current in the outer coil. Since the magnetic field within a solenoid is proportional to the current (Eq. 29-6), there is a momentary build-up of *B*, and therefore a momentary increase of ϕ, through both coils because $\phi = BA$ (Eq. 30-8). At the moment that the switch is completely closed, the current in the outer coil has reached a constant value (maximum) at which time the magnetic flux is no longer changing, thereby bringing the induced voltage to zero. The same argument prevails while the switch is opened.

30-7 FARADAY'S LAW FOR INDUCTION

The magnitude of the induced emf in a coil of N turns is given by

$$\mathscr{E} = -N\frac{\Delta\phi}{\Delta t} \qquad (30\text{-}10)$$

where $\mathscr{E}$ is the average induced emf and is expressed in volts, and $\Delta\phi$ is the change in flux through the coil, expressed in webers, occurring in a time Δt, which is in seconds. Equation 30-10 is referred to as *Faraday's law for induction*.

The change in flux ($\Delta\phi$ of Eq. 30-10) can be achieved in two general ways. One way involves a *changing* flux density going through a loop of constant area as in Fig. 30-8*a* to *d*. The other way involves a *constant* flux density going through a loop whose area is changing, as, for instance, in Fig. 30-9. Equation 30-10 can therefore be extended to

$$\mathscr{E} = -N\frac{\Delta\phi}{\Delta t} = -N\left(B\frac{\Delta A}{\Delta t} + A\frac{\Delta B}{\Delta t}\right) \quad (30\text{-}11)$$

If the area remains constant, then $\Delta A = 0$ and the first term within the bracket vanishes; if B remains constant, then $\Delta B = 0$ and the second term within the bracket vanishes.

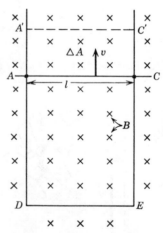

Figure 30-9 A conducting loop with a movable arm, *AC*. The flux density *B* is uniformly distributed throughout the loop. As the movable arm is pushed up, the flux through the closed loop increases because its area increases.

We can interpret Eq. 30-11 with the aid of Fig. 30-9. The conducting rectangular loop, *ACDE* is immersed in a constant field pointing into the page. During the interval Δt, during which the movable element *AC* is moved to *A'C'*, the area of the loop undergoes an increase, ΔA. Therefore, there is a $\Delta\phi$ (change in flux), $\Delta\phi = B\,\Delta A$. An induced emf, $\mathscr{E} = -B\,(\Delta A/\Delta t)$, results during the motion of the element. We refer to this as a *motional electromotive force*. An emf can also be induced in the loop *ACDE* without the motion of any of its parts by varying the magnitude or direction (or both) of B threading the loop. This will also give rise to a $\Delta\phi$ because $\Delta\phi = A\,\Delta B$. Equation 30-11 includes both possibilities.

If in Fig. 30-9 the element *AC* of length l is moved at a constant speed, v then in a time interval, Δt, this element has moved through a distance, $v\,\Delta t$. This motion brings about an area change, $\Delta A = lv\,\Delta t$ and a change in flux, $\Delta\phi = B\,\Delta A = Blv\,\Delta t$. The magnitude of the induced voltage during this interval is then, by Eq. 30-10, $\mathscr{E} = (Blv\,\Delta t)/\Delta t$ or

$$\mathscr{E} = Blv \qquad (30\text{-}12)$$

This equation applies to a wire of length l (in meters) that is moving at a constant velocity v (in m/sec) in a constant field B (in weber/m^2), and B, l, and v are mutually perpendicular. If the velocities direction forms an angle θ with the direction of the field, then

$$\mathscr{E} = Blv\sin\theta \qquad (30\text{-}13)$$

Example 30-6 The flux through a coil having 100 turns is changing at a constant rate from 0.5 to 0.7 weber in a time of 1 sec. The coil is connected to a small flashlight bulb, the combined resistance being 2 ohms. (a) Calculate the magnitude of the average induced voltage in the coil. (b) What is the average current through the filament of the bulb during this time interval?

Solution
(a) $\Delta\phi = (0.7 - 0.5)$ weber $= 0.2$ weber. By Eq. 30-10,

$$\mathscr{E} = \frac{-100 \text{ turns} \times 0.2 \text{ weber}}{1 \text{ sec}} = -20 \text{ volt}$$

(b) From Ohm's law,

$$i = \frac{\mathscr{E}}{R} = \frac{-20 \text{ volt}}{2 \text{ ohms}} = -10 \text{ amp}$$

The minus sign is related to the direction of the induced emf and current and will be discussed in the next section. ◀

Example 30-7 A constant magnetic field of flux density 0.03 weber/m^2 is directed normal to the plane of a square coil having 100 turns. The coil is then flipped so that in an interval of 0.5 sec its plane becomes parallel to the direction of the field. Calculate the average induced emf in the coil if its area is 0.25 m^2.

Solution. The area of the coil that is threaded (pierced) by the field changes from 0.25 m^2 to zero in a time of 0.5 sec. The change in area $\Delta A = 0.25$ m^2. From Eq. 30-11

$$\mathscr{E} = \frac{-NB\,\Delta A}{\Delta t}, \text{ since } \Delta B = 0$$

$$= -100 \text{ turns} \times 0.03\,\frac{\text{weber}}{\text{m}^2} \times \frac{0.25\text{ m}^2}{0.5 \text{ sec}}$$

$$= -1.5 \text{ volt}$$

The significance of the minus voltage will be explained in the next section. ◀

Example 30-8 The wire *AC* in Fig. 30-9 is 0.1 m long and moves at a constant velocity of 3 m/sec in a magnetic field of 0.35 weber/m^2. Calculate the induced emf.

Solution. By Eq. 30-12,

$$\mathscr{E} = 0.35\,\frac{\text{weber}}{\text{m}^2} \times 0.1\text{ m} \times 3\,\frac{\text{m}}{\text{sec}}.$$

$$\mathscr{E} = 0.105 \text{ volts}$$

30-8 CONSERVATION OF ENERGY; LENZ'S LAW

Figure 30-10a depicts a loop of wire being brought in from the right with the side *AC* just having entered the field. Equation 30-12 asserts that an emf ($\mathscr{E} = Blv$) is induced in *AC* and there is induced a potential difference between points *A* and *C*. Since the concept of potential difference is

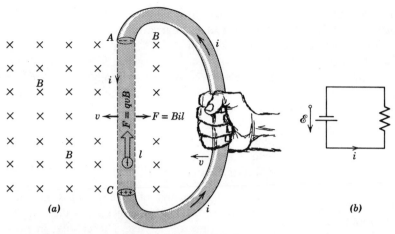

Figure 30-10 (a) As the loop is moved toward the left, the conduction electrons of the atoms of the loop element AC experience a force which is directed towards A (by the left-hand rule). This causes an accumulation of electrons near A and excess positive charges near C. This gives rise to a counterclockwise conventional current. (b) The situation in (a) is as if the wire element AC is replaced by a battery.

based on energy considerations (see Section 27-3), there arises a question concerning the conservation of energy. Because the wire is part of a closed loop, a current i will be induced in the loop. Now, the positive current can be directed along AC either from A to C or from C to A. Consideration of the conservation of energy principle leads to the conclusion that it must be from A to C. Therefore, the wire AC, while it is moved toward the left (Fig. 30-10), will experience a force toward the right, (Eq. 30-1). To continue moving the wire toward the left at a constant speed, an external agent (the hand) will have to exert an equal force toward the left. The external force will do work on the wire to transport it toward the left.

As the wire AC is brought into the field, the conduction electrons of its atoms will experience a force toward point A, $f = qvB$ (Eq. 30-3) causing the electrons to migrate toward point A, thereby leaving the region near C with an excess of positive charges. This situation suggests an analogy with the battery (Fig. 30-10b) where the chemical energy provides the work necessary to assemble the negative and positive charges, respectively, at the cathode and anode. The chemical energy is transformed into electric potential energy of the charges which, in turn, gives rise to the potential difference between anode and cathode. In the case of the wire AC in Fig. 30-10, the work necessary to assemble the negative charges in the region of point A, is provided by the mechanical work done by the hand in transporting the wire toward the left. This work is transformed into a gain in electric potential energy possessed by the charges that are assembled near the ends of the wire. Therefore, a potential difference is established between ends A and C.

Let us assume (incorrectly) that the induced current (conventional) moves from C to A. This situation would cause the wire AC to experience a force toward the left. This force alone, without the hand, would accelerate the wire toward the left, thereby increasing v. This, in turn, would increase the induced voltage which would then give rise to a bigger current. We would be getting electrical and mechanical energy for nothing. This violates the law of conservation of energy.

The direction of the induced current can be determined from Lenz's law:

> **The induced electromotive force and current must be in such a direction as to oppose the cause that gave rise to them.**

By means of this law, the conservation of energy principle is applied to induced voltages and currents. The minus sign in Faraday's law takes care of this consideration.

We can apply Lenz's law to the situation shown in Fig. 30-10a in the following manner. We construe the cause for the induced voltage in the wire AC to be its velocity toward the left. A way to oppose this cause would be to reduce v. A conventional current from A to C in AC would reduce v because it would give rise (by the left-hand rule) to a force toward the right. We can look at this in another way. As the loop moves toward the left, there is an increase in flux, which causes an emf. The current from A to C reduces this flux change, because this current would give rise to a field within the loop that is directed out of the page.

SUMMARY

A magnetic field exerts a force on a current-carrying wire: $F = Bil \sin \theta$.

The left-hand rule is used to determine the directions of F. The direction of the force is $\perp$ to the directions of both B and i.

A magnetic field exerts a force on a moving charge: $F = qvB \sin \theta$. The left-hand rule is used to determine the directions of F, v, or B. The force is $\perp$ to both v and B.

Charged particles move in circular orbits when the velocity of the particle is $\perp$ to B.

The cyclotron accelerates charged particles to very high speeds.

Relative motion between a magnet and a wire causes an induced emf in the wire.

If the wire is a closed loop, the induced emf causes an induced current.

Faraday's law is

$$\mathscr{E} = -N \frac{\Delta \phi}{\Delta t}$$

Lenz's law explains the conservation of energy principle when applied to induced voltages and currents.

QUESTIONS

1. Discuss electromagnetism by describing a few electromagnetic effects.

2. Can a cyclotron accelerate neutrons that have a zero net charge? Explain.

3. Deduce the magnitude and direction of the force on the lower wire in Example 30-2 by applying Newton's third law.

4. What is the direction of the force on the wire for each orientation in Fig. 30-11? The wire and magnet are on a horizontal plane.

5. A stream of moving electrons is substituted for the wire in Fig. 30-11, with their directions indicated by the arrows. What would be the direction of the force that the electrons experience?

6. The north end of a bar magnet is moved toward the center of the loop (see Fig. 30-12). Is the flux through the loop increasing or decreasing? Explain.

7. What is the direction of the induced current in the loop (as viewed from the left) in Fig. 30-12, while the magnet is in motion toward the right (apply Lenz's law)?

8. What is the direction of the induced current in the loop in Fig. 30-12, while the bar magnet is moved toward the right?

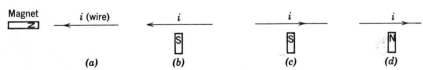

Figure 30-11

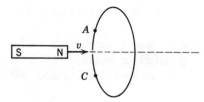

Figure 30-12

PROBLEMS

(A)

1. Calculate the magnitude of the force on a 1 m wire that is carrying a current of 10 amp toward the east in a region where the magnetic field is 10^{-2} weber/m² and pointing north.

2. What is the direction of the force in Problem 1?

3. Calculate the magnitude of the force on an electron ($e = 1.6 \times 10^{-19}$ C) that is moving at 10^8 m/sec toward the east in a region where the magnetic field is 10^{-2} weber/m² pointing toward the north.

4. What is the direction of the force in Problem 3?

5. A flux density of 10^{-2} weber/m² points in a direction that is normal to the plane of a loop of wire having an area of 100 cm². Calculate the flux in the loop.

6. Calculate the change in flux in the loop of Problem 5 if the flux density is reduced to 10^{-3} weber/m².

7. What is the induced emf in the loop of Problem 5 if the flux density is reduced to 10^{-3} weber/m² in a time of 10^{-2} sec.

8. An airplane having a wing spread of 50 m is moving in a horizontal plane at a speed of 350 km/hr and in a region where the vertical component of the earth's field is 10^{-5} weber/m². What is the induced voltage difference between the tips of the wing?

9. Assume that a proton (charge 1.6×10^{-19} C), which originated from a cosmic ray, is traveling vertically down at a speed of 3×10^5 m/sec in a region where the horizontal component of the earth's magnetic field is 10^{-5} weber/m². Calculate the magnitude and direction of the force on the proton.

10. Calculate the force (including the direction) on an electron traveling along the positive y axis, at a speed of 3×10^7 m/sec, in a uniform magnetic field of 5×10^{-4} weber/m² pointing along the positive x axis. (The charge on an electron is 1.6×10^{-19} C.)

(B)

1. A horizontal wire, carrying a current toward the south, is in a magnetic field directed from east to west. In what direction will the electromagnetic force on the wire point?

2. A wire carrying a current is located in a magnetic field parallel to the current. What is the direction of the electromagnetic force on the wire?

3. A square coil, 50 cm on a side and consisting of 100 turns of wire, is clamped in a magnetic field of magnitude 0.17 webers/m², and directed perpendicular to the plane of the coil.
   (a) Calculate the average electromotive force induced in the coil if the field changes to 0.15 weber/m² in 0.25 sec.
   (b) If the ends of the coil are connected to a lamp, the combined resistance of the lamp and coil is 2 ohms, what average current will flow through the circuit?

4. An electron is moving in a direction that is perpendicular to a uniform magnetic field. The electron experiences a southward force of 10^{-13} N while it is moving toward the west at 10^8 m/sec. What is the magnitude of the magnetic field?

5. What is the direction of the magnetic field in Problem 4B?

6. A wire 50 cm in length is carrying a current, which is directed toward the north in a magnetic field, $B = 10^{-1}$ weber/m², which is perpendicular to the wire. The force on the wire is 5 N and directed vertically downward. Calculate the magnitude of the current.

7. What is the direction of the field in Problem 6B?

8. A wire that is 70 cm in length is carrying a current of 10 amp directed eastward in a region of a uniform magnetic field, $B = 10^{-2}$ weber/m², which is directed at 60° north of east. Calculate:
   (a) The magnitude of the force on the wire.
   (b) The direction of this force.

9. Calculate the angle between the direction of the current in a wire and the field in order for the force on the wire to be 0.707 times the maximum value of the force?

10. A horizontal wire, carrying a 10-amp current directed westward, having a mass of 10 g and a length of 50 cm, lies on a table. What is the minimum flux density that will just lift the wire from the table? In what direction should this field be directed?

11. A proton is projected in a direction that is perpendicular to a constant magnetic field. Describe the shape of the line that the proton will move in. Why?

12. A proton has a mass of 1.67×10^{-27} kg. It is projected horizontally at a speed of 10^8 m/sec into a region in which there exists a constant vertical magnetic field, $B = 0.75$ weber/m². Calculate the radius of the proton's orbit.

13. In Problem 12(B), if the proton is projected in a westerly direction and the field is directed upward, what would be the direction of the proton's motion? Clockwise or counterclockwise as viewed from above?

14. What would the speed of an emerging proton be from a cyclotron having a maximum working radius of 1 m and a field of 0.5 W/m²?

15. What would be the energy of the emerging proton in Problem 14(B) expressed in joules and in electron volts.

16. The current in the upper wire of Fig. 30-13 is 5 amp and directed toward the right; the current in the lower wire is 10 amp and directed toward the left. Calculate the magnitude and direction of the force per unit length on each wire.

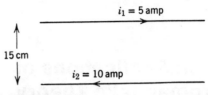

Figure 30-13

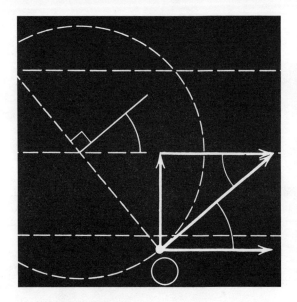

31.
Important Applications of Electromagnetic Theory

In Chapters 29 and 30 some of the fundamental principles of electromagnetic theory were discussed. In this chapter some important applications of those principles will be developed.

LEARNING GOALS

1. *To understand the principles governing the operation of a voltmeter and ammeter.*

2. *To understand the principles governing the operation of an electric generator.*

3. *To learn how an alternating-current generator can be converted to a direct current generator.*

4. *To understand the meaning of an alternating current.*

5. *To understand the electric transformer and its uses.*

31-1 GALVANOMETER, VOLTMETER, AND AMMETER

A *galvanometer* is an instrument that is used to detect small currents. With slight modifications, it can be made to measure larger currents and also voltages. The often-used laboratory or industrial galvanometer (Fig. 31-1) consists of a coil that is pivoted on low friction bearings, with the coil free to rotate about an axis perpendicular to the page and between the two pole pieces, N and S, of a permanent magnet. An iron core C is so shaped and positioned that the magnetic field has radial symmetry. A pointer P is attached to the coil so that it rotates with it, over a dial that is graduated in units of volts, or amperes, or both. The electric leads come from both ends of the coil.

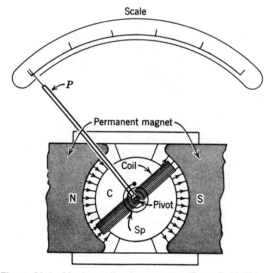

Figure 31-1   Moving coil galvanometer. (From D. Halliday and R. Resnick, *Fundamentals of Physics*, John Wiley & Sons, Inc. New York, 1970).

31-2 PRINCIPLE OF OPERATION

The principle for the operation of a galvanometer (or voltmeter) is based on the force exerted on a current-carrying wire located in a magnetic field, namely, $F = Bil \sin \theta$ (Eq. 30-1). A rectangular coil carrying a current i is depicted in Fig. 31-2. The coil is pivoted on an axis X-X' and the plane of the coil is parallel to a uniform magnetic field B. In this configuration sides MN and OP will not experience a force because the sine of the angle between i and B is zero. Side PM will experience an upward force (out of the page, by the left-hand rule) and side NO will experience a force into the page, the magnitude of each being $F = Bil$. These forces will cause a torque about the axis X-X', and the coil will rotate in a clockwise direction as viewed from the right. A moment later, the plane of the coil will make an angle α with B as in Fig. 31-2b. The forces on side NO and PM are still Bil. These forces will produce a torque that will continue the rotation. The magnitude of the torque can be computed by noting that the lever-arm distance for each of these forces is $w/2 \cos \alpha$. The total torque M is therefore

$$M = 2\,Bil\,\frac{w}{2}\cos\alpha = BiA\cos\alpha \quad (31\text{-}1)$$

where $A = lw$ is the area of coil. When the plane of the coil is perpendicular to the field, the torque is zero (because $\alpha = 90°$, $\cos 90° = 0$); when it is parallel to the field the torque is maximum (because $\alpha = 0°$, $\cos 0° = 1$).

If the coil consists of N closely wound turns, Eq. 31-1 becomes

$$M = NBiA\cos\alpha \quad (31\text{-}2)$$

Although this equation was derived for a rectangular coil, it is valid for a coil of any shape. In Fig. 31-2b, the line N is normal to the plane of the loop and θ is the angle between N and the flux density B. Since α and θ are complementary, Eq. 31-2 can also be expressed as

$$M = NBiA\sin\theta \quad (31\text{-}3)$$

In a given galvanometer, the number of turns, the flux density, and the effective area of the coil is constant. Therefore, the torque on the coil (by Eq. 31-3) is proportional to the two variable quantities, the current and the angle θ between the normal to the plane of the coil and the field.

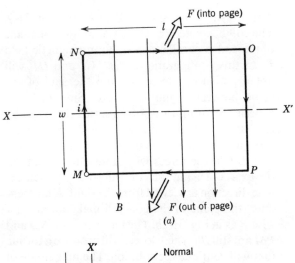

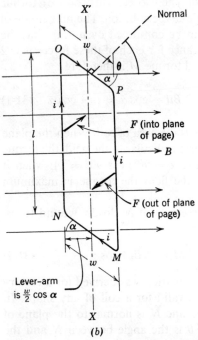

Figure 31-2 (a) The plane of the current-carrying loop is parallel to the magnetic field. (b) Left-end view of the loop of (a) a moment later.

To act as a current-reading instrument it is desirable that a given galvanometer be proportional to only one variable quantity, the current. This is accomplished by properly curving the magnets, which causes the field to be radial so that θ is kept constant at all the working angular orientations of the coil. For a given current, the coil and pointer will rotate until the restoring torque of the hairspring is equal and opposite to M. When this occurs, the coil is in equilibrium and the pointer settles down at the appropriate number on the dial.

Example 31-1   A coil consisting of 100 turns, each 5 cm × 10 cm, is pivoted as in Fig. 31-3. The flux density is 10^{-3} weber/m^2, and the current is 5 amp. Calculate the magnitude and direction of the force on each side of the coil at an instant when the plane of the coil is (a) parallel to B, and (b) perpendicular to B. (c) Calculate the torque on the coil at the two instants referred to in (a) and (b) above. (d) What torque is required to hold the coil in a position where the plane of the coil makes an angle of 30° with the field?

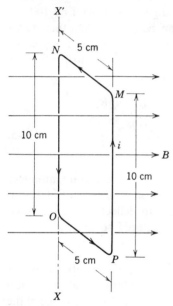

Figure 31-3

Solution.   The force on each wire is calculated from $F = Bil \sin \theta$.

(a) For wires MN and OP,

$$\theta = 0, \quad \sin \theta = 0, \quad F = 0$$

On wires NO and PM,

$$\theta = 90°, \quad \sin 90° = 1$$
$$F = 10^{-3} \times 5 \times 5 \times 10^{-1}$$
$$F = 5 \times 10^{-4} \text{ newton}$$

This is the force on each long wire. For 100 turns,

it is equal to

$$100 \times 5 \times 10^{-4} \text{ newton} = 5 \times 10^{-2} \text{ newton}$$

To find the direction of the force, apply the left-hand rule. The force on side MP is directed into the page. This force will cause a clockwise rotation about the axis $X\text{-}X'$ as viewed from below. The force on NO is directed out of the page.

(b) For wires MN and OP,

$$\theta = 90°, \quad \sin \theta = 1$$
$$F = 10^{-3} \times 5 \times 5 \times 10^{-2} \times 1$$
$$= 2.5 \times 10^{-4} \text{ newton}$$

On wires NO and PM, F is the same as in the case when the coil is parallel to the field.

For the coil of 100 turns, the force on each short sides is

$$100 \times 2.5 \times 10^{-4} \text{ newton} = 2.5 \times 10^{-2} \text{ newton}$$

and on the long sides

$$100 \times 5 \times 10^{-4} \text{ newton} = 5 \times 10^{-2} \text{ newton}$$

The directions of the forces are out of the page on side NO, into the page on side PM, in the positive y direction on side OP, and the negative y direction on side MN.

(c) $M = NBiA \cos \alpha$

where α is the angle that the plane of the coil makes with the field.

For the plane of the coil parallel to the field, $\alpha = 0°$, $\cos \alpha = 1$, $A = 5 \times 10^{-3}$ m^2

$$M = 100 \times 10^{-3} \times 5 \times 5 \times 10^{-3}$$
$$= 25 \times 10^{-4} \text{ newton-meter}$$

For the plane of the coil perpendicular to the field, $\alpha = 90°$, $\cos \alpha \doteq 0$, and therefore $M = 0$.

(d) When the coil makes an angle of 30° with the field, $\alpha = 30°$, $\cos 30° = 0.866$.

$$M = 100 \times 10^{-3} \times 5 \times 5 \times 10^{-3} \times 0.866$$
$$= 21.7 \times 10^{-4} \text{ newton-meter clockwise as viewed from below.}$$

Therefore, a counterclockwise torque of 21.7 × 10^{-4} newton-meter will hold the coil in this position.   ◀

31-3 AMMETER

A galvanometer can measure currents and voltages. When it measures currents, it is also called an *ammeter* and when it measures voltages it is called a *voltmeter*. The principle of operation in both cases involves the torque produced on a current-carrying loop in a magnetic field, which was described in Section 31-2. The resistance R_g of the coil and its leads in a galvanometer is low. As an example, let us say that it is 1 ohm in a given galvanometer. Suppose that the mechanical design of this instrument is such that a current of 0.01 amp ($= 10$ milliamperes $= 10$ mA) in its coil will swing the pointer to its maximum deflection (see Fig. 31-4). This means that half this current, or 5 mA will swing the pointer halfway and 1 mA will swing the pointer one-tenth of full scale, etc. This particular milliammeter can measure currents up to 10 mA.

This instrument can be converted to measure higher currents by connecting an appropriate resistor, called a *shunt* resistor in parallel with the coil. Suppose that we desire to convert this particular instrument to a 1-amp meter, i.e., a meter that can measure currents up to 1 amp. For full scale deflection of the pointer, a current of 0.01 amp must still be established in the coil. The remainder, or 1 amp–0.01 amp $= 0.99$ amp passes through the shunt, R_s. Since R_s is in parallel with

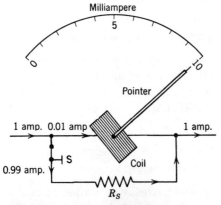

Figure 31-4   Full scale deflection on the galvanometer is 10 mA.

the coil resistance R_c, the voltages across these resistors are equal or

$$i_c R_c = i_s R_s \qquad (31\text{-}4)$$

where i_c is the current in the coil and i_s is the current in the shunt. Substituting the values and solving for R_s, we get

$$R_s = \frac{0.01 \text{ amp} \times 1 \text{ ohm}}{0.99 \text{ amp}} \approx 0.01 \text{ ohm}$$

This means that when a resistor of 0.01 ohm is connected in parallel with the coil of this instrument, the current range is multiplied by 100.

Equation 31-4 can be put into another useful form by noting that $i_c + i_s = i_d$, where i_d is the desired range of the instrument. Substituting for i_s, $i_d - i_c$ in Eq. 31-4, we get

$$i_c R_c = (i_d - i_c) R_s$$

or
$$R_s = \frac{i_c R_c}{i_d - i_c} \qquad (31\text{-}5)$$

Example 31-2   The coil of a certain ammeter has a resistance of 0.5 ohm. A current of 1 amp will cause its pointer to deflect full scale. What shunt resistance is required to convert it to a 10-amp meter?

Solution.   We can solve this problem by the use of Eq. 31-5. The current for full scale deflection (fsd) of the coil is $i_c = 1$ amp, the resistance of the coil $R_c = 0.5$ ohm, and the desired range $i_d = 10$ amp.

$$R_s = \frac{1 \text{ amp} \times 0.5 \text{ ohm}}{10 \text{ amp} - 1 \text{ amp}} = 0.056 \text{ ohm} \qquad \blacktriangleleft$$

31-4 VOLTMETER

A voltmeter is a galvanometer that is graduated in units of volts. If the resistance R_c of the coil of a given galvanometer is 1 ohm and the meter is designed to give a full scale deflection if the current in the coil, $i_c = 10^{-3}$ amp, then the voltage across the coil V_c can be deduced from Ohm's law. Thus

$$V_c = i_c R_c$$
$$= 10^{-3} \text{ amp} \times 1 \text{ ohm} = 10^{-3} \text{ volt}$$

This galvanometer can be used to measure voltages up to 10^{-3} volt ($=1$ millivolt).

This instrument can be converted to measure voltages that are higher than 10^{-3} volt by connecting an appropriate resistor R_v in series with the coil (see Fig. 31-5). Suppose that we want to convert this meter to read up to 100 volt. Then the voltage across the series combination of the

coil and R_v is

$$100 \text{ volt} = 10^{-3} \text{ amp} (R_v + 1 \text{ ohm})$$

$$R_v = \frac{100 \text{ volt}}{10^{-3} \text{ amp}} - 1 \text{ ohm} \approx 10^5 \text{ ohm}$$

The desired voltage range V_d is given by

$$V_d = i_c(R_c + R_v) \qquad (31\text{-}6)$$

Example 31-3 It is desired to convert a galvanometer with a coil resistance of 1 ohm and an f.s.d. of 0.01 amp to a voltmeter of 100-volt range. Calculate the resistance, R_v required. How should this resistance be connected?

Solution. $V_d = 100$ V, $i_c = 10^{-2}$ amp, $R_c = 1$ ohm. From Eq. 31-6,

$$R_v = \frac{100}{10^{-2}} - 1 = 9999 \text{ ohm}$$

This resistor is connected in series with the coil. ◀

31-5 THE ELECTRIC GENERATOR

The electric generator is a machine that converts mechanical energy into electrical energy. The essential features are indicated in Fig. 31-6. A coil,

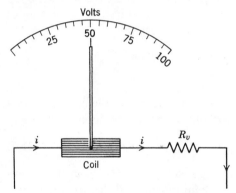

Figure 31-5 An appropriate series resistor, R_v, will convert the 1 millivolt meter to a 100-volt meter.

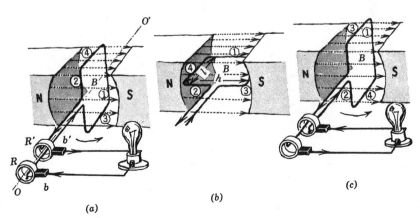

Figure 31-6 Rotating coil in a uniform magnetic field. The arrows in the lead-in wires to the coil indicate the direction of the induced current. The current reverses its direction every half-cycle.

located in a magnetic field, is rotated at a constant angular velocity by an outside agent. The outside agent can be a person or the steam produced by coal, oil, or a nuclear reactor. As the coil rotates, the flux through it changes from moment to moment. This condition fulfils the requirement for a voltage induction in the wires of the coil. To calculate the magnitude of the induced emf, the four sides of the rectangular coil are considered separately and the equation for the induced voltage in a wire, $\mathscr{E} = Blv \sin \theta$, (Eq. 30-13) is applied to each.

As side 1 in Fig. 31-6 rotates about the axis 0-0', half of it rotates in one direction with respect to the field and the other in the opposite direction. The net induced emf in this wire at any instant of time is therefore zero. The same argument holds true for side 2. The direction of the *instantaneous* velocity vector v of side 3 keeps changing from moment to moment (see Fig. 31-7). At the instant when the *velocity* of side 3 makes an angle of θ with the field, the induced emf in this side is given by

$$\mathscr{E}_{inst} = Blv \sin \theta \qquad (31\text{-}7)$$

where $v \sin \theta$ is the component of v that is perpendicular to the magnetic field.

The instantaneous velocity is associated with the angular velocity, ω of the coil by the equation

$v = \omega r$, (Eq. 11-3) where r is the radius of the circle swept out by wire 3. From Fig. 31-6 it is evident that $r = h/2$. In addition, the constant angular velocity of the coil is by definition θ/t. Making use of these facts, Eq. 31-7 takes the form

$$\mathscr{E}_{inst} = Blv \sin \theta = Bl\omega \frac{h}{2} \sin \theta \qquad (31\text{-}8)$$

for the instantaneous emf induced in wire 3. Identical arguments hold for wire 4. The emf in sides 3 and 4 reinforce in the sense that if the leads to coil are connected to an outside load, e.g., the lamp in Fig. 31-6, the direction of the induced current in these wires at any instant will either be clockwise or counterclockwise for both. The induced emf in the entire coil is the sum of the emf for the four sides:

$$\mathscr{E}_{inst} = 0 + 0 + 2\left(\frac{Bl\omega h}{2} \sin \theta\right) = BA\omega \sin \theta$$

where $l \times h = A$, the area of the coil. If the coil is comprised of N turns, the last equation becomes

$$\mathscr{E}_{inst} = NBA\omega \sin \theta \qquad (31\text{-}7)$$

It is emphasized that Eq. 31-7 gives the instantaneous value for the induced electromotive force in the coil of a generator. That this is so is evident

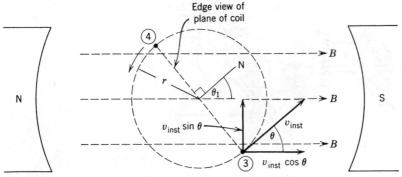

Figure 31-7   An end view of the sides ③ and ④. The instantaneous velocity vector, v_{inst}, is changing from moment to moment. The component $v_{inst} \sin \theta$ contributes to an induced emf and the component $v_{inst} \cos \theta$ does not, since the latter is parallel to B. The angles θ_1 and θ are equal.

from the sine factor in this equation. When $\theta = 0$, i.e., when the angle between the tangential velocity of sides 3 and 4 and the field is zero, $\mathscr{E}_{inst} = 0$. The induced emf is zero when the plane of the coil is perpendicular to the field. The maximum induced voltage occurs when the factor $\sin \theta$ is maximum, i.e., when $\theta = 90°$. The induced emf is maximum when the plane of the coil is parallel with the field.

Equation 31-7 can be put into another useful form by noting that $\theta = \omega t$, where ω is measured in radians/sec, and θ in radians. Also, ω is related to the frequency f by the equation $\omega = 2\pi f$, and f is measured in cycles/sec or revolutions/sec. Therefore, $\theta = 2\pi f t$. Equation 31-7 can, therefore, be expressed as

$$\mathscr{E}_{inst} = NBA(2\pi f) \sin(2\pi f t) \qquad (31-9)$$

This seems like a complicated equation, but it is not too difficult to apply.

Example 31-4   The coil (armature) of a certain generator is comprised of 1000 turns, each having an area of 75 cm². If the armature is rotated in a steady magnetic field of flux density 10^{-3} weber/m² and at a constant rate of 3600 rev/min, calculate (a) the frequency of the voltage generated, and (b) the maximum value of the voltage generated. At what position of the coil is maximum voltage achieved? (c) What is the magnitude of the voltage generated at an instant when the plane of the coil makes an angle of 60° with the field? (d) What is the magnitude of the voltage $\frac{1}{4}$ sec after the coil passes through the zero voltage position?

Solution
(a)   The frequency in cycles/sec is

$$3600\,\frac{\text{rev}}{\text{min}} \times \frac{1\,\text{min}}{60\,\text{sec}} = \frac{60\,\text{rev}}{\text{sec}} \text{ or 60 cycles/sec}$$

(b) $\mathscr{E}_{inst} = NBA\omega \sin \theta$ $\qquad$ (31-7)

The maximum value for $\mathscr{E}$ occurs when $\sin \theta$ is a maximum, at $\theta = 90°$. This occurs when the plane of the coil is parallel to the field. The

angular velocity of the coil, $\omega = 2\pi f = 2\pi 60 = 120\pi\,\text{rad/sec}$; $A = 75 \times 10^{-4}\,\text{m}^2$.

$$\begin{aligned}\mathscr{E}_{max} &= 1000 \times 10^{-3} \times 75 \times 10^{-4} \times 120\pi \\ &= 2.83 \text{ volts}\end{aligned}$$

(c)   By Eq. 31-7

$$\mathscr{E}_{inst} = NBA\omega \sin 30 = \mathscr{E}_{max} \sin 30 = 1.42 \text{ volts}$$

(d)   Since the coil rotates at a constant 60 rev/sec, it makes 15 complete revolutions in $\frac{1}{4}$ sec. If we start at the zero voltage position, then in an interval of $\frac{1}{4}$ sec later, the coil will be back in the same position and the induced voltage is zero. We can get the same answer from Eq. 31-9. In this equation, the factor $\sin(2\pi f t) = \sin(2\pi \times 60 \times \frac{1}{4}) = \sin(30\pi\,\text{rad}) = 0$. Therefore, $\mathscr{E}_{inst} = 0$. ◀

31-6 THE ALTERNATING CURRENT CYCLE

The coil of a generator is wound on an iron core in order to concentrate the magnetic field. This assembly is called the *armature*. Each of the two lead wires from the armature (Fig. 31-6) is connected to a *slip ring R* and *R'*. While the armature, lead wires, and slip ring rotate as one unit, the rings brush against stationary *brushes b* and *b'* to which supply lines are connected.

We can use Lenz's law to determine the direction of the induced current in the coil. In Fig. 31-6a if the plane of the coil is perpendicular to the field then the flux through the coil is maximum (see Eq. 30-9). A moment later the flux will be reduced and, by Lenz's law, the direction of the induced current must be such as to oppose this flux decrease. The induced current i in segments 3 and 4 must therefore be directed as shown in Fig. 31-6a. This direction of i (by the right-hand rule) would increase B from left to right and, therefore, the flux through the coil. This direction of the current is maintained until a moment after the coil rotates 180° (one-half cycle) as shown in Fig. 32-6c. During its rotation for the next 180° the current direction is reversed. Therefore, the direction of the current in segments 3 and 4 and

Table 31-1

θ	$\sin \theta$	$\mathscr{E}_{\text{inst}}$ (from Eq. 31-7)
0°	0	0
90	1	Positive maximum
180	0	0
270	−1	Negative maximum
360	0	0

in the external appliances connected to the generator changes every one-half circle.

Table 31-1 gives the extreme values of $\mathscr{E}_{\text{inst}}$ as a function of the angle.

Figure 31-8 is a graph of the values from Table 31-1, using the values of $\mathscr{E}$ as the ordinate and θ as the abscissa. The resulting graph is termed *sinusoidal* because it has the same shape as a graph of the sine of an angle versus the angle. It can be seen at a glance that for half of a cycle (180°) the voltage is positive; for the other half it is negative. The words *positive* and *negative* refer to the direction of the induced current. If the combined resistance of the coil and load, (the lamp) in Fig. 31-6 is R, then by Ohm's law, the instantaneous current, $i_{\text{inst}} = \mathscr{E}_{\text{inst}}/R$. The current in the coil of the generator therefore also varies sinusoidally. The alternating current supplied to most homes in the United States has a frequency of 60 cycles/sec. This means that the direction of the current in the appliances changes 120 times every second.

31-7 EFFECTIVE VALUES OF ALTERNATING VOLTAGE AND CURRENT

Figure 31-8 indicates that the magnitude of an alternating current changes from moment to moment. The *effective value** of an alternating current, i_{eff} is defined as

$$i_{\text{eff}} = 0.707 \, i_{\text{max}} \qquad (31\text{-}10)$$

where i_{max} is the maximum value of the current. The effective value is derived from a comparison of the heat developed in a resistor by a direct current (dc) and an alternating current (ac). It can be shown that a steady direct current of a magnitude equal to the effective value of a given alternating current will develop the same heat in the same time as this alternating current. As an example, suppose that a toaster is operated on alternating current having a maximum value of 10 amp. This toaster will develop the same heat, say, in 20 sec as it would if operated for 20 sec on a direct current having a value of 7.07 amp. The effective value of an alternating voltage, V_{eff} is given by

$$V_{\text{eff}} = 0.707 \, V_{\text{max}} \qquad (31\text{-}11)$$

where V_{max} is the maximum value of the alternating voltage.

The readings on ordinary ac ammeters and voltmeters indicate the effective values of currents and voltages, respectively.

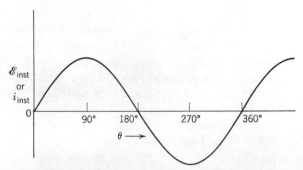

Figure 31-8 The voltage of an alternating voltage and current.

* The effective value is also referred to as the root-mean-square or r ms value.

31-8 THE DC GENERATOR

An ac generator can be converted to a dc generator by merely substituting a *split-ring commutator* (*C* in Fig. 31-9a) for the slip rings in Fig. 31-6. The commutator maintains a unidirectional induced current for the entire cycle of the armature. Notice that sides 3 and 4 of the coil are each permanently connected to one-half of the split ring and that the entire commutator assembly rotates with the armature. While these sides rotate through 180° (starting from the position of Fig. 31-9a) their respective commutator segments make separate contact with one of the brushes, *b*, *b'*. For the next half of the armature cycle, the commutator segments reverse the brush contact (Fig. 31-9c), causing the direction of the current in the outer circuit to remain constant during the entire cycle. One important appli-

cation of the dc generator is in the automobile, where it is used to recharge the storage battery.

31-9 THE DC MOTOR; BACK EMF

A motor is a machine that converts electrical energy into mechanical energy. In a motor, the electric current from an outside supply gives rise to the armature torque. In a generator, the torque supplied from an outside source to the armature gives rise to an induced current in it. In fact, a dc generator can be used as a motor, and a motor as a dc generator by merely reversing the kind of energy input. The generator in Fig. 31-6a would act as a motor if the lamp were replaced by a current source. The principle of a motor is based on the force experienced by a current-carrying wire residing in a magnetic field. These forces on

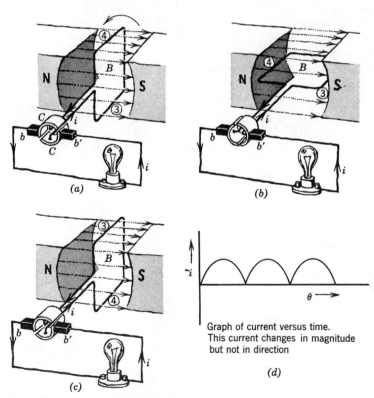

(a)

(b)

(c)

(d)

Graph of current versus time.
This current changes in magnitude
but not in direction

Figure 31-9 Direct current generator.

the armature wires give rise to a torque on the armature assembly (Sec. 31-2) which, in turn, provides a torque to an outside piece of equipment like an electric fan.

An emf will be induced in the rotating armature of a motor because its coils, rotating in a magnetic field, will experience a change in flux. By Lenz's law, the direction of the emf will oppose its cause, which is the current in the coil. The induced emf, called a *counter* or *back emf* will reduce the current in the armature. When the motor of an electric fan, for instance, is just starting up, the current in its armature is at a maximum; when it gains speed, the $\Delta\phi/\Delta t$ is larger, making the back emf larger, thereby reducing the current in the armature.

Two common types of dc motors are the series and shunt motors (Fig. 31-10). In the series motor, the resistance R_F of the electromagnet producing the magnetic field is in series with the armature resistance and with the supply line. Series motors are used when large starting torques are required. In the shunt motor, (Fig. 31-10b) the field resistance is in parallel with the armature. The shunt-wound motor has the disadvantage of low starting torque but the advantage that it develops constant torque at full speed.

Example 31-5   A 110-volt dc source is connected to a series wound motor; the combined field coil and armature resistance is 100 ohms, and the back emf $\mathscr{E}$ is 80 volts when the motor reaches full speed. (a) Calculate the initial current drawn from the supply line. (b) Calculate the current drawn at full speed. (c) Determine the power input to the motor and the power output by the motor at full speed.

Solution. (a) At the start, no back emf is developed in the armature. The current drawn is by Ohm's law

$$\frac{110 \text{ volts}}{100 \text{ ohms}} = 1.10 \text{ amp}$$

(b) At full speed the maximum value for the back emf $\mathscr{E} = 80$ volts. An equivalent circuit for full speed operation is shown in Fig. 31-11, where $\mathscr{E}$ can be imagined as being supplied by a battery, connected so as to buck the applied voltage from the supply. Applying Kirchhoff's loop theorem, starting from point A in Fig. 31-11, and going in a clockwise direction around the loop, we get

$$-iR - \mathscr{E} + V = 0$$

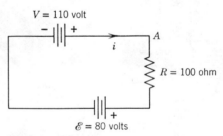

Figure 31-11

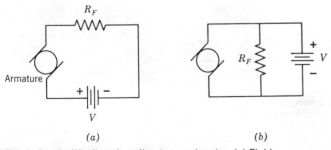

(a)   (b)

Figure 31-10   Windings in a direct current motor. (a) Field coil R_F in series with armature (series-wound motor) (b) Field coil R_F is in parallel with armature (shunt-wound motor).

or

$$i = \frac{V - \mathscr{E}}{R} = \frac{110 \text{ volts} - 80 \text{ volts}}{100 \text{ ohms}} = 0.3 \text{ amp}$$

(c) power in $= V \times i = 110 \text{ volts} \times 0.3 \text{ amp}$

$\qquad\qquad\quad = 33 \text{ watts}$

power out $=$ power in $- i^2 R$

$\qquad\qquad\quad = 33 \text{ watts} - (0.3 \text{ amp})^2$

$\qquad\qquad\qquad\quad \times 100 \text{ ohms}$

$\qquad\qquad\quad = 24 \text{ watts}$

The $i^2 R$ term represents the power lost in heating. Alternatively, we could take

power out $=$ back emf $\times$ armature current

$\qquad\qquad\quad = \mathscr{E}i$

$\qquad\qquad\quad = 80 \text{ volts } (0.3 \text{ amp}) = 24 \text{ watts}$

A dc motor may have its field wound with two separate coils. One of these coils is in series with the armature and the other is in shunt or parallel with the armature. The effect of this combination is a series and shunt-wound motor in one unit. This arrangement is called a compound motor. This type combines the advantages of the high starting torque of the series motor and the constant torque at full speed of the shunt motor. ◀

31-10 THE TRANSFORMER

One important practical advantage of alternating current over direct current is that the former lends itself more easily to increases or decreases of supply voltages. Certain instruments, like radios and sound systems, may require low supply voltages for some of its components and higher voltages for other of its components. A transformer connected to a 110-volt ac outlet can *step up* or *step down* the voltage to the proper value for a given appliance. Energy is conserved and, consequently, the total energy output of a transformer never exceeds the energy input.

A *transformer* (Fig. 31-12) consists essentially of two coils of wire, not connected electrically, wound near each other on a soft iron core. The coil that is connected to the ac voltage supply, say, the wall outlet, is called the *primary*, and the other coil is the *secondary*. Since the current in the primary is constantly changing in magnitude and direction, a changing magnetic flux is established within the core of the transformer. The same change in magnetic flux also passes through the secondary coil because the flux is almost entirely confined within the soft iron core.

From Faraday's law of induction, namely, $\mathscr{E} = -N \, \Delta\phi/\Delta t$, it follows that the voltage $\mathscr{E}_p$ induced in the primary is given by $\mathscr{E}_p = -N_p \, \Delta\phi/\Delta t$, and that induced in the secondary, $\mathscr{E}_s$ is given by $\mathscr{E}_s = -N_s \, \Delta\phi/\Delta t$, where N_p and N_s are, respectively, the number of turns in the primary and secondary. Since $\Delta\phi/\Delta t$ in the last two equations are the same, it follows that

$$\frac{\mathscr{E}_p}{\mathscr{E}_s} = \frac{N_p}{N_s} \qquad (31\text{-}12)$$

The efficiency of a transformer is defined as the ratio of the power output to the power input. Since

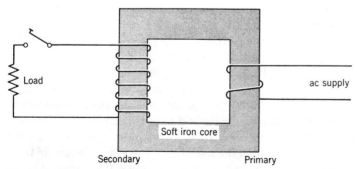

Figure 31-12 Transformer.

electrical power is given by the product of voltage and current, the efficiency of a transformer can be expressed by the equation

$$\text{efficiency} = \frac{\mathscr{E}_s i_s}{\mathscr{E}_p i_p} \qquad (31\text{-}13)$$

where i_s and i_p are, respectively, the currents established in the secondary and primary coils. Well-designed transformers have efficiencies higher than 95%, very large ones as high as 99%. Assuming 100% efficiency, i.e., a condition where the output power is equal to the power input, then, $\mathscr{E}_s i_s = \mathscr{E}_p i_p$ or

$$\frac{i_p}{i_s} = \frac{\mathscr{E}_s}{\mathscr{E}_p} = \frac{N_s}{N_p} \qquad (31\text{-}14)$$

In addition to many industrial and house appliances, transformers are used in power transmission from electric power houses to the consumers. The power losses involved in electrical transmission stem from the heat losses in the power lines. These can be reduced by decreasing the resistance of the lines. This would involve power lines of large diameter, since the resistance of a wire is inversely proportional to its cross-section area. However, practical considerations set an upper limit to the diameter of transmission lines that are miles in length. Since the power lost in a transmission wire is proportional to i^2R, and a lower limit on R is set by this consideration, less power is lost if somehow the current i can be reduced in the wire. This can be accomplished by a step-up transformer at the power plant and a step-down transformer at the destination. Assume that the generators of a given power plant can produce a terminal voltage of 10,000 at 10 amp. If the resistance of the transmission lines is 100 ohms, then $i^2R = (10)^2 \times 100 = 10,000$ watts would be lost in the transmission process. This represents a loss of 10% of the total power produced by the generators. If a step-up transformer at the power plant with a turn ratio of $N_s/N_p = 10$ is used, then from Eq. 31-14 the current transmitted will be reduced to 1 amp (discounting transformer losses) and the transmitted voltage increased to 100,000 volts, while the power trans-

mitted will remain the same. However, the transmission power loss will be reduced to $(1)^2 \times 100$ watts, or 1% of the total power output of the generator. Of course, a step-down transformer is required at the consumer end in order to reduce the voltages to safe values.

Example 31-6   In order to step-up the voltage from a 110-volt line to 2200 volts, a transformer having a 100-turn primary is used. Calculate the number of turns in the secondary and the current drawn from the line if the secondary draws $\frac{1}{2}$ amp. Assume the transformer to be 100% efficient.

Solution (from Eq. 31-14)

$$\frac{\mathscr{E}_p}{\mathscr{E}_s} = \frac{N_p}{N_s}; \qquad N_s = 100 \text{ turns} \times \frac{2200 \text{ volts}}{110 \text{ volts}}$$

$$= 2000 \text{ turns}$$

$$\frac{i_p}{i_s} = \frac{\mathscr{E}_s}{\mathscr{E}_p}; \qquad i_p = 20 \times \frac{1}{2} \text{ amp} = 10 \text{ amp} \blacktriangleleft$$

31-11 EDDY CURRENTS

Suppose that a conductor like a chunk of iron in Fig. 31-13 is moved to the right toward the north pole of a bar magnet. The iron will then experience a change of magnetic flux that is a sufficient condition for an emf to be induced in it. Since iron is a conductor, the emf will give rise to an induced current throughout its volume, in a direction (by Lenz's law) that will oppose this motion. An induced north pole at the right of the metal chunk will oppose its motion to the right (why?). The induced currents will therefore be in a plane perpendicular to the motion and in a direction shown by the current loop on plane A. Such current loops are called *eddy currents*.

In general, eddy currents will occur in conductors that experience a change of flux. This flux change can be due either to the mechanical motion in a magnetic field as in the armature of a motor, or to a changing current in the wires wound around the soft iron core of a transformer. In both cases, eddy currents cause i^2R losses due to

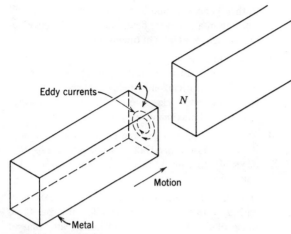

Figure 31-13 Eddy currents. Chunk of iron being moved toward north pole of magnet.

heating. These can be minimized by laminating the iron core, i.e., building it up with thin sheets covered by a thin coating of insulating varnish. This reduces the eddy currents.

SUMMARY

An alternating current generator produces a voltage that is changing in magnitude from moment to moment; when an electrical appliance is connected to the generator, the current changes in magnitude from moment to moment.

The operation of a voltmeter, ammeter, and motor is based on forces produced by a current in a wire that is located between the poles of a magnet. These forces produce torques.

The range of a voltmeter and ammeter can be extended by inserting a resistor in series or parallel, respectively.

Alternating-current voltmeters and ammeters give effective values that are 0.707 times the maximum values.

An electrical load on a generator produces a back torque in its armature. The greater the load, the greater must be the mechanical work required for turning the armature.

A mechanical load on a motor produces a back emf in its armature. The greater the load, the greater must be the electrical work required to turn the armature.

The transformer is a device that steps-up or steps-down alternating currents and voltages.

Eddy currents cause electrical losses in machines.

QUESTIONS

1. A voltmeter is "driven" by current. Explain.

2. Why is the internal resistance in a voltmeter high?

3. What is meant by an alternating current?

4. One often sees the house lights dimming for a moment when the motor of a refrigerator starts. Why?

5. A generator is essentially an ac device. Why?

6. Can both the voltage and current in the secondary of a transformer be stepped-up simultaneously? Why?

7. The primary and secondary in a transformer are electrically insulated from each other. Then how is energy transferred from the primary to the secondary?

8. It takes a mechanical force to rotate the metallic ring in Fig. 31-14(*a*). Does it take a somewhat larger force if the ring is in a magnetic field. Why? (See Fig. 31-14(*b*).)

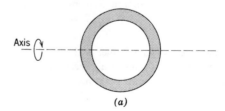

(*a*)

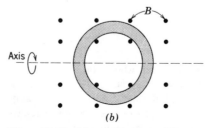

(*b*)

Figure 31-14 (*a*) and (*b*).

9. What is an eddy current?

10. Why is the efficiency of a transformer increased by constructing the iron core of separate thin layers?

PROBLEMS

(A)

1. An ac ammeter indicates the current in a certain circuit to be 5 amp. What is the maximum value of this current?

2. The maximum value of the ac voltage across a certain resistor is 50 V. What would be the reading on an ac voltmeter connected across the resistor?

3. The full scale deflection of a certain dc ammeter, having a coil resistance of 1 ohm, is 50 mA. What shunt resistance will enable this instrument to indicate a maximum current of 10 amp?

4. If the frequency of an alternating current is 60 cycles/sec, how long a time elapses between two successive zero values of the current?

5. What is the net force on the loop in Fig. 31-3 at any moment?

6. Is the net torque on the loop in Fig. 31-3 ever zero? If so, when?

7. At what orientation of the loop in Fig. 31-3 is the net torque a maximum?

8. The maximum value of the emf of a certain generator is 100 V when the armature turns at 60 rev/sec. What would be the maximum emf if the armature's speed is increased to 100 rev/sec?

9. Assume that a certain transformer is 100% efficient, the turns ratio of the primary to secondary is 5, and calculate the current ratio in the secondary to primary.

10. If the voltage across the secondary in the transformer of Problem 9 is 60 V, what is the voltage input?

(B)

1. A certain galvanometer, whose coil including electrical connections has a resistance of 0.2 ohms, is designed to give full scale deflection at 0.1 amp. What range of voltage can this instrument read if no additional resistors are connected to it?

2. Both a generator and a galvanometer have a coil rotating in a magnetic field. State the difference between their principles of operation.

3. A certain generator produces 60 cycles/sec voltage. Why is this an alternating voltage? What is the duration of 1 cycle? How many degrees of armature rotation does $\frac{1}{4}$ cycle represent? How many voltage minima and maxima occur per cycle?

4. The resistance of a galvanometer is 0.1 ohms and it has a fsd of 5 mA. Calculate the series resistance necessary to convert this instrument into a 100-V voltmeter.

5. It is desired to measure the value of a resistor X with a voltmeter and ammeter. The internal resistance of the voltmeter is 1000 ohms, and that of the ammeter is negligible. Determine the value of X if the voltmeter reads 50 volts and the ammeter 0.5 amp. (See Fig. 31-15.)

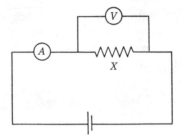

Figure 31-15

6. A step-up transformer has a turns ratio of 1:30 and is operated on a 110-volt line, which supplies a 10 amp current. Calculate the voltage and the current in the secondary, and the power output, assuming 100% efficiency.

7. A step-down transformer has 100 turns in the secondary winding. How many turns should it have in the primary in order to step-down a 2310-V supply line to 100 V?

8. (a) Calculate the maximum value for the emf produced by a generator having 5000 turns, each turn having an area of 750 cm², when the coil rotates at 100 rev/sec in a steady magnetic field of 0.5 weber/m².

   (b) What is the instantaneous voltage produced at an instant $1\frac{1}{2}$ sec after the minimum value?

9. Calculate the magnetic flux density required to give a coil of 500 turns a torque of 0.7 N-m when its plane is parallel to the field. The dimensions of each turn are 15 cm × 25 cm and the current is 10 amp.

10. Assume the horizontal component of the earth's magnetic field at a certain spot on the earth's surface to be 2.5×10^{-5} W/m². How many turns of wire, each turn having a 1.5 m² area, are required to exert a maximum torque of 1 N-m if $i = 2$ amp?

11. A ball park that has night games consumes electrical energy at a rate of 100,000 watts. If the terminal voltage across the two power lines at the park is 2000 V and the total resistance of these lines between the park and commercial generator is 10 ohms calculate:
    (a) The current drawn while the park lights are on.
    (b) The power loss in the lines.
    (c) The power supplied, assuming no transformer losses.

12. The rectangular motor coil in Fig. 31-16 has 100 turns and carries a 10-amp dc in the direction indicated. The flux density is 5×10^{-2} weber/m² and the coil can rotate about the axis X-X'. Calculate
    (a) The magnitude and direction of the force in sides 1, 2, 3, and 4.

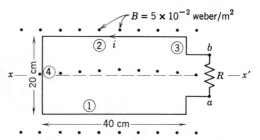

Figure 31-16

(b) The resultant force on the coil.
(c) The resultant torque on the coil at the moment that B is ⊥ to the plane of the coil.

13. Assume now that no external current is provided to the coil of Problem B12, that the coil rotates at 2400 rev/min and that $R = 5$ ohms. Calculate
    (a) The maximum induced voltage in the coil.
    (b) The maximum induced current.
    (c) The direction of the current in R a moment after the configuration shown if the coil rotates in a counterclockwise direction as viewed from the right.

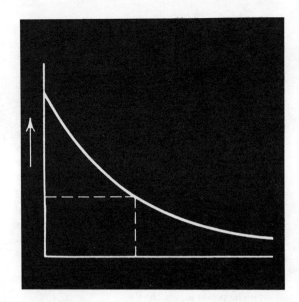

32.
Elements of an
Alternating Current

Most electrical appliances in the home and industrial installations are operated on an alternating current. An alternating current changes its direction from moment to moment while a direct current maintains the same direction. This chapter develops some principles related to the ac circuit.

LEARNING GOALS

1. *To learn the meaning and equation for inductance.*
2. *To learn the meaning and equation for capacitance.*
3. *To understand the concept of a time-constant.*
4. *To learn about phase relations between voltage and current in an ac circuit.*
5. *To understand and learn how to calculate the impedance of a circuit.*
6. *To learn how to calculate the power dissipated in an ac circuit.*

32-1 INDUCTANCE

The important element (external to the battery) in a dc circuit is the resistor. Such circuits involve currents and voltage drops across various combinations of resistors. The important elements in the external ac circuit, are the *inductor*, the *capacitor*, and the resistor.

An inductor is a loop of wire shaped in the form of a helix and is indicated as ᴍᴍ in a circuit diagram. When a steady current (dc) is established in an inductor, a magnetic field is set up within the volume formed by its loops, as described in Section 29-9. However, when an inductor is connected to an ac generator, the magnitude of the field within it changes from moment to moment because the magnitude of the current is constantly changing. Therefore, the magnetic flux within the volume of the inductor also changes and an induced emf results.

The *self-inductance L* of an inductor is constant and depends on its geometry (this is analogous to resistors). Self-inductance is defined by the equation

$$L = -\frac{\mathscr{E}}{\Delta i/\Delta t} \tag{32-1}$$

A given inductor has a value of one henry (named after the nineteenth century American scientist Joseph Henry) if one volt is induced in it when the current through it *changes* at the constant rate of one ampere per second. The minus sign in the equation refers to the direction of the induced emf, which can be deduced from Lenz's law.

Solving Eq. 32-1 for $\mathscr{E}$ yields

$$\mathscr{E} = -L\frac{\Delta i}{\Delta t} \tag{32-2}$$

The meaning of this equation is demonstrated by Fig. 32-1a, which shows an inductor connected to a battery through a switch S. As the blade of the switch begins to make contact, the resistance in the circuit decreases (compared with an open circuit) and the current in the circuit begins to increase. As the blade continues to be pushed down, the resistance further decreases and the current increases. Finally, after the blade of the switch is completely depressed, the current in the circuit increases toward a maximum value as long as the blade of the switch remains in the closed position.

During the short interval from the time when the blade begins to make contact to the moment that it is fully depressed, there is an *increase* in current through the circuit, which means that i is momentarily increasing and Δi is not equal to zero. According to Eq. 32-2, there is an emf $\mathscr{E}$ induced in the inductor L. To determine the direction of $\mathscr{E}$, we apply Lenz's law. The cause for the induced emf in Fig. 32-1a is a momentary current increase in L pointing toward the left so that the voltage induced must point to the right in L (Fig. 32-1b). The induced emf will prolong the time for the current to reach a maximum value.

Consider now what happens during the short interval when the switch is pulled up, as in Fig. 32-2a. The current in L is decreasing and Δi is not

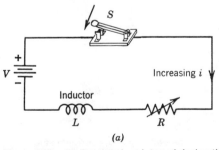

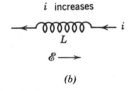

(a)   *(b)*

Figure 32-1 (a) For the short interval during the closing of the switch, the current in the inductor increases toward the left. (b) The direction of the induced voltage across the inductor is toward the right.

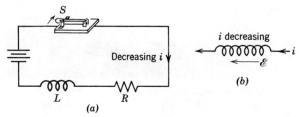

Figure 32-2   (a) The current is decreasing in the inductor. (b) The direction of the induced emf is such as to retard the current increase.

zero. This will cause an induced emf across the inductor directed toward the left, (Fig. 32-2b). This induced voltage will prolong the time for the current in L to fall to zero.

Example 32-1   (a) When the switch is moved down (see Fig. 32-1) the current through the 20-millihenry coil rises steadily from 0 to 5 amp in 0.01 sec. Calculate the induced emf in the coil during this interval. (b) After the switch has been down for some time, the resistance R is increased so that the current drops from 5 to 3 amp in 0.25 sec. What is the average induced emf in this interval?

Solution
(a) $L = 20 \times 10^{-3}$ henry, $\Delta t = 0.01$ sec, $\Delta i = 5$ amp.

$$\mathscr{E} = -L\frac{\Delta i}{\Delta t} = \frac{-(20 \times 10^{-3} \text{ h} \times 5 \text{ amp})}{0.01 \text{ sec}}$$

$$\mathscr{E} = -10 \text{ volt}$$

The minus sign in the above answer indicates the direction of $\mathscr{E}$ to be opposite to the current.
(b) $L = 20 \times 10^{-3}$ henry, $\Delta i = -2$ amp, $\Delta t = 0.25$ sec.

$$\mathscr{E} = \frac{-(20 \times 10^{-3})(-2)}{0.25} = +0.16 \text{ volt}$$

The plus indicates that $\mathscr{E}$ points in the same direction as the current.   ◀
An emf is generally associated with energy. The emf of a battery involves chemical energy. The induced emf involves the energy supplied to the inductor by the generator which, in this case, is

a battery. It can be shown that the total energy W supplied to an inductor in a circuit to establish the maximum current i_{max} is

$$W = \tfrac{1}{2}Li_{max}^2 \qquad (32\text{-}3)$$

This energy is stored by the inductor as magnetic energy as long as the circuit in Fig. 32-1a remains completed. When the switch is pulled up, this energy is released, often causing a visible spark. Actually, all inductors have some resistance. The energy dissipated by the resistive part is treated separately.   ◀

32-2 TIME CONSTANT OF AN L-R CIRCUIT

Suppose that L is not in the circuit of Fig. 32-3. A graph of the current established in the circuit versus time would look like Fig. 32-4 where Δt is the short time that it takes to close the switch.

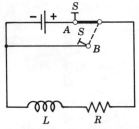

Figure 32-3   With the switch at A, the current in the inductor builds up to its maximum value. If the switch is then quickly turned to B, the current in L discharges through R.

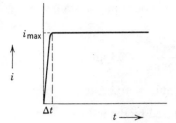

Figure 32-4 It takes a very short time for the current to reach the maximum value in a circuit containing resistors only.

The counter emf produced when an inductor is introduced, will stretch out the time that it takes for the maximum current $i_{max} = V/R$ to be established (Fig. 32-5).

In a time τ (Greek letter, tau) the current in the circuit will reach a value that is 0.631 of the maximum current. This time span is called the *inductive time constant* and is equal to the inductance divided by the resistance.

$$\tau = \frac{L}{R} \qquad (32\text{-}4)$$

In a time of two time constants, $2\tau = 2L/R$, the current reaches a value of 86.47% of i_{max}; in six time constants, $6\tau = 6L/R$, the current is 99.8% of i_{max}. Theoretically, the current never quite reaches i_{max}.

Suppose now that the switch in Fig. 32-3 has been in contact with point *A* for a long time and the maximum current has been established. By quickly turning the switch to point *B*, the battery is removed from the circuit and the current in the inductor starts falling. The emf induced across the inductor will retard this fall. In a time of one *time constant*, $\tau = L/R$, the magnitude of the current will have fallen approximately by 63% of i_{max}, or its value will be

$$100\% \ i_{max} - 63\% \ i_{max} = 37\% \ i_{max}$$

Theoretically, the current never goes to zero. See Fig. 32-6.

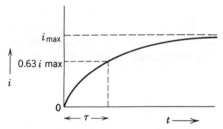

Figure 32-5 Graph of the current growth in a circuit containing an inductor. Theoretically, it takes an infinite time for the current to reach its maximum value.

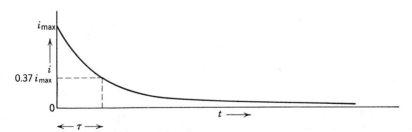

Figure 32-6 Decaying current in an inductor.

Example 32-2   Show that the time constant τ is measured in seconds.

Solution

$$\tau = \frac{L}{R}$$

A dimensional analysis of Eq. 32-1 gives

$$L = \frac{\text{volt-sec}}{\text{amp}} = \text{ohm-sec}$$

Substituting this value for L in $\tau = L/R$, we get

$$\tau = \frac{\text{ohm-sec}}{\text{ohm}} = \text{sec} \quad \blacktriangleleft$$

Example 32-3   If in Fig. 32-1 the battery voltage $V = 100$ volts and $L = 10$ henries. (a) What should be the value of R in order for the current to reach 63% of its maximum value in 10^{-2} sec? (b) What is the value of the current 10^{-2} sec after the switch is pushed down? (c) Compute the energy stored by the inductor.

Solution   (a) It takes one time constant for the current to reach $0.63i_{max}$. One time constant $= \tau = L/R$, $R = L/\tau$

$$R = \frac{10 \text{ henries}}{10^{-2} \text{ sec}} = 1000 \text{ ohms}$$

(b) The current i after one time constant is given by $i = 0.63i_{max}$.

$$i_{max} = \frac{V}{R} = \frac{100 \text{ volts}}{1000 \text{ ohms}} = 0.1 \text{ amp}$$

$$i = 0.63 \times 0.1 \text{ amp} = 0.063 \text{ amp} = 63 \text{ mA}$$

(c) The energy stored by the inductor after the maximum current is established in the circuit is (by Eq. 32-3)

$$W = \tfrac{1}{2}Li_{max}^2 = \tfrac{1}{2} \times 10(0.1)^2 = 5 \times 10^{-2} \text{ joule} \quad \blacktriangleleft$$

32-3 CAPACITORS

Circuit elements called capacitors also constitute an important part of an ac circuit. *Capacitors* consist of two conductors, usually in the form of parallel plates or concentric cylinders, which are separated by a highly insulating material, called a *dielectric*. Consequently, when capacitors are placed in a dc circuit, the current in that circuit will be zero. In a circuit diagram, capacitors are indicated by two parallel lines, as in Fig. 32-7. Capacitors are widely used in electronic equipment, as in radios and television receivers. The principles associated with capacitors can be brought out by their connections to a battery. In

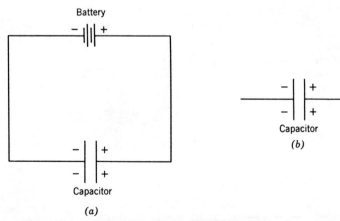

Figure 32-7   (a) A capacitor is charged by a battery. (b) The capacitor retains its charge after the battery is disconnected.

later sections their action in an ac circuit will be considered.

When the plates of a capacitor are connected to a battery, the negative charges will flow from the negative battery terminal to the capacitor plate connected to it (Fig. 32-7a). These negative charges will induce an equal number of positive charges on the other plate. If the contacts to the battery are then removed, the capacitor will maintain its charge (barring electrical leaks to the air). Oppositely charged plates of a capacitor give rise to a potential difference V between the plates. It is found that the quantity of charge q accumulated is proportional to the potential difference or

$$q \propto V$$

Transforming this proportionality statement into an equation yields

$$q = CV \qquad (32\text{-}5)$$

where C is the proportionality constant and is defined as the *capacitance*. The units for C are coulomb/volt, which is called a *farad* (after Michael Faraday).

> **A capacitor that develops a potential difference of one volt when a charge of one coulomb is accumulated on either plate has a capacitance of one farad.**

In practice, capacitances of the order of 10^{-6} farad ($1 \mu F$) and 10^{-12} F ($1 \mu\mu F$) are common.

The capacitance of a given capacitor depends on its geometry and on the dielectric material. For a parallel plate capacitor the capacitance C, in farads, is given by

$$C = \kappa\varepsilon_0 \frac{A}{d} \qquad (32\text{-}6)$$

where κ is called the *dielectric constant* and is constant for a given dielectric material; ε_0 is the permittivity of free space, expressed in mks units; A is the area of one plate, expressed in m^2, and d is the plate separation, in m. The dielectric constant, κ (Greek letter kappa), is a pure number, and is equal to approximately 1 for air and 7 for mica.

Example 32-4   What must be the area of each plate of a parallel plate capacitor if its capacitance is 10^{-8} farad? Take the plate separation to be 0.25 cm, the dielectric material to be mica, and $\varepsilon_0 = 8.85 \times 10^{-12}$ C^2/N-m^2.

Solution

$$C = \kappa\varepsilon_0 \frac{A}{d}, \qquad A = C \frac{d}{\kappa\varepsilon_0},$$

$$C = 10^{-8} \text{ farad}, \kappa = 7, d = 25 \times 10^{-4} \text{ m}$$

$$A = \frac{10^{-8} \times 25 \times 10^{-4}}{7 \times 8.85 \times 10^{-12}}$$

$$A = 0.4 \text{ m}^2$$

This is a large area for a portable radio, for instance. This area can be folded by an arrangement as in Fig. 32-8. ◀

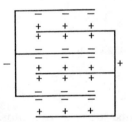

Figure 32-8   "Folded" capacitor.

32-4 CAPACITORS IN SERIES

When more than one capacitor is connected in a circuit, these can be replaced by a single capacitor having an equivalent capacitance, C_{equiv}. Figure 32-9a depicts three capacitors of capacitance C_1, C_2, and C_3 connected in series. A moment after the switch is pushed down, a negative charge q^- will accumulate on plate 1 which will, in turn, induce an equal but opposite charge q^+ on plate 2; q^+ will, in turn, induce an equal but opposite charge q^- on plate 3, etc., until the *plates of each capacitor in the series circuit carry equal charges* (in magnitude). The total charge drawn from the battery by the entire group of capacitors is $q = q^-$.

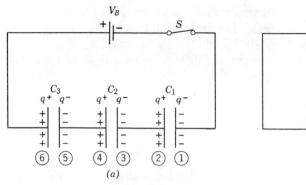

Figure 32-9   (a) and (b)

The equivalent capacitor will draw the same charge q from the battery as is drawn by the combination of capacitors. From Eq. 32-5 and Fig. 32-9b, $C_{equiv} = q/V$, where V is the voltage across the equivalent capacitor, which is equal to the battery voltage V_B. Thus

$$C_{equiv} = \frac{q}{V_B}, \qquad \text{or} \qquad V_B = \frac{q}{C_{equiv}} \quad (32\text{-}7)$$

The voltage across each capacitor is (from Eq. 32-5)

$$V_1 = \frac{q}{C_1}, \quad V_2 = \frac{q}{C_2}, \quad V_3 = \frac{q}{C_3} \quad (32\text{-}8)$$

where the q's are the same for each capacitor. Because this is a series circuit, $V_B = V_1 + V_2 + V_3$. Combining this equation with Eqs. 32-7 and 32-8 to

$$\frac{q}{C_{equiv}} = \frac{q}{C_1} + \frac{q}{C_2} + \frac{q}{C_3}$$

$$\frac{1}{C_{equiv}} = \frac{1}{C_1} + \frac{1}{C_2} + \frac{1}{C_3} \quad (32\text{-}9)$$

(Compare this equation with that governing *resistors in parallel*.) The equivalent capacitor has a smaller capacitance than any of the individual capacitors in the series circuit.

Example 32-5   Two capacitors are connected to a 100-volt source as in Fig. 32-10. Calculate (a) the equivalent capacitance of the two capacitors, (b) the charge on each plate, and (c) the voltage developed across each capacitor.

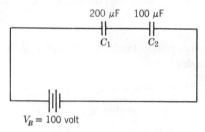

Figure 32-10

Solution.   (a) In a series circuit

$$\frac{1}{C_{equiv}} = \frac{1}{C_1} + \frac{1}{C_2}$$

$$= \frac{1}{(200 \times 10^{-6})\,\text{farad}} + \frac{1}{(100 \times 10^{-6})\,\text{farad}}$$

$$C_{equiv} = 66.6 \times 10^{-6}\,\text{farad}$$

(b) $C_{equiv} = \dfrac{q}{V_B}$

$$q = C_{equiv}V_B = 66.6 \times 10^{-6} \times 100$$
$$= 66.6 \times 10^{-4}\,\text{coulomb}$$

(c) $V_1 = \dfrac{q}{C_1} = \dfrac{66.6 \times 10^{-4}\,\text{C}}{200 \times 10^{-6}\,\text{V}} = 33.3\,\text{volts}$

$$V_2 = \frac{q}{C_2} = \frac{66.6 \times 10^{-4}}{100 \times 10^{-6}} = 66.6\,\text{volts}$$

The sum of V_1 and V_2 should add up to the total battery voltage.   ◀

32-5 CAPACITORS IN PARALLEL

The circuit diagram in Fig. 32-11a depicts three capacitors connected in parallel. The negative plate of the battery will supply a charge $-q_1$ to the plate of C_1 that is connected to it. This charge will induce an equal and opposite charge $+q_1$ on the other plate of C_1. The battery will also supply a charge $-q_2$ and $-q_3$ on the plates of the other two capacitors, which will induce equal and opposite charges on their respective positive plates. In general, $q_1 \neq q_2 \neq q_3$. The total charge, Q supplied by the battery is (in magnitude)

$$Q = q_1 + q_2 + q_3$$

An *equivalent* capacitor is one that will draw the same charge Q when connected to the same battery.

$$Q = q_1 + q_2 + q_3$$
$$C_{equiv}V = C_1V + C_2V + C_3V$$

Note that V, the voltage across each capacitor, is the same and equal to the battery voltage.

Dividing the last equation by V yields

$$C_1 + C_2 + C_3 = C_{equiv} \qquad (32\text{-}10)$$

This equation should be compared with the equation for *resistors connected in series*.

Example 32-6   Four capacitors are connected to a 200-volt battery as in Fig. 32-12a. Determine the value of one equivalent capacitor that could replace all four.

Solution.   The two 20-μF capacitors are in series. The equivalent capacitor for them, $C_{1\ equiv}$, is from Eq. 32-9

$$\frac{1}{C_{1\ equiv}} = \frac{1}{(20 \times 10^{-6})\ F} + \frac{1}{(20 \times 10^{-6}\ F)}$$

$$C_{1\ equiv} = 10 \times 10^{-6}\ F$$

The 40- and 10-μF capacitors are in parallel; $C_{2\ equiv}$ is simply their sum, or 50 μF. We now have an equivalent circuit of Fig. 32-12b. The equivalent capacitance, C for this circuit is obtained from

$$\frac{1}{C} = \frac{1}{(10 \times 10^{-6}\ F)} + \frac{1}{(50 \times 10^{-6}\ F)}$$

$$C = 8.33\ \mu F \qquad \blacktriangleleft$$

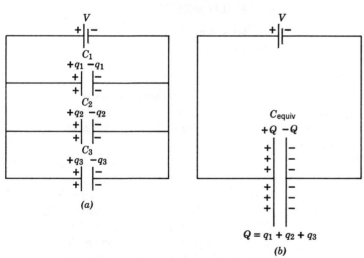

(a)

(b)

$Q = q_1 + q_2 + q_3$

Figure 32-11   (a) Three capacitors in parallel. The charge on each is, in general, different. The total charge drawn from the battery is $Q = q_1 + q_2 + q_3$. (b) The equivalent capacitor C_{equiv} must draw a charge Q from the battery.

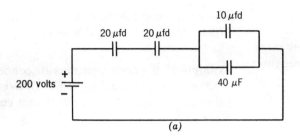

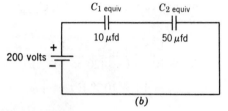

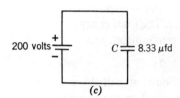

Figure 32-12

32-6 ENERGY REQUIRED TO CHARGE A CAPACITOR

After the first negative charge is delivered to the plate of a capacitor, it will become more difficult to accumulate additional negative charges. This is so because like charges repel, and the force of repulsion depends on the magnitude of the charge already at the plate. The battery, therefore, has to do work against this force. The energy W, to fully charge a capacitor, is given by

$$W = \frac{1}{2}\left(\frac{Q^2}{C}\right) \tag{32-11}$$

where Q is in coulombs, C in farads, and W in joules. Two other useful expressions for W are obtained by suitable substitutions. Substituting the quantity CV for Q in Eq. 32-11 yields

$$W = \frac{1}{2}(CV^2) \tag{32-12}$$

where V is the voltage across the capacitor. Substituting Q/V for C in Eq. 32-12 yields

$$W = \frac{1}{2}(QV) \tag{32-13}$$

A dimensional analysis of the last equation will show that W is in joules because V is in volts, which is the same as joule/coul.

Example 32-7 How much work was expended by the battery in order to charge the capacitors in Example 32-6?

Solution

$$\begin{aligned} W &= \tfrac{1}{2}CV^2 \\ &= \tfrac{1}{2} \times 8.33 \times 10^{-6} \text{ farad} \times (200)^2 \text{ volt}^2 \\ &= 0.17 \text{ joule} \end{aligned}$$

This work is not lost; it is stored as energy by the capacitor. ◀

32-7 TIME FOR CHARGING AND DISCHARGING A CAPACITOR; THE CAPACITIVE TIME CONSTANT

Consider a circuit containing a capacitor and resistor, as in Fig. 32-13a. When the switch is down, negative charges will start to flow to the plate of the capacitor but not beyond. As more negative charges accumulate on the capacitor plate, a voltage will develop across the capacitor that will retard the electron flow. A graph of the charge accumulated versus time is given in Fig. 32-13b. The graph indicates that it takes an infinite time for the charge to build up to the max-

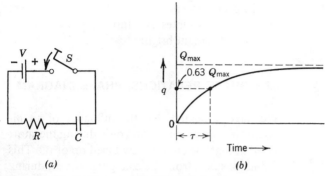

Figure 32-13 (a) The capacitor will start charging when the switch S is depressed. (b) Curve of the charge q on the capacitor vs. time. The charge approaches but never quite reaches Q_{max}.

imum value, $Q_{max} = CV$. A mathematical analysis shows that in a time of RC seconds the amount of charge accumulated on the plates is approximately 63% of the maximum charge. This time is called the *capacitive time constant*, τ (tau) of the circuit,

$$\tau = RC \qquad (32\text{-}14)$$

and τ is measured in seconds when R is in ohms and C in farads.

It also takes time for a fully charged capacitor to discharge. A graph of the charge q remaining after a time t, when the discharge has started, is given by Fig. 32-14b. After one time constant, i.e., in a time of $\tau = RC$, approximately 37% of its initial charge remains on the plates, or the capacitor has discharged 63% of its original value. It takes an infinite time for the capacitor to fully discharge.

Example 32-8 A series circuit consists of a 5-μF capacitor, a 5×10^4-ohm resistor, and a 12-volt battery. (a) What is the maximum charge on the capacitor? (b) How long does it take for the charge to reach Q_{max}? (c) What is the time constant of the circuit? (d) What will be the charge on the plates at a time of one time constant? (c) What charge will remain on the plates one time constant after this capacitor is discharged from a fully charged condition?

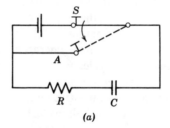

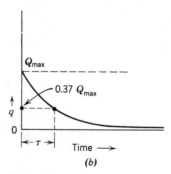

Figure 32-14 (a) After the capacitor is fully charged, the switch is pushed down to make contact with A. This short-circuits the battery and discharges the capacitor through R. (b) The remaining charge approaches but never quite reaches zero.

Solution

(a) $Q_{max} = CV = 5 \times 10^{-6}$ F $\times$ 12 volts
$$= 60 \times 10^{-6} \text{ coulomb}$$

(b) Theoretically, an infinite time. Practically, a time equivalent to about five time constants, or

$$5\tau = 5RC = 5 \times 5 \times 10^4 \text{ ohms} \times 5 \times 10^{-6} \text{ F}$$
$$= 1.25 \text{ sec}$$

(c) $\tau = RC = 5 \times 10^4 \text{ ohms} \times 5 \times 10^{-6} \text{ F}$
$$= 0.25 \text{ sec}$$

(d) After one time constant,

$$q = 0.63 Q_{max} = 0.63 \times 60 \times 10^{-6} \text{ couloumb}$$
$$= 37.8 \times 10^{-6} \text{ couloumb}$$

(e) $0.37 \times 60 \times 10^{-6} \text{ C} = 22.2 \times 10^{-6} \text{ C}$ ◀

A current cannot be maintained in a circuit consisting of a capacitor and battery. This is so because of the very high resistance of the air or the dielectric material between the capacitor's plates. However, a current is maintained in a circuit containing a capacitor and an ac generator. For one-half the generator's cycle, a negative charge is impressed on one of the capacitor's plates, thereby inducing an equal positive charge on the other plate. During the next half-cycle, the negative charges flow back to the generator while it impresses a negative charge on the other plate. This back and forth motion of the charges constitute a continuous alternating current in the circuit even though there is no current between the capacitor's plates. A lamp introduced in this circuit will remain bright.

32-8 PHASE RELATIONS, PHASE DIAGRAM

We have indicated that the influence of an inductor in a dc circuit occurs only during the short interval that the circuit is switched on or off. This influence stems from the changing current during this interval and the momentary induced emf. If an inductor is connected to the terminals of an ac generator (instead of a battery), the current is constantly changing and the inductor's influence occurs at every moment.

First, let us consider an ac generator and a pure resistor, Fig. 32-15a. At any instant

$$\mathscr{E}_{inst} = i_{inst}R = V_{R\ inst} \qquad (32\text{-}15)$$

where $V_{R\ inst}$, i_{inst} and $\mathscr{E}_{inst}$ are, respectively, the instantaneous values for the voltage across the resistor, the current through the resistor, and the generator emf. Since R is a constant in Eq. 32-15, when $i_{inst} = 0$, $V_{R\ inst} = 0$, and when i_{inst} is a maximum, $V_{R\ inst}$ is a maximum. When the maximum values of voltage and current occur at the same instant, and when the minimum values occur

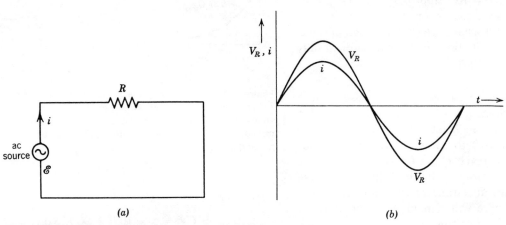

(a)   *(b)*

Figure 32-15 (a) Pure resistance in an ac circuit. (b) Voltage and current are in phase in a pure resistor.

at the same instant, the voltage and current are said to be in phase. A plot of the voltage and current in a pure resistor is given in Fig. 32-15b.

We consider next the phase relation between the current and voltage across a pure inductor (Fig. 32-16a). The voltage V_L across the inductor is porportional to $\Delta i/\Delta t$ (from Eq. 32-2). The quantity $\Delta i/\Delta t$ represents the slope of the current versus time curve (see Fig. 32-16b). At the instant when the magnitude of i is maximum (points 1 and 2 in

the figure), $\Delta i/\Delta t = 0$, therefore, $V_L = 0$. At the instant when $i = 0$ (points 0 and 3 in the figure), the slope $\Delta i/\Delta t$ and V_L are, respectively, positive maximum and negative maximum.

In a pure inductor, the voltage leads the current by 90°

Figure 32-17b demonstrates the phase relation between the voltage V_C across a capacitor and the current in a pure capacitive circuit.

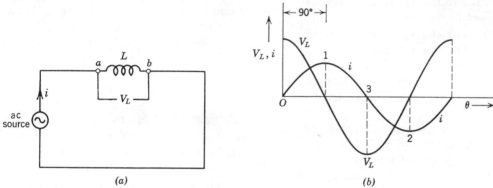

Figure 32-16   (a) Inductor in an ac circuit. (b) Voltage leads the current by 90° in a pure inductive circuit.

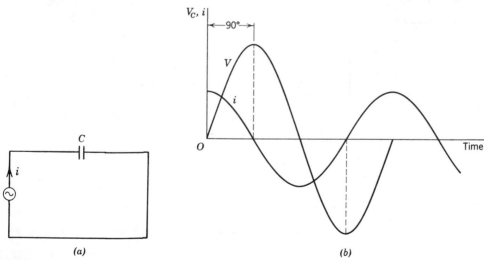

Figure 32-17   (a) A pure capacitive circuit. (b) In a pure capacitive circuit current leads the voltage by 90°.

The voltage lags behind the current by 90° in a pure capacitor.

The ac series circuit in Fig. 32-18 contains a resistor, capacitor, and inductor. Recall that the generator emf at a given instant in a dc series circuit is the sum of the voltage drops across each element. This is also the case in an ac series circuit and total energy is conserved. Let us assume in Fig. 32-18 that the effective values $V_R = 40$ volts, $V_C = 10$ volts, and $V_L = 30$ volts. This can be indicated by Fig. 32-19a, which is called a *phase diagram*. The voltage across the resistor V_R is drawn horizontally, pointing to the right. The voltage across the inductor V_L is drawn vertically, pointing up, and the voltage across the capacitor V_C is drawn vertically, pointing down.

In this and in the following sections, the quantities V, V_L, V_R, V_C, and i, refer to their respective effective values.

The effective value of voltage across any combination of elements is given by the *vector sum* of the voltages across each element. Thus the magnitude of the voltage across the resistor and

inductor combination is, from Fig. 32-19b,

$$\sqrt{V_L^2 + V_R^2} = \sqrt{900 + 1600} = 50 \text{ volts}$$

By the same method the effective voltage across the entire circuit, V, (Fig. 32-19c) is given by

$$V = \sqrt{V_R^2 + (V_L - V_C)^2} \qquad (32\text{-}16)$$
$$V = \sqrt{1600 + (30 - 10)^2}$$
$$= 44.7 \text{ volts}$$

Since in this example $V_L > V_C$, the circuit is said to be inductive, and the voltage will lead the current. The angle ϕ by which the voltage leads the current can be computed from Fig. 32-19c as

$$\tan \phi = \frac{V_L - V_C}{V_R} \qquad (32\text{-}17)$$

$$\tan \phi = \frac{20}{40}, \qquad \phi = 27°$$

Example 32-9 An inductor and resistor are connected in series to a 60 cycles/sec line (Fig. 33-20a). An ac voltmeter connected across each of these elements reads 100 volts for the inductor and 50 volts for the resistor. (a) Draw a phase diagram. (b) What is the phase difference between the voltage and current? Which leads? (c) What is the effective voltage across the entire circuit?

Solution. (a, b). In a circuit containing inductance and resistance, the voltage leads the current

$$\tan \phi = \frac{V_L}{V_R} = \frac{100}{50} = 2$$

$$\phi = 64°$$

The voltage leads the current by 64°.

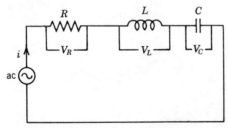

Figure 32-18   Inductance, capacitance, and resistance in series in an ac circuit.

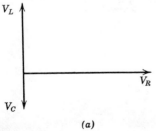

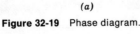
(a)

Figure 32-19   Phase diagram.

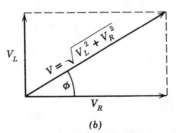
(b)

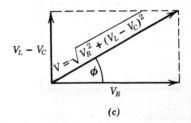
(c)

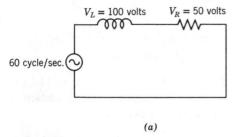

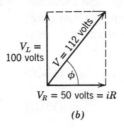

(a)   (b)

Figure 32-20

(c) A voltmeter put across the entire circuit will render an effective value

$$V = \sqrt{(100)^2 + (50)^2} = \sqrt{12,500 \text{ volts}}$$
$$= 112 \text{ volts} \qquad \blacktriangleleft$$

Example 32-10 An inductor, resistor, and capacitor are connected in series to a 60 cycle/sec line. A voltmeter, connected across each of the circuit elements, gives the following readings: $V_R = 50$ volts, $V_L = 100$ volts, and $V_C = 150$ volts (see Fig. 32-21a). (a) Draw a phase diagram.

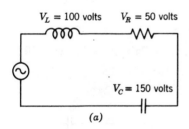

(a)

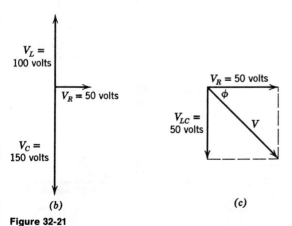

(b)   (c)

Figure 32-21

(b) What is the phase difference between the voltage and current? Which leads? (c) What is the effective voltage across the entire circuit?

Solution. (a) The voltage across the inductor *and* the capacitor is the vector sum of the two, or 150 volts down + 100 volts up = 50 volts down (see Fig. 32-21b,c).

(b) The voltage lags the current by an angle ϕ, because $V_C > V_L$.

$$\tan \phi = -50/50 = -1$$
$$\phi = -45°$$

(c) $V = \sqrt{(50)^2 + (50)^2} = 70.7$ volts $\qquad \blacktriangleleft$

32-9 REACTANCE AND IMPEDANCE

A pure inductor exhibits a resistance to alternating current. This resistance, called the *inductive reactance*, is symbolized as X_L and is given by

$$X_L = 2\pi f L \qquad (32\text{-}18)$$

where X_L is measured in ohms when the generator frequency f is in cycles/sec and the inductance L in henries. Note that the inductive reactance is not a constant but depends on the frequency.

A capacitor also offers resistance to alternating current. Its resistance is called *capacitive reactance*, X_C, and is given by

$$X_C = \frac{1}{2\pi f C} \qquad (32\text{-}19)$$

where X_C is measured in ohms when f is in cycles/sec and the capacitance C in farads.

The voltage across the inductor and capacitor can be expressed in the same form as Ohm's law, namely, as the product of the current and resistance. Thus

$$V_L = iX_L = i2\pi fL \qquad (32\text{-}20)$$

$$V_C = iX_C = \frac{i}{2\pi fC} \qquad (32\text{-}21)$$

and $$V_R = iR$$

Equation 32-16 can now be written as

$$V = \sqrt{i^2R^2 + i^2(X_L - X_C)^2} = i\sqrt{R^2 + (X_L - X_C)^2}$$

or $$i = \frac{V}{\sqrt{R^2 + (X_L - X_C)^2}} \qquad (32\text{-}22)$$

The denominator of Eq. 32-22 is known as the *impedance* of the circuit, Z. The higher the impedance of a circuit, the lower is the current in it.

$$Z = \sqrt{R^2 + (X_L - X_C)^2} \qquad (32\text{-}23)$$

and from Eq. 32-22

$$i = \frac{V}{Z} \qquad (32\text{-}24)$$

It is evident from Eq. 32-23 that the impedance is measured in ohms because R, X_L, and X_C have dimensions of ohms.

Example 32-11   A 30-millihenry inductor, having a resistance of 100 ohms, is connected to a 110-volt, 60 cycle-sec line. Determine (a) the inductive reactance of the inductor, (b) the impedance of the circuit, (c) the effective value of the current in the circuit, (d) the phase difference.

Solution

(a) $X_L = 2\pi fL$ (Eq. 32-18)

$$= 2\pi \times 60 \text{ cycles/sec} \times 30 \times 10^{-3} \text{ henry}$$

$$X_L = 11.3 \text{ ohms}$$

(b) $Z = \sqrt{R^2 + (X_L - C_C)^2}$ (Eq. 32-23)

but $X_C = 0$ because the circuit does not have a

capacitor, so

$$Z = \sqrt{R^2 + X_L{}^2} = \sqrt{(100)^2 + (11.3)^2}$$

$$Z = 101 \text{ ohms}$$

(c) By Eq. 32-24 $V = iZ$, $i = V/Z = 110/101$ amp $= 1.09$ amp. The current at any instant is the same in all parts of the circuit.

(d) By Eq. 32-17 $\tan\phi = \dfrac{V_L}{V_R} = \dfrac{iX_L}{iR} = \dfrac{11.3}{100} =$

0.113

$$\phi = 6.5° \qquad \blacktriangleleft$$

Example 32-12   A series circuit comprising a 100-millihenry inductor, 100-μF capacitor, and a 200-ohm resistor has impressed on it a potential difference having an effective value of 200 volts at 60 cycles/sec (Fig. 32-22). Calculate (a) the inductive reactance, (b) the capacitive reactance, (c) the impedance of the entire circuit, (d) the effective value of the current, (e) the phase difference, and (f) the voltage across each element.

Solution

(a) $X_L = 2\pi fL$ (Eq. 32-18)

$$= 2\pi \times 60 \text{ cycles/sec} \times 10^2 \times 10^{-3} \text{ henry}$$

$$= 37.6 \text{ ohms}$$

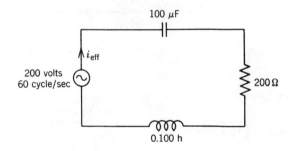

Figure 32-22

(b) $X_C = \dfrac{1}{2\pi fC}$ (Eq. 32-19)

$= \left(\dfrac{1}{2\pi \times 60 \text{ cycles/sec} \times 100 \times 10^{-6} \text{ F}}\right)$

$= 26.6$ ohms

(c) $Z = \sqrt{R^2 + (X_L - X_C)^2}$ (Eq. 32-23)

$= \sqrt{(200)^2 \text{ ohm}^2 + (37.6 - 26.6)^2 \text{ ohm}^2}$

$= 200$ ohms

(d) By Eq. 32-24, $i = \dfrac{V}{Z} = \dfrac{200 \text{ volts}}{200 \text{ ohms}} = 1$ amp

(e) $\tan \phi = (X_L - X_C)/R = \dfrac{11 \text{ ohms}}{200 \text{ ohms}} = 0.055$

$\phi = 3.2°$

(f) $V_R = iR = 1$ amp $\times 200$ ohms $= 200$ volts

$V_L = iX_L = 1$ amp $\times 37.6$ ohms $= 37.6$ volts

$V_C = iX_C = 1$ amp $\times 26.6$ ohms $= 26.6$ volts

From the phase diagram it can be seen that the effective voltage V across the entire circuit is

$V = \sqrt{V_R^2 + (V_L - V_C)^2} = \sqrt{(200)^2 + (37.6 - 26.6)^2}$
$= \sqrt{40,000 + 121} = 200$ volts

This turns out to be the same as the effective value of the generator voltage. ◄

32-10 RESONANCE

An ac circuit containing resistance, inductance, and capacitance is said to be in a state of resonance when the current through the circuit is maximum. From Eq. 32-22 it can be seen that i is at a maximum when the denominator $\sqrt{R^2 + (X_L - X_C)^2}$ is a minimum. This is so when

$$X_L = X_C \qquad (32\text{-}25)$$

for then $X_L - X_C = 0$ and the denominator goes to a minimum value R. Combining Eq. 32-19

with Eqs. 32-18 and 32-25 we get

$$2\pi fL = \dfrac{1}{2\pi fC}$$

Solving this equation for f, we get

$$f = \dfrac{1}{2\pi \sqrt{LC}} \qquad (32\text{-}26)$$

where f is called the *resonant frequency*.

Example 32-13 How can we make the circuit in Example 32-12 a resonant circuit?

Solution. The condition for resonance is

$$X_L = X_C$$

The solution for these quantities in Example 32-12 gave $X_L = 37.6$ ohms and $X_C = 26.6$ ohms. Either X_C must be increased or X_L decreased. Let us solve this problem by increasing X_C to 37.6 ohms.

$$X_C = \dfrac{1}{2\pi fC} = 37.6 \text{ ohms}$$

Solving for C, we obtain

$$C = \dfrac{1}{2\pi fX_C} = \dfrac{1}{2\pi \times 60 \times 37.6} \text{ F} = 7.05 \times 10^{-5} \text{ F}$$

$= 70.5 \times 10^{-6}$ farad $= 70.5 \ \mu$F

Thus a reduction of the capacitance in the circuit to 70.5 μfarad will render the circuit resonant. This change will, of course, increase the value of the current. ◄

32-11 POWER AND POWER FACTOR

The power dissipated (lost) in an ac circuit stems from the power losses in each of the elements of that circuit. Part of the electrical energy that is supplied to the resistor elements is lost by joule heating of the resistor. This gives rise to the usual i^2R or (Vi) loss. The pure inductor and capacitor elements of an ac circuit suffer no power loss. The inductor converts the electrical energy fed to it by the ac generator into a momentary magnetic field for half a cycle. During the other half of the cycle, this magnetic field collapses, and the magnetic energy is fed back to the generator without any net energy loss during one cycle. The capacitor, also, does not dissipate energy. Work is done to charge the capacitor, but this energy is fed back to the circuit when the capacitor discharges.

In order to obtain an expression for the power dissipated, we refer to a phase diagram of a general RLC series circuit (see Fig. 32-19). As was brought out in the above paragraph, the power P, dissipated in the entire circuit, stems only from the power dissipated in the resistor. Therefore,

$$P = V_R i$$

But $V_R = V \cos \phi$. Therefore, the average power,

$$P = iV \cos \phi \qquad (32\text{-}27)$$

Power is measured in the usual power unit, the watt. The factor $\cos \phi$ is called the *power factor* of the circuit.

Example 32-14 The circuit of Example 32-11 is shown in Fig. 32-23. What is the average power consumed in the circuit and the power factor?

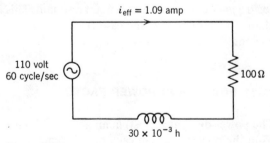

Figure 32-23

Solution

$P = iV \cos \phi$

$= 110 \text{ volts} \times 1.09 \text{ amp} \times \cos 6.5° = 119 \text{ watts}$

Since the power in an ac circuit is dissipated in the resistor only, $P = i^2 R$ should also yield the correct value.

$$P = (1.09)^2 \text{ amp}^2 \times 100 \text{ ohms} = 118 \text{ watts}$$
$$\text{(difference in rounding off)}$$

The power factor $\cos \phi = \cos 6.5° = 0.994$. ◀

SUMMARY

The induced voltage across an inductor depends on the time rate of change in current.

The inductive time constant is the time it takes for the current in an inductor to build up to 63% of its maximum value.

The capacitance depends on the geometry of a capacitor and the dielectric material between the plates.

The reciprocal of the equivalent capacitance for capacitors in series is given by the sum of the reciprocals of the individual capacitances. This is similar to resistors in parallel.

The equivalent capacitance of capacitors in parallel is given by the sum of the individual capacitances. This is equivalent to resistors in series.

The voltage and current are out of phase in an ac circuit.

The voltage leads the current in an inductive circuit.

The current leads the voltage in a capacitive circuit.

QUESTIONS

1. Is there a steady current in a dc circuit when a capacitor is part of it? Explain.

2. Is there a steady current in a dc circuit when an inductor is part of it? Explain.

3. Explain the electrical oscillations in an ac circuit containing an inductor and capacitor.

4. What is meant by the phase difference?

5. Explain the meaning of *impedance*. On what does it depend?

6. Does the maximum current in an ac circuit depend on the frequency of the generator? Explain.

7. Compare the capacitive reactance of a capacitor at 60 cycles/sec with the reactance at 120 cycles/sec.

8. The "apparent" (i^2R) power in an *R-L* circuit is 2000 watts. Is the real power dissipated greater or smaller? Explain.

9. There is no power loss in an ideal inductor. Is there a power loss in a real inductor? Explain.

What does a power factor of 1 indicate?

PROBLEMS

(A)

1. Calculate the magnitude of the induced voltage in a 2-millihenry inductor in which the current is changing at a steady rate of 0.2 amp per 0.1 sec.

2. What should be the inductance if it is desired to induce 20 V across the inductor when the current in it changes at 0.2 amp/sec?

3. A good capacitor, $C = 10^{-10}$ F, is connected to a 100-V battery. Calculate the maximum charge on the capacitor.

4. How much charge remains at a time of one time constant after the capacitor in Problem 3 has been disconnected from the battery and discharged through a resistor?

5. A certain *L-R* circuit has a time constant of 0.001 sec. What is the inductance of this coil if the resistance has a value of 3500 ohms?

6. An *L-R* circuit has a time constant of 10^{-3} sec. Calculate the resistance if $L = 10$ henrys.

7. What is the time constant when $L = 1500$ henrys and $R = 3000$ ohms?

8. The plates of a capacitor receive a charge of 3.5×10^{-3} C when put across a potential difference of 150 volts. Calculate the capacitance.

9. What voltage is developed across a 6 μF capacitor when a charge of 10^{-10} C is deposited on its plates?

10. Calculate the capacitive time constant when $C = 100$ μF and $R = 10^6$ ohms.

11. Calculate the effective capacitance for three capacitors connected in series: $C_1 = 5$ μF, $C_2 = 10$ μF, and $C_3 = 15$ μF.

12. What is the voltage developed across each of the capacitors in Problem 11 if the charge on each capacitor is 10^{-12} C?

13. Repeat Problem 11 when the same capacitors are connected in parallel.

(B)

1. A certain capacitor, when charged to 50 V, accumulates a charge of 2×10^{-3} C. How much energy was required to deliver this charge? What happened to this energy?

2. The inductance of a certain coil, having 300 turns, is 10^{-2} henry. Calculate the induced emf if $\Delta i/\Delta t = 4$ amp/sec. Sketch the direction of $\mathscr{E}$.

3. When a capacitor has mica as its dielectric ($\kappa = 7$), its capacitance is 30 $\mu\mu$F. What is its capacitance with glycerin as a dielectric ($\kappa = 56$)?

4. Compute the charge on the plates of the capacitor in Problem B3 when it is put across a potential difference of 50 V (for both cases).

5. A 5-henry inductor is connected in series to a 100-ohm resistor and a switch, and the combination is connected to a 50-V battery.
(a) What is the maximum current in the circuit?
(b) How long does it take to achieve this current?
(c) What is the value of the current at a time of 0.05 sec after the switch is closed?
(d) What is the time constant of this circuit?

6. Calculate the average induced voltage in the coil of Problem B5, under the same conditions, in the first 0.05 sec after the switch is closed.

7. Show that the time constant $\tau = RC$ is measured in seconds.

8. Compute the resonant frequency of a $100 = \mu$F capacitor and a 0.5 H inductor.

9. Given the circuit in Fig. 32-24 in which all numbers are in μF. Calculate the capacitance between points *A* and *B*.

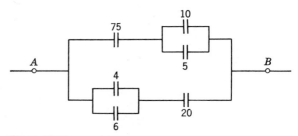

Figure 32-24

10. A 0.5-H inductor is connected in series to a 5000-ohm resistor and a 100-V battery. After this circuit has been connected for a long time, the battery is short-circuited. Discuss the current as a function of time.

11. Calculate the value of the current after two time constants in a 1.5-H coil connected to a 50-ohm resistor in series with a 100-V battery.

12. Show that the reactances (inductive and capacitive) each have dimensions of ohms.

13. A 50-mH coil having a resistance of 50 ohms is connected to a 60-cycle, 110-V line forming a series circuit. Compute the effective values of:
    (a) The voltage across each element.
    (b) The current in the circuit.
    (c) Construct a phase diagram.
    (d) What is the phase difference between voltage and current? Which leads?

14. A 15-μF capacitor is connected in series to a 100-ohm resistor on a 220-V, 100-cycle line. Calculate:
    (a) The reactance.
    (b) The impedance of the circuit.
    (c) The power consumed.
    (d) The power factor.

33.
Electronics

In this chapter we discuss several of the electronic innovations that have revolutionized the field of communications during the past 100 years. In the forefront of those discoveries which contributed the most to the technological advances society has made during this period are the development of *vacuum tubes* and the subsequent invention of the *transistor*.

LEARNING GOALS

The following learning objectives should be kept in mind during the study of this chapter.

1. *Learning the Edison effect and the role its discovery played in the development of the electronics industry.*

2. *Learning the functions of the different parts of a diode vacuum tube and understanding the operations of the diode as a rectifier.*

3. *Learning the parts of the triode vacuum tube and understanding the operation of the triode as an amplifier.*

4. *Learning how to calculate the amplification ratio of a triode.*

5. *Understanding the distinction between amplitude modulation (AM) and frequency modulation (FM) radio signals.*

6. *Learning about the development of transistors, which utilize semiconducting materials, and recognizing the similarity in function between the operation of the transistor and vacuum tubes.*

7. *Understanding why the transistor revolutionized the electronics industry.*

33-1 EDISON EFFECT

The success that science has enjoyed in revamping the field of electronics, especially in the area of radio and television, began with an accidental discovery attributed to Thomas A. Edison in the 1880s (Fig. 33-1). While searching for a way to perfect the electric light bulb, which he had invented, Edison happened upon an amazing curiosity. He introduced a metal plate into the electric bulb at some distance from the filament (Fig. 33-2). The metal plate is connected in series to a galvanometer and battery and back to one side of the filament. The filament is heated separately by a low voltage battery V_1. When the plate is maintained at a negative potential with respect to the filament, no charge flows through the galvanometer. If, however, the plate is maintained at a positive potential with respect to the filament, a current is detected by the galvanometer. This phenomenon of electron flow from a heated filament across a vacuum to a metal plate is called the *Edison effect*. Current is only detected when the filament is heated. Only then are electrons produced near the filament and attracted to the positive plate. The filament may be heated in several ways. In Fig. 33-2, an electric current produced by a battery, V_1, is used to heat the filament. A flame or nuclear energy could also be used to heat the filament (Fig. 33-3). An electron emitter such as a filament is often called a

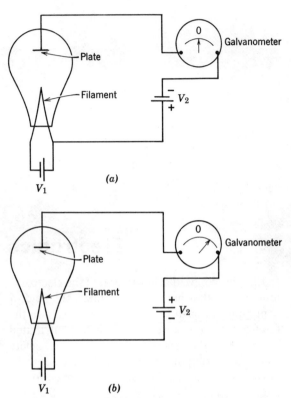

Figure 33-2   The Edison effect. In (*a*) the plate is maintained at a negative potential with respect to the filament and the galvanometer shows no current. In (*b*) a current is present when the plate is maintained at a positive potential with respect to the filament.

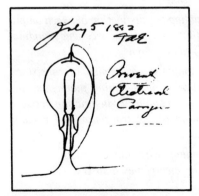

Figure 33-1   Original Edison sketch of discovery of Edison effect. (Courtesy Edison Foundation.)

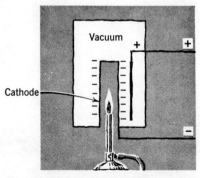

Figure 33-3   Model of thermionic generator.

cathode. The plate that receives the electrons is called the *anode*. A cathode is often heated indirectly by having it surround a heated metal wire. As the wire heats the cathode, the latter emits electrons (Fig. 33-4). Electrons emitted from hot metallic surfaces are called *thermoelectrons*.

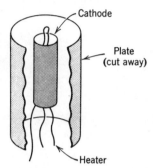

Figure 33-4 Heater-type diode tube.

33-2 THE DIODE

A vacuum tube with only a cathode and an anode is called a *diode*. If the cathode is heated, electrons will move only from the cathode to the anode. A diode allows electrons to move in only one direction, much like a check valve. When the plate is connected to a source of alternating current rather than direct current, the plate will have its polarity changed each half-cycle. When the plate is positive, current will pass from the filament to the plate. However, no current passes from the filament to the plate during the half-cycles when the plate is negative. Consequently, the diode changes alternating current impressed upon the plate to a pulsating direct current. We say that the alternating current has been *rectified* by the diode. For this reason the diode is often referred to as a rectifier (Fig. 33-5). When only half of the ac wave is changed to direct current, the process is called *half-wave rectification. Full-wave rectification* can be achieved with a double-diode vacuum tube. Such a tube has two plates; each one allows electron flow during half a cycle (Fig. 33-6). The resulting pulsating direct current can be "smoothed out" into a more constant direct current by passing the current through specially arranged sequences of capacitors and inductors.

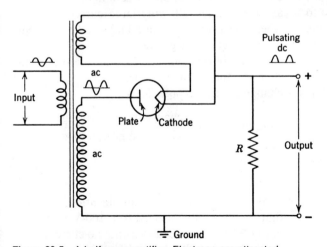

Figure 33-5 A half-wave rectifier. Electrons are attracted to the plate, only when the plate is more positive than the cathode. The input is alternating current and the output is a pulsating direct current.

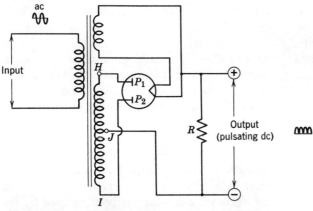

Figure 33-6 Full-wave rectification produced by a double diode. When end *H* of coil *HI* is positive, *P₁* is positive, *I* is negative, and *P₂* is negative. Electrons flow from filament to *P₂*. During the next half-cycle, *I* is positive, *P₂* is positive, *H* is negative, and *P₁* is negative. The electron flow is now from filament to *P₂*. In both cases the electrons flow through point *J* to the load *R*. The current through *R* is pulsating direct current.

33-3 THE TRIODE

In 1907 a young American engineer, Lee DeForest, added a third part to the diode. Because it now has three parts, it is called a *triode* (Fig. 33-7). The additional part is called a *grid*, since it looks like a piece of window screen. The grid is used to control the electron flow from cathode to plate. A negatively charged grid will repel some or all of the electrons emitted from the cathode. A positively charged grid will attract more electrons from the cathode. The current through the plate is a function of the grid voltage. Not only does the grid play the role of a rectifier by permitting current flow only in one direction, but it also acts as an amplifier, allowing large currents to flow when the grid is positive and smaller currents when the grid is negative.

The amplification ratio for a triode tube is the ratio of the change in plate voltage per unit change in grid voltage. Thus

$$\text{amplification} = \frac{\text{change in plate voltage}}{\text{change in grid voltage}}$$

$$= \frac{V_p}{V_g} \tag{33-1}$$

As an illustration of amplification, suppose that the plate is connected to a 100-V battery as in Fig. 33-8. Suppose further that no current flows when the grid voltage is −4 V. When no current flows, the plate voltage is 100 V relative to the cathode. When the grid voltage is 0 V, current

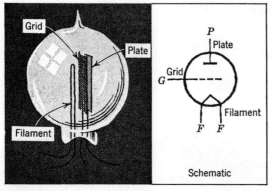

Figure 33-7 Triode tube.

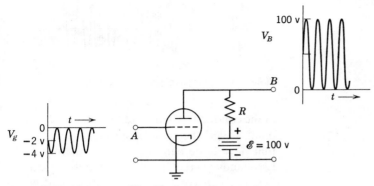

Figure 33-8 A triode amplifier. The input voltage on the grid is amplified by a factor of 25.

flows and the plate voltage relative to the cathode is essentially zero. Therefore, a change of 4 V on the grid produces a change of 100 V on the plate. The amplification ratio will be

$$\text{amplification ratio} = \frac{V_p}{V_g} = \frac{100 \text{ V}}{4 \text{ V}} = 25$$

In the amplifier of a hi-fi, radio, or TV, the amplification is provided by a triode. In a radio receiver, the signal that reaches the grid from an antenna is high frequency (radio frequency) A.C. The rectification or detection of the radio signal is accomplished by means of a triode. The triode changes high frequency A.C. to audio frequency if the grid circuit contains the correct capacitance and resistance.

33-4 MODULATION AM AND FM

Radio frequency signals must be modulated to carry information. In radio transmission the generation of a sine-wave high frequency signal produces what is called the *carrier signal*. The carrier transports no information other than its wavelength and frequency. The information, voice, music, or other data must be superposed on the carrier. This affects the amplitude of the carrier wave as in Fig. 33-9. This is called *modulation*. In the case under discussion, the wavelength and frequency are constant, but examination of

Fig. 33-9 shows that it is the amplitude that is varied by the modulation. Therefore this process is called *amplitude modulation* (AM). However, static and man-made electrical discharges can also affect the amplitude and interfere with the received signal. In the tropics where atmospheric static is great, the amplitude of the static signal can be so great as to interfere with the modulated signal. Therefore, for static-free reception, a system in which the amplitude is constant is required.

In frequency modulation (FM) the amplitude of the carrier wave is constant, as shown in Fig. 33-10. The modulation signal is made to vary the frequency of the carrier wave as indicated in Fig. 33-10. Note that the amplitude is constant, but the frequency and wavelength vary. Thus, at all times, the amplitude is at a maximum, and there is little or no interference from man-made static.

33-5 SEMICONDUCTORS AND TRANSISTORS

The modern explanation of the different electrical properties of solids depends upon the distribution of the electrons in a solid into energy levels, called *bands*. The highest energy levels in a solid comprise the *conduction band*. The energy levels immediately below this band comprise the *valence band*. The properties of a material depend on the size of the energy separation between the bottom

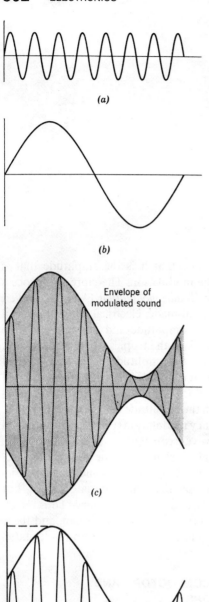

Frequency modulated signal

Figure 33-10   Frequency modulated signal.

of the conduction band and the top of the valence band, called the *band* or *energy gap*. Materials that have large band gaps are insulators. If the two bands overlap, there is no band gap and the material is a conductor. When the band gap is small, the material is classified as a semiconductor (Fig. 33-11).

Because the band gap in semiconductors is small, it is possible for a small quantity of energy, supplied by light, heat or a low voltage, to cause electrons to jump from the valence band to the conduction band. Electrons reaching the conduction band become free electrons, in the classical sense of the phrase. The empty position left behind by the electron in the valence band is called a *hole*. A nearby valence electron can jump into this hole, leaving behind a different hole. Thus a series of holes flow toward the negative terminal while electrons flow in the opposite direction. The holes are referred to as positive charge carriers. Semiconductors, therefore, have both negative and positive charge carriers (Fig. 33-12).

The properties of semiconductors can be used to construct circuit elements that have characteristics similar to electron tubes but with certain advantages. These circuit elements are called *transistors*. The transistor is much smaller than a vacuum tube, so circuits can be made more compact. There is no need for a heater to heat a filament, and the lack of a filament generally improves stability and increases the lifetime of the element.

The first transistors were of the point contact variety (Fig. 33-13). This transistor consists of a small block of semiconductor, usually germanium, to which has been added an impurity which makes the resultant material rich in either electrons (*n*-type germanium) or holes (*p*-type

(a)

(b)

Envelope of modulated sound

(c)

(d)

Figure 33-9   (*a*) Radio-frequency carrier wave.
(*b*) Audio-frequency voice wave (modulating signal).
(*c*) Amplitude-modulated carrier wave. (*d*) Rectified signal in receiver.

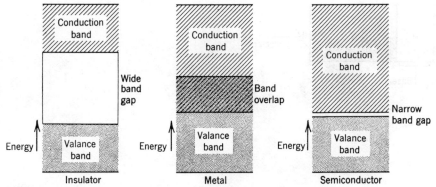

Figure 33-11 In the insulator, the wide gap between bands prevents the electrons from making the jump from the valence band to the conduction band. In metal, the bands usually overlap, allowing free motion of the electrons. Semiconductors have a small band gap.

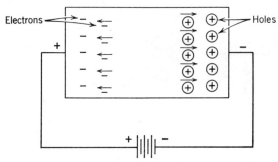

Figure 33-12 Movement of electrons (negative charges) and holes (positive charges) of a semiconductor material.

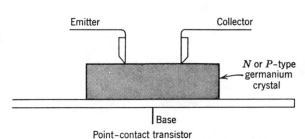

Point–contact transistor

Figure 33-13 Point-contact transistor. The first transistors were point-contact. Most transistors today are junction transistors.

germanium), and three electrodes. One electrode, the *base*, makes close contact with the semiconductor. The other two are fine points making

very light contact. One contact is called the *emitter*, the other the *collector*. The emitter is kept at a low positive voltage by a battery connecting it to the base. The collector is maintained negative by a higher voltage battery connecting it to the base. With this arrangement, small changes in the emitter current will produce larger changes in the collector current. The transistor acts, therefore, as an amplifier of current, just as the electron tube amplified voltage.

A second type of transistor is the *junction* type, composed of two *n*-type semiconductors surrounding a *p*-type semiconductor (*n-p-n* type), or two *p*-type semiconductors surrounding an *n*-type semiconductor (*p-n-p* type), as shown in Fig. 33-14. Junction transistors are better amplifiers than point-contact transistors.

Transistors can replace vacuum tubes in oscillator circuits and modulating circuits. Vacuum tubes require appreciable filament current and fairly high plate voltages. As a result of these power requirements, tubes dissipate large amounts of heat. On the other hand, transistors require no filament or heater current and operate at very low voltages and current. A typical radio receiver transistor consumes only $\frac{1}{25}$ of 1 watt. The low power requirements and minimum heat dissipated by transistors make it possible to produce compact and economical circuitry.

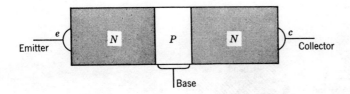

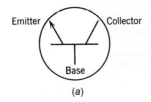

(a)

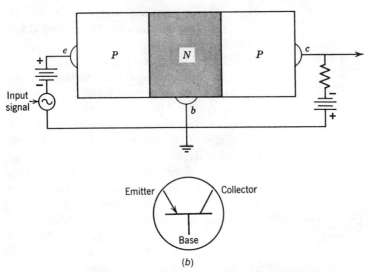

(b)

Figure 33-14 (a) *N-P-N* transistor showing contacts and schematic symbol used in circuit diagrams. (b) *P-N-P* circuit and *P-N-P* transistor symbol.

SUMMARY

A *diode* is a vacuum tube containing two elements called the cathode and the anode.

A diode is used as a *rectifier* to change alternating current to direct current.

A *triode* contains an additional element called a *grid* that enables this tube to amplify as well as

rectify. The amplification ratio of a triode is the ratio of the change in plate voltage to the corresponding change in grid voltage.

The superposition of information on a carrier wave is called *modulation*. Modulation can be achieved by varying the amplitude of the carrier wave (AM), or the frequency of the carrier wave (FM).

Semiconductors have small energy gaps between the top of their valence band and the bottom of the conduction band.

Semiconducting materials are used as the basic component of electronic devices called *transistors,* which can perform the same functions as electronic tubes.

QUESTIONS

1. Explain how the diode is used as a rectifier.

2. What is the Edison effect?

3. Explain why slight changes in grid voltage causes large changes in the plate voltage of a triode.

4. What is meant by pulsating direct current? Draw a sketch of such a current as a function of time.

5. Explain the difference between amplitude modulation and frequency modulation.

6. Why is there very little static in frequency modulation reception, compared with amplitude modulation?

7. Distinguish between conductors and insulators in terms of their conduction and valence bands.

8. Explain the difference between holes and electrons in a semiconductor.

9. Why is the flow of holes equivalent to a positive current opposite to the electron flow?

10. What are the functions of the emitter, base, and collector of a transistor?

PROBLEMS

1. Why does a negatively charged plate prevent the passage of current in a diode?

2. Explain how a space charge of electrons around a cathode may prevent electrons from reaching the plate.

3. What is the amplification ratio of a triode if a 2-V change in the grid circuit causes a 48-V change in the plate voltage?

4. In a certain triode a change of 180 V in the plate voltage occurs when the grid voltage changes by 6 V. What is the amplification ratio for this tube?

5. Consider the triode in Problem 4. If the grid changed by 9 V, what would be the change in plate voltage?

6. A triode has an amplification ratio of 15. What change in plate voltage occurs when the grid voltage is changed by 5 V?

7. A triode has an amplification ratio of 40. What change in grid voltage produces a 60-V change in plate voltage?

8. What change in grid voltage produces a 120-V change in plate voltage in a triode whose amplification ratio is 25?

9. Why does it require little energy to move electrons across the band gap of a semiconductor?

10. What are some advantages that transistors have when compared with electron tubes?

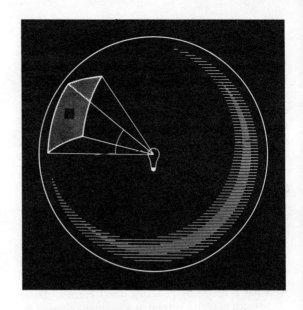

34.
Light and Photometry

Light is a necessary ingredient for seeing. As a consequence, all instruments used for seeing, from the eye itself and eyeglasses to compound microscopes and telescopes are based on the principles governing the behavior of light. The subject of light extends into the field of electromagnetic theory and its applications. In fact, light is an electromagnetic wave similar to a radio wave, the only difference being that the wavelength of the latter is much longer. The principles of light behavior, therefore, find many applications in radio and television broadcasting and radar. In addition to having many practical applications, the subject of light is prominent in theoretical physics. An important hypothesis of the theory of relativity is that the speed of light in a vacuum is constant and is the highest speed that can be attained.

LEARNING GOALS

1. *To learn about the nature of light.*
2. *To achieve an understanding of the electromagnetic spectrum.*
3. *To learn about the corpuscular and wave nature of light.*
4. *To learn about photometry, which deals with the measurement of the quantities associated with visible light.*

34-1 THE ELECTROMAGNETIC SPECTRUM

Electromagnetic waves can be represented by an electric and magnetic wave, each at right angles to the other, and both at right angles to the direction of travel (see Fig. 34-1a). Visible light, radio and television waves, heat waves, x rays and gamma rays are such waves. These are produced within atoms and radiated out into space at a frequency that is characteristic of each. A certain shade of green light, for instance, has a frequency of approximately 6×10^{14} cycles/sec ($= 6 \times 10^{14}$ hertz). Radio waves have a much lower frequency. Radio station WCBS in New York City has a carrier frequency of 880×10^3 hertz; i.e., alternating currents set up in the antenna of its transmitter send out electromagnetic waves of this frequency. The human eye is sensitive to only a very narrow region in the electromagnetic spectrum (Fig. 34-1b). Different wavelengths in the visible region are perceived as different colors. The approximate range of wave-

length perceived by the human eye is from $\lambda = 4 \times 10^{-5}$ cm, which corresponds to the violet color, to $\lambda = 7 \times 10^{-5}$ cm, which corresponds to the color red.

After these waves are produced, they are transmitted into space and travel at a speed that is characteristic of the space. All electromagnetic waves travel through a vacuum at a speed of 2.998×10^8 m/sec (usually rounded off as 3×10^8 m/sec) and through other materials, such as water or glass, at lower speeds. The association between the speed of light, c, its wavelength, λ, and its frequency, f, is given by the wave equation

$$c = \lambda f \qquad (34\text{-}1)$$

Example 34-1   A certain dental x-ray unit produces X rays at a frequency of 3×10^{19} hertz. Calculate their wavelength in air.

Solution.   X rays, like light are electromagnetic waves. Their speed in air is very nearly the same as in vacuum and can be taken as 3×10^{10} cm/sec.

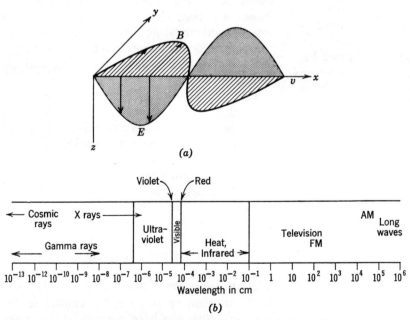

(a)

(b)

Figure 34-1   (a) Electromagnetic wave. Magnetic flux density B and the electric intensity E are perpendicular to each other and to the velocity vector v. (b) The electromagnetic spectrum.

From Eq. 34-1

$$\lambda = \frac{c}{f} = \frac{3 \times 10^{10} \text{ cm/sec}}{3 \times 10^{19} \text{ hertz}}$$

$$\lambda = 10^{-9} \text{ cm} \qquad \blacktriangleleft$$

Visible light has a frequency range from approximately 4.3×10^{14} cycles/sec (red color) to approximately 7.5×10^{14} cycles/sec (violet color). By simple computation from the wave equation, the wavelength in air for these frequencies can be shown to be, respectively, 7000×10^{-8} cm and 4000×10^{-8} cm. Since wavelengths for visible light are so small, two other practical units are usually used. These are the *angstrom*, abbreviated as Å and the *millimicron*, abbreviated as mμ.

$$1 \text{ Å} = 10^{-8} \text{ cm} = 10^{-10} \text{ m}$$
$$1 \text{ m}\mu = 10^{-9} \text{ m} = 10 \text{ Å} \qquad (34\text{-}2)$$

34-2 LIGHT WAVES AND CORPUSCLES

The production of light takes place within the atoms of the light source. If atoms are somehow energized, as by heat, some orbital electrons may be caused to jump into a higher orbit (see Fig. 34-2). At some later time these electrons may jump back into orbits closer to the nucleus, thereby releasing, in the form of light, some, or all, of the energy imparted to them.

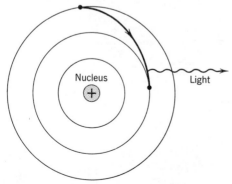

Figure 34-2 Emission of light by an atom.

But what is the nature of light as it travels through space? We can pose this question in another way: How is light energy transferred through empty space from the emitter to the receiver? For instance, when light shines on the surface of metals, it can transfer to the atomic electrons of the metal enough kinetic energy for the electrons to be torn away and leave the metal. This phenomenon is called the *photoelectric effect* and has wide applications in electronics. There are many less sophisticated experiments which indicate that light is a carrier of energy. The light coming from the sun will cause a body on earth to gain heat energy, etc.

In the game of billiards the moving mass of the Q-ball will transfer energy to the 8-ball when these collide. Similarly, when light strikes an object (as in the photoelectric effect) it behaves as if it consists of tiny particles called *photons* or *corpuscles*. Now, there is a mass associated with even the smallest of particles, for instance the electron. But the mass of a photon is zero. Since a photon cannot exist at zero speed, its mass, called the *rest mass* when it is at rest, is zero. At its singular speed, c, in a vacuum the photon has masslike characteristics in the sense that it can transfer energy and momentum to ordinary particles, but it has no mass in the usual sense that can be measured in grams.

That light behaves as a particle seems reasonable when we compare its behavior with that of larger particles, like small steel spheres. Both will be reflected in the same manner when they hit a hard surface, and both will change their direction of travel when they pass from one medium into another (refract), as from air into water. A theory of light called the *corpuscular theory* maintains that light consists of photons having masslike characteristics.

Another theory concerning the nature of light is called the *wave theory*. This theory also considers the question of how light energy is transferred through space from the emitter to the receiver. Most of us have experienced the energy of a water wave by being pushed or knocked down by one while standing in the ocean or a lake.

Waves also can transfer energy from one point in space to another. But there is a fundamental difference between waves and particles. Consider a small wave in a calm lake that is caused by a chunk of wood that is dropped straight down into the lake. The circular wave will move out for many meters from the point of impact between the wood and water while the wood will remain floating more or less near the point where it first landed. The wave hitting another object, say, a piece of paper floating 10 m from the wood, will transfer energy to the paper and cause it to move. But the important thing to realize is that the water particles did not move from the point in the lake where the wood landed to the paper. Energy was transferred not by the particles but by a wave. According to the wave theory, light energy is transported through space by waves.

Convincing evidence exists that light is a wave. One of the great physicists, James Clerk Maxwell, in the mid-nineteenth century proved on a theoretical level that light is an electromagnetic wave and predicted that other such waves, not necessarily visible, could be produced. A few years later, Heinrich Hertz succeeded in producing and propagating invisible electromagnetic waves. These waves are what we call radio waves today. Convincing arguments for the wave nature of light will be presented in Chapter 38, where interference and diffraction of light are discussed.

In spite of the fact that the corpuscular and wave theories differ on the nature of light, there exists ample experimental evidence to support both; both theories are equally accepted. There seems to be a duality in nature concerning light. That branch of physics which investigates this duality is called *quantum physics.*

34-3 STRAIGHT-LINE PROPAGATION OF LIGHT; SHADOWS

In the sections that follow, the behavior of light is discussed without necessarily analyzing the duality concerning the nature of light. Convincing evidence that light travels in a straight line can be obtained by a simple experiment, as depicted in Fig. 34-3. The point source of light, S, issues rays in all directions. No rays will penetrate the barrier because it is opaque. The shadow will appear sharply defined and entirely dark, indicating that no light has curved around the barrier. Such a shadow is called an *umbra.*

Another type of shadow results when the source is an extended one, rather than a point (Fig. 34-4). It is to be noted from the figure that the area on the screen marked A_1 has no light rays incident on it. This area again is called the umbra. In the areas marked A_2, rays from some part of the source appear but not from all parts. This area is

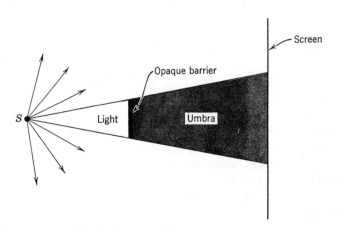

Figure 34-3 Shadow from a point source.

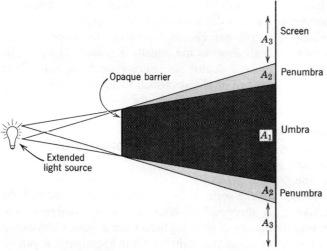

Figure 34-4   Shadows from an extended light source.

called the *penumbra*. The areas marked A_3 appear bright.

34-4 THE SPEED OF LIGHT

The speed of light in vacuum, denoted by the letter c, is one of the most important constants in physics. It is the same for all colors traveling in vacuum; in fact, it is the same for all electromagnetic radiations. One of the first accurate measurements of its speed was made by the French scientist Armand Fizeau in 1849. His experiment involved a rotating toothed wheel, having the same number of teeth as voids (see Fig. 34-5), a narrow light beam coming from a source located behind the wheel, and a plane (flat) mirror mounted a few miles in front and in line with the wheel's axle. The wheel and mirror are mounted at high elevations to avoid obstruction between them. By sighting through the void of the wheel, it is possible to see the reflections of the light from the mirror.

Assume the wheel to rotate at a low constant frequency. Light from the source passing through the void will travel to the mirror, reflect from it,

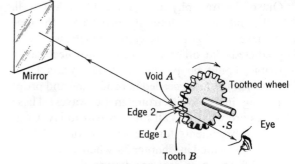

Figure 34-5   Schematic of Fizeau's apparatus for measuring the speed of light. Light from a light source is transmitted to the mirror through the voids and stopped by teeth. The eye can see only the reflected light from the mirror that returns through a void.

and return to the wheel. If the rotational frequency is low enough, the edge of tooth B will not have moved over far enough to block the returning light from the eye. The eye, behind void A, will then see a light spot.

If the rotational frequency is slowly increased, a constant frequency is just reached when no light appears to the eye as the voids continue to pass it. Under these circumstance, the time that it takes for edge 1 to move to edge 2, a distance equal to the width of the void, or of a tooth (they are the same), is exactly the time it takes for the light to make a round trip to the mirror.

Example 34-2   Fizeau used a wheel with 720 teeth and placed the mirror at a distance of 8630 m. He found that the lowest wheel frequency at which he could see no light through the void was 12.6 rev/sec. Calculate the speed of light from these data.

Solution.   Since the frequency was 12.6 rev/sec, the period was 1/12.6 sec, i.e., it took 1/12.6 sec for the whole wheel, and therefore for 720 teeth and 720 voids, to make one revolution. The time it takes for the outside of the wheel to move a distance corresponding to the width of one tooth (or one void) is

$$\frac{1}{12.6} \times \frac{1}{2 \times 720} = \frac{1}{18{,}144} \text{ sec}$$

This is the time it takes for the light to make one round trip, a distance of 2×8630 m $= 17{,}260$ m. Therefore, the speed of light (as determined by Fizeau) is

$$c = \frac{17{,}260 \text{ m}}{1/18{,}144 \text{ sec}} = 3.13 \times 10^8 \text{ m/sec}$$

The value for c that is presently accepted is 2.998×10^8 m/sec.   ◀

34-5 PHOTOMETRY; LUMINOUS FLUX, THE LUMEN

A given light source may emit electromagnetic energy distributed over many wavelengths in the visible and invisible region. The subject of photometry deals with light measurements *in the visible region only*, i.e., in the region approximately between 4000 and 7000 Å. Light energy is expressed in the usual energy units, the joule and the erg. The process of seeing involves visible light energy per unit time, called *luminous flux*. Lumi-

nous flux, therefore, is measured in joules/sec or watts.

The human eye is not equally sensitive to all colors. A green light source, producing light energy at a rate of 1 W, will evoke a greater sensation of brightness than a blue or red source producing the same light power. A graph depicting the response of the eye to various wavelengths is called the *spectral luminous efficiency curve* and is given in Fig. 34-6. The wavelength corresponding to the maximum visual sensitivity, which is 5550 Å, a yellow-green color, is assigned the value of 100%; the other wavelengths are expressed as fractions of the maximum. From the luminosity curve it can be seen that at the same power a color corresponding to a wavelength of approximately 5000 Å (a green color) and 6200 Å (a red color) will produce 30% of the visual sensation produced by light of 5550-Å wavelength. The luminosity curve is based on the visual response of many individuals to controlled light sources. Since it is difficult to express any human response with great quantitative accuracy, the luminosity curve

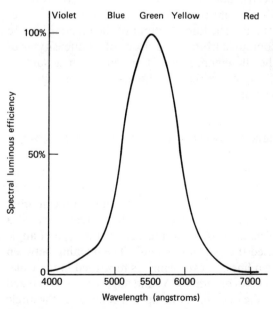

Figure 34-6   Spectral luminous efficiency curve.

is only approximate and applies to the "average" eye.

The visual sensation produced by visible light, then, depends on two factors: the light energy per unit time that is incident on the eye and the wavelength of this energy. The *lumen* is a unit of luminous flux that takes both factors into account. The lumen is equivalent to $\frac{1}{680}$ watt of light power of wavelength 5550 Å.

Example 34-3 A monochromatic light source produces light energy in the 5000-Å region at a rate of 2 W. Calculate the luminous flux in lumens produced by this source.

Solution. From the luminosity curve, the visual response in the 5000-Å region is about 30% of that in the 5550-Å region for which the lumen is defined. Since one lumen is equivalent to $\frac{1}{680}$ W of light power at 5550 Å, 2 W are equivalent to 1360 lumens at the same wavelength. Two watts of light power in the 5000-Å region will produce 30% of 1360 lumens, or 408 lumens. ◀

The usual light source, like the incandescent electric bulb, is not monochromatic. It produces visible light that is distributed over many wavelengths. The lumen output of such a lamp can be computed from a knowledge of its intensity or of the illuminance that it produces on a surface. These quantities are discussed in the following sections.

34-6 LUMINOUS INTENSITY—THE CANDLE

The usual light source emits light energy in many directions. The "strength" of a given light source, referred to as its *intensity*, depends on the number of lumens that it emits within a region called a *solid angle*. A solid angle is not the type of angle used thus far in this book. The opening between two lines, or two planes, is measured in the usual degrees or radians. A solid angle can be visualized as the opening at the tip of a cone, or as the angle subtended at the center of a sphere by an element of area on its surface. Solid angles are measured in

units of *steradians*. One steradian is the solid angle subtended at the center of a sphere by an area A on its surface that is equal to the square of its radius R (Fig. 34-7). The angle Ω, in steradians, is given by

$$\Omega = \frac{A}{R^2} \tag{34-3}$$

Since the total surface area of a sphere is $4\pi R^2$, it follows that the total solid angle of a sphere is $4\pi R^2/R^2$ or 4π steradians. (This should be compared with the definition of a radian, and the total angle of a circle, expressed in radians.)

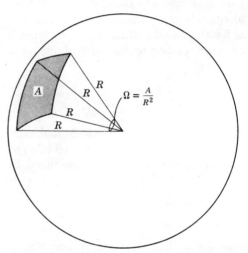

Figure 34-7   Solid angle Ω subtended by area A at the center of the sphere.

Example 34-4   What solid angle is subtended at the center of a 12-ft-diameter sphere, by a 2.5 ft² are on its surface?

Solution.   From Eq. 34-3

$$\Omega = \frac{2.5 \text{ ft}^2}{(6 \text{ ft})^2} = 0.07 \text{ steradians}$$

It should be noted that solid angles like plane angles are pure numbers. ◀

We are now in a position to further our discussion of luminous intensity. The intensity I of a given light source is defined as the luminous flux, F (i.e., number of lumens) it emits per unit solid angle Ω, or

$$I = \frac{F}{\Omega} \tag{34-4}$$

The unit for intensity is lumens/steradian, which is called a *candela* or *candle*. Thus a light source has an intensity of one candle if it emits one lumen per steradian. The usual light source does not have the same intensity in all directions. It is therefore more proper to speak of the *intensity in a specified direction*. If a point light source of I candela emits uniformly in all directions, i.e., if every part of the inner surface of a sphere receives the same flux with the source at its center, then it follows that the source must emit a total of $4\pi I$ lumens. *A uniform, one-candela point source therefore emits 4π lumens.*

The internationally accepted standard candle or candela (Fig. 34-8) is the intensity of light originating from a surface of platinum that is $\frac{1}{60}$ cm^2 in area maintained at its melting point 1769°C. By way of comparison, a 40-W electric bulb is equivalent to about 35 candelas or candles.

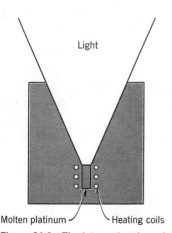

Molten platinum ⎯   ⎯ Heating coils

Figure 34-8   The international candela (candle).

Example 34-5   A 12-cm × 20-cm page is illuminated by a lamp hung directly above its center, at a perpendicular distance of 2 m. (a) If the intensity

of the lamp is 250 candelas, in the direction of the page, compute the flux falling on the page. (b) What is the total luminous flux produced by the lamp, assuming that its intensity is 250 candelas in all directions?

Solution. (a) The solid angle Ω subtended by the page on the lamp, assuming that the lamp is small compared with its distance to the page, can be computed by imagining a sphere of 2-m radius drawn around the lamp. From Eq. 34-3,

$$\Omega = \frac{A}{R^2} = \frac{0.12 \text{ m} \times 0.2 \text{ m}}{4 \text{ m}^2}$$

$$= 6 \times 10^{-3} \text{ steradians}$$

From Eq. 34-4

$$F = I\Omega = 250 \text{ candles} \times 6 \times 10^{-3} \text{ steradians}$$

$$= 1.5 \text{ lumens}$$

(b) A uniform source produces a total of $4\pi I$ lumens. Therefore, this lamp produces a total of $4\pi \times 250$ candles = 3142 lumens.   ◀

34-7 ILLUMINANCE

Thus far two quantities in photometry have been discussed. One describes the strength of a given light source, or its intensity, and the other describes the luminous flux emitted by the source. Another important quantity, called *illuminance*, describes the illumination of surfaces. What, for instance, is the illumination, or, more properly, the illuminance, of the page you are now reading? The average illuminance E of a given surface is defined as the total flux F falling on it, divided by its area, or

$$E = \frac{F}{A} \tag{34-5}$$

Illuminance is measured in units of lumen/m^2, called the *meter-candle* and in lumen/ft^2, called the *foot-candle*. The unit of meter-candle is sometimes referred to as *lux*.

Illuminance can also be expressed in terms of the intensity, I of a point source. In Fig. 34-7 the

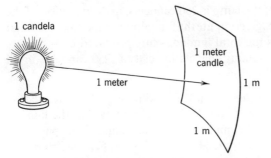

Figure 34-9 A surface having an area of 1 m² will have an illuminance of 1 meter-candle when it is 1 m away from a small 1 candela light source.

solid angle, Ω subtended by area A at the source (provided the distance R from every point on the area to the source is the same) is $\Omega = A/R^2$.

Substituting this value for Ω in Eq. 34-4 we get

$$I = \frac{F}{\Omega} = \frac{F}{A/R^2} \quad \text{and} \quad F = \frac{IA}{R^2} \quad (34\text{-}6)$$

But, by Eq. 34-5, $E = F/A$. Substituting for F in this equation its value given by Eq. 34-6 we get

$$E = \frac{F}{A} = \frac{IA}{R^2 A}$$

$$E = \frac{I}{R^2} \quad (34\text{-}7)$$

Therefore, a 1 m × 1 m surface will receive an illuminance of 1 meter candle from a small light source having an intensity of 1 candela placed at a perpendicular distance of 1 m from the center of the surface (Fig. 34-9). Equation 34-7 indicates that if the source of light is small (a point source) the illuminance on a surface is inversely proportional to the square of the distance. If the source is distributed along a line, like a fluoresent lamp, the illuminance is inversely proportional to the first power of the distance that is, the illuminance does not decrease as rapidly with distance as it does with a point source.

Example 34-6 A 10 cm × 20 cm page is illuminated by an electric bulb at a perpendicular distance of 1.5 m from its center. If the light intensity

of the bulb is 150 candelas (roughly equivalent to 150-W bulb) in the direction of the page, calculate (a) the luminous flux incident on the page and (b) the illuminance of the page.

Solution
(a) The solid angle subtended by the page at the lamp is given approximately by Eq. 34-3,

$$\Omega = \frac{0.1 \text{ m} \times 0.2 \text{ m}}{(1.5 \text{ m})^2} = 8.9 \times 10^{-3} \text{ steradian}$$

By Eq. 34-4,

$$F = I\Omega$$

$$F = \frac{150 \text{ lumens}}{\text{steradian}} \times 8.9 \times 10^{-3} \text{ steradian}$$

$$= 1.34 \text{ lumens}$$

(b) The distance is large enough for every point on the page to be considered as about equidistant to the small light source. By Eq. 34-7, $E = I/R^2$

$$E = \frac{150 \text{ candelas}}{(1.5 \text{ m})^2} = 66.7 \text{ lux}$$

An alternate method involves Eq. 34-5.

$$E = \frac{F}{A} = \frac{1.34 \text{ lumens}}{0.1 \times 0.2 \text{ m}} = 67 \text{ lux} \quad \blacktriangleleft$$

Consider next a small light source of intensity I, illuminating a surface of area A at a distance of R from the source as in Fig. 34-10. Notice that the light from the source is not perpendicular to the surface. The solid angle Ω, subtended by the surface at the source, is

$$\Omega = \frac{A \cos \theta}{R^2} \quad (34\text{-}8)$$

where $A \cos \theta$ is the projected area of the surface along a plane perpendicular to the rays and θ is the angle between the normal to the surface and the line connecting the source and center of the surface. Substituting this value for Ω in Eq. 34-4, we get

$$F = I\Omega = \frac{IA \cos \theta}{R^2} \quad (34\text{-}9)$$

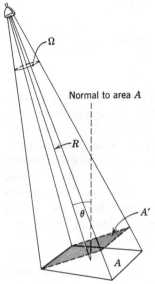

Figure 34-10 The component of the area A perpendicular to the rays is $A^1 = A \cos \theta$.

But the defining equation for illuminance, Eq. 34-5, is $E = F/A$. Substituting $IA \cos \theta/R^2$ for F, in Eq. 34-5, we finally get

$$E = \frac{I \cos \theta}{R^2} \qquad (34\text{-}10)$$

Example 34-7 A lamp having a uniform intensity of 100 candles rests on a flat table (no lampstand) at a distance of 50 cm from the center of a flat page resting on the same table. Compute the illuminance on the page.

Solution. The angle θ between the normal to the page and the line connecting the source and the center of the page is 90°. From Eq. 34-10

$$E = 0$$

Thus the page receives no illumination directly from the lamp. Actually, some illumination would reach it, since the lamp filament is at a small height above the table surface. ◀

Example 34-8 The lamp in Fig. 34-10 is 1.5 m away from the center of surface, which has an area of 0.15 m². The intensity of the lamp is 200 candles in the direction of the surface, and $\theta = 45°$. Calculate (a) the illuminance of the surface, and (b) the luminous flux incident on the surface.

Solution
(a) $I = 200$ candles, $R = 1.5$ m, $\cos \theta = 0.707$, $E = ?$ Substituting in Eq. 34-10 gives

$$E = \frac{200 \text{ candles} \times 0.707}{(1.5 \text{ m})^2} = 63 \text{ meter-candles}$$

(b) $F = I\Omega, \Omega = \dfrac{A \cos \theta}{R^2}$

from Eqs. 34-4 and 34-6, respectively.

$$\Omega = \frac{0.15 \text{ m}^2 \times 0.707}{(1.5 \text{ m})^2} = 4.7 \times 10^{-2} \text{ steradian}$$

Substituting this value for Ω in Eq. 34-4 results in

$$F = 200 \text{ candles} \times 4.7 \times 10^{-2} \text{ steradian}$$
$$= 9.4 \text{ lumens}$$

The same answer would be obtained if Eq. 34-5 were used. ◀

34-8 INSTRUMENT PHOTOMETRY

The preceding sections on photometry dealt with the definitions of photometric terms and with equations relating these terms. The illustrative examples assumed a knowledge of lamp intensities or one other photometric characteristic, from which the others were computed from appropriate equations. One of the usual problems confronting an illumination engineer is to determine such photometric characteristics of a given light source as intensity, flux output, and illuminance produced by it, without a previous knowledge of any of these. In order to determine these, he must have available a *standard light source* and an instrument called a *photometer*. A standard light source is usually an incandescent lamp of known intensity (candles) that has been calibrated directly or indirectly against an internationally defined standard light source. Having a lamp of known intensity in a given direction, the engineer can determine the intensity of an unknown lamp by means of a photometer and Eq. 34-8.

One of the often-used photometers, called a Bunsen photometer (Fig. 34-11), consists essentially of a translucent grease spot S, mounted on each side of an opaque plate, with two mirrors M so attached that they concurrently reflect an image of both grease spots to the eye, placed at the opening O. By moving the photometer back and forth until a position is finally reached where the illuminance produced on the two grease spots is the same, as judged by the eye, we can find the place where

$$E_s = E_x \qquad (34\text{-}11)$$

Here E_s is the illuminance produced by the standard lamp of known intensity I_s, and E_x is the illuminance produced by the unknown lamp of intensity, I_x. Substituting for the values of E in Eq. 34-7 gives

$$\frac{I_s}{R_s{}^2} = \frac{I_x}{R_x{}^2} \qquad (34\text{-}12)$$

where R_x and R_s are, respectively, the distances from the unknown lamp and known lamp to the grease spots.

The eye is a reliable indicator of matched illuminances to within a few percent, provided that the colors are the same. If the colors are different, other means have to be adopted.

Example 34-9 Balanced illuminance is achieved in a photometer when it is placed 10 cm from a lamp of unknown intensity and 20 cm from a standard lamp of 150 candles in the direction of the photometer. What is the intensity of the unknown lamp in the direction of the photometer?

Solution. From Eq. 34-12,

$$I_x = \frac{I_s R_x{}^2}{R_s{}^2} = 150 \text{ candles} \times \frac{(10 \text{ cm})^2}{(20 \text{ cm})^2}$$

$$= 37.5 \text{ candles} \qquad \blacktriangleleft$$

Photometers in general are not direct-reading instruments. They require the eye to make judgments of equal illuminances. There exist light meters (Fig. 34-12) which, when exposed to illumination, render direct readings of illuminance in meter-candles or foot-candles. Photographic exposure meters are of this type. Such meters consist essentially of an iron disk coated with a fine layer of the element selenium, and a galvanometer. When luminous flux is incident on the selenium side of the disk, an electric current between the selenium and iron results, which is proportional to the flux. For a disk of a given area, the galvanometer can be calibrated in units of meter-candles or foot-candles. No outside

Figure 34-12 Weston illumination meter.

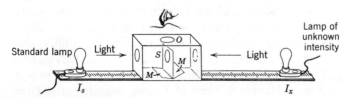

Figure 34-11 Bunsen photometer.

source of potential is needed for this type of meter. Such instruments are not highly accurate because the current produced is color-dependent.

Another type of light-measuring device is the phototube. Light incident on its photocathode causes a current, which is proportional to the luminous flux, to be established in the circuit consisting of the tube, a battery, and a galvanometer. The galvanometer is usually calibrated to to read in flux units (lumens). Photocells of various spectral sensitivities are manufactured commercially.

SUMMARY

Light is an electromagnetic wave.

The corpuscular theory and the wave theory advance different methods by which light energy is transported through space. Each of these theories is verified by different experiments. Both are considered correct.

The speed of light in vacuum is approximately 3×10^8 m/sec = 3×10^{10} cm/sec. The speed of light is an important constant of nature.

Different wavelengths of light are perceived as different colors.

Only a small portion of the electromagnetic spectrum can be perceived by the human eye.

The subject of photometry deals with measurements of light in the visible region.

The "average" human eye is most sensitive to the color yellow-green in the region of 5500 Å.

The luminous output of a light source is measured in lumens.

The luminous intensity of a source is measured in candelas.

The amount of the illumination on a surface is measured in meter-candles, or foot-candles.

QUESTIONS

1. Cite some everyday experiences that indicate that light is a carrier of energy.

2. Is external energy required to produce light? Why?

3. What is the difference between radio waves and the waves in the visible region?

4. Name some other electromagnetic waves.

5. Explain the meaning of a photon.

6. What is meant by rest mass? Does a photon have a rest mass?

7. Explain a crucial differences between the corpuscular and wave theories of light.

8. You can produce a water wave by inserting a pencil point into water in a tray. The wave will travel across the water in the tray. How can you determine experimentally that the water did not actually travel across the tray?

9. In what units is luminous flux measured?

10. Explain the photometric units lumen, candle, and meter-candle.

PROBLEMS

(A)

1. Assume that a certain radio station sends out radio waves at a carrier frequency of 770×10^3 cycles/sec.
   (a) Express this frequency in units of kilo hertz.
   (b) Calculate the wavelength of these waves.

2. Compare the frequency of the radio waves of Problem 1 with that of red light having a wavelength of 7000 Å.

3. The wavelength corresponding to the color red, $\lambda = 7000$ Å is approximately the longest to which the average human eye is sensitive. Compare this wavelength with that of the radio wave in Problem 1.

4. Television stations broadcast at much higher frequencies than radio stations. Compare the wavelengths of radio waves and television waves.

5. When a gas of hydrogen atoms is energized by an electric spark, this gas emits light having a wavelength of 656.3 millimicrons.
   (a) Express this wavelength in angstroms and in centimeters.
   (b) Is this light visible to the human eye?

6. The human eye can detect electromagnetic radiation in the wavelength range between 4000 to 7000 Å. Calculate the frequency range to which the human eye is sensitive.

7. (a) At what speed do x rays travel in air?
   (b) Does the speed of electromagnetic waves depend on frequency?

8. A certain medical x-ray unit produces x rays having a wavelength of 0.5 Å. Determine the frequency of such x rays.

9. How long does it take for light to travel from the sun to the earth when the sun is 148.8×10^6 km away?

10. What is the time delay in the radio communication between an astronaut on the moon and the space-center in Houston, Texas when the moon is 240×10^3 miles from the earth?

(B)

1. A monochromatic source, $\lambda = 550$ mμ produces light energy at a rate of 0.5 W. Calculate the luminous flux produced by this source.

2. Infrared rays are associated with the sensation of heat. Assume a wavelength of 8×10^3 Å in the infrared region of the electromagnetic spectrum and calculate the frequency of this wave.

3. Define and explain the lumen.

4. A monochromatic light source, $\lambda = 5500$ Å emitting light of wavelength 5500×10^{-8} cm, produces light energy at a rate of 1.5 joules/sec. Calculate the luminous flux in lumens produced by this source.

5. A monochromatic light source emitting light of wavelength 6000 Å produces light energy at a rate of 1.5 W. Calculate the luminous flux in lumens produced by this source (see Fig. 34-6).

6. A camera lens is located at a perpendicular distance of 15 ft from the center of a 1 ft $\times$ 2 ft window. Calculate the solid angle subtended by the window on the lens, assuming that every point on the window is about equidistant from the lens.

7. A flashlight projects a circle of light that is 30 in. in diameter on a wall 15 ft away. Calculate the solid angle projected by the circle at the flashlight.

8. Calculate the luminus flux in the light beam produced by the flashlight in Problem B7 if the luminous intensity of the flashlight is 50 candelas in the direction of the wall.

9. At what perpendicular distance should a small 40-W lamp be placed from the center of a desk to give it an illuminance of 100 foot-candles? Assume the lamp to have an intensity of 30 candelas.

10. Calculate the solid angle subtended at the center of a sphere of radius 20 cm by an area of 30 cm^2 on the surface of the sphere.

11. A small lamp is hung at a height of 2 m directly above the center of a 10 cm $\times$ 20 cm photographic film. If the lamp intensity is 500 candles in the direction of the film, compute the luminous flux incident on the film.

12. Calculate the total luminous flux of the lamp in Problem B11, assuming that its intensity is 250 candelas in all directions.

13. The small lamp in Fig. 34-10 is 2 m away from the center of the surface, which has an area of 0.2 m^2. The intensity of the lamp is 300 candelas in the direction of the surface and $\theta = 60°$. Calculate:
   (a) The solid angle subtended by the surface at the lamp.
   (b) The illuminance of the surface.

14. Assume that the area A in Fig. 34-10 is a page that has an illuminance of 200 foot-candles when the lamp is directly overhead ($\theta = 0°$). What would be the illuminance if the page is turned so that
   (a) $\theta = 30°$,
   (b) $\theta = 45°$ and
   (c) $\theta = 90°$?

15. A flashlight, having an intensity of 20 candelas is viewed through a camera lens at a distance of 1 km.
   (a) Calculate the number of lumens that enter the lens having an aperture of 10 cm^2.
   (b) the illuminance at the lens.

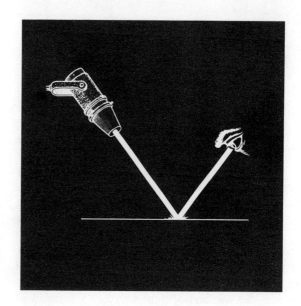

35.
Light Reflection and Mirrors

LEARNING GOALS

1. *To learn the Law of Reflection for light.*
2. *To understand how to apply this law to flat and curved mirrors.*
3. *To learn the mirror equation and its applications.*
4. *To understand the meaning of real and virtual images.*
5. *To learn Huygen's principle and understand how the law of reflection is derived from it.*

35-1 RAY OPTICS

The study of the behavior of light is enhanced by diagrams involving rays. A pinhole in a piece of cardboard, placed in front of a light source like a lamp, will produce a *beam of light* in the shape of a cone. By allowing the light to pass through a second pinhole, as in Fig. 35-1 a very narrow bundle, called a *pencil of light* is produced. An infinitessimal narrow pencil of light is a *ray*. According to the wave theory of light, point sources emit expanding spherical waves (see Fig. 35-16*b*) and lines (constructed) perpendicular to the wave front are called *rays*. An infinite number of rays can, therefore, be associated with even the smallest of light sources. A method of ray diagrams is helpful in describing reflection and refraction. This method is commonly referred to as *ray optics* or *geometrical optics*. The reason for the latter name stems from the fact that the principles of geometry are used extensively, as will become evident from the succeeding sections.

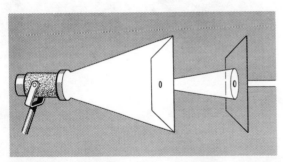

Figure 35-1 Production of a very narrow beam or pencil of light.

Point sources of light are often considered. A point source, practically speaking, is a small object that radiates light in all directions, its size being small compared with the distances considered. Stars, for instance, are considered as point sources with respect to an observer on the earth. A small, frosted light bulb can be considered as a point source when viewed from a distance of meters. A source of light is either luminous or illuminated. A *luminous* body produces its own light, as the sun or a glowing lamp filament. An *illuminated* body is one that gives off light by reflection, as the moon, a necktie, or any body that receives light from other sources. A point source of light, then, is a small luminous or illuminated body that gives off light in all directions. The subject of optics is partially concerned with tracing the path of this light energy, sometimes by means of rays and sometimes by means of waves, whichever is more convenient, through air and other materials. It will become apparent in the sections that follow that from a knowledge of the paths of a few selected rays, the position and size of mirror and lens images can be deduced.

35-2 THE LAW OF REFLECTION

When a light beam strikes a smooth surface, it is reflected from that surface in a predictable manner. An experiment shown in Fig. 35-2*a* depicts a light source issuing a narrow beam that is intercepted by a mirror. The light beam will be reflected from the mirror at a definite angle. Notice that the ray diagram of Fig. 35-2*b* reduces the oncoming and reflected beams, each to a single ray.

The perpendicular line to the mirror at the point of intersection between the ray and mirror is called the *normal*. The angle between the *incident ray* and the normal is the *angle of incidence*; the angle between the *reflected ray* and the normal is called the *angle of reflection*. The law of reflection states that:

1. The angle of incidence equals the angle of reflection, or

$$\angle i = \angle r \qquad (35\text{-}1)$$

2. The incident ray, reflected ray, and the normal lie in the same plane.

Example 35-1 A ray of light strikes a mirror at an angle of 35° with the mirror. Compute the

THE LAW OF REFLECTION

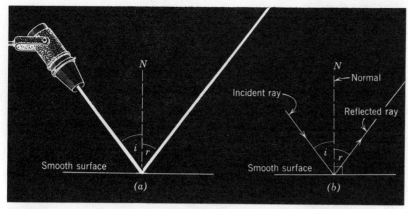

Figure 35-2 (a) Reflection of a narrow beam of light from a smooth surface. A beam consists of many rays, one of which is shown in (b). (b) For a given ray $\angle i = \angle r$ and the incident ray, reflected ray, and the normal lie in the same plane.

angle that the reflected ray will make with the mirror. What are the angles of incidence and reflection?

Solution

$$35° + \angle i = 90°$$
$$\angle i = 55°$$

From Eq. 35-1 $\angle i = \angle r = 55°$

Therefore, the reflected ray makes an angle of $90° - 55° = 35°$ with the mirror. ◀

If a narrow bundle of parallel rays is incident on a highly polished surface such as a mirror, the reflected rays will also be parallel. An observer will see a bright area ("hot spot") corresponding to the area where the beam hit the surface (see Fig. 35-3a).

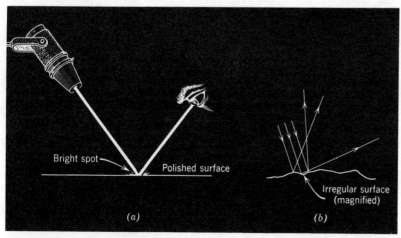

Figure 35-3 (a) Specular reflection. The narrow bundle of reflected rays will give a sensation of a bright spot. If the eye is moved away, the bright spot will disappear. (b) Diffuse reflection. The reflected rays are diffused over a wide region. The surface will appear toned down.

Such a surface is said to be a *specular reflector*. When a narrow bundle of rays is incident on a rough surface (Fig. 35-3*b*), the incident light scatters over wider areas and less light from a given part of the surface reaches an observer's eyes. Such a surface is called a *diffuse* or *mat reflector* (e.g., blotter, pitted wall).

35-3 IMAGE FORMATION

Every point of a luminous or an illuminated object sends out light rays in many directions. An observer locates these points in space by "following" the rays that strike his eyes, along a straight line and in a backward direction, to their point of intersection (Fig. 35-4). When an object is placed in front of a plane (flat) mirror, some of the rays emanating from its points will strike the mirror, Fig. 35-5. By applying the law of reflection to these rays, the direction of each reflected ray can be deduced. Notice that the reflected rays are *divergent* (separate), and that if continued to the left of the mirror, they will *converge* (come together) at a single point, *I*. If an observer's eye is placed so as to intercept some of the reflected rays, his eye will "follow" the reflected rays back, as in Fig. 35-4, and perceive the image to be located at the point where the rays *seem* to con-

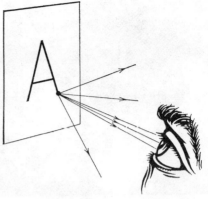

Figure 35-4 Every point on the "A" sends out rays. The eye locates a point on the "A" by intercepting at least two rays and "following" them back in a straight line, to the point of intersection.

verge, point *I*. This point *I* is said to be the *virtual image* of the object point, *O*. It is emphasized that no light comes from the point *I*; it only *seems* to come from there, which is the reason this type of image is called virtual.

35-4 IMAGE LOCATION

The location of the image in a plane mirror can be determined either by the construction method

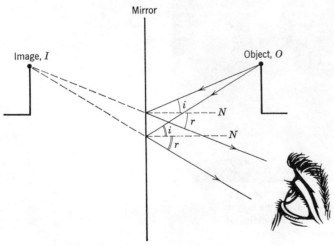

Figure 35-5 Image formation in a plane mirror.

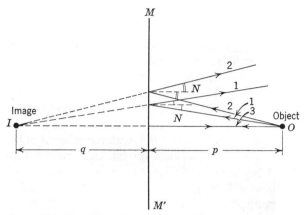

Figure 35-6   Image location by construction.

or analytically. In Fig. 35-6, the point O represents an object point located at a distance p from a plane mirror, MM'. Since all reflected rays from an object point will converge at a single image point, it is necessary to draw only two rays from O in order to locate the image point. If any two such rays are drawn and the reflected rays constructed with the aid of a protractor, the latter rays will lead to a point of convergence I, a distance q from the mirror. If the construction is performed carefully, it will be found that $p = q$.

Analytically, the image distance q can be deduced from the following considerations: In

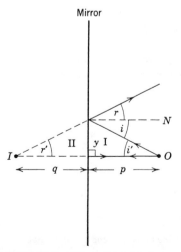

Figure 35-7   Object distance p is equal to the image distance q because triangles I and II are congruent.

Fig. 35-7 right triangles I and II are congruent because side y is common to both and $\angle r' = \angle i'$. Therefore, $p = q$, or

in a plane mirror, the object distance is equal to the image distance.

35-5 IMAGE SIZE AND MAGNIFICATION

In the preceding section the image distance of a point object was considered. An examination of Fig. 35-7 will reveal that the image lies on a straight line going through the point object and perpendicular to the mirror. If the object considered is now an extended shape, as the letter L in Fig. 35-8, the shape and size of its mirror image can be deduced from a consideration of a few well-chosen points. The top point 1 of the object will be imaged at point 1′, which is located on the perpendicular line from point 1 through the mirror and at the same distance behind the mirror. Similarly, the bottom points 2, 3, 4 are imaged, respectively, at 2′, 3′, and 4′. Since, for every point on the object, there is a corresponding image point that is on line with it and equidistant from the mirror, it follows that for a plane mirror:

  1. The image size is the same as the object size, and
  2. The image is erect (not upside down as is possible with curved mirrors).

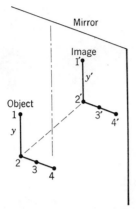

Figure 35-8 Height of the image is the same as the height of the object.

The magnification M for any mirror is defined as the negative ratio of the image height y' to the object height y, or

$$M = -\left(\frac{y'}{y}\right) \qquad (35\text{-}2)$$

Since in a plane mirror $y' = y$, Eq. 35-2 yields a magnification of -1, the minus indicating a virtual image.

Example 35-2 A 5-ft, 8-in. student stands erect at a distance of 4 ft from a full-length plane mirror. How tall is his image? (How tall does he see himself in the mirror?) What is his image distance? Describe the image.

Solution. Since in a plane mirror, the object size is equal to the image size, the magnification is -1, and he sees himself as 5 ft, 8 in. tall. Also, since the object distance is equal to the image distance, he sees his image 4 ft behind the mirror. His image is erect, and there is a right-left reversal. ◀

35-6 SPHERICAL MIRRORS

It was mentioned in Section 35-4 that a plane mirror diverges the light from a point object. A consequence of this is that plane mirrors cannot

be used to concentrate or *focus* light into a small spot. Curved mirrors, on the other hand, have the ability to collect rays emanating from a point object into a more or less small region, depending on the shape of the mirror and on the object distance. The spherical mirror (Fig. 35-9) will converge into a point the rays from a point object (e.g., an illuminated grain of sand). If two separate point objects, A and B, are placed in front of the concave side (inside) of a spherical mirror, with B on top (Fig. 35-10), two points of converged light will result, A' and B', with A' on top. The point A' is formed by the rays from point A, and B' by the rays from point B. If an extended object, such as an upright arrow, is placed in front of the mirror (Fig. 35-11), the rays from each of its points are converged into image points, this array of points describing an upside-down arrow is called the image of the object arrow. This is a *real image* in the sense that light rays actually exist at the image region, as contrasted with the virtual image given by a plane mirror.

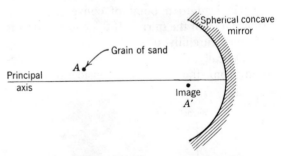

Figure 35-9 A concave mirror will render a point image A' for a point object A if the rays from A are paraxial.

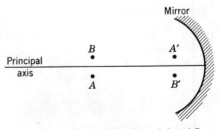

Figure 35-10 Two grains of sand A and B are imaged at A' and B'.

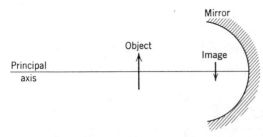

Figure 35-11 Image inversion in a spherical concave mirror.

Concave spherical mirrors can converge the rays from a point object into an effective image point, but only those rays that strike it near the central axis of the mirror, called the *principal axis* (see Fig. 35-12). Such rays are called *paraxial rays.* Rays from the object O that strike the mirror at outer regions will converge at different image points. A spherical mirror therefore renders more than one image point for a given object point. This condition is undesirable because it results in a distorted image, a defect of spherical mirrors referred to as *spherical aberration.* By making use of only the central portion of the mirror, called

the *paraxial region*, this aberration is minimized. The discussion on spherical mirrors in the following sections pertains to the paraxial region.

Concave mirrors shaped as a paraboloid, called *parabolic* mirrors, do not suffer from spherical aberration, thus making them capable of producing better images. Reflecting telescopes use parabolic mirrors. If a small lamp is placed at the focus of a parabolic mirror, the rays hitting the mirror will be reflected out parallel to each other. Automobile headlights make use of this principle and thus produce an even and parallel forward beam.

35-7 IMAGE LOCATION BY THE CONSTRUCTION METHOD

The location and characteristics of images formed by spherical concave and convex mirrors can be determined by careful ray construction. In Fig. 35-13, the arrow $O–O'$ represents an illuminated object whose image we want to locate. The point C represents the center of curvature

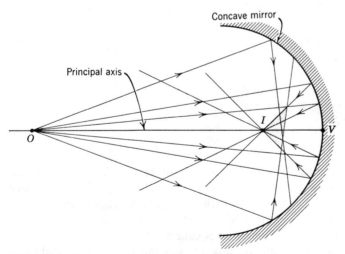

Figure 35-12 Spherical aberration. Rays starting from the point object O and at small angles with the principal axis are called paraxial rays. These rays are imaged at point I. Nonparaxial rays, as the two outer rays, are imaged somewhere within the shaded area.

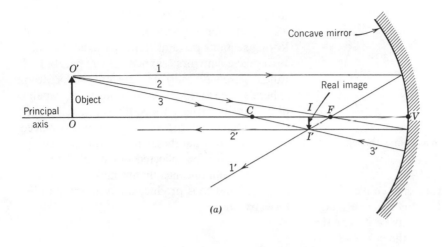

(a)

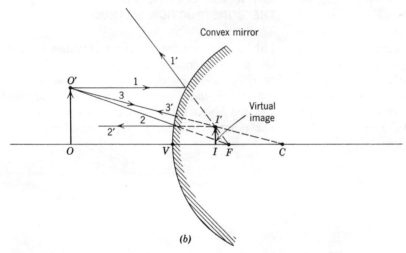

(b)

Figure 35-13 Image location by ray construction. The principal focus *F* is halfway between points *C* and *V*. (*a*) A real image formed by a concave mirror. (*b*) Convex mirrors form virtual images only.

of the mirror, which is at a distance of one radius from any point of the mirror. The principal axis, goes through the points *C* and *V*.

In order to locate the image of any object point, we need to trace any two rays from that point to the mirror. The point of convergence of the reflected rays is the image point. Three rays, rays 1, 2, and 3 in Fig. 35-13, are especially convenient. Ray 1 is parallel to the principal axis. Careful construction, conforming with the law of reflec-

tion (Eq. 35-1) will reveal that rays that are parallel to the principal axis (provided only that they fall within the paraxial region) intersect the principal axis at a single point after reflection. This point is called the *principal focus*, *F*, of the mirror. Ray 2 passes through the principal focus. All such rays will reflect along a line that is parallel to the principal axis. Rays passing through the center of curvature *C*, in Fig. 35-13a, will reflect back on themselves. (Why?) By drawing any two of the above rays from a few points of the object, the entire image can be constructed.

Example 35-3 An electric bulb, 2-in. high, is placed 15 in. in front of a concave spherical mirror, having a radius of curvature of 10 in. Determine, by the construction method, the location and size of the image formed. The focus is on the principal axis, 5 in. from the mirror.

Solution. Every dimension in Fig. 35-14 is scaled down to one-quarter size. Thus the mirror is an arc formed by a compass extended $2\frac{1}{2}$ in. The three rays drawn from the top of the bulb are the ones discussed above. Only two are necessary. The image distance q comes out by measurement with a ruler to be approximately 1.9 in. which, after scaling up, becomes 7.6 in. The image height is 1 in. ◀

35-8 IMAGE LOCATION; ANALYTICAL METHOD

The construction method gives approximate answers. While the approximations are good, provided that the construction is carried out carefully, the analytical method gives better answers. In Fig. 35-15, an object in the shape of an arrow is placed on the principal axis of a concave mirror at a distance p from the mirror. Two rays are traced from the top of the arrow, one ray passing through the center of curvature C, and the other heading toward the vertex V of the mirror. Right triangles $OO'V$ and $II'V$ are similar, since $\angle i = \angle r$ (from the law of reflection, Eq. 35-1). Therefore,

$$\frac{y}{y'} = \frac{p}{q} \tag{35-3}$$

Right triangles $OO'C$ and $II'C$ are also similar. Therefore,

$$\frac{y}{y'} = \frac{p - r}{r - q}$$

Since the right sides of the last two equations are equal to y/y', they are also equal to each other, or

$$\frac{p}{q} = \frac{p - r}{r - q}$$

Cross multiplication and rearrangement of terms yields

$$2pq = pr + rq$$

Dividing both sides of the equation by pqr, we get

$$\frac{2}{r} = \frac{1}{p} + \frac{1}{q} \tag{35-4}$$

This equation is a useful one, but it does not contain the important quantity, called the focal length of the mirror. Spherical mirrors are designated by

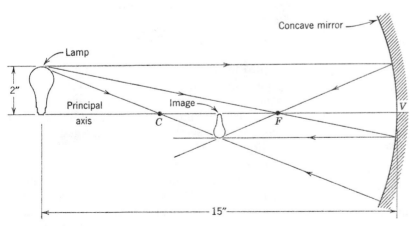

Figure 35-14

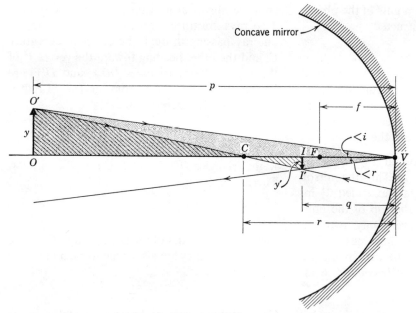

Figure 35-15   Construction for the derivation of the mirror equation.

either their focal length f or their radius of curvature. These quantities are related. The *focal length* is defined as the image distance for an infinitely far object. In other words, if p in Eq. 35-4 is allowed to approach infinity, $1/p$ approaches zero and q approaches f. Making these substitutions in Eq. 35-4 gives

$$\frac{2}{r} = 0 + \frac{1}{f}$$

Solving the last equation for f, we get

$$f = \frac{r}{2} \quad \text{or} \quad r = 2f \quad (35\text{-}5)$$

The radius of curvature of a spherical mirror is twice its focal length.

Substituting $2f$ for r in Eq. 35-4 leads to

$$\frac{1}{f} = \frac{1}{p} + \frac{1}{q} \quad (35\text{-}6)$$

This equation is referred to as the mirror equation.

35-9 SIGN CONVENTION

Concave mirrors form real or virtual images depending on the object distance. Convex mirrors form virtual images only. Real images are always inverted and are located on the same side of the mirror as the object. Virtual images are always erect with respect to the object and are on the opposite side of the mirror. The mirror equation, Eq. 35-6, can be used for both convex and concave mirrors, provided that appropriate signs are assigned to f, q, and r. The focal length f and the radius of curvature r are taken as *positive* for *concave* mirrors and *negative* for *convex* mirrors. The image distance q is taken as positive for real images, and negative for virtual images.

Equation 35-6 is also appropriate for plane mirrors because of the following considerations. A plane surface has an infinite radius of curvature. Therefore, by Eq. 35-5, the focal length of a plane mirror is also infinite and $1/f \rightarrow 0$. By substituting zero for $1/f$ in Eq. 35-6, we get

$$0 = \frac{1}{p} + \frac{1}{q} \quad \text{or} \quad p = -q$$

The image distance is equal to the object distance. The negative value for the image distance indicates that a plane mirror yields a virtual image.

Example 35-4 A lighted candle, 6 in. high, is placed at a distance of 20 in. from a concave spherical mirror having a radius of curvature of 5 in. Compute the image position and image height. Is it a real or virtual image?

Solution

$$p = 20 \text{ in.}, \, r = 5 \text{ in.}, \, y = 6 \text{ in.}, \, y' = ? \, q = ?$$

The focal length is one-half of the radius of curvature from Eq. 35-5:

$$r = 2f, \quad 5 \text{ in.} = 2f, \quad f = 2.5 \text{ in.}$$

From the mirror equation, Eq. 35-6,

$$\frac{1}{f} = \frac{1}{p} + \frac{1}{q}$$

$$\frac{1}{q} = \frac{1}{f} - \frac{1}{p} = \frac{1}{2.5} - \frac{1}{20}$$

$$q = +2.86 \text{ in.}$$

The plus sign indicates that the image is real and therefore inverted. From Eq. 35-3

$$\frac{y'}{y} = \frac{q}{p}$$

$$y' = \frac{yq}{p} = 6 \text{ in.} \times \frac{2.86 \text{ in.}}{20 \text{ in.}}$$

$$y' = 0.858 \text{ in.}$$

The image is smaller than the object. ◀

35-10 LATERAL MAGNIFICATION

The phrase, lateral magnification, refers to the ratio of image height, (y') to object height y. To be able to differentiate between real and virtual images in the expression for the magnification, M a minus is used.

$$M = -\left(\frac{y'}{y}\right) \tag{35-7}$$

Real images are always inverted with respect to the object. By adopting a sign convention which makes y' in Fig. 35-15 negative (it points down) and y positive (it points up) the magnification given by Eq. 35-7 will turn out positive for real images. Since virtual images are erect with respect to the object, by assigning the same sense, plus or minus to both y' and y, the magnification given by Eq. 35-7 will turn out to be negative for virtual images.

By Eq. 35-3, $y'/y = q/p$. This gives another equation for the magnification.

$$M = \frac{q}{p} \tag{35-8}$$

No minus is needed in this equation since by the sign convention discussed in Section 35-9, the image distance, q is taken as positive for real images and negative for virtual images. The object distance, p, is always taken as positive for a single mirror.

Example 35-5 A shaving mirror is usually a concave spherical mirror. It renders a magnified and erect image when the object, a portion of the face, is at a distance which is smaller than the focal length. If a certain shaving mirror has a focal length of 15 in., at what distance should the mirror be held from the face in order to give a magnification of 2? Locate the image.

Solution

$$f = 15 \text{ in.}, \, p = ? \, q = ? \, M = -2$$

Since the image is erect, it is virtual. By the sign convention, the magnification M is taken as negative. Therefore, $M = -2$ and by Eq. 35-8,

$-2 = q/p$ or $q = -2p$. From the mirror equation (Eq. 35-6)

$$\frac{1}{f} = \frac{1}{p} - \frac{1}{2p}; \quad \frac{1}{15 \text{ in.}} = \frac{1}{p} - \frac{1}{2p}$$

$$p = 7.5 \text{ in.}$$

Thus the part of the face that is to be magnified at twice its size should be placed at 7.5 in. from the center of the mirror.

Since $q = -2p$, $q = -15$ in. The minus indicates that the image is virtual, erect, and located 15 in. behind the mirror. ◀

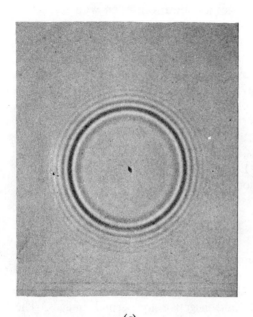

(a)

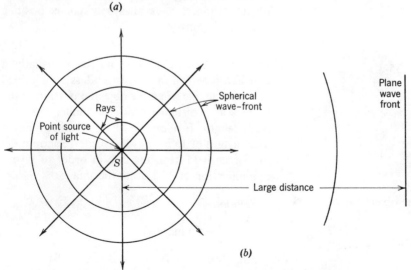

(b)

Figure 35-16 (a) Circular wave fronts in water. (b) Spherical light waves. At a large distance from the source, a small section of the wave front is plane (straight).

35-11 LAW OF REFLECTION-WAVE METHOD; HUYGENS' PRINCIPLE

One major test of the validity of the Wave Theory of light is its ability to explain the observed facts concerning light reflection. The law of reflection was experimentally deduced in Section 35-2. This law is derived in this section by means of waves.

A small pebble dropped in a still lake will cause a circular water wave to radiate out in an increasing diameter. Water waves can also be studied by dipping a small object like a marble at a constant frequency into a level tray of water. This will result in an array of concentric circles (waves) generated by the disturbance in the water from the point where the marble keeps hitting the water (see Fig. 35-16a). In an analogous manner, a point source of light issues spherical waves (see Fig. 35-16b), each wave front traveling away from the source and getting larger the farther it gets. At a large enough distance from the source (Fig. 35-16b), the wave front is so large that a small portion of it can be considered as a plane, which is represented as a line in the figure. The line representing the wave is called a *wave front*, and lines perpendicular to wave fronts are called *rays*.

A method based on Huygens' principle (Christian Huygens, seventeenth-century Dutch scientist) is followed in the construction of wave fronts. There are two parts to this principle. The first states in effect that every point on a wave front can be considered as a source for little, circular (really spherical) wavelets. Thus, in Fig. 35-17 the little arcs swung from any point on the big wave front represent the Huygens' wavelets. The second part of Huygens' principle deals with the new position of the wave front after it has traveled for a time t. The principle states, in effect, that a line tangent to the wavelets represents the new wave front.

Figure 35-18 depicts a plane wave front, AB, moving toward a mirror. The wave front comes from a point source of light S, issuing spherical waves that are pictured as circles in the diagram. At a great distance from S, a given wave front

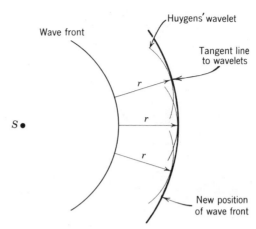

Figure 35-17   Huygen's construction.

has so expanded that a small portion of it is almost a straight line, depicted by AB in the figure. The three lines parallel to AB depict the same wave front as it approaches the mirror. We seek the position and shape of this wave front after it is reflected.

Point A of this front strikes the mirror before point B. In the time that it takes for B to reach the mirror, point A will have already been reflected, and will have traveled the same distance from the mirror as point B traveled toward the mirror, a distance of BD. Point A will be somewhere on the arc of radius BD (which represents a Huygens' wavelet) drawn from point A. In the meantime, the other points of the wave front will have advanced the same distance. Point E, for instance, will also have traveled a total distance of BD. Part of the distance, EG, is traveled toward the mirror, while the remainder, GF, has traveled away from the mirror. Point E then is located somewhere on the arc of radius GF drawn from point G. The line drawn from point D and tangent to the arcs represents the reflected wave front (by Huygens' principle).

In Fig. 35-18 the angle of incidence *for wave fronts* is the angle between the wave front and the mirror, angle i, and the angle of reflection is the angle between the reflected wave front and the mirror, angle r. These angles have the same

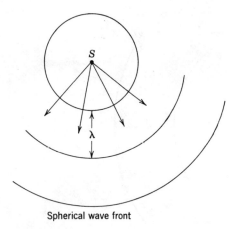

Spherical wave front

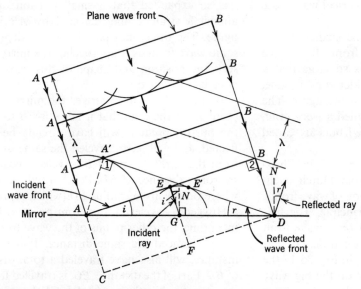

Figure 35-18   Law of reflection by Huygen's construction. The angles *i* and *r* in this diagram are, respectively equal to the angles of incidence and reflection as defined in Section 35-2.

value as the angles of incidence and reflection for rays as defined in Section 35-2.

Triangles $AA'D$ and BDA are congruent because lines AA' and BD are equal; line AD is common to both, and angles 1 and 2 are right angles. Angle 1 is 90° because it is between a

radius and a tangent line. Angle 2 is 90° because it is between a wave front and a ray. Therefore,

$$\angle i = \angle r$$

This is the law of reflection, as stated in Section 35-2.

SUMMARY

The angle of incidence equals the angle of reflection:

$$\angle i = \angle r$$

A real image of an object is real in the sense that light from the object is focused (assembled) at the vicinity of the image.

A virtual image of an object is virtual in the sense that *no* light from the object reaches the image.

Plane mirrors render virtual images only.

Concave mirrors render real and virtual images of objects depending on the object distance.

Convex mirrors render virtual images only.

The mirror equation

$$\frac{1}{f} = \frac{1}{p} + \frac{1}{q}$$

used with the appropriate sign convention is applicable to plane, concave, and convex mirrors.

The magnification of an image is the ratio of image height to object height.

Spherical aberration is a defect of spherical mirrors. This defect causes "fuzzy" images.

Parabolic mirrors correct for spherical aberration.

A point source of light issues expanding concentric spherical waves with the point source at the center. The law of reflection can be derived by Huygens' method of waves. This reinforces the wave theory of light.

QUESTIONS

1. In what sense is an image virtual or real?

2. If a screen is held at the position of a virtual image, can a person standing near the screen see this image? Explain.

3. If a screen is held at the position of a real image, can this image be seen on the screen?

4. What is meant by spherical aberration?

5. Is it possible to use a spherical mirror instead of a lens to take photographs? Explain.

6. Under what circumstances will a concave mirror render a virtual image?

7. What is meant by the focal length of a mirror?

8. How can the focal point of a certain concave mirror be experimentally determined?

9. What is the focal length of a plane mirror?

10. How does the wave theory of light explain the behavior of light as far as reflection is concerned?

PROBLEMS

(A)

1. Define the angles of incidence and reflection.

2. State and explain the law of reflection.

3. Explain the meaning of the mirror equation in words,

$$\frac{1}{f} = \frac{1}{p} + \frac{1}{q}$$

4. The angle of incidence of a ray incident on a plane mirror is 35°. Carefully sketch the incident ray, the normal ray, and reflected rays. Use a straight-edge and protractor.

5. An incident ray makes an angle of 60° with the plane of a mirror. Calculate:
   (a) The angle of incidence.
   (b) The angle of reflection.
   (c) The angle that the reflected ray makes with the plane of the mirror.

6. A person 70 in. tall stands 4 ft in front of a full-length plane mirror.
   (a) At what distance behind the mirror will he see his image?
   (b) How tall is his image?
   (c) Is the image real or virtual? Explain.
   (d) Explain the right-left reversal.

7. Calculate the image magnification in the preceding problem.

8. The radius of curvature of a certain spherical mirror is 1.5 m. Calculate the focal length of this mirror.

9. Check out the results obtained in Example 35-3 by the analytical method.

10. An upright arrow, 20 cm high, is placed on the principal axis and 1.5 m in front of a concave mirror

of 80-cm focal length. Determine by the construction method, the image
(a) height,
(b) location, and
(c) magnification.

11. Check your results of the answers to Problem 10 by the analytical method.

12. Repeat Problem 10 (by the analytical method) if the arrow is placed at a distance of 160 cm from the mirror.

13. Repeat Problem 10 (by the analytical method) if the arrow is placed at a distance of 50 cm from the mirror.

(B)

1. The focal length of a concave mouth mirror used by a dentist is 2.5 cm. What will be the magnification of a cavity's image when he places the mirror 2 cm from the cavity?

2. The filament of a clear electric lamp is 8 cm high. At what distance from a concave mirror should a screen be placed so that a 24-cm-high filament image appears on it. The focal length of the mirror is 40 cm.

3. Construct a ray diagram for Problem B2.

4. A straight arrow 5 cm high is placed upright on the principal axis and 50 cm from the vertex of a convex spherical mirror of 30 cm focal length. Construct a ray diagram for this setup.

5. Calculate for Problem B4:
(a) the image distance, and
(b) the image height.
(c) Describe the image.

6. Can a convex mirror yield a real image? Explain with the aid of a diagram.

7. The image of a 2-in. lamp filament on a wall 8 ft away is 5 in. high. Calculate the focal length of a concave mirror and the object distance for this to occur.

8. What size image will be produced by a concave mirror of 2-ft focal length when the object is an experimental balloon, 50 ft in diameter, at a height of 45 miles?

9. At what distance from a concave mirror should a candle flame be placed in order for its image to be the same size? What will then be the image distance?

10. An image of the moon is seen on a sheet of paper and is produced by means of a spherical-concave mirror having a radius of curvature of 24 in. Take the distance to the moon as 240,000 miles. How far from the vertex of the mirror should the paper be placed? (Do not go through extensive mathematics).

11. A girl holds a 2-in. mirror 12 in. from her eyes. The image of a boy standing at the entrance to the physics lab, 35 ft behind her, just covers the length of the mirror. How tall is he?

12. What type of mirror is needed in order to magnify a transparent picture seven times on a screen 12 ft from the mirror?

13. Assume the rear-view mirror of a convertible car, top down, is 6 in. wide and on line with, and one foot away from, the eyes of the driver. What angle of rear view does he have?

14. The angle of incidence for a particular ray striking a plane mirror is 15°. Through what angle will the reflected ray be turned if the mirror is rotated 20°?

15. A strong radioactive source is located at the bottom of a vertical hole. In order not to look straight down, a technician attaches a plane mirror at the top of the hole into which he can look horizontally. At what angle with the vertical should the plane of the mirror be attached so that he can see the source?

16. What is the minimum height of a plane mirror in order for a person to see his full height in it. *Hint*: Draw a plane mirror, with its top on a horizontal line with the person's eyes (assumed at the normal hairline for simplicity). Suitable rays entering the person's eyes and simple geometry should give the answer.

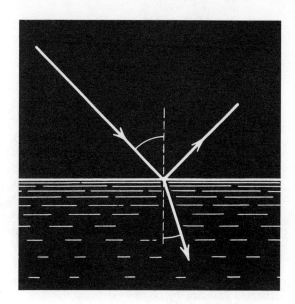

36.
Refraction

Light travels in a straight line when in one medium, like air or water or glass. When light moves from one medium into another, it bends, (refracts). This chapter discusses some of the principles associated with the refraction of light.

LEARNING GOALS

1. *To learn the meaning of the index of refraction.*
2. *To learn Snell's law and its applications.*
3. *To learn the equations associating the index of refraction with the speed of light, and wavelength.*
4. *To understand the meaning of the dispersion of light by a prism.*

36-1 INDEX OF REFRACTION; SNELL'S LAW

Light bends when it goes from one medium to another. In Fig. 36-1 a narrow beam is depicted as originating in air and directed toward a body of water. While in air, it travels in a straight line; while in the water it also travels in a straight line. Notice, however, that its path in the water is not a continuation of that in the air, and at the point of entrance into the water, the light bends abruptly. The bending of light when it goes from one medium into another is referred to as *refraction*. Notice that light also reflects from the boundary of two media. The direction of the reflected ray can be computed from the law of reflection, Section 35-2.

In Fig. 36-1, the perpendicular line N at the point of intersection between a ray and the two media is called the *normal*. The angle between the incident ray and normal is called the *angle of incidence* and the angle between the refracted ray and the normal is the *angle of refraction*. In general, when light goes from an optically rarer medium into a denser medium it bends toward the normal; when it goes from an optically denser medium into a rarer medium, it bends away from the normal.

The angle of incidence can be computed from a knowledge of the angle of refraction, or vice versa, provided a constant, called the absolute index of refraction, is known. The equation

relating these quantities is

$$\frac{\sin \theta_v}{\sin \theta_W} = n_W \qquad (36\text{-}1)$$

where θ_v is the angle between the ray in vacuum and the normal θ_W is the angle between the ray in the water and the normal, and n_W is a constant for the vacuum-water combination, called the *absolute index of refraction*, for water. The word *absolute* refers to the index of refraction of water with respect to vacuum, i.e., light traveling between the two media vacuum and water. If light were traveling between a vacuum and another material, e.g., glass, Eq. 36-1 would be expressed as

$$\frac{\sin \theta_v}{\sin \theta_g} = n_g \qquad (36\text{-}2)$$

where n_g is the absolute index of refraction for glass. Equations 36-1 and 36-2 can be generalized to cover light traveling between vacuum and any other medium by

$$\frac{\sin \theta_v}{\sin \theta} = n \qquad (36\text{-}3)$$

where θ is the angle between the normal and the ray in any medium, and n is the absolute index of refraction for that medium. Since the speed of light is almost the same in air as in vacuum, Eq. 36-3 can be used (with small error) for *vacuum or air* and any other medium. Equation 36-3 is referred to as Snell's law for refraction. Table 36-1 gives the refractive index for some substances.

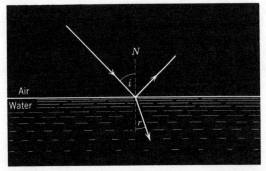

Figure 36-1   Light bends toward normal in going from air to water.

Table 36-1   Refractive indices of some materials (for $\lambda = 5893$ Å)

Vacuum	1.00 (exactly)
Air	1.00029
Alcohol, ethyl	1.36
Canada balsam	1.53
Carbon disulfide	1.64
Diamond	2.47
Glass, crown	1.51
Glass, flint	1.57–1.89
Water	1.33

Example 36-1   The angle between the beam in the air and the normal in Fig. 36-1 is measured to be 30°. The absolute index of refraction for water is 1.33. Calculate the angle between the beam in water and the normal.

Solution.   Since air is involved, Eq. 36-3 can be used.

$$\frac{\sin \theta_{air}}{\sin \theta_{water}} = n_{water}$$

or $$\sin \theta_{water} = \frac{\sin \theta_{air}}{n_{water}} = \frac{0.5}{1.33} = 0.375$$

By using trigonometric tables, θ_{water} is determined to be approximately 22°.   ◀

Example 36-2   A light source, submerged in water, issues a narrow beam going into the air above. Calculate the angle of refraction in the air, if the angle of incidence in the water is 22° and the absolute index of refraction of the water is 1.33.

Solution.   Notice that in this example, the light goes from water into air, which is the reverse of the situation in Example 36-1. Equation 36-3 applies to this situation as well.

$$\theta_{water} = 22°, \qquad n_{water} = 1.33, \qquad \theta_{air} = ?$$

By Snell's law,

$$\sin \theta_{air} = n_{water} \sin \theta_{water}$$
$$\sin \theta_{air} = 1.33 \times 0.375 \approx 0.5$$

By the use of trigonometric tables, θ_{air} is determined as 30°. Examples 36-1 and 36-2 illustrate the reversibility of the path of light when it undergoes refraction. In going from water to air as in Example 36-2, the beam will follow the same path as in going from air to water as in Example 36-1.   ◀

36-2 THE INDEX OF REFRACTION AND THE SPEED OF LIGHT

The index of refraction, n, of a medium is related to the speed of light v in that medium by the

equation

$$v = \frac{c}{n} \qquad (36-4)$$

where c is the speed of light in vacuum, 3×10^8 m/sec. Thus the speed of light in a type of glass of absolute index 1.5 is

$$v = \frac{3 \times 10^8 \text{ m}}{1.5 \text{ sec}} = 2 \times 10^8 \text{ m/sec}$$

In Fig. 36-2 a light ray goes from a medium M_1 having an absolute index of n_1 into a medium M_2 of index n_2, neither medium being a vacuum necessarily. The wave nature of light, as interpreted by Huygens' principle (refer to Section 35-11) leads to two important relations:

$$\frac{\sin \theta_1}{\sin \theta_2} = \frac{v_1}{v_2} \qquad (36-5)$$

and $$n_1 \sin \theta_1 = n_2 \sin \theta_2 \qquad (36-6)$$

These equations are developed in Section 36-7. Notice that Eq. 36-6 applies to light traveling between *any two* media, whereas Snell's law as stated by Eq. 36-3 applies to a pair of media, one of which must be a vacuum (or air). Equation 36-6 is a more general expression of Snell's law.

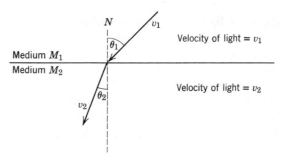

Figure 36-2 $V_1 > V_2$ since $\theta_1 > \theta_2$.

Equations 36-5 and 36-6 can be combined to yield

$$\frac{v_1}{v_2} = \frac{\sin \theta_1}{\sin \theta_2} = \frac{n_2}{n_1} \qquad (36-7)$$

The fraction n_2/n_1 is referred to as the *relative index of refraction* of medium M_2 with respect to medium M_1. If M_1 happens to be a vacuum, then the relative index of M_2 with respect to M_1, i.e., n_2/n_1, is equal to n_2, since n for the vacuum is equal to 1, by definition. In common usage, the phrase "index of refraction" implies the absolute index of refraction. This usage will be followed in the following sections.

Let us arbitrarily assume that the medium, M_1 in Fig. 36-2 is air and medium M_2 is glass. Experiments indicate that light bends toward the normal when entering glass from air. Therefore, $\theta_1 > \theta_2$, $\sin \theta_1 > \sin \theta_2$ and $\sin \theta_1/\sin \theta_2 > 1$. By Eq. 36-5

$$\frac{\sin \theta_1}{\sin \theta_2} = \frac{v_1}{v_2}$$

Therefore, $v_1/v_2 > 1$, from which it follows that $v_1 > v_2$. Therefore, the velocity of light in medium M_1, which is air, is greater than in M_2, which is glass. We can generalize Fig. 36-2 to depict any two media. If the angle θ_1 is larger than θ_2, the velocity of light is greater in medium 1 than in medium 2. Also the index of refraction of medium 1 is smaller than that of medium 2.

> **The lower the index of refraction of a medium the higher is the speed of light in that medium.**

Example 36-3 A light beam originating in air (Fig. 36-3) is incident on water contained in a thick-bottomed glass vessel, at an angle of incidence of 70°. After continuing through the water and glass bottom, the beam finally emerges into air again. Take the index of refraction of water as 1.33 and that for the glass as 1.5, and compute the angle at which the beam emerges.

Solution. We first calculate the angle of refraction in the water, θ_2.

$$n_1 \sin \theta_1 = n_2 \sin \theta_2, \text{ from Eq. 36-6, and}$$
$$n_1 = 1, \quad \theta_1 = 70°, \sin \theta_1 = 0.94,$$
$$n_2 = 1.33, \quad \theta_2 = ?$$

$$\sin \theta_2 = \frac{1 \times 0.94}{1.33} = 0.707$$

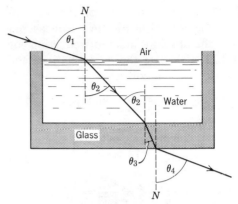

Figure 36-3 Refraction in media having parallel surfaces. The angles θ_1 and θ_4 are equal no matter how many intervening media exist.

From the trigonometric tables, θ_2 is determined as 45°. The two angles θ_2 in the water (Fig. 36-3) are equal because they are alternate interior angles of the two normals, which are parallel. We next calculate θ_3, the angle of refraction in the glass.

$$n_2 \sin \theta_2 = n_3 \sin \theta_3, \text{ from Eq. 36-6, and}$$
$$n_2 = 1.33, \quad \theta_2 = 45°; \quad \sin \theta_2 = 0.707,$$
$$n_3 = 1.5, \quad \theta_3 = ?$$

$$\sin \theta_3 = \frac{1.33}{1.5} \times 0.707 = 0.63$$

Therefore, $\theta_3 = 39°$ (approximately)

Finally, to calculate θ_4, we again use Eq. 36-6, and $n_3 = 1.5$,

$$\theta_3 = 39°, \quad \sin \theta_3 = 0.63, \quad n_4 = 1,$$
$$\text{and} \qquad \theta_4 = ?$$

$$\sin \theta_4 = \frac{1.5}{1} \times 0.63 = 0.94$$

$$\theta_4 = 70°$$

This example could have been solved by a shortcut method:

$$n_1 \sin \theta_1 = n_2 \sin \theta_2$$
$$n_2 \sin \theta_2 = n_3 \sin \theta_3$$
$$n_3 \sin \theta_3 = n_4 \sin \theta_4$$

Therefore, $n_1 \sin \theta_1 = n_4 \sin \theta_4$

The angle θ_4 can be solved for in the last equation, since the other quantities are given. Note that the ray in the above example leaves from the bottom surface at the same angle that it enters the top surface, i.e., the emergent ray from the bottom surface is parallel to the incident ray at the top surface. This is true no matter how many layers of different media are traversed, provided that all surfaces of the media are parallel and the initial and final media are the same.   ◀

Example 36-4   (a) Calculate the velocity of light in the water and in glass in Example 36-3 from the data given in that example. (b) Calculate the relative index of refraction of the glass with respect to the water, and of the water with respect to the glass.

Solution

(a) For the air–water combination, by Eq. 36-7,

$$\frac{v_1}{v_2} = \frac{n_2}{n_1}$$

The index of refraction of air (n_1) is 1 and for water (n_2) is given as 1.33. The velocity of light in air is v_1, which is taken as c, or 3×10^8 m/sec. Therefore,

$$v_2 = \frac{1 \times 3 \times 10^8 \text{ m/sec}}{1.33}$$

$$= 2.26 \times 10^8 \text{ m/sec   (in the water)}$$

For the water-glass combination:

$$\frac{v_2}{v_3} = \frac{n_3}{n_2} \quad \text{(by Eq. 36-7)}$$

$$v_3 = \frac{1.33 \times 2.26 \times 10^8 \text{ m/sec}}{1.5}$$

$$= 2.0 \times 10^8 \text{ m/sec   (in the glass)}$$

(b) The relative index of glass with respect to water is, by definition,

$$\frac{n_{\text{glass}}}{n_{\text{water}}} \quad \text{or} \quad \frac{1.5}{1.33} = 1.13$$

The relative index of water with respect to glass is, by definition,

$$\frac{n_{\text{water}}}{n_{\text{glass}}} \quad \text{or} \quad \frac{1.33}{1.5} = 0.886 \quad \blacktriangleleft$$

36-3 INDEX OF REFRACTION AND WAVELENGTH

A given wavelength of light becomes shorter in an optically denser media i.e., in a media of higher refractive index. This statement can be proved by assuming monochromatic light wavelength λ_A in air to travel into an optically denser medium of refractive index n_m. From Eq. 36-4, the velocity v_m of the light in the denser medium is given by $v_m = c/n_m$. But from the wave equation, $v_m = f\lambda_m$ and $c = f\lambda_A$, where λ_m is the wavelength of this light in the denser medium and f is the frequency of the light, which is the same for both media because f is a function of the light emitter only. Substituting the expression for v and c given by the last two equations into Eq. 36-4 gives

$$f\lambda_m = \frac{f\lambda_A}{n_m}$$

or

$$\lambda_m = \frac{\lambda_A}{n_m} \tag{36-8}$$

Thus λ_m is a smaller number than λ_A because n_m is larger than one.

The wavelength of monochromatic light is longest in vacuum (or air); it is shorter in media having a higher index of refraction.

Example 36-5   The wavelength in air of a certain shade of red light is 6800 Å. What is its wavelength in glass of refractive index 1.5?

Solution   From Eq. 36-8, $\lambda_{\text{glass}} = 6800\,\text{Å}/1.5 = 4533\,\text{Å} = 4533 \times 10^{-10}$ m. When this light gets back into air again, its wavelength returns to 6800 Å.   ◀

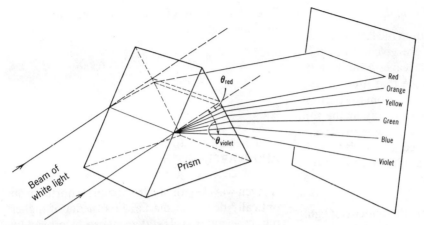

Figure 36-4   Dispersion of white light by a glass prism.

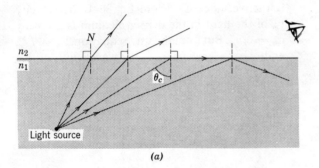

(a)

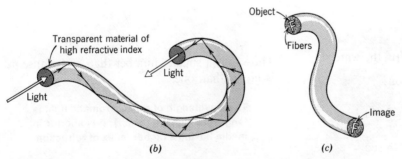

(b)   (c)

Figure 36-5   (a) The critical angle is that angle of incidence for which the angle of refraction is 90°. (b) The light "pipe", multiple internal reflections "pipe" the light from one end to the other through some plastic or glass rods. Some plastic fibers exhibit the same property. (c) A bundle of interwoven fibers transmits an image from one end to the other. One medical application of fiber optics involves the seeing of parts of the stomach through a long flexible fiber-bundle that is inserted through the mouth.

36-4 DISPERSION OF LIGHT BY A PRISM

White light entering a clear prism will emerge separated into its component colors (Fig. 36-4). The separation of light into its component wavelengths is called *dispersion*. The phenomenon of dispersion indicates that the index of refraction of a certain transparent material depends on the material *and on the wavelength of light traversing it*. This can be shown to be so by noting in Fig. 36-4 that the angle of bending from the original path, angle θ, increases for decreasing wavelengths. For instance, θ_{red} within the glass is smaller than θ_{violet}. Since the more a ray is bent upon entering a medium, the larger is the index of refraction of the medium, the index for the wavelength corresponding to the color violet is greater than that for red. Also, since the ratio of the velocities of light depends inversely on the indexes of refraction, $v_1/v_2 = n_2/n_1$ (Eq. 36-7), the velocity of red light is greater in glass than the velocity of violet light in the same glass. This does not negate the fact that the velocities of all colors of light are the same in vacuum.

36-5 TOTAL REFLECTION—CRITICAL ANGLE

Figure 36-5a depicts a submerged light source in a transparent medium of refractive index n_1 issuing light rays in all directions, some of which head toward the upper medium of refractive index n_2, n_2 being smaller than n_1. Not all the rays heading toward the upper medium will emerge into that medium. Those rays having an angle of incidence equal to, or greater than, a particular angle of incidence, called the *critical angle*, will not get out. The eye in Fig. 36-5 will not see the submerged light source. Notice from this figure that the angle of refraction in the upper medium increases as the incident angle in the lower medium increases. The critical angle is defined as that angle of incidence for which the refracted angle is 90°. Letting θ_1 in Eq. 36-6 equal the

critical angle, θ_c, and θ_2 equal 90°, we get
$$n_1 \sin \theta_c = n_2 \times 1$$
or
$$\sin \theta_c = \frac{n_2}{n_1} \qquad (36\text{-}9)$$

Rays having an angle of incidence in the optically denser medium that is larger than θ_c will be reflected from the boundary of the two media as from a mirror, according to the law of reflection.

Example 36-6   Calculate the critical angle for crown glass. Take the index of refraction for the glass as 1.5.

Solution.   By Eq. 36-9
$$\sin \theta_c = \frac{n_{air}}{n_{glass}} = \frac{1}{1.5} = 0.667$$
$$\theta_c = 42° \quad \text{(approximately)} \qquad \blacktriangleleft$$

36-6 APPARENT DEPTH OF SUBMERGED OBJECTS

A submerged object in water, or glass, will appear less deep than it really is. The reason can be understood from the ray diagram of Fig. 36-6, which depicts two light rays emanating from a speck of dust embedded at a depth h in glass of refractive

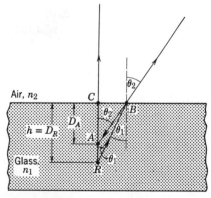

Figure 36-6   Apparent depth.

index n_1. The normal ray in this figure will emerge into the air undeviated. The other ray will bend away from the normal, since it goes into a medium of lower refractive index. An eye, intercepting these rays, will "follow" them backward to their point of intersection and interpret the speck of dust to be situated at that point, which seems to be less submerged than it really is. We can compute the apparent depth D_A from the following considerations. Since the eye can only intercept the rays leaving the speck of dust at small angles from the vertical, θ_1 and θ_2 are very small. For small angles, $\sin \theta = \tan \theta$. Therefore,

$$\sin \theta_1 = \tan \theta_1 = \frac{BC}{D_R}$$

$$\sin \theta_2 = \tan \theta_2 = \frac{BC}{D_A}$$

provided that θ_1 and θ_2 are small. Substituting these values in Eq. 36-6,

$$\frac{n_1 BC}{D_R} = \frac{n_2 BC}{D_A}$$

or
$$D_A = D_R \times \frac{n_2}{n_1} \qquad (36\text{-}10)$$

The apparent depth D_A must be smaller than the real depth D_R, because the fraction n_2/n_1 is less than one. Also, since submerged object seem less submerged than they really are, they seem nearer to the eye and, therefore, larger. The diameter of a penny submerged in water will seem larger than it really is.

Example 36-7   A dime is submerged in water at a known depth of 2 ft. How deep does it seem when viewed from directly above? Take the refractive index of the water to be 1.3.

Solution.   From Eq. 36-10 the apparent depth D_A is given by

$$D_A = 2 \text{ ft} \times \frac{1}{1.3} = 1.54 \text{ ft} \qquad \blacktriangleleft$$

36-7 HUYGENS' CONSTRUCTION FOR REFRACTION

The important equations concerning refraction (Eq. 36-6) are derived from considerations that light behaves as a wave. As in Fig. 35-18, a point source of light S, which is located in *any* medium M_1 issues expanding spherical waves advancing toward any other medium M_2. At a great distance from the source, a small portion of the wave front can be considered straight, which is indicated as line AB in Fig. 36-7a. It is desired to determine the position of this portion of the wave front at a time t that it takes for point B on it to travel to point C, a distance of BC. This is accomplished by drawing Huygens' wavelets according to Huygens' principle, as discussed in Section 35-11 above. A straight line tangent to these wavelets represents the new position of the wave front.

Point A in Fig. 36-7a is about to enter medium M_2. Assuming that the velocity of light v_2 in M_2 is smaller than the velocity v_1 in medium M_1, point A will travel a distance AA' in M_2, which is shorter than the distance BC. The distance BC can be expressed as $v_1 t$, and the distance AA' as $v_2 t$. Some consideration will indicate that the distance $AA' = v_2 t$ is also equal to $BC \times v_2/v_1$. An arc of this radius is drawn from point A, this arc representing a Huygens' wavelet, originating from point A.

Similarly, any other point, like point F on the wave front AB, will travel to point Q, and then start slowing up. If the second medium were not there, it would reach point S. The partial distance, QS, that it would have traveled without the second medium is reduced to $QS \times v_2/v_1$. An arc of this radius is drawn from point Q, this arc representing another Huygens' wavelet. We now have two wavelets. A straight line from point C, tangent to the two arcs gives, by Huygens' principle, the new wave front $A'C$ after a time t.

In right triangle ABC,

$$\sin \theta_1 = \frac{v_1 t}{AC}$$

In right triangle $AA'C$,

$$\sin \theta_2 = \frac{v_2 t}{AC}$$

Dividing the first equation by the second equation, we find that

$$\frac{\sin \theta_1}{\sin \theta_2} = \frac{v_1}{v_2} \qquad (36\text{-}5)$$

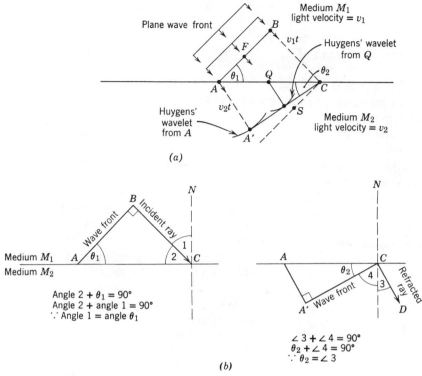

(a)

(b)

Figure 36-7 Derivation of Snell's law from Huygen's principle. (a) A plane wave front AB from a distant point source (not shown) is about to enter another medium, M_2. This wave front becomes $A'C$ in M_2. (b) The angles θ_1 and θ_2 are respectively equal to the angles of incidence and refraction because $\angle \theta_1 = \angle 1$ and $\angle 1$ is the angle between an incident ray and normal; $\angle \theta_2 = \angle 3$ and $\angle 3$ is the angle between a refracted ray and normal.

Notice that θ_1 is the angle between the wave front and the second medium. This angle is shown by Fig. 36-7b to be the angle of incidence, as defined in Section 36-1. Similarly, θ_2 is shown to be angle of refraction in medium M_2.

Since medium M_1 can be any medium, let it be a vacuum. In that case, the last equation above becomes

$$\frac{\sin \theta_{\text{vacuum}}}{\sin \theta_2} = \frac{v_{\text{vacuum}}}{v_2} = \frac{c}{v_2} = n_2 \text{ where } n_2 \text{ is a constant}$$

called the absolute index of refraction of medium M_2. From these equations, we get an important relation between the absolute index of refraction n of any medium and the speed of light v in that medium,

$$v = \frac{c}{n}$$

Substituting for v_1, c/n_1, and for v_2, c/n_2, in Eq. 36-5, we get

$$\frac{\sin \theta_1}{\sin \theta_2} = \frac{c/n_1}{c/n_2}$$

or

$$n_1 \sin \theta_1 = n_2 \sin \theta_2 \qquad (36\text{-}6)$$

SUMMARY

Light bends when it travels from one medium into another. Snell's law predicts the amount of bending.

The index of refraction for a transparent material is larger the more the light bends when entering the material from a vacuum (or air).

The velocity of light is smaller in materials having a higher index of refraction.

Dispersion refers to the separation of white light into its component colors (wavelengths).

The velocity of light in a medium other than vacuum is not a constant but depends on wavelength.

Submerged objects in water or glass appear less deep than they really are.

The important equations in refraction are derived from Huygens's principle, which treats light as waves. This lends strong support for the wave theory of light.

QUESTIONS

1. Under what circumstances will light, coming from water, not get into the air?

2. Is the speed of light in a medium (excluding a vacuum) a constant? Explain.

3. Why can the sun be seen for a minute or so after it sets?

4. Explain a mirage in a hot desert?

5. Explain why a straight stick, partially dipped obliquely in water, will appear "broken."

6. Why does an object that is submerged in water appear at a lesser depth than it really is?

7. Suppose that the situation is reversed and the eye is submerged in a pool of water viewing an object in air directly above. Will the object seem nearer than it really is?

8. Compare the speed of red light with that of green light in vacuum.

9. Compare the speed of red light with that of green light in glass.

10. Try this experiment: Place a cube of glass on a table and a penny near it. The penny cannot be seen through the top surface of the cube no matter where you place your eye. Why not?

PROBLEMS

(A)

1. A narrow beam of light enters obliquely into air from water. Will the beam bend toward or away from the normal?

2. A narrow beam of light enters obliquely into glass ($n = 1.5$) from water ($n = 1.33$). Will the beam bend toward or away from the normal?

3. At what angle must a beam enter any medium in order for it not to bend at all?

4. A pencil of light enters from air into water. The angle in air between the light pencil and the surface of the water is 37°. Calculate:
   (a) The angle of incidence.
   (b) The angle of refraction. Take the index of refraction of water to be 1.33.

5. An underwater lamp in a swimming pool projects a narrow beam that gets out into the air above. If the angle of incidence is 37°, what is the angle of refraction? The refractive index for water is 1.33.

6. Calculate the speed of light in diamond. The index of refraction of diamond is 2.47.

7. A narrow beam of light enters material A from material B having a higher index of refraction.
   (a) Will the light bend toward or away from the normal.
   (b) In which material is the speed of light greater?

8. The index refraction is 1.33 for water and 1.45 for benzene. In which of these materials is the speed of light greater?

9. A certain material has an index of refraction of 1.8. What is the ratio of the speed of light in this material to the speed of light in vacuum?

10. A certain shade of red light has a wavelength of 7000 Å in air. What is its wavelength while moving in glass with an index of refraction of 1.5.

11. A narrow beam of sunlight enters one side of a thick vertical window, ($n = 1.5$) at an angle of 30° with

the vertical. At what angle does the beam emerge from the other side?

12. A beam of light enters a container of carbon disulfide at an angle of incidence of 60°. What is the angle of refraction? Take the index for CS_2 as 1.63.

The speed of light in a certain type of glass is 120,000 miles/sec. Find the index of refraction for that type of glass.

(B)

1. The angle of incidence of a light beam going from water into quartz glass is 65°. Find the speed of light in this type of glass if the angle of refraction in the glass is 51° and the velocity of light in water is 2.26×10^8 m/sec.

2. Light, in going from carbon disulfide ($n = 1.63$) into glass ($n = 1.5$) has an angle of incidence of 35°. Find the angle of refraction in the glass.

3. Yellow light enters a triangular crown glass 60-60-60 prism, at an angle of incidence of 30°. What angle will the emerging light make with a side of the prism? Use Table 36-1 to determine the index.

4. Calculate the critical angle for light passing from glass of refractive index 1.5 to a liquid of refractive index 1.3.

5. What is the apparent depth of an object that is submerged 10 ft in a lake? The object is viewed vertically.

6. Compare the velocities of monochromatic light in media of different indices. The frequencies. The wavelengths.

7. The wavelength of a certain color is 5000 Å in air. Calculate the wavelength in glass ($n = 1.5$).

8. A plane mirror, face up, rests at the bottom of a water-filled tank, 15 in. deep. A submerged small object is 5 in. immediately above the mirror. Calculate the distance of the object's image from the water level, as seen by an observer directly above.

9. A front-surfaced plane mirror lies flat under a layer of clear water $\frac{1}{2}$ in. thick. An observer looks down at the mirror image of a fingertip $\frac{1}{4}$ in. within the water. Calculate his image distance.

10. A layer of water ($n = 1.33$) in a glass jar ($n = 1.5$) is separated from a layer of carbon disulfide ($n = 1.63$) by a layer of glass, with the water at the bottom. A ray of light traveling up in the jar strikes the glass layer at an angle of incidence of 20°.
    (a) Calculate the angle with which the light entered the jar from the bottom.
    (b) Trace a ray of light through the jar.
    (c) What is the angle of the ray of light in the CS_2 layer.
    (d) What is the exit angle?

11. Light passes from air to oil ($n = 1.5$) in a glass beaker ($n = 1.4$), through the bottom of the beaker, and out into the air. If the angle of incidence on the oil is 15°, calculate the angle of emergence.

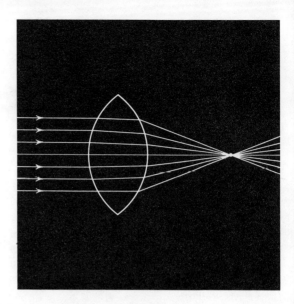

37.
Thin Lenses—
Optical Instruments

All optical instruments make use of either mirrors or lenses, and sometimes both. The underlying principle of instruments involving lens systems such as the camera, microscope, or indeed the eye itself, is refraction.

LEARNING GOALS

1. *To learn the ray-diagram method for determining the location of images formed by convex and concave lenses.*

2. *To learn the lens equations and their applications.*

3. *To gain an understanding of how some principles in optics apply to the eye.*

4. *To understand how the principles in optics are applied to important optical instruments such as the camera, magnifying glass, compound microscope, and telescope.*

5. *To learn about an important scientific instrument, the spectroscope, and its uses.*

37-1 SPHERICAL LENSES

Lenses can be better understood by first considering triangular prisms. Light rays incident on one side of a transparent prism will emerge on the other side bent toward the base of the prism (Fig. 37-1). Let I in this figure refer to an incident ray having an angle of incidence θ_1. Remembering that light will bend toward the normal in going from air to glass, we see that the ray within the glass is bent toward the base. Upon emerging into the air again, the ray will bend some more toward the base because it is going from an optically denser to a rarer medium. A quantitative determination of the direction of the emergent ray can be obtained from $n_1 \sin \theta_1 = n_2 \sin \theta_2$ (Eq. 36-6).

In Fig. 37-1 we considered the path of a ray through a single prism. If we allow many rays to be incident on a pair of prisms, placed base to base and arranged as in Fig. 37-2, two things will be accomplished. First, more rays from an object will be bent toward the line (actually the surface) joining the prism pair. In addition, these rays are caused to become concentrated into a relatively small area I. The small area can be considered as an image of the point O on the object but not a very good image. The reason why it is not very good is because it does not depict a point, while the object is a point. Imagine a photographic film placed in the vicinity of I and the point source replaced by two grains of sand near each other. A good image on the film should show *two* spots near each other. But the film will show only one spot. The relatively large images of both grains merge into a single image. Thus the two prisms would not be suitable as a camera lens.

By shaping the outer surfaces of the prisms into smooth spherical surfaces (Fig. 37-3a) a spherical lens results, which images point objects as almost point images. A spherical lens, which is thicker in the center than at its upper and lower edges, will form real images. Such a lens is called either a *convex, converging,* or a *positive* lens. A lens that is thinner at the center than at its upper and lower edges (Fig. 37-3b) will form virtual images only. Such a lens is called a *concave, diverging,* or a *negative* lens.

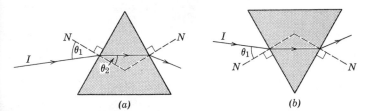

Figure 37-1   Emergent ray from a prism bends toward its base.

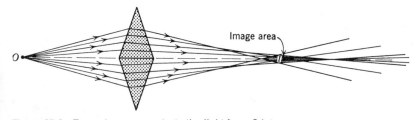

Figure 37-2   Two prisms concentrate the light from O into a small image area.

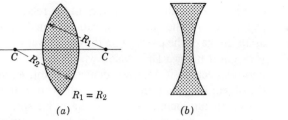

Figure 37-3 (a) Convex lens with surfaces of equal curvature. (b) Concave lens. (c) Convex lens with surfaces of unequal curvature.

37-2 FOCAL POINT; FOCAL LENGTH

Lenses are identified by their *focal length*, which can be explained with the aid of Fig. 37-4. A spherical lens has two outer spherical surfaces and the *principal axis* is the straight line going through the center of curvature of both surfaces. Rays parallel to the principal axis striking a convex lens at one side will converge at a point F on the other side called the *focal point*. The distance, f, from the center of the lens to the focal point is the *focal length* of the lens. Since the bundle of parallel rays must originate from an infinitely distant (very far) object, the focal length of the lens can be taken as the distance along the principal axis, from the center of the lens to

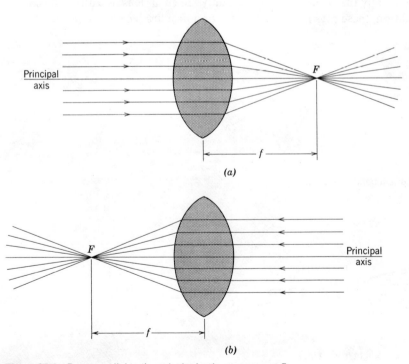

Figure 37-4 Rays parallel to the principal axis converge at F.

the image it forms of an infinitely distant object on the principal axis. The focal length of a given *thin* convex lens can be experimentally determined by forming a sharp image on a piece of cardboard of a distant body like the moon. The measured distance from the lens, along the principal axis, to the image can be taken as the focal length of this lens. A lens can be considered to be thin if its thickness is small compared with its focal length.

37-3 LOCATION OF IMAGES BY RAY DIAGRAMS

The position of lens images can be determined by construction. In Fig. 37-5a three *principal rays* emanating from the top point of the upright arrow, which is considered as the object, are traced through the thin convex lens. *The analyses that follow apply to thin lenses.* Ray 1 is drawn

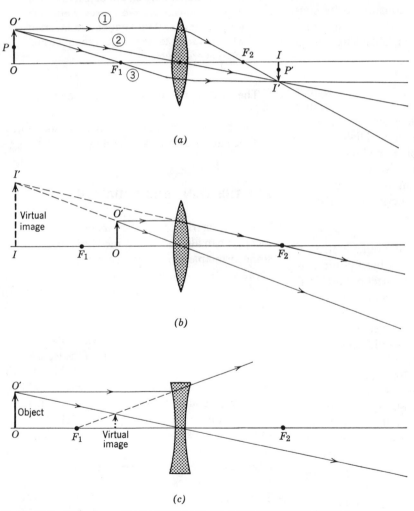

Figure 37-5 Image formation by thin lenses. (*a*) A converging lens forms a real image if the object distance is greater than the focal length and (*b*) a virtual image if object distance is shorter than the focal length. (*c*) A diverging lens forms virtual images only.

parallel to the principal axis. All rays parallel to the principal axis on the object side of the lens converge at the focal point F_2 on the image side. The point F_2 is referred to as the second focal point, to distinguish it from the first focal point F_1, on the other side of the lens. Both F_1 and F_2 are at the distance of one focal length from the lens, showing that it does not matter which side of the lens is exposed to the object. The rays from a given point on an object at F_1 will emerge parallel on the other side of the lens and will not converge. Since an image is constituted by convergent rays, a lens will not form an image of a body placed at a distance of one focal length from it.

Ray 2 in Fig. 37-5a is drawn through the center point of the lens. All rays going through the center of a thin lens emerge unbent. This is so because the two small elements of lens surfaces near its center are parallel to each other, and light rays go through parallel surfaces of a medium unbent (refer to Ex. 36-3). Ray 3 is drawn through the first focal point F_1. Rays going through the focal point on one side of the lens emerge parallel to the principal axis on the other side.

The intersection of any two rays from an object point forms the image of that point. Notice in Fig. 37-5a that the image of the object point O' is at I' and that the image of the object point O is at I. We can construct image points of other points on the object. For instance, if any two of the principal rays mentioned above, were traced from point P on the object, these would converge at point P'. By joining such image points as I' and P', an inverted arrow I-I' results, which represents the inverted image of the object O-O'.

In Fig. 37-5b the object is placed at a distance from the convex lens that is smaller than its focal length. Notice that the rays from a point on the object such as O' diverge as they emerge from the lens. To determine the image location of this point, the divergent rays are traced backwards to their point of intersection, I'. This point is the *virtual image* of the point O' because the lens does not actually project any rays to I'. An eye placed

at the right side of the lens in the Fig. 37-5b will see the magnified virtual image, I'-I.

In Fig. 37-5c the object is placed in front of a concave lens. Rays from any point on the object diverge after passing through a concave lens. The image of O' is located by tracing backward any two divergent rays to their point of intersection. A concave lens forms only virtual images.

Convex lenses yield either real or virtual images, depending on the object distance from the lens. Concave lenses yield only virtual images. Real images are inverted with respect to the object and virtual images are erect.

The size of the image and its height can be determined if the ray diagram is drawn to scale, i.e., if the focal length, object height, and object distance are drawn to a suitable scale.

37-4 THE THIN LENS EQUATION

The characteristics of the image, its size, and location can also be determined analytically from some principles in plane geometry. Figure 37-6 depicts a thin spherical convex lens, having radii of curvature that are not necessarily equal. An object O-O', in the form of an upright arrow, is placed a distance p from the lens. Ray 1, which is parallel to the principal axis, will pass through the focal point on the image side of the lens. Ray 2, which is drawn through the center of the lens, will emerge unbent. The point of intersection of these rays, at a distance q from the lens, constitutes the image point I' of the object point O'. The lens is so thin that it does not practically matter from which side of it p or q is measured. The heights of the object and image are, respectively, h and h'.

Right triangles AOO' and AII' are similar. Therefore,

$$\frac{h'}{h} = \frac{q}{p}$$

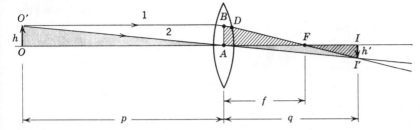

Figure 37-6   Construction for derivation of thin lens equation. The image height h' is taken as negative because it is directed in the negative y direction.

Triangles $ABF*$ and FII' are similar. Therefore,

$$\frac{h'}{h} = \frac{q - f}{f}$$

Combining the last two equations, we get

$$\frac{q}{p} = \frac{q - f}{f} = \frac{q}{f} - 1$$

Dividing by q and rearranging the terms, we get the *thin lens equation,*

$$\frac{1}{p} + \frac{1}{q} = \frac{1}{f} \qquad (37\text{-}1)$$

The lens equation can be used for both concave and convex lenses, provided that the following sign convention is followed:

1. The focal length f is positive for converging lenses and negative for diverging lenses.

2. The object distance p is positive when a single lens is used.

3. The image distance q is positive for real images and negative for virtual images.

Example 37-1   A 6-in.-high object is placed 24 in. from a 6-in.-focal-length converging lens. Find the position and height of the image.

* Actually, the line BI' is not straight because it bends as it emerges from the lens. But since this is a thin lens, points B and D are very near each other and line BI' can be considered as straight.

Solution

$f = +6$ in., $\quad p = +24$ in., $\quad h = 6$ in., $\quad q = ?$

$$\frac{1}{f} = \frac{1}{p} + \frac{1}{q} \quad \text{(lens equation)}$$

$$\frac{1}{q} = \frac{1}{6 \text{ in.}} - \frac{1}{24 \text{ in.}} = \frac{3}{24 \text{ in.}}$$

$$q = +8 \text{ in.}$$

The plus indicates that the image is real. From Fig. 37-6

$$\frac{h'}{h} = \frac{q}{p}$$

$$h' = \frac{6 \text{ in.} \times 8 \text{ in.}}{24 \text{ in.}} = 2 \text{ in.}$$

The height of the image is 2 in.; since it is a real image, it is inverted with respect to the object.   ◀

37-5 LATERAL MAGNIFICATION

The lateral magnification m of a lens is given by the negative ratio of the image height h' to the object height h or

$$m = -\left(\frac{h'}{h}\right) \qquad (37\text{-}2)$$

If h, in Fig. 37-6, is taken as positive (it points up) and h' is taken as negative, (it points down), the magnification will be positive for real images and negative for virtual images. Also, from Fig. 37-6

$h'/h = q/p$. The magnification can, therefore, also be expressed in terms of the object distance, p and the image distance q by the equation

$$m = \frac{q}{p} \qquad (37\text{-}3)$$

Since, by the sign convention, q is positive if the image is real and negative if the image is virtual, the magnification, as expressed by Eq. 37-3 will also turn out to be positive for real images and negative for virtual images. When a combination of lenses is used, the total magnification of the combination is equal to the product of the individual magnifications.

Example 37-2   An object in the form of an illuminated arrow is located at a distance of 10 cm from a thin converging lens having a focal length of 20 cm. (a) Calculate the distance of the arrow's image from the lens. Describe the image. (b) what is the image height if the height of the arrow is 8 cm?

Solution

(a) $p = 10$ cm and positive (sign convention)

$f = 20$ cm and positive (sign convention)

Substituting in Eq. 37-1, we get

$$\frac{1}{q} = \frac{1}{20} - \frac{1}{10}, \qquad q = -20 \text{ cm}$$

The minus sign indicates that the image (1) is virtual, (2) falls on the object side of the lens, and (3) is erect.

(b) From Eq. 37-3, $m = q/p$ where $p = 10$ cm and $q = -20$ cm.

$$m = \frac{-20}{10} = -2$$

The minus magnification indicates that the image is virtual. The virtual image is twice as high as the object and, therefore, 16 cm high. ◀

Example 37-3   The glowing vertical filament of a hanging incandescent lamp is 3 cm long and at a horizontal distance of 50 cm from a diverging lens having a focal length of 20 cm. (a) Locate the filament's image. (b) Determine the magnification and the size of the image. Describe the image.

Solution

(a) Since the lens is diverging, its focal length is taken as negative and $f = -20$ cm. By Eq. 37-1

$$-\frac{1}{20 \text{ cm}} = \frac{1}{50 \text{ cm}} + \frac{1}{q}$$

$$\frac{1}{q} = -\frac{1}{20 \text{ cm}} - \frac{1}{50 \text{ cm}}$$

$$q = -14.3 \text{ cm}$$

(b) The magnification is given by Eq. 37-3,

$$m = \frac{q}{p} = \frac{-14.3 \text{ cm}}{50 \text{ cm}} = -0.29$$

By Eq. 37-2, $m = -(h'/h)$; $-0.29 = -h'/3$ cm; $h' = +0.87$ cm. The minus image distance or minus magnification indicates that the image is virtual, erect, and located on the same side as the lamp. ◀

Example 37-4   Two positive thin lenses are 60 cm apart, and arranged as in Fig. 37-7. Lens I has a focal length of 10 cm, and the focal length of the other lens is 15 cm. Locate the final image of an object positioned 20 cm to the left of lens I.

Solution.   Applying Eq. 37-1 to the left lens, we get

$$\frac{1}{q_1} = \frac{1}{10} - \frac{1}{20}, \qquad q_1 = +20 \text{ cm}$$

The plus indicates that the image is real and lies to the right of lens I. This image now serves as a real object for lens II, or $p_2 = +40$ cm. Applying Eq. 37-1 to lens II, we get

$$\frac{1}{q_2} = \frac{1}{15} - \frac{1}{40}$$

$$q_2 = +24 \text{ cm}$$

The plus sign indicates that the final image is real and 24 cm to the right of lens II or 84 cm to the right of lens I.

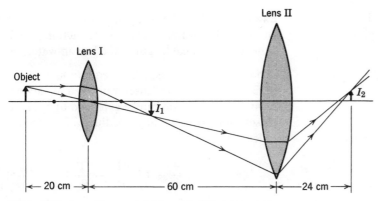

Figure 37-7 Lens I forms an image at I_1. This image acts as a real object for lens II, which forms a real image I. Notice that I_2 is inverted with respect to I_1 and erect with respect to the object.

The magnification, M, of a combination of lenses is the product of the individual magnifications. From Eq. 37-3 the magnification, m, produced by lens I is

$$m_1 = \frac{q_1}{p_1} = \frac{20}{20} = 1$$

The magnification, m_2 produced by lens II is

$$m_2 = \frac{q_2}{p_2} = \frac{24}{40} = 0.6$$

The total magnification, M produced by both lenses is

$$M = 1 \times 0.6 = 0.6$$

The final image is erect with respect to the original object and is 0.6 times its height. ◄

37-6 THE LENSMAKER'S EQUATION

The focal length f, of a particular thin lens can also be determined from a knowledge of the radii of curvature of the two lens surfaces, R_1 and R_2, and the index of refraction n of the lens from the equation

$$\frac{1}{f} = (n - 1)\left(\frac{1}{R_1} - \frac{1}{R_2}\right) \qquad (37\text{-}4)$$

This equation is referred to as the *lensmaker's equation*. The radius of curvature of a lens surface is taken as positive when a ray strikes it on the convex side; it is taken as negative when a ray strikes it on the concave side. By following this sign convention, the focal length will be positive when the lens is thicker at the center than at its edges. Equation 37-4 indicates that lenses with smaller radii of curvature have shorter focal lengths.

Example 37-5 The radius of curvature of each surface of the spherical thin lens in Fig. 37-3a is 3 in. Calculate the focal length of the lens if the index of refraction of the lens material is 1.6.

Solution. If we assume that light from an object approaches the lens from the left, it would then strike the lens at the convex side. By the sign convention. $R_1 = +3$ in. Upon continuing through the lens, the light will strike the right surface of the lens from the concave side, making $R_2 = -3$ in. From the lensmaker's formula,

$$\frac{1}{f} = (1.6 - 1)\left(\frac{1}{3} - \frac{1}{-3}\right) = (0.6) \times \frac{2}{3} = 0.4$$

$$f = +2.5 \text{ in.} \qquad ◄$$

Example 37-6 A given glass lens is concave on both sides, each surface having a radius of curvature of 3 in. Calculate the focal length of this lens if the index of refraction of the glass is 1.6.

Solution

$$n = 1.6, \qquad R_1 = -3 \text{ in.}, \qquad R_2 = +3 \text{ in.}$$

From the lensmaker's formula,

$$\frac{1}{f} = (1.6 - 1)\left(\frac{1}{-3} - \frac{1}{3}\right) = -0.4$$

$$f = -2.5 \text{ in.} \qquad \blacktriangleleft$$

Example 37-7 The curved surface of the plano-concave lens in Fig. 37-8 has a radius of curvature of 3 in. Take the glass to be the same as in Example 37-6 and calculate the focal length of this lens.

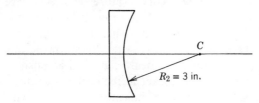

Figure 37-8

Solution. The radius of curvature of a plane surface is infinite and $1/R$ approaches 0. By the lensmaker's formula,

$$\frac{1}{f} = (1.6 - 1)\left(0 - \frac{1}{3 \text{ in.}}\right) = -0.2 \text{ in.}^{-1}$$

$$f = -5 \text{ in.} \qquad \blacktriangleleft$$

Example 37-8 The lens of Example 37-5 is submerged in water of index of refraction of 1.3. Calculate the focal length of the lens in water.

Solution. The lensmaker's equation, as given in Eq. 37-4 applies to lenses immersed in vacuum or air. This equation can be extended to apply to lenses immersed in any medium of refractive index n' by

$$\frac{1}{f} = \left(\frac{n}{n'} - 1\right)\left(\frac{1}{R_1} - \frac{1}{R_2}\right) \qquad (37\text{-}5)$$

Applying this equation, we get,

$$\frac{1}{f} = \left(\frac{1.6}{1.3} - 1\right)\left(\frac{1}{3 \text{ in.}} + \frac{1}{3 \text{ in.}}\right)$$

$$f = 6.5 \text{ in.} \qquad \blacktriangleleft$$

37-7 LENS ABERRATIONS

The purpose of a lens is to form a good image. If, for instance, the image of a point object consists of multiple points or a line the image is distorted. A spherical lens will form multiple image points of a point object, Fig. 37-9. The rays entering the lens near its edges form an image at I_1 while those rays entering the lens near its central region, called *paraxial rays*, form an image at I_2. This undesirable characteristic of a spherical lens is called *spherical aberration*. This condition is minimized by placing an opaque barrier called a *lens stop* in front of the lens to exclude the extreme rays and admit the paraxial rays.

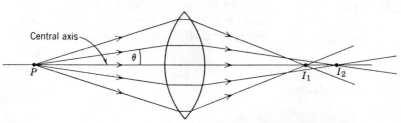

Figure 37-9 Spherical aberration is caused by rays that make a large angle with the central axis.

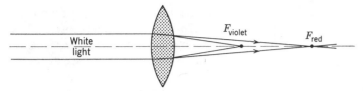

Figure 37-10 Chromatic aberration.

Another defect of a lens, called *chromatic aberration* is caused by the dispersion of white light by the lens. Since the index of refraction of a lens material varies with the wavelength of light (Sec. 36-4) and the focal length of a lens depends on the index (Eq. 37-4), a given lens has a different focal length for different colors, Fig. 37-10. This means that white light coming from a point object will be focused at different image points with the shorter wave-lengths converging nearer the lens. This defect is corrected to a great extent by an *achromatic doublet*, which consists of a converging and a diverging lens cemented together. Achromatic doublets can be designed to focus any two colors at a single point.

As was mentioned earlier, a good image is composed of single image points corresponding to points on the object. But an additional criterion has to do with the spacial orientations of the image points. If the object lies in a plane, say the flat wall of a house, the lens image will consist of image points that are on a curved surface. This defect of a lens is referred to as *curvature of field*. Also, the magnification of a lens varies with the angle that the object deviates from the principal axis of the lens. This defect, called *distortion*, causes different magnifications for different parts of an object, yielding a disproportionate composite image.

37-8 THE EYE

Our primary optical instrument is the eye. The eyeball, which is of approximate spherical shape and about one inch in diameter, comprises an ingenius optical system. Light enters the eye by first passing through a tough outer membrane called the *cornea*, Fig. 37-11. It then continues to the fibrous crystalline lens through an opening, in the *iris* called the *pupil* (P). In the normal eye, the lens forms an image on the *retina*, which consists of very small light-sensitive nerve endings called *rods* and *cones*. These transmit the image to the brain through the *optic nerve*. The rods are much more sensitive to light than are the cones. It is by means of the cones that we perceive color, whereas the rods yield images that are in shades of gray. The amount of light entering the eye is determined by the diameter of the pupil. The tiny ciliary muscles in the iris cause the pupil to dilate to about 7 mm in diameter in situations of dim light and to contract to a diameter of about 2 mm when intense light is incident on the eye. This automatic response is called *adaptation*.

The distinct seeing of objects depends on the ability of the eye lens to form a good image on the retina. The relaxed eye will form images on

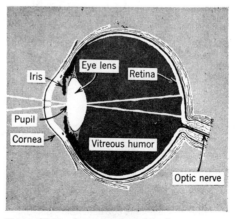

Figure 37-11 Cross section of the human eye.

the retina of objects at a large distance. To see nearer objects distinctly, the eyeball becomes more curved. This is accomplished by tiny muscles attached to the lens. The normal eye can thus *accommodate* to distinctly see objects all the ways from infinity to an object distance of about 25 cm, called the *near point*. However, as a person grows older, the eye muscles responsible for accommodation become weaker and the near point recedes further and further from the eye. At the age of 60 or so the near point can be as large as 2 m. The increase of the near-point distance with age is referred to as *presbyopia*. This condition is corrected by eyeglasses having convex lenses of appropriate radius of curvature that can bring the distinct vision of objects as close as 25 cm from the eye.

Two common defects of the eye, *nearsightedness* (myopia) and *farsightedness* (hyperopia) have to do with the inability of the eye to form distinct images on the retina of far objects. In the former, the eyeball is too long and the image is formed in front of the retina. This condition is corrected by eyeglasses having concave lenses, Fig. 37-12a. The condition of farsightedness is caused by too short an eyeball (with respect to the curvature of the

eye lens) and the image is formed beyond the retina, Fig. 37-12b. This condition is corrected by a convex lens. Another common eye defect is called *astigmatism*. The astigmatic eye may see distinctly the lines *AB* in Fig. 37-13 while the lines *CD* may appear blurred. Correction for astigmatism is accomplished by eyeglasses having cylindrical rather than spherical surfaces.

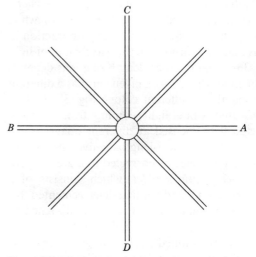

Figure 37-13

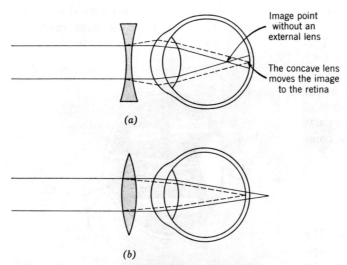

Figure 37-12   (a) A concave lens corrects nearsightedness. (b) A convex lens corrects farsightedness.

37-9 THE CAMERA

The lens of a camera forms a sharp image of the object photographed on a flat film that is coated with a light-sensitive emulsion. The inside of a camera is light-tight and no light falls on the film while in the camera except for the fraction of a second when the picture is taken. During this moment, the metallic diaphragm shutter springs open to an aperture of the desired diameter that allows light from the object to be focused on the film. A reaction takes place at the spots on the emulsion that were struck by light that, after subsequent treatment with a chemical called the *developer*, will turn dark. The degree of darkness depends on the intensity of the light. More intense light will cause darker spots.

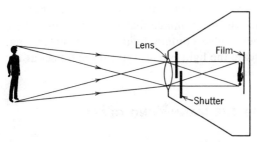

Figure 37-14 A camera. During the "taking" of a picture the shutter springs open and the light from the subject is focused on the film. The lens can be moved closer or farther from the film (in moderately priced cameras).

Since the composite lens image on the film consists of spots that receive bundles of rays of varying intensity from the object, the developed image will consist of varying shades of black and white. This film image is called a *negative* because it depicts as darker areas the areas of the object that are brighter. After an additional procedure in the photographic processing laboratory, which involves the exposure of the negative to light that is imaged by a lens onto another film, and its subsequent development, the usual black and white photographic picture results.

The taking of a good photograph involves the proper exposure on the film of the light from the object to be photographed. This entails the control of the light intensity (light energy per unit time and area) falling on the film. The intensity, I, depends on the area, A of the aperture in the camera shutter through which the light from the object reaches the film and the duration called the *exposure time* that the shutter remains open. In moderately expensive cameras the aperture size and exposure time can be varied. These variations must be matched to the *film speed* or the degree of the light sensitivity of a given film emulsion. Instructions on the proper exposure time and aperture for given lighting conditions are usually given by the film manufacturer.

The intensity of light from a given object falling on the film depends on the square of the aperture diameter, D^2 because the area of a circle, $A = \pi D^2/4$. The intensity is also inversely proportional to the square of the focal length of the lens. This dependence can be understood by considering the lens to be a small light source. The image distance (lens to film distance) is equal or very nearly equal to the focal length for objects distances varying from infinity to a few feet from the lens. Since the intensity at a point is inversely proportional to the square of the distance from the point to the light source, the intensity of the light falling on the film in a camera is inversely proportional to the square of the focal length, f^2, of the lens. These considerations can be summarized by

$$I \propto \frac{D^2}{f^2} \qquad (37\text{-}6)$$

The ratio of the focal length to the *maximum* aperture diameter, f/D is a quantity that is used to describe a given lens. This ratio is called the *focal ratio* or *f-number* and is usually marked on the side of the lens. Thus a lens having a focal ratio of 4, which is designated as $f/4$, is one whose maximum aperture diameter is $\frac{1}{4}$ of its focal length; an $f/8$ lens is one whose maximum diameter is $\frac{1}{8}$ of the focal length. Therefore the maximum aperture diameter of an $f/4$ lens is twice as large as that of an $f/8$ lens provided the focal lengths are the same. Since the intensity of light on the film depends on the square of the diameter,

(Eq. 37-6) the $f/4$ lens is capable of forming an image on the film that is four times as intense as the image formed by the $f/8$ lens of the same object. A *lower* focal ratio indicates a *higher* intensity that the lens is capable of producing on the film.

If a picture is taken of a dimly lighted object, say a person on a very cloudy day, it is desirable to use the maximum aperture diameter, that is, set the focal ratio on the lens at its minimum value. If a picture is taken on a bright sunny day, it is desirable to *stop-down* the aperture diameter, that is, increase the focal ratio setting on the lens. The focal ratio can be varied by moving a pointer to the desired value, Fig. 37-15. For instance, the focal length of a lens usually used in a 35 mm camera (a camera that uses 35 mm-wide film) is 50 mm. If this is an $f/2.8$ lens, its maximum working diameter is

$$2.8 = \frac{50}{D}, \qquad D = 17.86 \text{ mm}$$

The next higher focal ratio setting is $f/4$. The aperture diameter for this setting is

$$4 = \frac{50}{D}, \qquad D = 12.5 \text{ mm}$$

Since the intensity on the film is proportional to the square of the diameter, the ratio of the intensities produced on the film by these settings is

$$\left(\frac{17.86}{12.5}\right)^2 \approx 2$$

The $f/2.8$ setting produces an intensity that is twice as great as that produced by the $f/4$ setting when those settings are used to photograph the same object. Similar calculations will reveal that each f/stop setting produces an intensity which is about twice as great as that produced by the next higher setting.

The cost of a given camera depends to a large extent on the type of lens. The more expensive cameras have lenses consisting of a few elements that correct for chromatic aberration, curvature of field and distortion. The *stopping down* of a lens to a small aperture corrects for spherical aberration. In addition, the costly cameras have elaborate view-finders that allow the photographers to see the exact image that the film "sees"; these cameras also contain built-in light meters that indicate the light intensity from the object at the camera.

37-10 THE MAGNIFYING GLASS

The *magnifier*, which is sometimes referred to as a *magnifying glass* or *simple microscope*, is a convex lens that helps the eye to read very small print or

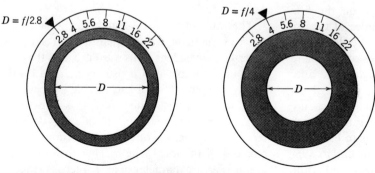

Figure 37-15 Various focal-ratio settings on a moderately priced camera. At a given setting, the intensity of light from an object at the film is twice that at the next higher setting.

to distinctly see very small objects such as the fine components of the inside of a watch. The magnifier produces a virtual image that the eye can see more distinctly because it is larger than the object.

Objects appear bigger when they are closer to the eye. Since the average eye cannot accommodate to object distances that are closer than 25 cm, a small object is most distinct to the unaided eye when it is 25 cm from it. Even at this distance the eye may not distinctly see the object if it is too small. If a convex lens is placed near the eye and at a distance of less than one focal length from the small object, an enlarged virtual image will result, Fig. 37-16. The maximum useful height of this image is obtained when the image distance, q is about 25 cm and the eye is held close to the lens.

By Eq. 37-3, the magnification, m is given by $m = q/p$, where $q = -25$ cm (the minus because of the sign convention for virtual images). Therefore $m = -25/p$. From the thin lens equation,

Eq. 37-1, $1/p = 1/f - 1/q$. By combining the last two equations we get

$$m = -25\left(\frac{1}{f} - \frac{1}{(-25)}\right)$$

$$m = -\left(\frac{25 \text{ cm}}{f} + 1\right) \tag{37-7}$$

This expresses the maximum magnification obtained by a magnifier with the minus indicating a virtual image. The magnification increases as the focal length of the lens decreases. Since the focal length of any spherical lens is dependent on the radius of curvature (Eq. 37-4), the magnifier must have a small radius of curvature (must be sharply curved) to obtain a large magnification. Magnifications of about 50 are obtained by the jewelers' magnifiers, which means that these have a focal length of about $\frac{1}{2}$ cm.

37-11 THE COMPOUND MICROSCOPE

The compound microscope is an optical instrument consisting of two lenses that in combination can produce a greater magnification than the single-lens magnifier. The microscopic body to be examined is placed at a distance p_o from a convex lens called the *objective*, which is slightly greater than its focal length, Fig. 37-17. The objective therefore forms a real image within the tube containing the lenses. This image is formed at a distance from a second convex lens called the *eyepiece*

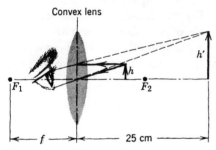

Figure 37-16   The magnifier.

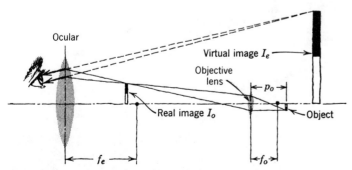

Figure 37-17   The compound microscope.

or *ocular*, which is slightly less than its focal length, f_e. The eye placed above the ocular will see an enlarged virtual image of the real image produced by the objective. Both lenses are highly corrected for aberrations.

The magnification, m_e produced by the eyepiece is from Eq. 37-7

$$m_e = -\left(\frac{25 \text{ cm}}{f_e} + 1\right)$$

The magnification, m_o produced by the objective lens is

$$m_o = \frac{q_o}{p_o}$$

where q_o is the image distance of the image formed by the objective. The total magnification, M is given by the product $m_e m_o$.

$$M = -\left(\frac{q_o}{p_o}\right)\left(\frac{25 \text{ cm}}{f_e} + 1\right) \qquad (37\text{-}8)$$

with the minus indicating a virtual image.

Example 37-9   The objective and eyepiece of a certain microscope each have a focal length of 2.5 cm. If the object to be examined is placed at a distance of 2.8 cm from the objective lens, calculate the magnification.

Solution.   The image distance q_o can be determined from the lens equation.

$$\frac{1}{f_o} = \frac{1}{q_o} + \frac{1}{p}$$

$$\frac{1}{q_o} = \frac{1}{2.5} - \frac{1}{2.8}, \qquad q_o = 23 \text{ cm}$$

By Eq. 37-8

$$M = -\left(\frac{23}{2.8}\right)\left(\frac{25 \text{ cm}}{2.5 \text{ cm}} + 1\right) = -90$$

The minus in the answer is usually omitted. Each linear dimension of the object will appear 90 times larger.   ◀

37-12 THE OPTICAL TELESCOPE

The purpose of a telescope is to make distant objects appear closer than when viewed by the unaided eye. There are two general types of optical telescopes. The *refracting telescope* utilizes a lens while the reflecting telescope utilizes a concave, mirror, Fig. 37-18. Notice that the image of a refracting telescope is formed near the end of the telescope tube. This allows for the possibility of the image to be examined by the eye with the

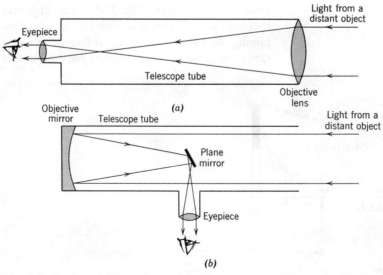

Figure 37-18   (a) Refracting telescope; (b) Reflecting telescope.

aid of a suitable eyepiece, placed at the end of the tube. In the reflecting telescope the image is formed within the tube. To make it accessible to the eye or some other sensing device as a photographic plate, the image can be deflected by a plane mirror to the side of the tube where it can conveniently be examined. This is called a *Newtonian* arrangement.

One important function of an astronomical telescope is to produce bright images of celestial bodies like stars, which appear as points. This is accomplished by lenses or curved mirrors with large diameters, Fig. 37-19. The larger the diameter the more light it can gather from the celestial body and thus produce a brighter image. The largest reflecting telescope is the 200-inch.

(200 in. diameter) Hale Telescope on Palomar Mountain in California. One advantage of reflecting telescopes over refractors is that the former does not exhibit chromatic aberration. Reflecting telescopes are usually parabolic (rather than spherical), thus making them free of spherical aberrations as well. The largest refracting telescope is the 40-in. telescope at the Yerkes Observatory in Wisconsin. The lenses usually have two or three elements, which means that four or six glass surfaces have to be carefully polished and prepared. The mirror of a reflector requires the careful preparation of only one surface.

The words "magnification" or "lateral magnification" as used so far in this book refer to the ratio of image height to object height. This ratio

Figure 37-19 The 120-inch reflecting telescope at the Lick Observatory on Mount Hamilton, California.

involves linear dimensions. An object appears larger when it is nearer to the eye because it subtends a larger angle at the eye. For an example, the sun's visible diameter, D is 1.39×10^6 km and its mean distance to the earth R is 148.8×10^6 km. The angular size, that is, the angle α that the sun's diameter subtends on the eye of an observer on earth is

$$\alpha = \frac{D}{R} \text{ radians}$$

$$\alpha = \frac{1.39 \times 10^6 \text{ km}}{148.8 \times 10^6 \text{ km}}$$

$$= 9.34 \times 10^{-3} \text{ radians} \approx 0.5 \text{ degree}$$

The diameter of the moon is 3476 km and its mean distance from the earth is 384404 km. The angle, β, that the full moon subtends on the eye of an observer on the earth is

$$\beta = \frac{3476 \text{ km}}{384404 \text{ km}} \approx 9 \times 10^{-3} \text{ rad} \approx 0.5 \text{ degree}$$

To an observer on the earth the moon appears to be the same size as the sun even though the sun's diameter is about 400 times larger (Fig. 37-20).

The *magnifying power* of a telescope is defined as the ratio of the angle subtended on the eye by

the telescopic image of a body to the angle subtended by the body on the unaided eye. In Fig. 37-21, I' is the image formed by the objective as seen through the eyepiece of a telescope. This image subtends an angle β on the eye. The angle α is the angle subtended by the same distant object on the objective lens, which is also the same angle that it would subtend on the unaided eye. The magnifying power, MP is defined as

$$MP = \frac{\beta}{\alpha} \qquad (37\text{-}9)$$

The angles β and α are small and their values in radians are $\beta = AB/f_e$ and $\alpha = AB/f_o$ where f_o and f_e are, respectively, the focal lengths of the objective and eyepiece.

$$MP = \frac{AB}{f_e} \div \frac{AB}{f_o} = \frac{f_o}{f_e}$$

$$MP = \frac{f_o}{f_e} \qquad (37\text{-}10)$$

The larger the focal length of the objective lens or mirror the higher is the magnification power of the telescope. The focal length of the Hale Telescope is 16.7 m. If an eyepiece having a focal length of 5 cm is used, the resulting magnifying power comes to $MP = (16.7 \text{ m})/(5 \times 10^{-2} \text{ m}) = 334$.

Magnification is not the only criterion of a good image. A good image should also exhibit fineness of detail in the object. For instance, a good microscopic image of a cell should exhibit the nucleus; a good telescopic image of twin-stars (two stars that are separated by a small angle) should show two small points and not a single larger point. The ability of an optical system to produce images with fine detail is called *resolu-*

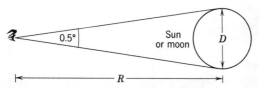

Figure 37-20 The sun and moon appear to be about the same size because they subtend approximately the same angle at the earth.

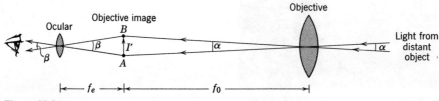

Figure 37-21

tion. The resolution given by a camera, microscope, or telescope is improved by the use of lenses or mirrors that are corrected for the various aberrations.

But there exists a limit to the highest resolution given by an optical instrument even if its components were perfectly corrected. This has to do with a phenomenon called diffraction (see Section 38-1) that is exhibited by light waves. Because of the wave nature of light the image formed of a point object by even the best optical system consists of a small central bright circle with concentric alternate dark and bright rings, Fig. 37-22. When two object points are very close together, the diffraction patterns of each image point overlap and the image depicts one fuzz spot. This phenomenon sets a fundamental limit on resolution.

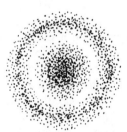

Figure 37-22   An image formed by a telescope of a point object like a star consists of a central bright spot followed by a number of alternate dark and less bright rings.

The resolution of a lens or mirror image depends on the wavelength of the light used in forming the image and on the diameter of the mirror or lens. The *resolving power* of a telescope is measured by the smallest angle, α, that a telescope can resolve. This angle (in radians) is given approximately by

$$\alpha = \frac{\lambda}{D} \qquad (37\text{-}11)$$

where λ is the wavelength of light and D is the diameter of the objective. The shorter the wavelength the smaller is the angle that can be resolved. The human unaided eye cannot resolve object points that are closer than about one-minute of arc (which equals $1°/60$) apart. Telescopes having large diameters can resolve objects that are a fraction of a minute of arc apart. This equation sets a limit on the resolving power of any instrument. The resolution cannot be improved no matter how many times the image is magnified by other instruments.

37-13 THE SPECTROSCOPE

When white light is passed through a prism, the prism will disperse (separate) the light into its component colors (wavelengths). Actually, the light need not be white. The light from a Bunsen burner or from any other light source, regardless of its apparent color will also be separated into its component wavelengths by a prism. When a chemical element or compound is heated to a high enough temperature it will emit light whose wavelengths is characteristic of the particular element or compound. An instrument called a *prism spectroscope* displays the wavelengths in an array called a *spectrum*, which can directly be viewed and examined by the eye. If the spectroscope also has a calibrated scale of wavelengths superimposed on the spectrum, it becomes a *spectrometer* and the wavelengths, in angstroms, emitted by the source can be directly determined. If the spectrometer has a recording device such as a photographic plate on which the spectrum can be recorded, it is called a *spectrograph*.

The essential components of a spectrograph are sketched in Fig. 37-23. The light to be examined passes through a narrow rectangular slit to an achromatic lens, *L*. The lens is placed at a distance of one focal length from the slit, thereby *collimating* the light or causing bundles of parallel rays to emerge from the lens and head toward the prism. The different wavelengths are separated by the prism and converged by the camera lens on photographic film. Multiple images of the slit will appear side by side on the film, each image corresponding to a given wavelength emitted by the source, Fig. 37-24.

When a few grains of table salt (sodium chloride) are sprinkled on a cooking-gas flame

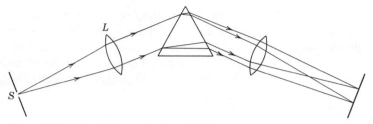

Figure 37-23 Essential components of a prism spectroscope.

the color of the flame will briefly turn yellow. Of course, the unaided eye cannot precisely determine the wavelength of this color. But, if subjected to a spectrographic analysis, the film would show this yellow light as two bright lines of wavelengths 5889.95 and 5895.92 Å, called the *sodium D lines*, which the eye perceives as yellow. This particular yellow light is characteristic of the sodium in the sodium chloride.

The film resulting from a spectrographic analysis of the light emitted by heated sodium vapor will appear black except for the sodium D lines and a few other less-bright lines in the red and blue-green regions, Fig. 37-24. This shows that sodium selectively emits light in narrow regions of wave-lengths that show up as bright lines on a spectrograph. Such a spectrum is called a *bright-line* spectrum. In general, heated vapors of the metallic elements and monatomic gases emit bright-line spectra. The molecular gases yield *band spectra* and heated metals yield *continuous* spectra.

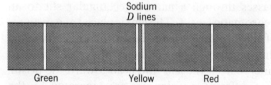

Figure 37-24 Some bright lines in the emmission spectrum of sodium.

When white light is passed through a monatomic gas at room temperature, the gas will absorb precisely those wavelengths that it emits when excited by intense heat. Dark lines will appear on

the spectrographic film at precisely those positions where the bright lines would appear if the same gas were heated to the point of emitting light. Such a spectrum is called a *dark-line* spectrum or an *absorption* spectrum.

Each of the elements and compounds exhibit a unique spectrum. The spectra thus provide an identifying signature. Spectroscopic analysis help chemists identify the presence of minute amounts of a given element or compound in a material. Spectroscopy is also widely used by astronomers. Spectroscopic analyses of the light from stars can yield important information on the elements present on the stars. Spectroscopy has also provided important information that has led to a better understanding of the structure of the atom. For a further discussion on this see Section 39-5.

SUMMARY

Convex lenses yield real and virtual images depending on the object-distance.

Concave lenses yield virtual images only.

The focal length of a thin lens is the distance from the lens to the image it forms of an infinitely distant object.

The thin lens equation: $1/f = 1/p + 1/q$. This equation is appropriate for convex and concave lenses provided the appropriate sign convention is used.

The lensmaker's equation: $1/f = (n - 1)(1/R_1 - 1/R_2)$. This equation is also appropriate for both concave and convex lenses with the appropriate sign convention.

The lateral magnification of a lens is given by the ratio of image height to object height.

Spherical aberration is a defect of spherical lenses that is exhibited by multiple image points for a single object point.

Chromatic aberration is a defect of lenses that is exhibited by multiple image points for a multi-colored object.

Seeing involves the formation of a sharp image on the retina.

Various optical instruments such as the camera, magnifying glass, compound microscope, and the refracting telescope are based on the principles governing the formation of images by lenses.

The spectroscope is based on the dispersion of light going through a prism. Minute quantities of a given element or compound in a material can be detected by spectrographic analysis.

QUESTIONS

1. What are some of the criteria of a good image?
2. Why is spherical aberration a lens defect?
3. Photographers often "stop-down" the lenses of their cameras in certain light conditions. By this is meant that they adjust the camera to block out light from the outer portions of the lens. How does this improve the picture?
4. Can a camera equipped only with a concave lens take a photograph? Explain.
5. Can a lens produce an image that is larger than the object? In what circumstances?
6. Differentiate between focal length and focal point.
7. How can the focal length of a given convex thin lens be determined by a simple experiment?
8. What is meant by the magnification of a lens?
9. Why is chromatic aberration a lens defect?
10. What causes chromatic aberration?
11. What is meant by focusing a camera?
12. What is meant by the resolution of a lens?
13. Explain the principles on which the prism spectrograph is based?
14. What is an emission spectrum?
15. Explain the meaning of an absorption spectrum.

PROBLEMS

(A)

1. Explain with the aid of a ray diagram the meaning of a real image that is formed by a convex lens.
2. Explain with the aid of a ray diagram the meaning of a virtual image that is formed by a concave lens.
3. A thin convex lens has a focal length of 100 mm. How far from the lens will the image of the full moon appear if the lens is aimed at the moon?
4. If the lens in Problem 3 is used to form an image of a person 2 m away, how far from the lens will the sharp image appear?
5. Explain the reason for the feature of a moderately priced camera that provides the ability to move the lens nearer or further from the film.
6. A 6-cm long pencil is stood up 20 cm in front of a convex lens whose focal length is 10 cm. Determine the image height and its location by means of a ray diagram drawn to scale.
7. (a) Verify your answer to Problem 6 analytically by use of appropriate equations.
   (b) Is the image erect or inverted and real or virtual?
8. Repeat Problem 6 for a 10-cm focal length concave instead of a convex lens.
9. (a) Verify your answer to Problem 8 analytically by use of the appropriate equations.
   (b) Is the image erect or inverted and real or virtual?
10. An object 30 cm high is placed at a distance of 8 cm from a thin positive lens having a focal length of 20 cm.
    (a) Calculate the image distance.
    (b) Describe the image.
11. A camera lens forms an image of a person 2 m tall. If the height of the image is 2 cm, find the magnification of the lens.
12. A thin convex lens forms a sharp image of an object that is 3 m from the lens. If the image distance is 7 cm. Determine:
    (a) The focal length of the lens.
    (b) The magnification.

(B)

1. (a) If the lens in Problem A12 forms an image of an incandescent lamp that is 6 m away, how far from the lens will a sharp image appear?

(b) What is the magnification of this image?

(c) What is the image height, if the lamp height is 4 in.?

2. Assume that in Fig. 37-3a $R_1 = 15$ cm and $R_2 = 10$ cm and the index of refraction of the lens material is 1.7. Calculate the focal length of this lens.

3. Repeat Problem B2 with $R_2 = 15$ cm instead of 10 cm.

4. Assume that in Fig. 37-3b $R_1 = 10$ cm and $R_2 = 10$ cm and the index of refraction of the lens material is 1.5. Calculate the focal length of this lens.

5. Calculate the focal length of a lens that will form an image of a lamp filament placed 4 in. away on a screen 20 ft from the lens.

6. A student uses a 15-mm focal length magnifying glass to examine an ant.
   (a) Draw a ray diagram for a 12-mm object distance.
   (b) What magnification will he achieve?

7. A microscope objective ($n = 1.5$) is in water. If its focal length in air is 5 mm, determine its focal length in water.

8. A double-convex lens has radii of curvature of 20 cm and 10 cm, respectively. If the lens material has an index of 1.5, determine its focal length. Is the lens converging or diverging?

9. A camera lens has a focal length of 5 in. How far from the infinity setting does the lens have to be moved out in order to photograph an object 15 ft away?

10. A 10-cm focal length camera lens is focused on a person 6 ft tall at a distance of 20 ft.
    (a) How far from the lens, assuming it is thin, does the $2\frac{1}{4}$-in. $\times$ $2\frac{1}{4}$-in. film have to be moved out?
    (b) What size image will be formed on the film?

11. What is the minimum film size that is needed to depict the full image of a person 6 ft tall and 7 ft from the camera lens of focal length 2 in.?

12. A lens of 15-cm focal length is to form a real image that is four times the size of the object. How far from the lens should the object be placed?

13. An illuminated arrow is placed on an optical bench 26 cm to the left of a 10-cm focal-length converging lens. Another converging lens that has a focal length of 10 cm is placed 30 cm to the right of the first lens.
    (a) Construct a ray diagram leading to the final image.
    (b) Calculate the position of the final image; is it real?

14. Substitute for the second lens in Problem B13 a concave mirror of 15-cm focal length.

15. Two thin lenses are placed 40 cm apart with their principal axis coinciding, as in Fig. 37-7. Lens I is converging and has a focal length of 15 cm; lens II is diverging, with a focal length of -10 cm.
    (a) Calculate the final image position of an object placed 40 cm to the left of lens I.
    (b) Describe the final image.
    (c) What is the total magnification of the system?

16. An illuminated object 10 cm high is placed in front of a thin double-convex lens, having radii of curvature of 15 cm and 20 cm. The lens has a refractive index of 1.4.
    (a) Locate the image position for an object distance of 25 cm.
    (b) Determine the magnification and height of the image.
    (c) Describe the image.

17. Show that an object distance that is less than the focal length of a converging lens will yield a virtual image.

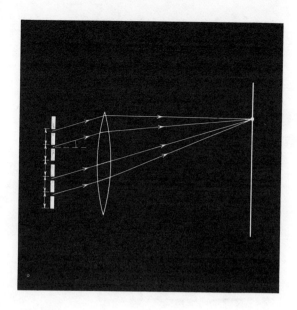

38.
The Wave Nature of Light

It is an interesting observation that while light is an indispensable ingredient for seeing, we cannot see light itself. When we see something, it is either because it emits or reflects light of a wavelength that falls within the range of approximately 4000 to 7000 Å. But we cannot see light itself in the sense that we cannot see the "stuff" of which it is composed. So far in our treatment of light we have been discussing its behavior. Lenses, for instance, bring out the behavior of light when it passes from one medium into another (the lens) and then out again. *But what is it that is actually traveling through the lens?* This chapter describes some of the experimental evidence for the wave theory of light.

LEARNING GOALS

1. *To understand more about the wave nature of light.*

2. *To learn about diffraction and interference of light waves.*

3. *To understand Young's experiment and how it supports the wave nature of light.*

4. *To learn about how the interference of light waves explains the color patches observed on soap bubbles and thin films of oil on the ground.*

5. *To learn about the polarization of light.*

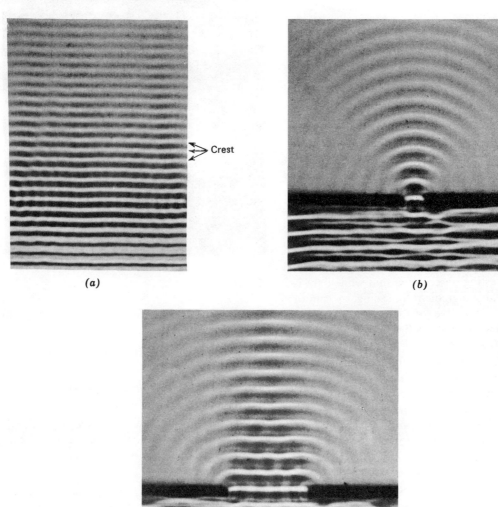

Figure 38-1 (*a*) Water waves produced by dipping the long edge of a ruler into the water at a constant frequency. (*b*) The water waves diffract or bend around the edges of an opening through which they pass. (*c*) Diffraction of the waves is less noticeable if the width of the opening is much larger than the wavelength. (From Education Development Center, Inc., Newton, Massachusetts.)

38-1 DIFFRACTION

An understanding of the wave nature of light is strengthened by a study of water waves, which are easier to visualize. If a straight-edge, like the edge of a ruler is dipped at regular intervals (constant frequency) into a tray of water called a ripple tank a regular pattern of straight waves will move away from the ruler along the surface of the water, Fig. 38-1a. The ripples represent small rises of the water surface called *crests*, and the distance between crests is one wavelength. Halfway between crests there is a depression in the water called a *trough*. Figure 38-1b depicts these waves as they pass through the opening in a barrier having a width that is approximately the same as the wavelength of the wave. Notice that the waves bend around the edges of the opening; *the bending of waves around edges is referred to as diffraction of waves.* Figure 38-1c depicts the water waves passing through an opening which is much wider than the wavelength. Notice that the diffraction of the waves is less in this situation.

Light waves also diffract. Figure 38-2 depicts light passing through an opening in an opaque barrier. The shadow of the barrier's edge is fuzzy and not sharply defined. If the aperture is wide in comparison with the wavelength of light, the

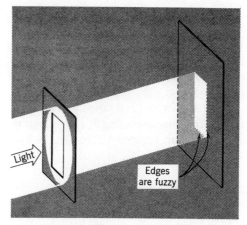

Figure 38-2 Diffraction causes the edges of shadows not to be sharply defined.

diffraction effects are minimized and the well-defined umbra of Fig. 34-3 results. Sound waves also diffract. A listener can hear the sound coming from around the corner of a sound barrier.

38-2 INTERFERENCE AND SUPERPOSITION OF WAVES

Figure 38-3 shows the multiple bands of light (more than 2) on a screen resulting from the light

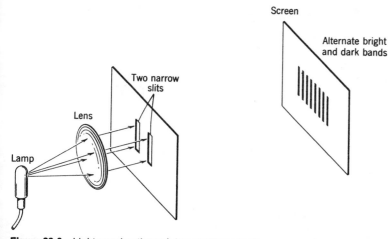

Figure 38-3 Light passing through two narrow openings can produce multiple (more than two) bright bands on a screen.

passing through the two narrow slits in the opaque barrier. Considerations resulting from the straight-line motion of light would lead one to conclude that there should appear only two bands of light on the screen. But this is not what happens.

To understand why, we first consider two light waves having the same wavelength and traveling in the direction from A to point P in Fig. 38-4a.

Notice that at point P, the crest of one wave meets the crest of another wave. What would happen to the water level at point P of Fig. 38-4a if the waves were water waves instead of light waves? The wave marked (1) would by itself produce a short rise, called a crest, in the water at point P while the wave marked (2) would also produce a crest at P. The combined effect caused by the *superposition* of the two waves would

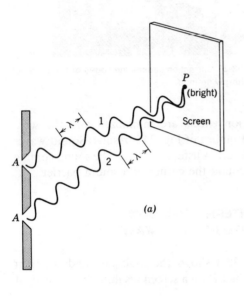

(a)

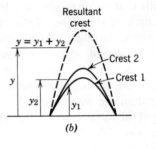

(b)

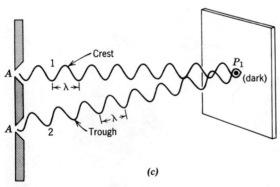

(c)

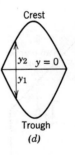

(d)

Figure 38-4 (a) The crests of two light waves arriving at the same time at point P. (b) The superposition of the two crests causes a bright spot at P. (c) The superposition of a crest of one wave on a trough of another wave at P. (d) The superposition of a crest of one wave with a trough of another wave causes a dark spot.

result in a crest that is twice as high (Fig. 38-4*b*). *The arrival of two crests (or two troughs) at the same moment at a point causes a reinforcement or bright spot at that point.* Such a reinforcement is also sometimes called *constructive interference.*

Consider next what happens at point P_1 in Fig. 38-4*c* where a crest meets a trough. If these were water waves, their superposition would result in a flat region of the water in the region of *P*. Analogously, the resulting superposition of a light trough and a crest at P_1 causes a dark spot

or *cancellation* of light at that point. This is sometimes referred to as *destructive interference.*

38-3 YOUNG'S EXPERIMENT

In 1801 Thomas Young, a British scientist, first considered the experiment sketched in Fig. 38-5. It is now referred to as *Young's double-slit experiment.* A front view of the screen of Fig. 38-5*a* is shown in Fig. 38-5*b*. The regions marked

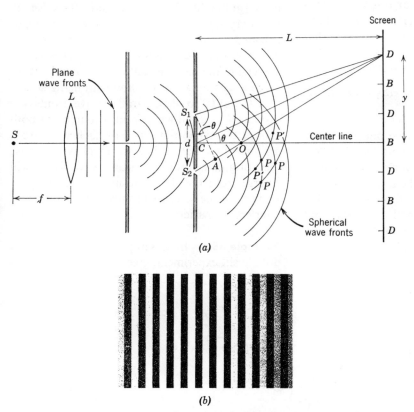

(a)

(b)

Figure 38-5 (a) Waves of light, originating at the monochromatic source, *S* arrive at the two narrow slits, S_1 and S_2. These split a given wave into wavelets which interfere with one another as they move toward the screen. Points *B* appear bright because they are located on a crest from S_1 and S_2. Points *D* appear dark because these lie at a trough from S_2 and on a crest from S_1. (b) Alternate bright and dark bands appear on the screen.

D are dark. In order for these to be dark, a crest of the light wave from one of the slits must have arrived at the same time as a trough from the other, thus superposing and causing cancellation. Since the wavelength λ of the light from both slits is the same, the path-length difference between the two rays, represented by the distance AS_2 in Fig. 38-6a, must be equal to one half-wavelength, or three half-wavelengths, or five half-wavelengths, etc. This can be summarized by

$$\text{path-length difference} = AS_2 = n\left(\frac{\lambda}{2}\right), \quad (38\text{-}1)$$

$$n = 1, 3, 5, \ldots$$

In order for a spot to appear bright, points B in Fig. 38-5 must receive either crests or troughs from both slits at the same time. This means that the path-length differences must be either zero or integral numbers of whole wavelengths, or

$$\text{path-length difference} = AS_2 = n\lambda, \quad (38\text{-}2)$$

$$n = 0, 1, 2, 3, \ldots$$

The last two equations can be put in a more useful form by expressing the path-length difference in terms of the distance of a given bright or dark line from the center of the screen. Since the screen distance L is very large compared with the slit separation d and θ is small, lines DS_1 and DS_2 are essentially parallel, so that $AS_2 = d \sin \theta$ where angle $S_1 AS_2 \approx 90°$. Finally, since θ is small, $\sin \theta$ is approximately equal to $\tan \theta$. Equations

38-1 and 38-2 can now be rewritten, respectively, as

$$d \sin \theta = \frac{d(y)}{L} = n\left(\frac{\lambda}{2}\right), \quad (38\text{-}3)$$

$$n = 1, 3, 5, \ldots \text{(cancellation)}$$

and

$$d \sin \theta = \frac{d(y)}{L} = n\lambda, \quad (38\text{-}4)$$

$$n = 0, 1, 2, 3, \ldots \text{(reinforcement)}$$

For the regular pattern of bands to appear on the screen, the light passing through the two slits must be monochromatic and arrive in step (in phase) at the slits. Such light is said to be *coherent*. Ordinary light sources, even if they are monochromatic, do not produce coherent light (see the discussion on lasers in Section 39-6). One way of achieving the regular pattern of bands is to place a lens between the monochromatic source and the slits so that parallel wave fronts are produced. A given wave front will arrive at both slits at the same moment. For each wave front arriving, two waves that are in phase (because they originated from the same wave front) will emerge from the slits. The interference in their journey to the screen will thus cause the regular interference patterns.

Example 38-1   In a setup similar to Young's double-slit experiment, green light of wavelength

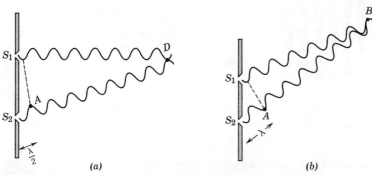

(a)   (b)

Figure 38-6   The path length difference for the two waves arriving at the screen is one-half wavelength in (a), and one wavelength λ in (b).

540 mμ (= 5400 Å = 5400 × 10⁻⁸ cm) is used with a slit separation of 0.05 mm. If the screen is 2 m away, calculate the distance from the center of the screen to (a) the first bright band, (b) the third dark band, and (c) the distance between the two first-order bright bands.

Solution
(a) We let $n = 1$ in Eq. 38-4 and solve for y.

$$y = \frac{1 \times \lambda \times L}{d} = \frac{5400 \times 10^{-8} \text{ cm} \times 200 \text{ cm}}{5 \times 10^{-3} \text{ cm}}$$

$$y = 2.16 \text{ cm}$$

(b) We let $n = 3$ in Eq. 38-3

$$y = \frac{3 \times L}{d}\left(\frac{\lambda}{2}\right) = \frac{3 \times 200}{5 \times 10^{-3}} \frac{(5400 \times 10^{-8})}{2}$$

$$y = 3.24 \text{ cm}$$

(c) The first bright band on each side of the central bright band is called the first-order bright band. The center-to-center distance between the two first-order bright bands is 2×2.16 cm = 4.32 cm. ◀

38-4 THE DIFFRACTION GRATING

The diffraction grating is used to determine wavelengths of light and to analyze light spectra.

The transmission grating consists of a glass plate upon which many equidistant narrow lines are ruled. The lines act as opaque barriers to the light to be analyzed, while the spaces act as slits. Some of the better gratings have as many as 30,000 lines per inch.

The principle of the grating is similar to that discussed in Section 38-3. Parallel rays of the light to be analyzed are incident on the grating from the left (Fig. 38-7). By Huygens' principle, each spacing in the grating acts as a source for wavelets that spread out toward the right. The lens L converges the light on the screen. Rays leave the grating spaces in many directions. However, for a given direction θ, the rays from all spacings will arrive on the screen in phase, i.e., they will reinforce so as to cause a maximum at that point. The condition for reinforcement is

$$d \sin \theta = n\lambda \qquad n = 0, 1, 2, 3, \ldots \quad (38\text{-}5)$$

where d is the distance between spaces. By measuring θ and counting the order n of a particular dark or bright fringe, the wavelength λ can be deduced.

Example 38-2 A transmission grating, having 20,000 lines/in., produces a first-order bright line at an angle of 25° when monochromatic light of unknown wavelength is passed through it. Calculate the wavelength.

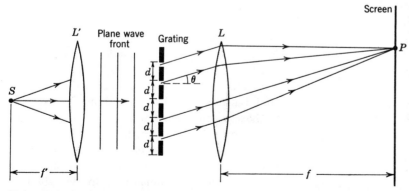

Figure 38-7 The diffraction grating (transmission type).

Solution. Since the grating has 20,000 lines/in., the distance between lines is

$$\frac{1}{20,000}\frac{\text{in.}}{\text{line}} \times 2.54\,\frac{\text{cm}}{\text{in.}} = 1.27 \times 10^{-4}\ \text{cm}$$

For the first order, $n = 1$ so that by Eq. 38-5

$$\lambda = 1.27 \times 10^{-4}\ \text{cm} \times \sin 25°$$
$$= 1.27 \times 10^{-4}\ \text{cm} \times 0.423$$
$$= 0.537 \times 10^{-4}\ \text{cm} = 5370\ \text{Å} \quad \blacktriangleleft$$

38-5 INTERFERENCE IN TRANSPARENT THIN FILMS

A thin layer of oil on the ground will exhibit many colors when viewed from above. In general, thin transparent objects like patches of oil or soap bubbles will display color patterns when viewed by reflected white light, or dark and bright patterns when viewed by reflected monochromatic light. This phenomenon can be explained by the interference of light waves.

Figure 38-8 depicts the appearance of a film of soap when viewed by reflected light; Fig. 38-9 shows a small portion of the film wall facing the

Figure 38-8   A thin transparent film such as a soap bubble will exhibit dark and brighter patches.

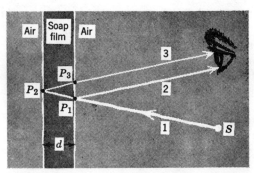

Figure 38-9   A narrow beam is incident on a soap film of thickness *d*. It is partially reflected by the front surface and partially by the rear surface of the film.

eye. Wave 1 leaving the source S will, upon reaching the film, be reflected by its front surface at P_1 and by its rear surface, at P_2. The two reflected waves, 2 and 3, will interfere with one another and will cause the eye to see a bright or dark spot depending on the thickness d of the film. It might be expected that cancellation would occur when the path-length difference between waves 2 and 3, which is about $2d$ for near-normal incidence, is an odd multiple of half wavelengths (Eq. 38-1). Actually, another consideration must be made. This consideration has to do with the phase change that light waves undergo upon reflection.

Figure 38-10 depicts two ropes that are joined by a knot, the mass per unit length of each rope being different. When a transverse pulse is sent down the heavier rope, a transmission of the pulse to the lighter rope will occur at the knot, but a reflection of the wave will also occur. Notice that the reflected pulse at the knot is in phase; i.e., a crest approaching the knot will reflect as a crest. In Fig. 38-11 the situation is reversed by sending a pulse from the lighter to the heavier rope. Here a reflection also takes place at the knot, but this time, the reflected pulse is 180° out of phase with the incident one; i.e., an oncoming crest is reflected as a trough.

This change of phase occurs in light waves as well. Wave 2 in Fig. 38-9 is reflected from the air-to-soap interface (analogous to the lighter to heavier rope junction in Fig. 38-11*b*) and will

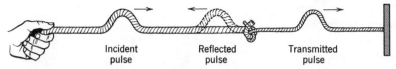

Figure 38-10   A transverse pulse in a heavy rope is traveling to the right, toward the juction with a lighter rope. The reflected pulse from the juction is in phase with the incident pulse.

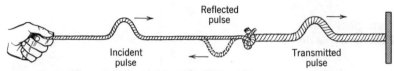

Figure 38-11   A transverse pulse in a rope is traveling toward a junction with a heavier rope. The reflected pulse will be out of phase with the incident pulse.

suffer $180°$ phase change. Wave 3, on the other hand, is reflected from the soap-air interface (analogous to the heavier to lighter rope junction), and will suffer no phase shift. If the film is infinitesimally thin and if, say wave 2 at point P_1 is reflected as a crest, wave 3 will arrive from point P_2 which is very near to P_1 as a trough (because of the $180°$ phase shift and the negligible path-length difference) and the two will cancel, causing the eye to see a dark spot. However, if the thickness of the film is $\lambda/4$, making the path-length difference $\lambda/2$, wave 3 will arrive at point P_3 one half of a wavelength later, and the two waves will reinforce.

The mathematical expressions for cancellation and reinforcement for thin films when viewed by almost normal incident monochromatic light are

$$\text{path-length difference} = 2d = m\lambda_m \quad (38\text{-}6)$$

$$m = 0, 1, 2, \ldots \text{(cancellation)}$$

and

$$\text{path-length difference} = 2d = (m + \tfrac{1}{2})\lambda_m, \quad (38\text{-}7)$$

$$= 0, 1, 2, \ldots \text{(reinforcement)}$$

The λ_m in the two preceding equations refers to the wavelength of the light while in the re-

flecting medium. Recalling from Section 36-3 that λ_m in a medium is given by

$$\lambda_m = \frac{\lambda_{\text{air}}}{n_m} \quad (36\text{-}8)$$

where n_m is the index of refraction of the medium, Eqs. 38-6 and 38-7 become

$$2dn_m = m\lambda_{\text{air}} \quad (38\text{-}8)$$

$$m = 0, 1, 2, \ldots \text{(cancellation)}$$

and

$$2dn_m = (m + \tfrac{1}{2})\lambda_{\text{air}} \quad (38\text{-}9)$$

$$m = 0, 1, 2, \ldots \text{(reinforcement)}$$

Example 38-3   A soap bubble is illuminated normally by monochromatic light having a wavelength of 500 mμ in air. What minimum wall thickness must it have in order to be bright? Take the index of refraction for the bubble wall to be 1.5.

Solution.   In order to be bright, reinforcement must occur. By Eq. 38-9

$$2d \times 1.5 = \tfrac{1}{2} \times 5000 \times 10^{-8} \text{ cm}$$

$$d = 8.333 \times 10^{-6} \text{ cm} \quad \blacktriangleleft$$

Immediately after a soap bubble is formed, the soap fluid will start moving down the bubble from its top surface, causing the thickness of its walls to increase toward the bottom. The different thicknesses will favor the reinforcement of different wavelengths. Therefore, when viewed by white light (which consists of many wavelengths) the eye will perceive the bubble to be multicolored. It is for the same reason that thin patches of oil appear multicolored.

The films of air caught between glass plates placed on top of each other will also give rise to interference patterns. The irregular fringes in Fig. 38-12 indicate air films of different thicknesses, or depressions in one of the plates when the other plate is flat. These patterns are useful insofar as they indicate areas on the plates that require polishing in order to render the plate optically flat. Newton first observed this phenomenon by placing a convex lens on a flat piece of glass, convex side down as in Fig. 38-13a. The interference patterns known as Newton's rings are caused by interference of the light in passing through the air space between the lens and the plate, Fig. 38-13b.

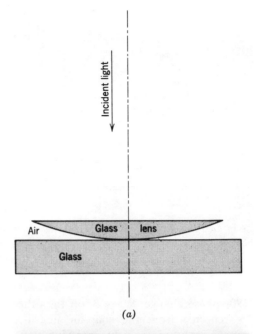

(a)

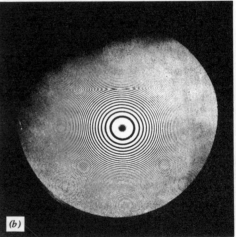

(b)

Figure 38-13 (a) Various thicknesses of air exist between the lens and flat glass surface. (b) These air thicknesses produce symmetrical interference patterns known as Newton's rings (Courtesy Bausch and Lomb, Inc.).

38-6 POLARIZATION

Figure 38-14 shows a transverse wave in a rope heading toward two narrow slots. The wave is

Figure 38-12 Air films of uneven thickness between glass plates.

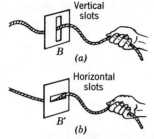

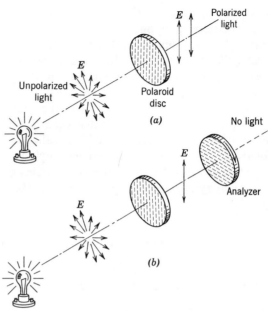

Figure 38-14 (a) The wave train will emerge from the narrow vertical slot at B. (b) The wave will not emerge from the narrow horizontal slot at B'.

Figure 38-15 (a) A polarizer passes light whose electric vibrations are directed along one direction. (b) A polaroid disc used as an analyzer. By turning the analyzer, one can detect whether polarized light is incident on it.

transverse because the direction of the motion of the wave (from right to left in the figure) is perpendicular to the vibrations of the rope. In Fig. 38-14a the wave will emerge from the narrow vertical slot, while in Fig. 38-14b, it will not emerge as a wave.

The analogous behavior of light is called polarization. Light is also a transverse wave. In the case of waves in a rope it is the vibration of the particles of the rope that constitutes the wave; in the case of a water wave it is the material particles of the water that constitute the wave. There are no material particles associated with a light wave. Then what is it that is "waving" in light? It is the magnitude of the electric vector E and the magnetic vector B that vary at a given point in space at a given moment (see Fig. 34-1a). Light is an electromagnetic wave. It is a transverse wave because both E and B are perpendicular to the direction of the light's motion. Transverse waves exhibit a characteristic called *polarization*. Longitudinal waves (sound waves) do not.

Light waves originate in the huge number of atoms that produce the light. Even a single atom of a given light emitter, say the tungsten filament of a lamp, emits waves having different frequencies and energies. Also, the direction of the electric vector E, is, in general, different for different waves. Such waves are said to be *unpolarized*, Fig. 38-15.

The transmission of light through transparent materials such as glass or water is accomplished by the interaction of the electric vibrations of the light wave with the electric charges constituting the transmitting material. In general, unpolarized light entering such a transparent material, will emerge unpolarized. But the alignment of the regular arrangement of the atoms of certain crystals like Iceland spar ($CaCO_3$) is such that the crystals will transmit the light waves whose electric vibrations E are only in one line. Such light is referred to as *line-polarized*. It is possible to properly embed such tiny crystals into very thin sheets of transparent plastic. Such sheets, called *polaroids*, will transmit polarized light when ordinary, unpolarized light strikes it.

When unpolarized light (ordinary light) is reflected from a transparent surface, like water or glass, some of the reflected light is polarized. The direction of the polarization is along lines that are parallel to the reflecting surface. This means that a person who is wearing eyeglasses made of polaroid material, which usually passes light with its electric component in the vertical direction, will not see some of the light reflecting from a lake. Polaroid glasses reduce the glare from horizontal surfaces. A reflecting medium (non-conducting) will polarize unpolarized light that is incident on the medium such that the angle between the reflected and refracted beam is 90° (Fig. 38-16). This angle of incidence, is called the *polarizing angle*, $\angle p$. The polarizing angle can be determined from a knowledge of the index of refraction, by consideration of Snell's law (Section 36-1),

$$n = \frac{\sin \theta_{\text{air}}}{\sin \theta_m} = \frac{\sin p}{\sin \theta_m}$$

But from Fig. 38-16 the angles p and θ_m are complementary, and $\sin \theta_m = \cos p$. Therefore,

$$n = \frac{\sin p}{\cos p}$$

$$n = \tan p \qquad (38\text{-}10)$$

Equation 38-10 is referred to as Brewster's law. As an example, the index of refraction for water is 1.33. By Brewster's law, 1.33 = tan p, and p = 53°. This means that water will reflect as polarized light the unpolarized light that strikes it at an angle of incidence of 53°. Some of the incident light will also enter the water.

SUMMARY

The diffraction of light refers to its bending around edges.

Diffraction effects are noticed when waves pass through narrow openings having a width that is about as small as the wavelength.

Two waves arriving at the same point at the same time interfere with each other. Interference can cause a reinforcement or cancellation of waves.

Diffraction and interference are characteristic of all waves—water waves, sound waves, light waves, radio waves. Polarization effects are exhibited by transverse waves only.

Young's experiment renders important evidence for the wave nature of light.

The multiple colors displayed by thin films like soap bubbles are caused by the interference of the reflected light waves.

QUESTIONS

1. What is the difference between diffraction and refraction?

2. How does the diffraction of light reinforce the wave nature of light?

3. Would you expect particles to diffract? Explain.

4. Explain why refraction does not necessarily indicate that light is a wave.

5. Can radio waves be polarized? Why?

6. Explain the polarization of light.

7. Explain the bright and dark bands in Young's experiment.

8. Would it be reasonable to expect the alternate bright and dark bands in Young's experiment if light consisted of particles? Explain.

9. What is the difference between a transverse and a longitudinal wave?

10. Explain the cause of the bright and dark bands

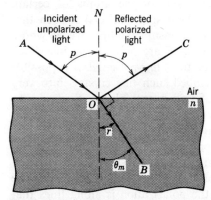

Figure 38-16 A reflecting medium-like water will at the same time reflect and transmit a beam of light. The polarizing angle is that angle of incidence at which the reflected and refracted rays are 90° apart.

exhibited by thin films when viewed by monochromatic light.

11. Sharply defined alternate bright and dark bands will not be formed on a screen when monochromatic light from two separate lamps is used to illuminate the two slits in Young's experiment. Why?

12. Both a glass prism and a thin film of a soap bubble will separate white light into different colors but for different reasons. Explain.

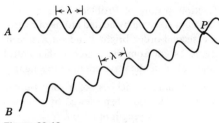

Figure 38-18

9. Will the superposition of the two waves at point P in the preceding figure cause a cancellation or reinforcement? Explain.

10. Explain the cause of the dark bands and bright bonds in Fig. 38-5.

PROBLEMS

(A)

1. Determine the wavelength of the water waves in Fig. 38-1a by measurement with a ruler.

2. If the waves in Fig. 38-1a were produced by dipping the straight-edge into the water at a steady frequency of 20 times per second, what would their speed be?

3. What is meant by the "crest" and "trough" of a water wave?

4. What is the distance between two successive crests in a train of waves?

5. What is the distance between the lowest point on a trough and the highest point of the next crest?

6. The crest of one water wave arrives at a small region in still water at the same time as a crest from another wave.
   (a) Describe the water level where the two crests meet.
   (b) Describe the water level where a crest of one wave meets a trough of another wave.

7. What is the path-length difference, expressed in number of wavelengths, between the wave trains A and B in Fig. 38-17?

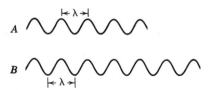

Figure 38-17

8. What is the path-length difference between the wave trains A and B in Fig. 38-18?

(B)

1. If the distance between the two slits is doubled in Young's experiment what will be the change in the distance between centers of
   (a) two successive bright bands and
   (b) two successive dark bands?

2. Red light, $\lambda = 695$ mμ passes through two narrow slits 4 mm apart. Calculate the position of the first dark and third bright bond on a screen 1 m away.

3. Calculate the width of the central bright band for Problem B2.

4. Blue light of wavelength 4600 Å is incident on two slits, 1 mm apart. Calculate the distance between the central bright band and the second dark band on a screen 1 m away.

5. Do Problem B4 with the slit separation doubled.

6. Find the distance between the two first-order bright bands in Problem B4.

7. What is the angular deviation from the center-line of the third bright band produced by green light $\lambda = 5200$ Å in a double-slit experiment, where the separation between slits is 0.1 mm and the screen distance is 1.5 m?

8. A double slit produces a central bright image that is 1.5 mm wide. What is the slit separation if sodium light of 5890 Å wavelength is used and the screen distance is 1.5 m?

9. A certain diffraction grating has 5000 equidistant lines/cm. What is the distance between two adjacent lines?

10. The diffraction grating of Problem B9 is used to analyze the light given off by heated atoms of hydrogen. If the second-order bright band on a screen, 1 m away, is 2.4 cm from the center of the central bright band, calculate the wavelength of this light.

11. What is the minimum thickness of a soap bubble $[n = 1.5]$ that looks red when viewed by reflected red light having a wavelength of 6850 Å?

12. What is the next larger thickness of soap bubble in Problem B11 that will also appear red.

13. Determine the minimum thickness of the part of a soap bubble that looks bright when viewed by reflected green light having a wavelength of 5461 Å.

14. Determine the next two larger thicknesses of the soap bubble in Problem B13 that will also appear bright green.

15. Calculate the polarizing angle for flint glass having an index of refraction of 1.7.

16. What must be the angle of incidence of a beam of light falling on ethyl alcohol having an index of refraction of 1.4 for the beam to be reflected as polarized light?

39.
Modern Physics

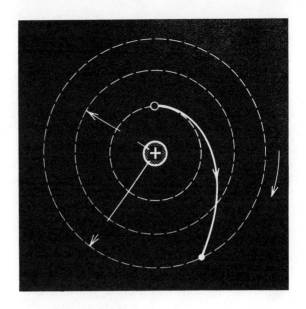

The phrase *modern physics* refers to important investigations and discoveries in physics that were made from about the year 1900. It was about that time that the first important discoveries began on the submicroscopic building blocks of matter—the atom and its components. These discoveries led to other important questions about the structure of matter which, in turn, led to other important discoveries, etc. Another profound advancement in physics was made by the theory of relativity that was formulated by Albert Einstein in 1905. Relativity theory deals with such topics as the conversion of matter to energy, energy to matter, the ultimate speed of particles, to mention but a few. Applications of this theory enhance the understanding of the smallest aspects of nature–atoms and nuclei and the very largest–the formation of stars and galaxies. Modern physics emphasizes atomic and nuclear physics and the theory of relativity. Pre-twentieth century physics and its methods are referred to as *classical physics*.

LEARNING GOALS

1. *To learn some important elements of modern physics.*

2. *To understand the planetary model of the atom.*

3. *To learn about the particle nature of waves and the wave nature of particles.*

4. *To understand how the Bohr model of the atom explains the wavelengths of light emitted by heated atoms.*

5. *To gain an understanding of the nucleus and nuclear transformations.*

6. *To learn about the nuclear reactor.*

7. *To learn about radioactivity.*

39-1 THE ATOM—ATOMIC MASS

Since a given chemical element exhibits certain unique chemical properties, there must exist a smallest "piece" of an element that still exhibits such properties. This smallest piece is called an *atom*. The first quantitative and important information on the atom was advanced in 1810 by the British chemist John Dalton in his atomic theory. Quantitative experiments with many chemical elements and compounds led him to a generalization referred to as the *law of definite proportions*, which asserts that the ratio of the masses of elements in a given compound is always the same. Thus a sample of the compound water always exhibits the ratio of 7.94 g of oxygen to 1 g of hydrogen. If 7.94 g of oxygen is chemically combined with 5 g of hydrogen, only 8.94 g of water would result, with 4 g of hydrogen left over. At the time that Dalton advanced his atomic theory, there were about 20 different elements identified by chemists. Today, there are approximately 100, which means that there are 100 different types of atoms.

How do atoms differ? As a result of Dalton's atomic theory, later discoveries of the formulas for molecules were discovered; e.g., the formula for the water molecule is $H_2O = 2$ atoms of hydrogen and 1 atom of oxygen. A combined knowledge of the molecular formula and the masses of definite proportions leads to the knowledge of the relative masses of atoms. Thus if 7.94 g of oxygen combine with 1 g of hydrogen to form water, the number n_o of oxygen atoms times the mass m_o of one atom is equal to 7.94 g or

$$n_o m_o = 7.94 \text{ g}$$

Similarly, $$n_h m_h = 1 \text{ g}$$

where n_h and m_h, respectively, represent the number of hydrogen atoms and the mass of one atom of hydrogen. Upon dividing the first equation by the second and noting that there are twice as many hydrogen atoms as oxygen atoms in a molecule of water, $n_h = 2n_o$, we get

$$\frac{n_o m_o}{2n_o m_h} = \frac{7.94 \text{ g}}{1 \text{ g}}$$

or $$m_o = 15.88 \, m_h$$

The oxygen atom is 15.88 times as massive as the hydrogen atom. From a knowledge of the molecular formulas and definite proportions of other compounds, the relative masses of the other elements were determined.

The actual mass of atoms, in grams, was deduced in the latter half of the nineteenth century from the hypothesis that was advanced in 1811 by the Italian scientist Avogadro. According to this hypothesis, 22.4 liters of any atomic gas at standard temperature and pressure contains the same number of atoms. This number is called Avogadro's number, N_0 which was experimentally determined in the latter half of the nineteenth century to be

$$N_0 = 6.02 \times 10^{23} \frac{\text{molecules}}{\text{mole}}$$

The mass of a molecule of a given gas can be determined by weighing a known volume of this gas and a knowledge of the number of molecules in this volume, which can be deduced from Avogadro's number. From a knowledge of the number of atoms in this molecule, the mass in grams of the atoms can be determined. The mass of one atom of the most abundant type of carbon was determined to be 19.92×10^{-24} g. Exactly one-twelfth of this mass, or 19.92×10^{-24} g/12 = 1.6600×10^{-24} g, is called one *atomic mass unit* (abbreviated as 1 u) and is used as a unit for expressing the masses of small particles.

$$1 \text{ u} = 1.6600 \times 10^{-24} \text{ g} \qquad (39\text{-}1)$$

The masses of atoms differ because they contain different numbers of the *fundamental particles*: the electron, proton, and neutron. These particles are fundamental in the sense that they are the building blocks of all atoms and, therefore, of all matter. The mass and charge of the electron and proton were experimentally determined in the early part of this century. The neutron and its mass were discovered in 1932. The neutron carries a zero electrical charge. All matter on this earth is electrically neutral in the sense that its atoms contain the same number of electrons as protons.

Table 39-1   Atomic masses of some of the lighter elements and the masses of the three stable subatomic particles

Element	Isotope Symbol	Atomic Number (Z)	Atomic Mass (u)	Atomic Weight (u)	Percent Natural Abundance
Hydrogen	^1_1H	1	1.007825	1.00797	99.985
Hydrogen	^2_1H	1	2.014103		0.015
Helium	^3_2He	2	3.016030	4.0026	0.0001
Helium	^4_2He	2	4.002604		99.9999
Lithium	^6_3Li	3	6.015126	6.939	7.52
Lithium	^7_3Li	3	7.016005		92.48
Beryllium	^9_4Be	4	9.012186	9.012186	100
Boron	$^{10}_5B$	5	10.012939	10.811	18.6
Boron	$^{11}_5B$	5	11.009305		81.4
Carbon	$^{12}_6C$	6	12.000000	12.01115	98.892
Carbon	$^{13}_6C$	6	13.003354		1.108
Uranium	$^{235}_{92}U$	92	235.04393	238.03	0.75
Uranium (heaviest natural atom)	$^{238}_{92}U$	92	238.05076		99.25

Subatomic Particles

Electron	$^0_{-1}e$		0.000549		
Proton	^1_1p		1.007276		
Neutron	^1_0n		1.008665		

The *atomic number* column in Table 39-1 indicates the number of protons (or electrons) in the atoms. The chemical properties of an element depend on this number. There are, however, different types of a given element found in nature. As an example, the most abundant type of hydrogen, occurring in 99.985% of the natural samples of hydrogen, consists of atoms containing one electron and one proton. The atoms of 0.015% of the natural hydrogen contain one electron, one proton, and one neutron. This heavy form of hydrogen, called *deuterium*, has the same chemical properties as the more abundant type of hydrogen because it has the same Z number. Deuterium is an *isotope* of hydrogen. *All isotopes of a given element have the same chemical properties and the same atomic number but a different number of neutrons* (see Fig. 39-1). The total number of

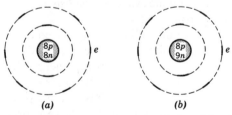

Figure 39-1   A schematic representation of the atoms of two isotopes of oxygen. (a) $^{16}_8O$, and (b) $^{17}_8O$. The electron structure and chemical properties are the same for both.

protons and neutrons in a given atom is called the *mass number* and is denoted by A. The various isotopes are abbreviated by the chemical symbol of the element with the mass number as a superscript and the atomic number as a subscript. Thus, $^{13}_6C$ indicates a carbon atom having 6 protons, 6 electrons and $13 - 6 = 7$ neutrons; $^{238}_{92}U$ indicates an atom of a uranium isotope having 92 protons, 92 electrons and $238 - 92 = 146$ neutrons.

The atomic weight column in Table 39-1 indicates the average masses of the elements found on earth without regard to the fact that the elements may consist of a mixture of different isotopes of the element. The atomic mass column gives the atomic masses of the separate isotopes in atomic mass units. As an example, a given sample of the element boron found on earth may consists of two different isotopes of boron. The mass of an average atom is given in the atomic weight column as 10.811 u, but the mass of an atom of the isotope $^{11}_{5}B$ is 11.009305 u.

39-2 THE PLANETARY MODEL OF THE ATOM

Although the masses of individual atoms were quite well determined by the beginning of the twentieth century, the size of the atom and the manner in which the electrons and protons are distributed within it was not discovered until 1911. It was the English physicist Ernest Rutherford who devised an experiment which established that the protons of an atom are packed in a relatively small volume called the *nucleus*, around which the electrons orbit.

We can deduce the radius of a hydrogen atom in a manner that is similar in principle to the one used to determine the radius of a planet's orbit around the sun (see Section 10-7). The electron moving in a circular orbit must have a centripetal force, $F_c = mv^2/r$ acting on it. This force is provided by the coulomb force F of attraction between the electron and proton, as in Fig. 39-2.

$$F = F_c$$

$$\frac{ke^2}{r^2} = \frac{mv^2}{r}$$

or $\qquad \frac{1}{2}mv^2 = \frac{ke^2}{2r} = K$

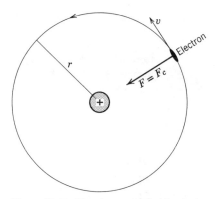

Figure 39-2 Planetary model of the hydrogen atom. The electron revolves in a circular orbit around the nucleus that consists of a proton.

where r is the radius of the hydrogen atom, m and e are, respectively, the electron's mass and charge, and k is the electrostatic constant. The total energy E of the atom (assuming that the proton is at rest) is the sum of the electron's kinetic energy K and electrical potential energy U, where $U = -ke^2/r$ (see Eq. 27-6)

$$E = K + U$$

$$E = \frac{ke^2}{2r} + \frac{(-ke^2)}{r}$$

$$E = \frac{-ke^2}{2r} \qquad (39\text{-}2)$$

The total energy of the hydrogen atom is negative, signifying an attractive force. This force binds the electron to the proton in the atom. These are "tied" together with a *binding energy*, E_B given by

$$E_B = -E = \frac{ke^2}{2r} \qquad (39\text{-}3)$$

To tear the electron and proton apart to a large distance of separation, an external energy, equal to the binding energy, must be provided. This energy is also called the *ionization energy*.

Careful experiments have been performed on binding energy. Results indicate that it takes 13.6 electron volts to ionize the hydrogen atom.

Recalling from Section 27-5 that 1 eV = 1.6 × 10^{-19} joule, we find that

$$13.6 \text{ eV} \times 1.6 \times 10^{-19} \frac{\text{joule}}{\text{eV}}$$

$$= 21.76 \times 10^{-19} \text{ joule}$$

By putting this value for E_B into Eq. 39-3, we can determine the radius of a hydrogen atom. Thus, from the equation $r = ke^2/2E_B$

$$r = \frac{9 \times 10^9 (1.6 \times 10^{-19})^2}{2(21.76 \times 10^{-19})} = 0.53 \times 10^{-10} \text{ m}$$

Since the radius of a nucleus, determined by other experiments, is approximately 10^{-14} m, the ratio of the atomic radius to the nuclear radius is about $10^{-10}/10^{-14} = 10^4$. The radius of an atom is 10,000 times larger than that of a nucleus. Since an electron is much smaller than a nucleus, we conclude that matter consists mostly of empty space.

It is an amazing fact that Newton's laws which explain the motions of such large objects as the planets in the solar system, could also aid in explaining the tiny atom. The planetary model, sometimes referred to as *Rutherford's model* or the *classical model* (because it has been superseded by more modern models) correctly predicts the size of the atom. A model of the atom, called the Bohr model, is a better one because it also explains other occurrences involving the atom. But this model involves two other principles in modern physics, which are discussed below.

39-3 PLANCK'S QUANTUM THEORY

The extension of the planetary model of the atom involves two important quantitative interpretations of the wave-particle duality of light discussed in Section 34-2. The first was advanced by the German physicist Max Planck in 1900. In spite of the fact that light travels through space as a wave, light is emitted and interacts with matter as a particle. Light is emitted as individual bursts of particle-like bundles, called *photons* or light *quanta*, and the energy E of each quantum is expressed by

$$E = hf \qquad (39\text{-}4)$$

where f represents the frequency of the light and h is called *Planck's constant*. The units for h can be determined from Eq. 39-4 to be joule sec. Many careful experiments show the value for h to be

$$h = 6.625 \times 10^{-34} \text{ joule sec}$$

Since the speed of any wave is given by the product of the wavelength λ and frequency, Eq. 39-4 can also be expressed as

$$E = \frac{hc}{\lambda} \qquad (39\text{-}5)$$

As an example, the energy associated with a single photon of light in the violet region, having a wavelength 4000 × 10^{-10} m can be computed from Eq. 39-5 as

$$E = \frac{6.6 \times 10^{-34} \times 3 \times 10^8}{4000 \times 10^{-10}} = 4.95 \times 10^{-19} \text{ joule}$$

or

$$E = 4.95 \times 10^{-19} \text{ joule} \times \frac{1 \text{ eV}}{1.6 \times 10^{-19} \text{ joule}}$$

$$= 3.1 \text{ eV}$$

A single photon associated with violet light will transfer 3.1 eV to an electron when the two collide. When light shines on metals, it can transfer enough energy to the electrons to tear them away from the atoms constituting the metal. These electrons constitute a photocurrent that is utilized in many applications in the field of electronics.

The right sides of Eqs. 39-4 and 39-5 contain the quantities wavelength and frequency, both of which are associated with the wave nature of light. We can interpret the energy E in these equations as the energy transferred to material particles (like electrons) by other "particles." These equations therefore relate the particle nature of light to its wave characteristics. In the game of billiards, when the 8-ball, initially at rest, is struck elastically head-on by the Q-ball, the latter stops and transfers all its kinetic energy to the 8-ball. When a photon strikes an electron, the photon's speed is reduced to zero and the photon ceases to exist. It gives up its entire energy to the electron. The energy of light is said to be *quantized* because it can only exist in whole-number multiples of hf. Also, when light interacts with electrons, it transfers all or none of its energy; it does not transfer a fraction of its energy.

Planck's quantum theory has been greatly extended since its inception in 1900 and has developed into a branch of physics called *quantum physics*. Planck's constant is considered as a fundamental constant of nature.

Example 39-1　Infrared light of wavelength 7500 Å, incident on 1 g of water, causes the temperature of the water to rise 1°C. (a) What is the energy associated with 1 photon of this light? (b) Calculate the number of photons absorbed by the water, assuming that all of the light energy is transformed to heat energy of the water.

Solution

(a) $\lambda = 7500$, Å $= 7500 \times 10^{-10}$ m, $c = 10^8$ m/sec, $h = 6.6 \times 10^{-34}$ joule sec. By Eq. 39-5

$$E = \frac{hc}{\lambda}$$

$$E = \frac{6.6 \times 10^{-34} \times 3 \times 10^8}{7500 \times 10^{-10}} = 26.4 \times 10^{-20} \text{ joule}$$

(b) It takes 1 calorie ($=4.186$ joule) of heat energy to raise the temperature of 1 g of water by 1 C°.

number of photons (n) × energy of 1 photon (E)

$$= 4.186 \text{ joule}$$

$$n \times 26.4 \times 10^{-20} = 4.186$$

$$n = \frac{4.186}{26.4 \times 10^{-20}} = 1.6 \times 10^{19} \text{ photons} \blacktriangleleft$$

39-4 THE WAVE NATURE OF PARTICLES

Section 39-3 dealt with the particle nature of light. In 1924, Louis de Broglie, the French physicist,

hypothesized that *a moving particle also behaves as a wave*. The de Broglie wavelength λ for a particle of mass m and speed v is given by

$$\lambda = \frac{h}{mv} \qquad (39\text{-}6)$$

where h is Planck's constant. As an example, a 1 g ($=10^{-3}$ kg) marble, moving at a speed of 10^3 m/sec, has a de Broglie wavelength of

$$\lambda = \frac{6.6 \times 10^{-34}}{10^{-3} \times 10^3} = 6.6 \times 10^{-34} \text{ m}$$

But how can such a small wavelength be measured? A good experimental test for waves involves their diffraction and interference, as discussed in Section 38-1. It was stated in that section that in order to show up the wave nature of water waves or light waves, slit widths are required that are of the same approximate size as a single wavelength. To construct a slit width as narrow as 10^{-34} m, which is required to prove the wave nature of the moving marble, is practically impossible. However, it can be seen from Eq. 39-6 that if the mass is decreased, its de Broglie wavelength is increased.

Example 39-2　Calculate the de Broglie wavelength of an electron that is accelerated through a potential difference, $V = 100$ volts.

Solution.　From Eq. 27-15 $Ve = \frac{1}{2}mv^2$, where e is the charge on an electron ($=1.6 \times 10^{-19}$ C) and m is its mass (9.1×10^{-31} kg)

$$v = \sqrt{\frac{2Ve}{m}}$$

By Eq. 39-6

$$\lambda = \frac{h}{mv}$$

$$\lambda = \frac{h}{m\sqrt{2Ve/m}} = \frac{h}{\sqrt{2Vem}} = \frac{6.6 \times 10^{-34}}{54 \times 10^{-25}}$$

$$= 1.2 \times 10^{-10} \text{ m} \qquad \blacktriangleleft$$

The above example indicates that an electron having a kinetic energy of 100 eV can be expected to exhibit a wavelength of about 1 Å. In 1927 two American physicists, C. S. Davisson and L. H. Germer, performed an experiment involving the reflection of 54 eV electrons from a nickel crystal. They reasoned that the natural and regular spaces between the atoms of some crystals is about 1 Å, and that such crystals can provide the small "slit" widths required to bring out the wave nature of electrons. The resulting electron diffraction pattern indicated that there is a wave associated with electrons and that the wavelength is correctly given by the de Broglie relation, Eq. 39-6. Other experiments involving larger particles like neutrons and atoms indicated that these also behave as waves. The de Broglie hypothesis suggests a certain symmetry to nature. Waves exhibit particle characteristics and particles exhibit wave characteristics.

An important application of the de Broglie hypothesis is the *electron microscope*, which is used extensively in biological and medical research. The ability of the lenses of any microscope to render fine detail depends on the wavelength of the light used. The shorter the wavelength the better is the resolution (see Eq. 37-11). The reason for this stems from the diffraction effects which become more prominent when the wavelength is of the same approximate size as the specimen to be examined by the microscope. Therefore, an optical microscope, when using even violet light, which has the shortest visible wavelength, $\lambda = 4000$ Å becomes unsuitable to examine small specimens of about 10^{-6} m in size. The de Broglie wavelength of 100-V electrons is about 1 Å. These are much more suitable for the "illumination" and high magnification of very small specimens, which have a size of 10^{-6} m or smaller. Electrons cannot be focused by glass lenses. The focusing in electron microscopes is accomplished magnetically, and a magnification of 50,000 is achieved.

39-5 THE BOHR ATOM

As was stated above, the classical model of the atom correctly predicts that the protons in an atom are concentrated in a relatively small volume called the nucleus and that the radius of a hydrogen atom is of the order of 0.5×10^{-10} m. But there are other occurrences involving atoms that this model must explain. These have to do with the production of light.

The production of light takes place in the atom. Thus, for instance, the light furnished by an electric lamp is produced by the atoms of its filament. The electric energy from the battery is somehow converted by the filament's atoms into light energy, which is radiated out. When hydrogen atoms are excited by intense heat or by the energy resulting from an electric spark, the atoms will emit visible light, the wavelengths of which can be separated and identified by a spectrometer (see Section 37-13). The resulting emission spectrum is called a *line spectrum* because it is composed of separated and distinct lines which are in the visible, infrared, and ultraviolet regions. For instance, the H_α and H_β lines in Fig. 39-3 indicate that hydrogen atoms emit light of wavelengths 6563 and 4863 Å, respectively. Why this selectivity? Why, for instance, do hydrogen atoms not emit light of wavelength 5000 Å?

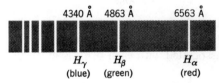

Figure 39-3 Emission spectrum of atomic hydrogen. The lines represent distinct colors having the given wavelengths.

Such questions were not satisfactorily answered until the Danish physicist Niels Bohr proposed in 1913 an extension of the planetary model of the atom considered in Section 39-2. According to Bohr, electrons orbiting a nucleus do not emit light. However, when the atom is somehow energized as by intense heat, the orbiting electron is caused to jump into a higher orbit having a larger radius. At the larger radius, the electron's energy is larger. At a small time later, the electron will

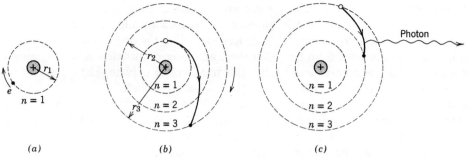

Figure 39-4 The hydrogen atom. (a) The smallest radius, r_1, of the electron's circular orbit around the nucleus. (b) As the atom is energized, the electron jumps out to a higher orbit. (c) A moment later, the electron will jump back to an inner orbit, radiating light in the process.

spontaneously jump back to some lower orbit, giving up some of its energy in the form of light, which is radiated out into space (Fig. 39-4).

The Bohr model of the hydrogen atom allows only certain radii, r_n, of the orbiting electron given by

$$r_n = n^2 r_1, \qquad n = 1, 2, 3 \ldots \qquad (39\text{-}7)$$

and the lowest radius, r_1 is

$$r_1 = \frac{h^2}{4\pi^2 kme^2} \qquad (39\text{-}8)$$

where h is Planck's constant and m and e are the mass and charge of the electron, respectively. The energy E_n associated with each orbit is given by

$$E_n = \frac{1}{n^2} E_1, \qquad n = 1, 2, 3 \ldots \qquad (39\text{-}9)$$

$$E_1 = \frac{-ke^2}{2r_1} \qquad (39\text{-}10)$$

where k is the electrostatic constant. Notice that the energy can have only certain values, $E_1/1$, $E_1/4$, $E_1/9$, etc. The energy of an electron in an atom is said to be *quantized*. The quantity n in Eqs. 39-9 is called the *principal quantum number*, which indicates the ring or shell in which the electron is orbiting. The value $n = 1$ indicates the closest ring to the nucleus, which is also called the K shell; $n = 2$ indicates the next larger ring and is called the L shell, etc.

Example 39-3 (a) Calculate the smallest and next larger radius at which the electron can orbit in the hydrogen atom. (b) What are the electron's energies at these orbits?

Solution
(a) By Eq. 39-8

$$r_1 = \frac{(6.6 \times 10^{-34})^2}{4\pi^2 \times 9 \times 10^9 \times 9.1 \times 10^{-31}(1.6 \times 10^{-19})^2}$$

$$= 0.528 \times 10^{-10} \, \text{m}$$

This is the smallest radius of the hydrogen atom. When the atom is energized, the electron can jump to the next higher orbit, which by Eq. 39-7 is given as

$$r_2 = 4r_1 = 2.112 \times 10^{-10} \, \text{m}$$

(b) By Eq. 39-10

$$E_1 = \frac{-9 \times 10^9 (1.6 \times 10^{-19})^2}{2 \times 0.528 \times 10^{-10}}$$

$$= -21.82 \times 10^{-19} \, \text{joule}$$

$$= -13.64 \, \text{eV}$$

This represents the electron's energy in the lowest orbit. Letting $n = 2$ in Eq. 39-9, we get

$$E_2 = \frac{E_1}{(2)^2} = \frac{-13.64 \, \text{eV}}{4} = -3.41 \, \text{eV}$$

which is the electron's energy while moving in the

second orbit. Notice that $E_2 > E_1$ because E_2 is less negative than E_1. ◀

Figure 39-5 is called an energy-level diagram and depicts the electron's energies while at the different electron shells, as calculated from Eq. 39-9. In jumping from a higher to a lower shell, the electron releases a photon whose energy can be computed from the difference in the electron's energy in the two shells. For instance, the H_B line in Fig. 39-5 consists of light that has a wavelength of 4863 Å, as measured by a spectrometer. This corresponds to the light emitted by an electron when it jumps from the fourth to the second orbit. Thus, by Eq. 39-9

$$E_4 = -\frac{1}{(4)^2} E_1 = -\frac{13.64}{16} \text{ eV}$$

$$= -0.85 \text{ eV}$$

$$E_2 = -\frac{13.64}{(2)^2} \text{ eV} = -3.41 \text{ eV}$$

$$E_4 - E_2 = -0.85 \text{ eV} - (-3.41 \text{ eV})$$

$$= 2.56 \text{ eV}$$

$$2.56 \cancel{\text{eV}} \times 1.6 \times 10^{-19} \frac{\text{joule}}{\cancel{\text{eV}}} = 4.1 \times 10^{-19} \text{ joule}$$

This is the energy of the photon that is radiated out. Its wavelength can be computed from Eq. 39-5

$$\lambda = \frac{hc}{(E_4 - E_2)} = \frac{6.6 \times 10^{-34} \times 3 \times 10^8}{4.1 \times 10^{-19}}$$

$$= 4.829 \times 10^{-7} \text{ m} = 4829 \text{ Å}$$

This is in pretty good agreement with the measured wavelength. The other spectral lines can be similarly computed.

Did Bohr's model of the atom completely describe the atom? Soon after his theory was advanced, better spectroscopes were developed which identified other spectral lines that the Bohr model did not explain. It was also determined that the spectral lines change when atoms are subjected to an external magnetic field. To account for these, Bohr's model was extended to include elliptical orbits of the electrons in addition to the original circular orbits. Other analyses included the motion of the nucleus, which the original Bohr model did not take into account. A more general analysis of atomic structure and the radiation of light involves mathematical solutions of the *wave equation*, first advanced by the German physicist Erwin Schrödinger in 1926. The Schrödinger equation has given rise to a branch of physics called *wave mechanics*.

It should be realized that any fundamental experiment involving atomic structure must use instruments that are fine enough to yield measurements associated with the smallest particle in nature, the electron. Since such instruments are impossible to design, we resort to "thought experiments." We imagine a microscope powerful enough to locate an electron in its orbit around the nucleus. To see this electron through the microscope, the electron would have to be illuminated by photons of light that would bounce off from it into the microscope. But the collision between photon and electron would alter the latter's

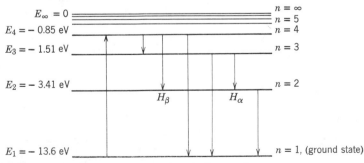

Figure 39-5   Energy level diagram for a hydrogen atom.

original position. The position and velocity of the electron would be changed by the very instrument designed to measure them. Such ideas are expressed by the *uncertainty principle*, first advanced by the German physicist Werner Heisenberg in 1925. The uncertainty principle, quantum theory, wave equation, and de Broglie waves are involved in present-day analyses of atomic structure.

39-6 THE LASER

An instrument called the *laser*, which stands for light amplification by stimulated emission of radiation, is based on a fascinating and useful application of the principles associated with the Bohr model of the atom and quantum theory. According to the Bohr model, when a hydrogen atom absorbs a quantum of light of proper energy, its electron is caused to jump to a higher orbit and the atom is in an excited state. A short time after, the electron will spontaneously jump back into an orbit of lower energy and in the process emit a photon. This is the explanation of how light is produced, Fig. 39-4. In the case of the usual light source (say the filament of an electric lamp) a huge number of atoms are involved in producing light in this manner, which is called *spontaneous emission*.

An excited atom can be made to emit two identical photons and in the same direction when the atom is struck by a photon from an external source, Fig. 39-6. This process is called *stimulated*

emission and can occur if the energy of the incident photon is the same as the energy difference of the atom in its excited state and its energy after the photon emission. The two photons emitted are identical in energy and direction and the light waves associated with the photons are coherent and in phase, that is, move through space with the crests and troughs in line. This is unlike the usual light source where the spontaneous emission produces waves that are out of phase and that move out in many directions.

In a laser, the two emitted photons will strike two similar excited atoms, with each struck atom emitting two more identical photons, resulting in four identical photons. These can strike four other excited atoms resulting in eight excited atoms, etc. This chain reaction produces a huge number of identical photons in a very short time. Because of the coherence of the light waves associated with the photons, and their single direction the intensity (energy per unit time per unit area) of the light beam emerging from the laser is very high. Short bursts of energy at a rate of 10^{13} watts have been produced! In addition, the laser beam is collimated into a narrow pencil of light that remains narrow even after traveling through a very large distance. A beam emitted from the usual light source spreads out as it moves away from the source, thereby causing the light energy per unit area to decrease with increased distance from the source.

In order to produce the huge number of identical photons, there must be a supply of a large number of excited atoms at the proper energy

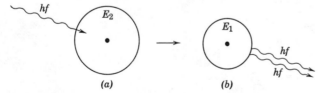

Figure 39-6 Stimulated emission of photons. (*a*) A photon of energy *hf* strikes an excited atom in an energy state E_2. (*b*) If $hf = E_2 - E_1$, the atom emits another photon of the same energy. The emitted and incident photons are radiated out in phase.

level. The excited atoms of a lamp filament are at many energy levels and, therefore, are not suitable. Atoms of a material called ruby, which is aluminum oxide with traces of chromium, are raised to a narrow energy band when excited by white light. Continuous *optical pumping* of the ruby atoms to the desired energy level is accomplished by the photons produced by an ordinary lamp, Fig. 39-7. The photons with an energy that is favorable for the production of many identical photons are given many opportunities to continue the chain reaction by their multiple reflections from the two flat sides of the cylinder made of ruby. The laser beam emerging has a single wavelength that appears red.

Because of its high intensity and narrowness the laser beam has wide applications. It has been used to precisely determine the distance between the moon and earth. A laser beam, directed at the moon, is reflected back to earth by a reflector left on the moon by the astronauts. By measuring the time of travel for the round trip and from a knowledge of the speed of light, the moon-earth distance can be determined to an accuracy of a few meters. The laser beam is also applied in surgical procedures, some involving the removal of unwanted tissue and other types of surgery involving the eye. It is also used to deliver the

high energy required to start controlled nuclear fusion in experiments that hopefully will lead to cheaper and cleaner nuclear energy (Section 39-9).

39-7 THE NUCLEUS AND THE THEORY OF RELATIVITY

The classical and Bohr models of the atom held that the atom consists of planetary electrons revolving around a nucleus containing protons. These models were extended in 1932 to include another fundamental particle, the *neutron*, which also resides in the nucleus. The mass of the neutron is slightly greater than that of the proton, and the neutron carries a zero electrical charge. Since the nucleus is a compact volume having a radius of about 10^{-14} m, the question arises as to why the repelling forces of the protons do not cause these particles to fly apart and shatter the nucleus. Experimental evidence indicates that at the small nuclear distances, protons attract other protons, protons attract neutrons, and neutrons even attract other neutrons. These are referred to as *nuclear forces*, which are much stronger than electrical forces. These nuclear forces cause a *nuclear binding energy* that holds the nucleus together.

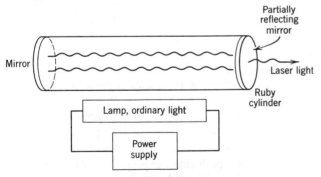

Figure 39-7   Sketch of a ruby laser. The ruby atoms are raised to an excited energy state by the light from the lamp. These will spontaneously release photons of different energies. Those having a favorable energy will stimulate the excited ruby atoms to emit other identical photons that are radiated out in the narrow laser beam.

The nuclear binding energy is computed from a principle in a most fundamental theory in physics called the *special theory of relativity*, which was advanced by Albert Einstein in 1905. This principle involves the equivalence between mass and energy that is given by the famous equation,

$$E = mc^2 \qquad (39-11)$$

The mass m in this equation refers to the moving mass of a particle which is different from the mass, m_0, called the *rest mass*, when the particle is at rest. The relation between the two masses is given by

$$m = \frac{m_0}{\sqrt{1 - v^2/c^2}} \qquad (39-12)$$

where v and c, respectively, represent the speed of the particle and the speed of light in vacuum. For example, if the speed of a particle is 98% of the speed of light, $v = 0.98c$, the particle's moving mass is

$$m = \frac{m_0}{\sqrt{1 - [(0.98c)^2/c^2]}} = \frac{m_0}{0.2} = 5\,m_0$$

A person who is at rest with respect to this moving particle would measure its mass to be 5 times its rest mass. For speeds that we ordinarily encounter or even for the speeds associated with airplane jets or space vehicles, the ratio of v^2/c^2 is so very small that Eq. 39-12 reduces to $m = m_0$. However, electrons and other particles are accelerated in large research accelerators to speeds which approach c. The moving mass of these particles is experimentally determined to be larger than their rest mass. The theory of relativity also maintains that the speed of light in vacuum is the ultimate speed and that a material particle can only approach this speed but never quite reach or exceed it. This has also been verified by the modern particle accelerating machines.

If the value for m in Eq. 39-11 is replaced by its equivalent value given by Eq. 39-12, we get

$$E = \frac{m_0 c^2}{\sqrt{1 - v^2/c^2}}$$

For low speeds, $v^2/c^2 \rightarrow 0$, giving

$$E_0 = m_0 c^2 \qquad (39-13)$$

where E_0 and $m_0 c^2$ are referred to as the *rest energy* of a particle and m_0 is called the *rest mass*. In addition to the other types of energies associated with a body at rest, which were discussed in earlier chapters, such as gravitational and electrical potential energies, a body also possesses a rest energy given by Eq. 39-13. This equation also asserts an equivalence between mass and energy, in the sense that mass can be created out of an equivalent amount of energy or that energy can be created out of an equivalent amount of mass. It is the latter type of mass–energy equivalence that is responsible for the tremendous energy release in atomic and hydrogen bombs.

That mass can be created out of energy is verified in experiments involving *pair production*. The energy of a photon (having zero rest mass) can be converted to an electron and another particle, called a *positron*, both having a non zero rest mass. A positron is a particle having the same mass as an electron but carrying a positive charge. The minimum energy that the photon must have to accomplish this can be obtained from Eq. 29-13 as $2m_0 c^2$, where $m_0 = 9.1 \times 10^{-31}$ kgm is the rest mass of an electron which is the same for the positron:

$$2m_0 c^2 = 2 \times 9.1 \times 10^{-31}(3 \times 10^8)^2$$
$$= 163.8 \times 10^{-15} \text{ joule}$$
$$= 1.02 \text{ MeV}$$

Photons possessing this energy are called gamma rays which are spontaneously emitted from some radioactive isotopes. Figure 39-8 is a photograph showing the destruction of a gamma ray and the creation of an electron positron pair.

We now return to the binding energy of nuclei. Carefully performed experiments reveal that the mass of a given nucleus is less than the combined masses of its constituent particles, measured individually. For example, the mass of a helium nucleus consisting of 2 neutrons + 2 protons is measured as 4.001506 atomic mass units. The

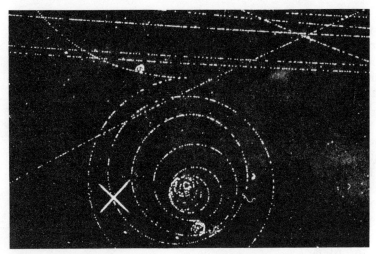

Figure 39-8   Pair production. This bubble chamber photograph shows the production of an electron-positron pair. The paths in a pair are indicated by two curved lines in opposite directions starting from a single point. (Photograph courtesy of Brookhaven National Laboratory.)

mass of a single neutron is 1.008665 u and that of a proton is 1.007276 u. Therefore,

mass of 2 free protons + 2 free neutrons

$$= 4.031882 \text{ u}$$

mass of helium neucleus $\quad = 4.001506 \text{ u}$

mass difference $\quad\quad\quad = \overline{0.030376 \text{ u}}$

There is a mass loss, called a mass defect of

$$0.030376 \,\text{u} \times \frac{1.66 \times 10^{-27} \,\text{kg}}{\text{u}} = 5.042 \times 10^{-29} \,\text{kg}$$

This loss of mass represents a gain in energy given by Eq. 39-12 as

$$E = m_0 c^2$$
$$= 5.042 \times 10^{-29} \text{ kg} \times (3 \times 10^8 \text{ m/sec})^2$$
$$= 45.38 \times 11^{-13} \text{ joule}$$

$$45.38 \times 10^{-13} \text{ joule} \times \frac{1 \text{ eV}}{1.6 \times 10^{-19} \text{ joule}}$$

$$= 28.3 \times 10^6 \text{ eV} = 28.3 \text{ MeV}$$

The four particles in the helium nucleus are bound by this energy. To shatter this nucleus into its four constituent separated particles would require an external energy of 28.3 MeV. The binding energy of heavier nuclei is even greater. It is interesting to compare the atomic and nuclear binding energies. In Section 39-2 it was stated that the binding energy of a hydrogen atom is about 13.6 eV. This is extremely low compared with the 28.3 MeV for the helium nucleus. Even the highest energies associated with atoms are much smaller than the energies "locked up" in nuclei.

39-8 RADIOACTIVITY

The very strong forces among the nuclear particles cause the nuclei of some isotopes to be unstable and to naturally disintegrate. Such isotopes are called *radioactive*. The disintegration of radioactive nuclei is accompanied by ejections of energetic particles and electromagnetic rays, thus leaving the remaining nucleus more stable. There is no clear understanding of the reasons for the stability or instability of nuclei. The stable lighter elements up to mass number 40 have about the same number of neutrons and protons, as exemplified by $_2^4$He, $_6^{12}$C and $_{20}^{40}$Ca. The stable heavier nuclei have a preference for an excess of neutrons over protons, as exemplified by the elements molybdenum ($_{42}^{92}$Mo), cesium ($_{58}^{136}$Ce), and others.

Radioactivity was discovered quite accidentally by the French physicist Henri Becquerel in 1896. He happened to leave photographic film wrapped in a light-tight cover in the same drawer with some uranium salts that he used for his experiments on fluorescence. After developing the film he observed its being fogged in the same way that it would be if it were exposed to ordinary light. Evidently, the rays from the uranium salts were strong enough to penetrate the paper in which the film was wrapped, which ordinary light cannot do. Pierre and Marie Curie pursued this discovery by their investigations of the various ores of uranium in order to isolate the elements that emit these strong radiations. In 1898 they discovered the two new elements radium and polonium, both of which are radioactive. The latter was named after Madame Curie's native country, Poland.

The nuclei of some radioactive elements disintegrate by their emission of a chunky particle, called an alpha (α) particle. This particle really consists of four particles, two protons and two neutrons which makes the alpha particle identical with a helium nucleus. As an example, the isotope of radium, $_{88}^{226}$Ra consists of atoms whose nuclei contain 88 protons and $226 - 88 = 138$ neutrons. This is a radioactive isotope because each of its nuclei spontaneously emits an alpha particle. This disintegration can be summarized by the nuclear reaction.

$$_{88}^{226}\text{Ra} \rightarrow \, _{86}^{222}\text{Rn} + \, _2^4\text{He} \qquad (39\text{-}14)$$

where Rn is the chemical symbol for the element radon and $_2^4$He is the symbol for an alpha particle (2 protons and $4 - 2 = 2$ neutrons). Thus this isotope of radium is unstable and naturally *transmutes* into a nucleus of another chemical element, radon. Alpha particles are ejected at high speeds from some isotopes (Fig. 39-9). This makes these alphas suitable as projectiles in experiments involving the bombardment and the shattering of other stable nuclei. Such experiments have yielded important information on the nucleus and its component particles.

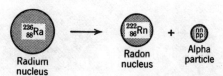

Figure 39-9 Nuclei of the isotope radium-226 spontaneously eject chunky particles called alpha particles.

Some radioactive elements disintegrate by emissions of *beta* (β) *particles* (Fig. 39-10). The β particle carries the same charge and is of the same mass as an electron. For instance, the disintegration of radioactive carbon $_6^{14}$C is summarized by

$$_6^{14}\text{C} \rightarrow \, _7^{14}\text{N} + \, _{-1}^0\beta \qquad (39\text{-}15)$$

where $_{-1}^0\beta$ is the symbol for a beta particle and $_7^{14}$N represents the nucleus of the element nitrogen. At some time later, this nucleus will gather seven electrons from its surroundings and become a nitrogen atom. The production of a beta particle by a nucleus is achieved by the disintegration in the nucleus of a neutron into a proton and a β particle, with the latter ejected. This causes the carbon nucleus to gain one proton and thus become a nitrogen nucleus. The ejection of betas is sometimes accompanied by the emission from the nucleus of a *gamma* (γ) *ray*. The γ ray has no mass, carries no electrical charge, and is similar to photons of ordinary light or X rays, except the γ is more energetic.

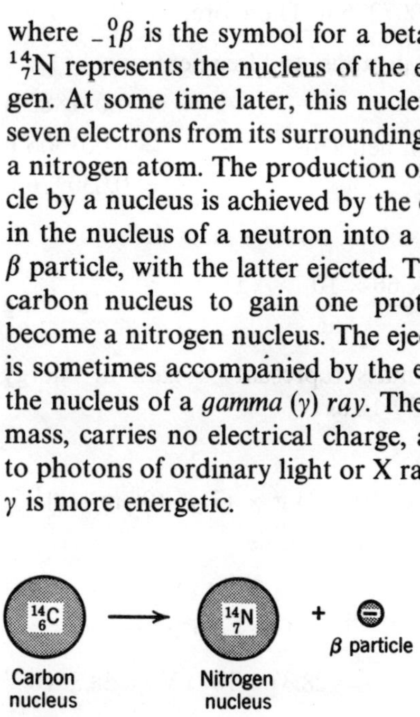

Figure 39-10 Beta decay.

It should be noticed in Eqs. 39-14 and 39-15 that the sum of the subscripts on the left side of the arrow is the same as that on the right side. This indicates that in a nuclear reaction the net charge remains the same. Also, the sum of the superscripts is the same on both sides of the arrow. This indicates that the combined sum of protons and neutrons before a reaction is the same as the sum of these particles after the reaction.

The time for decay of radioactive nuclei is measured by a time called the *half-life*. The half-life of a given isotope is the time that it takes for half the number of its nuclei to disintegrate. For example, the half-life of cobalt–60 (^{60}Co) is 5.26 years. This means that in this time the number of its radioactive nuclei will drop to one-half of the original number. The half-lives of different isotopes vary over a wide range, from a fraction of a second to billions of years. Also, the amount of a given isotope is measured in *curies*, where one curie represents an activity of 3.7×10^{10} disintegrations per second.

Example 39-4 The radioactive isotope sodium, $^{24}_{11}$Na, is a beta emitter and has a half-life of 15.0 hours. Suppose that the activity of a sample of this isotope is now measured to be 5 curies. How much of it will remain at a time of 75 hours ($= 5$ half-lives) later?

Solution. In a time of one half-life (15 hr), $\frac{5}{2}$ curies remain; in a time of 2 half-lives, $5/(2 \times 2)$ curies remain; in a time of 5 half-lives, $5/2^5$ curies $= 0.156$ curies remains. ◀

The energies of the emitted α and β particles and γ rays can be as high as several MeV. Consequently, these energies will penetrate and move through matter with which they happen to come in contact. The chunky α particle will cause much ionization because of its collisions with the atomic electrons of the matter that it moves through. This is also true, although to a lesser extent, of the betas and gammas. The ionization of human tissue can cause certain chemical changes within the body that are extremely harmful. At the same time, radioactive isotopes are widely used for the diagnosis and treatment of certain diseases. One example involves the radioisotope sodium, $^{24}_{11}$Na. When injected into the bloodstream, this chemical will be distributed to all parts of the body within a few minutes. The emitted betas are energetic enough to exit the body and be "counted" by an instrument called a Geiger–Mueller counter. By placing this instrument near a given part of the body, say the foot, physicians can determine whether a proper amount of sodium has reached the foot. In this way, the proper or improper mode of blood circulation can be determined by the β particles, which act as tracers of the blood in its movement in the body.

39-9 THE NUCLEAR REACTOR

A nuclear reaction that involves the shattering of a heavy nucleus when it is bombarded by neutrons, into two lighter nuclei, is called *nuclear fission*. The fissioning of nuclei of the uranium ($^{235}_{92}$U) isotope is accompanied by a release of energy that is harnessed by a machine called a *nuclear power reactor*. The nuclear energy is converted into heat energy of water, producing steam. The steam is made to drive electric generators, thus producing electrical energy, which is transmitted by power lines to consumers.

The nuclear fuel used to run most reactors is the uranium isotope, $^{235}_{92}$U, which constitutes about 0.75% of the mined uranium. This isotope has to be separated from the remainder, or 99.25% of the mined uranium, which is the isotope $^{238}_{92}$U. The energy resulting from fission stems from a mass loss. A slow moving neutron will be captured by a nucleus of ^{235}U causing the nucleus to become excited. In a short interval thereafter, the excited nucleus will split (fission) into a pair of lighter nuclei and also eject a few free neutrons. In a typical fission, the resulting pair of fragments are the nuclei of the elements molybdenum and xenon plus two free neutrons. The combined mass of the fragment nuclei plus the two free neutrons is less by about 0.22 atomic mass units than the combined mass of the ^{235}U nucleus plus the neutron that caused its fission. By the equivalence of mass–energy principle, the energy release per fission is

$$0.22 \text{ u} \times 934 \frac{\text{MeV}}{\text{u}} = 205 \text{ MeV}$$

This released energy is distributed in the kinetic energies of the fission fragments and neutrons and the energies associated with the released beta particles and gamma rays. Reactors are designed with surrounding shields in order to protect personnel from the possible harmful effects of the β particles and γ rays.

Example 39-5 How many fissions per second are required to produce 1000 watts of power in a ^{235}U reactor?

Solution. The energy release per single fission is about

$$200 \frac{\text{MeV}}{\text{fission}} \times 1.6 \times 10^{-13} \frac{\text{joules}}{\text{MeV}}$$

$$= 320 \times 10^{-13} \text{ joule/fission}$$

$$1000 \text{ watt} = 1000 \frac{\text{joule}}{\text{sec}} \times \frac{1 \text{ fission}}{320 \times 10^{-13} \text{ joule}}$$

$$= 3 \times 10^{13} \text{ fission/scc} \quad \blacktriangleleft$$

The fissioning of different uranium nuclei in a reactor can yield different pairs of fragment nuclei which are accompanied by the release of different small numbers of free neutrons. On the average, 2.4 free neutrons are released per fission. The free neutrons strike neighboring uranium nuclei, causing them to fission and expel more free neutrons. This *chain reaction* causes a huge buildup of the number of free neutrons and the number of fissions in a short time. This process is controlled by the insertion of *control rods* into a reactor, the rods being made of a material that absorbs the free neutrons, thus reducing the number of fissions. See Fig. 39-11.

The probability for the fission of a ^{235}U nucleus is greatly increased if the speed of the neutron striking it is much lower than the speed with which the neutrons are ejected after a fission. To lower the neutron speeds, the fuel in a reactor is usually immersed in ordinary water, which is called a *moderator*. The collisions between the fast neutrons and the nuclei of the water slow the former down to the required low speeds, called *thermal speeds*, which are most favorable for the fission process. The slow neutrons are called *thermal neutrons*; the reactors requiring thermal neutrons are classified as *thermal reactors*. The circulating water moderator also acts as a *coolant* of the regions near and at the fuel rods, where a large amount of heat is produced. In some reactors the water is pressurized to allow it to reach higher than the normal boiling temperature, thus making the heat transfer process more efficient.

The three nuclear fuels that can be used in thermal fission reactors are $^{235}_{92}$U, another isotope of uranium, $^{233}_{92}$U, and plutonium ^{239}Pu. Of these fuels, only ^{235}U occurs naturally while the other two can be produced artificially in *breeder reactors*. Uranium-233 is produced by the nuclear reaction involving the combination of a neutron with a nucleus of the element thorium, which is naturally more abundant than uranium. Plutonium is produced by a combination of a neutron and a nucleus of uranium-238. When "aprons" of thorium or uranium-238 are inserted into a reactor, it is possible to obtain more new fuel of ^{233}U or Pu than the uranium ^{235}U used up.

In 1976 approximately 8% of the total commercial electrical power produced in the United States was derived from nuclear fission. A national debate continues on the advisability of the construction of more nuclear power plants. On the one hand, nuclear energy can diminish our energy problem, which is caused by the diminishing supply and high price of the conventional oil or coal that fuels the conventional electrical power plants. Also, the exhausts produced by the burning of oil or coal pollute the atmosphere by releasing toxic gases like sulfur dioxide. Nuclear power plants are "cleaner" in this respect. At the same time there are other types of risks associated with nuclear power plants. These involve the intense radioactivity produced in the components of a nuclear reactor. The longer a nuclear plant is in operation the more radioactive scrap

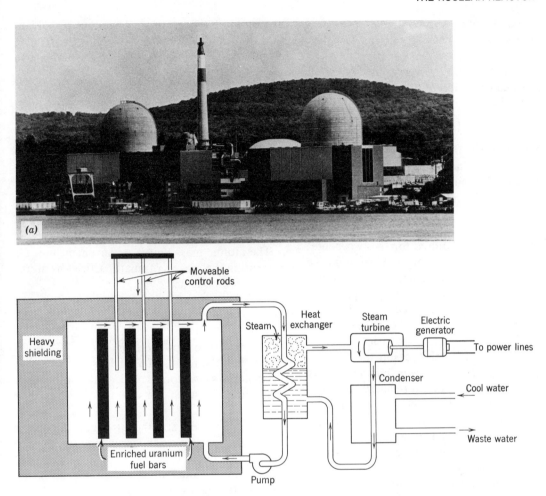

Figure 39-11 (a) View of the Con Edison Nuclear plant at Indian Point, N.Y. Courtesy of Consolidated Edison Company of New York, Inc. (b) Schematic diagram.

it accumulates, which has to be disposed of in some way. Since the scrap is contaminated by the radioactive by-products of fission having half-lives of hundreds and thousands of years, its disposal could in time pollute our land and waters. In addition, it is possible to hypothesize a major nuclear accident in a nuclear power plant that could result in a release of radioactive contaminants to the atmosphere and spread to the surrounding heavily populated areas.

The present debate is also accompanied by a research effort on the part of our government

and the nuclear power industry on safe ways of the disposal of nuclear wastes, and on ways to minimize to a probability approaching zero, the chances for a major accident in a nuclear power plant.

A research effort in this and in other countries is now in progress on the harnessing of the energy resulting from a nuclear reaction called *fusion*. Although the first fusion reactor is projected to a time of about a few decades or so from now, its operation will greatly reduce the problems associated with the radioactive by-products. A nuclear reaction that involves the combination of two lighter nuclei into a heavier nucleus is called *fusion*. One such reaction that has been achieved in laboratories for only a very short duration involves the combination of two hydrogen nuclei of the isotope deuterium, $_1^2H$.

$$_1^2H + _1^2H \rightarrow _2^3He + _0^1n + 3.3 \text{ MeV}$$

where $_1^2H$ represents the nucleus of the hydrogen isotope deuterium, the $_0^1n$ represents a neutron and $_2^3He$ is the nucleus of the isotope of helium having 2 protons and 1 neutron. This reactions is accompanied by the indicated energy release that stems from an equivalent mass loss.

External energy is required for the two deuterium nuclei to fuse because of the large repulsive forces between these nuclei at small distances (but greater than nuclear distances) of separation. This energy is provided by heating them to a very high temperature, about 10^7 degrees, thereby causing the nuclei to smash into each other at high speeds and fuse. For this reason, the reaction is also called a *thermonuclear reaction*. It is believed that the huge amount of energy released by the sun is produced in this way. This is also the reaction that causes the very large energy release produced by a hydrogen bomb.

The harnessing of the energy involves the containment of the very high speed deuterium nuclei and the controlled conversion of the nuclear energy into electrical energy. One avenue taken by researchers is to use external magnets to confine the nuclei into the desired volume. Since the nuclei carry a net electrical charge, their direction of motion can be controlled and confined by the direction of the field. Researchers are also involved in a method for attaining the high speeds required for a fusion reaction. One method involves condensing the nuclear fuel to small pellets which are then bombarded by high energy laser beams. The fusion fuel, deuterium, can be extracted from a virtually limitless supply in the waters of the ocean. This is one large advantage over fission reactors, which rely on the mined uranium or thorium, which is estimated to exist in quantities that will be enough to supply us for only 100 years or so. The other big advantage of fusion reactors over fission is the relative cleanliness of the former as far as radioactive wastes are concerned.

SUMMARY

All atoms of a given chemical element have the same number of electrons and protons.

Different isotopes of a given element exhibit the same chemical characteristics and are composed of atoms having the same number of electrons and protons but different numbers of neutrons.

The atomic mass unit is a convenient unit for measuring the masses of small particles as atoms and its components.

The protons and neutrons of an atom are concentrated in a small dense volume called the nucleus.

The radius of an atom is about 10^4 times larger than the radius of a nucleus.

Coulomb's law alone is inadequate for the small separations between the charged particles in a nucleus.

Nuclear forces are the strongest forces in nature and give rise to the large binding energies of nuclei.

The Bohr model of the atom explains the color of the light emitted by energized atoms of hydrogen.

The theory of relativity correctly predicts that:

1. The speed of light in vacuum is the ultimate speed that any particle can attain.

2. The moving mass of a body is greater than its rest mass.

3. The equivalence between mass and energy.

A nuclear fission power reactor harnesses nuclear energy so that it can be converted into electrical energy that can be distributed to consumers.

Radioactivity refers to the spontaneous ejection by the nuclei of radioactive isotopes of alpha particles, beta particles, or gamma rays.

QUESTIONS

1. Advance some arguments for the existence of atoms.

2. Is a photon a particle? Compare it to an electron.

3. What is meant by the equivalence of mass and energy?

4. Explain the meaning of $E_0 = m_0 c^2$.

5. What is meant by matter waves?

6. Why does the electron in a hydrogen atom not crash into its proton-nucleus? Compare with the earth–sun situation.

7. What is meant by a "thought experiment"?

8. Discuss the theoretical possibility of constructing a powerful microscope through which one can locate the position of an electron in its motion in an atom at a given moment of time.

9. With our modern high energy machines, why is it not possible to make a stable nucleus consisting of, say, 500 protons and neutrons?

10. The filament of an electric lamp is energized by the current from a battery. How are the various colors of the light produced by the filament, according to the Bohr model of the atom?

PROBLEMS

1. How many protons, electrons and neutrons are there in an atom of the lithium isotope, ^7_3Li?

2. What is the difference between the two isotopes of oxygen, $^{16}_8\text{O}$ and $^{17}_8\text{O}$? How do their chemical characteristics compare?

3. The mass of 1 atom of the carbon isotope, $^{12}_6\text{C}$ is 1 u. Express this mass in grams.

4. A neutron has a rest-mass of 1.009 u. What is its rest-mass in kilograms?

5. Calculate the equivalent rest energy of a neutron in joules?

6. What is the rest energy of a neutron expressed in electron volts and million electron volts?

7. Compute the rest energy in million electron volts of a proton. The rest mass of a proton is 1.007 u.

8. Compare the radii of a hydrogen atom when the electron is in the K shell and M shell.

9. How much external energy is required to ionize a hydrogen atom when its electron is in the K shell?

10. Compare the energies of a hydrogen atom when the electron is in the K shell and M shell.

11. Calculate the wavelength of the light that is produced by the atom of hydrogen when its electron jumps from the M shell to the K shell.

12. What is the energy carried by a single photon of *red light having a* wavelength of 7000×10^{-8} cm?

13. What is the energy of a single photon of x rays having a wavelength of 1×10^{-8} cm?

14. Compare the energies of a single photon of violet light whose wavelength is 4000×10^{-8} cm with the energy of a photon of red light, $\lambda = 7000 \times 10^{-8}$ cm.

15. How many photons of red light, $\lambda = 7000$ Å must strike a surface every second in order to illuminate the surface with a light power of 10^{-6} watt?

16. At what speed must a body travel in order for its mass to be three times its rest mass?

17. The radioactive isotope of iodine, ^{131}I which is used in medicine, emits a gamma ray having an energy of 0.364 MeV.
   (a) Express this energy in joules.
   (b) What is the wavelength of this γ ray?
   (c) What is its frequency?

18. How many atoms are disintegrating per sec in a sample of 1 microcurie ($=10^{-6}$ curies) of ^{131}I?

19. The half-life of the phosphorous radioisotope, $^{32}_{15}\text{P}$ is 14.3 days. If we start with a quantity having an activity of 10 curies, how many curies remain after an elapsed time of 4 half-lives ($=57.2$ days)?

20. The nucleus of the thorium radioisotope, $^{230}_{90}\text{Th}$ decays into a radium (Ra) nucleus by emitting an alpha particle. How many neutrons and protons remain in the radium nucleus? Write down the equation for this nuclear reaction.

21. The nucleus of the radioisotope of carbon, $^{14}_6\text{C}$ decays into a nitrogen (N) nucleus by emitting a β particle. How many neutrons and protons remain in the nitrogen nucleus? Write down the equation for this reaction.

22. Assuming an energy yield of 200 MeV per fission of a $^{235}_{92}\text{U}$ nucleus, calculate the number of fissions per seconds in a power reactor in order to produce a steady power of 1 megawatt ($=10^6$ watts).

Appendix

Three-Place Table of Sines, Cosines, Tangents, and Cotangents

Angles	Sines	Cosines	Tangents	Cotangents	
0°00′	0.000	1.00	0.000		90°00′
30	0.009	1.00	0.009	115	30
1°00′	0.018	1.00	0.018	57.3	89°00′
30	0.026	1.00	0.026	38.2	30
2°00′	0.035	0.999	0.035	28.6	88°00′
30	0.044	0.999	0.044	22.9	30
3°00′	0.052	0.999	0.052	19.1	87°00′
30	0.061	0.998	0.061	16.4	30
4°00′	0.070	0.998	0.070	14.3	86°00′
30	0.078	0.997	0.079	12.7	30
5°00′	0.087	0.996	0.088	11.4	85°00′
30	0.096	0.995	0.096	10.4	30
6°00′	0.104	0.994	0.105	9.51	84°00′
30	0.113	0.994	0.114	8.78	30
7°00′	0.122	0.992	0.123	8.14	83°00′
30	0.130	0.991	0.132	7.60	30
8°00′	0.139	0.990	0.140	7.12	82°00′
30	0.148	0.989	0.150	6.69	30
9°00′	0.156	0.988	0.158	6.31	81°00′
30	0.165	0.986	0.167	5.98	30
10°00′	0.174	0.985	0.176	5.67	80°00′
30	0.182	0.983	0.185	5.40	30
11°00′	0.191	0.982	0.194	5.14	79°00′
30	0.199	0.980	0.204	4.92	30
12°00′	0.208	0.978	0.213	4.70	78°00′
30	0.216	0.976	0.222	4.51	30
	Cosines	Sines	Cotangents	Tangents	Angles

Three-Place Table of Sines, Cosines, Tangents, and Cotangents
(*Continued*)

Angles	Sines	Cosines	Tangents	Cotangents	
13°00'	0.225	0.974	0.231	4.33	77°00'
30	0.233	0.972	0.240	4.17	30
14°00'	0.242	0.970	0.249	4.01	76°00'
30	0.250	0.968	0.259	3.87	30
15°00'	0.259	0.966	0.268	3.73	75°00'
30	0.267	0.964	0.277	3.61	30
16°00'	0.276	0.961	0.287	3.49	74°00'
30	0.284	0.959	0.296	3.38	30
17°00'	0.292	0.956	0.306	3.27	73°00'
30	0.301	0.954	0.315	3.17	30
18°00'	0.309	0.951	0.325	3.08	72°00'
30	0.317	0.948	0.335	2.99	30
19°00'	0.326	0.946	0.344	2.90	71°00'
30	0.334	0.943	0.354	2.82	30
20°00'	0.342	0.940	0.364	2.75	70°00'
30	0.350	0.937	0.374	2.67	30
21°00'	0.358	0.934	0.384	2.61	69°00'
30	0.366	0.930	0.394	2.54	30
22°00'	0.375	0.927	0.404	2.48	68°00'
30	0.383	0.924	0.414	2.41	30
23°00'	0.391	0.920	0.424	2.36	67°00'
30	0.399	0.917	0.435	2.30	30
24°00'	0.407	0.914	0.445	2.25	66°00'
30	0.415	0.910	0.456	2.19	30
25°00'	0.423	0.906	0.466	2.14	65°00'
30	0.430	0.903	0.477	2.10	30
26°00'	0.438	0.899	0.488	2.05	64°00'
30	0.446	0.895	0.499	2.01	30
27°00'	0.454	0.891	0.510	1.96	63°00'
30	0.462	0.887	0.521	1.92	30
28°00'	0.470	0.883	0.532	1.88	62°00'
30	0.477	0.879	0.543	1.84	30
29°00'	0.485	0.875	0.554	1.80	61°00'
30	0.492	0.870	0.566	1.77	30
30°00'	0.500	0.866	0.577	1.73	60°00'
30	0.508	0.862	0.589	1.70	30
31°00'	0.515	0.857	0.601	1.66	59°00'
30	0.522	0.853	0.613	1.63	30
32°00'	0.530	0.848	0.625	1.60	58°00'
30	0.537	0.843	0.637	1.57	30
33°00'	0.545	0.839	0.649	1.54	57°00'
30	0.552	0.834	0.662	1.51	30
	Cosines	Sines	Cotangents	Tangents	Angles

Three-Place Table of Sines, Cosines, Tangents, and Cotangents (*Continued*)

Angles	Sines	Cosines	Tangents	Cotangents	
34°00′	0.559	0.829	0.674	1.48	56°00′
30	0.566	0.824	0.687	1.46	30
35°00′	0.574	0.819	0.700	1.43	55°00′
30	0.581	0.814	0.713	1.40	30
36°00′	0.588	0.809	0.726	1.38	54°00′
30	0.595	0.804	0.740	1.35	30
37°00′	0.602	0.799	0.754	1.33	53°00′
30	0.609	0.793	0.767	1.30	30
38°00′	0.616	0.788	0.781	1.28	52°00′
30	0.622	0.783	0.795	1.26	30
39°00′	0.629	0.777	0.810	1.23	51°00′
30	0.636	0.772	0.824	1.21	30
40°00′	0.643	0.766	0.839	1.19	50°00′
30	0.649	0.760	0.854	1.17	30
41°00′	0.656	0.755	0.869	1.15	49°00′
30	0.663	0.749	0.885	1.13	30
42°00′	0.669	0.743	0.900	1.11	48°00′
30	0.676	0.737	0.916	1.09	30
43°00′	0.682	0.731	0.932	1.07	47°00′
30	0.688	0.725	0.949	1.05	30
44°00′	0.695	0.719	0.966	1.04	46°00′
30	0.701	0.713	0.983	1.02	30
45°00′	0.707	0.707	1.00	1.00	45°00′
	Cosines	Sines	Cotangents	Tangents	Angles

Logarithms to Base Ten

Numbers	0	1	2	3	4	5	6	7	8	9
10	0000	0043	0086	0128	0170	0212	0253	0294	0334	0374
11	0414	0453	0492	0531	0569	0607	0645	0682	0719	0755
12	0792	0828	0864	0899	0934	0969	1004	1038	1072	1106
13	1139	1173	1206	1239	1271	1303	1335	1367	1399	1430
14	1461	1492	1523	1553	1584	1614	1644	1673	1703	1723
15	1761	1790	1818	1847	1875	1903	1931	1959	1987	2014
16	2041	2068	2095	2122	2148	2175	2201	2227	2253	2279
17	2304	2330	2335	2380	2405	2430	2455	2480	2504	2529
18	2553	2577	2601	2625	2648	2672	2695	2718	2742	2765
19	2788	2810	2833	2856	2878	2900	2923	2945	2967	2989
20	3010	3032	3054	3075	3096	3118	3139	3160	3181	3201
21	3222	3243	3263	3284	3304	3324	3345	3365	3385	3404
22	3424	3444	3464	3483	3502	3522	3541	3560	3579	3598
23	3617	3636	3655	3674	3692	3711	3729	3747	3766	3784
24	3802	3820	3838	3856	3874	3892	3909	3927	3945	3962
25	3979	3997	4014	4031	4048	4065	4082	4099	4116	4133
26	4150	4166	4183	4200	4216	4232	4249	4265	4281	4298
27	4314	4330	4346	4362	4378	4393	4409	4425	4440	4456
28	4472	4487	4502	4518	4533	4548	4564	4579	4594	4609
29	4624	4639	4654	4669	4683	4698	4713	4728	4742	4757
30	4771	4786	4800	4814	4829	4843	4857	4871	4886	4900
31	4914	4928	4942	4955	4969	4983	4997	5011	5024	5038
32	5051	5065	5079	5092	5105	5119	5132	5145	5159	5172
33	5185	5198	5211	5224	5237	5250	5263	5276	5289	5302
34	5315	5328	5340	5353	5366	5378	5391	5403	5416	5428
35	5441	5453	5465	5478	5490	5502	5514	5527	5539	5551
36	5563	5575	5587	5599	5611	5623	5635	5647	5658	5670
37	5682	5694	5705	5717	5729	5740	5752	5763	5775	5786
38	5798	5809	5821	5832	5843	5855	5866	5877	5888	5899
39	5911	5922	5933	5944	5955	5966	5977	5988	5999	6010
40	6021	6031	6042	6053	6064	6075	6085	6096	6107	6117
41	6128	6138	6149	6160	6170	6180	6191	6201	6212	6222
42	6232	6243	6253	6263	6274	6284	6294	6304	6314	6325
43	6335	6345	6355	6365	6375	6385	6395	6405	6415	6425
44	6435	6444	6454	6464	6474	6484	6493	6503	6513	6522
45	6532	6542	6551	6561	6571	6580	6590	6599	6609	6618
46	6628	6637	6646	6656	6665	6675	6684	6693	6702	6712
47	6721	6730	6739	6749	6758	6767	6776	6785	6794	6803
48	6812	6821	6830	6839	6848	6857	6866	6875	6884	6893
49	6902	6911	6920	6928	6937	6946	6955	6964	6972	6981
50	6990	6998	7007	7016	7024	7033	7042	7050	7059	7067
51	7076	7084	7093	7101	7110	7118	7126	7135	7143	7152
52	7160	7168	7177	7185	7193	7202	7210	7218	7226	7235
53	7243	7251	7259	7267	7275	7284	7292	7300	7308	7316
54	7324	7332	7340	7348	7356	7364	7372	7380	7388	7396

Numbers	0	1	2	3	4	5	6	7	8	9
55	7404	7412	7419	7427	7435	7443	7451	7459	7466	7474
56	7482	7490	7497	7505	7513	7520	7528	7536	7543	7551
57	7559	7566	7574	7582	7589	7597	7604	7612	7619	7627
58	7634	7642	7649	7657	7664	7672	7679	7686	7694	7701
59	7709	7716	7723	7731	7738	7745	7752	7760	7767	7774
60	7782	7789	7796	7803	7810	7818	7825	7832	7839	7846
61	7853	7860	7868	7875	7882	7889	7896	7903	7910	7917
62	7924	7931	7938	7945	7952	7959	7966	7937	7980	7987
63	7993	8000	8007	8014	8021	8028	8035	8041	8048	8055
64	8062	8069	8075	8082	8089	8096	8102	8109	8116	8122
65	8129	8136	8142	8149	8156	8162	8169	8176	8182	8189
66	8195	8202	8209	8215	8222	8228	8235	8241	8248	8254
67	8261	8267	8274	8280	8287	8293	8299	8306	8312	8319
68	8325	8331	8338	8344	8351	8357	8363	8370	8376	8382
69	8388	8395	8401	8407	8414	8420	8426	8432	8439	8445
70	8451	8457	8463	8470	8476	8482	8488	8494	8500	8506
71	8513	8519	8525	8531	8537	8543	8549	8555	8561	8567
72	8573	8579	8585	8591	8597	8603	8609	8615	8621	8627
73	8633	8639	8645	8651	8657	8663	8669	8675	8681	8686
74	8692	8698	8704	8710	8716	8722	8727	8733	8739	8745
75	8751	8756	8762	8768	8774	8779	8785	8791	8797	8802
76	8808	8814	8820	8825	8831	8837	8842	8848	8854	8859
77	8865	8871	8876	8882	8887	8893	8899	8904	8910	8915
78	8921	8927	8932	8938	8943	8949	8954	8960	8965	8971
79	8976	8982	8987	8993	8998	9004	9009	9015	9020	9025
80	9031	9036	9042	9047	9053	9058	9063	9069	9074	9079
81	9085	9090	9096	9101	9106	9112	9117	9122	9128	9133
82	9138	9143	9149	9154	9159	9165	9170	9175	9180	9186
83	9191	9196	9201	9206	9212	9217	9222	9227	9232	9238
84	9243	9248	9253	9258	9263	9269	9274	9279	9284	9289
85	9294	9299	9304	9309	9315	9320	9325	9330	9335	9340
86	9345	9350	9355	9360	9365	9370	9375	9380	9385	9390
87	9395	9400	9405	9410	9415	9420	9425	9430	9435	9440
88	9445	9450	9455	9460	9465	9469	9474	9479	9484	9489
89	9494	9499	9504	9509	9513	9518	9523	9528	9533	9538
90	9542	9547	9552	9557	9562	9566	9571	9576	9581	9586
91	9590	9595	9600	9605	9609	9614	9619	9624	9628	9633
92	9638	9643	9647	9652	9657	9661	9666	9671	9675	9680
93	9685	9689	9694	9699	9703	9708	9713	9717	9722	9727
94	9731	9736	9741	9745	9750	9754	9759	9763	9768	9773
95	9777	9782	9786	9791	9795	9800	9805	9809	9814	9818
96	9823	9827	9832	9836	9841	9845	9850	9854	9859	9863
97	9868	9872	9877	9881	9886	9890	9894	9899	9903	9908
98	9912	9917	9921	9926	9930	9934	9939	9943	9948	9952
99	9956	9961	9965	9969	9974	9978	9983	9987	9991	9996

Wire Table, Standard Annealed Copper
(American Wire Gauge (B. & S.) English Units)

Gauge Number	Diameter in mils at 20°C	Cross-Section at 20°C		Ohms per 1000 ft	
		Circular mils	Square inches	0°C (32°F)	20°C (68°F)
0000	460.0	211600	0.1662	0.04516	0.04901
000	409.6	167800	0.1318	0.05695	0.06180
00	364.8	133100	0.1045	0.07181	0.07793
0	324.9	105500	0.08289	0.09055	0.09827
1	289.3	83690	0.06573	0.1142	0.1239
2	257.6	66370	0.05213	0.1440	0.1563
3	229.4	52640	0.04134	0.1816	0.1970
4	204.3	41740	0.03278	0.2289	0.2485
5	181.9	33100	0.02600	0.2887	0.3133
6	162.0	26250	0.02062	0.3640	0.3951
7	144.3	20820	0.01635	0.4590	0.4982
8	128.5	16510	0.01297	0.5788	0.6282
9	114.4	13090	0.01028	0.7299	0.7921
10	101.9	10380	0.008155	0.9203	0.9989
11	90.74	8234	0.006467	1.161	1.260
12	80.81	6530	0.005129	1.463	1.588
13	71.96	5178	0.004067	1.845	2.003
14	64.08	4107	0.003225	2.327	2.525
15	57.07	3257	0.002558	2.934	3.184
16	50.82	2583	0.002028	3.700	4.016
17	45.26	2048	0.001609	4.666	5.064
18	40.30	1624	0.001276	5.883	6.385
19	35.89	1288	0.001012	7.418	8.051
20	31.96	1022	0.0008023	9.355	10.15
21	28.45	810.1	0.0006363	11.80	12.80
22	25.35	642.4	0.0005046	14.87	16.14
23	22.57	509.5	0.0004002	18.76	20.36
24	20.10	404.0	0.0003173	23.65	25.67
25	17.90	320.4	0.0002517	29.82	32.37
26	15.94	254.1	0.0001996	37.61	40.81
27	14.20	201.5	0.0001583	47.42	51.47
28	12.64	159.8	0.0001255	59.80	64.90
29	11.26	126.7	0.00009953	75.40	81.83
30	10.03	100.5	0.00007894	95.08	103.2
31	8.928	79.70	0.00006260	119.9	130.1
32	7.950	63.21	0.00004964	151.2	164.1
33	7.080	50.13	0.00003937	190.6	206.9
34	6.305	39.75	0.00003122	240.4	260.9
35	5.615	31.52	0.00002476	303.1	329.0
36	5.000	25.00	0.00001964	382.2	414.8
37	4.453	19.83	0.00001557	482.0	523.1
38	3.965	15.72	0.00001235	607.8	659.6
39	3.531	12.47	0.000009793	766.4	831.8
40	3.145	9.888	0.000007766	966.5	1049

Some Useful Conversion Factors

	Meter^2	cm^2	ft^2	in.^{2}
1 square meter =	1	10^4	10.76	1550
1 square centimeter =	10^{-4}	1	1.076×10^{-3}	0.1550
1 square foot =	9.290×10^{-2}	929.0	1	144
1 square inch =	6.452×10^{-4}	6.452	6.944×10^{-3}	1

Volume

	Meter^3	cm^3	1	ft^3	in.^{3}
1 cubic meter =	1	10^6	1000	35.31	6.102×10^4
1 cubic centimeter =	10^{-6}	1	1.000×10^{-3}	3.531×10^{-5}	6.102×10^{-2}
1 liter =	1.000×10^{-3}	1000	1	3.531×10^{-2}	61.02
1 cubic foot =	2.832×10^{-2}	2.832×10^4	28.32	1	1728
1 cubic inch =	1.639×10^{-5}	16.39	1.639×10^{-2}	5.787×10^{-4}	1

1 U.S. fluid gallon = 4 U.S. fluid quarts = 8 U.S. pints = 128 U.S. fluid ounces = 231 in.^{3}

Mass

	g	kg	slug	u
1 gram =	1	0.001	6.852×10^{-5}	6.024×10^{23}
1 kilogram =	1000	1	6.852×10^{-2}	6.024×10^{26}
1 slug =	1.459×10^4	14.59	1	8.789×10^{27}
1 u =	1.660×10^{-24}	1.660×10^{-27}	1.137×10^{-28}	1

Mass Density

	slug/ft^3	kg/meter^3	gm/cm^3
1 slug per ft^3 =	1	515.4	0.5154
1 kilogram per meter^3 =	1.940×10^{-3}	1	0.001
1 gram per cm^3 =	1.940	1000	1

Time

	yr	day	hr	min	sec
1 year =	1	365.2	8.766×10^3	5.259×10^5	3.156×10^7
1 day =	2.738×10^{-3}	1	24	1440	8.640×10^4
1 hour =	1.141×10^{-4}	4.167×10^{-2}	1	60	3600
1 minute =	1.901×10^{-6}	6.944×10^{-4}	1.667×10^{-2}	1	60
1 second =	3.169×10^{-8}	1.157×10^{-5}	2.778×10^{-4}	1.667×10^{-2}	1

1 year = 365.24219879 days

Force

	dyne	newton	lb
1 dyne =	1	10^{-5}	2.248×10^{-6}
1 newton =	10^5	1	0.2248
1 pound =	4.448×10^5	4.448	1

Pressure

	atm	dyne/cm^2	cm Hg	newton/meter^2	lb/in.^{2}	lb/ft^2
1 atmosphere =	1	1.013×10^6	76	1.013×10^5	14.70	2116
1 dyne per cm^2 =	9.869×10^{-7}	1	7.501×10^{-5}	0.1	1.450×10^{-5}	2.089×10^{-3}
1 centimeter of mercury at 0°C^a =	1.316×10^{-2}	1.333×10^4	1	1333	0.1934	27.85
1 newton per meter^2 =	9.869×10^{-6}	10	7.501×10^{-4}	1	1.450×10^{-4}	2.089×10^{-2}
1 pound per in.^{2} =	6.805×10^{-2}	6.895×10^4	5.171	6.895×10^3	1	144
1 pound per ft^2 =	4.725×10^{-4}	478.8	3.591×10^{-2}	47.88	6.944×10^{-3}	1

Energy, Work, Heat

	Btu	erg	ft-lb	hp-hr	joules	cal	kW-hr	eV	MeV
1 British thermal unit =	1	1.055×10^{10}	777.9	3.929×10^{-4}	1055	252.0	2.930×10^{-4}	6.585×10^{21}	6.585×10^{15}
1 erg =	9.481×10^{-11}	1	7.376×10^{-8}	3.725×10^{-14}	10^{-7}	2.389×10^{-8}	2.778×10^{-14}	6.242×10^{11}	6.242×10^{5}
1 foot-pound =	1.285×10^{-3}	1.356×10^{7}	1	5.051×10^{-7}	1.356	0.3239	3.766×10^{-7}	8.464×10^{18}	8.464×10^{12}
1 horsepower-hour =	2545	2.685×10^{13}	1.980×10^{6}	1	2.685×10^{6}	6.414×10^{5}	0.7457	1.676×10^{25}	1.676×10^{19}
1 joule =	9.481×10^{-4}	10^{7}	0.7376	3.725×10^{-7}	1	0.2389	2.778×10^{-7}	6.242×10^{18}	6.242×10^{12}
1 calorie =	3.968×10^{-3}	4.186×10^{7}	3.087	1.559×10^{-6}	4.186	1	1.163×10^{-6}	2.613×10^{19}	2.613×10^{13}
1 kilowatt-hour =	3413	3.6×10^{13}	2.655×10^{6}	1.341	3.6×10^{6}	8.601×10^{5}	1	2.247×10^{25}	2.270×10^{19}
1 electron volt =	1.519×10^{-22}	1.602×10^{-12}	1.182×10^{-19}	5.967×10^{-26}	1.602×10^{-19}	3.827×10^{-20}	4.450×10^{-26}	1	10^{-6}
1 million electron volts =	1.519×10^{-16}	1.602×10^{-6}	1.182×10^{-13}	5.967×10^{-20}	1.602×10^{-13}	3.827×10^{-14}	4.450×10^{-20}	10^{6}	1

Periodic chart of the elements

IA	IIA	IIIB	IVB	VB	VIB	VIIB	VIII	VIII	VIII	IB	IIB	IIIA	IVA	VA	VIA	VIIA	
1 H 1.0080																	2 He 4.003
3 Li 6.940	4 Be 0.013											5 B 10.82	6 C 12.010	7 X 14.008	8 O 16.0000	9 F 19.00	10 Ne 20.183
11 Na 22.997	12 Mg 24.32											13 Al 26.98	14 Si 28.09	15 P 30.975	16 S 32.066	17 Cl 35.457	18 A 39.944
19 K 39.100	20 Ca 40.08	21 Sc 44.96	22 Ti 47.90	23 V 50.95	24 Cr 52.01	25 Mn 54.93	26 Fe 55.85	27 Co 58.94	28 Ni 58.69	29 Cu 63.54	30 Zn 65.38	31 Ga 69.72	32 Ge 72.60	33 As 74.91	34 Se 78.96	35 Br 79.916	36 Kr 83.8
37 Rb 85.48	38 Sr 87.63	39 Y 88.92	40 Zr 91.22	41 Nb 92.91	42 Mo 95.95	43 Tc (99)	44 Ru 101.7	45 Rh 102.91	46 Pd 106.7	47 Ag 107.880	48 Cd 112.41	49 In 114.76	50 Sn 118.70	51 Sb 121.76	52 Te 127.61	53 I 126.91	54 Xe 131.3
55 Cs 132.91	56 Ba 137.36	57–71 Rare Earths	72 Hf 178.6	73 Ta 180.88	74 W 183.92	75 Re 186.31	76 Os 190.2	77 Ir 193.1	78 Pt 195.23	79 Au 197.2	80 Hg 200.61	81 Tl 204.39	82 Pb 207.21	83 Bi 209.00	84 Po (210)	85 At (210)	86 Rn 222
87 Fr (223)	88 Ra 226.05	89– Acti- nides															

Atomic Number
Element Symbol
Atomic Mass
(u)

Rare earths (Lanthanide series)

57 La 138.92	58 Ce 140.13	59 Pr 140.92	60 Nd 144.27	61 Pm (145)	62 Sm 150.43	63 Eu 152.0	64 Gd 156.9	65 Tb 159.2	66 Dy 162.46	67 Ho 164.94	68 Er 167.2	69 Tm 169.4	70 Yb 173.04	71 Lu 174.99

Actinide series

89 Ac (227)	90 Th 232.12	91 Pa 231	92 U 238.07	93 Np (237)	94 Pu (242)	95 Am (243)	96 Cm (245)	97 Bk (249)	98 Cf (249)	99 Es (253)	100 Fm (253)	101 Md (256)	102 No	103 Lw

609

Answers to Problems

Chapter 2

(A)
1. 200 lb to the right.
3. 20 lb to the left.
5. a) 4 mph, upstream b) 16 mph, downstream.
7. a) 600 knots b) 440 knots.
9. $F_x = 100$ lb; $F_y = -173$ lb.
11. a) 500 lb; 37° NE b) 500 lb; 37° SW.
13. 100 lb; 53° SW.

(B)
1. 224 lb; 26.5° NE
3. 575 lb; 31°.
5. 120°.
7. 217 lb; 134° west of north.
9. a) 52.5° b) 81°.
11. 155 lb; 19.5° with 60 lb force.
15. a) 171° west of north b) 4.6 hr.
17. $R_x = 5$ lb; $R_y = -190$ lb; 88.5° to the right and below the horizontal.
19. a) 128 N; 38.7° to the right and below the horizontal
    b) 128 N; 38.7° to the left and above the horizontal.

Chapter 3

(A)
1. 100 lb; 53° to the right and above the horizontal.
3. 200 lb.
5. b) 280 lb c) 180 lb.
7. 312.5 lb.
9. b) $C = 1000$ lb; $T = 600$ lb.
11. 1879 lb.
13. 219.3 lb.

(B)
1. $T_A = 1000$ lb; $T_B = T_C = 577.4$ lb.
3. 1000 lb.

5. $F_x = -10$ lb; $F_y = 86.6$ lb.
7. 250 lb.
9. 166.7 lb.
11. a) 13.2 ft; 40° b) 88.2 lb.
13. $C = 2732$ lb; $T = 1932$ lb.
15. a) 200 lb b) $PA = 447.2$ lb; $\theta = 63.4°$.
17. a) 125 lb b) 75 lb.

Chapter 4

(A)
1. a) -300 lb ft b) -260 lb ft c) -150 lb ft
   d) 0.
3. a) $+240$ lb ft b) -60 lb ft.
5. 5.6 ft.
7. a) 80 lb b) 130 lb.
9. 200 lb.
11. a) 125 lb b) 75 lb.
13. 4 ft to right of 10 lb weight.
15. 100 lb and 600 lb ft.
17. a) 4091 lb b) 909 lb.

(B)
1. $x = -0.56$ ft, $y = 1.78$ ft
3. 0.286 lb.
5. a) 375 lb b) $H = 225$ lb; $V = 100$ lb.
7. a) 120.4 lb toward left; 59.7 lb downward.
   b) 16.0 ft to right of A.
9. a) 250 lb
   b) 30 lb toward right; 400 lb upward.
11. 222.8 lb.
13. a) 25 lb at A; 75 lb at B b) 86.6 lb.

Chapter 5

(A)
1. 20 lb.
3. 0.33.
5. 1600 lb.

7. a) 0.09 b) 0.11.
9. 0.574.
11. 0.436.
13. 30 lb ft.

(B) 1. a) 20 lb b) 40 lb c) 35 lb d) 35 lb.
3. a) 23.1 lb b) 0.231.
5. a) 175 lb b) 4.29 ft.
7. $N = 135$ lb, $P = 96.7$ lb.
9. 55.6 lb.
11. 63.4.

Chapter 6

(A) 1. 19.2 lb/ft.
3. 10 lb.
5. 4800.
7. 11,781 lb.
9. 1200 lb.
11. 3.84×10^{-3} in.
13. a) 2×10^{-4} b) 5000 lb/in.^{2}

(B) 1. 16×10^2 lb/in.^{2}
3. a) 100 lb
   b) K_2 stretches 5 in., and K_1 stretches 10 in.
5. a) 4584 lb/in.^{2} b) 1.83×10^{-2} in.
7. a) 50% b) 128 lb/ft^3.
9. a) 250 lb/in.^{2} b) 6000 lb/ft.

11. a) $\dfrac{1}{900}$ b) $\dfrac{1}{90}$ gal.

13. a) 4000 lb b) 0.19 in. c) 0.4 in.^{2}
    d) 8.6×10^{-4} rad $= 0.049$ deg.

Chapter 7

(A) 1. 30 ft/sec.
3. a) 1382 ft/min b) 23 ft/sec c) 15.7 mph.
5. a) 7.8 m/sec b) 25.6 ft/sec.
7. 5.41 m/sec.
9. a) 48 ft/sec b) 32.7 mph.
11. a) 100 ft/sec b) 500 ft c) 50 ft/sec
    d) 10 ft/sec^2.
13. b) 12.5 ft/sec^2 c) 625 ft.
15. 2.4×10^6 cm/sec^2.
17. a) 3.39×10^{-7} sec b) 29.5 cm.
19. 20 m/sec^2.
21. 10 ft/sec^2.

(B) 1. a) 3.35 hr b) 47.8 mph.
3. b) 25 sec c) 1250 ft.
5. a) 75 m/sec b) 300 m/sec c) 200 m/sec
   d) 3,687 m e) 245.8 m/sec.
7. b) from 100 m to 150 m c) 1.43 m/sec.
9. 26.7 mph.
11. a) $t = 1$ sec, 5 sec b) $t = 1$ sec, 3 sec, 5 sec
    c) 15.7 cm/sec.

Chapter 8

(A) 1. a) 22.5 ft b) 15 ft/sec.
3. a) 330 ft b) 8.8 ft/sec^2.
5. a) 4 ft/sec^2 b) 80 ft/sec.
7. a) 0.8 m/sec^2 b) 4 m/sec.
9. 87.6 ft/sec.
11. a) 50 m/sec^2 b) 9×10^4 m.
13. a) 3.35 sec b) 32.8 m/sec.
15. a) 4.4 ft/sec^2 b) 385 ft.
17. a) 400 ft b) 5 sec.
19. a) 20 ft/sec^2 b) 136.4 mph.
21. a) 31.3 m b) 50 m/sec c) 10.2 sec.

(B) 1. a) 6 sec b) 300 m.
3. a) 600 ft b) 200 ft/sec.
5. 2674 ft.
7. a) 442.7 m/sec b) 2×10^4 m.
9. a) 218.5 sec b) 806 mi.
11. a) 4×10^4 ft b) 1.2×10^5 ft
    c) $x = 7.2 \times 10^4$ ft; $y = 3.84 \times 10^4$ ft
    d) 1240 ft/sec.
13. 136.7 ft/sec.

Chapter 9

(A) 1. a) 2 slug b) 20.4 kg c) 2.55 gm.
3. 3.125 slug.
5. a) 160 lb b) 98 N c) 2.94×10^4 dynes.
7. 80 N.
9. a) 5 slug b) 5 slug c) 26.6 lb.
11. a) 4 ft/sec^2 b) 1 lb c) 100 ft.
13. a) 75 slug b) 600 lb.
15. a) 8 ft/sec^2 b) 0.3125 slug c) 2.5 lb.
17. a) 2.5 ft/sec^2 b) 3.125 lb.
19. 8 ft/sec^2.

(B) 1. 1.33 ft/sec^2.
3. a) 8×10^3 dynes b) 1.8×10^3 dynes.
5. a) 468 g b) 4.56×10^5 dynes.

7. 48 ft/sec².
9. 2.51×10^{-13} N.
11. a) 7400 lb   b) 6400 lb.
13. a) 192.4 lb   b) 33.3 ft/sec².
15. a) 98 N   b) 98 N   c) 98 N   d) 178 N.
   e) 18 N.
19. 5.96×10^{24} Kg.

Chapter 10

(A) **1.** 50 ft/sec².
   3. 38.7 ft/sec².
   5. a) 1/6 sec   b) 113.1 ft/sec   c) 4,260 ft/sec².
   7. a) 3.13×10^{16} m/sec²   b) 28.5×10^{-15} N.
   9. a) 26.2 ft/sec   b) 821.7 ft/sec².
   11. 18.6 ft/sec.
   13. 7.6 rpm.
   15. 668 rpm.
   17. a) 1225 lb   b) 0.34.
   19. a) 80 ft/sec   b) 382 rpm

(B) **1.** a) 0.07 sec   b) 70.7 ft/sec   c) 23.6 ft/sec
   d) 6662 ft/sec².
   3. 8067 ft.
   5. a) 6 ft/sec   b) 0.25 lb   c) 0.45 lb.
   7. a) 2.36 ft/sec   b) 0.02 lb.
   9. 7.8 ft/sec.
   11. a) 11.5 mi   b) 43.2 sec.
   13. 0.61 ft.
   15. a) $AB = 32.7$ N, $BC = -26.2$ N
   b) $AB = 32.7$ N, $BC = 52.8$ N.

Chapter 11

(A) **1.** 0.06 sec.
   3. 314 rad/sec
   5. a) 95.5 rpm   b) 0.63 sec.
   7. a) 50 rad/sec   b) 477 rpm.
   9. a) 0.79 rad/sec²   b) 94.2 rad/sec.
   11. a) 1257 cm/sec   b) 628.5 cm/sec.
   13. 10 kg-m².
   15. 8.38 rad/sec².
   17. 1000 N.
   19. 12.6 rad/sec.
   21. a) 5 rev/sec   b) 0.2 sec.

(B) **1.** 1.4 rad/sec².
   3. a) 5 rad/sec²   b) 1.5 m/sec²
   c) 1.69×10^3 m/sec².

5. 62.8 lb.
7. a) 25.1 rad/sec²   b) 1200 rpm   c) 3.98 sl-ft².
9. a) 2 rad/sec   b) 25.1 sec   c) 31.4 sec.
11. 414 lb-ft.
13. a) 38 N   b) 15 rad/sec   c) 1.01 kg-m².
15. a) $v_x = 4$ ft/sec; $v_y = -2.7$ ft/sec
   b) $a_x = 11.4$ ft/sec², $a_y = -10$ ft/sec².
17. a) 37.0 ft/sec²; $\alpha = 12.3$ rad/sec²   b) 15 lb.

Chapter 12

(A) **1.** 2000 ft-lb.
   3. 4.8×10^{-6} erg.
   5. 10^4 joules.
   7. 20.9 ft-lb.
   9. 8 N-m.
   11. 4×10^{-14} N.
   13. 450 ft-lb.
   15. 300 ft-lb.
   17. 1.885×10^4 ft-lb.
   19. 50 ft-lb.
   21. a) 120 ft-lb   b) 280 ft-lb   c) 30 ft/sec.

(B) **1.** a) 86.6 ft-lb   b) 50 ft-lb   c) 36.6 ft-lb.
   3. a) 1.7×10^5 ft-lb   b) 1.7×10^3 lb.
   5. a) 260 joules   b) 100 joules   c) 260 joules.
   7. a) 4.9×10^3 dynes/cm   b) 9.8×10^5 ergs.
   9. 8.5 in.
   11. a) 4.93×10^4 ft-lb   b) 78.5 lb-ft.
   13. a) $W_A = 10^3$ joules; $W_B = -900$ joules;
   $W_C = -600$ joules
   b) -500 joules.
   15. a) 8 ft-lb   b) 11.3 ft/sec.
   17. 1.67×10^5 joules.
   19. 1.25×10^7 joules.

Chapter 13

(A) **1.** 9.1 hp.
   3. a) 1400 W   b) 1.88 hp.
   5. 90.9 hp.
   7. 3.64 hp.
   9. a) 4000 ft-lb   b) $400 \dfrac{\text{ft-lb}}{\text{sec}}$; 0.73 hp.
   11. 1568 W.
   13. 0.545 ft from weight.
   15. 48%.
   17. a) $1.26 \times 10^5 \dfrac{\text{ft-lb}}{\text{sec}}$ b) 229 hp.

19. a) $6.6 \times 10^6 \dfrac{\text{ft-lb}}{\text{min}}$ b) 2.64×10^7 ft-lb

c) 149.2 kW.

21. 80 hp.

23. a) 2000 ft-lb   b) $500 \dfrac{\text{ft-lb}}{\text{sec}}$.

(B)   **1.** a) 4.84×10^5 ft-lb   b) 4.84×10^5 ft-lb

c) 88 hp.

3. a) $200 \dfrac{\text{ft-lb}}{\text{sec}}$ b) 2.5   c) 3   d) 83.3%.

5. 28.6 N-m; 21.1 lb-ft.

7. a) 303.3 hp   b) 505.5 hp.

9. a) $400 \dfrac{\text{ft-lb}}{\text{sec}}$ b) 0.73 hp.

11. a) 40 lb-ft   b) 1.52 hp.

13. a) 100   b) 192   c) 31.25 ft-lb   d) 28.75 ft-lb

e) 52.1%.

15. a) 0.327 hp   b) 83.3%.

Chapter 14

(A)   **1.** 4 lb-sec.

3. a) $120 \dfrac{\text{kg-m}}{\text{sec}}$ b) 30 N.

5. 4000 lb-ft-sec.

7. 8 m/sec².

9. a) 1.17 lb-sec   b) 117 lb.

11. a) 75 ft/sec   b) 1.41×10^8 ft-lb

c) 3.52×10^6 ft-lb.

13. a) 5 m/sec

b) $V_{10} = -10$ m/sec; $V_{30} = 10$ m/sec.

15. a) 5 ft/sec

b) $V_{10} = 10$ ft/sec, left; $V_{30} = 10$ ft/sec, right.

17. $V_{16} = 7.78$ ft/sec; $V_2 = 17.78$ ft/sec.

19. a) 2513 lb-ft-sec   b) 1257 lb-ft-sec

c) 125.7 lb-ft.

(B)   **1.** a) $V = 10$ cm/sec, left; 2.7×10^4 ergs

b) $V_{10} = 70$ cm/sec, left; $V_{20} = 20 \dfrac{\text{cm}}{\text{sec}}$, right.

3. a) 500 ft/sec   b) 5.8×10^3 ft-lb.

5. a) -2 m/sec   b) 60 joules   c) 0.8.

7. 50 ft/sec.

9. a) 942 N-m-sec   b) 7.85 N-m.

11. a) 2.5 sl-ft²   b) $785 \dfrac{\text{sl-ft}^2}{\text{sec}}$ c) 1.23×10^5 ft-lb.

13. 6×10^4 N.

15. a) 25 ft/sec   b) 6.25×10^6 ft-lb

c) 1.56×10^5 ft-lb   d) 7.8×10^4 lb

e) 1.1×10^6 lb.

17. a) 10^4 gm-cm²; 3.77×10^7 dyne-cm-sec

b) 1.96×10^6 dyne-cm; $0.035 \dfrac{\text{rad}}{\text{sec}}$.

Chapter 15

(A)   **1.** a) 8 lb/ft   b) 0.32 Hz.

3. a) 48 lb/ft   b) 96 ft/sec²   c) 6.93 ft/sec

d) 0.45 sec.

5. 25 gm.

7. 987 cm/sec².

9. 3.14 sec.

11. 1.05 sec.

13. a) 15.7 ft/sec   b) 37.0 ft-lb   c) 49.3 ft-lb.

15. a) 9 cm

b) $v = 14.1$ cm/sec, $a = 22.2$ cm/sec²

c) $v = \pm 11.8$ cm/sec, $a = 12.3$ cm/sec²

d) at midpoint $x = 0$ and $F = 0$

e) 3.46×10^4 dynes.

17. a) 1.8 sec   b) 2.67 ft.

(B)   **1.** 1.0 sec.

3. a) 0.2 m   b) 24.5 N/m   c) 9.8 m/sec².

5. 1.3 sec.

7. 4.9 cm.

9. a) $2k$ b) $4k$.

11. a) 1.6 Hz   b) 1 m/sec   c) 10 m/sec²

d) 4 N.

13. a) 1.64 ft/sec²   b) 1.05 ft/sec   c) 1.16 ft/sec²

d) 4 inches.

15. a) 1.97×10^4 ft/sec²   b) 616 lb.

Chapter 16

(A)   **1.** 29.4 cm³.

3. 0.66 g/cm³; 0.66.

5. 900 g.

7. 1600 lb/in.²

9. 104 ft.

11. 1.04×10^4 dyne/cm².

13. 500 N.

15. 0.33 g/cm³ or 20.8 lb/ft³.

17. 0.85 ft.

19. 145.6 dynes/cm^2.

21. 5 cm.

(B) **1.** 187.2 lb/ft^3.

3. 1050 lb.

5. 34.6 ft^2.

7. 1.54 g/cm^3.

9. 100 cm^3.

11. 1476 gal/hr.

13. $v_1 = 20.8$ ft/sec, $v_2 = 52$ ft/sec.

15. gauge pressure $= 35.7$ lb/in.^{2}

17. $v = 25$ ft/sec.

Chapter 17

(A) **1.** 37.0°C.

3. 2,795°F; 5,432°F.

5. 99.6°R.

7. 1482°R.

9. 24.4°C; -126.1°C; 237.8°C.

11. 243°K; 351°K; 233°K.

13. 10,832°F; 11,292°R; 6,273°K.

15. 10,521°F.

17. 0.039 ft.

19. 88°C.

(B) **1.** 35°C.

3. 50°F.

5. 303°K.

7. 606.12 l.

9. 8.77 g/cm^3.

11. 48.2 ft^2.

15. 210.4 cm^3.

17. 1.00006 sec.

Chapter 18

(A) **1.** 10,549 J.

3. 12,558 W.

5. 16,000 cal.

7. 1056 cal.

9. 18.4°C.

11. 6200 cal.

13. 79,240 Btu.

(B) **1.** 7653 Btu.

3. 105.3 Btu.

5. 7300 cal.

7. 45.2°C.

9. 63°F.

11. 54,000 Btu.

13. 15,000 Btu/lb.

15. 1.6 g steam and 50.4 g of water at 100°C.

17. 121°C.

Chapter 19

(A) **1.** 1.2×10^8 J.

3. 20 l-atm.

5. 700 cal.

7. 66.7%.

9. 1200 cal.

11. 2000°K.

13. 5/3.

(B) **1.** 1.03 Btu.

3. 201 Btu.

5. 2.34 cal.

7. No; violates first law.

9. 8.33×10^5 N/m^2.

11. a) 360 W b) 3440 cal.

13. 39%.

15. 433.8°K; 5/3.

17. 61.8%.

19. 109,883 J.

Chapter 20

(A) **1.** 1.47 atm.

3. 0.46 mole.

5. 0.18 mole or 5.76 g.

7. 1.05 atm.

9. 1.98×10^{-21} cal/molecule.

11. 15,288 cal; 10,920 cal.

13. 3750 cal; 4500 cal.

15. 8 atm.

(B) **1.** 1.33 atm.

3. 0.446 moles.

5. 5.83×10^5 g.

7. 470 g.

9. $44.6 \dfrac{\text{ft-lb}}{\text{lb-°R}}$.

11. 0.96×10^6 dynes/cm^2.

13. -83.7°C.

15. 39.2 g.

17. a) 250°K b) 500°K; 1000°K; 500°K

c) 750 cal d) 2500 cal e) 500 cal f) 0.

Chapter 21

(A) 1. 29%.
3. 25.36 mm of Hg.
5. 4.88 mm of Hg; 0.8°C.

(B) 1. 8.55 g/m^3; 47°F.
3. 54.5%.
5. 72.9%.
7. 11.1 mm of Hg.

Chapter 22

(A) 1. 40 C°/m.
3. a) 10.6 cal/sec   b) 636 cal.
5. 12.5 cm.
7. 3840 Btu.
9. 5.67 × 10^4 W/m^2.

11. 5.54 × 10^3 $\dfrac{\text{cal}}{\text{cm}^2\text{-sec}}$.

(B) 1. 1.17 × 10^5 cal.
3. 2.3 × 10^5 cal.
5. a) 116°F   b) 3.6 × 10^5 Btu.
7. 946 cal.

Chapter 23

(A) 1. 338 m/sec.
3. 700 m/sec.
5. 5 × 10^5 cm/sec.
7. 17.3 ft.
9. 3.5 × 10^5 cm/sec.
11. 560 Hz.
13. 632 m/sec.
15. 72 N.

(B) 1. 1.4 × 10^5 cm/sec.
3. 2.68 × 10^4 cm/sec.
5. 4800 m/sec.
7. 3200 dynes.
9. 5.49 sec.
11. 11 beats/sec.

Chapter 24

(A) 1. a) 1400   b) 1650 m.
3. a) 308 Hz   b) 256 Hz.

5. 273 Hz.
7. 1.32 m/sec.
9. 1500 mph.
11. 15.8.
13. 60.

(B) 1. v_A = 22.6 m/sec, v_B = 37.8 m/sec.
3. 7.92 m/sec.
5. 23.6°.

Chapter 25

(A) 1. 8.1 × 10^{-5} N; attractive.
3. 1.44 × 10^{-1} N; attractive.
5. 1.024 × 10^{-10} N.
7. 10^{14} electrons.
9. a) 4.1 × 10^{-47} N   b) 9.2 × 10^{-8} N   c) Yes.
11. 0.28 m.

(B) 1. 22.3 N; 79.1° with line joining $-2\mu C$ and $-6\mu C$ charges.
3. 2.67 cm from smaller charge.
5. 1.3 × 10^{-3} N.
7. 9.2 × 10^{-8} N; 7.2 × 10^{15} Hz.
9. 6.8 m.

Chapter 26

(A) 1. 3 × 10^5 N/C.
3. 9 × 10^3 N/C; 1.44 × 10^{-15} N.
5. 4 × 10^2 N/C.
7. 2.45 m.
9. 72.7 cm to the left of q_1.
11. 1.77 × 10^{-10} C.

(B) 1. 5.2 cm from the smaller charge and between the charges.
3. 6.4 × 10^{-18} J; 3.75 × 10^6 m/s.
5. 9.8 × 10^{-8} C.
7. a = 5.76 × 10^{10} m/s compared to 9.8 m/s.
9. 1.17 × 10^8 N.

Chapter 27

(A) 1. 216 × 10^{-15} joule.
3. 24 × 10^3 volts.
5. $V_A - V_B$ = 12 × 10^3 volts;
$V_B - V_A$ = -12 × 10^3 volts.

7. $V_A - V_B = 17.28$ volts.
9. 3×10^4 volts/m.

(B)   1. 90 eV; 144×10^{-19} joule.
3. 10^5 eV; 1.6×10^{-14} joule.
5. -72.003×10^{-7} joule; No.
7. a) -38.4×10^{-20} joule
   b) -38.4×10^{-20} joule
   c) -11.52×10^{-20} joule.
9. 10^{-7} joule.

Chapter 28

(A)   1. 3.33 amp.
3. 20.8×10^{18} electrons.
5. a) 2.5 ohms   b) 1.25 ohms.
7. a) 150 ohms   b) 16.67 ohms.
9. a) 15×10^3 joules by 20-ohm resistor;
      7.5×10^3 joules by 40-ohm resistor.
   b) 500 watts by the 20-ohm resistor;
      250 watts by the 40-ohm resistor.
11. a) 1.8×10^{-2} ohm   b) 0.45×10^{-2} ohm.

(B)   1. 333.3 ohms; 1000 volts across the 333.3-ohm
      resistor; 750 volts across the 250-ohm resistor.
3. 18.8 ohms.
5. 7 volts.
7. 22 ohms.
9. $P_5 = 235$ watts; $P_{10} = 209$ watts;
   $P_{20} = 104.5$ watts.
11. a) 1.47 amp   b) 441 coul
    c) 4.862×10^4 watt-sec.
13. 26,122 joules in the 2-ohm resistor;
    13,061 joules in the 4-ohm resistor;
    6530.5 joules in the 8-ohm resistor.
15. a) 3 amp   b) 1 amp   c) 3 V.

Chapter 29

(A)   1. 900 dynes.
3. 6.7×10^{-6} weber/m^2.
5. 6.7×10^{-6} weber/m^2.
7. a) Toward the east   b) toward the north
   c) toward the west   d) toward the south.
9. Straight up.

(B)   1. 62.8×10^{-4} weber/m^2.
5. 1.34×10^{-5} weber/m^2 (south).
7. a) 32,800 dynes (toward the left)
   b) 32,800 dynes (toward the right).

9. a) Force on left magnet is 68.7×10^3 dynes
      toward the right.
   b) Force on right magnet is 68.7×10^3 dynes
      toward the left.
11. 80.

Chapter 30

(A)   1. 0.1 newton.
3. 1.6×10^{-13} newton.
5. 10^{-4} weber.
7. 9×10^{-3} volt.
9. 4.8×10^{-19} newton (east).

(B)   1. Straight down.
3. a) 2 volts   b) 1 amp.
5. Straight up.
7. Toward the east.
9. 45°.
11. Circular orbit.
13. Clockwise.
15. 19.16×10^{-13} joule; 11.97×10^6 eV.

Chapter 31

(A)   1. 7.07 amp.
3. 5.025×10^{-3} ohm.
5. Zero.
7. When the plane of loop is parallel to the field.
9. 5.

(B)   1. 0–20 millivolts.
3. $\frac{1}{60}$ sec; 90°; 2 each.
5. 111 ohms.
7. 2310 turns.
9. 3.7×10^{-3} weber/m^2.
11. a) 50 amp   b) 25,000 watts
    c) 125,000 watts.
13. a) 100.5 volts   b) 20.1 amp
    c) from a to b in R.

Chapter 32

(A)   1. 4×10^{-3} volt.
3. 10^{-8} coulomb.
5. 3.5 henry.
7. 0.5 sec.
9. 16.7×10^{-6} volt.

11. 2.73 μF.

13. 30 μF.

(B) **1.** 50 $\times$ 10^{-3} joule; it is converted to potential energy.

3. 240 $\mu\mu$F.

5. a) 0.5 amp

b) theoretically a very long time; practically, in a time of 6 time constants, 99.8% of the maximum current is achieved.

c) 0.05 sec   d) 0.315 amp.

9. 19.17 μF.

11. 1.73 amp.

13. a) $V_R = 103$ V; $V_L = 38.8$ V   b) 2.06 amp

c) $\phi = 20.6°$; voltage leads the current.

Chapter 33

3. 24.

5. 270 V.

7. 1.5 V.

Chapter 34

(A) **1.** a) 770 kilo-hertz   b) 389.6 m.

3. 7 $\times$ 10^{-7} m.

5. a) 6563 Å; 6563 $\times$ 10^{-8} cm   b) yes.

7. About 10^8 m/sec   b) no.

9. 496 sec.

(B) **1.** 340 lumens.

5. About 408 lumens.

7. 22 $\times$ 10^{-3} steradian.

9. 0.55 ft.

11. 2.5 lumens.

13. a) 0.025 steradian   b) 37.5 meter-candles.

15. a) 2 $\times$ 10^{-8} lumen

b) 20 $\times$ 10^{-6} meter-candle.

Chapter 35

(A) **5.** a) 30°   b) 30°   c) 60°.

7. -1.

9. image distance = 7.5 in.; image height = 1 in.

11. a) 22.9 cm   b) 171.4 cm   c) 1.145.

13. a) 53.3 cm (erect)

b) -133.3 cm (behind mirror)

c) -2.7 (virtual image).

(B) **1.** -5; virtual image.

5. a) -18.75 cm   b) 1.875 cm (erect)

c) image is virtual, erect and smaller than the object.

7. 3.79 ft (focal length); 5.3 ft (object distance).

9. Two focal lengths; twice the focal length.

11. 74 in.

13. 28°.

15. 45°.

Chapter 36

(A) **1.** Away from the normal.

3. 90° from the surface.

5. 52.9°.

7. Away from the normal; material A.

9. $\dfrac{1}{1.8} = \dfrac{5}{9}$

11. 30° from the vertical.

(B) **1.** 1.94 $\times$ 10^8 m/sec.

3. 10°

5. 7.5 ft.

7. 3333 Å.

9. 0.56 in.

11. 15° from the normal.

Chapter 37

(A) **3.** 100 mm.

7. a) The image is 6 cm in height and is located 20 cm behind the lens

b) The image is real and inverted.

9. a) The image is 2 cm high and is located 6.7 cm on the object side of the lens

b) erect and virtual.

11. 0.01.

(B) **1.** a) 6.92 cm   b) 0.01   c) 0.04 in.

3. $+10.7$ cm.

5. $+3.93$ in.

7. $+19.5$ mm.

9. 0.143 in.

11. 1.76 in.

13. b) The final image is real and 92.7 cm to the right of the object arrow.

15. a) 6.15 cm to the right of lens I

b) virtual and inverted with respect to the object

c) −0.23.

Chapter 38

(A) 5. $\dfrac{\lambda}{2}$.

(B) 1. a) $\dfrac{\lambda L}{2d}$ b) $\dfrac{\lambda L}{4d}$.

3. 868.75×10^{-5} cm.
5. 23×10^{-3} cm.
7. $0.9°$.
9. 2×10^{-4} cm.
11. 1141.7×10^{-8} cm.

13. 910×10^{-8} cm.
15. $59.5°$.

Chapter 39

1. 3 electrons, 3 protons, 4 neutrons.
3. 1.66×10^{-24} gm.
5. 15.07×10^{-11} joule.
7. 940 MeV.
9. 13.64 eV.
11. 1020.85 Å.
13. 12.375×10^3 eV.
15. 35.3×10^{14} photons.
17. a) 5.8×10^{-14} joule b) 3.4×10^{-12} m
c) 8.8×10^{19} hertz.
19. 0.625 curie.
21. 7 protons, 7 neutrons; $^{14}_{6}C \rightarrow {}^{14}_{7}N + {}^{0}_{-1}\beta$.

Index

Absolute humidity, 327, 333
Absolute pressure, 249
Absolute temperature scale, 273
Absolute zero, 273, 308, 314
Acceleration, 82-84, 108-110, 115, 117, 121-123, 133-138
  angular, 145-147, 159
  centripetal, 121-123
  (g), 93-94, 100, 110, 115-116, 124, 129, 134
  due to gravity, 92, 95, 100, 112-113, 117
  in SHM, 231, 233-234
Action, 111
Action-at-a-distance, 387
Actual mechanical advantage, 191, 197, 199, 203, 204
Adaptation, 555
Adiabatic constant, 303, 304, 349
Adiabatic demagnetization, 307
Adiabatic process, 296, 302, 312, 323
Air columns, 353
Air resistance, 92, 98-100
Alpha particle, 594
Alternating current-cycle, 469
AM, 497, 501
Ammeter, 466
Ampere, definition of, 451
Amplification ratio, 497, 500
Amplitude, 349, 355, 368
Amplititude modulation, 497, 501
Angle of incidence, 520
  of reflection, 520
  of refraction, 536
  of repose, 58-59, 61
  of shear, 72
  of uniform slip, 59
Angular acceleration, 145-147, 149-150, 159
Angular displacement, 144, 159

Angular impulse, 217-220, 226
Angular momentum, 217-226
Angular momentum vector, 222-223
Angular motion, 147, 159
Angular velocity, 145, 159
Anode, 499
Antinodes, 345, 351
Apparent depth, 541
Apparent weight, 133-135, 138
Archimedes' principle, 245, 253
Artificial gravity, 134
Atmosphere, 71, 73
Atom, 584, 585
Atomic binding energy, 581
Atomic mass unit, 582
Atomic number, 583
Atwood's machine, 115-116
Average speed, 79, 84, 137
Average velocity, 79-80, 84
Avogadro's number, 316

Banking, 126-129, 138
Bearings, 53
Beats, 345, 356, 358
Bels, 368
Belt drive, 200-201
Bernoulli's equation, 245, 260
  derivation, 258
  and work-energy principle, 258
Beta particle, 594
Bimetallic strip, 276
Black body, 337, 341
Block and tackle, 194-195
Bohr atom, 587
Boiling, 327, 331

Boiling point, 330, 332
  table, 330
Boltzmann's constant, 317
Boom and mast problem, 26-27, 43
Boyle's law, 312, 313, 328
Breeder reactor, 597
Brewster's law, 578
British thermal unit, 284
Btu, 284
Bulk modulus, 70, 71
Buoyancy, 245
Buoyant force, 253

Calorie, 284
Calorimeter, 287
Camera, 557
Candela, 513
Candle, 513
Capacitance, 482
Capacitive time constant, 481
Capacitors, in parallel, 485
  in series, 484
Capillarity, 245, 265
Cardan, 2
Carnot cycle, 300, 308
  efficiency, 293, 301
Cathode, 499
Celsius, 272
Center of curvature, 137, 139
Center of gravity, 39-42, 44
Center of percussion, 219-220
Centrifuge, 135-136
Centripetal acceleration, 122-125, 136-138
Centripetal force, 125-127, 135, 138-139
Chain drive, 200-201
Chain reaction, 596
Change of phase, 288, 327, 574
Charles' law, 312, 314
Chromatic aberration, 555
Clausius, 298
Coefficient of, friction, 113-114, 126-127
  kinetic friction, 55, 59-61
  linear expansion, 275
  performance, 301, 307
  rigidity, 73
  sliding friction, 54
  starting friction, 54-55, 61
  static friction, 54, 58-59
  volume expansion, 277
Coil spring, 67-68
Collector, 503
Collision, elastic, 214-215, 226
  inelastic, 211-213, 226
  partially elastic, 215-216, 226
Color patches, 579

Combustion, 290
Commutator, 471
Components of vectors, 14-15, 18
Composition of vectors, 9-18
Compound microscope, 559
Compressibility, 71, 73
Compression, 26-27
Compressions, 347, 355
Compressive stress, 70
Concave lens, 549
Concave mirror, 524
Concurrent forces, 17, 22-24, 27
Conduction band, 501
Conduction, heat, 337, 338
Conductor, electrical, 375, 379, 394
Conservation of, angular momentum, 220-222, 226
    energy, 173-174, 179, 181, 187
    momentum, 208, 210-211, 217, 226
Constructive interference, 571
Continuity equation, 258
Control rods, 596
Convection, 337, 339
  forced, 340
Convergent rays, 522
Convex lens, 547
Convex mirror, 526
Corpuscular theory of light, 508
Cosecant, 6
Cosine, 7
Cotangent, 7
Coulomb's law, 2, 375, 380, 388
Critical angle, 541
Critical point, 329
Critical values, 329
Critical velocity, 131, 132, 138
Cryogenics, 307
Curie (unit), 595
Current, 411
Curvilinear motion, 121-139
Cyclotron, 452

Dalton's law, 312, 320
De Broglie wavelength, 586
Deceleration, 83
Decibels, 368
Density, 107, 246
  and expansion, 278-279
  table for common liquids and gases, 246
Destructive interference, 571
Deuterium, 583
Dew point, 327, 333, 335
Diamagnetic materials, 442
Diesel engine, 303
Differential hoist, 197-198
Differential pulley, 197-198

Diffraction, 569
Diffraction grating, 573
Diffuse reflector, 521
Diode, 497, 499
Discharge rate in fluid, 257
Dispersion of light, 540
Displacement, 77-78, 80-81, 83-84, 89-91, 94, 98, 100
    in SHM, 231-234
Displacement-time diagram, 80-82, 84
Divergent rays, 522
Doppler effect, 363, 364
Dry ice puck, 106
Dynamics, 105, 116
Dynamometer, 199-200
Dyne, 109-110, 117

Earth satellites, 132-133
Eddy current, 474
Edison effect, 497, 498
Effective values, 470
Efficiency, 187, 189-190, 191, 199, 203
    Carnot engine, 293, 301
    diesel engine, 303
    jet engine, 304
    rotating machines, 199-203
Efflux, velocity of, 260
Elastic collisions, 214-216, 226
Elastic deformation, 66-68, 73
Elastic forces, 66-70, 73
Elastic limit, 67, 69, 73, 232
Elastic modulus, 66, 69-70, 72-73
Elastic potential energy, 171-172, 181
Electric charge, 375, 376, 380, 383, 396
Electric field, 386, 387, 388, 393
Electric force, 375, 380, 387
Electric potential, 403
Electric potential energy, 400
Electrolysis, laws of, 2
Electromagnetic induction, 455
Electromagnetic spectrum, 507
Electromotive force, 412
Electron, 376, 383, 396
Electronics, 497
Electronic shells, 376
Electron-volt, 406
Electroscope, 375, 378
Emissivity, 342
Emitter, 503
Energy, 164, 168, 170-182, 189
    kinetic, 168, 170, 172-175, 176-178, 181-182
    potential, 164, 170-175, 178-182
    in rotation, 176-178, 181
    in space mechanics, 178-181
    thermal, 283, 288
Energy gap, 502

Energy-level diagram, 589
English engineering system, 77-78, 109-110, 165-166
Enthalpy, 297
Entropy, 298
Equal arm balance, 107-108
Equation of state, 317, 318, 320, 323, 329, 333
Equilibrant, 22, 23, 27
Equilibrium, 22, 23-25, 27
Equivalence of mass and energy, 592
Erg, 165-166, 181
Evaporation, 327, 331
Expansion, area, 276
    linear, 275
    volumetric, 277
    water, 270, 278
Exposure time, 557
Eye, 555

Fahrenheit, 272
Faraday, Michael, 2, 395
    law of induction, 456
Farsightedness, 556
Ferromagnetic materials, 447
Filament, 498, 503
Film speed, 557
Final velocity, 90, 98
First condition of equilibrium, 23-24, 27
First law of motion, 104-106, 108, 116-117
    of thermodynamics, 293-294, 296, 321
Flexure, 73
Floatation, law of, 266
Flow rate, 257
Fluid dynamics, 245, 257
Fluid mechanics, 245
FM, 497, 501
Focal length, 548
Focal point, 548
Focal ratio, 557
Foot candle, 513
Foot-pound, 165, 181
Force, 10, 17-18, 22, 105-106, 108-110, 117
    and displacement, 169-170
    of gravity, 109-110, 111-113, 117, 132-134, 138
    on moving charges, 451
    table, 33-34
    on a wire, 449
    and work, 165-166, 181
Forced vibrations, 355
Forces, 10, 17, 22, 27
    adhesive, 265
    cohesive, 265
    rolling, 157-158
    thermal, 270, 279
Fourier analysis, 369
Franklin, Benjamin, 2, 376

Free-body diagram, 25, 27, 35-37, 113-116, 130-131, 133-134
Free expansion, 296
Freezing, 327, 331
Frequency, 124, 138, 231, 234, 345, 369
Frequency modulation, 497, 501
Friction, 52-62, 113-116, 126-127, 138
F-s diagram, 169-170
Fundamental frequency, 345, 351, 359
Fusion, 597
  heat of, 288, 289

Galileo, 111
Galvanometer, 463
Gamma rays, 594
Gasoline engine, 302
Gas turbine, 304
Gauge pressure, 249
Gauss (unit), 433
Gaussian surface, 392, 393, 394
Gauss' law, 386, 390, 391
Gear, spur, 201-202
  train, 202-203
Generator, ac, 468
  brushes, 468
  commutator, 471
  dc, 471
  slip-ring, 468
  split-ring, 471
Geometric optics, 520
Gilbert, 2
Governor, 129
Gram, 5, 107, 110, 117
Gram-molecular weight, 316
Gravitation, 111-113, 117
Gravitational mass, 107
Grid, 500
Grounding, 378
Gyroscope, 223-226

Half-life, 595
Harmonic, 345, 360
Harmonic motion, *see* Simple harmonic motion
Heat, defined, 270
  and energy transfer, 271, 283
  latent, 283, 289
  specific, 285, 321
  transfer, 337
Heat engine, 293, 298, 302, 303
Heat transfer, 337
Heisenberg, Werner, 590
Hertz, Heinrich, 509
High fidelity, 363, 370
Hooke's law, 66, 67, 69, 72, 73, 236

Horizontal range, 96-98, 99-100
Horsepower, 187-189
Humidity, 327, 333
  absolute, 327, 333
  relative, 327, 334
Huygen's principle, 531
Hydraulic jack, 249, 251
Hydraulic press, 249
Hydraulic systems, 249
Hydrodynamics, 245
Hydrostatic paradox, 249
Hydrostatic pressure, 247, 254
Hydrostatics, 245, 260
Hygrometer, 335
Hyperopia, 556
Hysteresis, 443

Ideal gas, 315
  equation of state, 312, 317
  properties, 312, 317, 320
Ideal mechanical advantage, 191-196, 199, 201, 203-204
Illuminance, 479
Impulse, 208-211, 226
Inclined axis, 15
Inclined plane, 24, 56-58, 178, 192-193
Index of refraction, 536, 537, 539
Inductance, 489
Induction, 375, 378
Inductive circuit, 489
Inductor, 479
Inelastic collision, 208, 211-213, 215-216, 226
Inertia, 105
Inertial mass, 106, 135
Initial velocity, 90, 92, 93
Instantaneous velocity, 80, 81, 84
Insulation, heat, 342
Insulator, electrical, 375, 379
Intensity, electric field, 386, 387, 388, 393
Intensity level, 363
  sound waves, 363, 368
Interference, 345, 356
  constructive, 357, 370
  destructive, 358, 370
  in films, 574
  of waves, 569
Internal energy, 271, 294
International Bureau of Weights and Measures, 107
Isobaric process, 297
Isothermal process, 296, 312
Isovolumic process, 296

Jet engine, 304
Jets, conservation of momentum, 217
Joule, James Prescott, 165
  experiment, 284

(unit), 165-166
Joule's law, 424

Kelvin, 273, 301, 315
Kelvin-Planck, 298
Kepler, Johannes, 133
Kilogram, 107, 110
Kinematics, 77-84, 105
Kinetic energy, 168-170, 172-178, 181
Kinetic friction, 55-56, 61, 193-194
Kirchhoff's rules, 416, 419
Kundt's tube, 351, 359

Laminar flow, 257
Laser, 590
Latent heat, 283, 289
Law of cosines, 6
Law of gravity, 111-113, 117, 132
Law of inertia, 105-106, 108, 116-117
Law of partial pressures, 312, 320
Law of reflection, 520
Law of sines, 6
Lawrence, Ernest, 452
Laws of thermodynamics, 293-297
Left-hand rule, 450
Lens, concave, 549
  convex, 547
Lens aberrations, chromatic, 555
  curvature of field, 555
  distortion, 555
  spherical, 554
Lens images, 550
Lenz's law, 457
Lever, 36, 46, 192
Lever arm, 37, 44
Light-"pipe," 540
Light-quantum, 585
Linear acceleration in rotation, 148–149, 159
Linear coefficients of expansion, 275
Linear displacement, 78–79, 83–84
Linear motion, 77–117
Linear velocity in rotation, 147–148, 159
Lines of force, 386, 390
Liquefaction, 327, 328
Liquid drop, 261, 263
Lodestone, 2
Lorentz, 2
Low temperature physics, 307
Lumen (unit), 511
Luminous body, 520
Luminous flux, 511
Luminous intensity, 512
Lux (unit), 513

Mach number, 366

Magnet, 431
Magnetic, angle of dip, 435
  declination, 435
  flux, 433, 454
  flux density, 449
  induction, 449
Magnetic field, 434
  of circular loop, 439
  of long wire, 439
  of solenoid, 440
Magnetism of earth, 435
Magnetomotive force, 444
Magnification, 524
Magnifying glass, 557
Magnifying power, 562
Mass, 104, 106-108, 110-112, 117, 125, 149-150, 152-153, 159
Mass density, 107
Mass of earth, 112, 117
Maxwell, James Clerk, 2, 509
Measurement, 3-5
Mechanical advantage, 190-204, 250
Mechanical energy, 170, 173, 174-175, 181
Mechanical equivalent of heat, 284
  Joule's experiment, 284
Melting point, table, 289
Metric system of units, 5, 78, 109, 110, 117, 165-166, 188
Millikan, 2
  experiment, 382, 386, 395
Million electron-volts, 406
Mirror equation, 528
Moderator, 596
Modulation,
  AM, 497, 501
  FM, 497, 501
Modulus, bulk, 70-71
  elastic, 66, 69-70, 72-73
  shear, 71-73
  Young's, 279, 349
Modulus of a spring, 236
Mole, 316
Moment, 17
Moment of a couple, 38, 44
Moment of a force, 35, 37-39, 42, 44
Moment of inertia, 150-154, 159, 218-219, 221, 226
  physical pendulum, 240
  torsion pendulum, 240
Momentum, 106, 108, 169, 214-217, 226
  angular, 217-226
Muzzle velocity, 91, 96-97, 168

Natural frequency, 351
Near-point, 556
Net work, 166-167, 181
Neutral equilibrium, 42

Neutron, 591
Newton, Sir Isaac, 105, 111, 117
  (unit), 109, 110, 117, 165-166
Newton's first law, 105-106, 117
  law of gravitation, 104, 111-112, 117
  laws of motion, 104, 105-106, 108-109, 110-111, 116-117
  second law, 108-109, 117
  third law, 111, 117
Nodes, 345, 350, 351
Noncurrent forces, 17-18, 35-44
Normal acceleration, 137-138, 139
Normal force, 24, 53-55, 61, 126-127, 129, 138
Normal-line, 520
Nuclear, binding energy, 591
  chain reaction, 595
  fission, 595
  fusion, 597
Nucleus, 376, 591

Ohm's law, 411
Optical "pumping," 591
Orbits, 132-134, 180-181
Oscillatory motion, *see* Simple harmonic motion
Otto cycle, 302
Overtone, 345, 351

Parabolic mirror, 525
Parabolic trajectory, 98-99
Parallel forces, 36-37, 38-39, 44
Parallelogram method, 9, 12-13, 18
Paramagnetic materials, 442
Paraxial rays, 525
Partially elastic collision, 215-216
Partial pressure, 312, 320
Pascal's principle, 245, 247, 249
Path length, 78-79, 84
Pendulum, physical, 231, 240, 242
  simple, 231, 238, 241
  torsion, 231, 240, 241
Penumbra, 510
Period, 123, 138
Periodic motion, 231-242
Period of satellite, 133
Permanent magnets, 442
Permeability constant, 439
Permittivity, 391
Phase changes, 288, 327, 574
Photoelectric effect, 508
Photometry, 511
Photons, 508
Physical pendulum, 231, 240, 242
Physics, definition, 1-5
Planck's quantum theory, 585
Plane mirror, 523
Plane wave-front, 532

Point-source, 520
Polarization of light, 576
Polarized light, 577
Polarizing angle, 578
Polygon method, 9, 11-12, 18
Positron, 592
Potential difference, 403
Potential energy, 164, 170-175, 178-182
  of spring, 171-172, 181
Potentiometer, 423
Pound of force, 109, 110, 117
Power, 187-190, 198-204
  factor, 493
  output and torque, 199, 203
  rotating machines, 199-203
  sound waves, 368
  steering, 250, 251
Precession, 224-226
Presbyopia, 556
Pressure, 70-71, 73
  absolute, 249
  atmosphere, 71
  definition, 248
  gauge, 249
  hydrostatic, 247, 254
Principal axis, 525
Principal focus, 526
Principal quantum number, 588
Principle of moments, 17, 37-38, 44
Projectiles, 92-98, 99, 100
Proton, 376
Pulley, 33, 116, 193-196, 198

Quality of sound, 351, 363, 370
Quantum electrodynamics, 2

Radian, 144
Radiation, 337, 341
Radioactivity, 593
Radius of curvature, 125-127, 137, 138
Rankine, 273, 315
Rarefactions, 348
Ray optics, 520
Reactance, 491
Reaction, 111, 133, 135, 138
Real image, 522
Rectification, 499
Reference circle, SHM, 231, 233, 241
Reflection of light, 520
Refraction, 536
Refrigerator, 293, 306
Regelation, 332
Relative index of refraction, 537
Relative permeability, 442
Reluctance, 444
Resistors in parallel, 419

in series, 419
Resolution, 562
Resonance, 345, 355, 493
  closed pipes, 345, 353
  fixed string, 345
  open pipes, 345, 354
  rods, 345, 359
Rest-energy, 592
Rest-mass, 592
Restoring force, 67-68, 73
Resultant, 9, 10-17, 18, 23
Resultant force, 108-109, 113, 117, 127, 167
Reversible engine, 300, 308
Rocket engine, 304
Rockets, conservation of momentum, 217, 226
Rolling friction, 53, 157-158
Rolling and rotation, 155-158, 160
Rotary engine, 305
Rotary motion, 144
Rotation, 143-155, 159
Rotational energy, 176, 181
Rotational inertia, 150, 159
Rotation and rolling, 155-158, 160
Rowland ring, 442
Rutherford atom, 375, 377, 585

Satellites, 92, 112-113, 132-133, 180-181
Scalars, 10, 11, 18
Screw-rule, 450
Secant, 6
Second condition for equilibrium, 42, 44
Second law of motion, 108-109, 117
Second law for rotating object, 150-151, 157, 159-160
Second law of thermodynamics, 293, 297
Semiconductors, 501, 503
Series circuit, 419
Shearing strain, 72
Shearing stress, 71-73
Shear modulus, 71-73
Shunt resistor, 466
Simple harmonic motion, acceleration, 231, 232, 233, 234
  displacement, 231, 233, 234
  graphs, 231, 234
  period, 231, 234, 237, 238, 239, 240
  physical pendulum, 231, 240, 242
  reference circle, 233, 241
  simple pendulum, 231, 238, 241
  spring, 231, 236
  torsion pendulum, 231, 240, 241
  velocity, 231, 233, 234
Simple machines, 190-198, 203-204
Simple microscope, 557
Simple pendulum, 231, 238, 241
Sliding friction, 53-55, 61-62
Slip-ring, 471

Slug, 109-110, 117
Soap bubble film, 262
Sodium D lines, 564
Solid angle, 512
Sonar, 371
Sound, db levels, 368
  intensity, 363, 368
  power, 368
Space capsule, 111, 132, 134
Space mechanics, 132-134, 178-181
Specific gravity, defined, 246
Specific heat capacity, 284
  of gases, 321
  molar, 321
  table, 285
  water, 340
Spectroscope, 563
Spectrum, absorption, 564
  bright-line, 564
  dark-line, 564
Specular reflector, 521
Speed, 79-80, 84
Speed of light, 510
Spherical mirror, 524
Spontaneous emission of light, 590
Spring balance, 108
Spring constant, 66, 67-68, 73, 231, 236
Sprocket, 200-201
Spur gear, 201-202
Stable equilibrium, 42, 44
Standing waves, 345, 351, 352
Starting friction, 54-55, 58, 61
Static friction, 54, 58, 61
Statics, 23, 61, 78
Stationary waves, 345, 351, 352
Stefan-Boltzmann law, 337, 341
Steradian, 512
Stimulated emission of light, 590
Strain, 66, 69, 70, 72, 73
Streamline flow, 257
Stress, 66, 69-70, 71-72, 73
Sublimation, heat of 288, 332
Superelevation, 128, 129
Superfluidity, 308
Superposition, 350, 569
Supersonic, 366, 371
Surface tension, 245
  and liquid drop, 261, 263
  soap bubble, 262
  table, 264
Sympathetic vibrations, 345, 355

Tangent, 6
Tangential acceleration, 136-137, 139
Tangential velocity, 147-148, 159

Telescope, magnifying power, 562
  optical, 560
  reflecting, 560
  refracting, 560
Temperature, 270, 271
  absolute scales, 273
  and average kinetic energy, 271, 318
  coefficient of resistivity, 415
  and expansion, 270, 275
  and thermometers, 270
  units, 272
Temperature gradient, 337, 338
Tensile stress, 70
Tension, 26, 27, 125, 129, 131-132
Terminal velocity, 92
Terminal voltage, 421
Tesla (unit), 449
Thales, 2
Thermal conductivity, 338, 339
Thermal energy, 283
  and mechanical energy, 283
  and phase changes, 283, 288
  and temperature changes, 283
Thermal neutrons, 596
Thermal stress, 270, 279
Thermodynamics, first law of, 293, 294, 296, 321
  second law of, 293, 297
Thermoelectron, 499
Thermometer, 272
  constant-volume gas, 272, 274
  liquid-in-glass, 272, 274
  resistance, 274
  optical, 274
Thermometry, 270
Thermonuclear reaction, 597
Thomson, 2
Thought experiment, 589
Throttling process, 297, 306
Timbre, 370
Time constant, capacitive, 481
  inductive, 480
Torque, 17, 149-151, 157-159, 176, 198-199, 203-204
  on current loop, 463
  resultant, 149-150, 159
  in rolling, 157-158
Torque-time diagram, 218
Torque transmission, 198-199, 203
Torricelli's theorem, 261
Torsional stress, 73
Torsion pendulum, 231, 240, 241
Traction, 53

Transfer of heat, 337
Transformer, 473
  efficiency, 474
Transistor, 497, 501, 502, 504
Translation, 158-159, 160
Traveling waves, 345, 349
Triangle method, 12, 13-14, 18, 24, 25, 27
Trigonometric functions, 6
Triode, 497, 500
Triple point, 332

Ultrasonic, 363, 371
Umbra, 509
Uncertainty principle, 590
Uniform angular acceleration, 146-147, 159
Uniform circular motion, 122-124, 136
Uniformly accelerated motion, 89-100
Units of measurement, 3-5
Universal gas constant, 316
Unpolarized light, 577
Unstable equilibrium, 42, 44
U.S. National Bureau of Standards, 5, 107
Vacuum tube, 497
Valence band, 501
Van der Waal's equation, 319, 329, 330
Vaporization, heat of, 288
  table, 289
Vapor pressure, 330, 331, 334
  saturated, 334
Vector addition, 9-18
Vectors, 9-18
Vector sum, 10
Velocity, 77, 79-83, 84, 89, 93-94, 98, 100, 122-123, 131, 133, 136, 138, 147-148, 159
  angular, 143, 145-148, 159, 198-199, 200, 203, 204
  efflux, 260
  rolling, 155, 160
  root-mean-square (rms), 319
  simple harmonic motion, 231, 233, 234
  sound, 348, 349
  strings, 348
  wave, 347
Velocity-time curve, 77, 82-83, 84
Venturi tube, example, 259
Vertical circle, 130-132, 138
Vertical loop, 130
Vibrations, 345, 353-354, 359
Virtual image, 522
Volt (unit), 403
Voltaic pile, 2
Voltmeter, 466

Wankel engine, 305
Water, expansion of, 278
  table of densities, 279
Waterproofing, 266
Watt (unit), 424
Wavelength, 346
Wave mechanics, 589
Wave motion, 345
  longitudinal, 345, 346
  transverse, 345, 346
Wave-nature of particles, 586
Waves, crest, trough, 570
  interference of, 569
  light, 569
  superposition of, 569
  water, 569
Weight, 39, 44, 109-110, 112-113, 117, 133-135, 138
Weightlessness, 133-134

Wheatstone bridge, 423
Wheel and axle, 193
Work, 164-182, 188-190, 203
  from F-s diagram, 169-170
  input, 190, 198-199, 203
  net, 166-167, 181
  output, 189-190, 198-199, 203
  and power, 187, 188-189, 203
  in rotation, 175-177
  space mechanics, 178-180
  in thermodynamics, 293
Work-energy principle, 164, 172-175, 181

X-ray tube, 405

Young's experiment, 571
Young's modulus, 69-70, 73, 279, 349